CONTROL OF CELLULAR DIVISION AND DEVELOPMENT
Part A

PROGRESS IN CLINICAL AND BIOLOGICAL RESEARCH

Series Editors
Nathan Back
George J. Brewer

Vincent P. Eijsvoogel
Robert Grover
Kurt Hirschhorn

Seymour S. Kety
Sidney Udenfriend
Jonathan W. Uhr

RECENT TITLES

Vol 52: **Conduction Velocity Distributions: A Population Approach to Electrophysiology of Nerve,** Leslie J. Dorfman, Kenneth L. Cummins, and Larry J. Leifer, *Editors*

Vol 53: **Cancer Among Black Populations,** Curtis Mettlin and Gerald P. Murphy, *Editors*

Vol 54: **Connective Tissue Research: Chemistry, Biology, and Physiology,** Zdenek Deyl and Milan Adam, *Editors*

Vol 55: **The Red Cell: Fifth Ann Arbor Conference,** George J. Brewer, *Editor*

Vol 56: **Erythrocyte Membranes 2: Recent Clinical and Experimental Advances,** Walter C. Kruckeberg, John W. Eaton, and George J. Brewer, *Editors*

Vol 57: **Progress in Cancer Control,** Curtis Mettlin and Gerald P. Murphy, *Editors*

Vol 58: **The Lymphocyte,** Kenneth W. Sell and William V. Miller, *Editors*

Vol 59: **Eleventh International Congress of Anatomy,** Enrique Acosta Vidrio, *Editor-in-Chief*. Published in 3 volumes:
  Part A: **Glial and Neuronal Cell Biology,** Sergey Fedoroff, *Editor*
  Part B: **Advances in the Morphology of Cells and Tissues,** Miguel A. Galina, *Editor*
  Part C: **Biological Rhythms in Structure and Function,** Heinz von Mayersbach, Lawrence E. Scheving, and John E. Pauly, *Editors*

Vol 60: **Advances in Hemoglobin Analysis,** Samir M. Hanash and George J. Brewer, *Editors*

Vol 61: **Nutrition and Child Health: Perspectives for the 1980s,** Reginald C. Tsang and Buford Lee Nichols, Jr., *Editors*

Vol 62: **Pathophysiological Effects of Endotoxins at the Cellular Level,** Jeannine A. Majde and Robert J. Person, *Editors*

Vol 63: **Membrane Transport and Neuroreceptors,** Dale Oxender, Arthur Blume, Ivan Diamond, and C. Fred Fox, *Editors*

Vol 64: **Bacteriophage Assembly,** Michael S. DuBow, *Editor*

Vol 65: **Apheresis: Development, Applications, and Collection Procedures,** C. Harold Mielke, Jr., *Editor*

Vol 66: **Control of Cellular Division and Development,** Dennis Cunningham, Eugene Goldwasser, James Watson, and C. Fred Fox, *Editors*. Published in 2 Volumes.

Vol 67: **Nutrition in the 1980s: Constraints on Our Knowledge,** Nancy Selvey and Philip L. White, *Editors*

Vol 68: **The Role of Peptides and Amino Acids as Neurotransmitters,** J. Barry Lombardini and Alexander D. Kenny, *Editors*

Vol 69: **Twin Research 3, Proceedings of the Third International Congress on Twin Studies,** Luigi Gedda, Paolo Parisi and Walter E. Nance, *Editors*. Published in 3 Volumes:
  Part A: **Twin Biology and Multiple Pregnancy**
  Part B: **Intelligence, Personality, and Development**
  Part C: **Epidemiological and Clinical Studies**

Vol 70: **Reproductive Immunology,** Norbert Gleicher, *Editor*

Vol 71: **Psychopharmacology of Clonidine,** Harbans Lal and Stuart Fielding, *Editors*

See pages 613–614 for previous titles in the series.

CONTROL OF CELLULAR DIVISION AND DEVELOPMENT

Part A

Proceedings of the ICN-UCLA Symposium
held at Keystone, Colorado
March 2-8, 1980, Part A

Editors

DENNIS CUNNINGHAM
University of California
Irvine, California

EUGENE GOLDWASSER
University of Chicago
Chicago, Illinois

JAMES WATSON
University of California
Irvine, California

C. FRED FOX
University of California
Los Angeles, California

Alan R. Liss, Inc. • New York

Address all Inquiries to the Publisher
Alan R. Liss, Inc., 150 Fifth Avenue, New York, NY 10011
Copyright © 1981 Alan R. Liss, Inc.

Printed in the United States of America.

Under the conditions stated below the owner of copyright for this book hereby grants permission to users to make photocopy reproductions of any part or all of its contents for personal or internal organizational use, or for personal or internal use of specific clients. This consent is given on the condition that the copier pay the stated per-copy fee through the Copyright Clearance Center, Incorporated, 21 Congress Street, Salem, MA 01970, as listed in the most current issue of "Permissions to Photocopy" (Publisher's Fee List, distributed by CCC, Inc.) for copying beyond that permitted by sections 107 or 108 of the US Copyright Law. This consent does not extend to other kinds of copying, such as copying for general distribution, for advertising or promotional purposes, for creating new collective works, or for resale.

Library of Congress Cataloging in Publication Data

Main entry under title:

Control of cellular division and development.

   (Progress in clinical and biological research;
v. 66)
   Includes bibliographical references and index.
   1. Cellular control mechanisms — Congresses.
2. Cell division — Congresses. 3. Developmental cytology — Congresses. I. Cunningham, Dennis.
II. ICN Pharmaceuticals, inc. III. University of California, Los Angeles. IV. Series. [DNLM:
1. Cell division — Congresses. 2. Cell differentiation — Congresses. W1 PR668E v. 66/QH 605 C764 1980]

QH604.C654   574.87'6   81-8434
ISBN 0-8451-0066-1 (2-Volume Set)   AACR2
ISBN 0-8451-0156-0 (pt. A)
ISBN 0-8451-0157-9 (pt. B)

Pages 1–472 of this volume are reprinted from the Journal of Supramolecular Structure, Volumes 13 and 14, and the Journal of Supramolecular Structure and Cellular Biochemistry, Volume 15. The Journal is the only appropriate literature citation for the articles printed on these pages. The page numbers in the table of contents, contributors list, and subject index of this volume correspond to the page numbers at the foot of these pages.

The table of contents does not necessarily follow the pattern of the plenary sessions. Instead, it reflects the thrust of the meeting as it evolved from the combination of plenary sessions, poster sessions, and workshops, culminating in the final collection of invited papers, submitted papers, and workshop summaries. The order in which articles appear in this volume does not follow the order of citation in the table of contents. Many of the articles in this volume were published in the Journal of Supramolecular Structure (now Journal of Supramolecular Structure and Cellular Biochemistry), and they are reprinted here. These articles appear in the order in which they were accepted for publication and then published in the Journal. They are followed by papers which were submitted solely for publication in the proceedings.

Contents

Contents of Part B xi
Contributors to Part A xiii
Preface
    Dennis Cunningham, Eugene Goldwasser, James Watson, and C. Fred Fox xxi

I. OVERVIEWS ON THE CONTROL OF CELL DIVISION AND DEVELOPMENT

Transforming Growth Factors (TGFs): Properties and Possible Mechanisms of Action
    George J. Todaro, Joseph E. De Larco, Charlotte Fryling, Patricia A. Johnson, and Michael B. Sporn 445

Normal and Neoplastic Lymphocyte Maturation
    I.L. Weissman, M.S. McGrath, E. Pillemer, N. Hollander, R.V. Rouse, L. Jerabek, S.K. Stevens, R.G. Scollay, and E.C. Butcher 461

Control of Hemopoietic Cell Proliferation and Differentiation
    Donald Metcalf 473

II. EFFECTORS OF DNA SYNTHESIS OR DEVELOPMENT

Polypeptide Growth Factors: Some Structural and Mechanistic Considerations
    Ralph A. Bradshaw and Jeffrey S. Rubin 193

Inhibition of Cell Proliferation and Protease Activity by Cartilage Factors and Heparin
    Victor B. Hatcher, Grace Tsien, Martin S. Oberman, and Peter G. Burk 153

Regulation of B Lymphocyte Activation by the Fc Portion of Immunoglobulin
    Edward L. Morgan and William O. Weigle 211

The Role of Colony-Stimulating Factor in Granulopoiesis
    Richard K. Shadduck, Giuseppe Pigoli, Abdul Waheed, and Florence Boegel 265

Evidence for a Low-Molecular-Weight Plasma Peptide Which Stimulates Chick Chondrocyte Metabolism
    Loren Pickart 99

Multiplication Stimulating Activity (MSA) From the BRL 3A Rat Liver Cell Line: Relation to Human Somatomedins and Insulin
    Matthew M. Rechler, S. Peter Nissley, George L. King, Alan C. Moses, Ellen E. Van Obberghen-Schilling, Joyce A. Romanus, Alfred B. Knight, Patricia A. Short, and Robert M. White 411

Contents

Activation of T Lymphocytes by the Fc Portion of Immunoglobulin
Marilyn L. Thoman, Edward L. Morgan, and William O. Weigle 131

Erythropoietin and Red Cell Differentiation
Eugene Goldwasser 487

On Deciding Which Factors Regulate Cell Growth
Arthur B. Pardee, Paul V. Cherington, and Estela E. Medrano 495

The Platelet-Derived Growth Factor: A Perspective
Russell Ross 503

Plasma Components in Growth and Development: Workshop Report
S.P. Nissley and W.J. Pledger 507

Lymphocyte Triggering: Workshop Report
C.W. Parker and J.D. Watson 511

III. PRODUCTION AND PURIFICATION OF GROWTH/ DIFFERENTIATION FACTORS AND THEIR CARRIERS

Erythropoietic Enhancing Activity (EEA) Secreted by the Human Cell Line, GCT
Camille N. Abboud, John F. DiPersio, James K. Brennan, and
Marshall A. Lichtman 9

Purification of Human Interleukin 1
Lawrence B. Lachman, Stella O. Page, and Richard S. Metzgar 109

Human T-Cell Growth Factor: Parameters for Production
Francis W. Ruscetti, James W. Mier, and Robert C. Gallo 39

Release of Erythropoietin From Macrophages by Treatment With Silica
I.N. Rich, V. Anselstetter, W. Heit, E. Zanjani, and B. Kubanek 387

Purification and Characterization of Multiplication-Stimulating Activity (MSA) Carrier Protein
Daniel J. Knauer, Fred W. Wagner, and Gary L. Smith 395

Stimulatory Activity of PHA-LCM for Normal Human Hemopoietic Progenitors and Leukemic Blast Cell Precursors: Separation by Isoelectric Focusing
A.A. Fauser and H.A. Messner 365

Purification of Murine Helper T cell-Replacing Factors
James Watson, Diane Mochizuki, and Steven Gillis 513

IV. CELL SURFACE RECEPTORS

Modulation of EGF Binding and Action by Succinylated Concanavalin A in Fibroblast Cell Cultures
Kurt Ballmer and Max M. Burger 219

Chicken Tissue Binding Sites for a Purified Chicken Lectin
Eric C. Beyer and Samuel H. Barondes 29

Visualization of Thrombin Receptors on Mouse Embryo Fibroblasts Using Fluorescein-Amine Conjugated Human α-Thrombin
Darrell H. Carney 119

Polyclonal Activation of Ts Cells With Antiserum Directed Against an IGH-1 Linked Candidate for a T-Cell Receptor Constant Region Marker
Frances L. Owen 185

Cell Surface Receptors for Endogenous Mouse Type C Viral Glycoproteins and Epidermal Growth Factor: Tissue Distribution In Vivo and Possible Participation in Specific Cell-Cell Interaction
U.R. Rapp and Thomas H. Marshall 255

Selective Protein Transport: Identity of the Solubilized Phosvitin
Receptor From Chicken Oocytes
    John W. Woods and Thomas F. Roth 315
Insulin Receptor Synthesis and Turnover in Differentiating 3T3-L1
Preadipocytes
    M. Daniel Lane, Brent C. Reed, and Peter R. Clements 523
Cell Surface Receptors: Workshop Report
    H.R. Herschman and J.F. Perdue 543
Coated Pits and Vesicles in Intracellular Protein Transport: Workshop
Report
    Ralph A. Bradshaw, Stanley N. Cohen, and Thomas F. Roth 549

V. BIOCHEMICAL ALTERATIONS OF CELL SURFACE RECEPTORS

Proteolytic Domains of the Epidermal Growth Factor Receptor of
Human Placenta
    Edward J. O'Keefe, Teresa K. Battin, and Vann Bennett 339
Cleavage of Cell Surface Proteins by Thrombin
    Martin Moss and Dennis D. Cunningham 373
Controlled Proteolysis of EGF Receptors: Evidence for
Transmembrane Distribution of the EGF Binding and Phosphate
Acceptor Sites
    Peter S. Linsley and C. Fred Fox 303
Direct Linkage of EGF to Its Receptor: Characterization and Biological
Relevance
    Peter S. Linsley and C. Fred Fox 283
Photoaffinity Labeling of the Insulin Receptor in H4 Hepatoma Cells:
Lack of Cellular Receptor Processing
    Cecilia Hofmann, Tae H. Ji, Bonnie Miller, and Donald F. Steiner 325
Epidermal Growth Factor–Receptor–Protein Kinase Interactions
    Stanley Cohen, Graham Carpenter, and Lloyd King, Jr. 557
Role of Proteases in Growth and Development: Workshop Report
    J.F. Perdue and D. Rifkin 569

VI. CELL-CELL AND CELL-SUBSTRATUM INTERACTIONS

Effects of a Serum Spreading Factor on Growth and Morphology of
Cells in Serum-Free Medium
    David Barnes, Richard Wolfe, Ginette Serrero, Don McClure, and
    Gordon Sato 167
Cell-Cell Contact and Growth Regulation of Pinocytosis in 3T3 Cells
    Peter F. Davies 21
The Cell Substratum Modulates Skeletal Muscle Differentiation
    Hannah Friedman Elson and Joanne S. Ingwall 225
The Extracellular Matrix and the Control of Proliferation of Vascular
Endothelial and Vascular Smooth Muscle Cells
    D. Gospodarowicz, I. Vlodavsky, and N. Savion 53
Interaction of Serum and Cell Spreading Affects the Growth of
Neoplastic and Non-Neoplastic Fibroblasts
    R.W. Tucker, C.E. Butterfield, and J. Folkman 353
Cell-Mediated Lympholysis in *H-2K/D* Identical Congenic Strain
Combinations
    Dorothee Wernet and Jan Klein 573

x   Contents

**Cellular Interaction and the Environment in Lymphocyte Development:
The Roles of Antigen, Histocompatibility, and Growth Factors in
T cell-Dependent B cell Stimulation**
   Fritz Melchers, Jan Andersson, Waldemar Lernhardt, and Max H. Schreier 579
Density Inhibition of Growth: Workshop Report
   Robert W. Tucker and Charles W. Boone 589

VII. THE CELL CYCLE
Control of Animal Cell Proliferation
   Robert W. Holley 1
**Inhibition of Lymphocyte Mitogenesis by and Arachidonic Acid
Hydroperoxide**
   Michael G. Goodman and William O. Weigle 87
Mapping the Mitotic Clock by Phase Perturbation
   R.R. Klevecz, G.A. King, and R.M. Shymko 241
   See also **Errata** 611
Regulation of the Balb/c-3T3 Cell Cycle-Effects of Growth Factors
   C.D. Stiles, W.J. Pledger, R.W. Tucker, R.G. Martin, and C.D. Scher 141
Random Transitions and Cell Cycle Control
   Robert F. Brooks 593
Control of the Cell Cycle: Workshop Report
   R. Baserga, R.F. Brooks, and A.B. Pardee 603
Subject Index 605

Contents of Part B

INTRACELLULAR EVENTS IN THE TRIGGERING OF CELL DIVISION AND DEVELOPMENT

Hormonal Regulation of the Adrenocortical Cell / *Gordon N. Gill, Peter J. Hornsby, and Michael H. Simonian*
Total and Exchangeable Calcium in Lymphocytes: Effects of PHA and A23187 / *Andrew H. Lichtman, George B. Segel, and Marshall A. Lichtman*
Specific Protein Phosphorylation During Cyclic AMP-Mediated Morphological Reversion of Transformed Cells / *George S. Bloom and Arthur H. Lockwood*
Interleukin 2 in Cell-Mediated Immune Responses / *Verner Paetkau, Jennifer Shaw, Barry Caplan, Gordon B. Mills, and Kwok-Choy Lee*
Alterations in the Responsiveness of Diabetic Fibroblasts to Insulin / *Mohan K. Raizada and Robert E. Fellows*
Inhibition of Adipose Conversion in 3T3-L2 Cells by Retinoic Acid / *Thomas Murray and Thomas R. Russell*
Induction of an Intracellular Mitogenic Messenger by Epidermal Growth Factor / *Manjusri Das*
Intracellular Events in Growth and Development: Role of Ions as Intracellular Transducers: Workshop Report / *H. Rubin*

ORGANIZATION AND EXPRESSION OF GENES

Glucocorticoid-Induced Lymphocytolysis: State of the Genetic Analysis / *Suzanne Bourgeois*
Hybrid Ia Antigens: Genetic, Serologic, and Biochemical Analyses / *William P. Lafuse, Shumpei Yokota, and Chella S. David*
Expression of Red Cell Membrane Proteins in Erythroid Precursor Cells / *Peter D. Yurchenco and Heinz Furthmayr*
Restricted Expression of an MHC Alloantigen in Cells of the Erythroid Series: A Specific Marker for Erythroid Differentiation / *B.M. Longenecker and T.R. Mosmann*
In Vitro Maintenance of Differentiation Marker Synthesis by Subpopulations of Mouse Thymocytes / *Ellen Rothenberg and Dennis Triglia*
Effect of Heme on Globin Messenger RNA Synthesis in Spleen Erythroid Cells / *O. Fuchs, P. Ponka, J. Borova, J. Neuwirt, and M. Travnicek*
Neonatal Imprinting and Hepatic Cytochrome P-450 II. Partial Purification of a Sex-Dependent and Neonatally Imprinted Form(s) of Cytochrome P-450 / *Leland W.K. Chung, Mark Colvin, and Haiyen Chao*
The T/t-Complex in the Mouse: Mutations That Impair Differentiation / *Dorothea Bennett*
Definition of the Transcription Unit of the Natural Ovalbumin Gene / *Dennis R. Roop, Sophia Y. Tsai, Ming-Jer Tsai, and Bert W. O'Malley*
Some New Developments in Genetic Analysis of Somatic Mammalian Cells / *Theodore T. Puck*
Regulation of Expression of Genes for Specialized Proteins: Actin, Insulin, Globins: Workshop Report / *G. Stamatoyannopoulos and P.A. Marks*

TUMOR CELL BIOLOGY AND TERATOCARCINOMA

Alterations in Growth Requirements of Kidney Epithelial Cells in Defined Medium Associated With Malignant Transformation / *Mary Taub, Ben Ü, Lorraine Chuman, Michael J. Rindler, Milton H. Saier, Jr., and Gordon Sato*
Plate-Induced Tumors of BALB / 3T3 Cells Exhibiting Foci of Differentiation Into Pericytes, Chondrocytes, and Fibroblasts / *Charles W. Boone and Robert E. Scott*
H1 Histone and Nucleosome Repeat Length Alterations Associated With the In Vitro Differentiation of Murine Embryonal Carcinoma Cells to Extra-Embryonic Endoderm / *Robert Oshima, Diana Curiel, and Elwood Linney*

xii Contents of Part B

The Effects of Laminin on the Growth and Differentiation of Embryonal Carcinoma Cells in Defined Media / *Angie Rizzino, Victor Terranova, David Rohrbach, Craig Crowley, and Heather Rizzino*
Restricted Infectivity of Ecotropic Type C Retroviruses in Mouse Teratocarcinoma Cells: Studies on Viral DNA Intermediates / *Wen K. Yang, Luc d'Auriol, Den-Mei Yang, James O. Kiggans, Jr., Chin-yih Ou, Jorge Périès, and Rodica Emanoil-Ravicovitch*
The Regulation of SV40 Gene Expression in Nonpermissive Cells / *S. Segal and G. Khoury*
Murine Teratocarcinoma: A Model for Virus-Cell Interaction in a Differentiating Cell System / *Thomas D. Friedrich and John M. Lehman*

CLONAL SELECTION OF CELL POPULATIONS

Promotion of Hematopoietic Stem Cell Differentiation In Vitro by a Soluble Mediator, Allogeneic Effect Factor / *Amnon Altman, Thomas D. Gilmartin, and David H. Katz*
An In Vitro Assay for T Lymphocyte Progenitors (CFU-preT) / *Bonita R. Acuff and J. John Cohen*
The Role of Cells and Their Products in the Regulation of In Vitro Stem Cell Proliferation and Granulocyte Development / *T.M. Dexter, E. Spooncer, D. Toksoz, and L.G. Lajtha*
Immunoregulation by Thymopoietin / *Gideon Goldstein and Catherine Y. Lau*
Self-Renewal of Factor-Dependent Hemopoietic Progenitor Cell-Lines Derived From Long-Term Bone Marrow Cultures Demonstrates Significant Mouse Strain Genotypic Variation / *Joel S. Greenberger*
Control of Mouse Myoblast Commitment to Terminal Differentiation by Mitogens / *Thomas A. Linkhart, Christopher H. Clegg, and Stephen D. Hauschka*
H-2 I Alloantigens and Recall of Memory Cytotoxic Responses / *Charles G. Orosz, Stuart Macphail, and Fritz H. Bach*
Enhanced Lymphoid and Decreased Myeloid Reconstituting Ability of Stem Cells From Long-Term Cultures of Mouse Bone Marrow / *R.A. Phillips*
Characterization of Monolayer and Organ Cultures of Cloned and Enriched Lymphohematopoietic Stromal Cell Populations / *R.W. Anderson and J.G. Sharp*
Induction and Long-Term Maintenance of Thy-1 Positive T Lymphocytes: Derivation From Continuous Bone Marrow Cultures / *Gérard Tertian, Yee Pang Yung, and Malcolm A.S. Moore*
Hemopoietic Stem Cells: Workshop Report / *J. Byron, C.J. Eaves, G.R. Johnson, B. Lord, and H.A. Messner*
Lymphocyte Effector Molecules and Clonal Growth of Lymphoid Cells: Workshop Report / *Steven B. Mizel, Markus Nabholz, and Kendall A. Smith*

IMPROVED CELL CULTURE SYSTEMS

Extracellular Regulation of Fibroblast Multiplication: A Direct Kinetic Approach to Analysis of Role of Low Molecular Weight Nutrients and Serum Growth Factors / *Wallace L. McKeehan and Kerstin A. McKeehan*
Long-Term Maintenance of "Cloned" Human PLT Cells in TCGF With LCL Cells as a Feeder Layer / *Jacquelyn A. Hank, Hiroo Inouye, Lynn A. Guy, Barbara J. Alter, and Fritz H. Bach*
Culture of Hematopoietic Cells: Workshop Report / *A. Axelrad, A. Burgess, N. Iscove, and G. Wagemaker*
Defined Culture Media for Anchorage-Dependent Cells: Workshop Report / *W.L. McKeehan*

Contributors to Part A

Camille N. Abboud [9]
Department of Medicine, University of Rochester School of Medicine, Rochester, NY 14642

Jan Andersson [579]
Biomedicum, University of Uppsala, Uppsala, Sweden

V. Anselstetter [387]
Department of Inner Medicine III, University of Ulm, Federal Republic of Germany

Kurt Ballmer [219]
Department of Biochemistry, Biocenter of the University of Basel, CH-4056 Basel, Switzerland

David Barnes [167]
Department of Biological Sciences, University of Pittsburgh, Pittsburgh, PA 15260

Samuel H. Barondes [29]
Department of Psychiatry, University of California, San Diego, La Jolla, CA 92093 and Veterans Administration Hospital, San Diego, CA 92161

R. Baserga [603]
Temple University, Philadelphia, PA 19140

Teresa K. Battin [339]
Department of Dermatology, University of North Carolina, Chapel Hill, NC 27514

Vann Bennett [339]
Department of Cell Biology and Anatomy, The Johns Hopkins School of Medicine, Baltimore, MD 21205

Eric C. Beyer [29]
Department of Psychiatry, University of California, San Diego, La Jolla, CA 92093 and Veterans Administration Hospital, San Diego, CA 92161

Florence Boegel [265]
Department of Medicine, Montefiore Hospital, University of Pittsburgh School of Medicine, Pittsburgh, PA 15213

Charles W. Boone [592]
National Cancer Institute, National Institutes of Health, Bethesda, MD 20205

Ralph A. Bradshaw [193,549]
Department of Biological Chemistry, Washington University School of Medicine, St. Louis, MO 63110

James K. Brennan [9]
Department of Medicine, University of Rochester School of Medicine, Rochester, NY 14642

Robert F. Brooks [593,603]
Cell Proliferation Laboratory, Imperial Cancer Research Fund Laboratories, London WC2A 3PX, England

The boldface number in brackets following each contributor's name is the opening page number of that contributor's article.

Contributors to Part A

Max M. Burger [219]
Department of Biochemistry, Biocenter of the University of Basel, CH-4056 Basel, Switzerland

Peter G. Burk [153]
Albert Einstein College of Medicine, Montefiore Hospital and Medical Center, Bronx, NY 10467

E.C. Butcher [461]
Laboratory of Experimental Oncology, Stanford University School of Medicine, Stanford, CA 94305

C.E. Butterfield [353]
The Children's Hospital Medical Center, Harvard Medical School, Boston, MA 02114

Darrell H. Carney [119]
Department of Human Biological Chemistry and Genetics, University of Texas Medical Branch, Galveston, TX 77550

Graham Carpenter [557]
Departments of Biochemistry and Medicine (Dermatology), Vanderbilt University School of Medicine, Nashville, TN 37232

Paul V. Cherington [495]
Laboratory of Tumor Biology, Sidney Farber Cancer Institute, and Department of Pharmacology, Harvard Medical School, Boston, MA 02115

Peter R. Clements [523]
Department of Physiological Chemistry, The Johns Hopkins University School of Medicine, Baltimore, MD 21205

Stanley Cohen [557]
Department of Biochemistry, Vanderbilt University School of Medicine, Nashville, TN 37232

Stanley N. Cohen [549]
Departments of Genetics and Medicine, Stanford University School of Medicine, Stanford, CA 94305

Dennis D. Cunningham [373]
Department of Microbiology, College of Medicine, University of California, Irvine, CA 92717

Peter F. Davies [21]
Departments of Pathology, Peter Bent Brigham Hospital and Harvard Medical School, Boston, MA 02115

Joseph E. De Larco [445]
Laboratory of Viral Carcinogenesis, National Cancer Institute, National Institutes of Health, Bethesda, MD 20205

John F. DiPersio [9]
Department of Medicine, University of Rochester School of Medicine, Rochester, NY 14642

Hannah Friedman Elson [225]
Department of Biology, University of California, San Diego, La Jolla, CA 92093

A.A. Fauser [365]
Institute of Medical Science, University of Toronto, Toronto, Ontario, Canada M4X 1K9

J. Folkman [353]
The Children's Hospital Medical Center, Harvard Medical School, Boston, MA 02114

C. Fred Fox [283,303]
Molecular Biology Institute, University of California, Los Angeles, CA 90024

Charlotte Fryling [445]
Laboratory of Viral Carcinogenesis, National Cancer Institute, National Institutes of Health, Bethesda, MD 20205

Robert C. Gallo [39]
Laboratory for Tumor Cell Biology, National Cancer Institute, National Institutes of Health, Bethesda, MD 20205

Steven Gillis [513]
Fred Hutchinson Cancer Research Center, Seattle, WA 98104

Eugene Goldwasser [487]
Department of Biochemistry, University of Chicago, Chicago, IL 60637

Michael G. Goodman [87]
Department of Immunopathology, Scripps Clinic and Research Foundation, La Jolla, CA 92037

D. Gospodarowicz [53]
Cancer Research Institute and the Department of Medicine, University of California Medical Center, San Francisco, CA 94143

Victor B. Hatcher [153]
Albert Einstein College of Medicine, Montefiore Hospital and Medical Center, Bronx, NY 10467

W. Heit [387]
Department of Inner Medicine III, University of Ulm, Federal Republic of Germany

H.R. Herschman [543]
University of California, Los Angeles, CA 90024

Cecilia Hofmann [325]
Department of Biochemistry, University of Chicago, Chicago, IL 60637

N. Hollander [461]
Laboratory of Experimental Oncology, Stanford University School of Medicine, Stanford, CA 94305

Robert W. Holley [1]
Molecular Biology Laboratory, The Salk Institute for Biological Studies, San Diego, CA 92138

Joanne S. Ingwall [225]
Department of Medicine, University of California, San Diego, La Jolla, CA 92093

L. Jerabek [461]
Laboratory of Experimental Oncology, Stanford University School of Medicine, Stanford, CA 94305

Tae H. Ji [325]
Department of Biochemistry, University of Wyoming, Laramie, WY 82071

Patricia A. Johnson [445]
Laboratory of Viral Carcinogenesis, National Cancer Institute, National Institutes of Health, Bethesda, MD 20205

G.A. King [241,611]
Division of Biology, City of Hope Research Institute, Duarte, CA 91010

George L. King [411]
Section on Cellular and Molecular Physiology, Diabetes Branch, National Institute of

Arthritis, Metabolism and Digestive Diseases, National Institutes of Health, Bethesda, MD 20205

Lloyd King, Jr. [557]
Department of Medicine (Dermatology), Vanderbilt University School of Medicine, and Veterans Administration Hospital, Nashville, TN 37232

Jan Klein [573]
Abteilung Immungenetik, Max-Planck-Institut für Biologie, 7400 Tübingen, Federal Republic of Germany

R.R. Klevecz [241,611]
Division of Biology, City of Hope Research Institute, Duarte, CA 91010

Daniel J. Knauer [395]
School of Life Sciences, University of Nebraska, Lincoln, NE 68583

Alfred B. Knight [411]
Section on Biochemistry of Cell Regulation, Laboratory of Biochemical Pharmacology, National Institute of Arthritis, Metabolism and Digestive Diseases, National Institutes of Health, Bethesda, MD 20205

B. Kubanek [387]
Department of Transfusion Medicine, University of Ulm, Federal Republic of Germany

Lawrence B. Lachman [109]
Department of Microbiology and Immunology, Duke University Medical Center, Durham, NC 27710

M. Daniel Lane [523]
Department of Physiological Chemistry, The Johns Hopkins University School of Medicine, Baltimore, MD 21205

Waldemar Lernhardt [579]
Basel Institute for Immunology, CH-4058 Basel, Switzerland

Marshall A. Lichtman [9]
Department of Radiation Biology and Biophysics, University of Rochester School of Medicine, Rochester, NY 14642

Peter S. Linsley [283,303]
Molecular Biology Institute, University of California, Los Angeles, CA 90024

Don McClure [167]
Department of Biology, University of California, San Diego, La Jolla, CA 92093

M.S. McGrath [461]
Laboratory of Experimental Oncology, Stanford University School of Medicine, Stanford, CA 94305

Thomas H. Marshall [255]
Laboratory of Viral Carcinogenesis, National Cancer Institute, National Institutes of Health, Bethesda, MD 20205

R.G. Martin [141]
National Institutes of Health, Bethesda, MD 20205

Estela E. Medrano [495]
Laboratory of Tumor Biology, Sidney Farber Cancer Institute, and Department of Pharmacology, Harvard Medical School, Boston, MA 02115

Fritz Melchers [579]
Basel Institute for Immunology, CH-4058 Basel, Switzerland

H.A. Messner [365]
Ontario Cancer Institute, Department of Medicine, and Institute of Medical Science, University of Toronto, Toronto, Ontario, Canada M4X 1K9

Donald Metcalf [473]
Cancer Research Unit, Walter and Eliza Hall Institute of Medical Research, Royal Melbourne Hospital, Victoria 3050, Australia

Richard S. Metzgar [109]
Department of Microbiology and Immunology, Duke University Medical Center, Durham, NC 27710

James W. Mier [39]
Laboratory for Tumor Cell Biology, National Cancer Institute, National Institutes of Health, Bethesda, MD 20205

Bonnie Miller [325]
Department of Biochemistry, University of Chicago, Chicago, IL 60637

Diane Mochizuki [513]
Department of Microbiology, University of California, Irvine, CA 92717

Edward L. Morgan [131,211]
Department of Immunopathology, Scripps Clinic and Research Foundation, La Jolla, CA 92037

Alan C. Moses [411]
Endocrine Section, Metabolism Branch, National Cancer Institute, National Institutes of Health, Bethesda, MD 20205

Martin Moss [373]
Department of Microbiology, College of Medicine, University of California, Irvine, CA 92717

S. Peter Nissley [411,507]
Endocrine Section, Metabolism Branch, National Cancer Institute, National Institutes of Health, Bethesda, MD 20205

Edward J. O'Keefe [339]
Department of Dermatology, University of North Carolina, Chapel Hill, NC 27514

Ellen E. Van Obberghen-Schilling [411]
Section on Biochemistry of Cell Regulation, Laboratory of Biochemical Pharmacology, National Institute of Arthritis, Metabolism and Digestive Diseases, National Institutes of Health, Bethesda, MD 20205

Martin S. Oberman [153]
Albert Einstein College of Medicine, Montefiore Hospital and Medical Center, Bronx, NY 10467

Frances L. Owen [185]
Department of Pathology and Cancer Research Center, Tufts University School of Medicine, Boston, MA 02111

Stella O. Page [109]
Department of Microbiology and Immunology, Duke University Medical Center, Durham, NC 27710

A.B. Pardee [495,604]
Laboratory of Tumor Biology, Sidney Farber Cancer Institute, and Department of Pharmacology, Harvard Medical School, Boston, MA 02115

C.W. Parker [511]
Department of Microbiology, University of California, Irvine, CA 92717

Contributors to Part A

J.F. Perdue [543,569]
Lady Davis Institute for Medical Research, Montreal, Quebec, Canada H3T 1E2

Loren Pickart [99]
Virginia Mason Research Center, Seattle, WA 98101

Giuseppe Pigoli [265]
Department of Medicine, Montefiore Hospital, University of Pittsburgh School of Medicine, Pittsburgh, PA 15213

E. Pillemer [461]
Laboratory of Experimental Oncology, Stanford University School of Medicine, Stanford, CA 94305

W.J. Pledger [141,507]
University of North Carolina, Chapel Hill, NC 27514

U.R. Rapp [255]
Laboratory of Viral Carcinogenesis, National Cancer Institute, National Institutes of Health, Bethesda, MD 20205

Matthew M. Rechler [411]
Section on Biochemistry of Cell Regulation, Laboratory of Biochemical Pharmacology, National Institute of Arthritis, Metabolism and Digestive Diseases, National Institutes of Health, Bethesda, MD 20205

Brent C. Reed [523]
Department of Physiological Chemistry, The Johns Hopkins University School of Medicine, Baltimore, MD 21205

I.N. Rich [387]
Department of Transfusion Medicine, University of Ulm, Federal Republic of Germany

D. Rifkin [569]
New York University Medical School, New York, NY 10016

Joyce A. Romanus [411]
Section on Biochemistry of Cell Regulation, Laboratory of Biochemical Pharmacology, National Institute of Arthritis, Metabolism and Digestive Diseases, National Institutes of Health, Bethesda, MD 20205

Russell Ross [503]
University of Washington School of Medicine, Seattle, WA 98195

Thomas F. Roth [315,549]
Department of Biological Sciences, University of Maryland, Baltimore County, Catonsville, MD 21228

R.V. Rouse [461]
Laboratory of Experimental Oncology, Stanford University School of Medicine, Stanford, CA 94305

Jeffrey S. Rubin [193]
Department of Biological Chemistry, Washington University School of Medicine, St. Louis, MO 63110

Francis W. Ruscetti [39]
Laboratory for Tumor Cell Biology, National Cancer Institute, National Institutes of Health, Bethesda, MD 20205

Gordon Sato [167]
Department of Biology, University of California, San Diego, La Jolla, CA 92093

N. Savion [53]
Cancer Research Institute and the Department of Medicine, University of California Medical Center, San Francisco, CA 94143

C.D. Scher [141]
Harvard Medical School, Boston, MA 02115

Max H. Schreier [579]
Basel Institute for Immunology, CH-4058 Basel, Switzerland

R.G. Scollay [461]
Laboratory of Experimental Oncology, Stanford University School of Medicine, Stanford, CA 94305

Ginette Serrero [167]
Department of Biology, University of California, San Diego, La Jolla, CA 92093

Richard K. Shadduck [265]
Department of Medicine, Montefiore Hospital, University of Pittsburgh School of Medicine, Pittsburgh, PA 15213

Patricia A. Short [411]
Endocrine Section, Metabolism Branch, National Cancer Institute, National Institutes of Health, Bethesda, MD 20205

R.M. Shymko [241,611]
Division of Radiation Oncology, City of Hope Medical Center, Duarte, CA 91010

Gary L. Smith [395]
School of Life Sciences, University of Nebraska, Lincoln, NE 68583

Michael B. Sporn [445]
Laboratory of Chemoprevention, National Cancer Institute, National Institutes of Health, Bethesda, MD 20205

Donald F. Steiner [326]
Department of Biochemistry, University of Chicago, Chicago, IL 60637

S.K. Stevens [461]
Laboratory of Experimental Oncology, Stanford University School of Medicine, Stanford, CA 94305

C.D. Stiles [141]
Harvard Medical School, Boston, MA 02115

Marilyn L. Thoman [131]
Department of Immunopathology, Scripps Clinic and Research Foundation, La Jolla, CA 92037

George J. Todaro [445]
Laboratory of Viral Carcinogenesis, National Cancer Institute, National Institutes of Health, Bethesda, MD 20205

Grace Tsien [153]
Albert Einstein College of Medicine, Montefiore Hospital and Medical Center, Bronx, NY 10467

R.W. Tucker [141,353,589]
The Johns Hopkins Oncology Center, Baltimore, MD 21205

I. Vlodavsky [53]
Cancer Research Institute and the Department of Medicine, University of California Medical Center, San Francisco, CA 94143

Fred W. Wagner [395]
School of Life Sciences, and Laboratory of Agricultural Biochemistry, University of Nebraska, Lincoln, NE 68583

Abdul Waheed [265]
Department of Medicine, Montefiore Hospital, University of Pittsburgh School of Medicine, Pittsburgh, PA 15213

Contributors to Part A

James Watson [511,513]
Department of Microbiology, University of California, Irvine, CA 92717

William O. Weigle [87,131,211]
Department of Immunopathology, Scripps Clinic and Research Foundation, La Jolla, CA 92037

I.L. Weissman [461]
Laboratory of Experimental Oncology, Stanford University School of Medicine, Stanford, CA 94305

Dorothee Wernet [573]
Abteilung Immungenetik, Max-Planck-Institute für Biologie, 7400 Tübingen, Federal Republic of Germany

Robert M. White [411]
Endocrine Section, Metabolism Branch, National Cancer Institute, National Institutes of Health, Bethesda, MD 20205

Richard Wolfe [167]
Roche Institute of Molecular Biology, Nutley, NJ 07110

John W. Woods [315]
Department of Biological Sciences, University of Maryland Baltimore County, Catonsville, MD 21228

E. Zanjani [387]
Department of Medicine, Veterans Administration Medical Center, University of Minnesota, Minneapolis, MN 55417

Preface

This volume is part of the proceedings of a symposium which examined molecular, cellular, and tissue level mechanisms in the control of cell division and development. The symposium provided an opportunity for synergystic discussions among scientists who meet rarely as a single group. The four research areas represented were: 1) triggering of clonal expansion and development in lymphoid cells, 2) control of division and development in myeloid cells, 3) control of proliferation and development in anchorage-dependent cells, and 4) molecular control of developmental processes that lead to cells with specialized biosynthetic functions. These areas have been created somewhat arbitrarily by scientists themselves as they have communicated their findings in separate meetings and scientific journals. The symposium represented an attempt to undo these artificial boundaries.

The proceedings have been divided into Parts A and B along topical lines. Part A deals with regulation of cell division and development at the levels of the cell and its environment, emphasizing hormones and/or substratum interactions that initiate intracellular processes. Cell surface receptors and direct biochemical transactions involving them, for example, receptor internalization, proteolysis or phosphorylation, also are included in Part A, as are general aspects of ligand-mediated cell cycle regulation. The focus of Part B is directed more towards intracellular events which follow directly the interactions of cells with extracellular ligands or a substratum. Also considered are general organization and expression as they relate to triggering of cell proliferation or maturation processes. Part B concludes with contributions that address cell proliferation control mechanisms at the level of clonal selection of both normal and abnormal cell populations.

Each of the areas treated in the symposium is characterized by novel advances made with highly refined model systems. For example, studies on the mechanisms of triggering of clonal expansion and development in lymphoid cells have the advantage of well established genetic systems with well characterized mitogenic and developmental responses. A distinct feature of studies on myeloid cells is the extent to which information on their precursor stem cells and their stages of development has facilitated progress. The factors which control proliferation of anchorage-dependent cells is another highly active field. In particular, studies with purified peptide growth factors and homogeneous cell populations are providing insights into the properties of receptors for key regulatory molecules. The control of developmental processes at the molecular level has been most effectively studied in systems where control of synthesis of a particularly abundant molecule, such as hemoglobin or albumin, has been examined.

The organizing committee made a special effort to create interactions among investigators from these four areas to stimulate the development of new ideas and approaches. This interdisciplinary tone had particular emphasis in the plenary sessions where speakers were encouraged to present conceptual aspects of their studies. The first section of the proceedings contains three lectures delivered during the first plenary session. Each of these treated one of the three major cell systems represented in the symposium and emphasized the particular experimental advantages which characterize each cell system, as described in the preceding paragraph. The other sections contain articles contributed by both plenary session speakers and poster session participants, as well as summaries contributed by the workshop conveners. The symposium workshops and poster sessions provided an opportunity for highly focussed discussion in a setting where smaller groups of investigators exchanged ideas on topics of more limited scope. We are grateful to the workshop participants who overcame the temptation to show their "one miserable slide" and promoted stimulating conceptual discussions instead.

A number of articles in this volume were submitted for consideration by the Journal of Supramolecular Structure (now Journal of Supramolecular Structure and Cellular Biochemistry) and are reprinted here. We gratefully acknowledge the efforts of the editorial board members of the Journal and the referees who reviewed these articles. The assembly of this volume was administered by Betty Handy.

The formulation of a program with the topical breadth represented in the symposium was derived with the extensive participation of a small group of investigators who met for four consecutive days of data presentation and open discussion. These discussions occurred during two consecutive workshops, first at the Molecular Biology Institute of the University of California, Los Angeles, and then at the Batelle Memorial Institute conference center in Seattle, Washington. Most of the participants presented updates of research in their areas at an ICN-UCLA mini-symposium held in January 1979 in Los Angeles. The conference program was organized during the Seattle working sessions which commenced the day following the mini-symposium. Robert Holley, Lee Hood, Charles Scher, Gordon Sato, Russell Ross, George Todaro, John Lehman, and Richard Shadduck participated with us in one or both of these essential preliminary working sessions. The arrangements for the Seattle session were made by Russell Ross and the Center for Advanced Studies, University of Washington. Lodging and subsistence expenses in Seattle were borne by the ICN-UCLA Symposia and the University of Washington Center for Advanced Studies. The public mini-symposium held in Los Angeles was cosponsored by the Tumor Cell Biology and Molecular and Cellular Biology training programs of the University of California, Los Angeles.

We thank the National Cancer Institute, National Institute of Arthritis, Metabolism and Digestive Diseases, National Institute on Aging, National Institute of Child Health and Human Development, Fogarty International Center, March of Dimes Birth Defects Foundation, Burroughs Wellcome Co., Merck Sharp & Dohme, and the series sponsor, ICN Pharmaceuticals, for their financial contributions to the program.

Dennis Cunningham
Eugene Goldwasser
James Watson
C. Fred Fox

Control of Animal Cell Proliferation

Robert W. Holley

Molecular Biology Laboratory, The Salk Institute for Biological Studies, San Diego, California 92138

Present understanding of the control of animal cell proliferation is summarized briefly. Major gaps in present knowledge are listed. Models of growth control are discussed.

Key words: growth factors, growth inhibitors, models of growth control

Those interested in the control of growth of animal cells are concerned with two different but related questions. First, what are the factors outside the cell that control cell growth? Second, what happens inside the cell when the growth-controlling factors act? In this paper I plan to discuss both of these questions, placing the emphasis on problems that remain to be solved.

WHAT ARE THE FACTORS OUTSIDE THE CELL THAT CONTROL CELL GROWTH?

Many factors are known that can control the growth of animal cells in culture. The list of factors is striking both for its length and for its variety [1]. To simplify, the factors can be grouped into four general classes.

(1) Growth factors. There are many different growth-stimulating factors. Among them are a number of polypeptide growth factors, such as epidermal growth factor (EGF) and fibroblast growth factor (FGF), that are active at ng/ml concentrations [2, 3]. There are other, very different types of growth-stimulating factors, for example, prostaglandin $F_2\alpha$ stimulates the growth of some cells and is active at approximately 100 ng/ml [4].

(2) Nutrients. Various common nutrients stimulate or inhibit growth when their concentrations in the growth medium are raised or lowered. Included are amino acids [5], glucose [6], cations [7], and anions [8].

(3) Growth inhibitors. Many growth inhibitors have also been observed. These will be discussed later.

(4) Cell shape and surface area. The shape and surface area of cells often affect growth [9–11]. In general, growth is favored by increasing the amount of cell surface area exposed to the medium.

Received April 2, 1980; accepted May 8, 1980.

0091-7419/80/1302-0191$01.70 © 1980 Alan R. Liss, Inc.

As this summary suggests, enough is known at present to permit the culture of many cells in completely defined medium [12, 13]. The medium contains an appropriate set of growth factors, plus adequate concentrations of all of the necessary nutrients. Growth can be limited by growth inhibitors or by restriction of the surface area of the cells.

With this as background, let us consider the major gaps that remain in our knowledge of growth-controlling factors outside the cell. In my view they are the following.

Additional Polypeptide Growth Factors Remain to Be Identified

There is much evidence that additional polypeptide growth factors exist. Growth-promoting activity for many cells is found in biological fluids [1, 14] and in media conditioned by the growth of other cells [1, 15]. In many instances, the activity cannot be replaced by known growth factors. It seems likely that some of the unidentified growth factors have very important actions in vivo. The investigator is faced with the practical problem of deciding which growth-promoting activity to isolate, and also must choose a good source of the factor, as well as a good assay.

The Growth Factors That Act in Different Situations In Vivo Are Not Known

At present almost nothing is known about the identities of the growth factors that are required for the growth of the various tissues and organs in vivo. This is a very important problem. Until recently this problem has been difficult experimentally, but it seems to me that an approach is now available. The approach involves the use of monoclonal antibody prepared against a growth factor and injected into the animal to inactivate the growth factor in vivo. The approach is similar to that used by Levi-Montalcini and Booker [16] in their demonstration, in which they used rabbit antiserum to nerve growth factor (NGF), of an NGF requirement during the development of the sympathetic nervous system. The advantage of using monoclonal antibody instead of normal antibody is that a much higher concentration of antibody can be achieved. Arrest of growth of certain types of cells, as the result of inactivating a growth factor with monoclonal antibody, may limit growth or may cause developmental changes in a rapidly growing animal. By studying a variety of growth processes in the presence of monoclonal antibody, it should be possible to elucidate the action of a growth factor in different situations.

Many Growth Factor Interactions Remain to Be Studied

In cell culture, synergisms are often observed in the action of pairs of different growth factors [17, 18]. Some of these interactions have been studied, but many other possible interactions have not been investigated. One question of interest is whether specificity for control of growth of different cell types can result from highly specific interactions among low concentrations of different growth factors.

Interactions Between Different Growth Factors In Vivo Are Unknown

Based on observations in cell culture, it seems possible that specificity of growth factor action in vivo is achieved by the combined action of a specific set of growth factors, with a different set acting on each cell type. In principle, such interactions can be identified by inactivating combinations of growth factors in vivo with combinations of monoclonal antibodies. This will require a collection of different monoclonal antibodies.

Many Growth Inhibitors Remain to Be Purified

For technical reasons, growth inhibitors are less well studied than growth-stimulating factors. Nevertheless, there is evidence that suggests that growth inhibitors may be as numerous as growth factors, and it seems likely that growth inhibitors are important in vivo.

Recently we have purified a growth inhibitor that is produced by the BSC-1 cell line, a kidney epithelial cell line of African green monkey origin. Conditioned medium removed from crowded BSC-1 cells contains both low and high molecular weight growth inhibitors [19]. The low and high molecular weight inhibitors can be separated by ultrafiltration. The high molecular weight inhibitor has been concentrated 1,000-fold from serum-free conditioned medium, and has been purified by gel filtration followed by high-pressure liquid chromatography. This kidney epithelial cell growth inhibitor has the properties of a protein with a molecular weight of approximately 15,000. It arrests the growth of BSC-1 cells in the G_1 phase of the cell cycle. Present preparations give approximately a 50% inhibition of growth of 1 ng/ml. The action of the kidney epithelial cell growth inhibitor is reversible; that is, the cells resume growth after the inhibitor is removed. To the extent that it has been tested, it is highly specific for kidney epithelial cells. It has no interferon activity.

Preliminary evidence from this laboratory suggests that other epithelial cells also produce growth inhibitors.

The best studied growth inhibitors at present are ACTH [20] and interferon [21, 22]. They are growth inhibitory at approximately 1 ng/ml. A variety of other growth inhibitors of lower specific activity have been reported [23–28].

Thus far there is no information on the normal in vivo action of growth inhibitors. In principle, once inhibitors are available for the preparation of monoclonal antibodies, then these can be used to inactivate the endogenous growth inhibitors in vivo, and it will be possible to study the effect of the inhibitors on growth. It is possible that the growth inhibitors act as differentiation factors in vivo.

The Role of Cell Shape and Surface Area Require More Study

Much evidence indicates that increasing the surface area of cells favors growth. Decreasing the surface area inhibits growth [9–11]. Density-dependent regulation of growth is an example of this phenomenon. Cells become anchorage independent when they are able to grow with a minimum of surface area. Present evidence indicates that density-dependent regulation of growth has different causes in different situations [29], suggesting that the mechanisms by which cell shape and surface area act are complex. It is quite possible that different cells become anchorage independent for different reasons, depending on the factors that favor growth of the particular cell. Studies are needed of these phenomena with a number of different cells in a variety of growth situations.

WHAT HAPPENS INSIDE THE CELL WHEN THE GROWTH-CONTROLLING FACTORS ACT?

Turning to the question of how the growth factors act, it is clear that the polypeptide growth factors bind to specific cell surface receptors. Among the growth factors, EGF has been studied most extensively [2]. EGF binds to its specific receptor, the receptor–EGF complex is internalized, and the EGF is degraded [2]. With EGF, the process of binding, internalization, and degradation continues, repeatedly, for many hours before the stimulated cells are committed to initiate DNA synthesis.

What Happens Inside the Cell?

It is clear that EGF and other growth factors have effects as soon as they bind to the cell surface receptors. Effects on transport processes [30], ion fluxes [31, 32], membrane composition (phosphorylation [33], phospholipase A_2 activation [34]), and cyclic nucleotide concentrations [35] are observed within minutes after binding of the factor. Internalization of the growth factor does not appear to be required for these early effects. However, there is no proof that these early events lead to the initiation of DNA synthesis. Is internalization of the growth factor required for the initiation of DNA synthesis? Various types of evidence (immobilization of the growth factor [36], inhibition of internalization [37]) suggest that little, if any, internalization is required. Nevertheless, there seems to be no way to exclude the possibility that the growth factor (or a fragment of it or of the receptor) also acts internally. Even one molecule arriving at the nucleus could be sufficient, and there is no way to exclude this possibility.

How Do External Nutrient Concentrations Act To Influence Growth Control?

It seems likely that the external concentrations influence internal concentrations of the nutrients. Whether these then act directly or indirectly, by influencing the concentration of other effectors, is unknown.

How Do Growth Inhibitors Act?

High molecular weight growth inhibitors such as the kidney epithelial cell growth inhibitor presumably act by binding to specific cell surface receptors. With ACTH, interaction with the cells has been studied extensively [20], and it seems likely that ACTH inhibits DNA synthesis in adrenal cells in culture by increasing the intracellular concentration of cAMP. The intracellular effect of the kidney epithelial cell growth inhibitor is unknown. The inhibitory action of the kidney epithelial cell growth inhibitor is counteracted by EGF, and vice versa. It appears that this takes place intracellularly, since there is no indication that the inhibitor affects EGF binding. The interaction of EGF and the inhibitor is an illustration of the complexity of growth control; growth appears to be under the simultaneous control of many different external factors.

Control of growth by cell shape and surface area is another example of the complexity of interacting growth-controlling factors. Cell surface area probably influences a variety of cell membrane and transport effects.

MODELS OF GROWTH CONTROL

The observation of simultaneous control of growth by numerous external growth-controlling factors is more or less puzzling depending on one's model of the events inside the cell that lead to DNA synthesis. There are two very different simple models of these internal events. At one extreme, the various external growth-controlling factors are considered to act on the cell and influence, directly, a series of processes that lead to the initiation of DNA synthesis. Alternatively, at the other extreme, the various external factors are considered to act on normal cellular processes, such as energy production, protein synthesis, and RNA synthesis, among others, and it is the state of these cellular processes that controls the initiation of DNA synthesis. Present data can be explained with either model. Nevertheless, one's choice of model has a great influence on one's experimental approach to studies of internal events.

According to the first model, quiescent cells might be expected to be blocked at a specific place, a "restriction point" [38], corresponding to the first of the series of biochemical reactions, stimulated by the growth factor, that leads to the initiation of DNA synthesis. If one accepts this, studies of the early events after growth can be expected to lead to the identity of the "restriction point." According to the second model, the early events after growth stimulation affect general cellular processes, and the "restriction point" is, in a sense, an inactive metabolic state. The study of early events after growth stimulation will lead only to general cellular processes.

Though there seems to be no way to distinguish between the two models on the basis of present data, there is the possibility that the initiator of DNA synthesis can be identified directly and then pathways can be traced backward from it. Grummt has reported [39] that addition of Ap_4A to permeabilized cells leads to the initiation of DNA synthesis. Das [40] has reported the isolation of a protein fraction from stimulated cells that in turn stimulates the initiation of DNA synthesis in isolated Xenopus nuclei. Unfortunately, we do not know at this time whether either of these experiments represents the natural course of events. Nevertheless, the experiments do suggest that it may be possible to elucidate the series of internal events after growth stimulation by identifying the initiator of DNA synthesis and working backward from this.

Whatever model one favors, the series of internal events must be complicated, since there is typically a 12–15-hour period between the beginning of growth stimulation and the initiation of DNA synthesis. One's explanation of this long time period is influenced in turn by one's choice of model of the cell cycle.

In the classical model of the cell cycle, quiescent cells are considered to be in a special, G_0 state, which is in some way outside the normal cell cycle. The general explanation for the long delay before initiation of DNA synthesis in quiescent cells is that it takes a number of hours for a G_0 cell to return to the normal cell cycle.

In the Smith-Martin model [41] for the cell cycle, quiescent cells are primarily in an indeterminate, A, state, from which they leave with a very low transition probability. Stimulating quiescent cells with growth factors increases the transition probability, and it is the delay in changing the transition probability that causes the delay in changing the rate of initiation of DNA synthesis.

Both models of the cell cycle are in widespread use, and each has its appeals. The Smith-Martin model does have the advantage that it is consistent with the general observation that quiescent cell cultures usually have a significant labeling index and the quiescent cells are not strictly in a G_0 state. Also the Smith-Martin model might predict that there would be an additional 12–15-hour delay in increasing the transition probability when a culture that has had suboptimal growth stimulation is subjected to maximum growth stimulation.

An experiment of this type has been reported by Brooks [42]. He found that there is an additional 12-hour delay before the effect of the second increase in growth stimulation is observed. This experiment indicates that it takes approximately the same time to increase the rate of initiation of DNA synthesis whether the cell culture is already growing or not. This is consistent with general observations, but it is not what might be anticipated from simple theories of the cell cycle.

There are, however, experimental conditions that give different results. Jimenez de Asua et al report [18] that there is only a short delay after the second growth stimulation under some conditions with prostaglandin $F_2\alpha$ and with FGF. There may be experimental differences that lead to the differing results, or it may be that the pathways that lead to stimulation of the initiation of DNA synthesis in different situations are different. Whichever explanation one prefers, the observations suggest that the experimental situation is complex.

In summary, consideration of the various models, and the numerous unsolved problems, leads to an awareness that growth-control mechanisms are probably very complicated. Nevertheless, there is a strong drive to devise simple models. The dilemma is that the models may be misleading. Probably each of us is willing to tolerate being misled a little because we hope that our preferred model will be of more help than hindrance. Unfortunately, there is no way to know in advance which model will be a help and which will be a hindrance.

ACKNOWLEDGMENTS

This work was supported in part by grant CA11176, awarded by the National Cancer Institute, DHEW, and grant VC-304-F from the American Cancer Society. R. W. Holley is an American Cancer Society Professor of Molecular Biology.

REFERENCES

1. Gospodarowicz D, Moran JS: Annu Rev Biochem 45:531, 1976.
2. Carpenter G, Lembach KJ, Morrison MM, Cohen S: J Biol Chem 250:4297, 1975.
3. Gospodarowicz D: J Biol Chem 250:2515, 1975.
4. Jimenez de Asua L, Clingan D, Rudland PS: Proc Natl Acad Sci USA 72:2724, 1975.
5. Ley KD, Tobey RA: J Cell Biol 47:453, 1970.
6. Holley RW, Armour R, Baldwin JH: Proc Natl Acad Sci USA 75:339, 1978.
7. Hori C, Oka T: Proc Natl Acad Sci USA 76:2823, 1979.
8. Holley RW, Kiernan JA: Proc Natl Acad Sci USA 71:2942, 1974.
9. Castor LN: J Cell Physiol 72:161, 1968.
10. Folkman J, Moscona A: Nature 273:345, 1978.
11. Zetterberg A, Auer G: Exp Cell Res 67:260, 1971.
12. Bottenstein JA, Sato GH: Proc Natl Acad Sci USA 76:514, 1979.
13. Mather JP, Sato GH: Exp Cell Res 120:191, 1979.
14. Klagsbrun M: J Cell Biol 84:808, 1980.
15. Rheinwald JG, Green H: Cell 6:331, 1975.
16. Levi-Montalcini R, Booker B: Proc Natl Acad Sci USA 46:384, 1960.
17. Holley RW, Kiernan JA: Proc Natl Acad Sci USA 71:2908, 1974.
18. Jimenez de Asua L, Richmond KMW, Otto AM, Kubler AM, O'Farrell MK, Rudland PS: In Sato GH, Ross R (eds): "Hormones and Cell Culture." Cold Spring Harbor Conferences on Cell Proliferation. Cold Spring Harbor, New York: Cold Spring Harbor Laboratory, 1979, p 403.
19. Holley RW: In Shields R, Levi-Montalcini R, Iacobelli S, Jimenez de Asua L (eds): "Control Mechanisms in Cultured Animal Cells." New York: Raven Press (in press).
20. Weidman ER, Gill GN: J Cell Physiol 90:91, 1976.
21. Balkwill F, Taylor-Papadimitriou J: Nature 274:798, 1978.
22. Watanabe Y, Sokawa Y: J Gen Biol 41:411, 1978.
23. Kinders RJ, Johnson TC: J Cell Biol 83:68a, 1979.
24. Bertsch S, Marks F: Cancer Res 39:239, 1979.
25. Gaugas JM, Dewey DL: Br J Cancer 39:548, 1979.
26. Sekas G, Owen WG, Cook RT: Exp Cell Res 122:47, 1979.
27. Rytomaa T, Kiviniemi K: Eur J Cancer 4:595, 1968.

28. Richter H: Hoppe Seyler's Z Physiol Chem 359:1435, 1978.
29. Holley RW, Armour R, Baldwin JH: Proc Natl Acad Sci USA 75:1864, 1978.
30. Cunningham DD, Pardee AB: Proc Natl Acad Sci USA 64:1049, 1969.
31. Rozengurt E, Heppel LA: Proc Natl Acad Sci USA 72:4492, 1975.
32. Smith JB and Rozengurt E: Proc Natl Acad Sci USA 75:5560, 1978.
33. Carpenter G, King Jr L, Cohen S: Nature 276:409, 1978.
34. Shier WT: Proc Natl Acad Sci USA 77:137, 1980.
35. Seifert WE, Rudland PS: Nature 248:138, 1974.
36. Carney DH, Cunningham DD: Cell 15:1341, 1978.
37. Maxfield FR, Davies PJA, Klempner L, Willingham MC, Pastan I: Proc Natl Acad Sci USA 76:5331, 1979.
38. Pardee AB: Proc Natl Acad Sci USA 71:1286, 1974.
39. Grummt F: Proc Natl Acad Sci USA 75:371, 1978.
40. Das M: Proc Natl Acad Sci USA 77:112, 1980.
41. Smith JA, Martin L: Proc Natl Acad Sci USA 70:1263, 1973.
42. Brooks RF: Nature 260:248, 1976.

Erythropoietic Enhancing Activity (EEA) Secreted by the Human Cell Line, GCT

Camille N. Abboud, John F. DiPersio, James K. Brennan, and Marshall A. Lichtman

Departments of Medicine and Radiation Biology and Biophysics, University of Rochester School of Medicine, Rochester, New York 14642

Medium conditioned by the monocyte-like cell line GCT contains colony-stimulating activity (CSA), a mediator of in vitro granulopoiesis. Also, the conditioned medium (CM) contains erythroid-enhancing activity (EEA), which can be demonstrated in a system utilizing either nonadherent marrow or blood mononuclear cells, erythropoietin (1–2 units/ml), and 20 ml/dl fetal calf serum. Under these conditions, GCT CM enhances the growth of CFU-E and BFU-E. Attempts were made to characterize the molecular features of EEA. Serum-free GCT cell CM was fractionated on Sephacryl S200 and Ultrogel AcA54. EEA and CSA cochromatographed with apparent molecular weights of $\sim$ 40,000 daltons on Sephacryl and $\sim$ 30,000 daltons on Ultrogel. Fractionation on DEAE Sephacel led to an apparent separation of CSA from EEA; however, when diluted, the fractions containing CSA had EEA. Undiluted fractions containing potent CSA inhibited erythropoiesis; however, dilution of these fractions resulted in marked EEA. Diluted crude GCT CM and DEAE Sephacel fractions enriched in EEA were also capable of sustaining BFU-E in liquid culture and mediating erythropoietin-independent colony growth. CSA could not be unequivocally separated from EEA on concanavalin A-Sepharose, since the diluted void volume containing CSA also had EEA. EEA was present in CM boiled for 60 minutes, whereas CSA was markedly reduced but not abolished. The inverse relationship between CSA concentration and EEA mandates dilution of fractions when bioassayed for these two activities. Although CSA and EEA are similar in molecular weight, they appear to be partially separable by ion-exchange chromatography and heat stability.

Key words: erythropoiesis, granulopoiesis, colony stimulating factors, hematopoiesis, erythroid cell growth

Recently, a two-regulator model of erythropoiesis was described [1]. This model proposes that early stages of erythropoiesis are controlled by a factor other than erythropoietin, whereas the latter is required for the terminal maturation of erythroid precursors. In vitro studies of early erythropoiesis have shown that erythroid burst-forming units (BFU-E) require a mediator termed "erythroid enhancing activity" (EEA) [2], "burst-promoting activity" (BPA) [1], or "burst-feeder activity" [3]. This activity has been

Received March 7, 1980; accepted May 23, 1980.

0091-7419/80/1302-0199$02.30 © 1980 Alan R. Liss, Inc.

described for both mouse and human BFU-E, and it is present in serum [1, 4] and in medium conditioned by peripheral blood leukocytes [2], non-irradiated [5] or irradiated [3] bone marrow cells, mitogen-stimulated mononuclear cells [1, 6, 7], normal or transformed T lymphocytes [8, 9], and monocytes and macrophages [10]. All of these sources of EEA also contain colony-stimulating activity (CSA), which is necessary for the growth of granulocyte-monocyte progenitors in viscous culture [2, 11, 12].

The human monocyte-like cell line GCT elaborates potent CSA for man and other species [13, 14]. We have found that EEA is also present in medium conditioned by GCT cells, and in this report we describe our efforts at fractionation of this activity. The importance of diluting elution fractions in measuring the biological activity of EEA is also emphasized.

METHODS

Cell Preparation

Bone marrow cells were obtained by 0.5–1 ml aspirates from normal human volunteers. Red cells were sedimented in 4.5 g/dl dextran, and residual nucleated cells were exposed to plastic surfaces in a two-step manner as described by Messner [15]. The following modifications were made: cells were diluted to a 50 ml volume in McCoy's 5A with 10% (v/v) fetal calf serum and placed in 150 cm^2 Falcon plastic flasks for adherence. This adherence procedure resulted in no cluster or colony formation when 1×10^5 nonadherent cells were plated in the absence of CSA, even with 30% fetal calf serum.

Blood mononuclear cells were obtained from plateletpheresis residues and prepared as previously described [16]. Monocytes were depleted by iron filing adherence prior to Isopaque Ficoll density centrifugation and overnight adherence in 150 cm^2 Falcon plastic flasks. Blood cells were cultured at 2.5×10^5 ml.

Viscous Culture Technique

An 0.8 g/dl methylcellulose matrix was used for cell culture, as described previously [16], but with the following modifications: both granulocytic and erythroid progenitors were cultured with the same lot of fetal calf serum (20 ml/dl) and Iscove's modified Dulbecco's MEM without mercaptoethanol was used instead of Alpha MEM with mercaptoethanol. Bovine serum albumin (1.0 g/dl) was included in the cultures, and all incubations were at 5% CO_2, 37°C, 100% humidity. Bone marrow assays were done at 1×10^5 nonadherent cells per ml and blood cells at 2.5×10^5 per ml.

Erythropoietin, 2 units/ml (step III of Connaught), was used in all EEA assays unless otherwise specified. All CFU-C assays were done by adding 10% (v/v) of dialyzed, unconcentrated GCT cell-conditioned medium (CM) prepared as previously described [13]. Fractionation studies were made using serum-free GCT CM. The latter was prepared from GCT cells grown in McCoy's 5A medium containing 0.01 g/dl polyethylene glycol 6000 instead of calf serum as described previously [14].

CM fractions from DEAE and Ultrogel columns were usually diluted 1:3 in McCoy's 5A containing 5 ml/dl fetal calf serum prior to EEA and CSA assays.

CFU-E colonies were counted at day 7; BFU-E colonies were counted at day 14. Both were identified by the red color imparted by hemoglobinization. CFU-C colonies were scored at days 7 and 14, using 20 cells as minimum colony size.

Suspension Cultures

Nonadherent marrow cells were cultured in Falcon tissue culture tubes at 1×10^6/ml in McCoy's 5A containing 20 ml/dl fetal calf serum. Test fractions were added at 10 ml/dl to duplicate tubes, and incubation was carried out at 5% CO_2, 37°C for a period of 5–6 days. Prior to assay, cells were pelleted, supernates were discarded, and pellets were resuspended to the original volume of 1 ml. Aliquots of 0.1 ml were used for assay of EEA or CSA.

RESULTS

Effects of GCT Conditioned Medium on Erythroid Colonies

Medium conditioned by GCT cells consistently produced an enhancing activity for the human erythroid progenitors, CFU-E and BFU-E as well as potent CSA. The effect of GCT cell CM on blood BFU-E growth at different serum concentrations is shown in Figure 1. When blood mononuclear cells depleted on monocytes were plated at a cell density of 2.5×10^5 ml, enhancement of BFU-E growth was observed at each concentration of fetal

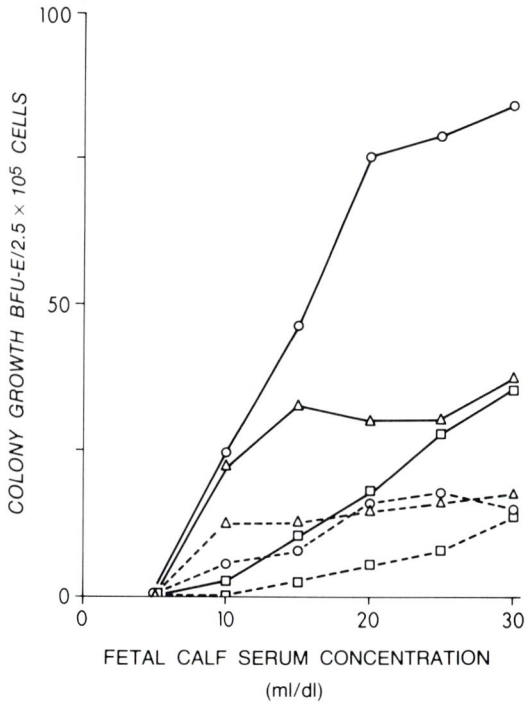

Fig. 1. Effect of dialyzed GCT conditioned medium on the fetal calf serum required for blood BFU-E growth. Solid lines represent cultures with 10% (v/v) CM; dashed lines, those without CM. The mean of duplicate observations for three experiments with and without CM are depicted by the paired symbols. Erythropoietin 2 units/ml was used in all cultures. Similar results were obtained with serum-free GCT conditioned medium.

calf serum tested. Bone marrow BFU-E growth also was enhanced by GCT CM. In all subsequent studies, the same fetal calf serum lot was used at a concentration of 20% (v/v) and erythropoietin at 2 units/ml in both CFU-E and BFU-E assays.

We examined the effect of GCT CM concentration on the growth of bone marrow CFU-E and BFU-E (Fig. 2a,b). The mean CFU-E in the absence of added GCT CM was 59 colonies. The optimum enhancement of CFU-E was at 2.5 ml/dl CM. Higher concentrations reduced, and eventually abolished, all enhancing activity. In some experiments growth was reduced below that of controls. A similar optimum was observed for marrow BFU-E. Thus, EEA was no longer seen when CM concentration was high.

GCT cells remain viable in serum-free medium fortified with polyethylene glycol [14]. Under these conditions they elaborate potent CSA. EEA was also present in serum-free GCT conditioned medium. Serum-free CM was used in the fractionation of EEA, which minimizes the interference of serum albumin in chromatographic analysis.

Chromatographic Fractionation of GCT Conditioned Medium

Ion-exchange chromatography. Initially, we placed concentrated serum-free CM through ion-exchange chromatography on DEAE Sephacel. The peak of the EEA eluted after the bulk (~90%) of CSA, when a linear gradient of sodium chloride was used (Fig. 3).

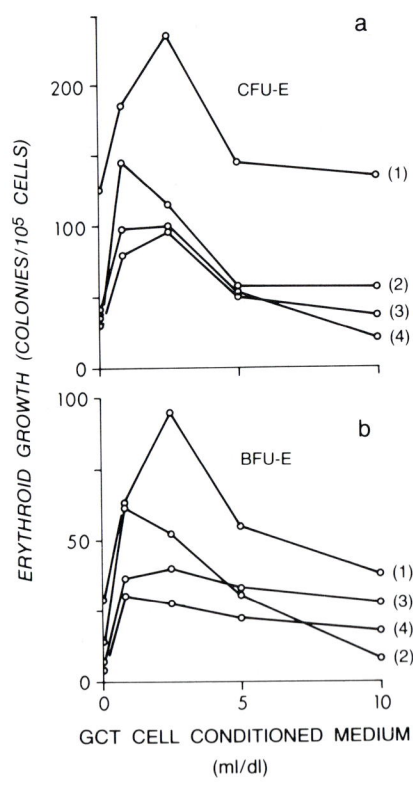

Fig. 2. Effect of dialyzed GCT conditioned medium on marrow CFU-E (a) and BFU-E (b). All cultures were made in 20% FCS with EPO 2 units/ml using 1×10^5 nonadherent cells. Numbers indicate paired experiments; mean of duplicate plates is shown.

CSA and EEA eluted between 0.1 and 0.2 M sodium chloride. We examined the peak CSA fraction (tube 155) and the peak EEA fraction (tube 170) for EEA at various dilutions (Table I). The fraction enriched for CSA contained EEA, which increased as the active fraction was diluted with control medium from 10% to 1%. The ratio of CSA/EEA was always higher on the ascending portion of the activity curves than on the descending limb, suggesting that the expression of EEA was more than simply a function of CSA concentration. The same elution profile was seen in four separate runs.

Gel filtration. In an effort to further discriminate between molecular species with weights in the 25,000–45,000 dalton range, we applied DEAE fractions containing CSA and EEA to Ultrogel AcA54 columns. DEAE fractions were pooled and concentrated by Amicon ultrafiltration with a PM 10 membrane. The elution profiles of CSA and EEA are shown in Figure 4. Both EEA and CSA appeared to co-chromatograph with a molecular weight of about 30,000 daltons. In studies not shown using Sephacryl S-200 chromatography, EEA and CSA eluted in similar fractions, with an apparent molecular weight of about 40,000 daltons.

In these experiments no EEA could be demonstrated in peak CSA fractions until dilution of samples was performed. EEA was tested, therefore, by adding 1.0% of the 1:3 dilutions of the Ultrogel fractions to cultures, whereas CSA was tested by adding 5%. When peak CSA fractions from the Ultrogel chromatography were tested at 10% no EEA could be demonstrated. Thus, EEA, which was present in the peak CSA fractions, could be detected only if dilutions were performed.

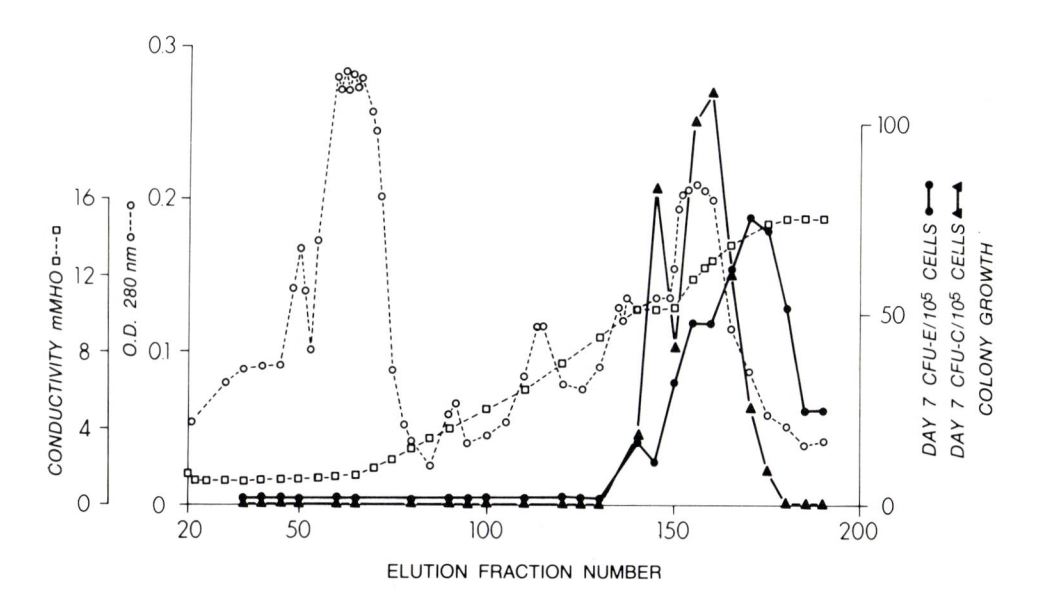

Fig. 3. DEAE Sephacel chromatography of a concentrated sample of serum-free GCT conditioned medium (3 liters) on a 5 × 80 cm column equilibrated in 0.03 M Tris-HCl buffer at pH 7.4 and eluted by a linear NaCl gradient from 0 to 0.25 M. All fractions were diluted 1:3 in McCoy's 5A with 5 ml/dl FCS and filtered prior to assay. CFU-C day 7 (closed triangles) and CFU-E day 7 (closed circles) were both cultured in 20% (v/v) FCS with a 10% (v/v) addition of each diluted fraction. Both CSA and EEA activities overlap, but the apparent peak of EEA elutes behind the peak of CSA. A similar elution profile was seen whether day 7 or day 14 colonies were assayed.

Affinity chromatography. In an effort further to separate EEA and CSA, we hoped to capitalize on a preferential binding to lectin by one or the other material. Therefore, we applied serum-free GCT CM to concanavalin A Sepharose 4B. As shown in Figure 5, CSA and EEA were partially bound to the lectin and were eluted by α-methyl mannoside; however, both CSA and EEA were present in the unbound fraction. When these fractions were pooled and rechromatographed on the same column, no further binding occurred, ruling out overloading as an explanation for the presence of both EEA and CSA in the unbound fractions. In other experiments we found that when the unbound fraction containing CSA was pooled prior to assay, EEA activity was present only when dilutions were carried out. Hence, in order to rule out the presence of EEA in unbound fractions, one had to test diluted fractions to ensure that inhibition did not occur from potent CSA.

Effect of Boiling on CSA and EEA in GCT Conditioned Medium

Golde et al [9] reported on the dissociation of EEA and CSA by boiling Mo cell line conditioned medium. We investigated the effect of varying times of boiling up to 60 min on EEA and CSA, both assayed in identical systems except for the EPO added to EEA assays (Fig. 6). EEA was present in GCT conditioned medium boiled for up to 60 min, and its apparent activity increased in tandem with a reciprocal decrease in CSA, both activities

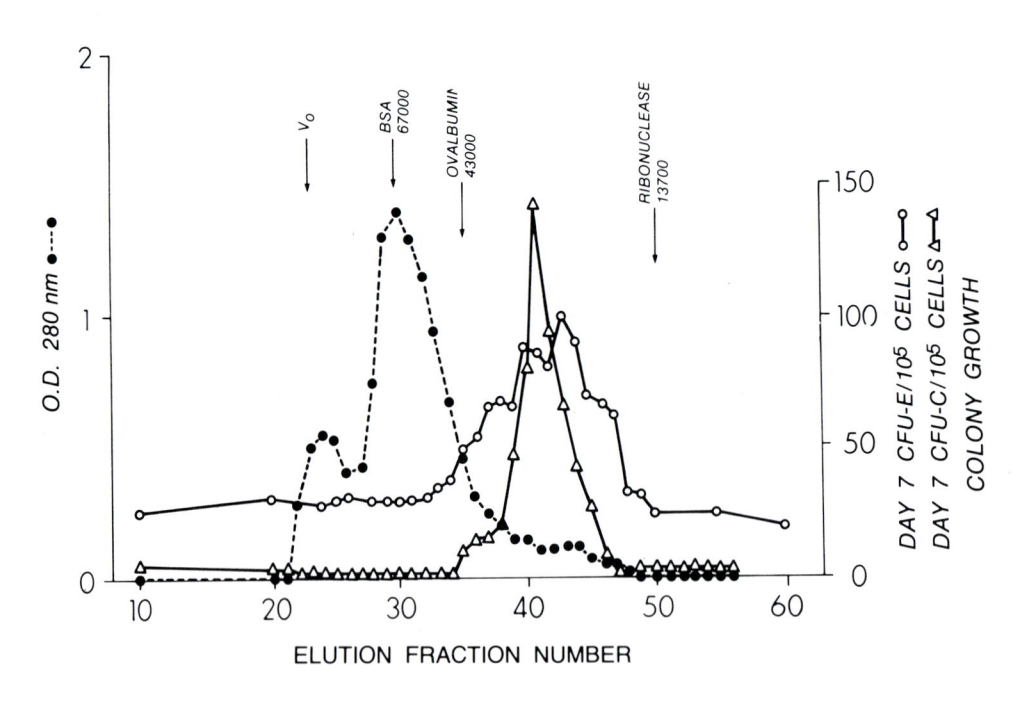

Fig. 4. Ultrogel AcA54 elution profile of serum-free GCT conditioned medium. DEAE Sephacel active CSA-EEA pools representing ~ 10 liters of medium were concentrated to 10 ml by ultrafiltration. A sample of 6 ml was applied to a 2.5 × 10 cm Ultrogel column in 0.05 M Tris HCl buffer, pH 7.4, 0.1 M NaCl, 0.01% PEG 6000, and eluted at 0.5 ml/min. Ten-ml fractions were diluted 1:3 prior to testing for EEA at 10 μl/ml EPO 2 units/ml, or CSA at 50 μl/ml in the same system without EPO. Both CSA and EEA peaks eluted with apparent weights of ~ 30,000 daltons as judged by assay of day 7 colonies. A similar profile was obtained when day 14 colonies were assayed.

reaching plateaus after 10 min. CSA was never completely abolished, since up to 25 small colonies/10^5 cells were obtained even after 60 min of boiling.

Effect of GCT CM on Erythropoietin-Independent Growth in Liquid Cultures

GCT conditioned medium was capable of sustaining BFU-E growth in viscous culture when erythropoietin addition was delayed for two days after cultures were initiated (Fig. 7). A similar sustaining action of GCT CM on marrow BFU-E was demonstrated when cells were preincubated in liquid suspension cultures for up to six days prior to colony assay (Table II). Enhancement of erythroid colonies occurred at the lowest dilutions of fractions containing CSA. This was associated with an increase in CFU-C as well. In control tubes BFU-E number dropped after the six days of incubation prior to assay.

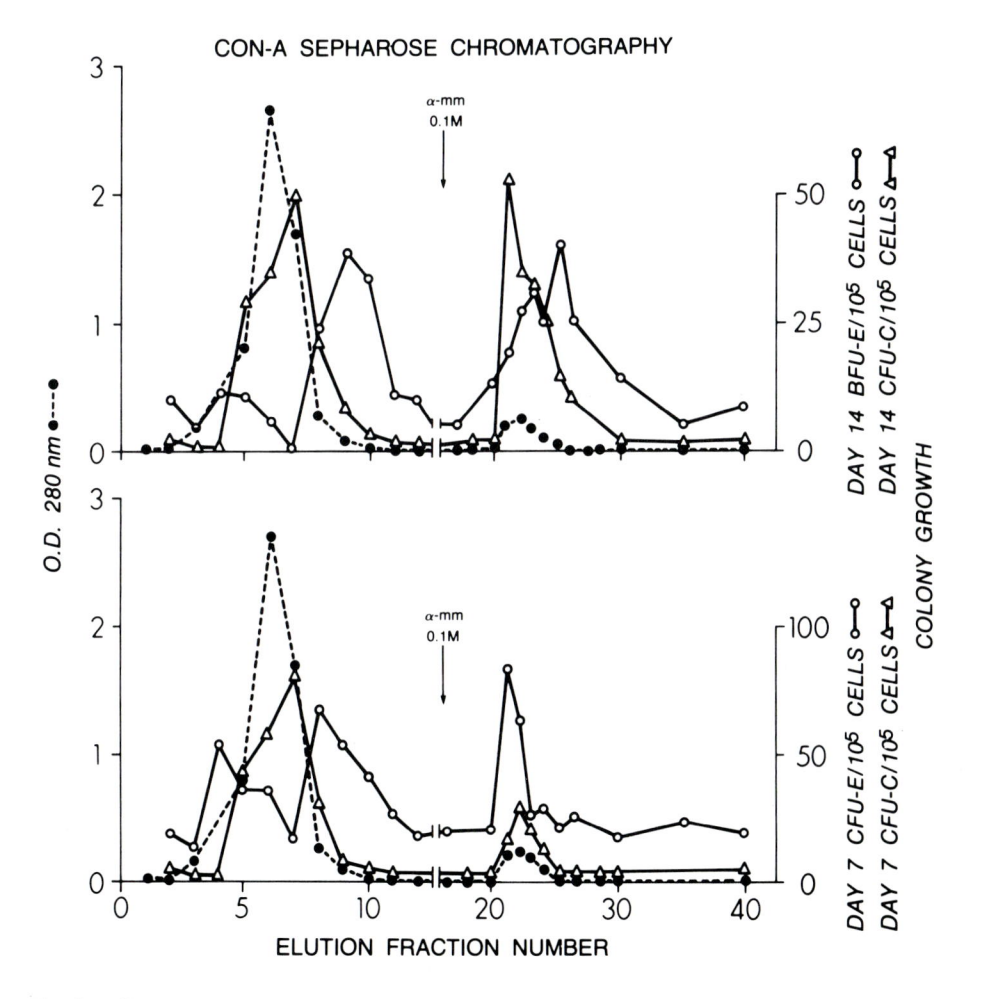

Fig. 5. Affinity chromatography of serum-free GCT conditioned medium. One liter was concentrated to 25 ml and applied to a 2.5 × 30 cm concanavalin A Sepharose column equilibrated in Tris-HCl, 0.05 M, pH 7.4; NaCl, 0.5M; polyethelene glycol, 0.05%; and $CaCl_2$, $MgCl_2$, $MnCl_2$ 0.1 mM at 4°C. α-methyl mannoside (0.1 M) in the same buffer was used for the elution. Fractions were dialyzed against 0.05 M Tris HCl pH 7.4, before assay, at 10% v/v. CSA and EEA were present in unbound fractions as was the fraction eluted with α-methyl mannoside.

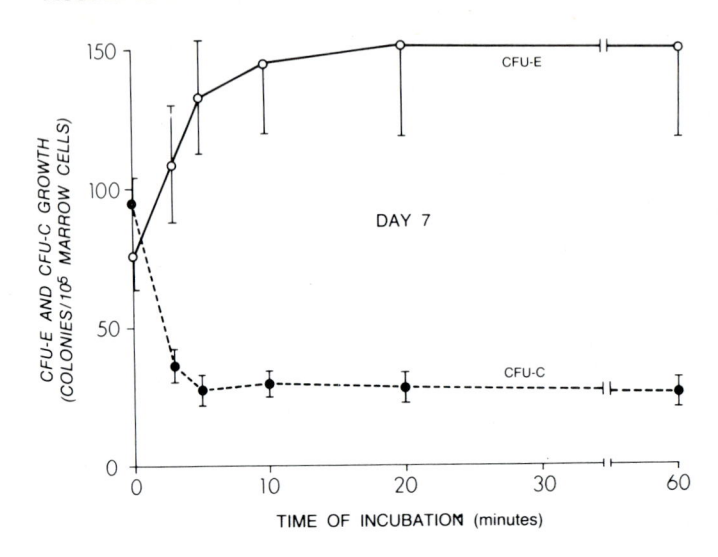

Fig. 6. Effect of boiling on CSA and EEA in dialyzed GCT conditioned medium. Test fractions were added at 10% (v/v) in cultures containing 20 ml/dl fetal calf serum, 2 units/ml erythropoietin, and 1 × 10⁵ nonadherent marrow cells. A rapid decrease in CSA occurred after three minutes of boiling coincident with an increase in the number of erythroid colonies at 7 days. Similar results were seen with day 14 colonies. Mean + SE are shown, erythroid colonies in triplicate, granulocytic in duplicate, for two different experiments.

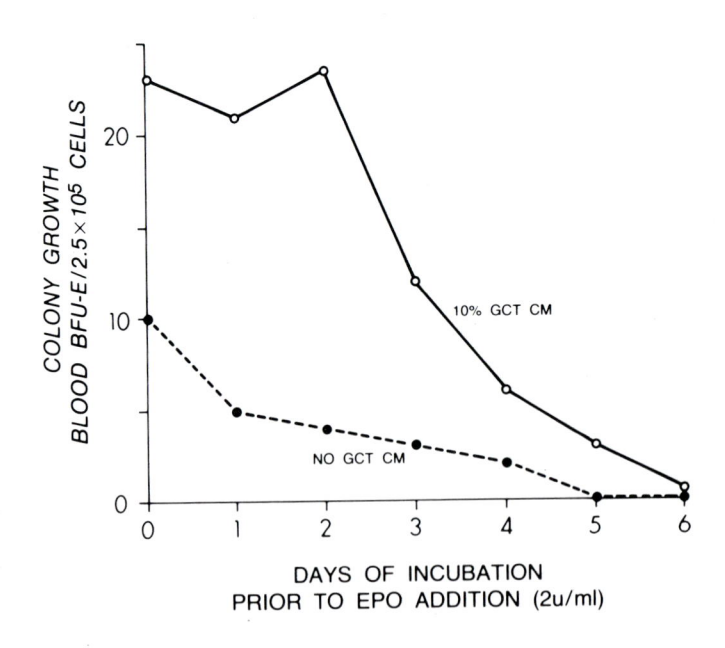

Fig. 7. Effect of GCT CM on BFU-E growth when erythropoietin addition is delayed. EPO 2 units/ml in 100 μl volume was pipetted gently over dish without disturbing the cultures. A similar effect was seen when marrow BFU-E were analyzed.

TABLE I. Effect of Dilution on EEA Activity in DEAE Fractions

DEAE Sephacel fractions	DEAE fractions added to culture (vol %)	Colonies/10^5 cells[a] CFU-E	BFU-E
CSA peak (tube 155)	10	68	18
	5	82	26
	2.5	100	32
	1	120	30
EEA peak (tube 170)	10	96	27
	5	91	18
	2.5	50	6
Control medium	10	40	4

[a] All cultures stimulated by 2 units/ml EPO, FCS 20% (v/v). Data represent means of duplicate plates. DEAE fractions were all added at volumes of 10% by addition of control medium.

TABLE II. Effect of GCT Conditioned Medium on BFU-E and CFU-C Growth in Liquid Suspension After 6 Days of Incubation*

Test fraction	Day 14 BFU-E	% Control	Day 14 CFU-C	% Control
Control medium	22	100	49	100
GCT CM 20%	20	91	72	147
10%	27	123	78	159
1%	43	195	83	169
DEAE CSA peak (fraction 155) 10%	38	175	77	157
DEAE EEA peak (fraction 170) 10%	42	193	68	139
DEAE CSA pool (52–59) 1%	57	259	72	147

*Mean values of duplicate plates. Cells were centrifuged and reconstituted to 1 ml final volume with 0.1 ml added per ml culture. EPO 2 units/ml and FCS 20% (v/v) were included in assays. EPO was omitted in CSA assays. The day zero control contained 28 BFU-E and 50 CFU-C per 10^5 cells (measured at day 14 of viscous culture).

DISCUSSION

The description of an erythroid-enhancing activity in leukocyte conditioned medium was made initially by Aye [2], who used human marrow and blood cells thoroughly depleted of adherent cells. In parallel, evidence was accumulating that erythropoietin was not the sole regulator of the early BFU-E compartment in vivo from studies of mice subjected to hypertransfusion [1] and phenylhydrazine-induced hemolysis [3]. Subsequently EEA has been described in serum and conditioned media from various sources using normal or transformed cells [1–10].

Our studies describe the properties of EEA secreted by the GCT cells, a permanent culture derived from a malignant fibrous histiocytoma [13]. These cells elaborate, in addition to CSA, a low molecular weight inhibitor (CIA), which inhibits erythroid growth at low cell densities [14]; plasminogen activator (unpublished); lymphocyte activating factor (unpublished); and a factor responsible for the growth of primitive macrophage progenitors [19]. Both classes of erythroid progenitors, BFU-E and CFU-E, appear responsive to this mediator, indicating effects on very early, as well as later, stages of erythroid development. This finding is in agreement with the results obtained by Golde et al with EEA elaborated by the Mo cell line [9]. The active component in GCT conditioned medium containing EEA and CSA has an apparent molecular weight of about 30,000 daltons by gel filtration on Ultrogel AcA54. This value is somewhat lower than other estimates. Human-active EEA derived from the Mo cell line [9] and eluted from Ultrogel AcA44, or from leukocyte and placenta conditioned medium [17] with Sephadex G-100, had an apparent molecular weight of about 45,000 daltons. The active human CSA peaks from these other sources were around 30,000 daltons [9, 17]. These differences in molecular weight estimates of EEA and CSA by gel filtration can be reconciled if they represent differences in biochemical characteristics of EEA from various sources. Another explanation may reside in the fact that these authors [9, 17] used different bioassay conditions for CSA and EEA, with different matrices and fetal calf serum concentrations. The apparent separation of EEA from CSA by gel filtration may also be due to the inverse relationship of EEA expression and CSA titer. Hence, the CSA peak region around 30,000 daltons may still have EEA when dilutions of these fractions are made. This is similar to the observations of Van Zant et al [18] that describe reduced erythroid colony growth by high titers of purified mouse CSF and enhancement at low CSF concentrations.

Complete dissociation of CSA from EEA in GCT conditioned medium was not achieved by ion-exchange chromatography or boiling. Furthermore, we have shown partial binding of CSA and EEA to concanavalin A Sepharose. Aye [20] has reported that EEA from leukocyte-conditioned medium, unlike CSA, bound to con A Sepharose, but dilutions of the CSA peak were not reported to ensure the absence of EEA. Nevertheless, the finding that EEA elutes always after the CSA by ion-exchange chromatography suggests that these two entities may be separable.

The role of EEA in early erythropoiesis in vitro is underscored by the enhanced survival of BFU-E in delayed EPO addition experiments and in liquid suspension cultures. The relationship of EEA to CSA is at present unclear. Iscove and co-workers [17] have reported the presence of delta colony potentiating activity (ΔCPA), an activity sustaining CFU-C growth in suspension cultures, in their EEA elution fractions. Our results in suspension culture show a sustained proliferation of both CFU-C and BFU-E in diluted CSA fractions. This may indicate that EEA may affect the proliferation of primitive stem cells, regardless of their terminal pathway of differentiation. Alternatively, this could indicate that a granulocyte-enhancing activity is present in EEA fractions. Although it may appear physiologically unlikely that a single mediator would influence distinct hemopoietic lineages, proof of this contention awaits further purification. In this regard, studies by Metcalf et al [21] have suggested that highly purified mouse lung CSF can directly stimulate multipotential and early erythroid precursor cells in fetal liver cell cultures. These observations, as well as those of Van Zant et al [18], suggest enhanced erythroid growth by purified CSF. To answer this question beyond doubt, it will be necessary to quantitate CSA and EEA in the same preparations, preferably using identical assays in vitro.

The availability of human cell lines such as Mo and GCT should facilitate the characterization and purification of EEA and CSA from cell sources that elaborate factors similar to monocytes and T lymphocytes.

ACKNOWLEDGMENTS

We wish to thank David Nemchick for his excellent technical help, Dr. Grant Barlow for his helpful discussions, and Ms Stase Mickys for typing the manuscript.

This work was aided by American Cancer Society grants CH-71A and IN-18T 223; and by USPHS grants CA-25512 and HL-18208.

REFERENCES

1. Iscove NN: In Golde DW, Cline MJ, Metcalf D, Fox CF (eds): "Hematopoietic Cell Differentiation." ICN-UCLA Symposia on Molecular and Cellular Biology. New York: Academic Press, 1978, vol 10, pp 37–52.
2. Aye MT: J Cell Physiol 91:69, 1977.
3. Wagemaker G: In Murphy MJ Jr (ed): "In Vitro Aspects of Erythropoiesis." New York: Springer Verlag, 1978, pp 44–57.
4. Nissen C, Iscove NN, Speck B: In Baum SJ, Ledney GD (eds): "Experimental Hematology Today." New York: Springer Verlag, 1979, pp 79–87.
5. Porter PN, Ogawa M, Leary AG: Exp Hematol 8:83, 1980.
6. Humphries RK, Eaves AC, Eaves CJ: Blood 53:746, 1979.
7. Meytes D, Ma A, Ortega JA, Shore NA, Dukes PP: Blood 54:1050, 1979.
8. Nathan DG, Chess L, Hillman DG, Clarke B, Breard J, Merler E, Housman DE: J Exp Med 147:324, 1978.
9. Golde DW, Bersch N, Quan SG, Lusis AJ: Proc Natl Acad Sci USA 77:593, 1980.
10. Zanjani ED, Kaplan ME: In Brown EB (ed): "Progress in Hematology, Vol XI." New York: Grune & Stratton, 1979, pp 173–191.
11. Burgess AW, Metcalf D, Nicola NA, Russell SHM: In Golde DW, Cline MJ, Metcalf D, Fox CF (eds): "Hematopoietic Cell Differentiation." ICN-UCLA Symposia on Molecular and Cellular Biology. New York: Academic Press, 1978, vol 10, pp 399–416.
12. Golde DW, Quan SG, Cline MJ: Blood 52:1068, 1978.
13. DiPersio JF, Brennan JK, Lichtman MA, Speiser RL: Blood 51:507, 1978.
14. DiPersio JF, Brennan JK, Lichtman MA: In Golde DW, Cline MJ, Metcalf D, Fox CF (eds): "Hematopoietic Cell Differentiation." ICN-UCLA Symposia on Molecular and Cellular Biology. New York: Academic Press, 1978, vol 10, pp 433–444.
15. Messner HA, Till JE, McCulloch EA: Blood 42:701, 1973.
16. Abboud CN, Brennan JK, Lichtman MA: Transfusion 20:9, 1980.
17. Hoang T, Iscove NN: International Society of Experimental Hematology, 8th Annual Meeting, 1979 (poster presentation).
18. Van Zant G, Goldwasser E, Pech N: Blood 53:946, 1979.
19. Bradley TR, Hodgson GS: Blood 54:1446, 1979.
20. Aye MT: Blood 50:122, 1977 (abstr).
21. Metcalf D, Johnson GR, Burgess AW: Blood 55:138, 1980.

Cell-Cell Contact and Growth Regulation of Pinocytosis in 3T3 Cells

Peter F. Davies

Departments of Pathology, Peter Bent Brigham Hospital and Harvard Medical School, Boston, Massachusetts 02115

In sub-confluent cultures of Balb/c-3T3 cells, pinocytosis rates were increased after exposure to specific growth factors (serum; platelet-derived growth factor, PDGF; epidermal growth factor, EGF). Conversely, as cells became growth-inhibited with increasing culture density, there was a corresponding decline in pinocytosis rate per cell. In order to test whether density-inhibition of pinocytosis was influenced either by the growth cycle or by cell contact independently of growth, cells were induced into a quiescent state at a range of subconfluent and confluent densities. Under such conditions, cell density did not significantly inhibit pinocytosis rate. When confluent quiescent cultures in 2.5% serum were exposed to 10% serum, the resulting round of DNA synthesis was accompanied by enhanced pinocytosis per cell, even though the cells were in contact with one another. Furthermore, in a SV40-viral transformed 3T3 cell line, both the growth fraction and the pinocytosis rate per cell remained unchanged over a wide range of culture densities. These studies indicate that density-dependent inhibition of pinocytosis in 3T3 cells appears to be secondary to growth-inhibition rather than to any direct physical effects of cell–cell contact.

Key words: pinocytosis, cell density, growth control, growth factors

The rate of fluid endocytosis (fluid pinocytosis) in a number of non-transformed and transformed cells in culture is related to the growth cycle [1–5]. We are particularly interested in vascular endothelium, where endocytosis is associated with transendothelial transport of molecules [6, 7] and where focal areas of arterial endothelial cell proliferation have been detected in vivo [8, 9]. Such areas are associated with increased permeability to a number of macromolecules [10, 11] and may be sites that are predisposed to atherogenesis. Recently, density-inhibition of arterial endothelial cell growth was demonstrated to correlate with decreased rates of pinocytosis in vitro [1, 4, 15]. In the case of density-inhibition of a membrane-associated event such as pinocytosis, however, the question arises as to whether cell–cell contact exerts direct influence upon pinocytosis independently of cell growth. This may be of importance with respect to endothelium, because in vitro removal of single cells results in spreading of adjacent cells to fill the "wound" without a round of

Received March 27, 1980; accepted June 4, 1980.

0091-7419/80/1302-0211$01.70 © 1980 Alan R. Liss, Inc.

cell division [12]. Thus, cell–cell contact is reestablished without growth. It is therefore of interest to test whether pinocytosis is stimulated by loss of cell–cell contact independently of growth. Unfortunately endothelial cells cannot be induced into a quiescent state (arrested in G_0/G_1) in the absence of known growth factors at sparse densities. The hypothesis that cell contact in the absence of growth can mediate pinocytosis was therefore tested in Balb/c-3T3 cells, a monolayer-forming, density inhibited cell line that, unlike endothelium, is responsive to growth factors. These experiments, conducted with quiescent serum-stimulated and SV40 viral transformed 3T3 cells, have led to the conclusions that cell contact in growth-arrested cells does not influence pinocytosis rate and that differences in pinocytosis rates at various cell densities are secondary to density-dependent inhibition of growth.

MATERIALS AND METHODS

Cell Culture

Mouse Balb/3T3, clone A31 (Balb/c) were obtained from the American Type Culture Collection and were maintained by serial passage. The culture medium was Dulbecco-Vogt's modification of Eagle's basal medium supplemented with 10% calf serum. For the studies with quiescent cells, cultures were trypsinised and plated at appropriate densities in 5% calf serum. After the cells were adherent to the plastic dish, the medium was replaced with 5% plasma-derived serum (PDS) prepared as described by Vogel et al [16]. The attainment of a quiescent state was determined by daily measurement of cell numbers (Coulter Electronics Inc., Hialeah, FL) and by autoradiographic labeling of cell nuclei with ^{3}H-thymidine. In quiescent cultures there were very few labeled nuclei (thymidine index range 0.03–0.08) after 48 h in the presence of PDS.

Quiescent cells were stimulated to divide either by addition of platelet-derived growth factor (PDGF, generously supplied by Dr. Ross and partially purified as previously described [16]), or by the addition of calf serum or, in some experiments, epidermal growth factor (EGF, Collaborative Research, Waltham, MA).

Pinocytosis Rate

Fluid phase pinocytosis was measured as the cellular uptake of (U-^{14}C)-sucrose (673 Ci/mole, New England Nuclear, Boston, MA) from the tissue culture medium by the method reported previously [1]. Radiolabeled sucrose was added in a small volume of BSS to a final activity of between 10 and 20 μCi per ml of tissue culture medium, depending upon the cell density. Following incubation at 4°C (control to inhibit endocytosis) or at 37°C, the cells were washed 5 times with ice-cold BSS containing 0.2% bovine serum albumen. Each culture was then rinsed twice with BSS, and the cells were dissolved in 1 ml 0.1% (aq) sodium dodecylsulphate (SDS, electrophoresis grade; Biorad, NY). The SDS lysate was transferred to a scintillation vial, 10 ml Ultrafluor cocktail (National Diagnostics, Somerville, NJ) was added, and the ^{14}C activities (dpm) were determined in a Beckman LS345 liquid scintillation counter. Counting efficiencies were determined from external standard ratios by reference to quenched (^{14}C)-toluene standards (Beckman). Rate of fluid endocytosis was expressed as the volume (nl) of fluid endocytosed per 10^6 cells per unit time [1, 17].

^{14}C-sucrose meets the criteria for a tracer of fluid phase endocytosis [18]. It is impermeable to the plasma membrane, does not significantly bind to the cell surface, and its uptake rate is directly proportional to its concentration in the culture medium. The rate of uptake of ^{14}C-sucrose remained constant for 18 h and essentially represented a cumulative measurement because of the absence of an intracellular degradative enzyme [19] and a slow rate of exocytosis ($t_{1/2} > 40$ h at 37°C) (Davies, unpublished).

Autoradiography

Cells were labeled for 24 h by addition of 0.3 µCi ^{3}H-thymidine per ml of medium. After washing with BSS, cultures were fixed in ethanol-acetic acid, further rinsed, then air dried. The dish bottom was cut out, mounted in halves on a glass slide with polyvinylpyrrolidone, sealed with lacquer and the slides dipped in Kodak NTB2 emulsion. Following exposure for 10 days, the autoradiographs were developed, stained with haematoxylin-eosin, and % labeled nuclei determined.

RESULTS

3T3 cells were maintained in a quiescent state of growth at a range of cell densities by the use of plasma-derived serum (PDS), from which platelets and platelet-derived growth factors (PDGF) had been removed. The cells were plated at various densities in 10% serum until adherent, at which time the medium was replaced by 10% PDS. Under these conditions the cells were limited to a single round of division before accumulation in a G_0/G_1 phase of the cell cycle [13, 14, 20]. When either serum or partially purified PDGF was added back to the medium, there was a stimulation of DNA synthesis dependent upon the coordinate control of growth factor and plasma components, as fully documented elsewhere [16, 21]. In 10% PDS, the effects of increasing concentrations of PDGF upon DNA synthesis and pinocytosis rates are as shown in Table I at subconfluent and confluent cell densities. As Vogel et al [16] have recently demonstrated, 3T3 cells require increasing concentrations of PDGF to elicit equal mitogenic response with increasing cell density. Table I shows that the rate of pinocytosis in sparse cultures is significantly enhanced in the presence of a lower concentration of PDGF than is the case for confluent cultures.

When 3T3 cultures became growth-inhibited in 10% serum, the rate of pinocytosis per cell declined to a basal level (Table II). In light of the data presented in Table I, the findings of Vogel et al [16], and previous studies relating pinocytosis to the growth cycle [1–5], it was hypothesised that the decline in pinocytosis rate was related to the attainment of quiescence. At confluence, however, there was considerable cell—cell contact,

TABLE I. Stimulation of DNA Synthesis and Pinocytosis in 3T3 Cells in Relation to Cell Density*

	7×10^3 cells/cm^2					3.6×10^4 cells/cm^2				
µg PDGF/ml:	0	5	10	15	25	0	5	10	15	25	50
Fraction labeled nuclei:	0.04	0.80	0.74	0.91	0.82	0.03	0.21	0.59	0.74	0.73	0.81
Pinocytosis rate at 24 h: (nl fluid/10^6 cells/h)	47	101	1111	151	147	44	48	86	71	79	136

*Cells were plated at nominal densities of 4×10^3/cm^2 and 2×10^4/cm^2 in 10% calf serum. The medium was changed to 10% PDS 6 h after plating, and the cells were maintained in this medium for 48 h. PDGF was then added at various doses with 0.3 µCi ^{3}H-thymidine/ml medium. Twenty hours later ^{14}C-sucrose was introduced into some dishes for measurement of pinocytosis rate. Finally, after 24 h continuous exposure to PDGF, all dishes were harvested and assayed for autoradiography and pinocytosis rates. Cell densities in the table were measured by electronic counting at 24 h.

TABLE II. Density-Inhibition of 3T3 Cell Pinocytosis in 10% Calf Serum*

Cell density ($\times 10^{-4}/cm^2$)	0.48	1.4	3.1	4.0
Rate of pinocytosis (nl fluid/10^6 cells/h)	163	119	61	72
Fraction labeled nuclei	0.91	0.87	0.14	0.09

*Cells plated at a density of $4 \times 10^3/cm^2$ in 10% calf serum. As cell numbers increased, rates of pinocytosis were measured during a 2 h period at each of the cell densities. Cultures were also incubated with ^{3}H-thymidine for 12 h before and 10 h after the period of measurement of pinocytosis rates. Thus the fraction of labeled nuclei represents an average around the cell density indicated.

which might directly influence plasma membrane flow and invagination by the imposition of direct physical restraints upon membrane mobility. In an attempt to separate the growth component from other aspects of this system, two approaches were used. First, 3T3 cells were made quiescent in PDS at a range of subconfluent and confluent densities, so that different degrees of cell contact occurred in the absence of growth. Second, in order to maintain the growth fraction maximal despite increasing cell—cell contact, a SV40-viral transformed 3T3 cell line was used. Figure 1 demonstrates that, over a range of cell densities, the rate of pinocytosis per cell remained unchanged in quiescent 3T3 cells (fraction of ^{3}H-thymidine-labeled nuclei, 0.03–0.06). In SV40/3T3 under conditions of constant, maximal growth (fraction of ^{3}H-thymidine-labeled nuclei effectively 1.0 at all cell densities), the pinocytosis rate also remained unchanged over a 20-fold range of cell densities. When quiescent 3T3 cells at different densities were exposed to serum, however, the pinocytosis rate was more greatly stimulated at lower cell densities than in confluent cultures (Fig. 1). These results indicate that growth, and not cell—cell contact by itself, mediates pinocytosis rates. Supporting this conclusion are the data in Figure 2, which demonstrate that confluent, growth-inhibited 3T3 cells in 2.5% serum responded to 10% serum by an enhanced rate of pinocytosis together with a round of DNA synthesis.

DISCUSSION

The rates of fluid pinocytosis in 3T3 cells [1], vascular smooth muscle cells [1, 5], endothelium [4], S774 macrophages [3], and hepatoma cells [1] appear to be related to the cell cycle. Variations of pinocytosis rates of up to 10-fold have been reported. The methods employed to synchronize cultures in such studies have included colcemid block [2], density-inhibition [1, 4], and arrest of the cell cycle at G_0/G_1 in growth-factor-depleted medium [1, 5]. Coincident with density-dependent inhibition of growth in vascular endothelium and 3T3 cells, however, is the establishment of significant contact between cells. Such contact has been proposed to inhibit the lateral mobility of ligand—receptor complexes in the plane of the plasma membrane of cultured endothelium [15]. The formation of interdigitated gap junctions in many cultured cell lines may also directly inhibit membrane-mediated events such as pinocytosis. Thus it was of interest to determine the relative contributions made by cell contact and cell growth to the inhibition of pinocytosis in confluent 3T3 cultures.

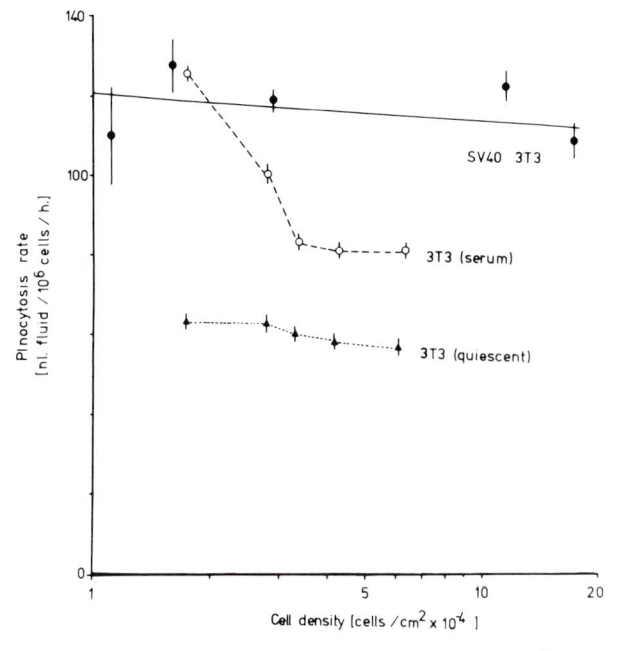

Fig. 1. Pinocytosis rate plotted against cell density for nontransformed Balb/c-3T3 cells and SV40 viral transformed 3T3 cells. Cells were plated at a range of densities in 10% calf serum After the nontransformed cells were adherent to the culture dish, the medium was changed to 10% PDS for 48 h. The 3T3 cells became quiescent at a range of cell densities. To half of the quiescent cultures was added 10% serum; the remainder received fresh 10% PDS. Twenty-four hours later, pinocytosis rate was measured during a 2 h period in the quiescent (▲) and serum-stimulated (○) 3T3 cells. Pinocytosis rates were also measured in the SV40/3T3 cells (●) at the various cell densities. Each point represents the mean of 4 determinations ± SD. The fraction ^{3}H-thymidine-labeled cells by autoradiography at lowest and highest density in each group was as follows: quiescent 3T3: 0.06 (low) 0.05 (high); serum stimulated 3T3: 0.86 (low) 0.11 (high); SV-40-3T3: 0.98 (low) 1.00 (high).

Plasma-derived serum (PDS), consisting of recalcified plasma from which platelets had been removed, was used to induce cells into a quiescent state of growth [13, 16, 22] at a range of cell densities. This occurs because platelets are removed from the plasma withoug degranulation and release of mitogenic platelet-derived growth factor [13]. Addition of purified growth factor to PDS results in growth responses that are comparable to serum. The measurements of pinocytosis rates in quiescent cells at different cell densities demonstrated that cell—cell contact in the absence of cell growth has no significant effect upon pinocytosis rates, suggesting that growth is the dominant mediator of density-dependent pinocytosis. This conclusion is supported by the observation that, in maximally growing transformed cells, the formation of a confluent monolayer, and even multilayers, did not significantly inhibit pinocytosis. The studies therefore suggest that the growth status of the cell most greatly influences pinocytosis rates at different cell densities, and that density inhibition of pinocytosis occurs secondary to density inhibition of growth.

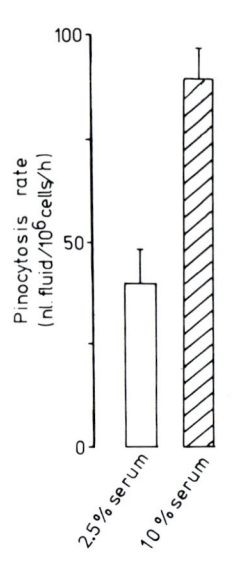

Fig. 2. Serum stimulation of pinocytosis in confluent 3T3 cells. 3T3 cells were plated in 2.5% calf serum at a density of $10^4/cm^2$. The cells grew to confluence (cell density $4 \times 10^4/cm^2$) and became quiescent. The medium was then changed in some dishes to 10% serum, and pinocytosis rate was measured over a 2 h period 22 h after serum-stimulation. Each column represents the mean of 4 determinations ± SD.

A number of mechanisms have been proposed for density-dependent inhibition of cell growth [23–25]. Most pertinent to the experiments described here are the recent studies of Vogel et al [26], who have provided evidence that a major regulator of density-dependent inhibition of growth in 3T3 cells is an increasing requirement for PDGF. Since we have previously demonstrated a close correlation between pinocytosis rate and entry into the cell cycle induced by specific growth factors, including PDGF, the present data are also consistent with the findings of Vogel et al.

Measurements of fluid pinocytosis rates indirectly reflect the internalization rate of plasma membrane, including ligand—receptor complexes at the cell surface. Since many such complexes are able to move laterally in the plane of the membrane, leading to patching, clustering, and capping [27–29], there may be a high degree of independence between fluid pinocytosis and ligand—receptor endocytosis, perhaps by the existence of separate populations of vesicles. Several studies, however, have provided evidence that the uptake of low-density lipoproteins (LDL) may be influenced by cell density in a pattern similar to that observed in the studies of pinocytosis. Stein and Stein [30] reported that uptake of LDL by cultured smooth muscle cells varied inversely with cell density, and recently Kruth et al [31] have demonstrated relationships between LDL receptor expression, lipid internalization, and fibroblast culture density. Their results were consistent with the possibility that LDL receptor-mediated uptake can be regulated via cell density. It will be of interest to apply methods of growth control such as used in the present study to investigate the specific endocytosis of macromolecules.

ACKNOWLEDGMENTS

I am most grateful for helpful discussion with Drs. Ross and Vogel of the University of Washington, Seattle, who also generously provided PDGF, and with Drs. Cotran and Gimbrone of the Harvard Medical School. I also thank Cathy Kerr for excellent technical assistance. The work was supported by USPHS grant HL-24612.

REFERENCES

1. Davies PF, Ross R: J Cell Biol 79:663, 1978.
2. Quintart J, Leroy-Houyet M-A, Trouet A, Baudhuin P: J Cell Biol 82:644, 1979.
3. Berlin RD, Oliver JM, Walter RJ: Cell 15:327, 1978.
4. Davies PF, Selden SC, Schwartz SM: J Cell Physiol 102:119, 1980.
5. Davies PF, Ross R: Exp Cell Res (in press).
6. French JE: Int Rev Exp Pathol 5:253, 1966.
7. Hüttner I, Boutet M, More RH: Lab Invest 28:678, 1973.
8. Payling-Wright H: Atherosclerosis 15:93, 1972.
9. Schwartz SM, Benditt EP: Am J Pathol 66:241, 1972.
10. Bell FP, Adamson IL, Schwartz CJ: Exp Mol Pathol 20:57, 1974.
11. Bell FP, Gallus AS, Schwartz CJ: Exp Mol Pathol 20:281, 1974.
12. Reidy MA, Schwartz SM: Fed Proc 39:1109, 1980.
13. Ross R, Glomset J, Kariya B, Harker L: Proc Natl Acad Sci USA 71:1207, 1974.
14. Pledger WJ, Stiles CD, Antoniades HN, Scher CD: Proc Natl Acad Sci USA 74:4481, 1977.
15. Vlodavsky I, Fielding PE, Johnson LK, Gospodarowicz D: J Cell Physiol 100:481, 1979.
16. Vogel A, Raines E, Kariya B, Rivest M-J, Ross R: Proc Natl Acad Sci USA 75:2810, 1978.
17. Williams KE, Kidston EM, Beck F, Lloyd JB: J Cell Biol 64:123, 1975.
18. Steinman RM, Brodie SE, Cohn ZA: J Cell Biol 68:665, 1976.
19. Cohn ZA, Ehrenreich BA: J Exp Med 129:201, 1969.
20. Ross R, Nist C, Kariya B, Rivest M-J, Raines E, Callis J: J Cell Physiol 97:497, 1978.
21. Pledger WJ, Stiles CA, Antoniades H, Scher CD: Proc Natl Acad Sci USA 75:2839, 1978.
22. Kohler N, Lipton H: Exp Cell Res 87:297, 1974.
23. Holley RW: Nature 258:487, 1975.
24. Dulbecco R, Stoker M: Proc Natl Acad Sci USA 66:204, 1970.
25. Todaro G, Lazar G, Green H: J Cell Comp Physiol 66:325, 1965.
26. Vogel A, Ross R, Raines E: J Cell Biol 85:377, 1980.
27. Schlessinger J, Shechter Y, Cuatrecasa P, Willingham MC, Pastan I: Proc Natl Acad Sci USA 75:663, 1978.
28. Ryan GB, Borysenko JZ, Karnovsky MJ: J Cell Biol 62:351, 1974.
29. Maxfield FR, Schlessinger J, Shechter Y, Pastan I, Willingham MC: Cell 14:805, 1978.
30. Stein O, Stein Y: Biochim Biophys Acta 398:377, 1975.
31. Kruth HS, Avigan J, Gamble W, Vaughan M: J Cell Biol 83:588, 1979.

Chicken Tissue Binding Sites for a Purified Chicken Lectin

Eric C. Beyer and Samuel H. Barondes

Department of Psychiatry, University of California, San Diego, La Jolla, California 92093 and Veterans Administration Hospital, San Diego, California 92161

A lactose-binding lectin previously purified from embryonic chicken muscle and adult chicken liver, and here referred to as chicken-lactose-lectin-I (CLL-I), was added to sections of various adult chicken tissues to detect available binding sites. Both the sites of binding of added CLL-I as well as the tissue distribution of endogenous CLL-I were determined by indirect immunofluorescence using a rabbit antibody to CLL-I followed by fluorescent goat anti-rabbit IgG. Some tissues such as intestine and kidney showed abundant extracellular binding sites for the lectin, primarily between cells, in basement membrane, and in material on the luminal surface. In contrast, adult heart showed no significant binding sites for CLL-I. Adult pancreas showed considerable endogenous CLL-I in an extracellular site surrounding exocrine lobules, but added CLL-I did not bind substantially. The distribution of CLL-I binding sites in intestine were mimicked by those of purpurin, another lactose-binding lectin. CLL-I binding sites were also detected on the surface of cultured chick embryo skin fibroblasts. The factors controlling the specific distribution of occupied and unoccupied CLL-I binding sites are not known.

Key words: lectins, lectin binding sites, cell surfaces, extracellular materials

A number of embryonic [1–3] and adult [4] chicken tissues contain an endogenous lactose-sensitive carbohydrate-binding protein which undergoes marked changes in activity with development. Because this protein may be assayed as a lactose-sensitive hemagglutinin it is referred to as a lectin. The role of this lectin in cellular differentiation or in other cellular functions has not yet been determined. The lectin has been purified from several embryonic and adult tissues [5–7]. It is a dimer of subunits with apparent molecular weight of 15,000. In this paper we will refer to this lectin as chicken-lactose-lectin-I (CLL-I) to distinguish it from another distinct lactose-binding lectin [7] from chicken tissues. This second lactose-binding lectin, like CLL-I, is found in adult chicken intestine, from which it has been purified [7].

Detailed comparisons of CLL-I purified from embryonic muscle and adult liver indicate that they are identical by all chemical and immunological tests that were used [7]. A potent antiserum has been raised by immunizing rabbits with this lectin [4]. Immuno-

Received April 9, 1980; accepted June 11, 1980.

0091-7419/80/1302-0219$02.00 © 1980 Alan R. Liss, Inc.

histochemical studies have shown that CLL-I is concentrated at different sites in different tissues. In developing muscle [5] and brain [8], some is detectable on the surface of myoblasts and neurons, but most is intracellular. Lectin levels in adult muscle and brain are low, but are high in several other adult tissues. CLL-I localization differs in three adult tissues studied [4]. It is concentrated in secretory granules of intestinal goblet cells, in the extracellular matrix surrounding pancreatic exocrine lobules, and in hepatic Kupffer cells.

The present study was designed to determine the location of binding sites for CLL-I in several adult chicken tissues and the relationship between these binding sites for added CLL-I and the localization of endogenous CLL-I. We found that in some tissues such as adult intestine and kidney there are many extracellular CLL-I binding sites not already occupied by endogenous CLL-I. Adult heart, in contrast, had neither endogenous lectin nor significant binding sites for CLL-I lectin, and the binding sites in adult pancreas were either absent or fairly well occupied by endogenous CLL-I which was abundant in the extracellular material surrounding lobules.

MATERIALS AND METHODS

Preparation of Lectins and Antisera

The lectin from adult chicken liver (CLL-I) was isolated and purified by affinity chromatography as described previously [4, 5]. This lectin is indistinguishable from the lactose-binding lectin previously purified from chick embryonic pectoral muscle [7].

Production of antiserum directed against this antigen and demonstration of its specificity have also been described [4]. Purpurin, the lactose-binding lectin from Dictyostelium purpureum, and its antiserum were obtained as described previously [9]. Concanavalin A and antiserum directed against it were purchased from Calbiochem.

Lectin Localization in Tissue Sections

Tissue fragments were fixed in 3% paraformaldehyde, 0.1% glutaraldehyde, and 0.5 to 1.0 μm frozen sections were cut as described previously [4]. Indirect immunofluorescent staining for endogenous lectin was done with immune gamma globulin followed by tetramethylrhodamine-conjugated goat anti-rabbit IgG (Cappel Labs) as described previously [4]. To determine binding of exogenous lectin to sections, they were incubated for 15 minutes with a small droplet of pure lectin (100 μg/ml). After unbound lectin was washed away, the sections were stained with anti-lectin gamma globulin and tetramethylrhodamine-conjugated goat anti-rabbit IgG as above. To verify specific binding of the lectin to sections, lectin was incubated in the presence of a high concentration (300 mM) of a hapten inhibitor.

Cell Culture

Primary cultures of chick embryo skin fibroblasts were grown on glass coverslips in Eagle's minimal essential medium supplemented with 10% fetal calf serum, as described previously [5, 10]. Cells fixed with 3% paraformaldehyde were stained by the same protocol as for the tissue sections.

Fluorescence Microscopy

Slides of sections or cells were examined with a Leitz Dialux epifluorescence microscope with a X40 (NA 1.3) oil immersion lens. Photomicrographs were taken with 30-sec exposures with Kodak Plus-X film and developed with Diafine developer. All photographs of each tissue were printed under identical conditions.

RESULTS

Lectin Binding to Tissue Sections

In adult chicken intestine, endogenous CLL-I was localized to the secretory granules of the mucus-secreting goblet cells (Fig. 1a) as observed previously [4]. Sections reacted with non-immune sera showed only very faint and diffuse labelling, and all specific staining could be adsorbed with purified CLL-I [4]. However, if 100 μg/ml CLL-I was incubated with the sections before antibody staining, much additional fluorescence was observed (Fig. 1b). This exogenous lectin bound to the epithelial basement membrane, the extracellular spaces between adjacent epithelial cells, the microvilli and their mucous coat, and extracellular materials in the lamina propria, especially surrounding smooth muscle bundles. CLL-I binding was all specifically abolished by incubation in the presence of the hapten inhibitor lactose (300 mM) (Fig. 1c). Endogenous CLL-I staining remained since this lectin was held in place by tissue fixation. It is possible that endogenous CLL-I in unfixed sections would also be solubilized by hapten, since lactose markedly facilitates extraction of this lectin from homogenates [5].

Since exogenous CLL-I bound so well to intestinal sections we compared the distribution of this bound lectin with that of two others. The distribution of bound Concanavalin A in intestinal sections differed somewhat from that of CLL-I in that it bound relatively more to diffuse cytoplasmic materials compared with intercellular substances (Fig. 2a). Concanavalin A also showed relatively less binding to the vesicles in goblet cells, although it did intensely stain materials on the intestinal luminal surface. In contrast, purpurin, a lactose-binding lectin from D purpureum [19] bound to the same structures as CLL-I, especially the goblet cell material and intercellular material between intestinal epithelial cells and muscle bundles (Fig. 2c). These results indicate that the receptors for a lectin with affinity for lactose residues are not highly specific for the chicken lactose lectin.

In adult chicken kidney, endogenous CLL-I was found in the cytoplasm and on the luminal and basal surfaces of tubule cells (Fig. 3a). Glomeruli were unstained. Nuclei in some kidney sections also bound antibody. Whether this indicates the presence of lectin or an immunologically cross-reactive material in the nuclei remains to be determined. Exogenously added CLL-I bound predominantly to the luminal surfaces of tubules and to peritubular and glomerular material (Fig. 3b).

In adult chicken pancreas, staining for endogenous lectin gave bright fluorescence which was confined to the extracellular matrix surrounding exocrine lobules (Fig. 4a). Exogenously added CLL-I (Fig. 4b) showed little additional staining, suggesting that there are few receptor sites not occupied by endogenous lectin.

Adult chicken heart contains little endogenous CLL-I as measured by hemagglutination [4]. This result was confirmed by the absence of significant staining for endogenous CLL-I (Fig. 5a). The adult heart also contained few binding sites for exogenously added CLL-I (Fig. 5b), except those associated with blood vessels.

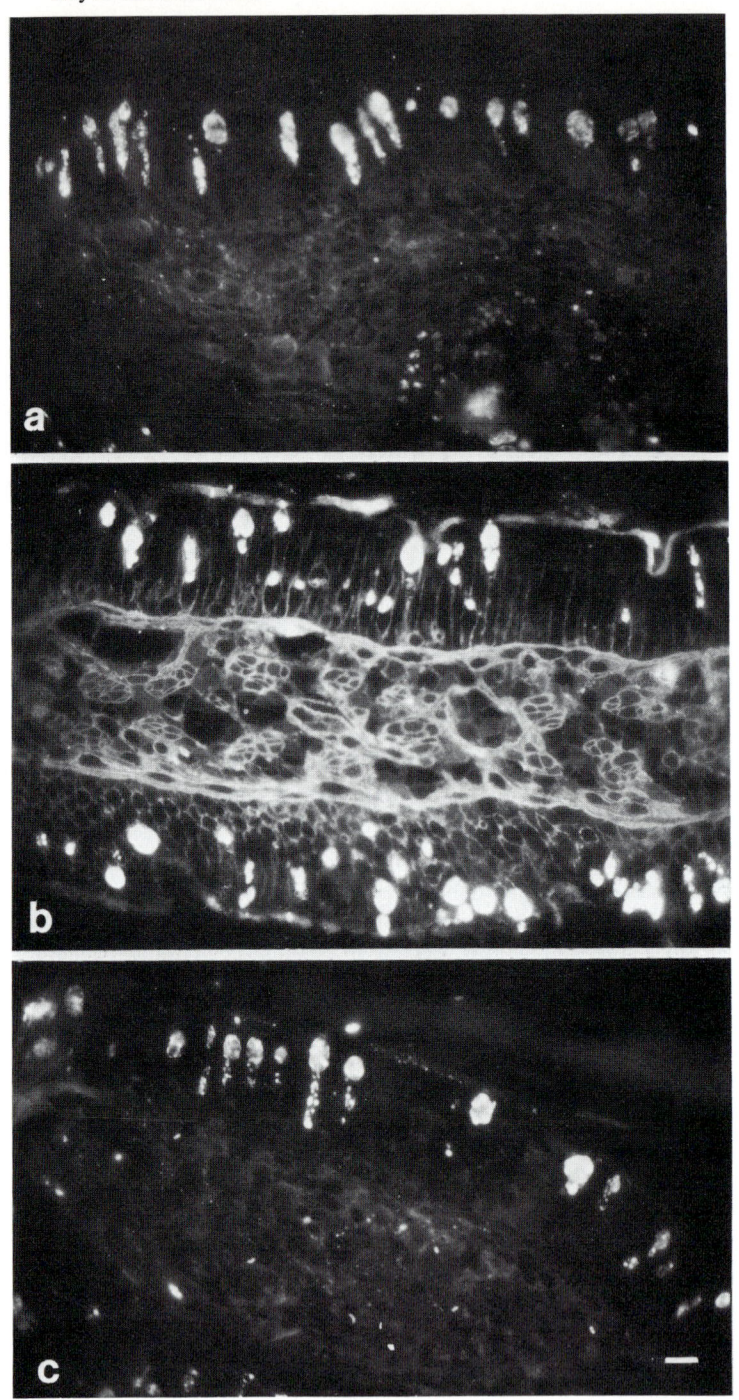

Fig. 1. Indirect immunofluorescence studies of sections of adult chicken intestine. a) Fixed intestinal tissue demonstrating localization of endogenous CLL-I. b) Staining after incubation of section with purified CLL-I demonstrating available binding sites for the added lectin as well as the endogenous lectin. c) Staining after incubation of section with purified CLL-I in the presence of 300 mM lactose which blocks the binding of the exogenously added lectin. Only the endogenous CLL-I fixed to the tissue is now detected by the antibody. Bar, 10 μm.

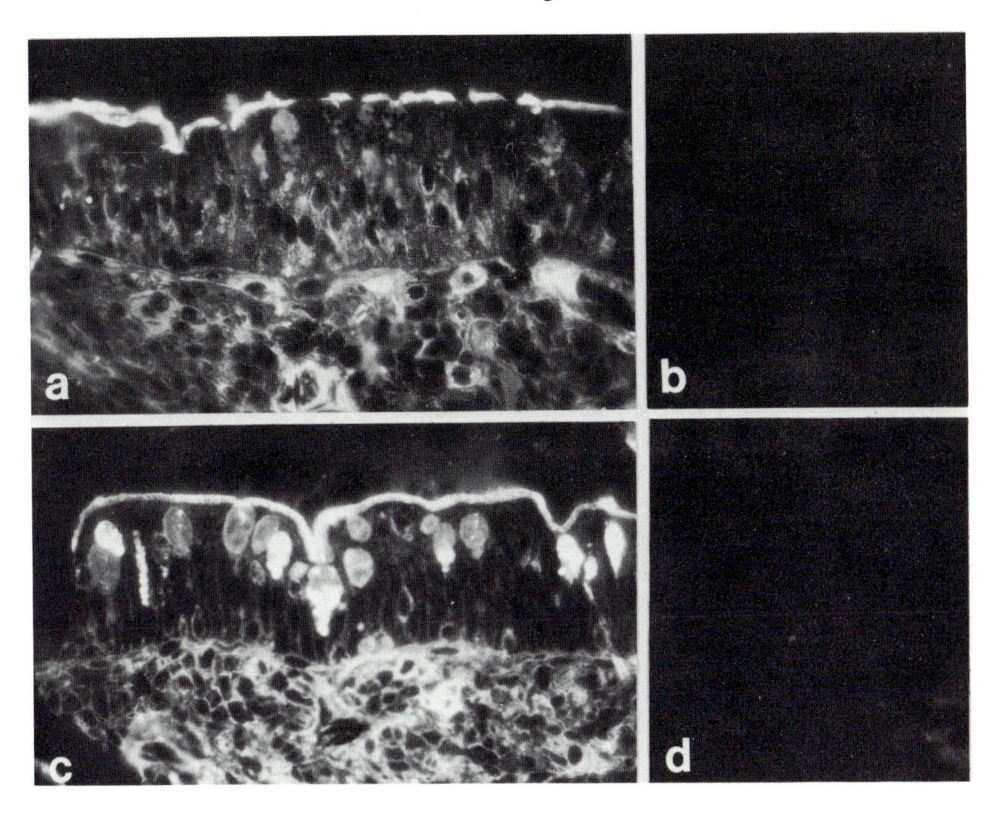

Fig. 2. Indirect immunofluorescence studies of sections of adult chicken intestine stained with different exogenous lectins. a) Fixed sections were incubated with 300 µg/ml of Concanavalin A, washed, then reacted with a rabbit immunoglobulin fraction containing antibodies directed against Concanavalin A followed by fluorescent goat anti-rabbit IgG. b) As in (a) but with incubation of Concanavalin A performed in the presence of 300 mM α-methyl mannoside. c) As in (a) but the exogenous lectin added was purpurin, a lactose-specific lectin from D purpureum and the first antibody was a rabbit immunoglobulin fraction raised against purpurin. d) As in (c) but purpurin was added to the sections in the presence of 300 mM lactose.

Lectin Binding to Cultured Cells

Chick embryo fibroblasts contain little or no endogenous CLL-I detectable by hemagglutination activity or radioimmunoassay [Beyer, Ceri, and Barondes, unpublished results]. When these cells were stained by indirect immunofluorescence, no surface lectin could be detected (Fig. 6a). However, fibroblast cell surfaces bound considerable exogenous CLL-I (Fig. 6b).

DISCUSSION

Lectins from plant and invertebrate sources which specifically bind to various carbohydrate moieties have been extensively used to examine the distribution of oligosaccharide-containing materials in cells and tissues. By using labelled lectins as histological reagents it

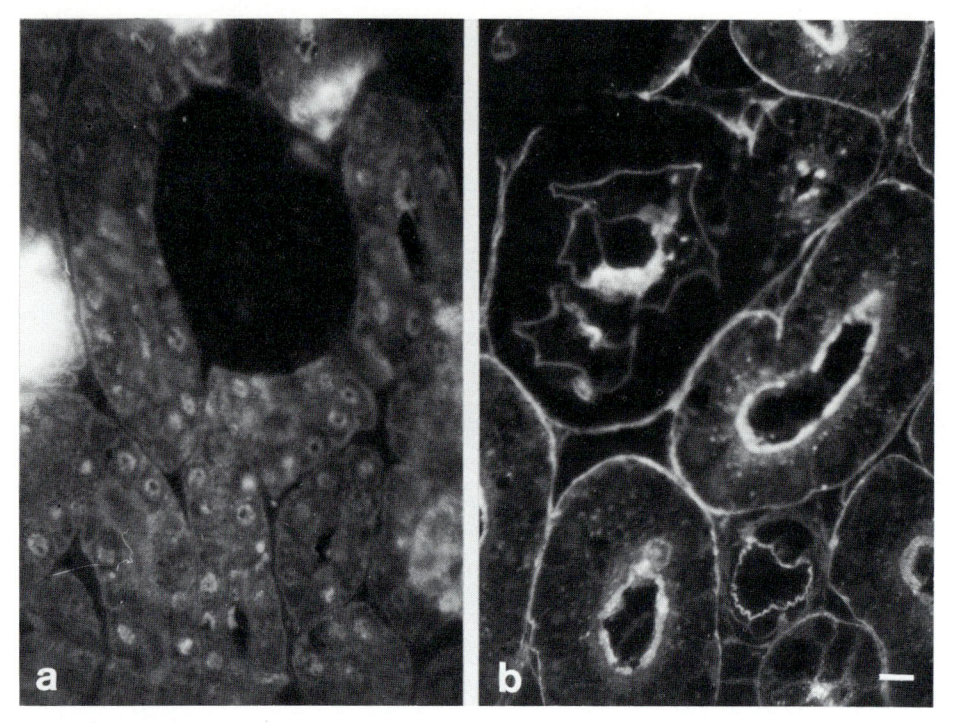

Fig. 3. Indirect immunofluorescence studies of sections of adult kidney cortex. a) Localization of endogenous CLL-I performed as in Figure 1a. b) Localization of both endogenous and added exogenous CLL-I as in Figure 1b. Controls in which the exogenous CLL-I was added in the presence of 300 mM lactose gave results indistinguishable from Figure 3a. Bar, 10 μm.

has been possible to show changes in embryonic [11] or tumor cell surfaces [12], differences in cell surface oligosaccharides of different cell types within an organ [13], and differential distribution of specific oligosaccharides within a tissue [14–16]. The present study describes the first attempt to localize binding sites for a vertebrate lectin within various tissues of the organism from which the lectin itself was derived.

In this report, we have shown that a lectin, CLL-I, detectable in a number of chicken tissues can bind to a variety of cellular and extracellular sites in the chicken. In some tissues, such as intestine, sites not normally occupied by the endogenous lectin can bind considerable exogenous lectin. These binding sites are found especially in intercellular spaces and basement membranes. The sites are not specific for the chicken lectin since another lactose-binding lectin, purpurin, also binds to them. In other chicken tissues such as pancreas, extracellular sites appear saturated by endogenous lectin; but similar sites surrounding kidney tubules are not saturated and bind considerable added CLL-I. Adult heart not only has little or no endogenous CLL-I but also few available binding sites.

The function of lectins in the tissues in which they are normally found is not presently understood. In the cellular slime molds where this problem has been studied extensively, lectins have been implicated in specific cell interactions including cell–cell adhesion [17]. Evidence that plant lectins play a role in cellular interactions has also been presented [18]. Although CLL-I is abundant at intracellular sites [5], its presence on cell surfaces [5, 8] and at extracellular sites [4] suggest that it may sometimes also function in this

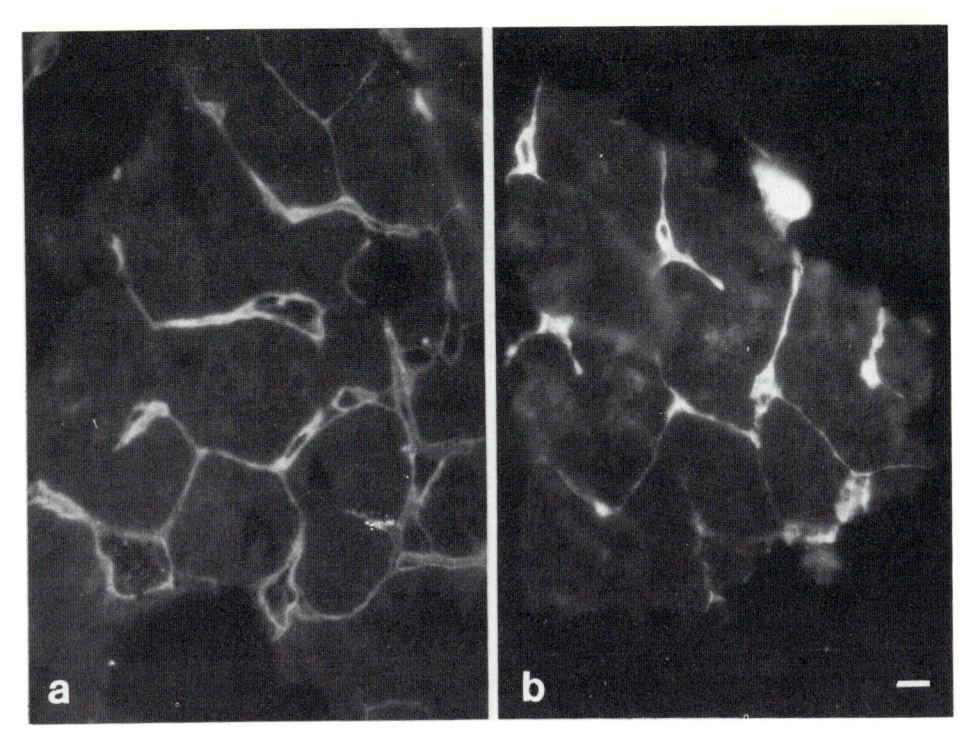

Fig. 4. Indirect immunofluorescence studies of sections of chicken pancreas. a) Localization of endogenous CLL-I performed as in Figure 1a. b) Localization of both endogenous and exogenously added CLL-I determined as in Figure 1b. Note that there was not much binding of exogenous CLL-I suggesting that there are few, if any, receptors for this lectin in this tissue or that they are already largely saturated by the endogenous lectin. Bar, 10 µm.

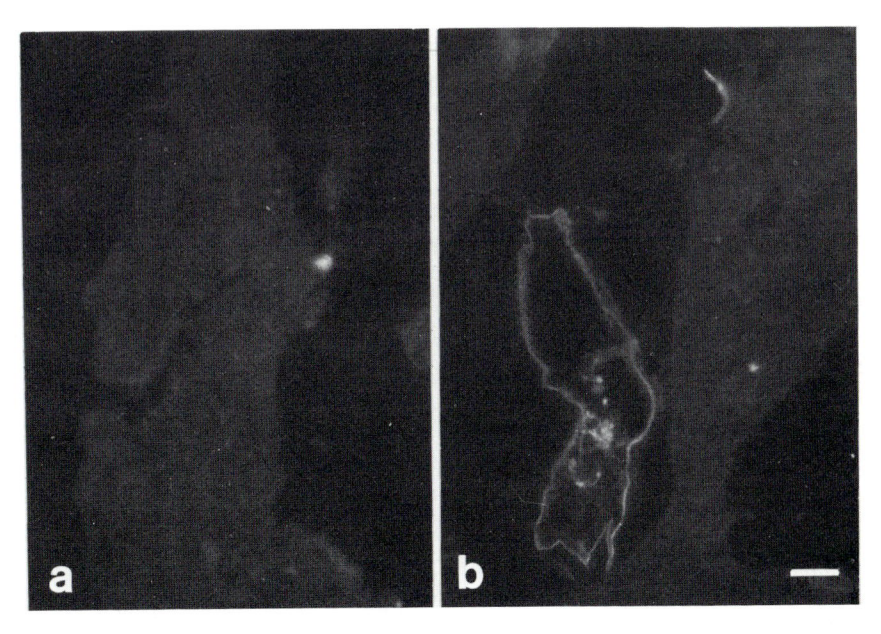

Fig. 5. Indirect immunofluorescence studies of adult heart. a) Localization of endogenous CLL-I as in Figure 1a showing very faint diffuse staining. b) Localization of exogenously added CLL-I determined as in Figure 1b. Note that there is no significant binding of the exogenous lectin to heart muscle or intercellular material. Bar, 10 µm.

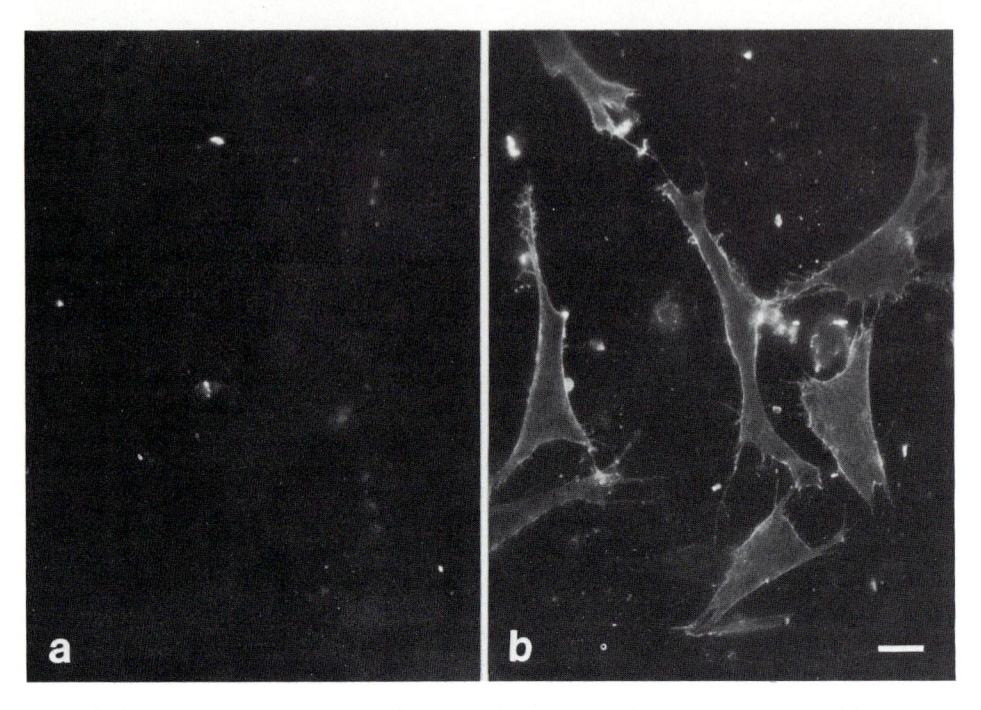

Fig. 6. Indirect immunofluorescence studies of fixed cultured chick embryo fibroblasts. a) Staining in the absence of exogenous CLL-I, as in Figure 1a. b) Staining after addition of exogenous CLL-I as in Figure 1b. Note that the added CLL-I binds to the surface of the fibroblasts. This binding was completely blocked by 300 mM lactose. Bar, 10 µm.

manner. Indeed, some evidence has been presented suggesting a role for this lectin in the fusion of vertebrate muscle cells [19, 20]. The present finding that receptors for added CLL-I are concentrated extracellularly as well as on fibroblast cell surfaces supports a possible role for this lectin in interactions between cells or between cells and extracellular materials.

The significance of the unoccupied binding sites for exogenous CLL-I in various chicken tissues is also not known. These sites are clearly not highly specific for CLL-I since purpurin, another lactose-binding lectin, will occupy them also. Specificity of receptors is also argued against by the finding that a mammalian lactose-binding lectin which is very similar to CLL-I binds to a large variety of substances with terminal β-galactose residues [21, 22] which are common in many cellular oligosaccharides. However, the finding that CLL-I, like other known lectins, does not have a very narrow range of binding specificities in no way precludes their possible role in biologically significant interactions with oligosaccharide receptors in the tissue sites where they are found. Rather, the specific role of lectins may be determined by the factors that determine their sites of deposition within and around cells and by the factors determining the deposition of the various oligosaccharide substances to which they can bind. The fact that specific binding sites for CLL-I are localized to certain regions of certain tissues indicates the potentiality for the interaction of such sites with cells containing CLL-I on their surface or with CLL-I secreted from cells.

ACKNOWLEDGMENTS

This research was supported by grants from the United States Public Health Service (HD13542), the McKnight Foundation, and the Veterans Administration. Eric Beyer is a graduate student supported by a Medical Science Trainee Award (GM07198). The assistance and invaluable advice of Dr. Howard Ceri, Dr. Kiyoteru Tokuyasu, and Patricia Haywood are gratefully acknowledged.

REFERENCES

1. Nowak TP, Haywood PL, Barondes SH: Biochem Biophys Res Commun 68:650, 1976.
2. Kobiler D, Barondes SH: Dev Biol 60:326, 1977.
3. Den H, Malinzak DA, Rosenberg A: Biochem Biophys Res Commun 69:621, 1976.
4. Beyer EC, Tokuyasu KT, Barondes SH: J Cell Biol 82:565, 1979.
5. Nowak TP, Kobiler D, Roel LE, Barondes SH: J Biol Chem 252:6026, 1977.
6. Kobiler D, Beyer EC, Barondes SH: Dev Biol 64:265, 1978.
7. Beyer EC, Zweig SE, Barondes SH: J Biol Chem 255:4236, 1980.
8. Gremo F, Kobiler D, Barondes SH: J Cell Biol 78:491, 1978.
9. Barondes SH, Haywood PL: Biochem Biophys Acta 550:297, 1979.
10. Ceri H, Shadle PJ, Kobiler D, Barondes SH: J Supramol Struc 11:61, 1979.
11. Oppenheimer SB: Curr Top Dev Biol 11:1, 1977.
12. Nicolson GL: Int Rev Cytol 39:89, 1974.
13. Maylie-Pfenninger MF, Jamieson JD: J Cell Biol 80:77, 1979.
14. Peters BP, Goldstein IJ: Exp Cell Res 120:321, 1979.
15. Bretton R, Bariety J: J Histochem Cytochem 24:1093, 1976.
16. Mazzuca M, Roche AC, Lhermitte M, Roussei P: J Histochem Cytochem 25:479, 1977.
17. Barondes SH: In Curtis and Pitts (eds): "Cell Adhesion and Motility." Cambridge: Cambridge University Press, 1980, pp 309–328.
18. Sequeira L: Ann Rev Phytopathol 16:453, 1978.
19. Gartner TK, Podleski TR: Biochem Biophys Res Commun 67:972, 1975.
20. Podleski TR, Greenberg I: Proc Natl Acad Sci USA 77:1054, 1980.
21. DeWaard A, Hickman S, Kornfeld S: J Biol Chem 251:7581, 1976.
22. Childs RA, Feizi T: FEBS Lett 99:175, 1979.

Human T-Cell Growth Factor: Parameters for Production

Francis W. Ruscetti, James W. Mier, and Robert C. Gallo

Laboratory for Tumor Cell Biology, National Cancer Institute, National Institutes of Health, Bethesda, Maryland 20205

Using conditioned media (CM) from phytohemagglutinin (PHA)-stimulated peripheral blood lymphocytes (PBL) we observed long-term selective growth of T-cells from normal human donors. This T-cell growth was continuously dependent on addition of a factor called T-cell growth factor (TCGF). The optimal method for preparing highly active CM from single donor PBL involves the addition of mitomycin C-treated B-lymphoblastoid cell lines to the mixture of PBL and PHA. A number of different cell lines greatly augmented the production of TCGF in 18/18 cases. Preparation of plasma membranes from the Daudi cell line could replace the intact cells in the production of TCGF but those from the cell line, Molt-4, could not. Since the cell surface of Daudi possesses HLA-D antigens but not HLA-A, B, and C, and Molt-4 has HLA-A and B and not HLA-D, it is possible that the Ia antigens (HLA-DR$_W$ in man) are important in the release of TCGF. Using this method for growth factor production, an analysis was made concerning the events necessary for lymphocyte activation and the requirements for production and release of TCGF. Removal of PHA 12 hr after incubation had no effect on lymphocyte transformation but decreased TCGF release by 90%. In addition, colchicine and cytosine arabinoside inhibited DNA synthesis but had no effect on TCGF release. Little or no TCGF activity was present after cellular protein synthesis was inhibited by puromycin and cycloheximide. These results suggest that TCGF production: a) requires protein synthesis; b) requires binding of the stimulating agent; c) can occur in a non-dividing cell, probably a terminally differentiated T-cell, without the need for cellular proliferation; and d) needs the assistance of an adherent cell which probably is a monocyte-macrophage. The ability to produce TCGF from single human donors will allow better understanding of the nature and action of TCGF.

Key words: T-cell growth factor, T-cell proliferation, cellular regulation, B-lymphoblastoid cell lines

The discovery of a soluble factor named T-cell growth factor (TCGF) in crude PHA-stimulated leukocyte conditioned media (Ly-CM), made it possible to selectively maintain normal human T-lymphocytes in continuous culture for the first time [1, 2]. This observation prompted attempts to maintain antigen-activated T-cells in continuous culture, and it

Received May 6, 1980; accepted June 24, 1980.

0091-7419/80/1302-0229$02.60 © 1980 Alan R. Liss, Inc.

is now evident that antigen-specific cytotoxic and helper T-cells may be grown for extended periods with retention of specificity using TCGF [3–10]. It now appears that the antigen stimulates the initial activation of the T-cell and confers specificity on the cell (ie, the first signal) [2, 11–14]. TCGF, while not involved in the activation process, is required as a second signal to cause the clonal expansion of activated T-cells [8, 12–14].

The method used in our laboratory [1, 2] for the prolonged culture of T-cells resulted in a saturation density of $1.5–2.0 \times 10^6$ cells/ml 4–5 days after an initial concentration of 5×10^5/ml. Subsequent reports by other investigators [8, 11] in both human and animal systems demonstrated that the factor-cell ratio is crucial for growth. In their systems, T-cell growth was optimal between 10^4 and 10^5 cells/ml, and their factor preparations would not sustain growth above 2×10^5 cells/ml. This emphasized the need for more defined methods for TCGF production and for some easy, reproducible way to quantitate factor levels or at least to compare the potency of different factor preparations in a more direct way than increase in cell number. Supernatants from mitogen or antigen-stimulated murine or rat spleen cells were potent sources of TCGF [7, 11]. However, attempts to produce highly active supernatants from mitogen-stimulated human peripheral blood leukocytes from a single donor were largely unsuccessful.

In the present study, procedures were developed that allowed an analysis of the cellular and molecular requirements for the production of human TCGF. In addition, it was found that a rapid assay for ^{3}H-TdR incorporation into cultured T cells, previously described for murine T-cells [11], could be used as a measure of human TCGF activity if used under precisely defined conditions.

MATERIALS AND METHODS

TCGF Production From Multiple Donors

Commercially collected heparinized whole blood was mixed with 0.3 ml of plasmagel (Associated Biomedics Systems, Buffalo, NY) per ml of whole blood and incubated without agitation at 37°C for 1 hour. The leukocyte-rich plasma was collected in such a way as to minimize red blood cell contamination and diluted in an equal volume of serum-free tissue culture medium. This plasma-culture medium mixture was then passed through nylon fiber filled columns (Dupont type-200 nylon, Dupont Chemical, Wilmington, DE) at a rate of 125 ml per min and the eluate from the column was centrifuged for 20 min at 1,000g. The cell pellet was resuspended in a small volume of culture media and mixed with 10–20 cell pellets from other ABO-compatible blood donors. To prepare conditioned media containing TCGF, a leukocyte concentration of 10^6 cells/ml was mixed with 1% human plasma and 0.1 ml PHA-P (Difco, Detroit, MI) per 100 ml of cell suspension and incubated for 72 hours. After 72 hr supernatant fluid was collected, filtered through sterile membranes, and stored at −20°C.

TCGF Production From Single Donors

Leukocyte-rich plasma samples from single donors were kindly provided by the staff of the NIH Blood Bank. Routinely, $5–10 \times 10^8$ leukocytes were obtained after ficoll-hypaque separation [15]. Supernatants to be assayed for TCGF activity were prepared by varying numerous parameters. First, 10^6 peripheral blood leukocytes/ml in RPMI 1640 medium and 5% v/v heat-inactivated fetal bovine serum (Reheis Chemical Co, Phoenix, AZ) were stimulated by PHA-M at 1% v/v (Difco, Detroit, MI) at 37°C. Culture supernatants were harvested cell-free after 72 hr of incubation and stored at $-20°C$ until assayed for TCGF activity. Within this basic scheme, the following parameters were changed: 1) cell concentrations (2×10^6, 5×10^6, 10^7 cells/ml); 2) media conditions (1% human plasma, 0.5% bovine serum albumin) (Sigma Chemical Co, St. Louis, MO), serum-free; 3) different mitogens at their optimal dose (Concanavalin A, protein A, leuco-agglutinin) (all from Pharmacia Chemicals, Uppsala, Sweden), Pokeweed mitogen, lipopolysaccharide (Difco, Detroit, MI); and 4) times of incubation (12, 24, 48, 72, 96, and 120 hr). Second, adherent cells in PBL were removed by nylon fiber filtration as described above. The resulting lymphocyte populations were greater than 99% pure as judged by esterase strains [16]. Third, 2×10^5 mitomycin C (25 μg/ml) treated cells of a B-lymphoblastoid cell line were added per ml of mixture of PBL and PHA. The cell lines used were Daudi, HSB-2, Molt-4, SB, and Raji (obtained from American Type Culture Collection, Rockville, MD). Finally, PBL were depleted of T-cells by binding to sheep erythrocytes (E-rosette) as previously described [17]. Also, T-cells were lysed using specific antisera and complement [18].

Cultured T-Cells

Long-term cultures of human T-cells were maintained as described previously [1, 2]. These cells were resuspended at 5×10^5 cells/ml in 50% v/v crude Ly-CM prepared from multiple donors. At day 4 or 5, the cells reached their saturation density of $1-2 \times 10^6$ cells/ml and were fed by dilution to 5×10^5 cells/ml. This process was repeated every 4–5 days.

TCGF Assay

TCGF-dependent long-term cultured T-cells were used as the target cells in the assay. The cells to be used as target were in culture for at least 20 days and were at their saturation density for 24 hours. These conditions are necessary to insure that only activated T-cells which need TCGF for growth remain in the cultures. The cells were washed free of growth media and resuspended in media containing 5% FCS. Cells were placed in 96-well microplates (No. 3596 Costar, Inc, Cambridge, MA) followed by serial dilutions of the TCGF sample to be assayed. After 48 hr of incubation, 0.5 μCi of ^{3}H-TdR (specific activity 3 Ci/mM, Schwartz/Mann, Inc, Orangeburg, NJ) were added to each well and cultured for an additional 4 hours. Cultures were harvested onto glass fiber filter strips and ^{3}H-TdR incorporation was determined as previously described [19]. CM were concentrated by Amicon filtration using a UM05 membrane (Amicon Corp, Lexington, MA) or fractionated by DEAE-Sepharose as previously described [14].

Plasma Membrane Isolation

The procedure was modified from one described by Jett et al [20]. Briefly, 5 grams of cells were extensively washed with Earle's balanced salt solution and then 90% glycerol was slowly added in small increments to the cells until a 30–40% final concentration was reached. Lysis buffer (10 mM Tris/HCL, pH 7.4, containing 1 mM $MgCl_2$ and 1 mM $CaCl_2$) equal to 1% of the original volume is rapidly added to the cell pellet and then gently homogenized. The cell-free lysate was centrifuged at 700g for 10 minutes. A crude membrane was obtained by sedimentation on a sucrose cushion at 30,000g for 90 minutes. The crude membrane preparation was then sedimented in a 25–50% sucrose gradient at 100,000g for 16 hours. The gradient fractions were assayed for thymidine 5'-phosphodiesterase [21], a cell membrane marker; β-glucuronidase [22], a lysosomal marker; succinate dehydrogenase [22], a mitochondrial marker; and glucose-6-phosphatase [23], a microsomal marker.

Using this procedure, there was a 15-fold increase in phosphodiesterase activity in the membrane preparation but there was still appreciable lysosomal contamination.

Metabolic Inhibitor Studies

The effect of colchicine, cycloheximide, dexamethasone (Sigma Chemical Co, St. Louis, MO), cytosine arabinoside (Upjohn Co, Kalamazoo, MI), and puromycin (Nutritional Biochemical Co, Cleveland, OH) on the production of TCGF were tested as described in Results. These substances were always used as freshly prepared solutions. Since most of these chemicals were toxic in the TCGF assay, the CM were thoroughly dialyzed against a Tris/HCL, pH 7.4, buffer before assaying for TCGF.

RESULTS

A typical growth factor assay is shown in Figure 1. Cultured T-cells in the presence of increasing concentrations of TCGF incorporated ^{3}H-TdR in a dose-dependent manner. In the crude Ly-CM prepared from multiple donors, inhibition of TCGF-mediated thymidine incorporation was seen at the highest doses of TCGF, but this effect was not observed with TCGF after it had been fractionated on an ion-exchange column [14]. This material then had a sigmoid dose-response curve which is characteristic of several growth factors. The observed inhibition with Ly-CM could be caused by PHA toxicity and, indeed,

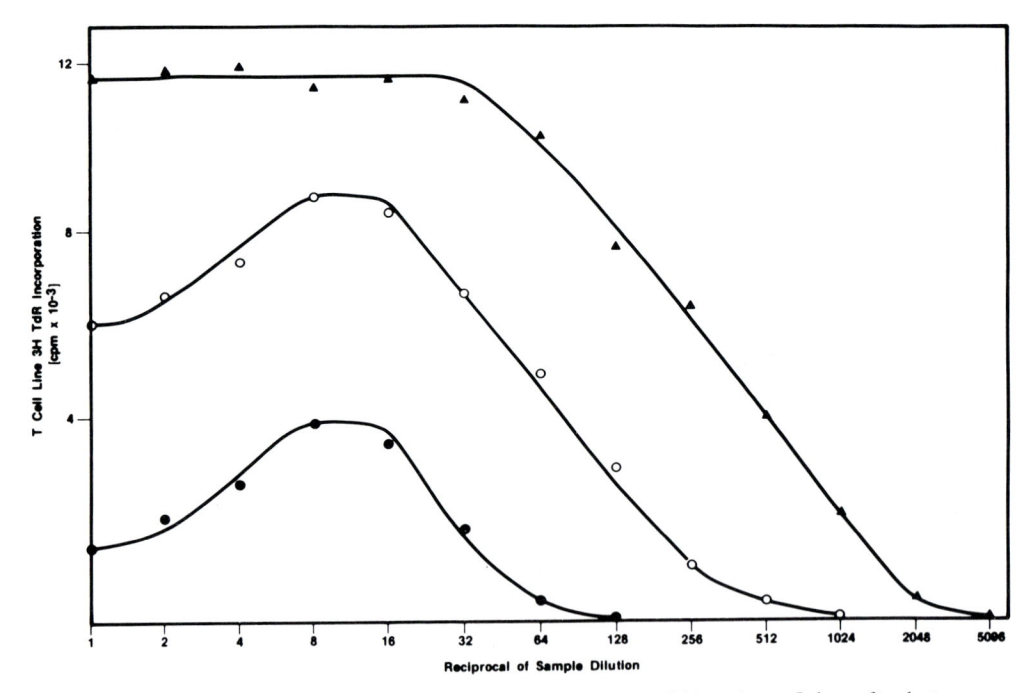

Fig. 1. The TCGF microassay. Human T-lymphocyte cell lines (day 56 in culture; 5 days after last addition of TCGF). ^{3}H-TdR incorporation plotted against the reciprocal of the sample dilution. The samples tested were from 72-hr conditioned media from PHA-stimulated pooled allogeneic peripheral blood leukocytes cultured at 10^6 cells/ml. Crude CM (solid circles), ammonium sulfate fractionated CM (open circles), and DEAE-Sepharose chromatographed TCGF (triangles) were assayed.

addition of optimal mitogenic concentrations of PHA-P to some TCGF-dependent T-cell lines resulted in a 30–60% inhibition of thymidine incorporation (data not shown). PHA was only weakly mitogenic (2–3-fold more thymidine incorporation into cultured T-cells than controls) in the absence of TCGF and was unable to sustain the growth of T-cells in culture.

Quantitation of TCGF Activity

The ^{3}H-TdR incorporation data generated by stimulating cultured T-cells with serial dilutions of a standard TCGF preparation was plotted on probability paper as a percentage of the maximum ^{3}H-TdR counts incorporated [11, 24]. The sample dilution that corresponded to the 50% level of the maximum incorporation could be defined as 1 unit/ml of TCGF activity. In a similar manner, other TCGF-containing media can be compared to the standard, and a value in units can be determined [11]. For example, the data displayed in Figure 1 was plotted on probability paper (Fig. 2). The factor preparations were calculated as 1 unit/ml, 7.8 units/ml, and 18.5 units/ml for the crude Ly-CM, the ammonium sulfate fractionated sample, and the TCGF after ion-exchange chromatography, respectively. The repeated use of the same TCGF standard allowed comparison of data obtained from different assays. All the remaining data presented is expressed in units/ml as compared to the same TCGF standard.

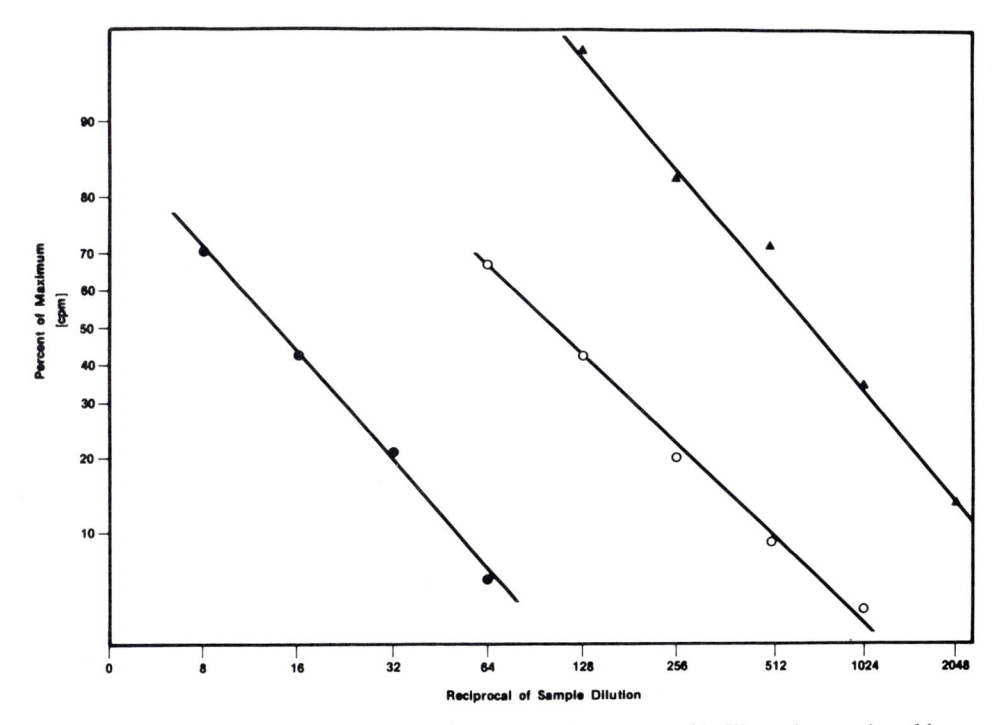

Fig. 2. Probit analysis of the TCGF microassay data. The data presented in Figure 1 were plotted by probit analysis (a value of 100% is given to the cpm of ^{3}H-TdR incorporated at the highest dose TCGF; all other data points are a percentage of the maximum). If the 50% value of the crude CM is assigned a value of 1 unit/ml, then the ammonium sulfate fractionated sample is 7.8 units/ml, and the chromatographed sample is 18.5 units/ml.

TCGF Production From Leukocytes From Single Donors

Attempts to generate CM containing TCGF activity from PBL obtained from individual donors are summarized in Table I. Using 18 different samples, little or no TCGF activity was produced. Varying the time of incubation did not appreciably alter the results. Although supernatants harvested after 24 and 48 hr of incubation stimulated the incorporation of more ^{3}H-TdR than 72 hr, they were not able to maintain the growth of T-cells even after concentration by Amicon filtration and could therefore not be considered positive for TCGF. The removal of adherent cells by nylon fiber filtration, use of other mitogens such as Con A and pokeweed, and the fractionation of inactive CM by ion-exchange chromatography to remove most of the mitogen [14] did not result in the recovery of TCGF activity (Table I). CM generated from leukocytes pooled from multiple donors were always much more active than CM from single donors.

Another approach to produce TCGF from leukocytes from a single donor was suggested by the work of Oliver and his colleagues [25]. They demonstrated that the requirement for the presence of pooled lymphocytes from many donors in assisting human T-cells to become cytolytic for autologous leukemic blast cells [26] could be replaced by using Daudi, a lymphoblastoid cell line. When PBL from single donors were stimulated with PHA in the presence of Daudi at a concentration of 2×10^5 cell/ml, TCGF production was almost as efficient as production using multiple donors (Table II). Addition of mitomycin-treated Daudi cells to a mixture of pooled lymphocytes from 15 donors and PHA led to a substantial increase in TCGF activity. PBL cultured with Daudi cells in the absence of PHA produced little TCGF. Daudi by itself or stimulated with PHA did not produce TCGF. Pokeweed mitogen could replace PHA-M but Con A could not.

TCGF Production Kinetics

To determine the time course of production and release of TCGF, conditioned media were harvested from replicate plates after various times of incubation and then assayed for TCGF (Table III). The kinetics of production were approximately the same for both the multiple and single donor methods. Only minimal activity was present after 12 hr, after which production rapidly increased, reaching a peak at 72 hr. The amount of TCGF in the CM significantly declined after 96 and 120 hr of incubation.

This decline in TCGF activity could be a result of the release of inhibitors of TCGF or nonspecific proteolytic enzymes. To examine this, CM obtained after 72 hr of incubation was diluted twofold with CM harvested after 120 hr incubation or with fresh tissue culture media and then assayed in a microassay for TCGF (Fig. 3). Both samples had equivalent amounts of TCGF, making the presence of an inhibitor in the 120 hr CM unlikely.

Role of Lymphoblastoid Cell Lines in the Stimulation of TCGF Production

In an attempt to better understand the nature of the stimulus supplied by Daudi cells, a number of lymphoblastoid cell lines were tested for their ability to stimulate production of TCGF (Table IV). Molt-4 and HSB-2 were negative and three B-cell lines, SB, Raji and Daudi were positive. Supernatants from these latter cell lines could not replace the need for the cells themselves in the stimulation of TCGF production, suggesting that cell-to-cell contact is necessary. A partial purification of plasma membranes from Daudi and Molt-4 was accomplished by hypotonic lysis of glycerol-loaded cells followed by differential centrifugation. In the case of Daudi cells, but not Molt-4, this crude membrane preparation could replace the need for intact cells in stimulating TCGF production (Table IV). The

TABLE I. T-Cell Growth Factor Production From Human Leukocytes

Cell source of PBL[a]	Samples tested	Mitogen[b] source	Time of incubation (hr)	TCGF Activity[c] (units/ml)
10–20 pooled donors	32	PHA	72	0.8–1.7
10–20 pooled donors	6	none	72	0.0–0.2
Individual donors	18	PHA	24	0.0–0.3
Individual donors	18	PHA	48	0.0–0.3
Individual donors	18	PHA	72	0.0–0.1
Individual donors	5	Con A	48	0.0
Individual donors	5	Pokeweed	48	0.0–0.2
Individual donors (macrophage depleted)	3	PHA	48	0.1–0.3

[a]Peripheral blood leukocytes were isolated as described in Materials and Methods. Macrophages were removed by nylon fiber columns.
[b]PHA-M (1% v/v), Con A (5 μg/ml), and Pokeweed mitogen (15 μg/ml) were used at their optimal mitogenic concentrations.
[c]Cell-free CM were harvested after the indicated time of incubation. The CM were assayed for TCGF activity as described in Materials and Methods.

TABLE II. Generation of Human T-Cell Growth Factor

Responding lymphocyte source[a]	Samples tested	Stimulation source[b]	TCGF Activity[c] (units/ml)
MNC from 15 pooled donors	3	PHA	0.89–1.20
MNC from 15 pooled donors	3	PHA + Daudi	1.30–1.55
MNC	18	PHA	0.00–0.20
MNC	18	PHA + Daudi	0.84–0.96
MNC	3	Pokeweed + Daudi	0.70–0.86
MNC	3	Con A + Daudi	0.00–0.18
MNC pretreated with anti-T-cell sera + C'	3	PHA + Daudi	0.20–0.35

[a]Mononuclear cells (MNC) were isolated as described in Materials and Methods.
[b]PHA-M (1% v/v), Con A (5 μg/ml), and Pokeweed mitogen (15 μl/ml) were used at optimal mitogenic concentrations. Daudi cells were mitomycin C-treated for 30 min and, after extensive washing, were added to the cultures at 2×10^5/ml.
[c]Cell-free CM were harvested after 72 hr of incubation. The CM were assayed for TCGF activity as described in Materials and Methods.

fact that the membrane preparation can be diluted tenfold while retaining 60% of maximal stimulation suggested that this activity is a membrane-bound activity. Daudi has HLA-D antigens but not HLA-A, B, and C on its cell surface [27], while Molt-4 has HLA-A and B but not HLA-D antigens [25]. This effect, whether related to Ia antigens or not, seemed to be directly involved in TCGF production, since mixing supernatants from Daudi and from PHA-stimulated PBL did not enhance TCGF activity.

Cellular Requirements for TCGF Production

The removal of T-cells using anti-T-cell antisera in the presence of complement markedly reduced TCGF production (Table II), as did the depletion of E-rosette positive cells from the PBL, suggesting that a mature T-cell is required (Fig. 4). E-rosette positive

TABLE III. Kinetics of TCGF Production From Human Leukocytes

Culture conditions[a]	Time of incubation[b] (hr)					
	12	24	48	72	96	120
MNC from 15 pooled donors + PHA	0.0–0.1	0.4–0.6	0.9–1.0	1.0–1.2	0.7–0.8	0.5–0.6
MNC + PHA + Daudi	0.0–0.2	0.6–0.7	0.8–0.9	0.9–1.0	0.5–0.6	0.2–0.3

[a]Mononuclear cells (MNC) were isolated as described in Materials and Methods. PHA-M (1% v/v) and Daudi cells, mitomycin C-treated for 30 min, were added to the cultures where indicated.
[b]Cell-free CM were harvested at the times indicated. The CM were assayed for TCGF activity as described in Materials and Methods and the units were calculated by probit analysis using a highly active CM as an arbitrary standard of 1 unit/ml.

cells were able to produce only about 30% of the activity produced by PBL, suggesting that another cell type is also required. To test for the role of adherent cells, nylon fiber columns were used to separate PBL into adherent and non-adherent fractions. Adherent cells (10^5/ml) did not produce detectable TCGF. The addition of 5×10^4 adherent cells to 10^6 E-rosette positive cells resulted in the reconstitution of all the activity found with stimulation of PBL. Human thymocytes, even with the addition of adherent cells, were unable to produce TCGF.

Macromolecular Requirements for TCGF Production

Studies to determine the relationship between DNA synthesis and TCGF production were performed using the mitotic inhibitors colchicine (1 μg/ml) and cytosine arabinoside (10 μg/ml) (Fig. 5). These inhibitors markedly reduced the incorporation of ^{3}H-TdR into DNA but had little effect on the TCGF production. The PHA-stimulated control cultures incorporated over 30,000 cpm of ^{3}H-TdR, while none of cultures containing inhibitors incorporated over 800 cpm; yet these cultures still produced 80% of the TCGF produced in the control. The use of inhibitors of protein synthesis, cycloheximide (50 μg/ml) and puromycin (10 μg/ml), virtually abolished the production of TCGF (Fig. 5). That this inhibition was due to a lack of protein synthesis and not to cellular toxicity was indicated by the finding that the cells could resume making TCGF after the removal of puromycin (data not shown). The removal of PHA after 12 hr of incubation markedly reduced the amount of TCGF production in a 72-hr incubation (Fig. 5).

DISCUSSION

The discovery that CM obtained from PHA-stimulated peripheral blood leukocytes could facilitate the long-term growth of T-lymphocytes [1, 2] has permitted the development of cloned T-cell lines which retained specificity after many months in culture [3–10]. It became apparent that the growth kinetics of cultured T-cells differed in several laboratories [8–11], suggesting that the TCGF-T-cell ratio was important in maintaining growth. It then became important to determine more precisely the cellular and molecular requirements for TCGF production.

First, it was necessary to develop an assay system that permitted a simple and rapid quantitation of TCGF levels in many preparations. A microassay based upon ^{3}H-TdR incorporation in factor-dependent cultured T-cells stimulated by CM containing TCGF, first

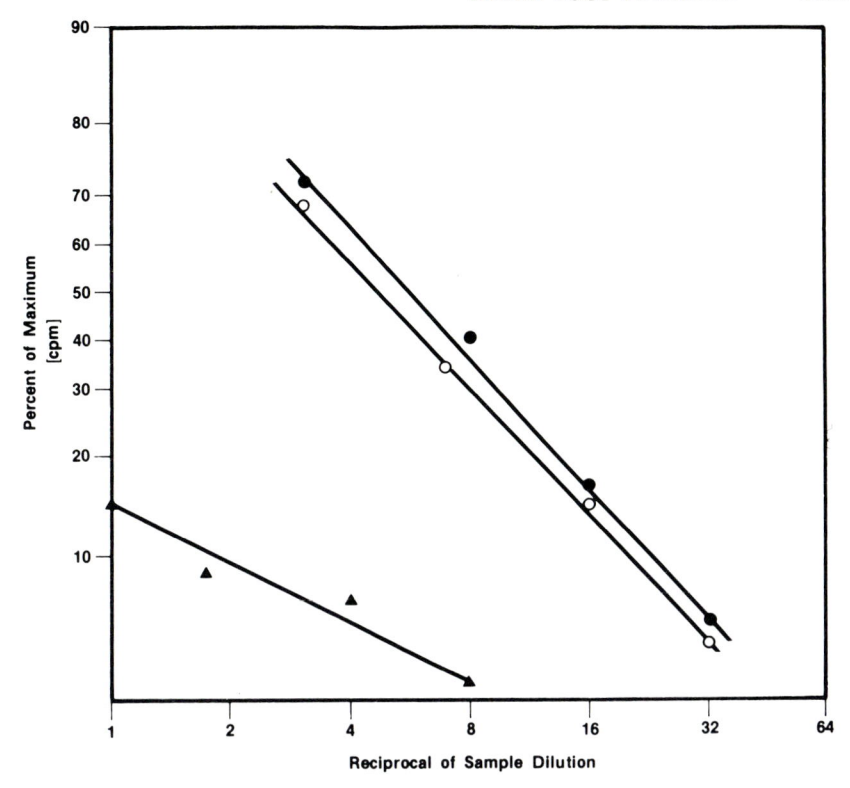

Fig. 3. Lack of inhibitors of TCGF in Ly-CM. Conditioned media for TCGF assay were prepared as described in Materials and Methods. Activity in units was determined by assaying against the standard in Figure 1. TCGF 1:1 dilution with RPMI (solid circles), TCGF 1:1 dilution with 120-hr Ly-CM (open circles), and 120-hr Ly-CM (triangles) by itself were assayed.

TABLE IV. Lymphoblastoid Cell Enhancement of Human TCGF Production

Addition to PBL[a]	HLA Antigens[b]	TCGF Activity[c] (units/ml) +PHA	−PHA
None	–	0.1–0.2	0.0
SB	A,B,D	0.7–0.9	0.0
HSB-2	A,B	0.1–0.3	0.0
Molt-4	A,B	0.0–0.3	0.0
Molt-4 plasma membrane	A,B	0.0	0.0
Daudi	D	0.8–1.3	0.2–0.3
Daudi plasma membrane	D	0.6–1.0	0.1
Daudi plasma membrane (1:10 dilution)	D	0.5–0.8	0.0

[a]Peripheral blood leukocytes were isolated as described in Materials and Methods. Plasma membranes were prepared by hypotonic lysis of glycerol loaded cells followed by differential centrifugation and were adjusted to a protein concentration of 100 μg/ml [27].
[b]The HLA antigens on the cell surface were taken from the literature for Daudi [28], Molt-4 [25], SB, and HSB-2 [29].
[c]PHA-M was used at 1% v/v. Cell-free CM were harvested after 72 hr of incubation at 37°C. The CM were assayed for TCGF activity as described in Materials and Methods and the units of activity were calculated by probit analysis using a highly active CM as an arbitrary standard of 1 unit/ml.

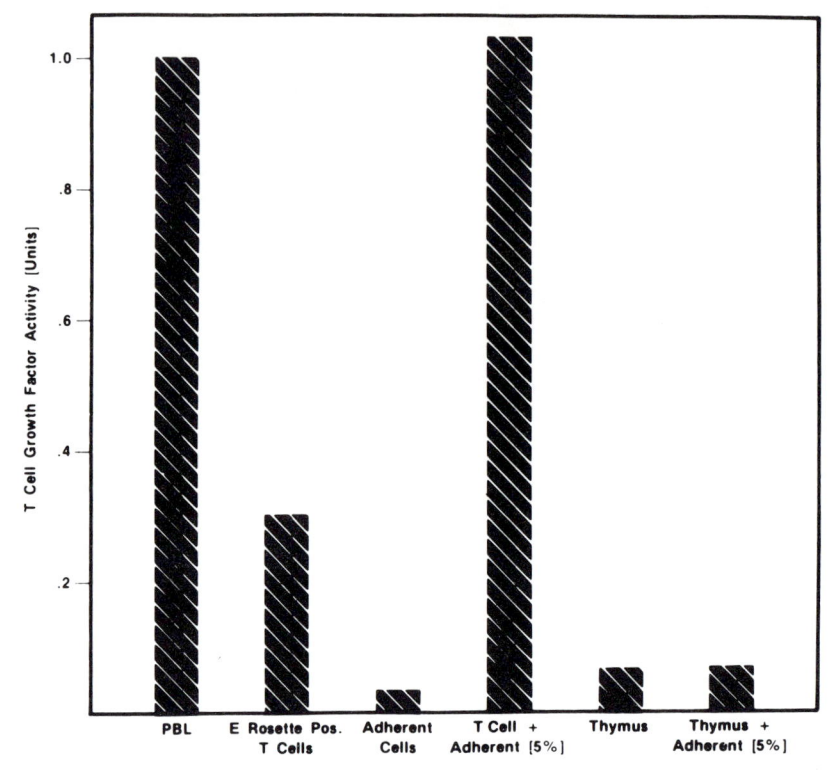

Fig. 4. Cellular requirements for TCGF production. Peripheral blood leukocytes from a single donor at 10⁶ cell/ml in the presence of PHA and Daudi cells as described in Materials and Methods. Ablation of T-cells with anti-T-cell serum was complement mediated. Adherent cells were removed by nylon wool, and mature T-cells were separated by sheep erythrocyte rosetting. Experiments were done in triplicate with the standard error less than 10%.

described for murine T-cells [11], was modified for use with human T-cells (Fig. 1). To ensure specificity of the assay, the T-cells need to be in culture for at least 20 days and require the addition of TCGF for further growth. The microassay requires only 0.2 ml of sample to generate a dose-response curve, which allows several samples to be conveniently tested in one day. The use of crude Ly-CM in the assay resulted in inhibition of TCGF-directed ^{3}H-TdR incorporation at high doses of Ly-CM. The addition of PHA at the optimal mitogenic dose to the assay also inhibited some but not all cultures by 30–60%. Inhibitors, however, could be removed by fractionation of Ly-CM by ion-exchange chromatography [14], permitting the generation of sigmoid dose-response curves (Fig. 2).

The statistical methods used to determine interferon levels [24] were employed to quantitate the amount of TCGF in a given CM. A standard lot of TCGF was arbitrarily assigned a value of one unit/ml. The data from a microassay were analyzed by assigning a value of 100% to the highest value of cpm of ^{3}H-TdR incorporated and expressing all other points as a percentage of the maximum. By plotting these values against the logarithm of the media dilution on probability paper, a straight line results where the points fall within two standard deviations of the mean (50% point) (Fig. 2). The TCGF activity of an unknown sample can then be determined from the standard curve.

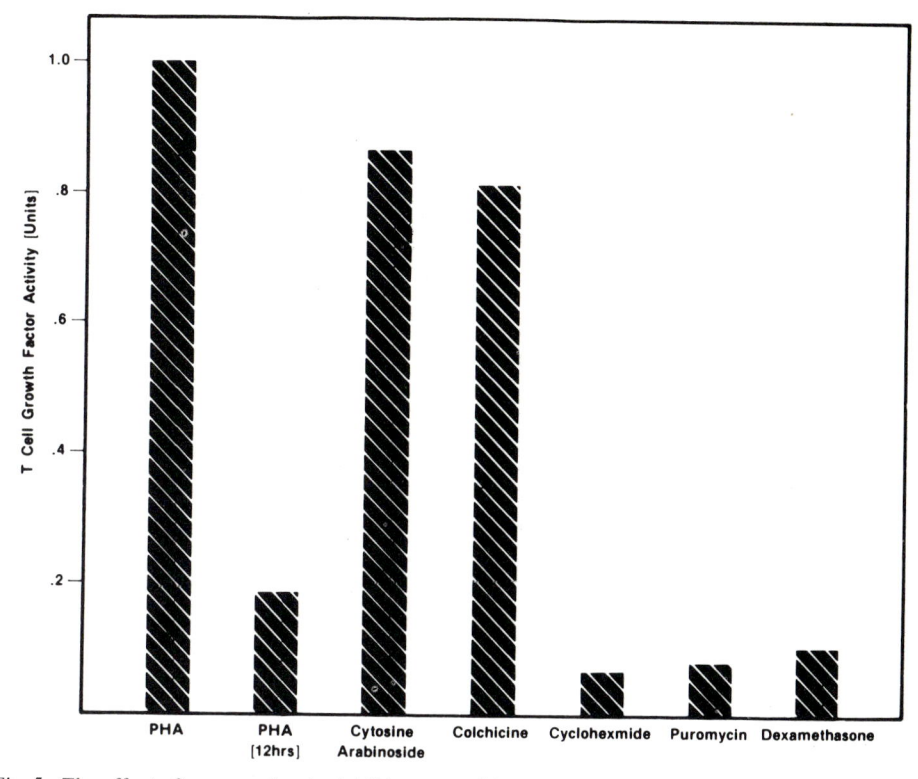

Fig. 5. The effect of macromolecular inhibitors on TCGF production. CM was prepared after 72 hr of incubation of single donor PHA-stimulated peripheral blood lymphocytes with mitomycin C-treated Daudi cells. The CM was dialyzed against 0.15 M phosphate buffered saline. Control CM, CM from cytosine arabinoside (10 μg/ml) treated cultures, CM from colchicine (100 μg/ml) treated cultures, CM from cyclohexamide (50 μg/ml) treated cultures, CM from puromycin (10 μg/ml) treated cultures, and CM from dexamethasone (10^{-6}M) treated cultures were assayed for TCGF activity. Experiments were done in triplicate with standard error less than 10%.

Production of TCGF by PHA-stimulated PBL from pooled multiple donors and from single donors was compared. Using a CM from PBL of multiple donors which was able to support the proliferation of T-cells for sixty days as the standard, we found that in all batches prepared from multiple donors' PBL gave comparable activity (Table I). In testing the PBL from 18 individual donors, most of the samples had less than 10% of the activity of the standard but some, in particular those CM harvested after 24 hr of incubation, stimulated as much as 30% of the incorporation of thymidine as standard (Table I). However, neither concentration by Amicon filtration nor fractionation by ion-exchange chromatography produced samples which were able to maintain growth of T-cells through two successive passages. We considered these samples negative, although development of a more sensitive assay, such as a radioimmune assay, may allow detection of low levels of TCGF in these CM. Use of other mitogens, removal of adherent cells, and use of different serum components did not alter these findings (Table I).

The addition of mitomycin C-treated Daudi cells to PHA-stimulated PBL from individual donors produced TCGF activity equal in amount to that produced by pooled multiple donors (Table II). This enhanced production has been observed with other lymphoblastoid cell lines [30]. The rationale for the use of B-cell lines was to provide a source of allogeneic stimulation and thus eliminate the need for mixing several donors. However, a recent study claimed that pooling allogeneic lymphocytes had no effect on TCGF production [31]. Whatever the basis for the effect, it is clear that not all lymphoblastoid cell lines are active (Table IV). Mixing supernatants of B-cell lines and of PHA-stimulated PBL did not increase TCGF production. This and the observation that crude plasma membrane preparations of Daudi cells could replace intact cells in production of TCGF shows that B-cell lines stimulate TCGF production rather than effect the action of TCGF.

It is possible that Ia-like antigens may be involved in the enhancing effect of lymphoblastoid cell lines in producing TCGF. Daudi cells, which stimulate TCGF production, have HLA-D antigens but no HLA-A, B, or C on their cell surface. Molt-4, a cell line which did not stimulate TCGF production, has HLA-A and B but not D antigens. It was found that over 90% of the cultured human T-cells possess HLA-DR$_w$ (Ia-like) antigens after 8 days of culture and that these cultured T-cells also express the same HLA-DR$_w$ phenotype as the fresh B-cells of the donor [18], suggesting that the Ia antigens were not absorbed onto the T-cells from the Ly-CM. Thus, it seems to us that Ia-like antigens play an important role in T-cell proliferation, and its relationship to TCGF is under study in our laboratory.

Mature T-cells are required for TCGF production. Depletion of E-rosette positive cells from mononuclear cells and the use of anti-T-cell sera, which is cytotoxic to T-cells, greatly reduced the production of TCGF (Fig. 4). Some evidence with murine cells indicate that the T-cell which releases TCGF belongs to a functionally distinct T-cell subset. Using specific antisera to lyse T-cell subsets, Wagner and Rollinghoff [32] found that the removal of Ly 2,3+ cells had no effect on TCGF production, whereas removal of the Ly 1+ cells completely abrogated its release. Interestingly, Ly 1+ murine T-cells have Ia antigens on their cell surface.

In addition to activation by a lectin or antigen, TCGF requires another signal derived from an adherent cell population, probably a macrophage (Fig. 4). It has been shown that in TCGF production by murine cells the adherent cell requirement can be replaced by soluble macrophage products and that TCGF production may be stimulated by macrophage-derived lymphocyte activating factor (LAF) [33, 34].

However, it is difficult to distinguish whether biologically active molecules like LAF, PHA, or Ia antigens act solely on TCGF production or also act synergistically with TCGF to amplify TCGF action. In addition, it is not known whether some TCGF producer cells survive in long-term cultures so that when these substances are put directly in the TCGF assay, the increase in ^{3}H-TdR incorporation observed is due to low level TCGF production during the assay period.

Studies on the regulation of production of TCGF with inhibitors of macromolecular synthesis showed that inhibition of DNA synthesis and cellular proliferation had no effect on TCGF release (Fig. 5). However, little or no TCGF activity was present if inhibitors of protein synthesis were used. Thus, TCGF production can occur in a non-dividing cell, probably in a terminally differentiated T-cell without the need for proliferation but requiring binding of a stimulating agent and protein synthesis. It is not clear whether the T-cell involved in production proliferates in response to TCGF or not. All the normal long-term T-cell lines that have been developed, whether they were helper, cytotoxic, or polyclonal in function, have not contained cells capable of producing enough TCGF to maintain growth of the cultures [2–10].

ACKNOWLEDGMENTS

The authors thank the staff of the NIH blood bank for providing leukocyte enriched plasma, and Ms S. Krishnan and Diana Linnekin for their excellent technical assistance.

REFERENCES

1. Morgan DA, Ruscetti FW, Gallo RC: Science 193:1007, 1976.
2. Ruscetti FW, Morgan DA, Gallo RC: J Immunol 119:131, 1978.
3. Gillis S, Smith KA: Nature 268:154, 1977.
4. Kasakura S: J Immunol 118:43, 1977.
5. Gillis S, Baker PE, Ruscetti FW, Smith KA: J Exp Med 148:1093, 1978.
6. Stausser JL, Rosenberg SA: J Immunol 121:1491, 1978.
7. Rosenberg SA, Swartz S, Spiess PJ: J Immunol 121:1951, 1978.
8. Kurnik JT, Gronvik KO, Kirmura KA, Lindbolm JB, Skoog VT, Sjoberg O, Wigzell H: J Immunol 122:1255, 1979.
9. Watson J: J Exp Med 150:1510, 1979.
10. Schrier MH, Tees R: Int Arch Allergy Appl Immunol 61:227, 1980.
11. Gillis S, Ferm MM, Ou W, Smith KA: J Immunol 120:2072, 1978.
12. Smith KA, Gillis S, Baker PE, McKenzie D, Ruscetti FW: Proc NY Acad Sci 332:423, 1979.
13. Bonnard GD, Yasaka K, Jacobson D: J Immunol 123:2704, 1979.
14. Mier JW, Gallo RC: Proc Natl Acad Sci USA (in press).
15. Boyum A: Scand J Clin Lab Invest 21 (Suppl. 97):224, 1968.
16. Yam LT, Li CY, Crosby W: Am J Clin Pathol 55:283, 1971.
17. Dean JH, Silva JS, McCoy JL, Leonard CB, Cannon GB, Herberman RB: J Immunol 118:1449, 1977.
18. Metzgar RS, Bertoglio J, Anderson JK, Bonnard GB, Ruscetti FW: J Immunol 122:949, 1979.
19. Oppenheim JJ, Schecter B: In: "Manual for Clinical Immunology," p 81, 1976.
20. Jett M, Seed TM, Jamieson GA: J Biol Chem 252:2134, 1977.
21. Koerner JF, Sinsheimer RL: J Biol Chem 228:1049, 1957.
22. Earl DCN, Korner A: Biochem J 94:721, 1965.
23. Ames BN: Enzymol 8:115, 1966.
24. Jordan GW: J Gen Virol 14:49, 1972.
25. Lee SK, Oliver RTD: J Exp Med 147:912, 1978.
26. Zarling JM, Raich PM, McKeough M, Bach FH: Nature 262:691, 1976.
27. Lowry OH, Rosebrough NJ, Farr AL, Randall RJ: J Biol Chem 193:265, 1951.
28. Jones EA, Goodfellow PN, Kennert RH, Bodmer WF: Somatic Cell Genet 2:483, 1976.
29. Royston I, Pitts RB, Smith RW, Graze PR: Transplant Proc 7:531, 1975.
30. Bonnard GB, Yasaka K, Maca RD: Cell Immunol 51:390, 1980.
31. Alvarez JM, Silva A, de Landazari MO: J Immunol 123:927, 1979.
32. Wagner H, Rollinghoff M: J Exp Med 148:1523, 1978.
33. Larsson EL, Iscove NN, Continuho A: Nature 283:664, 1980.
34. Smith KA, Lachman LB, Oppenheim JJ, Favata MF: J Exp Med 151:1551, 1980.

The Extracellular Matrix and the Control of Proliferation of Vascular Endothelial and Vascular Smooth Muscle Cells

D. Gospodarowicz, I. Vlodavsky, and N. Savion

Cancer Research Institute and the Department of Medicine, University of California Medical Center, San Francisco, California 94143

> In this short review we describe the observations which have led us to conclude that one of the most important components involved in modulating cell proliferation in vitro, and probably in vivo as well, may be the extracellular matrix upon which cells rest.

Key words: extracellular matrix, FGF, vascular endothelial cells, vascular smooth muscle cells, aging, differentiation

The extracellular matrix (ECM), or basal lamina, produced by cells is the natural substrate upon which cells migrate, proliferate, and differentiate in vivo. Although the exact nature and composition of ECMs are still to be elucidated, they are composed in large part of different types of collagen, glycosaminoglycans, proteoglycan [1, 2], and glycoproteins, among which is fibronectin [3], which has been shown to be a ubiquitous component of various types of extracellular matrices.

Numerous studies have dealt with the tissue differentiation induced by the extracellular matrices and basal lamina, but few studies have dealt with their effect on cell proliferation. Since in most organs cell proliferation precedes cell differentiation, it is likely that both proliferation and differentiation could be controlled directly or indirectly by the substrate upon which the cells rest [4]. This is particularly true of tissues which have only one developmental option. Early in vitro studies done on the control of cell proliferation by fibroblast growth factor (FGF) have indicated that FGF could modulate, either directly or indirectly, the rate of production of various cell surface proteins and extracellular matrix components. This was observed in at least two cell models, the vascular and corneal endothelia. In this short review we describe the observations which have led us to conclude that, in vitro, one of the most important components involved in modulating cell proliferation is the extracellular matrix produced by the cells and upon which cells rest.

Received March 27, 1980; accepted May 7, 1980.

0091-7419/80/1303-0339$05.60 © 1980 Alan R. Liss, Inc.

FGF AND THE PROLIFERATION OF CULTURED ENDOTHELIAL CELLS

Significant limitations on the culture of vascular and corneal endothelial cells have been imposed by the slow doubling time of these cultures, which can be passaged only at a high cell density if precocious senescence is to be avoided [5–7]. In contrast, if one adds FGF to the cultures, no such limitations are encountered. When primary cultures of cells from the bovine aortic endothelium are initiated with few cells (3 cells/cm^2), the development of a monolayer will depend on the presence of FGF in the culture medium [8, 9]. In 10% calf serum alone, small colonies develop from cell aggregates during the first few days but the cells look unhealthy and proliferate slowly, overlapping each other and becoming vacuolated and considerably enlarged. However, if FGF is added to the cultures, the population doubling time is reduced to as little as 18 hours and the cultures soon become confluent, exhibiting both the morphology and functions characteristic of the vascular endothelium in vivo. These include the formation of a highly contact-inhibited cell monolayer (Fig. 1) composed of closely apposed and flattened cells characterized by a nonthrombogenic upper cell surface and an active secretion of a fibrillar basement membrane composed mostly of fibronectin, proteoglycans, and collagen types III, IV, and V as well as by the synthesis of Factor VIII antigen and prostacyclins (Fig. 2), both of which are produced specifically by the vascular endothelium in vivo. Such cultures maintained and propagated in the presence of FGF divide with an average doubling time of 18 hours when seeded at either a high (up to 1:1000) or low split ratio (Fig. 3A). Upon reaching confluence, the cells adopt a morphological configuration similar to that of the confluent culture from which they originated (Fig. 4D).

In contrast, seeding of the same cells in the absence of FGF, even at a low (1:6) split ratio, results in a much longer doubling time (60–78 hr) (Figs. 3A,C) and in a strikingly different morphology. When seeded at a high split ratio (1:128) and in the absence of FGF, the cells proliferated poorly. The alterations in growth behavior were best demonstrated after 3 to 4 passages (15–20 generations) in the absence of FGF (Fig. 3A). The cells, by then 4- to 6-fold larger in size, failed to adopt a non-overlapping monolayer configuration even after being split at a 1:4 ratio. Instead, at sparse density they were flattened and highly spread (Fig. 4A) and at confluence grew on top of each other, leaving intercellular spaces (Fig. 4B). These cells exhibited a short lifespan, as reflected by vacuolization and cell degeneration after 30 generations.

Endothelial cells maintained for 3 passages in the absence of FGF and beginning to show a greatly increased average doubling time were still capable of responding to the mitogen, since addition of FGF resulted in a greatly increased growth-rate of the cultures (Fig. 3C) which at confluence adopted the characteristic endothelial configuration of a cell monolayer composed of small, highly flattened and closely apposed cuboidal cells, rather than their prior appearance as overlapping and large cells. In the case of cultured vascular endothelial cells, therefore, FGF is not only a potent mitogen which acts on cell plated at a clonal density but, by inducing the cultures to proliferate actively, it also serves as a survival factor, as is reflected by a substantial delay in the ultimate senescence of the cells.

VASCULAR ENDOTHELIAL CELLS AND AGING

There has been considerable discrepancy in the results of various laboratories regarding the number of doublings that vascular endothelial cells maintained in vitro will undergo. Schwartz [10], using Waymouth medium and endothelial cells derived from the adult

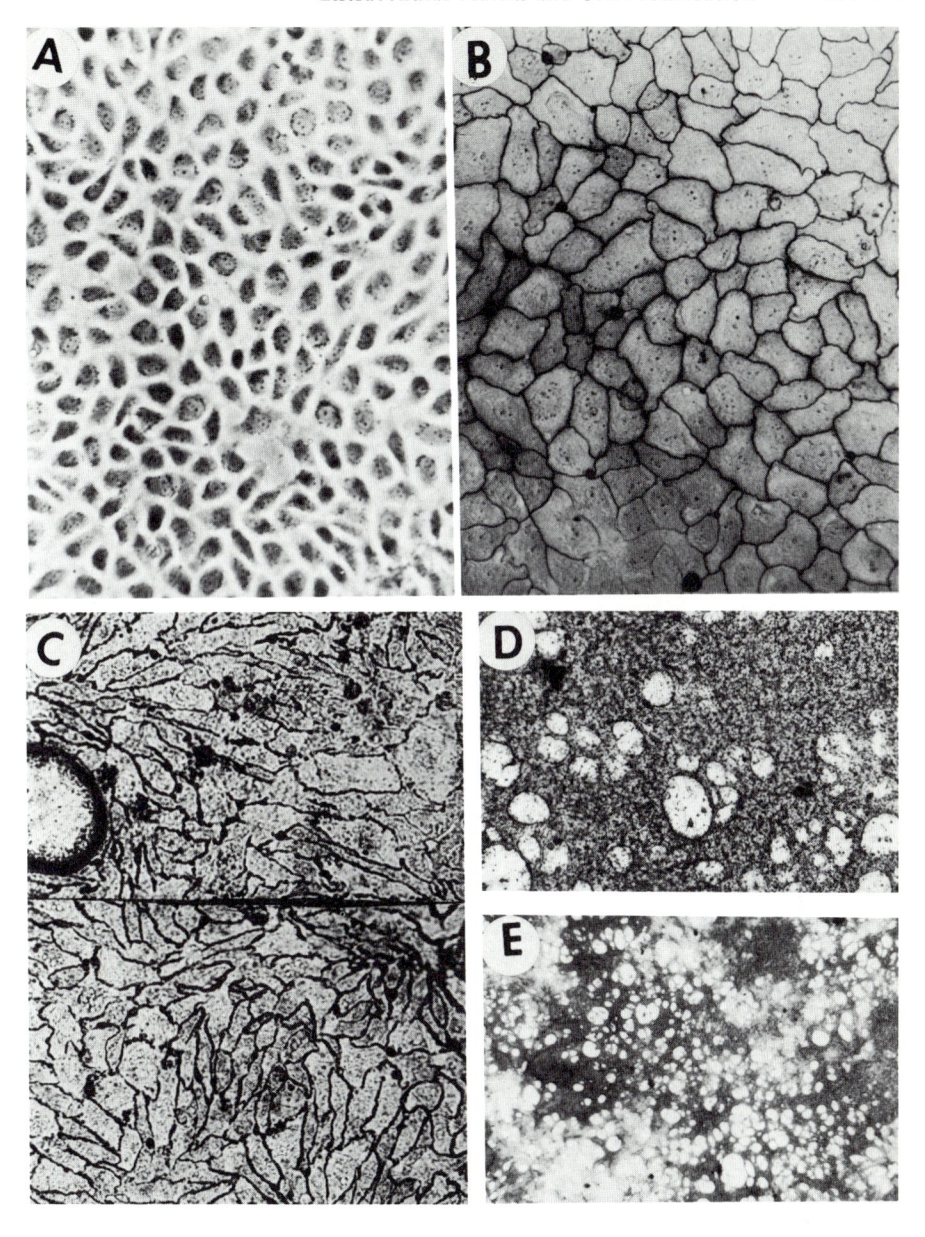

Fig. 1. A) Monolayer of vascular endothelial cells (13 passages, 65 generations) maintained in the presence of 10% calf serum and 100 ng/ml FGF. The cells are polygonal, closely apposed, and have an indistinct border (phase contrast, ×150). B) Same monolayer as in A, but stained with silver nitrate to show the cell borders (×210). The cells showed the same organization as did preparations of endothelium stained in situ. C) Adult bovine aortic endothelial cells maintained in culture for 3 weeks. The cultures were then exposed to alizarin red. The intercellular border stained bright red (×105). D) Adult bovine aortic endothelial cells were maintained in culture for a month and then exposed to 0.5% Triton X in PBS. After removal of the cell monolayer, the basement membrane was stained with alizarin red (×70). E) Same as D, but at a magnification of ×28; the basement membrane stained bright red.

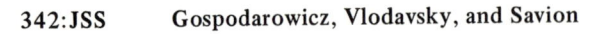

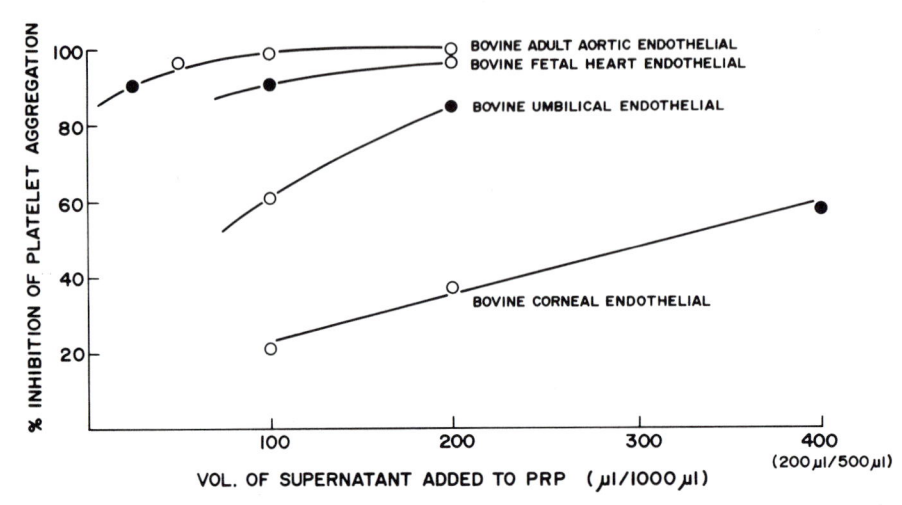

Fig. 2. Generation of PGI$_2$ activity by endothelial cell monolayer. Confluent vascular and corneal endothelial cell cultures were washed twice with DMEM, 15 mM Hepes and incubated in the presence of 1 ml of the same medium supplemented with arachidonic acid (1×10^{-4} M) for 5 min at 37°C before aliquots (200 μl, 25 μl) of the medium were added to stirred platelet-rich plasma (PRP). The PRP and medium were incubated together for 5 min at 37°C with stirring in a Payton Aggregation Module. Aggregation was induced by ADP ($5-8 \times 10^{-4}$ M) and was biphasic. The total reaction volume was 1 ml. Inhibition of the primary phase of the ADP-induced aggregation was monitored as an indication of the generation of prostacyclin-like activity. Aspirin (1×10^{-3} M) inhibited the generation of PGI$_2$-like activity by all four cell cultures. In addition, incubation of the cells with tranylcypromine (5×10^{-3} M), a known inhibitor of PGI$_2$ synthesis, also blocked the generation of inhibitory activity.

aortic arch, was able to passage cultures at a low split ratio (1:2). While the doubling time of early-passage cultures was 4 days, that of late-passage cultures was 7 days. Senescence was observed after 50 generations and could be speeded up if cultures were passaged at a split ratio of 1:4 [10]. Mueller et al [11], using MEM rather than Waymouth medium and cells derived from fetal rather than from adult aortic arch, have reported the development of clonal cell lines. However, the cloning efficiency was extremely low (approx. 1%) and could have led to great selective pressure. Cultures could be passaged at a split ratio of 1:11 for 80 generations, and early-passage cultures had an initial doubling time of 60 hours. This 60-hour doubling time is similar to that reported for ABAE cultures maintained in the absence of FGF [5]. In fact, when one corrects for the low split ratio used by Mueller et al, one discovers that the rate of proliferation of their cultures is the same as that reported by us earlier for vascular endothelial cells maintained in the absence of FGF (Fig. 5). Late-passage cultures had a doubling time of 72–96 hours. Similar results were claimed by McAuslan et al [12], who used medium 199 supplemented with thymidine to grow endothelial cells derived from adult bovine aortic arch. We have reported the cloning of various vascular endothelial cell lines using FGF. Using DMEM and FGF one can get a cloning efficiency as high as 50–75%. The risk of selecting special cell types is therefore greatly reduced. Such cultures maintained in the presence of FGF have been passaged at a high split ratio (1:64 every 5 days or 1:128 every 7 days) and can have average doubling time of 18–20 hours. Perfect contact inhibition in cultures derived from the adult aortic arch have been reported. Although the total generation number one can achieve with

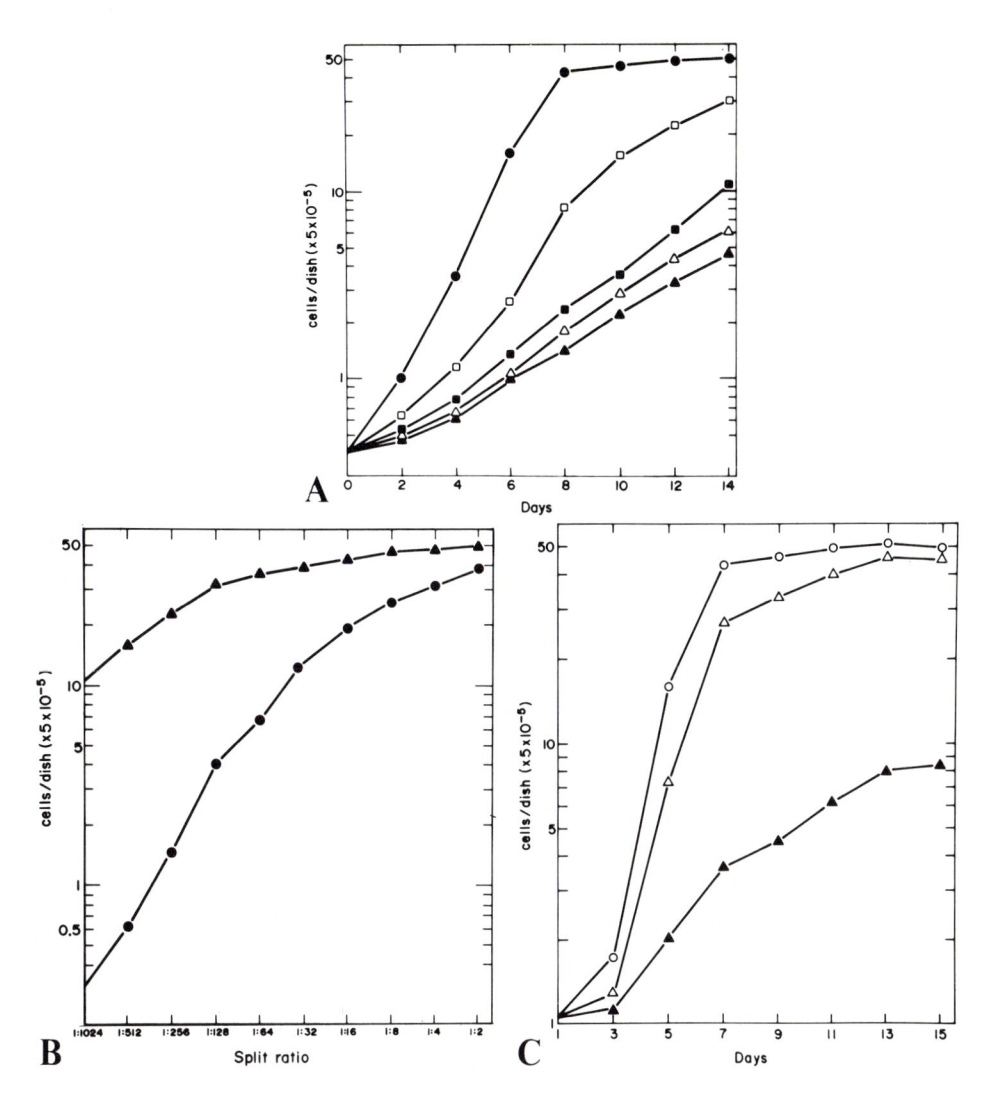

Fig. 3. Growth rate of vascular endothelial cells in the presence and absence of FGF. A) Bovine aortic endothelial cells (ABAE) maintained in the absence of FGF for 1 (□), 3 (■), 5 (△), and 6 (▲) cell passages (each passage representing 3–5 cell doublings and 6–8 days in culture) were seeded into 35-mm dishes (8 × 10^3 cells per dish) in DMEM supplemented with 10% bovine calf serum. The medium was replaced every four days, and duplicate dishes counted every other day. Control cultures (●) which were not subjected to FGF withdrawal were similarly seeded and counted, except that FGF was added every other day. These cells adopted at confluence a perfect monolayer configuration, whereas cells that were cultured (3 or more passages) in the absence of FGF grew on top of each other and formed an unorganized cell layer. B) Confluent endothelial cell cultures maintained with (16 passages, ▲) or without (4 passages, ●) FGF were split at various ratios and cultured in DMEM containing 10% calf serum in the presence and absence of FGF (added every other day), respectively. The medium was replaced after five days and triplicate dishes counted every other day. The number of cells after 9 days in culture is plotted as a function of the split ratio. The seeding level at a split ratio of 1:2 was 1.65 × 10^5 and 1.4 × 10^5 cells per 35-mm dish for cells maintained with and without FGF, respectively. C) Vascular endothelial cells derived from the adult aortic arch (32 passages; 160 generations) were plated into 35-mm dishes (2 × 10^4 cells per dish) in DMEM (H-16) supplemented with 10% calf serum. Duplicate cultures were counted every other day and the medium replaced every 4 days. ▲, Cells after 3 passages in the absence of FGF. No FGF was added. △, Cells after 3 passages in the absence of FGF. FGF was added on the third day after seeding and every other day thereafter. These cells adopted at confluence a perfect monolayer configuration indistinguishable from that of cells that were never subjected to FGF withdrawal. ○, Cells derived from endothelial cultures that were continuously maintained with FGF. FGF was added every other day.

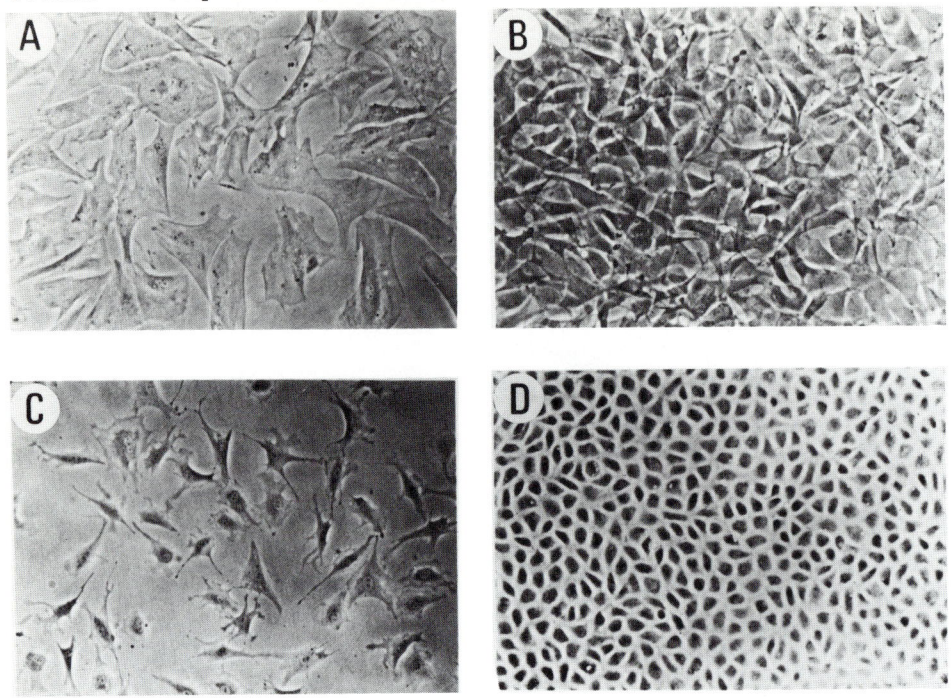

Fig. 4. Morphological appearance of sparse and confluent vascular endothelial cultures maintained with and without FGF (phase contrast, ×150). A) Sparse endothelial cells maintained in the absence of FGF. The cells are 4- to 6-fold larger and spread farther apart than sparse cells (C) seeded and maintained in the presence of FGF. B) Confluent endothelial cells after 4 passages (12 generations) in the absence of FGF. The cells grow on top of each other and in various directions. D) A confluent endothelial monolayer formed by cells (100 generations) maintained in the presence of FGF. The cells are highly flattened, closely apposed, and non-overlapping.

cultures originating from vascular territories as different in age as fetal versus adult and as different in origin as vein versus artery can itself differ greatly, our greatest success has been 390 generations. In all cases, cells maintained in the presence of FGF were capable of making Factor VIII antigen, as well as of producing high level of prostacyclins. Cultures were perfectly contact inhibited, as documented by their (^{3}H)thymidine and (^{35}S)methionine incorporation as a function of cell density and cell morphology. Therefore, in three different laboratories the number of generations that vascular endothelial cells will undergo is as different as 50 [10], 80 to 90 [11, 12], or 390 [13].

But is there really a difference? As shown in Table I, if one corrects for the average doubling time of the cultures, one arrives at the surprising result that, regardless of the number of generations the cells have undergone, the total time cultures have endured is very similar in all cases and averages 284 ± 29 days, despite the wide difference in the way the cultures were maintained (split ratio, medium, as well as vascular territory from which cells originated). Even more remarkable is that the average lifetime of the cultures is similar to that reported for WI-38 cultures passaged weekly at a split ratio of 1:2 for 50 weeks.

This simple observation seems to indicate that aging in culture is more a somatic problem than a fixed program of senescence imprinted in the DNA, since senescence of the cells correlates better with the *total lifetime* of the cultures than with the generation

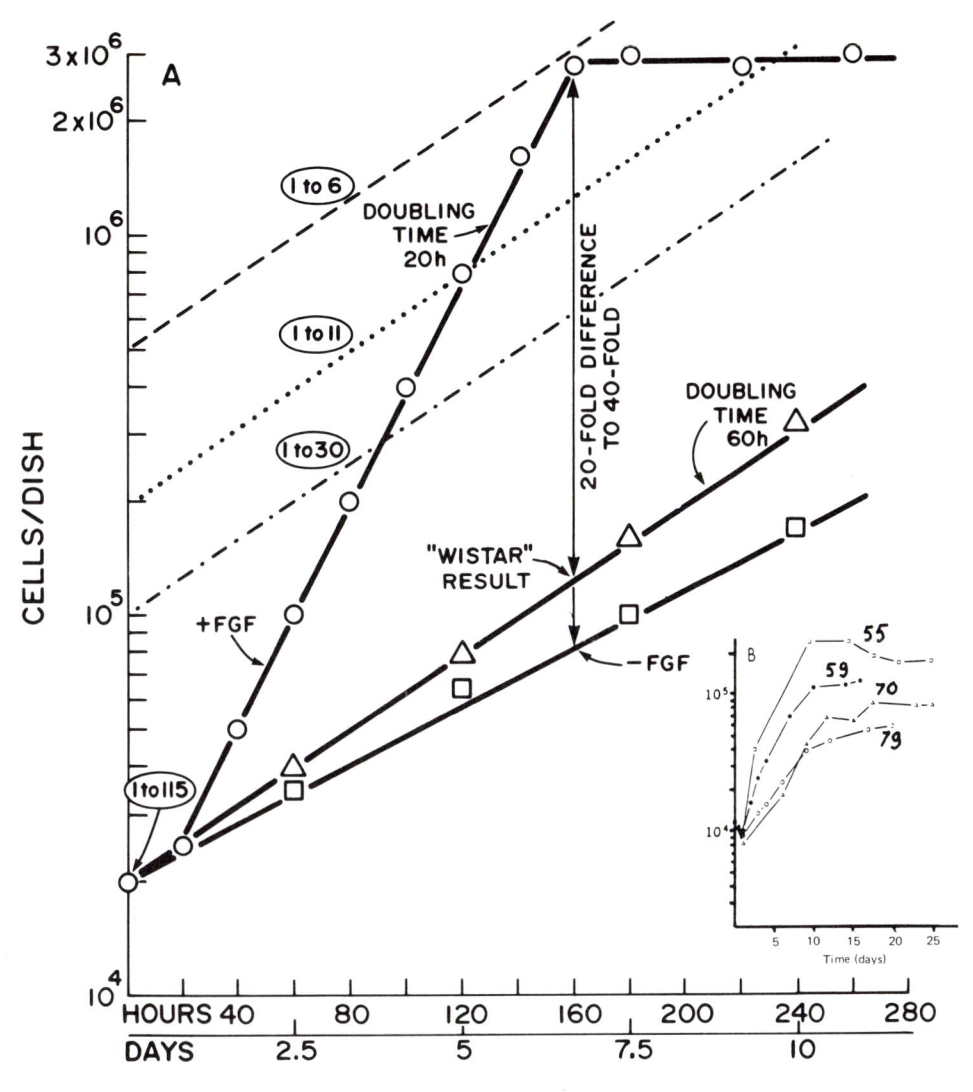

Fig. 5. Rate of proliferation of vascular endothelial cells maintained in the presence or absence of FGF. A) Vascular endothelial cells seeded at an initial cell density of 2×10^4 cells/6 cm dish and maintained in the presence of DMEM (H-16) supplemented with 10% calf serum. In the presence of FGF (○, average doubling time of the culture was 20 hours. In the absence of FGF, it was 72 hours (□). This average doubling time in the absence of FGF is similar to that reported by Mueller et al [11]. Their data are shown in the insert (B). When the optimal proliferative rate of their cultures was recomputed according to the split ratio at which we transferred our cultures (1:115) (Wistar result, △), there was little if any difference in the growth rate of control cultures maintained in the absence of FGF (□). The main difference, however, was that confluence in the presence of FGF was reached *after 6 days* (○). In the case of Mueller et al [11], it would not have been reached *before 17 days*. To reach confluence within 5 days, these cultures would have to be passaged at *a split ratio of 1:6* (– – –). At a split ratio of 1:15, confluence is reached by day 10 (·····), and at 1:30 it is reached by day 13. B) Growth curves as a function of the cumulative population doubling (CPDL) for endothelial cell clone BFA-1c developed by Mueller et al [11]. Cells were inoculated into 25-cm² flasks at densities of 1.0×10^4 to 1.2×10^4 cells/cm² (8 ml medium per flask). Flasks were incubated under standard conditions and given fresh culture medium every 6 or 7 days. At various times after inoculation, duplicate cultures were counted with a Coulter counter to determine density. The CPDLs studied were 55 (□), 59 (●), 70 (△), and 79 (○). Their results are expressed as cell density per cm² as a function of time [11].

TABLE I. Lifetime in Culture of Bovine Vascular Endothelial Cells Maintained in Different Laboratories

	Days in culture	Generation	Day of passage	Split ratio	Doubling time (hr)	Cell number generated from 1 cell
Mueller et al [12]	260	80–90	10–15	1:10	60–72	10^{24-27}
Schwartz [11]	280	50	15	1:2 or 1:4	84–168	10^{15}
Gospodarowicz et al [13]	325	390	5	1:64	18–20	10^{117}
McAuslan et al [12]	270	—	—	—	60–72	—
WI 38	200–250	50	4–5	1:2	96–120	10^{15}

Mean value of lifetime in culture of vascular endothelial cells ± standard error = 284 days ± 28.

number. In this regard, it is interesting to note that if senescence in culture is due to an oxidation problem (generation of free radicals), it makes more sense to consider it a somatic than a genetic problem. The effect of time in culture on mitochondrial replication and the ability of cells to perform metabolic functions adequately has not been studied in this context. Yet mitochondrial replication is as likely a candidate as nuclei replication as a locus for senescence.

If one were to predict the maximal number of generations cells could undergo in vitro based on a lifetime of the cultures of 300 days and the shortest cycle cells can achieve (16 hours), it would be 495 generations. How does this compare with the in vivo situation? Not very well. The basal cell layer of the corneal epithelium renews itself every week. Therefore, based on an average human lifespan of 70 years, each basal cell will go through 3850 generations in vivo. Yet, when put in culture this cell type will go through 50–100 generations at most. One should further point out that clonal proliferation, except in the case of the hematopoietic system, is a rare event in vivo. When a cell divides, in most cases it replaces a dead or a terminally differentiated cell. Usually one of the daughter cells resulting from the replication will lose its ability to proliferate while the other cells will retain it. This leads to an apparent zero growth rate, although cells in some organs such as the gut go through thousands of replications during the lifespan of the individual.

FGF AND THE DIFFERENTIATION OF CULTURED VASCULAR ENDOTHELIAL CELLS

As already pointed out, vascular endothelial cells grown in the presence of FGF adopt, upon reaching confluence, the morphology of a cell monolayer composed of tightly packed and flattened cells. This cell layer, as in vivo, shows an asymmetry of cell surfaces. While the apical cell surface is a nonthrombogenic surface to which platelets do not bind (Fig. 6), the basal cell surface is involved in the synthesis of a highly thrombogenic extracellular matrix which, when examined by immunofluorescence, is composed of collagen type III and, to a lesser extent, of collagen type IV [11] (Fig. 7). Also noteworthy is the redistribution or de novo appearance of new cell surface proteins which correlate with the organization of subconfluent cells not yet organized into a highly contact-inhibited cell monolayer. Sparse, actively growing vascular endothelial cells contain fibronectin associated with both their apical and basal cell surfaces as well in the areas of cell-cell contact. Upon reaching confluence, a dramatic redistribution of that cell surface protein occurs which results in its disappearance from the apical cell surface and its secretion into the basal cell surface, where it eventually becomes an integral component of the extracellular matrix [14, 15]. Parallel to the redistribution of fibronectin as the cells reorganize into a nonoverlapping cell monolayer, the appearance of a new cell surface protein called CSP-60 can also be observed [16]. This protein was found to be exposed to iodination by lactoperoxidase only in endothelial cells that have adopted a monolayer configuration. It was not detected either in actively growing or in unorganized endothelial cell cultures. Likewise, CSP-60 was no longer exposed for iodination in disorganized endothelial cell monolayers and was not present in sparse or confluent cultures of fibroblasts or vascular smooth muscle cells that grow in multiple layers [16]. CSP-60 has now been observed with all types of vascular endothelium studied to date, whether they be fetal or adult in origin or from territories as diverse as the endocardium, aortic arch, umbilical vein, or lung arteries (Fig. 8).

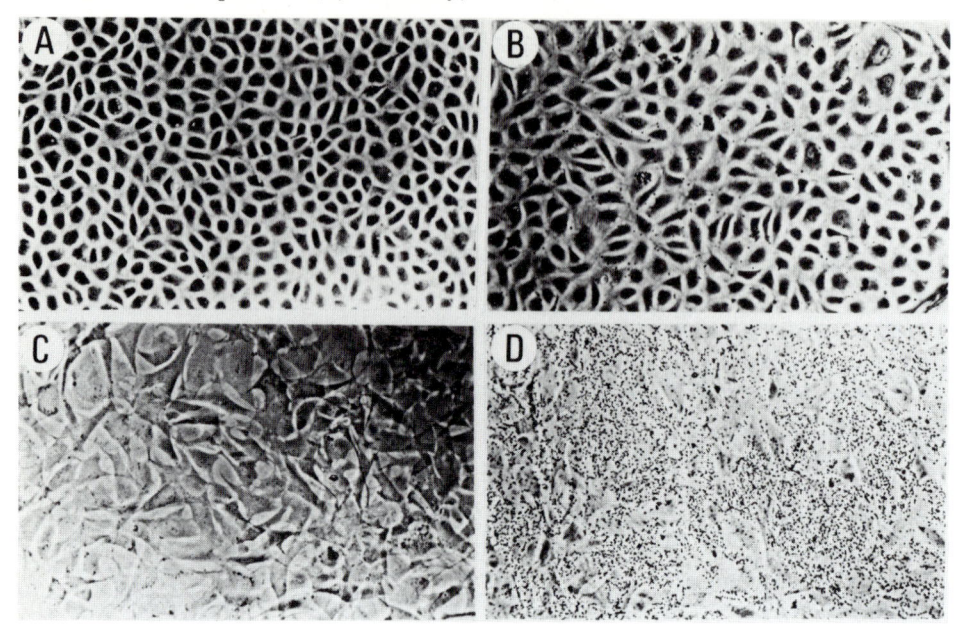

Fig. 6. Adherence of platelets to confluent vascular endothelial cultures maintained with or without FGF. Confluent endothelial cultures maintained in the absence (3 passages) or presence (during the phase of logarithmic growth) of FGF were incubated with human platelets (2×10^8/ml, 30 min, 37°C), washed, and observed by phase microscopy (×150). A, B) Cells maintained with FGF. Very little or no platelets can be seen (B) attached to the upper surface of cells that adopt a monolayer configuration as in (A). In contrast, most of the upper surface of unorganized cultures (C) maintained in the absence of FGF is covered with platelets which attached singly and do not form aggregates (D).

In contrast to cultures grown in the presence of FGF, cultures maintained in its absence lose within 3 passages their ability to form at confluence a monolayer of closely apposed and flattened cells. Instead, the cultures adopt a multilayer configuration consisting of large and overlapping cells which are no longer contact inhibited [5, 17]. Parallel to these changes in cell morphology, a loss of cell surface polarity is observed. Both the apical and basal cell surfaces are now covered by an extracellular matrix which fluoresces strongly when analyzed by immunofluorescence with purified antibodies against collagen types I, III, and IV [17] (Fig. 7). Parallel to this loss of orientation in the secretion of extracellular matrix, the apical cell surface becomes thrombogenic, as reflected by an increase in platelet-binding capacity [5, 17]. Likewise, marked changes in the distribution and appearance of cell surface proteins such as fibronectin and CSP-60 can be observed. Fibronectin, which in confluent and highly organized cultures grown in the presence of FGF is detected only in the basal cell surface, now appears in both basal and apical cell surfaces [5]; CSP-60 is no longer exposed for iodination, even late at confluence [5, 17].

The production of fibronectin by sparse and confluent endothelial cultures maintained with or without FGF has been further studied by exposing the cells to (^{35}S)methionine and subjecting both the cell layer and tissue culture medium to SDS slab gel electrophoresis before and after immunoprecipitation with anti-fibronectin antiserum. Although anti-fibronectin precipitated less than 3% of the total (^{35}S)-labeled proteins that were secreted into the culture medium of cells maintained and actively growing in the presence

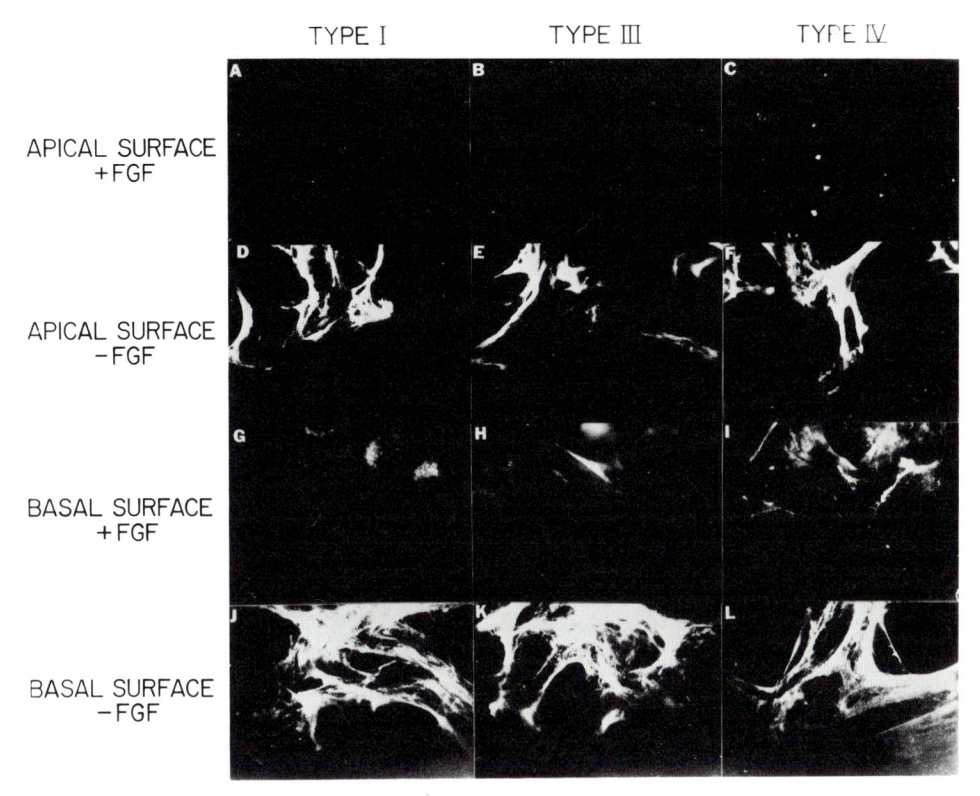

Fig. 7. Indirect immunofluorescent localization of collagen in vascular endothelial cells maintained in the presence (ABAE) or absence (ABAE-F) of FGF. Cultures were stained with affinity purified antisera (kindly provided by Drs. G. Martin and J. M. Foidart, Dental Institue, NIH, Bethesda, MD). ABAE cell monolayers were stained with antibodies directed against (A) collagen type I, (B) collagen type III, and (C) collagen type IV. Focusing was on the apical cell surface. ABAE-F cell monolayers were stained with antibodies directed against (D) collagen type I, (E) collagen type III, and (F) collagen type IV. Focusing was on the apical cell surface. The basal cell surface was studied by first permeabilizing the cells in acetone (10 min at $-20°C$) prior to staining. The ABAE basal cell surface was examined with antibodies directed against (G) collagen type I, (H) collagen type III, and (I) collagen type IV. The basal cell surface of ABAE-F was examined with antibodies directed against (J) collagen type I, (K) collagen type III, and (L) collagen type IV. In all cases, non-immune rabbit serum was non-reactive with either ABAE or ABAE-F cells (×200).

of FGF, 20–25% of the total (^{35}S)-labeled proteins were precipitated from the culture medium of cells that were maintained in the absence of FGF. When analyzed on SDS polyacrylamide gels, more than 90% of the immunoprecipitated radioactivity comigrated with fibronectin [5]. On the basis of the immunoprecipitation values, it can be calculated that sparse and confluent cultures maintained in the absence of FGF secreted into the medium 30- and 50-fold more fibronectin per cell than sparse and confluent cells cultured in the presence of FGF, respectively [5].

Endothelial cells previously maintained in the absence of FGF showed a normal proliferative response to readdition of FGF when reseeded at a low density (split ratio 1:16), as indicated by a 4-fold decrease in doubling time (18 hr versus 72 hr) and increased in final cell density at confluence (900 cells/mm^2) [5] (Fig. 3C). In terms of morphological

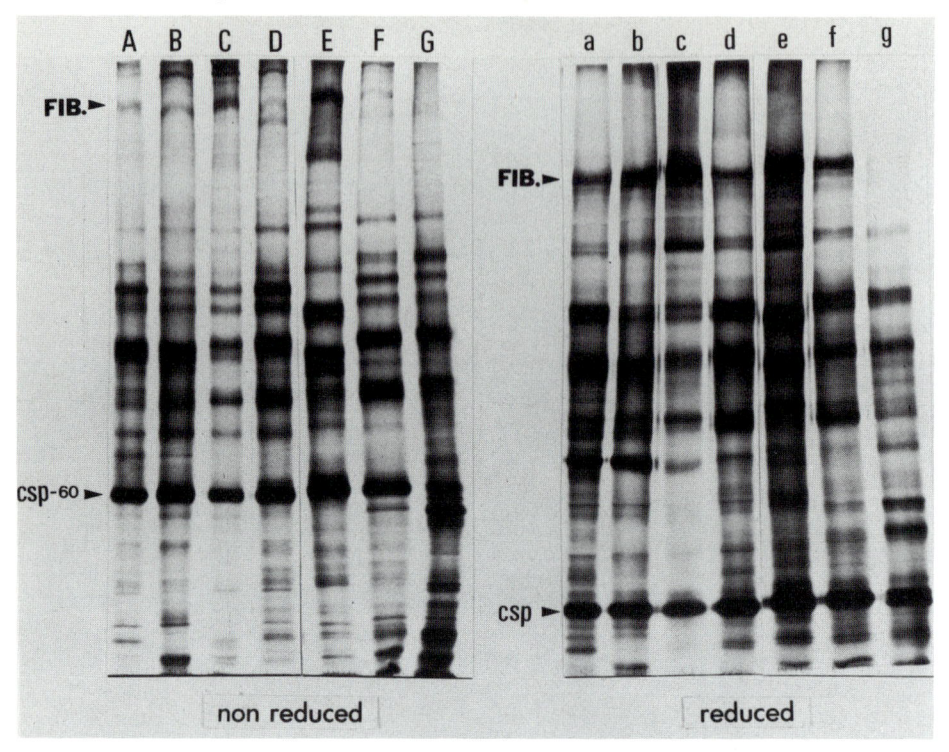

Fig. 8. Appearance of CSP-60 in various types of cultured vascular endothelial cells. Vascular endothelial cells of various origins were obtained, cloned, and maintained in culture as described [5, 21, 22]. Confluent cell monolayers (5 to 7 days after reaching confluence) were radioiodinated (lactoperoxidase/glucose oxidase) and analyzed by a gradient (5–16%) PAGE before (lanes A–G) or after (lanes a–g) reduction with 0.1 M DTT. Vascular endothelial cells of the following origins were studied: A, a) Bovine pulmonary artery; B, b) adult pig aorta; C, c) fetal bovine aorta; D, d) adult bovine aorta; E, e) fetal bovine heart; F, f) bovine umbilical vein; and G, g) calf bovine aorta. Arrows mark the positions of fibronectin (460K and 230 K) and CSP-60 (60K and 30K) before and after reducing the samples with DDT, respectively.

appearance, the cells became smaller and formed at confluence a highly organized monolayer in which little or no cell overlapping was observed. When the confluent cultures were tested for surface iodination pattern (Fig. 9D, I) [5], CSP-60 appeared as a major protein, whereas fibronectin was detected in a much smaller amount, as in a confluent monolayer of endothelial cells that were continuously maintained with FGF (Figs. 9A, F). Similarly, extracellular matrix was produced toward the basal part of the cells and no longer on top of the cell layer. The cells which reverted also showed the normal low rate of fibronectin synthesis and secretion as well as a nonthrombogenic apical surface to which platelets cannot adhere. Endothelial cells that were maintained for over 40 generations (7 passages) in the absence of FGF underwent cell senescence and hence no longer responded to FGF [5].

When FGF was added to sparse or subconfluent cultures previously maintained in its absence, it was found that as the cell density and degree of overlapping increased, the cells became less able to regain their normal differentiated properties. The reorganization of cells into a closely apposed monolayer of cells required at least 1 to 2 cell doublings and was always associated with a reappearance of CSP-60, a decreased synthesis and secre-

Extracellular Matrix and Cell Proliferation   JSS:351

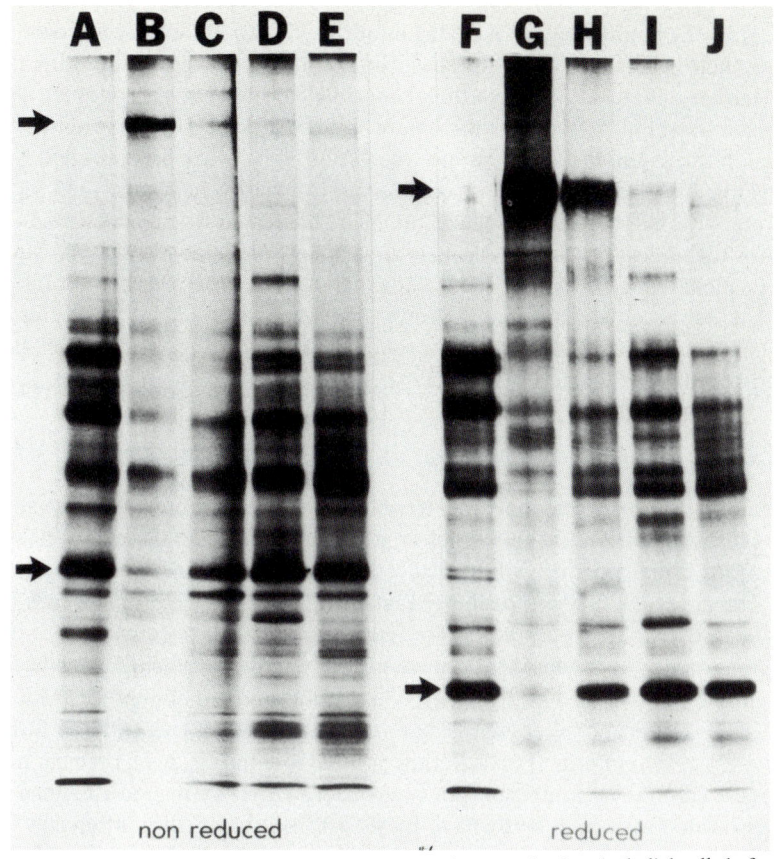

Fig. 9. Cell surface iodination pattern of overlapping and reorganized endothelial cells before and after being reexposed to FGF. Endothelial cells maintained for 3 passages in the absence of FGF were seeded at a split ratio of 1:8 and FGF was added (every other day) starting on day 2 or day 7 after seeding. Control cultures (not exposed to FGF) and reorganized cultures (reexposed to FGF) were iodinated at confluence (14 days after seeding) and analyzed by gradient (5–16%) polyacrylamide slab gel electrophoresis before (lanes A–E) and after (lanes F–J) reduction with 0.1 M DTT. A, F) A confluent monolayer of endothelial cells that were continuously maintained in the presence of FGF. CSP-60 appears as a major band. B, G) Confluent but unorganized endothelial culture maintained for 4 passages without FGF. Fibronectin appears as a major band; CSP-60 is missing. C, H) Same cells as in (B) and (G) exposed to FGF at a subconfluent density (7 days after seeding) and labeled 7 days later. Most of the cultures adopt a monolayer configuration, although some unorganized areas (about 20% of the culture) are still present. Both fibronectin and CSP-60 are exposed for iodination. D, I) The same cells as in (B) and (G) exposed to FGF at a sparse density (3 days after being split at a 1:8 ratio) and labeled 12 days afterwards, when the cells were highly organized and closely apposed. CSP-60 appears as a major band, whereas fibronectin is detected in small amounts as in cells that are maintained continuously with FGF. E, J) (-FGF) culture exposed when subconfluent to a medium conditioned by a confluent monolayer of endothelial cells and iodinated 7 days afterwards.

tion of fibronectin into the tissue culture medium, and the acquisition of non-thrombogenic properties. On the other hand, fibronectin already present on the apical cell surfaces disappeared only partially [5]. Limited reversion, if any, was obtained when unorganized cultures were reexposed to FGF late at confluence, suggesting that the spatial organization might, among other factors, determine whether or not and to what extent cells will respond to FGF. Unlike the limited degree of reversion obtained with FGF, total reversion of highly overlapping endothelial cell cultures was induced without promoting cell proliferation by a conditioned medium taken from confluent endothelial cells that have adopted a monolayer configuration (Figs. 10, 11). This medium, which has little mitogenic activity, is fully potent even in the absence of serum or after depletion of its fibronectin content by affinity chromatography on a gelatin-Sepharose column [17]. It is now being analyzed for the presence of proteolytic activity and various components of cellular origin which might be responsible for the induced changes in cellular morphology and growth characteristics. Therefore, endothelial cells that are seeded at a high density (1:2 to 1:10 split ratio), unlike cells that are seeded at a low clonal density, may, by conditioning the medium, adopt a cell monolayer configuration and the associated differentiated properties even in the absence of an added factor such as FGF [17].

THE EXTRACELLULAR MATRIX (ECM) AND CELL PROLIFERATION

The above results suggest that vascular endothelial cells maintained in the absence of FGF exhibit, in addition to a much slower growth rate, morphological as well as structural alterations which mostly involved changes in the composition and distribution of the ECM. This raises the further possibility that the ECM produced by these cells could have an effect on their ability to proliferate and to express their phenotype once confluent.

The importance of the ECM for normal growth and development in vivo has long been recognized [18]. It has been demonstrated by Dodson [19] and by Wessels [20] that the basal cell layers of the epidermis have to be in direct contact with the ECM upon which they rest in vivo in order to retain their normal orientation and to remain mitotically active in organ culture. This substrate can be produced either by the mesenchyme which is closely associated with most epithelia or by the epithelia themselves, after they interact with the ECM produced by other tissues. Such is the case with the isolated corneal epithelium which can recreate its own stroma if cultured in vitro on isolated lens capsule but not if cultured on a non-collagenous stroma [21]. Recent transfilter experiments have shown that direct contact by epithelial cells with a collagen substrate is required if they are to produce their own ECM and that the extent of cell surface area in contact with the substrate is directly proportional to the stimulation of stroma production [21–23]. This newly produced ECM could in turn be held responsible for the control of proliferation of the basal epithelial cell layer, possibly by affecting the cell shape [21]. Investigation of the role of extracellular materials at the epithelial-mesodermal interface has shown that glycosaminoglycans present as major molecular species at the junction of interacting tissues could be implicated in epithelial morphogenesis [24]. Likewise, evidence that the substrate upon which cells rest when maintained in tissue culture is important for their proliferation is now plentiful. The pioneering work of Ehrmann and Gey [25] has shown that various tissues demonstrate enhanced growth and differentiation when cultured on collagen gels. Recent studies have also shown that collagen is important in promoting cell attachment [26–30], cell migration [31], and cell proliferation [27, 32, 33]. Of particular interest is the study of Liotta and his colleagues [32] on the growth of fibroblasts in culture which indicates that, even when grown on plastic, the cells deposit a collagen substrate which is required for proliferation. This was demonstrated using the proline analog, cis-hydroxyproline, which decreases the amount of newly synthesized procollagen secretion when incorporated into collagen [34]. Cultures maintained on plastic and exposed to cis-hydroxyproline did not produce collagen and did not proliferate, while cultures exposed to cis-

Fig. 10. Restoration of the normal phenotypic expression of ABAE-F cells by conditioned medium (scanning electron microscopy). A, B, C) Confluent ABAE cells maintained with FGF do not overlap each other, are closely apposed, and have a prominent nucleus. D, E, F) ABAE-F cells maintained without FGF for 4 passages (1:8) are larger, disorganized, and form multiple layers. The cell nucleus is not prominent and the underlying extracellular matrix is exposed in some areas. G, H, I) When 600,000 ABAE-F cells are maintained for 8 days in conditioned medium, the non-overlapping monolayer is reestablished. The cells become smaller, organized, and non-overlapping, thus forming the typical "cobblestone pattern" (×2,000).

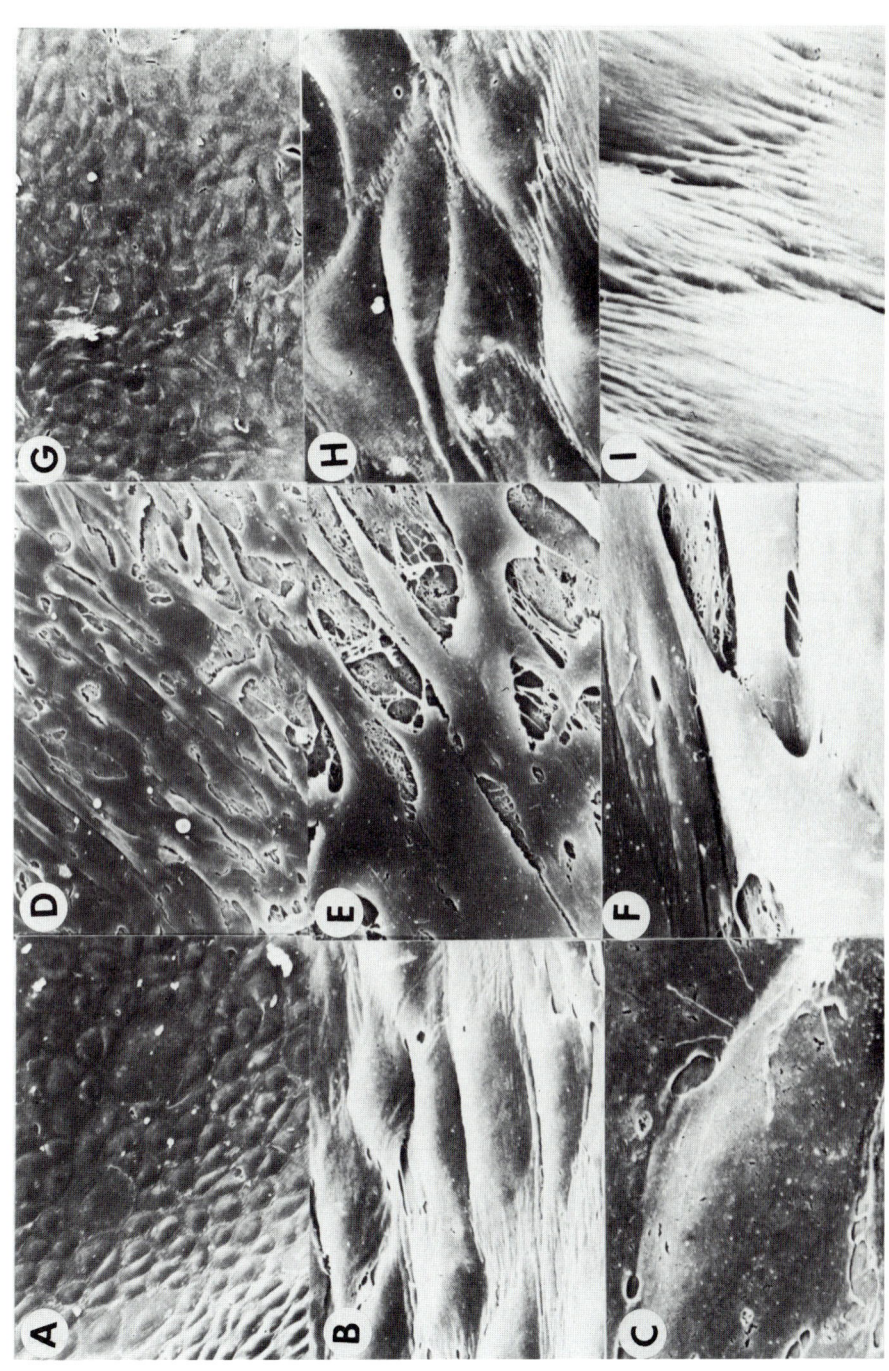

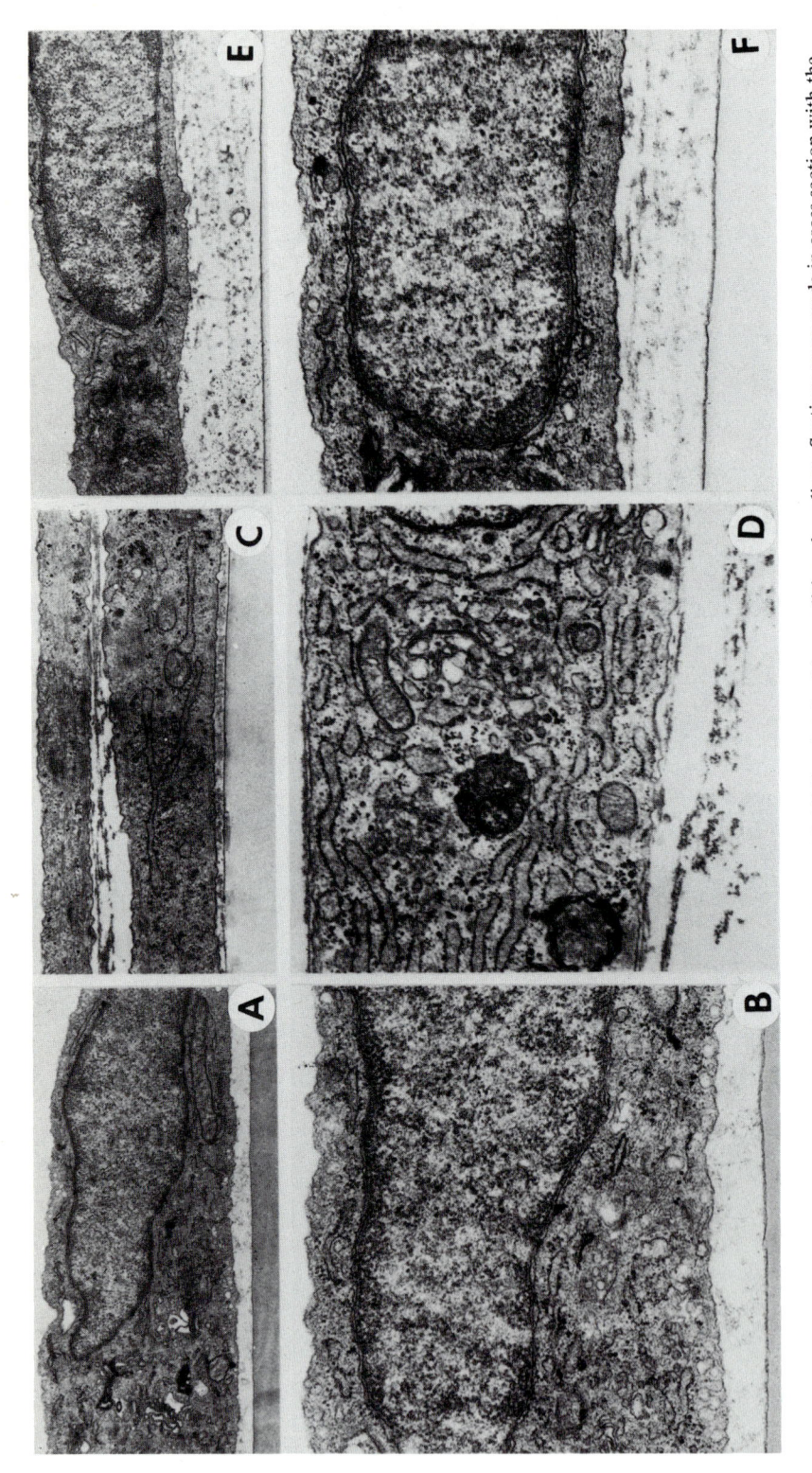

Fig. 11. Transmission electron microscopy of ABAE, ABAE-F, and ABAE-F maintained in conditioned medium. Sections were made in cross-section with the cell monolayer (A, C, E at ×20,000 and B, D, F at ×50,000). A, B) ABAE cell monolayer showing the presence of an amorphous basement membrane underlying the cell with no material on the apical cell surface. C, D) ABAE-F cells can be seen to overlap, and the presence of extracellular matrix on the basal and apical cell surfaces is illustrated. E, F) ABAE-F cells maintained for 8 days in conditioned medium have regained their characteristic morphology with the concurrent redistribution of extracellular matrix to the basal cell surface. No matrix can be seen on the apical cell surface.

hydroxyproline and provided with an artificial collagen substrate did proliferate [32]. This observation therefore linked collagen and ECM production to cell proliferation. It is not known, however, if the main effect of collagen is to promote cell attachment, thereby indirectly allowing the cells to proliferate, or if it has a direct effect on both cell attachment and cell proliferation. In either case, the possibility exists that cells which do not adapt and grow readily in tissue culture could be limited in their collagen production and that providing them with an ECM produced by other cell types could be one way to circumvent this limitation.

Among the other components of the ECM which have been studied in regard to cell attachment and proliferation in vitro is fibronectin. Like collagen, it has been shown to promote cell attachment, migration, and proliferation [35–37]. Whether fibronectin directly mediates these effects or acts by stimulating the production of ECM from cells which are exposed to it has not been analyzed.

The effect of the ECM on the proliferation of cells maintained in cultures has not been studied. This is mostly due to the fact that, with the exception of lens capsule, it is difficult in vivo to isolate such material from neighboring tissues. In vitro the reconstitution of an ECM from its separate elements (collagens, proteoglycans, glycosaminoglycans, and glycoproteins) may be difficult, if not impossible. Not only must the correct ratio of components constituting the ECM be respected, but they must also be linked in such a way that the resulting tri-dimensional structure will be like that of the extracellular scaffolding in vivo. The problem in reconstituting an ECM in vitro is made even more difficult by the fact that collagen types IV and V, which in vivo are the main components of basement membrane collagens, can only be extracted from tissue following proteolysis. This could result in structural alterations and prevent their proper polymerization in vitro. One must also consider that our knowledge of the ECM components is limited. Components such as laminin [38] have only been isolated recently, and the number of these which remain to be identified can only be guessed.

Corneal endothelial cells maintained in tissue culture retain their ability, in contrast to most cell types, to synthesize and secrete an ECM found underneath, but not on top of the cells [39]. As shown in Figure 12A, corneal endothelial cells upon reaching confluence form a monolayer of small, highly flattened, tighly packed (1,100 cells per mm^2), and non-overlapping cells. Secretion of an ECM takes place only underneath the cell layer [15, 40], and the underlying matrix is revealed after exposing the cell layer to 0.5% Triton X-100 and subsequent washing with PBS to remove remaining nuclei and cytoskeletons [41]. The matrix then appears as a uniform layer of amorphous material coating the entire area of the dish (Figs. 12B, C). The chemical composition of the ECM produced by corneal endothelial cells in vitro is currently being analyzed [C. Tseng, N. Savion, R. Stern, and D. Gospodarowicz, manuscript in preparation]. Based on immunofluorescence studies (Fig. 13), collagen types III and IV appear as the major collagen components of the matrix, forming an evenly distributed fibrillar meshwork (Fig. 13). Fibronectin and laminin are also present [15, 39]. This ECM, whose appearance has been shown to correlate with the acquisition by cultured corneal endothelial cells of their normal "in vivo" morphology, cell surface polarity, and function [15, 39], could substitute for the ECM produced by other cell types. The ability of corneal endothelial cells in tissue culture to produce an extensive ECM could, therefore, provide us with a tailor-made ECM with which to test the proliferation and response of other cell types to growth factors. We have, therefore, compared the rates of proliferation of vascular endothelial cells maintained on plastic versus an ECM.

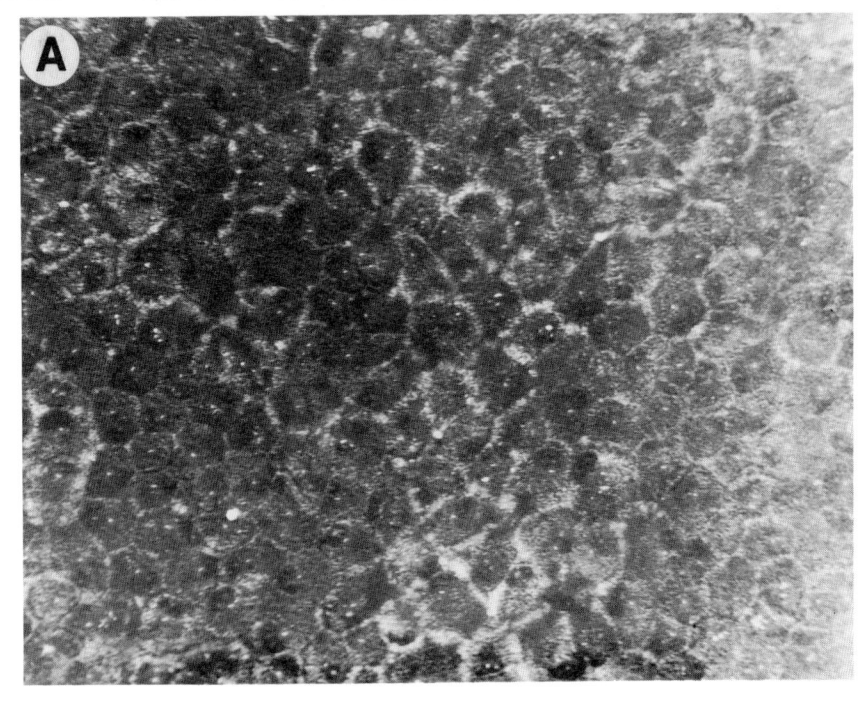

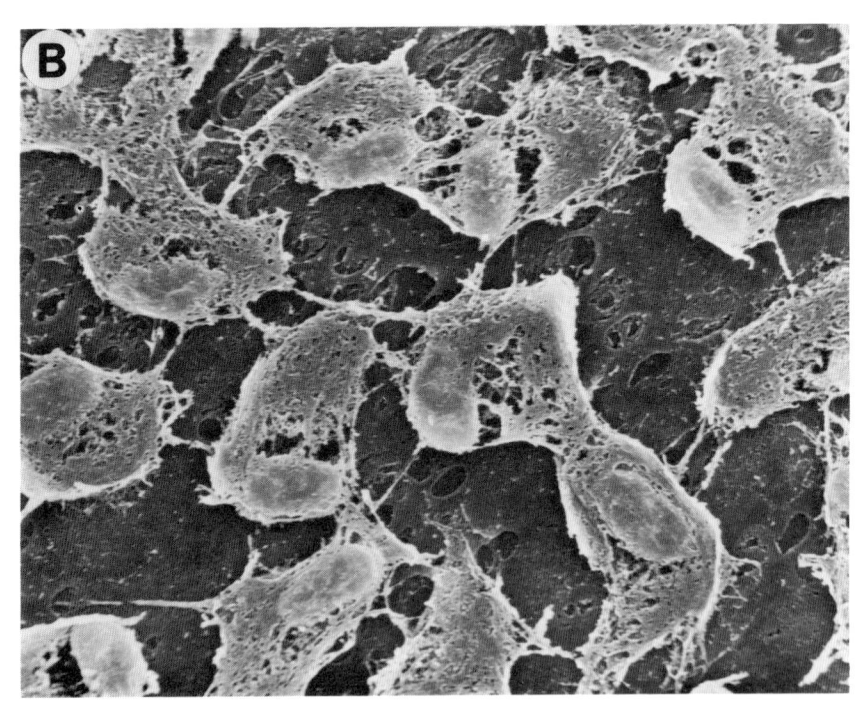

Fig. 12. Scanning electron microscopy of a monolayer of bovine corneal endothelial cells before and after exposure to Triton X-100. A monolayer composed of polygonal, highly flattened, and closely apposed cells can be seen in A (×600). After the monolayer has been treated with Triton X-100, it is composed of nuclei and cytoskeletons which no longer attach firmly to the extracellular matrix (B,

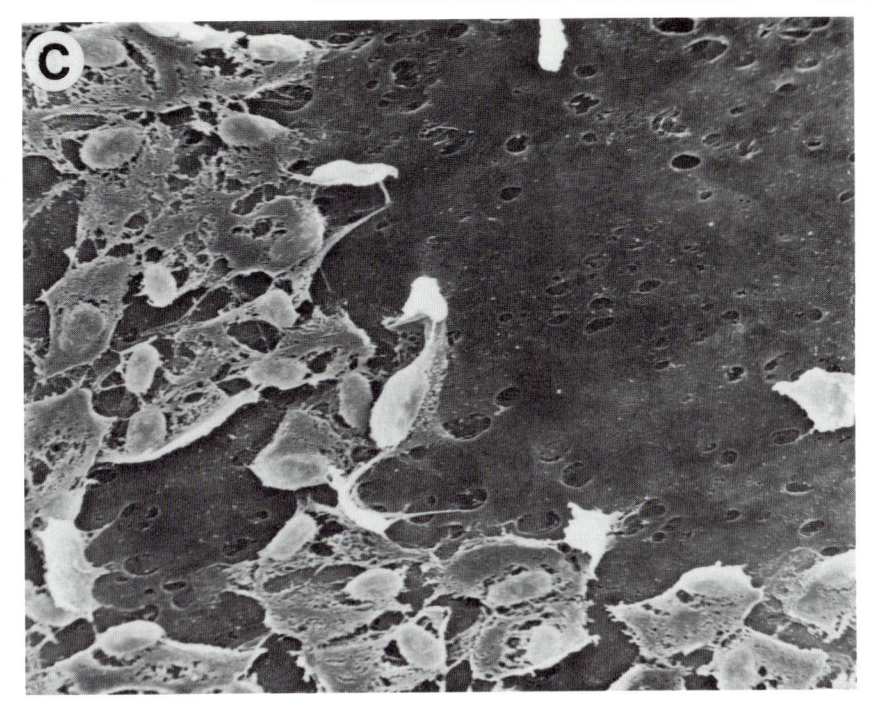

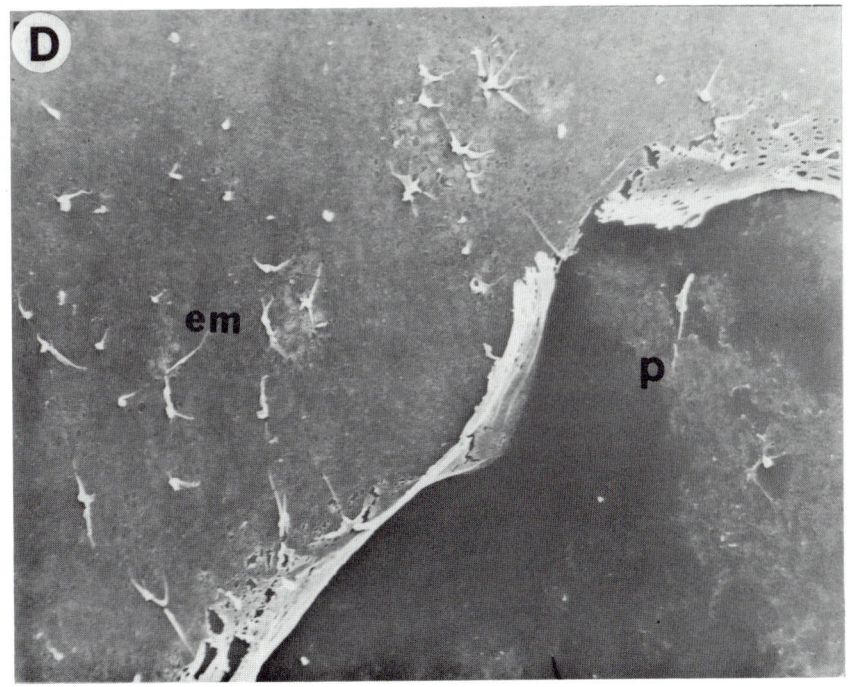

×200). In some areas the extracellular matrix has been exposed (C, ×600). Washing the dishes with PBS removed the cytoskeleton and exposed the extracellular matrix present underneath the cells (D, ×200). The plate has been scratched with a needle to expose the plastic (P) to which the extracellular matrix (em) strongly adheres.

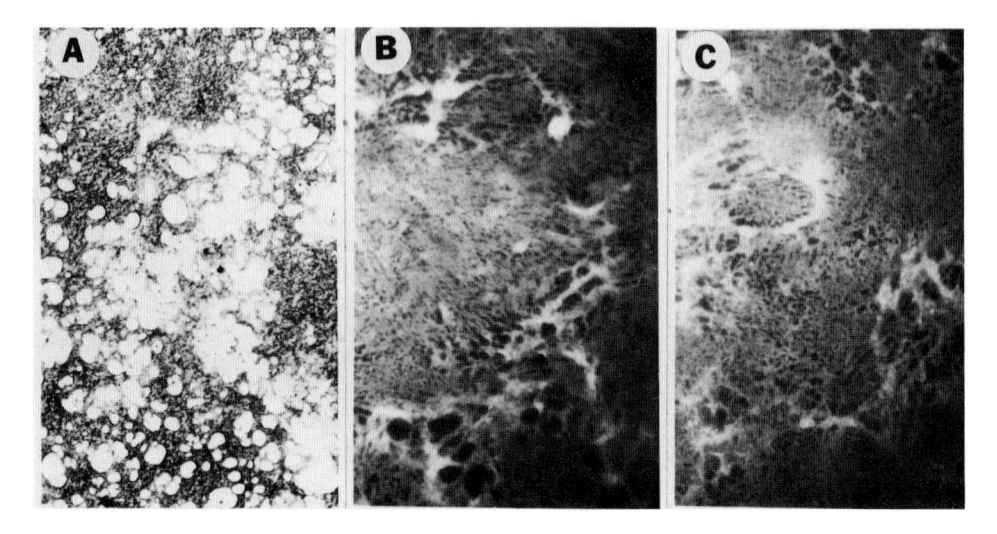

Fig. 13. Immunofluorescence of the extracellular matrix produced by corneal endothelial cell cultures. Confluent corneal endothelial cell cultures were treated (60 min, 24°C) with 0.5% Triton X-100 to remove the cell layer and to expose the underlying extracellular matrix shown in (A) stained with alizarin red. B–C) show the immunofluorescence pattern of the extracellular matrix when exposed to anti-collagen type III (B), or anti-collagen type IV (C) followed by incubation with fluorescein-conjugated anti-rabbit IgG. Collagen type IV (C) was present in lower amounts, as indicated by a 3- to 5-fold longer exposure time of the automatic camera.

GROWTH AND MORPHOLOGICAL APPEARANCE OF CULTURED BOVINE AND HUMAN VASCULAR ENDOTHELIAL CELLS MAINTAINED ON PLASTIC VERSUS ECM AND EXPOSED OR NOT TO FGF

When the growth of bovine vascular endothelial cells maintained on plastic versus ECM was compared, cells maintained on an ECM, regardless of whether they were exposed or not to FGF, reached a final cell density within 5 days which was 10- to 12-fold that of cultures maintained on plastic alone (Fig. 14A). Addition of FGF to cultures maintained on an ECM did not decrease their mean doubling time, which was already at a minimum (18–20 hr), nor did it result in a higher final cell density, which was already at a maximum (700–1000 cells/mm^2). When the morphology of confluent cultures of bovine vascular endothelial cells maintained on plastic and exposed to FGF was compared to that of confluent cultures maintained on ECM, it was found to be similar (Figs. 14C, D). In contrast to bovine vascular endothelial cells, human umbilical vein endothelial cell cultures did not proliferate when seeded at low cell density on tissue culture dishes (Fig. 14B). As previously reported [13, 42], addition of FGF to the cultures induced the cells to divide actively and within 11 days a 15-fold increase in cell number was observed. When cells were seeded on ECM instead of on plastic, a 6-fold increase in cell number was observed over the same period of time. In contrast, with bovine endothelial cell cultures, FGF was still required if cultures maintained on ECM were to reach confluence (Fig. 14B). Addition

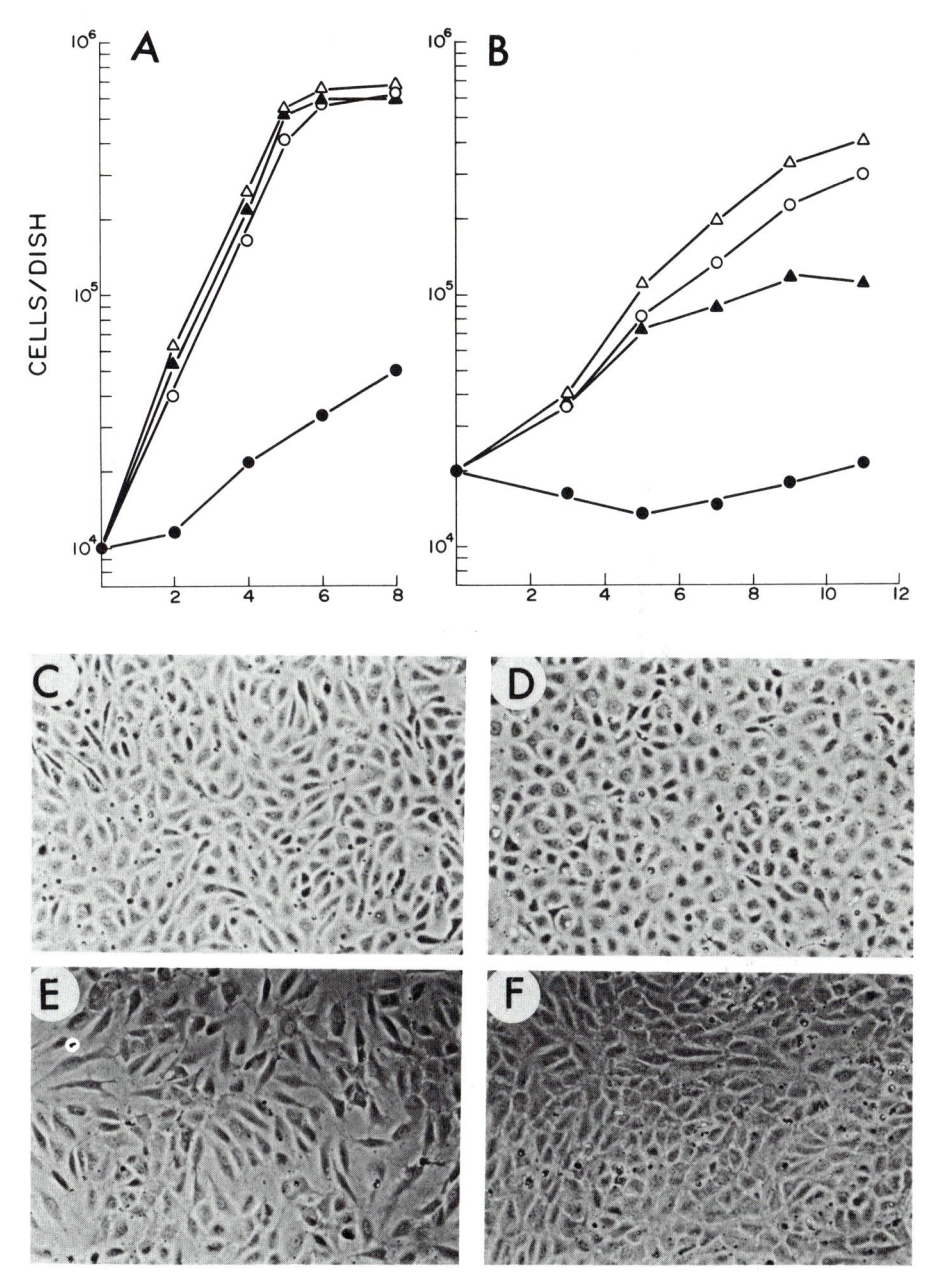

Fig. 14. Proliferation and morphological appearance of bovine aortic endothelial cells and human umbilical vein endothelial cells when maintained on plastic versus extracellular matrix (ECM) and exposed or not to FGF. A) Bovine vascular aortic endothelial cells were plated at an initial density of 1×10^4 cells per 35-mm dish coated (△, ▲) or not (○, ●) with an ECM. Cultures were maintained in the presence of DMEM supplemented with 10% calf serum and with (△, ○) or without (▲, ●) FGF (100 ng/ml) being added every other day. B) Human vascular endothelial cells were plated at an initial density of 2×10^4 cells per 35-mm dish coated (△, ▲) or not (○, ●) with an ECM. Cultures were maintained in the presence of DMEM supplemented with 10% calf serum and with (△, ○) or without (▲, ●) FGF (100 ng/ml) being added every other day. The morphological appearance of bovine vascular endothelial cells maintained on plastic and grown in the presence of FGF or maintained on an ECM and grown in the absence of FGF is shown in (C) and (D), respectively. The morphological appearance of human umbilical vein endothelial cells maintained on plastic or on an ECM and grown in both cases in the presence of FGF is shown in (E) and (F), respectively. Pictures were taken once the culture reached confluence (phase contrast, ×100).

of FGF to cultures maintained on ECM induced an optimal growth rate and the cultures became confluent within 11 days. The final cell density of the confluent cultures was 20-fold higher than that of cultures maintained on plastic and not exposed to FGF. When the morphological appearance of confluent cultures maintained on plastic (Fig. 14E) versus ECM (Fig. 14F) and exposed in both cases to FGF was compared, the cultures maintained on ECM were composed of cells more closely apposed and tightly packed than those maintained on plastic.

The rate of proliferation of bovine vascular endothelial cells maintained on plastic and exposed to FGF (Fig. 15B) or maintained on an extracellular matrix and not exposed to FGF (Fig. 15A) was a strict function of the serum concentration to which the cultures were exposed. While cells maintained on an extracellular matrix and exposed to serum concentration as low as 1% proliferated actively even in the absence of FGF (Fig. 15A), cells maintained on plastic, when exposed to a serum concentration as high as 10%, proliferated poorly (Fig. 15B). In contrast, if FGF was added to such cultures, then active proliferation resumed (Fig. 15B).

It can, therefore, be concluded that when the proliferation of bovine vascular endothelial cells maintained on plastic versus an ECM is compared, low-density cell cultures maintained on plastic proliferate poorly. FGF is, therefore, needed in order for the cultures to become confluent within a few days. In contrast, when the cultures are maintained on ECM, they proliferate actively and no longer require FGF in order to become confluent. In both cases (either maintained on plastic and exposed to FGF or maintained on an extracellular matrix), the rate of proliferation was a direct function of the serum concentration to which cultures were exposed. It is, therefore, likely that the effect of the ECM is more a permissive than a direct mitogenic one, since cells still required serum in order to proliferate.

To test the possibility that collagen or fibronectin alone could be the component of the ECM responsible for the increased rate of proliferation of cells plated on it, we have compared the growth of bovine vascular endothelial cells plated on dishes coated with purified collagen types I, II, III, and IV or with fibronectin. In no case did the cultures significantly increase their rate of growth when maintained on these different substrates (Fig. 16). In all cases, an aberrant morphological appearance was observed, the cultures being composed of large cells of which a high proportion are binucleated. Only cells maintained on an ECM proliferated actively (Fig. 16), reaching confluence within 5 days. This observation, therefore, excludes the possibility that the component in the ECM produced by corneal endothelial cells which could have a permissive effect on their proliferation is either collagen or fibronectin alone.

Proliferation of cells in culture is not only a function of the medium, serum, or growth factor(s) to which cells are exposed. It is also a function of cell density. While at high cell density, cells can rapidly condition their medium, thereby compensating for the nutrient deficiency of the medium when plated at clonal density they can no longer do so. Therefore, factors or nutrients required for cell survival and proliferation may be more readily apparent when cells are maintained at low rather than at high cell density. In particular, requirements for a proper substrate could become apparent. While cells maintained at high density could readily make a basement membrane, thereby facilitating further proliferation, at clonal density, even if every cell were to produce a basement membrane, it would be extremely difficult for them to cover the whole dish in a reasonable period of time. We have, therefore, analyzed the proliferation of vascular endothelial cells plated at clonal density on plastic versus plates coated with an extracellular matrix. As can be seen

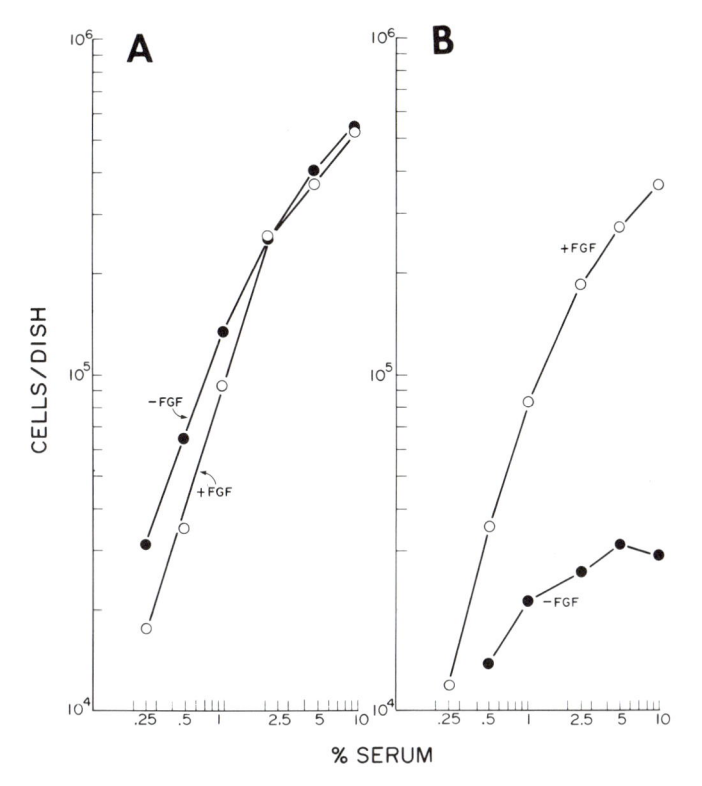

Fig. 15. Comparison of the rates of proliferation of bovine vascular endothelial cells maintained on plastic versus an extracellular matrix as a function of the serum concentration to which they are exposed. Bovine vascular endothelial cells were plated at 10^4 cells per 35-mm dish coated either with an extracellular matrix (A) or not (B). Cultures were maintained for 8 hr in the presence of DMEM, H-16 supplemented with 10% serum, 50 μg/ml Gentamycin, and 2.5 μg/ml Fungizone. After 8 hr, the medium was removed and the cultures washed once with DMEM, H-16. DMEM, H-16 supplemented with 2.5 μg/ml Fungizone, 50 μg/ml Gentamycin, and various concentrations of serum was then added to the dishes (○, ●). To half of the dishes FGF (100 ng/ml) was added every other day (○, +FGF). After 5 days the cultures were trypsinized and counted.

(Fig. 17), when cells were plated at low cell density on plastic, the plating efficiency was extremely poor or cells died rapidly, since no clones were visible after 10 days. If FGF was present in the medium, 25% of the cells gave rise to individual clones. In contrast, when cells were plated at low density on an ECM, not only was a 90% plating efficiency observed at all cell densities (from 0.012 cells/mm^2 to 1.2 cells/mm^2) but, in addition, *all* cells gave rise to clones even in the absence of FGF. This demonstrates that the substrate upon which cells rest, even at clonal density, is crucial to insure both their survival and proliferation in response to serum or plasma factors.

Since cultures maintained on ECM no longer require FGF in order to proliferate actively, we have also investigated its effect on the lifetime of vascular endothelial cells. Cultures which had been maintained on plastic in the presence of FGF for 50 generations and which had been shown to increase their average doubling time rapidly as soon as FGF was no longer added to the media [5] were maintained in the absence of FGF on dishes coated with an ECM (Fig. 18). When maintained on such a substrate the cells proliferated

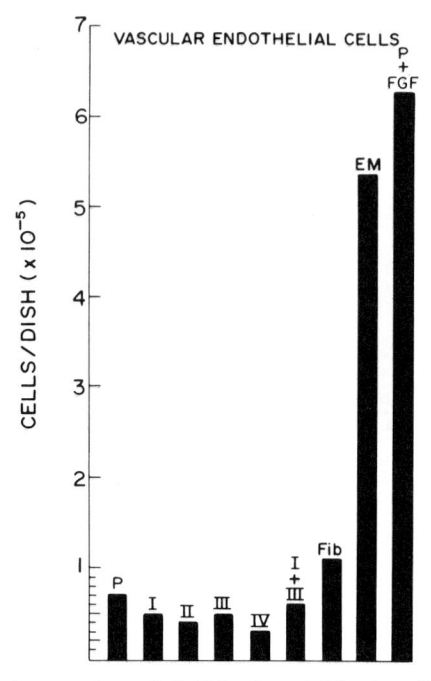

Fig. 16. Proliferation of bovine vascular endothelial cells maintained on different substrates. Bovine vascular endothelial cells (2×10^4 cells per 35 mm dish) were seeded on dishes coated with collagen types I, II, III, and IV, with collagen types I and III, with fibronectin (Fib), or with an extracellular matrix (EM). Cultures were maintained in the presence of DMEM, H-16 supplemented with 10% calf serum, 50 µg/ml Gentamycin, and 2.5 µg/ml Fungizone. After 5 days the cultures were trypsinized and the cells counted. The final cell density was compared to that of cultures maintained for the same period of time on plastic (P) and exposed or not to FGF (P + FGF).

actively and could be passaged weekly at a split ratio of 1:95. It was not until 50 generations later that a detectable increase in their average doubling time was observed (Fig. 18). It is, therefore, demonstrated that cells maintained on ECM have a much longer lifespan in culture than do cells maintained on plastic alone when passaged at a high split ratio (compare Figs. 3A and Fig. 18).

The present results, therefore, raise the possibility that although FGF is clearly mitogenic for a number of mesoderm-derived cells [43], its action on some of the cell types could be indirect. It could either replace the cellular requirement for a substrate such as the ECM and thereby make the cells fully responsive to growth factors present in serum and plasma even when the cells are maintained on plastic, or, alternatively, it could control the synthesis and secretion of the extracellular matrix produced by the cells. Such control could in turn make the cells sensitive to factors present in serum or plasma. That the latter alternative could occur finds support in our previous observation that FGF can control the production by vascular endothelial cells of extracellular and cell surface components such as fibronectin and various types of collagen [5, 17]. Since sparse cultures of endothelial cells proliferate poorly when maintained on plastic but not when maintained on an ECM, it may be that low density cell cultures maintained on plastic are unable to produce enough extracellular material to support further growth. The mitogenic effect of FGF on these cells could be the indirect result of an increased synthesis of the ECM by vascular endothelial cells. These possibilities are currently being tested.

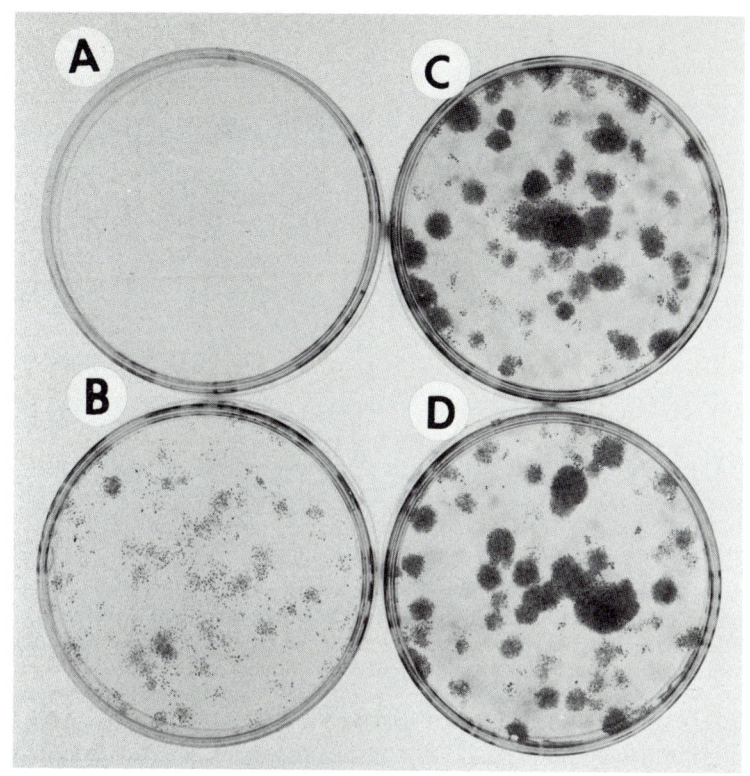

Fig. 17. Proliferation at a clonal density of bovine vascular endothelial cells maintained on plastic versus an extracellular matrix and exposed or not to FGF. Bovine vascular endothelial cells (50 cells per 35-mm dish) were seeded either on plastic dishes (A, B) or dishes coated with an extracellular matrix (C, D). Cultures were maintained in the presence of DMEM, H-16 supplemented with 10% calf serum, 50 µg/ml Gentamycin, and 2.5 µg/ml Fungizone. To half of the dishes (B, D) FGF (100 ng/ml) was added every other day. After 10 days (with a medium change on day 5), the medium was removed and the cultures were washed once and fixed with 10% formalin in PBS. Cultures were then stained with 0.1% crystal violet.

The ways in which the ECM exerts its permissive effect on cell proliferation can only be the object of speculation. One possible effect is, as pointed out earlier, to modify the cell shape in order to make it responsive to factor(s) to which the cells do not respond unless they adopt an appropriate shape. Recently, Folkman and Moscona [44], using vascular endothelial cells maintained on tissue culture dishes coated with an agent which modifies the adhesiveness of the cells to the dish, were able to control precisely the cellular shape in morphologies ranging from highly flattened to almost spheroidal. When the extent of cell spreading was correlated with DNA synthesis or cell growth, it was found to be highly coupled. Whereas highly flattened cells responded to serum factors, spheroidal cells no longer responded and intermediate degrees of response could be observed, depending on how flattened the cells were. Likewise, with corneal epithelial cells, changes in cell shape which depend on the substrate upon which the cells are maintained correspond to drastically altered sensitivities of the cells to EGF versus FGF [40, 45, 46].

The results presented above emphasize how drastically one can modify the proliferative response of a given cell type to serum factors depending on the substrate upon which the cells are maintained. It is possible that the lack of response of different cell types main-

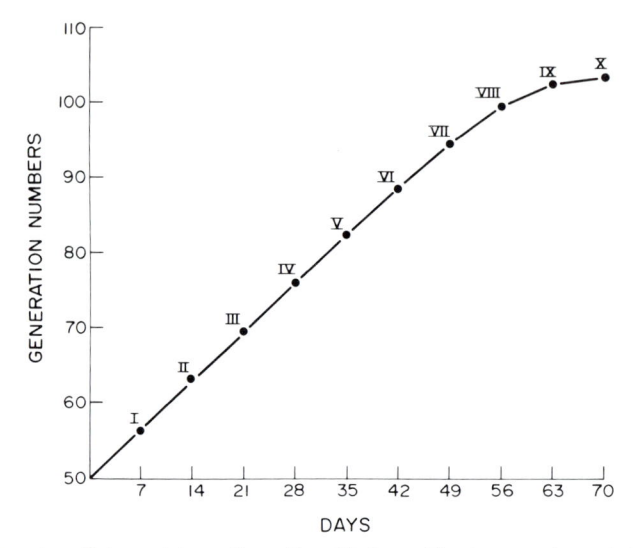

Fig. 18. Effect of the extracellular matrix on the culture lifetime of bovine vascular endothelial cells. Bovine vascular endothelial cells previously maintained on plastic tissue culture dishes and grown in the presence of FGF (100 ng/ml added every other day), as described in [8, 9, 13], for 50 generations were maintained and passaged weekly on dishes coated with an ECM. FGF was no longer added to the cultures. The number of generations was determined from the initial cell density 8 hours after seeding and the number of cells harvested at each transfer. Each point represents a single transfer. The average growth rate is given by the slope. Roman numbers indicate the passage number.

tained under tissue culture conditions to agents responsible in vivo for their proliferation and differentiation could be attributed to the artificial substrate, whether plastic or glass, upon which the cells rest and which limit their ability to produce an ECM.

PLASMA VERSUS SERUM. IS THERE A DIFFERENCE IN THEIR ABILITIES TO PROMOTE CELL GROWTH?

Culture of most cells in vitro requires the presence of serum [47]. Consequently, investigators have spent much effort in a search to identify the various factors in serum that stimulate cell growth in vitro. An important step in the search for serum growth factors has been the finding that one of the most potent mitogenic factors present in serum is derived from platelets. Such a possibility, first postulated by Balk [48], was based on studies of the growth of chick embryo fibroblasts in medium supplemented with plasma versus serum. While chicken fibroblasts do not proliferate in plasma-containing medium [48], when they are exposed to serum they proliferated actively. It was, therefore, concluded that serum contained growth-promoting activity which is lacking in plasma [49]. These studies were followed by reports which demonstrated that platelets are the source of a potent mitogen present in serum but not in plasma. While plasma was unable to support the growth of aortic smooth muscle cells [50] or that of BALB/c 3T3 cells [50], serum made from the same pool of blood stimulated their proliferation. Addition of a platelet extract to cell-free plasma-derived serum restored the growth-promoting activity [50–52]. One could, therefore, conclude that one of the principal mitogens responsible for the induction of DNA synthesis present in whole blood serum is derived from platelets [50–52]. The difference in the proliferative ability of cells exposed to plasma versus serum

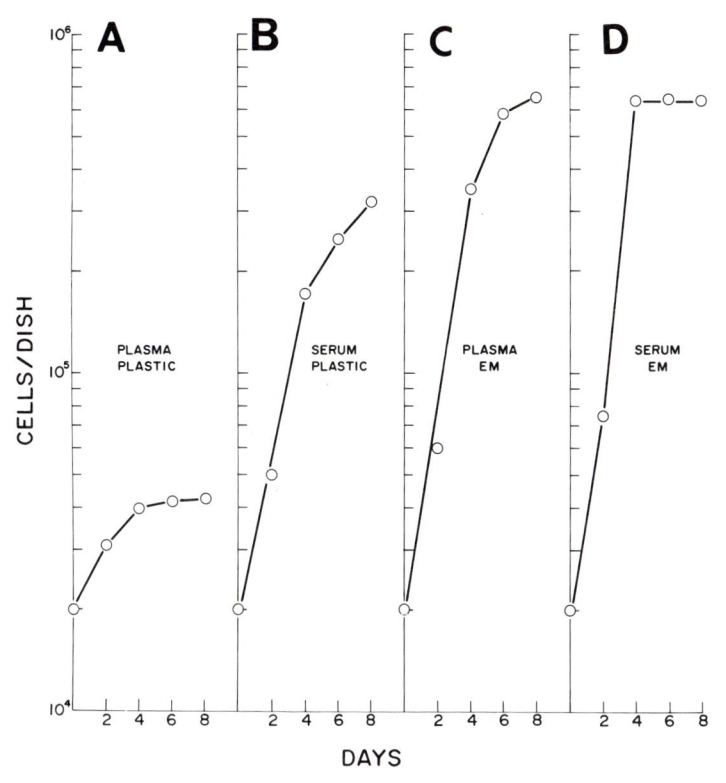

Fig. 19. Proliferation of bovine vascular smooth muscle cells when maintained on plastic versus extracellular matrix (ECM) and exposed to either plasma or serum. Vascular smooth muscle cells were seeded at 2×10^4 cells per 35-mm dish and maintained in the presence of DMEM supplemented with either 10% plasma (A, C) or serum (B, D). The cells were maintained on either plastic (A, B) or ECM (C, D).

results from the absence of the platelet factor in the former. However, all studies have, thus far, used cells maintained on plastic rather than on a basal lamina or an ECM. This difference in the substrate upon which the cells are maintained could have prevented their response to physiological factors present in plasma, thereby creating the difference in mitogenic activity between plasma and serum.

To explore the possibility that the serum factors to which cells maintained on ECM become sensitive are also present in plasma, we have compared the mitogenic activity of plasma versus serum, using as target cells vascular smooth muscle cells maintained on either plastic or an ECM.

Vascular smooth muscle cells maintained on plastic and exposed to plasma (10%) proliferate poorly. Within 4 days the cells go through one doubling and afterwards cease to proliferate (Fig. 19A). When the morphological appearance of such cultures was observed by phase contrast microcopy, the cells were considerably enlarged (Fig. 20A). When the same cultures were exposed to serum (10%) instead of plasma, the cells proliferated actively over a period of 6 to 8 days and underwent a 15-fold increase in cell number (Fig. 19B). During the logarithmic growth phase the mean doubling time of the cultures was 30 hours. These results, therefore, confirm previous results showing that when cells are maintained on plastic and exposed to plasma, they proliferate poorly or not at all,

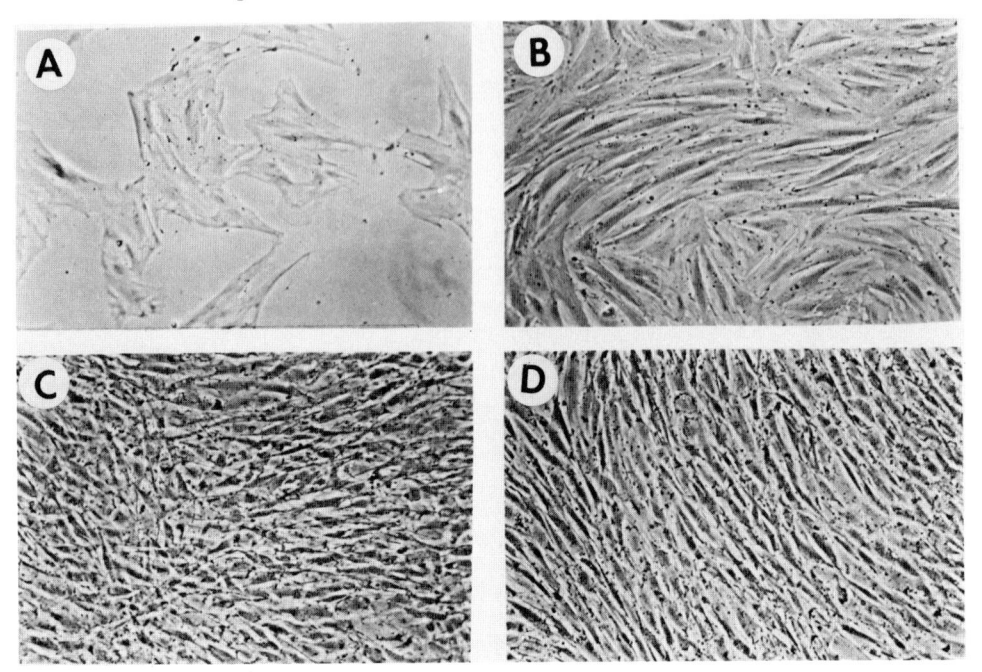

Fig. 20. Morphological appearance of vascular smooth muscle cells maintained on plastic (A, B) or an extracellular matrix (C, D). Cultures were plated and maintained as described in Fig. 19 and exposed to either 10% plasma (A, C) or 10% serum (B, D). Pictures were taken on day 8 with a phase contrast microscope (×100).

whereas they actively proliferate when exposed to serum [50, 51]. In contrast to the above results, when cells were maintained on an ECM and exposed to plasma, they proliferated actively (Fig. 19C). Within 6 days the cell number increased by 30-fold, and during the logarithmic growth phase the mean doubling time of the cultures was as low as 15 hours (Fig. 19C). Plasma was, therefore, even more mitogenic for cells maintained on an ECM than was serum for cells maintained on plastic. When the growth rate and the final cell density of cultures maintained on an ECM and exposed to either plasma (Fig. 19C) or serum (Fig. 19D) were compared, they were found to be the same. The differences between plasma and serum in their abilities to support cell growth, differences which are evident when the cells are maintained on plastic, therefore, vanish when the cells are maintained on an ECM. In Fig. 20, the morphological appearance of a culture maintained on plastic and exposed to serum (Fig. 20B) can be compared to that of a culture maintained on an ECM and exposed to either plasma (Fig. 20C) or serum (Fig. 20D). The difference between such cultures is apparent when one compares their respective average cell sizes. Cultures maintained on an ECM and exposed to either plasma or serum were composed of small, spindly, overlapping, and tightly packed cells which were on an average 3- to 5-fold smaller than the mean cell size of cultures maintained on plastic and exposed to serum (Figs. 20C, D). When the mean cell size of cultures maintained on ECM versus plastic and exposed in both cases to plasma was compared (Figs. 20A, C), cells maintained on plastic and exposed to plasma (Fig. 20A) had on an average a 10-fold larger size than cells maintained on an ECM (Fig. 20C).

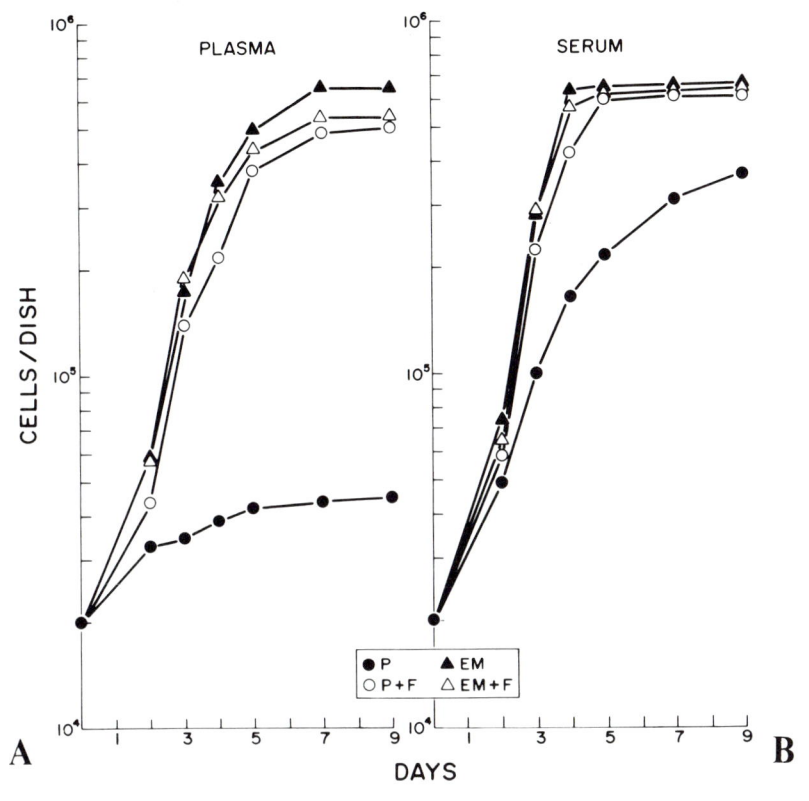

Fig. 21. Effect of FGF on the growth rate of vascular smooth muscle cell cultures maintained on either plastic or an extracellular matrix and exposed to 10% plasma (A) or 10% serum (B). Vascular smooth muscle cell cultures (2nd passage) were plated at 20,000 cells per 35-mm dish on plastic (o●, P) or an extracellular matrix (△▲, EM). The cultures were then maintained with DMEM, H-16 supplemented with 2.5 µg/ml Fungizone, 50 µg/ml Gentamycin, and with either 10% plasma (A) or 10% serum (B). FGF (100 ng/ml) was added every other day to half of the cultures maintained on plastic (o, P + F) or on an extracellular matrix (△, EM + F).

Earlier studies have shown that FGF, which has many similarities to the platelet-derived growth factor (same molecular weight and isoelectric point) [52], is a mitogen for vascular smooth muscle cells [9]. We have, therefore, compared its effect on cells maintained on plastic versus an ECM and exposed to either plasma or serum. As shown in Figure 21, cells maintained on plastic and exposed to plasma hardly proliferated. In contrast, when FGF was added to the cultures, the cells rapidly divided. After 7 days there was a 25-fold increase in cell number (Fig. 21A). The final cell density of cultures maintained on plastic and exposed to plasma plus FGF was higher than that of cultures maintained on plastic and exposed to serum alone. This demonstrates that the addition of FGF to the medium of cells maintained on plastic can make up for the difference in mitogenic activity between plasma and serum. When the growth rate of cultures maintained on plastic and exposed to plasma and FGF was compared to that of cultures maintained on an ECM and exposed to plasma but not to FGF, they were found to be similar. A noticeable but small difference was that the final cell density of cultures maintained on an ECM and exposed to plasma alone was 50% higher than that of cultures maintained on plastic and exposed to plasma and FGF. Addition of FGF to cultures maintained on an ECM and plasma (Fig. 21A) did not affect their growth rate. It resulted instead in a final cell density slightly lower than that observed with cultures not exposed to FGF (Fig. 21A). When the growth rate and final cell density of cultures exposed to serum and FGF were compared as a function of the substrate upon which cells were maintained, addition of FGF to cultures maintained on plastic and exposed to serum resulted in a decrease in the mean doubling time of the cultures (from 30–16 hr) as well as in a 2- to 3-fold increase in the final cell density (Fig. 21B). When FGF was added to cultures maintained on ECM, it did not affect their growth rate, which was already maximal (15-hr mean doubling time), nor did it affect their final cell density (Fig. 18B). Similar results were obtained when PDGF, instead of FGF, was used. One could, therefore, conclude that although FGF greatly increases the growth rate of cultures exposed to plasma and, to a lesser extent, that of cultures exposed to serum when the cultures are maintained on an ECM, it does not affect their growth rate since it become optimal when cultures are exposed to either plasma or serum.

The increased rate of proliferation of cells maintained on ECM and exposed to plasma or serum could either be the result of a direct mitogenic effect on the part of the ECM, the plasma or serum having a permissive role, or, conversely, the result of a direct mitogenic effect of plasma or serum, the ECM having a permissive role. To distinguish between these two possibilities, cells maintained on ECM were exposed to increasing concentrations of plasma or serum and the final cell densities were compared (Fig. 22). If the ECM should be the mitogen and the plasma or serum has a permissive effect, one would expect little difference in the rates of proliferation between high (10%) and low (0.5%) plasma or serum concentrations. When the final cell density of cultures maintained on an ECM and exposed to either plasma or serum was analyzed as a function of the serum or plasma concentration (Fig. 22) to which they were exposed, it was found to be a direct function of the serum or plasma concentration. It is, therefore, likely that the proliferation of vascular smooth muscle cells maintained on an ECM is controlled by factor(s) already present in plasma and that the ECM has a permissive role.

One could, therefore, conclude that, depending on the substrate upon which vascular endothelial cells are maintained, there are drastic differences in their requirements for growth factors. While cells maintained on plastic do not respond to plasma factor(s) [50, 53] and require serum [52], FGF [9], or PDGF [52] in order to proliferate when in close contact with an ECM the same cell types will respond to plasma factor(s) and no longer require serum for proliferation.

If in the case of vascular smooth muscle cells one were to extrapolate to the in vivo situation, maintaining them on an ECM and plasma is clearly a closer approximation to physiological conditions than exposing them to plastic and serum. Since vascular smooth muscle cells proliferate at a maximal rate when maintained on an ECM and exposed to plasma, it is likely that they are responding to mitogen(s) already present in plasma rather than to FGF or to mitogen(s) generated during the coagulation process.

THE COMMITMENT AND THE PROGRESSION FACTORS

Soluble factors which control the production of the ECM have not previously been reported. Until now, production of ECM was thought to be an automatic process which was mostly a function of the substrate upon which cells rest and of cell density. Although it is quite possibly that in vivo the production of an ECM is not under any control other

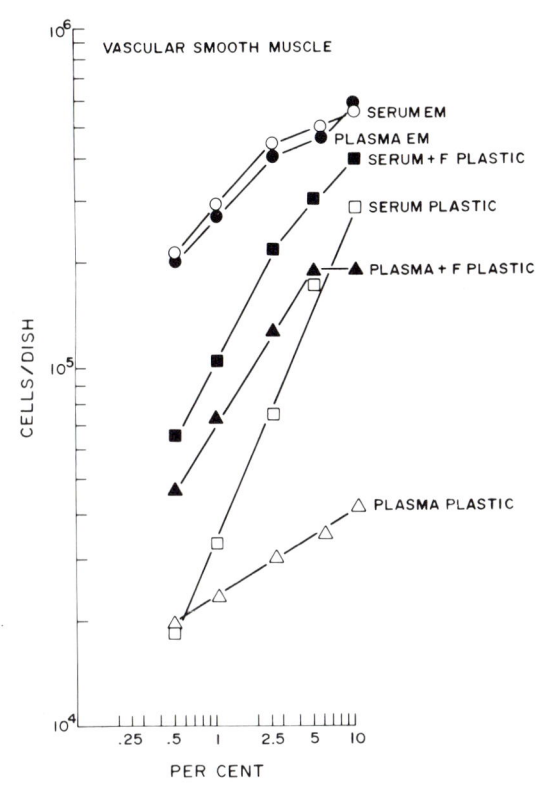

Fig. 22. Dependence for proliferation of vascular smooth muscle cells on plasma when the cells are maintained on an extracellular matrix. Vascular smooth muscle cells (2nd passage) were plated at 2 × 10^4 cells per 35-mm dish on plastic or on extracellular matrix (EM). Cultures were maintained for 8 hr in the presence of DMEM supplemented with either 10% plasma or serum. After 8 hr, the media were removed and the cultures washed once with DMEM, H-16. DMEM, H-16 supplemented with 2.5 μg/ml Fungizone and 50 μg/ml Gentamycin and containing different concentrations of serum or plasma (from 0.5% to 10%) was then added to the dishes. FGF (100 ng/ml) was added every other day to some of the dishes. After 5 days, the cultures were trypsinized and counted. Symbols are as follows. Cultures maintained on plastic and exposed to plasma (△) or to plasma and FGF (▲); cultures maintained on plastic and exposed to serum (□) or to serum and FGF (■); cultures maintained on an extracellular matrix and exposed to plasma (●) or to serum (○).

than that provided by the substrate upon which cells migrate during the early embryological phase, our results suggest that at least in vitro, factors such as FGF could influence the formation of an ECM scaffolding. We do not know the extent of this influence. Our observation that FGF could control the ECM production thereby making cells sensitive to plasma factor(s), is consistent with the findings of others that there could be two sets of growth factors in vitro, one of which, the commitment factor(s), could be involved in the formation of the ECM.

Earlier studies done on the control of cell cycle by growth factors have developed the concept that it could be controlled by 2 independent sets of factors present in serum, each of which controls a different phase of the cell cycle [54, 55]. One set is composed of heat-stable (100°C) factor(s) that are released into serum during the clotting process and induces BALB/c 3T3 cells to become capable of synthesizing DNA [56]. A second set of components, found in defibrinogenated platelet-poor plasma allows competent cells to progress through G_0G_1 and to synthesize DNA [54–56]. Stiles and his colleagues have further developed this observation by looking at the dual control of cell growth by the somatomedins, PDGF, and FGF [57]. Quiescent BALB/c 3T3 cells exposed briefly to PDGF or FGF become "competent" to replicate their DNA but do not "progress" into S phase unless exposed to somatomedin C, which is required for progression [57]. Since neither FGF nor PDGF is required to commit vascular smooth muscle cells to proliferation when they are maintained on an ECM and exposed to plasma, it is likely that competence factors would no longer be required if cells are maintained on an ECM.

THE IMPLICATIONS OF GROWING CELLS ON AN ECM INSTEAD OF ON PLASTIC

If one starts with a primary culture, it is likely that cells are selected which in the subsequent passages retain their ability to produce an ECM. Alteration in their phenotypic expression could be the direct result of an alteration in the type of ECM produced. This is best seen in the case of vascular endothelial cells, which express their normal phenotype when producing an ECM composed of collagen type III versus types IV and V at a ratio of 3:1 [58]. In contrast, when these cells no longer express their normal phenotype, the type of collagen produced is altered. Collagen type I begins to be synthesized, while the B and C chains of collagen type V are no longer produced [58]. This results in aberrant ECM production parallel with a loss of phenotypic expression [5, 17]. It is, therefore, possible that the widely acknowledged instability of the phenotypic expression of cultured cells could be due to their inability, when maintained on plastic, to continue to produce a normal ECM. If this should be the case, providing the cells with an artificial substrate closely resembling that produced in vivo should stabilize their phenotypic expression.

In the field of aging, it is generally recognized that senescent cells stop making an ECM [59]. Whether this is the result or the cause of senescence has not been investigated. It is to be suspected, however, that alterations in production or a loss of ability to produce a normal ECM could be directly linked to cell senescence, since this will result in a loss of their proliferative ability.

In the field of tumor cell biology, the growing recognition that the substrate upon which cells are maintained could modify their phenotypic expression is also important. One of the main characteristics of tumor cells grown in tissue culture is their loss of anchorage-dependence, as reflected by their ability to grow either in soft agar or in suspension when maintained on plastic. Yet in vivo, tumor cells from solid tumors, although they can adhere loosely to one another, can adhere tenaciously to the substrate which is provided by the host tissue or which they themselves produce. This is best reflected in the phenomenon of metastasis, where tumor cells which are carried away by the bloodstream can stick to basement membrane, infiltrate through it, and form secondary tumors in organs located far from the original tumor. If tumor cells were not anchorage-independent, one would expect them to proliferate freely in the bloodstream, but this rarely happens. It is, therefore, likely that if one provides tumor cells in culture with an adequate substrate, they could totally shift their pattern of growth. This is in fact what we have observed with hepatocarcinoma cells, Ewing's tumor cells or melanoma tumor cells [61]. It should also be pointed out that maintaining active proliferation of epithelial cell cultures of either normal or tumoral origin is a challenge. For example, in the case of carcinoma cells, less than 5% of the original cells put in culture give rise to cell lines [60]. Using an ECM as a natural substrate could change the figures and a higher percentage of either normal or tumor epithelial cells could be established in culture. It has also been shown by others that the ECM can control the morphogenetic and phenotypic expression of the tissue associated with it [60]. Use of ECM as a natural substrate for culturing the epithelial cells could, therefore, provide an opportunity to study not only their proliferation and the physiological factors controlling it but their differentiation as well. Although in the present study the ECM produced by corneal endothelial cells was used, other extracellular matrices produced by other cell types could also be used and would allow one to study the effects of various extracellular matrices on cell migration, proliferation, and differentiation.

Since cells maintained on an ECM now respond to plasma growth factors instead of to serum factors, one may wonder what these factors are. They could possibly be physiological agents such as trophic hormones which modulate cell proliferation in vivo but are inactive in vitro. Alternatively, they could be factors which have gone undetected because of the lack of sensitivity to them of cultured cells maintained on plastic. If this were so, one would now have an ideal substrate for restoring the normal growth response of many tissues to these naturally occurring factors. One might also suspect that conclusions concerning the mechanisms and controls of cell proliferation and cell migration of normal cells maintained on plastic may somehow differ from those which can be reached when cells are maintained on ECM.

ACKNOWLEDGMENTS

This work was supported by grants HL-23678 and EY-02186 from the National Institutes of Health. We wish to thank Mr. Harvey Scodel for his invaluable help in the preparation of this manuscript.

REFERENCES

1. Miller EJ: In Lash JW, Burger MM (eds): "Cell and Tissue Interactions," Society of General Physiologists Series, V 32, New York: Raven Press, 1977, pp 71–86.
2. Muir H: In Lash JW, Burger MM (eds): "Cell and Tissue Interactions," Society of General Physiologists Series, V 32, New York: Raven Press, 1977, pp 32–42.
3. Stenman S, Vaheri A: J Exp Med 147:1054, 1978.
4. Wessels NK: "Tissue Interaction and Development." San Francisco: WA Benjamin, 1977, pp 213–229.
5. Vlodavsky I, Johnson LK, Greenburg G, Gospodarowicz D: J Cell Biol 83:468, 1979.
6. Gimbrone MA: In Spaet T (ed): "Progress in Hemostasis and Thrombosis," V 3, New York: Green and Stratton, pp 1–28, 1976.
7. Booyse FM, Sedlak BJ, Rafelson ME: Thromb Diath Haemorrh 34:825, 1975.
8. Gospodarowicz D, Moran JS, Braun D: J Cell Physiol 91:377, 1977.
9. Gospodarowicz D, Moran JS, Braun D, Birdwell CR: Proc Natl Acad Sci USA 73:4120, 1976.
10. Schwartz SM: In Vitro 14:966, 1978.
11. Mueller SM, Rosen EM, Levine EM: Science 207:889, 1980.
12. McAuslan BR, Reilly W, Hannan GN: J Cell Physiol 100:87, 1979.
13. Gospodarowicz D, Greenburg G, Bialecki H, Zetter B: In Vitro 14:85, 1978.
14. Birdwell CR, Gospodarowicz D, Nicolson G: Proc Natl Acad Sci USA 75:3272, 1978.
15. Gospodarowicz D, Vlodavsky I, Greenburg G, Alvarado J, Johnson LK, Moran J: Rec Progr Hormone Res 35:375, 1979.
16. Vlodavsky I, Johnson LK, Gospodarowicz D: Proc Natl Acad Sci USA 76:2306, 1979.
17. Greenburg G, Vlodavsky I, Foidart JM, Gospodarowicz D: J Cell Physiol 103:333, 1980.
18. Grobstein C: Natl Cancer Inst Monogr 26:279, 1967.
19. Dodson JS: Exp Cell Res 31:233, 1963.
20. Wessels NK: Proc Natl Acad Sci USA 52:252, 1964.
21. Hay ED: In Littlefield JW, de Grouchy J (eds): "Birth Defects," Excerpta Medica, Int Cong Series 432, Amsterdam, Oxford, 1978, pp 126–140.
22. Hay ED: In Lash JW, Burger MM (eds): "Cell and Tissue Interactions," Society of General Physiologists Series, V 32, New York: Raven Press, 1977, pp 115–138.
23. Hay ED, Meier S: Dev Biol 52:141, 1976.
24. Bernfield MR: In Littlefield JW, de Grouchy J (eds): "Birth Defects," Excerpta Medica, Int Cong Series 432, Amsterdam: Oxford, 1978, pp 111–125.

25. Ehrmann RL, Gey GO: J Natl Cancer Inst 16:2:1375, 1956.
26. Gey GO, Svotelis M, Foard M, Band FB: Exp Cell Res 84:63, 1974.
27. Liu S-C, Karasek M: J Invest Dermatol 71:157, 1978.
28. Schar SL, Court J: J Cell Sci 38:267, 1979.
29. Sicha MS, Liotta LA, Garbioa S, Kidwell WR: Exp Cell Res 124:181, 1979.
30. Murray JC, Stingl G, Kleinman HK, Martin GR, Katz SI: J Cell Biol 80:197, 1979.
31. Fisher M, Solursh M: Exp Cell Res 123:1, 1979.
32. Liotta LA, Vembu D, Kleinman H, Martin GR, Boone C: Nature 272:622, 1978.
33. Yang J, Richards J, Bowman P, Guzman R, Ehami J, McCormick K, Hamamoto S, Pitelka D, Nandi S: Proc Natl Acad Sci USA 76:3401, 1979.
34. Vitto J, Prockop DY: Biochim Biophys Acta 336:234, 1974.
35. Yamada KJ, Yamada SS, Pastan I: Proc Natl Acad Sci USA 73:1217, 1976.
36. Ali IU, Mautner V, Lanza R, Hynes RO: Cell 11:115, 1977.
37. Orly J, Sato G: Cell 17:295, 1979.
38. Timpl R, Rohde H, Gehran-Robey P, Rennard SIL, Foidart JM, Martin GR: J Biol Chem 254:9933, 1979.
39. Gospodarowicz D, Greenburg G, Vlodavsky I, Alvarado J, Johnson LK: Exptl Eye Res 29:485, 1979.
40. Gospodarowicz D, Vlodavsky I, Greenburg G, Johnson LK: In Sato G, Ross R (eds): "Hormones and Cell Culture," Cold Spring Harbor Conferences on Cell Proliferation, New York: Cold Spring Harbor, pp 561–592, 1979.
41. Gospodarowicz D, Delgado D, Vlodavsky I: Proc Natl Acad Sci USA 77:4094, 1980.
42. Gospodarowicz D, Brown KS, Birdwell CR, Zetter BR: J Cell Biol 77:774, 1978.
43. Gospodarowicz D, Mescher AL, Birdwell CR: Natl Cancer Inst Monogr 48:109, 1978.
44. Folkman J, Moscona A: Nature 273:345, 1978.
45. Gospodarowicz D, Mescher AL, Brown K, Birdwell CR: Exp Eye Res 25:631, 1977.
46. Gospodarowicz D, Greenburg G, Birdwell CR: Cancer Res 38:4155, 1978.
47. Carrel AJ: J Exp Med 15:516, 1912.
48. Balk SD: Proc Natl Acad Sci USA 68:271, 1971.
49. Balk SD, Whitfield JF, Youdale T, Braun AC: Proc Natl Acad Sci USA 70:675, 1973.
50. Ross R, Glomset J, Kariya B, Harker L: Proc Natl Acad Sci USA 71:1207, 1974.
51. Kohler N, Lipton A: Exp Cell Res 87:297, 1974.
52. Ross R, Vogel A: Cell 14:203, 1978.
53. Ross R, Nist C, Kariya B, Rivest MJ, Raines E, Callis J: J Cell Physiol 97:497, 1978.
54. Vogel A, Raines E, Kariya B, Rivest MJ, Ross R: Proc Natl Acad Sci USA 75:2810, 1978.
55. Pledger WJ, Stiles CD, Antoniades HM, Scher CD: Proc Natl Acad Sci USA 75:2839, 1978.
56. Pledger WJ, Stiles CD, Antoniades HM, Scher CD: Proc Natl Acad Sci USA 74:4481, 1977.
57. Stiles CD, Capne GT, Scher CD, Antoniades HM, Van Wyk JJ, Pledger WJ: Proc Natl Acad Sci USA 76:1279, 1979.
58. Tseng S, Savion N, Stern R, Gospodarowicz D: J Biol Chem (in press).
59. Cristofalo VJ: Adv Gerontol Res 4:45, 1972.
60. Rafferty KA: Adv Cancer Res 21:249, 1975.
61. Vlodavsky I, Liu GM, Gospodarowicz D: Cell 19:607, 1980.

NOTE ADDED IN PROOF

Since submission of this manuscript, the factors present in plasma that regulate the proliferation of vascular endothelial cells and vascular smooth muscle cells have been identified. In the case of vascular endothelial cells, high density lipoprotein (HDL) is the major factor involved in the proliferation of vascular endothelial cells, since it can fully replace plasma [1, 2]. This is best shown by the observation that cells maintained in serum-free or plasma-free medium can be repeatedly passaged at low cell density and proliferate at an optimal rate if HDL is added to the medium. In the case of vascular smooth muscle cells, although HDL is also needed, somatomedin C and epidermal growth factor are required in order to induce optimal growth rate when cultures are maintained in plasma-free medium [3].

1. Tauber J-P, Cheng J, Gospodarowicz D: J Clin Inv 66:696–708, 1980.
2. Gospodarowicz D, Tauber J-P: Endocrine Reviews 1:201–277, 1980.
3. Gospodarowicz D, Hirabayashi K, Giguere L, Tauber J-P: J Cell Biol (in press).

Inhibition of Lymphocyte Mitogenesis by an Arachidonic Acid Hydroperoxide

Michael G. Goodman and William O. Weigle

Department of Immunopathology, Scripps Clinic and Research Foundation, La Jolla, California 92037

Incubation of murine spleen cells with the oxidation product of soybean lipoxidase-treated arachidonic acid results in profound inhibition of induction of proliferation and maturation of these cells. The active entity was shown to be the 15-hydroperoxide of arachidonic acid (15-HPAA). Inhibition of the enzymes of the cyclo-oxygenase pathway fails to disturb this effect, indicating that 15-HPAA is not a substrate for this series of enzymes. 15-HPAA produced in this manner interfered with RNA synthesis, DNA synthesis, and blastogenesis, while failing to exert cytotoxic effects on the cells themselves. A variety of lymphocyte subpopulations, distinguished by their responsiveness to a diverse group of mitogens, were all equally inhibited by the addition of 15-HPAA to culture. Addition of this agent even as late as 24 h after initiation of culture resulted in profound inhibition of the proliferative and differentiative responses of splenic B cells to bacterial lipopolysaccharide (LPS). Exposure of cells to 15-HPAA for 10–30 min was adequate to initiate inhibition, an event that exhibited marked temperature dependence. The effects of pre-incubation with 15-HPAA could not be reversed in its absence in recovery periods of up to 6 h prior to addition of LPS. The implications of these data with reference to cellular activation mechanisms are discussed.

Key words: hydroperoxide, mitogenesis, 15-HPAA, arachidonic acid, inhibition, lymphocyte activation

The metabolites of arachidonic acid (AA) generated via the cyclo-oxygenase pathway, ie, the prostaglandins (PG), thromboxanes, and prostacyclin, have been the object of a great deal of biological and biochemical research. However, a number of biological effects have been described that are dependent upon substances produced from arachidonic acid by lipoxidation. Such effects include the stimulation of human platelet guanylate cyclase [1], modulation of human lymphocyte responses to phytohemagglutinin [2], diminution in the adherence properties of baby hamster kidney cells to substrate [3], enhancement of mediator release from activated rat mast cells [4], and enhancement of release of slow-reacting substance from ionophore-stimulated rat mast cells [5]. One such oxidation product is the 15-hydroperoxide of arachidonic acid (15-HPAA), generated by treatment of substrate with soybean lipoxidase. This agent has been used as a probe in a number of mammalian systems, where it has been shown to augment the release of anaphylactic mediators

Received April 11, 1980; accepted June 24, 1980.

0091-7419/80/1303-0373$02.30 © 1980 Alan R. Liss, Inc.

from guinea pig lung [6], to inhibit the formation of 6-keto PGF$_{1\alpha}$ [7], to inhibit the generation of prostacyclin [8], and to stimulate human platelet guanylate cyclase activity [1]. Because the role that these compounds play in lymphocyte activation is not yet understood, experiments were undertaken to investigate the role of the soybean lipoxidation product of arachidonic acid on lymphocyte activation.

MATERIALS AND METHODS

Mice

C3H/St male mice, 8–16 weeks of age, were obtained from the mouse breeding facility at Scripps Clinic and Research Foundation, La Jolla, CA. All mice were maintained on Wayne Lab-Blox F6 pellets (Allied Mills, Inc., Chicago, IL) and chlorinated water acidified to a pH of 3.0 with HCl [9].

Mitogens

Bacterial lipopolysaccharide (LPS) 055:B5, extracted by the Boivin technique, was purchased from Difco Laboratories, Detroit, MI. Concanavalin A (Con A) was obtained from Miles-Yeda Laboratories, Rehovot, Israel. Polyinosinic-polycytidilic acid (Poly IC) double-stranded sodium salt was purchased from P-L Biochemicals, Inc., Milwaukee, WI. 2-Mercaptoethanol (2-ME), purchased from Matheson, Coleman and Bell, Los Angeles, CA, was dissolved in phosphate-buffered saline (PBS) and sterilized by filtration. Purified protein derivative (PPD), extracted from mycobacterium tuberculosis RT33, was obtained from Statens Serum-institut, Copenhagen, Denmark. Unless otherwise specified, all mitogens were sterilized by exposure to UV light and diluted in complete medium.

Culture Reagents

Dissociated spleen cells were cultured in serum-free medium, whose constituents have been detailed elsewhere [10]. Cultures were fed daily with nutritional cocktail as described previously [11]. Arachidonic acid was purchased from the Sigma Chemical Co., St. Louis, MO; ICN Pharmaceuticals, Inc., Irvine, CA; US Biochemical Corp., Cleveland, OH; and Vega Biochemicals, Tucson, AZ. Soybean lipoxidase and indomethacin were purchased from the Sigma Chemical Co., St. Louis, MO.

Lymphocyte Cultures

Spleen cell suspensions were prepared as described previously [10]. These cells were cultured in microculture plates (#3042, Falcon Plastics, Oxnard, CA) at a cell density of 5×10^6 viable cells/ml in a volume of 0.1 ml. Microculture plates were incubated at 37°C in a humidified atmosphere of 5% CO_2 in air. Cultures were fed daily with 8 μl of nutritional cocktail. Polyclonal activation of B cells was induced by LPS in cultures of 5×10^6 viable lymphocytes per well in a volume of 1.0 ml in tissue culture trays (#3008, Falcon Plastics). These cultures were incubated under conditions identical to those for microcultures and were fed with 60 μl of nutritional cocktail daily.

Measurement of DNA and RNA Synthesis

To evaluate DNA synthesis, cells were radiolabeled with 1.0 μCi of ^{3}H-thymidine (^{3}H-TdR) per culture (5.0 Ci/mmole, New England Nuclear, Boston, MA) during the final 24 h of culture. RNA synthesis was determined by pulsing cells with 1.0 μCi of ^{3}H-uridine

(^{3}H-UdR) per culture (5 Ci/mmole, New England Nuclear) for the final 4 hr of a 24 h culture period. Microcultures were harvested on a Brandel cell harvester, Model M24V (Biological Research and Development Laboratories, Rockville, MD) onto glass-fiber filter strips (Reeve Angel, Clifton, NJ). Filter discs were transferred to plastic scintillation vials (Kimball Products, Owens-Illinois, Toledo, OH), covered with liquid scintillation cocktail (Scintverse, Fisher Scientific Co., Fairlawn, NJ), and counted in a Beckman LS-230 liquid scintillation spectrometer.

Assay of Plaque-Forming Cells (PFC)

LPS-induced polyclonal B-cell activation was evaluated by determining the response to sheep red blood cells (SRBC). PFCs to SRBC were assayed at day 2 of culture [12], using a modification of hemolytic plaque assay of Jerne and Nordin [13]. Results are expressed as the arithmetic mean of triplicate cultures ± the standard error (SE).

Enumeration of Blast Cells

Histological preparations were generated on microscopic slides from individual lymphocyte cultures with the aid of a cytocentrifuge. Slides were stained with the methyl green-pyronin Y technique [14] to simplify the enumeration of pyroninophilic blast cells. Results are expressed as the arithmetic mean of triplicate cultures ± the standard error.

High Pressure Liquid Chromatography

Oxidation of AA by treatment with soybean lipoxidase was performed according to the protocol of Hamberg and Samuelsson [15]. Oxidation products were purified by HPLC using a Waters liquid chromatograph fitted with a 0.39 × 30 cm μ-Porasil column. The eluting solvent, n-hexane/propan-2-ol/acetic acid (994:5:1 [8]) was pumped through the column at 3 ml/min.

RESULTS

Effects of Arachidonic Acid Oxidation Product(s) on Lymphocyte Stimulation

Preliminary experiments demonstrated that neither arachidonic acid nor its lipoxidase-catalyzed oxidation product(s) were mitogenic for murine spleen cells, in spite of the reported ability of the latter product(s) to activate guanylate cyclase in human platelets [1]. Therefore, the ability of these substances to modify the mitogenic response of murine lymphocytes to LPS was evaluated. It was found that variable concentrations of arachidonic acid treated with soybean lipoxidase [15], when incubated with cultures in the presence of 100 μg/ml of LPS, led to inhibition of lymphocyte mitogenesis (Table I). Predominant conversion to 15-HPAA was verified by determining the absorbance at 234 nm [15] and by HPLC. The addition of lipoxidase alone to cultures of spleen cells incubated with LPS failed to alter the LPS response. Experiments undertaken to learn if 15-HPAA-mediated inhibition of mitogenesis was due to toxicity demonstrated that this substance caused no significant diminution of viability during the 3-day culture period (34% recovery without 15-HPAA, 37% with it).

Chromatography of 15-HPAA on HPLC produced one predominant peak; this substance reacted positively with the peroxide reagent ferrous thiocyanate. Assay of the HPLC fractions for inhibitory activity (Table II) indicated that most of the activity was attributable to 15-HPAA, eluting at 15 min. However, moderate inhibitory effects were seen with the shoulder eluting just prior to 15-HPAA, at 12.9 min.

TABLE I. Effect of 15-HPAA on LPS-Induced Mitogenesis

LPS[a]	[15-HPAA]	³H-TdR uptake (cpm/culture)[b]
−	−	3,830 ± 530
+	−	26,660 ± 3,830
+	10^{-7}	23,960 ± 1,200
+	3×10^{-7}	24,540 ± 840
+	10^{-6}	23,980 ± 800
+	3×10^{-6}	24,440 ± 520
+	10^{-5}	16,400 ± 1,000
+	3×10^{-5}	3,350 ± 590
+	10^{-4}	1,650 ± 220

[a] 5×10^5 viable C3H/St spleen cells were cultured with 100 µg/ml LPS in 0.1 ml of serum-free medium in the presence or absence of incremental concentrations of 15-HPAA.
[b] Cultures were pulsed with 1 µCi of ³H-TdR (5 Ci/mmole) for the final 24 h of the 3-day culture period. Results are expressed as the arithmetic mean of 5 replicate cultures ± the SE.

TABLE II. Effect of HPLC Separated AA Oxidation Products on the Mitogenic Response to LPS

Peak	Retention time (min)	Inhibition of LPS response (%)[a]
−	−	0
1	3.2	0
2	5.9	1
3	7.1	0
4	12.9	36
5	15.0	87

[a] 5×10^5 viable C3H/St spleen cells were cultured with 100 µg/ml LPS in 0.1 ml of serum-free medium in the presence or absence of the minimal concentration necessary for inhibition by any one component. Cultures were pulsed with 1.0 µCi ³H-TdR for the final 24 h of the 3-day culture period. Results are expressed as the percent inhibition of the mean LPS response of 5 replicate cultures. ³H-TdR uptake by LPS-stimulated cultures in the absence of peak fractions was 30,500 ± 750 cpm.

Inability of Indomethacin to Reverse the Effect of 15-HPAA

Prostaglandins are produced from arachidonic acid via the cyclo-oxygenase pathway, a pathway that can be blocked by the action of the "prostaglandin synthetase" inhibitor, indomethacin. Therefore, if 15-HPAA were acting as a substrate for any of the enzymes involved in synthesis of the prostaglandins, the thromboxanes, or prostacyclin, its inhibitory effect should be blocked when indomethacin is added to culture. The data presented in Table III are representative of experiments in which optimal amounts of 15-HPAA were added to cultures activated with 100 µg/ml of LPS, in the presence or absence of indomethacin. No significant difference can be seen in the presence or absence of several concentrations of this prostaglandin synthetase inhibitor, arguing that products of the cyclo-oxygenase pathway are not involved in this inhibitory phenomenon.

TABLE III. Inability of Indomethacin to Reverse the Effect of 15-HPAA

10^{-4} M 15-HPAA[a]	Indomethacin	^{3}H-TdR uptake (cpm/culture) (E-C)[b]
+	–	350 ± 90
–	–	8,540 ± 390
+	10^{-6}	470 ± 170
–	10^{-6}	9,340 ± 230
+	10^{-5}	420 ± 120
–	10^{-5}	8,620 ± 250

[a]5×10^5 viable C3H/St spleen cells were cultured with 100 μg/ml LPS in 0.1 ml of serum-free medium in the presence or absence of optimal concentrations of 15-HPAA in the presence of variable amounts of indomethacin.
[b]Cultures were pulsed with 1 μCi of ^{3}H-TdR (5 Ci/mmole) for the final 24 h of the 3-day culture period. Results are expressed as the arithmetic mean of 5 replicate cultures minus controls ± the SE.

15-HPAA Inhibits Lymphocyte Activation by a Spectrum of Mitogens

The ability of 15-HPAA to inhibit activation of a diversity of lymphocyte populations and subpopulations was investigated. It has previously been demonstrated that the B-cell subpopulations responsive to LPS and Poly IC are essentially the same, while each of those responding to PPD and 2-ME is distinct [10]. However, as shown in Table IV, activation of each of these B-cell subpopulations is completely inhibited by co-incubation with 15-HPAA. Moreover, activation of T cells by incubation of spleen cells with either the jack bean lectin Con A or with 2-ME (which stimulates both B and T cells separately) is similarly inhibited. It thus appears that 15-HPAA is capable of profoundly inhibiting the activation of a number of distinct B- and T-lymphocyte subpopulations.

Because 15-HPAA might theoretically interfere with ^{3}H-TdR uptake without otherwise altering mitogenesis, the effect of 15-HPAA on LPS-induced blastogenesis was investigated. Spleen cells from C3H/St mice were cultured with LPS under serum-free conditions in the presence or absence of 2.5×10^{-4} M 15-HPAA. After 1, 2, and 3 days of incubation, each culture was transferred to a slide by means of a cytocentrifuge, and stained as described in Materials and Methods. Enumeration of pyroninophilic blast cells under the microscope indicated that LPS-induced blastogenesis is completely inhibited by 15-HPAA regardless of the duration after which it is evaluated (data not shown).

Inhibition of RNA Synthesis by 15-HPAA

To this point, inhibition of cellular activation has been evaluated by assay of uptake of ^{3}H-TdR. Therefore, the effect of 15-HPAA on synthesis of cellular RNA was examined to determine whether earlier stages in the activation process were inhibited as well as later stages. In Table V, murine spleen cells were activated with optimal concentrations of LPS in the presence of incremental amounts of 15-HPAA. One μCi/culture of ^{3}H-UdR was added after 20 h of culture, and cells were harvested 4 h later. The data indicate that uridine uptake also is inhibitable by the action of 15-HPAA with a profile of similar shape to that for DNA synthesis.

Kinetics of 15-HPAA-Dependent Inhibition of Polyclonal B Cell Activation and Mitogenesis

The marked inhibitory effect of 15-HPAA on cellular proliferation led us to consider next whether cellular differentiation might be similarly affected. Therefore, taking

TABLE IV. Effect of 15-HPAA on Lymphocyte Activation by a Spectrum of Mitogens

Mitogen[a]	15-HPAA	^{3}H-TdR uptake (cpm/culture)[b]
–	–	1,395 ± 540
–	+	690 ± 120
LPS	–	50,650 ± 1,920
LPS	+	2,350 ± 420
PPD	–	66,150 ± 2,300
PPD	+	2,530 ± 1,190
Poly IC	–	23,790 ± 890
Poly IC	+	800 ± 180
2-ME	–	39,070 ± 2,080
2-ME	+	1,020 ± 530
Con A	–	285,180 ± 2,670
Con A	+	1,070 ± 500

[a] 5×10^5 viable C3H/St spleen cells were cultured in 0.1 ml of serum-free medium with either medium alone, 100 µg/ml LPS, 500 µg/ml Poly IC, 300 µg/ml PPD, 5×10^{-5} M 2-ME, or 1 µg/ml Con A. Each mitogen was tested in the presence and absence of 2.5×10^{-4} 15-HPAA.

[b] Cultures were pulsed with 1 µCi of ^{3}H-TdR (5 Ci/mmole) for the final 24 h of the 3-day culture period. Results are expressed as the arithmetic mean of 5 replicate cultures ± the SE.

TABLE V. Effect of 15-HPAA on Cellular RNA Synthesis

LPS[a]	15-HPAA	^{3}H-UdR uptake (cpm/culture)[b]
–	–	3,230 ± 385
+	–	10,401 ± 165
+	10^{-5}	9,049 ± 212
+	5×10^{-5}	9,906 ± 379
+	10^{-4}	8,757 ± 628
+	2.5×10^{-4}	1,061 ± 485
+	5×10^{-4}	574 ± 224

[a] 5×10^5 viable C3H/St spleen cells were cultured with 100 µg/ml LPS in 0.1 ml of serum-free medium in the presence of incremental concentrations of 15-HPAA.

[b] Cultures were pulsed with 1 µCi of ^{3}H-UdR (5 Ci/mmole) for the final 4 h of a 24-h culture period. Results are expressed as the arithmetic mean of 5 replicate cultures ± the SE.

advantage of the ability of LPS to activate lymphocyte cultures to polyclonal secretion of immunoglobulin (Ig), 15-HPAA was added at optimal inhibitory doses at various times after culture initiation. The data presented in Table VI demonstrate that this agent, added as late as 24 h after LPS, inhibits the polyclonal response as assayed against SRBC on day 2. However, at 45–48 h, when cellular differentiation to antibody secretion has occurred, 15-HPAA exerts only minimal inhibitory effects.

The ability of 15-HPAA to inhibit LPS-induced mitogenesis when added at sequential time-points after culture initiation was evaluated in experiments summarized in Table VII. The kinetic profile in this case is similar to that seen for polyclonal activation, except that the time of assay here is 72 h rather than 48 h. The data indicate that complete inhibi-

TABLE VI. Kinetics of 15-HPAA-Dependent Inhibition of Polyclonal B-Cell Activation

LPS[a]	15-HPAA added at	Anti-SRBC PFC/culture[b]
−	−	35 ± 6
+	−	427 ± 8
+	0 h	107 ± 11
+	24 h	103 ± 4
+	45 h	347 ± 12
+	Assay	348 ± 48

[a]5×10^6 viable C3H/St spleen cells were cultured in the presence or absence of 100 μg/ml LPS in 1.0 ml of 5% FCS-containing medium. At various times after culture initiation, cultures were brought to a final 15-HPAA concentration of 2.5×10^{-4} M.
[b]The direct PFC response to SRBC was assessed after 2 days of culture. Results are expressed as the arithmetic mean of triplicate cultures ± the SE.

TABLE VII. Kinetics of 15-HPAA-Dependent Inhibition of LPS-Induced Mitogenesis

LPS[a]	Time of 15-HPAA addition	^{3}H-TdR uptake (cpm/culture)[b]
−	−	3,400 ± 300
+	−	49,720 ± 1,290
+	0 h	240 ± 50
+	6 h	200 ± 90
+	12 h	770 ± 110
+	24 h	180 ± 90
+	48 h	13,350 ± 700

[a]5×10^5 viable C3H/St spleen cells were cultured with 100 μg/ml LPS in 0.1 ml of serum-free medium. At various times after culture initiation, cultures were brought to a final 15-HPAA concentration of 2.5×10^{-4} M.
[b]Cultures were pulsed with 1 μCi of ^{3}H-TdR for the final 24 h of the 3-day culture period. Results are expressed as the arithmetic mean of 5 replicate cultures ± the SE.

tion of mitogenesis occurs when 15-HPAA is added as late as 24 h after culture initiation. When added after 2 days of a 3-day culture period, 75% of the response can still be abolished. Thus, it appears that late events are important targets for the action of the inhibitory AA oxidation product.

Effect of Temperature on 15-HPAA-Mediated Inhibition

The profound nature of the inhibition of mitogenesis induced by 15-HPAA led us next to investigate whether this effect is temperature dependent. C3H/St spleen cells were incubated for variable periods of time with 15-HPAA at either 4°C or 37°C, following which they were washed in an excess of BSS 3 times, plated out in microculture wells in the presence or absence of optimal mitogenic concentrations of LPS, and incubated at 37°C for the remainder of the culture period. The results of these experiments are shown in Table VIII. It is apparent that at 37°C, a 10-min pre-incubation with 15-HPAA is sufficient to reduce the mitogenic effects of LPS to baseline levels. Pre-incubation longer than 30 min does not serve to promote additional inhibition. On the other hand, when pre-incubation is performed at 4°C, lymphocyte mitogenesis is inhibited by only approximately 50%. This inhibition also levels off after 10 min of exposure to 15-HPAA. In concert, these data suggest that the inhibitable entity is inactivated by exposure to 15-HPAA

TABLE VIII. Effect of 15-HPAA Pre-incubation on LPS-Induced Mitogenesis

LPS	Length of 15-HPAA pre-incubation[a]	^{3}H-TdR uptake (cpm/culture) (E-C)[b] 37°C	4°C
+	–	51,200 ± 2,400	51,200 ± 2,400
+	10 min	3,310 ± 1,480	23,320 ± 1,630
+	30 min	0	18,100 ± 2,670
+	1 h	0	24,730 ± 1,550
+	2 h	0	25,390 ± 3,540

[a]5×10^6 viable C3H/St spleen cells per ml were pre-incubated (in 12 × 75-mm test tubes) in the presence or absence of 2.5×10^{-4} M 15-HPAA for the time lengths shown, at either 4°C or 37°C. After pre-incubation, cells were washed 3 times in an excess of BSS. 5×10^5 viable cells were then cultured in the absence or presence of 100 μg/ml LPS in 0.1 ml of serum-free medium.
[b]Cultures were pulsed with 1 μCi of ^{3}H-TdR for the final 24 h of the 3-day culture period. Results are expressed as the arithmetic mean of 5 replicate cultures ± the SE.

very rapidly. Furthermore, this rapid inhibition includes a temperature-dependent phase resulting in a partial inhibition of cells pre-incubated at 4°C, which cannot be amended by subsequent incubation at 37°C for the duration of the culture period.

Inability of Cells to Recover From Inhibition by 15-HPAA

Finally, the possibility that cells inhibited by a brief pre-incubation with 15-HPAA might be able to regenerate 15-HPAA-sensitive function during a recovery period was tested. Therefore, cells were incubated with 15-HPAA at 37°C for 1 h, washed 3 times in an excess of serum-free medium, and cultured in the absence of inhibitor for variable periods of time before addition of LPS. As can be seen in Table IX, however, recovery periods up to 6 h were ineffectual in allowing cells to recover responsiveness to LPS. Because of the dimished stimulation obtained when addition of LPS is delayed 6 h or longer, the effects of later addition of LPS were not evaluated.

DISCUSSION

The specific introduction of a hydroperoxide group at the 15 position of arachidonic acid catalyzed by soybean lipoxidase [15] results in the creation of a compound with potent biological activities. Its abilities to enhance the release of histamine, SRS-A, and rabbit aortic-contracting substance (RCF) from guinea pig isolated perfused lungs [6], to inhibit the synthesis of 6-keto prostaglandin $F_{1\alpha}$ by aortic microsomes [7], and to inhibit conversion of prostaglandin endoperoxides to prostacyclin [8] have been well described. The current communication reports the further ability of this compound to inhibit lymphocyte mitogenesis as induced by a variety of B- and T-lymphocyte activators.

The inhibitory effect of 15-HPAA on murine lymphocytes is profound and exhibits marked dose-dependency. Inhibition does not appear to be attributable to processing by cyclo-oxygenase pathway enzymes. Indomethacin, a prostaglandin synthetase inhibitor, blocks the formation of the prostaglandins, the thromboxanes, and prostacyclin, via the cyclo-oxygenase pathway. However, this agent failed to interfere with the inhibition of lymphocyte activation mediated by 15-HPAA. Although there are differences in sensitivity of arachidonic acid metabolites to indomethacin, no effect was attributable to this agent over the entire range of concentrations that failed to alter the basic lymphocyte response itself.

TABLE IX. Inability of Cells to Recover from Inhibition by 15-HPAA

15-HPAA pre-incubation[a]	Length of recovery period[b]	LPS	^{3}H-TdR uptake (cpm/culture)[c]
−	0	−	990 ± 65
−	0	+	39,610 ± 3,130
+	0	+	2,400 ± 360
−	1 h	+	44,560 ± 1,970
+	1 h	+	3,100 ± 710
−	3 h	+	35,620 ± 5,090
+	3 h	+	3,650 ± 420
−	6 h	+	4,805 ± 530
+	6 h	+	3,900 ± 1,765

[a]5×10^6 viable C3H/St spleen cells per ml were pre-incubated (in 12 × 75-mm test tubes) in the presence or absence of 2.5×10^{-4} M 15-HPAA for 1 h at 37°C.
[b]Cells were washed 3 times in an excess of BSS, and 5×10^5 viable cells were cultured in 0.1 ml of serum-free medium in the absence of LPS. At the times shown, LPS was added to a final concentration of 100 μg/ml.
[c]Cultures were pulsed with 1 μCi of ^{3}H-TdR for the final 24 h of the 3-day culture period. Results are expressed as the arithmetic mean of quadruplicate cultures ± the SE.

The interference of 15-HPAA with cellular activities failed to result in the loss of cellular viability. Similarly, the diminution of tritiated thymidine uptake by cells was shown not to be an artifact of thymidine transport, insofar as blast transformation and RNA synthesis were similarly affected. Thus, the inhibitory effect of 15-HPAA appears to act by arresting the stimulation process without totally disrupting the function vital to cellular survival.

Activation of T lymphocytes, as well as a variety of B-cell subpopulations, were inhibited by 15-HPAA. Activation of these populations and subpopulations was elicited by the use of a number of different lymphocyte mitogens, whose cellular specificity has been previously verified [10]. These experiments, however, were carried out with whole spleen cells, and it is possible that 15-HPAA induces suppressor activity in one group of cells, ie, macrophages, resulting in an indirect mechanism of cellular inhibition.

The differentiative as well as the proliferative responses of murine spleen cells to LPS is subject to inhibition by 15-HPAA. It has been demonstrated by Andersson and Melchers [16] that maturation of resting B cells into 19S IgM-secreting plaque-forming cells (PFC) is able to occur even when DNA synthesis and proliferation are inhibited by the use of either hydroxyurea or cytosine arabinoside. Thus, differentiation can occur in the absence of proliferation, and in fact, these 2 activities have been dissociated by others [17]. Therefore, one would have to postulate that 15-HPAA exerts its effects upon a sequence of steps common to both the differentiative and proliferative pathways or that this agent interferes with a different step in each pathway. Addition of 15-HPAA to culture after LPS has effected the maturation of precursor B cells to IgM-secreting PFC failed to inhibit the development of PFC in a Jerne plaque assay. Thus, antibody secretion is not subject to the effects of 15-HPAA. Kinetic analysis of the inhibitory potential of 15-HPAA on polyclonal as well as proliferative responses to LPS indicated that this agent exerts its effects relatively late in the culture period. Others have noted that optimal activation, especially with B-lymphocyte mitogens, requires the presence of the mitogen for the first 24 h of culture [18]. Apparently, it is in this time, during which the activation signal is being received and processed by the target cell, that 15-HPAA is most effective.

Lymphocyte inhibition by 15-HPAA requires only a very brief exposure to this agent, as revealed by a kinetic analysis utilizing brief pre-incubation of spleen cells with 15-HPAA followed by extensive washing procedures. These kinetics closely resemble those described for 15-HPAA-mediated inhibition of 6-keto PGF$_{1\alpha}$ production by aortic microsomes [7]. Although the length of exposure is brief, the inhibitory event which occurs is temperature-dependent. Thus, if pre-incubation of the cells is carried out at 4°C, followed by washing and subsequent incubation at 37°C in the absence of 15-HPAA, the inhibition that follows is incomplete, averaging about 50%. This observation is taken to indicate that at least one part of the inhibitory process is dependent upon cellular metabolism while 15-HPAA is present, since later elevation of the temperature to 37°C does not correct the incomplete inhibition. However, an association of 15-HPAA with the inhibited entity seems to occur at 4°C, although one cannot determine from the data presented whether the actual process culminating in the partial inhibition observed occurs at this temperature or at the following incubation at 37°C.

Recovery periods of up to 6 h in length fail to result in the regeneration of LPS responsiveness by cells previously exposed to 15-HPAA. Further corroborating evidence for the irreversibility of this effect can be gleaned from the experiments in which cells were pre-incubated for brief periods of time in the presence of 15-HPAA followed by standard 2-day culture with LPS in the absence of this inhibitory agent. In no case was responsiveness recovered.

The reasons why inhibition of the endogenous lipoxygenase pathway with ETYA interferes with cellular activation [19], yet supplementation of cultures with exogenous products of AA lipoxidation inhibits activation, are not clear. While it is possible that the site of oxidation within the molecule may lead to the generation of products with opposing activities, that selective pathways are inhibited by a relatively large amount of exogenous hydroperoxide, or that initial exposure of these highly reactive compounds to the outside vs the inside of the cell results in very different effects, the actual cause remains to be elucidated. Further experiments, probing the mechanism by which this inhibitory action is exerted, are currently in progress.

ACKNOWLEDGMENTS

This is publication no. 2110 from the Department of Immunopathology, Scripps Clinic and Research Foundation, La Jolla, California. This work was supported by United States Public Health Service grant AI-07007, American Cancer Society grant IM-421, and Biomedical Research Support Program grant RRO-5514. M.G.G. is a recipient of United States Public Health Service grant AI-15284 and a fellowship from The Arthritis Foundation. The authors wish to thank Mr. Craig Gerard for assistance with the HPLC, Anne Zumbrun for excellent technical assistance, and Ms Janet Kuhns for her superb secretarial work in the preparation of the manuscript.

REFERENCES

1. Hidaka H, Asano T: Proc Natl Acad Sci USA 74:3657, 1977.
2. Kelly JP, Parker CW: J Immunol 122:1556, 1979.
3. Hoover RL, Lynch RD, Karnovsky MJ: Cell 12:295, 1977.
4. Sullivan TJ, Parker CW: J Immunol 122:431, 1979.
5. Yecies LD, Johnson SM, Wedner HJ, Parker CW: J Immunol 122:2090, 1979.
6. Adcock JJ, Garland LG, Moncada S, Salmon JA: Prostaglandins 16:163, 1978.

7. Salmon JA: Prostaglandins 15:383, 1978.
8. Salmon JA, Smith DR, Flauer RJ, Moncada S, Vane JR: Biochim Biophys Acta 523:250, 1978.
9. McPherson CW: Lab Anim Care 13:737, 1963.
10. Goodman MG, Fidler JM, Weigle WO: J Immunol 121:1905, 1978.
11. Mishell RI, Dutton RW: J Exp Med 126:423, 1967.
12. Coutinho A, Gronowicz E, Bullock WW, Möller G: J Exp Med 139:74, 1974.
13. Jerne NK, Nordin AA: Science 140:405, 1963.
14. Ling NR: "Lymphocyte Stimulation." New York: American Elsevier, 1968.
15. Hamberg M, Samuelsson B: J Biol Chem 242:5329, 1967.
16. Andersson J, Melchers F: Eur J Immunol 4:533, 1974.
17. Poe WJ, Michael JG: Immunology 30:241, 1976.
18. Sidman CL, Unanue ER: J Exp Med 144:882, 1976.
19. Kelly JP, Johnson MC, Parker CW: J Immunol 122:1563, 1979.

Evidence for a Low-Molecular-Weight Plasma Peptide Which Stimulates Chick Chondrocyte Metabolism

Loren Pickart

Virginia Mason Research Center, Seattle, Washington 98101

Plasma contains a number of insulin-like activities (ILA) of molecular weights 7,000 to 90,000 (somatomedins and insulin-like proteins) which stimulate cellular metabolism and may function as growth factors. We have found evidence for the presence of an 800 Dalton peptide in human plasma which markedly stimulates the metabolism of chick chondrocytes.

This peptide was extracted from human Cohn fraction IV-1 by procedures similar to those used for somatomedin isolations. At the Sephadex G-50 column separation step, the fraction with molecular weights of 300–1,000 was found to markedly stimulate chick chondrocyte metabolism. Rechromatography on Sephadex G-25 concentrated activity in peptides of molecular weight of about 800. An HPLC separation on a silica C-18 reverse phase column gave elution of the active peptide at 18% acetonitrile in water. This bioactivity appears to be a peptide which is free of lipids, carbohydrates, nucleic acids, metal ions, and immunoreactive insulin. This factor markedly increased the metabolism of cultured chick chondrocytes, but had only marginal activity on rat chondrocytes. When added at 1 µg/ml to chick chondrocytes cultured in F-12 medium plus 1.5% fetal calf serum, the HPLC-purified activity increased DNA synthesis 7.3-fold, lipid synthesis 10.2-fold, and lactate production 2.9-fold after 48 h incubation.

However, unlike somatomedins A and C, this factor did not displace insulin from placental membranes. These results suggest that low-molecular-weight peptides, which are smaller than the somatomedins, may contribute to the total ILA of human plasma.

Key words: insulin-like, somatomedin, chick chondrocytes, peptides, HPLC

Human plasma contains a number of bioactivities which express themselves in cell culture systems as "insulin-like" (eg, increased utilization of energy-producing substrates, stimulation of lipid synthesis) and stimulatory to growth [1–5]. The insulin-like activities (ILA) of plasma have been attributed to a variety of polypeptide fractions with molecular weights ranging from 500 to 90,000 [1–5]. These non-insulin factors account for 90%–95% of the ILA of plasma, the remainder being due to immunoreactive insulin [2]. The

Received April 7, 1980; accepted July 23, 1980.

0091-7419/80/1303-0385$02.00 © 1980 Alan R. Liss, Inc.

best characterized of these peptides, somatomedins A and C [2, 3] and IGF I and II [1], are of molecular weights of about 7,000. In addition, a non-suppressible insulin-like 88,000 Dalton protein (NSILP) has recently been isolated [5]. However, the contribution of these various factors to overall plasma ILA is uncertain since the losses of activity (>95%) incurred in the course of isolation of such factors [1–3, 5] leave open the possibility that significant ILA may reside in unidentified factors.

In the course of isolating somatomedin-C activities from human plasma, we found evidence for a low-molecular-weight (~800) peptide which markedly stimulates lipid and DNA synthesis, and lactate production, of cultured chick chondrocytes. Our results support the observation of Bala and co-workers [4] that human plasma contains a molecule, smaller than the somatomedins, which enhances chondrocyte metabolism.

MATERIALS AND METHODS

Puck's saline G (calcium and magnesium free) [6], trypsin, fetal calf serum, Ham's F-10 and F-12 medium were purchased from Grand Island Biologicals; collagenase, type CLS, #4194 from Worthington Biochemicals; chicken serum, #407, from International Scientific Industries; Lactate Determination Kit #826, Sephadex G-25 and G-50, blue dextran from Sigma Chemical Company; ^{3}H-thymidine (17.6 Ci/M) and 1-[^{14}C]-acetate (2 Ci/M) from New England Nuclear; ^{125}I-Insulin (114 μCi/μg) from Cambridge Nuclear Co.; Human Cohn plasma fractions III-0, IV-1, and V plus the final supernatant (Cutter SE-1) remaining after the removal of proteins by the Cohn process from 200 L of plasma, from Cutter Laboratories; thin-layer high-performance silica gel plates #5631, spray reagents for detection of compounds from E. M. Laboratories; HPLC grade solvents, from J. T. Baker; high-performance liquid chromatography (HPLC) column (RSil HL, C-18 reverse-phase, 10 micron, 25 × 0.4 cm), from Alltech.

Isolation of Chondrocyte-Stimulating Activity From Plasma

Cohn fraction IV-1 from 20 L plasma was extracted with an HCl-ethanol solvent by the method of Van Wyk et al [8] for the initial purpose of isolating somatomedin-C. Solubilized material was precipitated with acetone, the dried precipitate extracted 3 times over a 2 h period with 20% formic acid, and polypeptides in the resultant extract fractionated by gel filtration on Sephadex. For these studies, we followed the procedure as published by Van Wyk et al [8], omitting the G-75 column step, in order to separate low-molecular-weight chondrocyte-stimulating activities on a single column for purposes of comparison (Fig. 1). The effect of extending the formic acid extraction step to 24 h on recovery of bioactivity was also determined. Molecular weights of eluted fractions were estimated by using blue dextran, ^{125}I-insulin, bacitracin, H-gly-his-lys-OH, and NaCl as markers. It was assumed that the exclusion peak on G-50 represented proteins with molecular weights of greater than 20,000.

The formic acid extract from the acetone powder was fractionated on G-50 (100 × 2.5-cm column) in 1% formic acid at 3°C. Fraction H from the G-50 column was rechromatographed in 1% formic acid at 3°C on G-25 (100 × 0.9-cm column). Sephadex columns were used for this purpose at least 10 times before estimation of molecular weights to minimize errors due to polypeptide adsorption. Active fractions from G-25 were further purified by HPLC using a linear gradient of acetonitrile. All fractions were lyophilized, re-

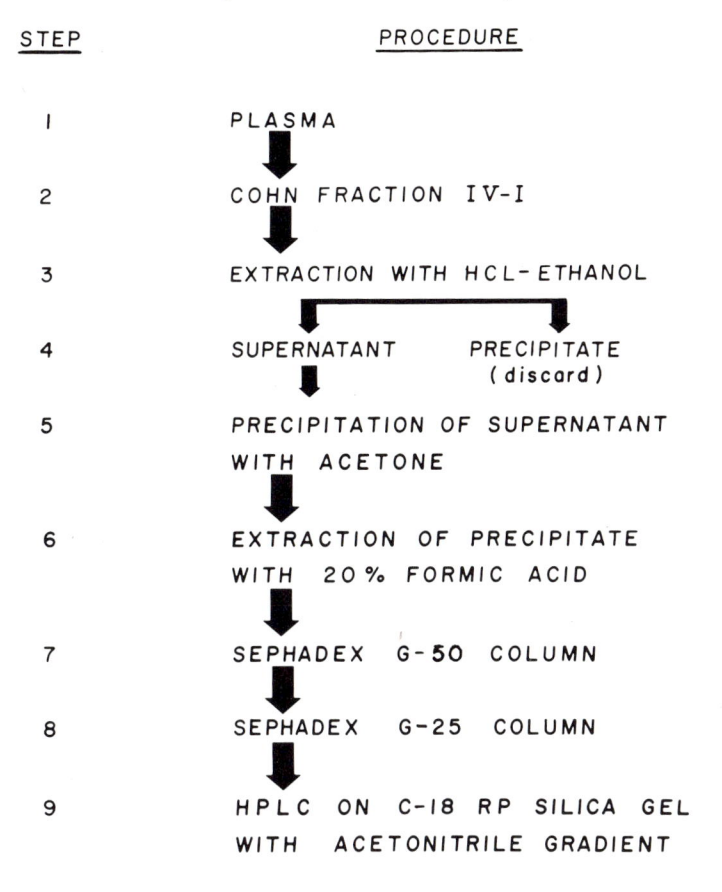

STEP	PROCEDURE
1	PLASMA ↓
2	COHN FRACTION IV-1 ↓
3	EXTRACTION WITH HCL-ETHANOL ↓
4	SUPERNATANT PRECIPITATE (discard)
5	PRECIPITATION OF SUPERNATANT WITH ACETONE ↓
6	EXTRACTION OF PRECIPITATE WITH 20% FORMIC ACID ↓
7	SEPHADEX G-50 COLUMN ↓
8	SEPHADEX G-25 COLUMN ↓
9	HPLC ON C-18 RP SILICA GEL WITH ACETONITRILE GRADIENT

Fig. 1. Purification process.

dissolved on 0.01 N HCl, and relyophilized. Samples were stored at −86°C prior to bioassay, dissolved in Dulbecco's phosphate-buffered saline (pH 7.4), their protein content measured by the Lowry reaction, and diluted to desired concentrations for use in experiments.

In ancillary experiments, Cohn fractions III-0 and V in amounts equivalent to 20 L starting plasma, and the final supernatant remaining in an amount equivalent to 200 L starting plasma, after the removal of proteins by the Cohn process (Cutter SE-1), were also extracted similarly to fraction IV-1 to assess their potential as sources of chondrocyte-stimulating activity.

The direct isolation of this activity from plasma by methods similar to those employed by Bala et al [4] was also attempted. Citrated human plasma (20 ml) was fractionated on a G-50 column (2.5 × 100 cm) either in 0.05 M phosphate buffer, pH 7.4, or in 1% formic acid, pH 3.3, in a manner similar to the extract from Cohn fraction IV-1. Fractions with elution volumes similar to fractions G and H (Fig. 2) were dialyzed with distilled H_2O on an ultrafiltration membrane (Amicon UM-05), then lyophilized and stored at −20°C prior to testing.

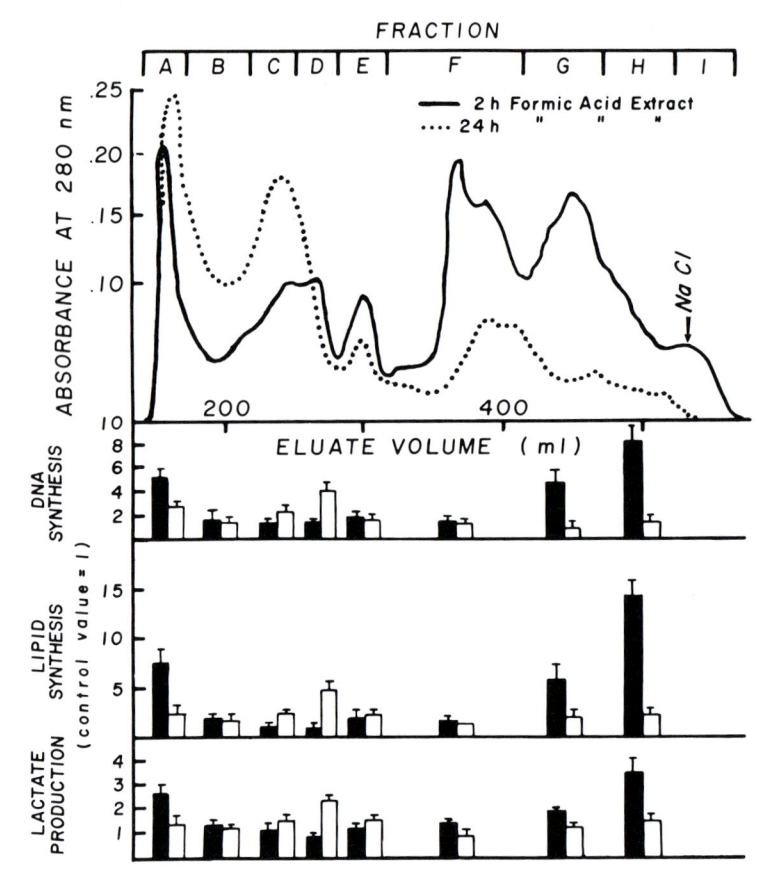

Fig. 2. Effect of fractions on G-50 column on DNA and lipid synthesis and lactate production by chick and rat chondrocytes. The elution pattern of the 2 h formic acid extract from plasma Cohn fraction IV-1 on a Sephadex G-50 column (2.5 × 100 cm) is shown by the solid line. Bar graphs represent stimulation of DNA and lipid synthesis and lactate production induced in chick chondrocytes (solid bars) and rat chondrocytes (open bars) induced by 5 μg/ml of proteinaceous material from the Sephadex G-50 fractions above each group of 3 bars. The elution pattern on G-50 obtained after extraction with formic acid for 24 h is given by the broken line. Fraction I could not be readily desalted and was not assayed.

Characterization of Bioactive Fraction From HPLC

Elemental analysis of the most active from HPLC was determined by x-ray photoelectron spectroscopy [9]. Specific tests were performed after drying this fraction on filter paper and staining with the following reagents: ninhydrin, triphenyl tetrazolium, diphenylamine, alkaline silver nitrate, iodine vapor, aniline phthalate, bromocresol green, 2′,7′-dichlorofluorescein, and dimethylaminobenzaldehyde. The ultraviolet spectrum of the fraction was determined.

Bioactive fractions from HPLC were chromatographed on thin-layer silica plates (solvent: chloroform/methanol/17% NH$_4$OH: 2/2/1 by volume), then stained with the detection reagents listed above.

Determination of Insulin-Like Activity and Immunoreactive Insulin in G-50 Fractions

Total insulin-like activity in fractions from the G-50 column was determined by radioactive receptor assay utilizing placental membranes and porcine ^{125}I-insulin, as described by Van Wyk et al [8]. Unlabeled porcine insulin (28.5 μU per nanogram) was used as a standard for insulin-like activity in the receptor titrations. Insulin concentrations were determined by standard radioimmunoassay [10].

Determination of DNA Synthesis, Lipid Synthesis, and Lactate Production

DNA synthesis was measured by incorporation of ^{3}H-thymidine into acid-precipitable protein by the procedure of Garland et al [11] for chondrocytes. Lipid synthesis was measured as the incorporation of ^{14}C-acetate into cellular lipids [12]. Lactate production was calculated from the lactic acid concentrations in the medium before and after incubation. Lactic acid was quantitated by a NAD-linked lactate dehydrogenase reaction [13].

Culture of Chick and Rat Chondrocytes

Chick chondrocytes were isolated by a modification of the method of Cahn et al [14]. Sterna from 12–15-day-old embryos were cleared of muscle tissue, excised, and submerged in ice-cold saline G containing 10% chicken serum. Fibroblasts and muscle cells were cleared from sterna by digestion in flasks (4 sterna per flask) containing 5 ml of saline G, pH 7.6, 0.2% collagenase, and 1.0% trypsin. After digestion for 15 min at 37°C on a rotary shaker (80 cycles/min), saline G (10 ml) was added and the solution decanted. Sterna were then washed and resuspended in saline G for examination under a dissecting microscope to select the cleanest fragments for isolation of chondrocytes by a repeat of the above digestion. The supernatant containing the isolated chondrocytes was decanted into 8 ml Ham's F-12 medium supplemented with 1.5 ml FCS. The digestion was repeated a third time and the final supernatant added to the previously isolated chondrocytes. Released cells and debris were sedimented by centrifugation at 1,000g for 10 min and resuspended in F-12 medium. Cellular debris was removed by centrifugation at 60g for 5 min, then the chondrocytes sedimented at 1,000g for 10 minutes. Chondrocytes were washed in F-12, then 2×10^5 cells incubated at 37°C in 2 ml Ham's F-12 containing 1.5% FCS in glass tubes (12 × 75 mm) on a roller drum (12 revolutions per hr) for 4 days. At the end of incubation, medium was removed and replaced with fresh F-12 medium containing 1.5% FCS, a peptide fraction to be tested, and 1 μCi of ^{3}H-thymidine or 1 μCi of ^{14}C-acetate, and incubation continued for an additional 48 h for determination of DNA and lipid synthesis and lactate production.

Rat chondrocytes were prepared from fragments of costal cartilage removed from 24-day-old male Sprague-Dawley rats. Adherent tissue was removed under a dissecting microscope and chondrocytes were isolated by the procedure used for chick chondrocytes. Experimental culture conditions and assays were similar to those employed with chick chondrocytes, except medium used was Ham's F-10 medium plus 1.5% FCS.

RESULTS

Chondrocyte-Stimulating Activity in Sephadex G-50 Fractions

The elution pattern of peptides in formic acid extracts chromatographed on Sephadex G-50 is shown in Figure 2. The approximate molecular weights of fractions containing major peaks were: A (exclusion peak), >20,000; C, 8,000–10,000; D, 5,000–7,000; E, 4,000–5,000; F, 2,000–4,000; G, 1,000–2,000; H, 300–1,000; I < 300.

TABLE I. Insulin-Like Activities of G-50 Fractions From Cohn IV-1 as Determined by Radioreceptor Assay and Radioimmunoassay

G-50 fractions	Insulin content (μU insulin/mg protein)	Insulin-like activity not due to insulin (μU ILA/mg protein)
A	60	450
B	30	1,470
C	120	6,080
D	530	12,080
E	470	5,230
F	ND[a]	650
G	ND	ND
H	ND	ND
I	ND	ND

Net ILA not due to insulin equals ILA determined by radioreceptor assay minus insulin content as measured by insulin radioimmunoassay. Details of procedure are given in text.
[a]ND, not detectable.

Optimal concentrations of bioactive protein fractions on cell cultures were determined in preliminary experiments. The effect of fractionated protein from Sephadex eluates on lipid synthesis in chick chondrocyte cultures was assayed at concentrations of 0.1 to 20 μg/ml incubation medium. It was found that most fractions gave nearly maximal stimulation at 5 μg/ml. When tested at this concentration, peptide fractions in the entire range of molecular sizes fractionated on Sephadex G-50 (500 to >20,000) stimulated DNA synthesis, lipid synthesis, and lactate production in chick and rat chondrocytes (Fig. 2). Chick chondrocytes were most responsive to fractions A, G, and H. In these cells, fraction A raised DNA synthesis 5.2-, lipid synthesis 7.7-, and lactate production 2.8-fold, while fraction H increased these processes 8.2-, 14.5-, and 3.5-fold, respectively. The effects of fraction G were intermediate between fractions A and H. In contrast with chick chondrocytes, rat chondrocytes responded maximally to fractions D, C, and A, in order of activity (Fig. 2). Fractions D and A increased DNA synthesis in rat chondrocytes 3.7- and 2.6-fold, lipid synthesis 4.8- and 2.5-fold, and lactate production 2.4- and 1.4-fold, respectively. Activities associated with fraction C were comparable to fraction A.

Insulin displacement activity was concentrated in fractions C, D, and E. Fraction D had the highest insulin-like activity (12,610 μU ILA/mg), as measured by ability to displace ^{125}I-labeled insulin from placental receptors (Table I). Displacement due to immunoreactive insulin was detectable in fractions A through D, ranging from approximately 2% of total displacement activity in fractions B and C, to 13% in fraction A. Fraction D contained the highest absolute amounts of immunoreactive insulin at concentrations of 530 μU ILA per mg protein or 19 ng insulin per mg protein, accounting for 4% of the total displacement activity in this fraction. However, the addition of insulin at concentrations ranging from 0.005 to 1.0 ng/ml had no detectable effect on DNA and lipid synthesis or lactate production in chick and rat chondrocytes.

Increasing the duration of the formic acid extraction to 24 h increased the recovery of peptides with molecular weights greater than 4,000 but had no effect on the production of smaller molecular species (Fig. 2). Thus, it is unlikely that the low-molecular-weight peptide fraction was generated by formic-acid-catalyzed proteolysis. Furthermore, the bioactivity of fractions G and H from G-50 was only marginally increased by the extended formic acid extraction.

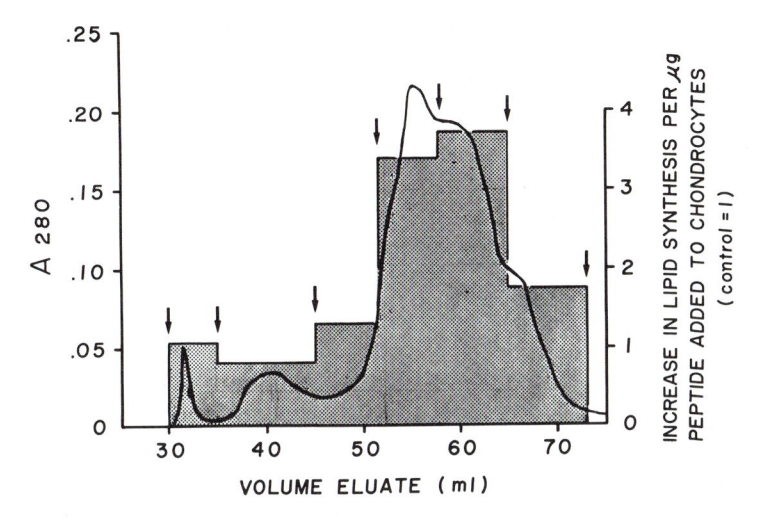

Fig. 3. Elution of G-50 fraction H on G-25. Bioactivity of fractions (stippled bars) measured as effect of peptide, when added at 1 µg/ml, on lipid synthesis in chick chondrocytes.

Cohn fractions III-0 and V, when processed in the manner of fraction IV-1, were found to possess similar bioactivities in G-50 fractions G and H, which stimulated chick chondrocyte metabolism. However, Cohn fraction III-0 yielded only 17% and fraction V only 46% of the comparable activity isolated from Cohn fraction IV-1. Extracts of Cutter fraction SE-1 were inactive.

G-50 fractions G and H from human plasma chromatographed at pH 3.3 possessed an activity which stimulated chick chondrocyte lipid and DNA synthesis, while similar fractions from plasma chromatographed at pH 7.4 was inactive. However, extraction of bioactive material by direct chromatography of plasma was impractical because of the low yield.

Chromatography of Sephadex G-50 Fraction H on Sephadex G-25

The most active low-molecular weight bioactive material isolated from Cohn IV-1 on G-50 (fraction H) was rechromatographed on a G-25 column and fractions lyophilized. The majority of activity associated with the main peak (Fig. 3). The addition of 1 µg/ml of the most active material to chick chondrocytes increased lipid synthesis approximately 3.5-fold. Comparison with standards gave a molecular weight of approximately 800.

Chromatography of Bioactive Fraction From Sephadex G-25 on HPLC

Material from the major peak of activity from G-25 was dissolved in 10% acetonitrile in H_2O, then chromatographed on a HPLC C-18 reverse-phase silica gel column into an acetonitrile gradient. Chick-chondrocyte-stimulating activity associated with a peak which eluted from the column in 18% acetonitrile (Fig. 4). The most active material, when added to chick chondrocytes at 1 µg/ml, increased lipid and DNA synthesis 10.2- and 7.3-fold, respectively, and lactate production 2.9-fold (Table II).

Elemental analysis, chemical tests, and ultraviolet absorbance spectra of the most active HPLC fraction indicated that it contained peptidic material which was relatively low in aromatic residues. Analysis by x-ray photoelectron spectroscopy found that hydro-

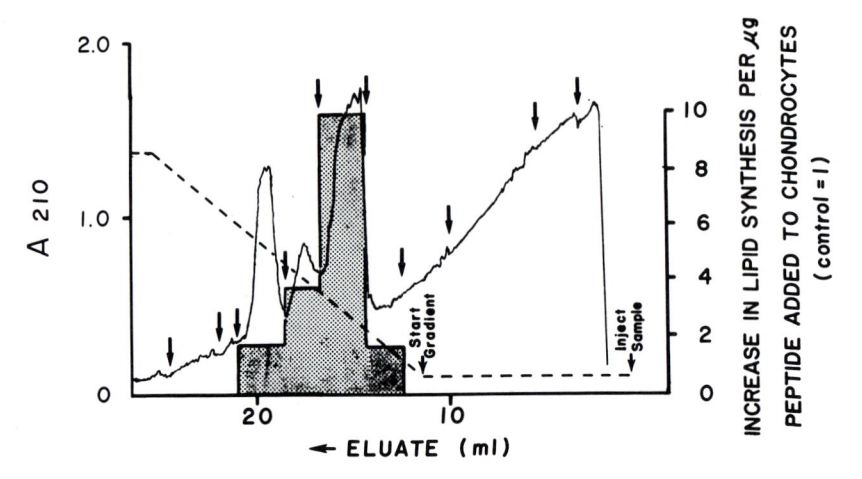

Fig. 4. HPLC of chick-chondrocyte-stimulating fraction from G-25. Bioactivity of fractions (stippled bars) measured as effect of peptide, when added at 1 μg/ml, on lipid synthesis in chick chondrocytes. Fractions without bars had no significant bioactivity. The dotted line represents the gradient which started at 10% acetonitrile and was increased to 70% acetonitrile.

TABLE II. Bioactivity of HPLC-Purified Isolates on Chick Chondrocytes

Peptide added, μg/ml	DNA synthesis	Lipid synthesis	Lactate production
0.0 (control)	1.0	1.0	1.0
0.1	1.9 ± 0.3	2.2 ± 0.2	1.3 ± 0.3
0.5	4.5 ± 0.7	6.4 ± 0.9	1.6 ± 0.2
1.0	7.3 ± 1.2	10.2 ± 2.3	2.9 ± 0.5

Values for control cultures normalized to unity for comparative purposes. Each determination is the average of 6 experiments.

gen, carbon, nitrogen, and oxygen in proportions characteristic of polypeptides comprised at least 99.5% of the sample. This fraction gave positive reactions to ninhydrin and iodine vapor; however, other staining reagents indicated that the fraction was free of carbohydrates, nucleic acids, reducing agents, steroids, and tryptophane. The ultraviolet spectra of the fraction had a typical peptide spectrum from 220–260 nanometers, with only a small increase in absorption at 280 nanometers, suggesting a low-content of aromatic residues (Fig. 5).

Chromatography of this fraction on TLC silica gel plates revealed three peptides as components. The fraction contained no amino acids. The major peptide migrated with an R_f = 0.30, gave a purple color with ninhydrin, and stained as a base with bromocresol green. The minor peptides, respectively, moved with R_f's of 0.09 and 0.37, stained orange and purple with ninhydrin, and gave acidic and neutral responses to bromocresol green. This isolate did not contain H-gly-his-lys-OH, a growth-modulating peptide previously isolated from human plasma [9]. Hydrolysis of this fraction in 6 N HCl for 24 h abolished the bioactivity and the three constituent peptides were replaced with a mixture of ninhydrin-staining spots characteristic of amino acids.

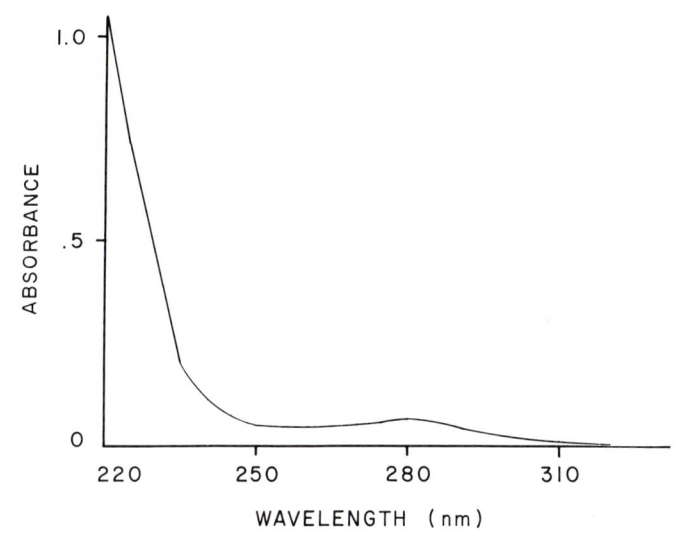

Fig. 5. Ultraviolet spectrum of chick-chondrocyte-stimulating activity from HPLC. The absorbance of bioactive material at a concentration of 0.2 mg/ml was determined in a 1-cm cell.

DISCUSSION

These studies demonstrate that diverse classes of peptides from human plasma possess insulin-like activities, as manifested by their stimulatory effects on chick and rat chondrocytes. When using plasma Cohn fraction IV-1 as a source, activities which stimulated chondrocyte metabolism were found to be associated with fractions of molecular weights greater than 20,000 (fraction A), approximately 6,000 (fraction D), and 300 to 2,000 (fractions G and H) (Fig. 2). These results are similar to the observations of Bala et al [4] who reported that sulfation of rat cartilage is stimulated by at least 6 polypeptide and protein fractions extracted from plasma which range in molecular size from 500 to 90,000. Insulin-like activities of molecular weights of approximately 7,000 isolated from acid-ethanol extracts of plasma have previously been attributed to a family of closely related peptides (somatomedins), while those of the high-molecular weight fractions to either a complex of binding proteins and somatomedins [1–3] or to a 90,000 Dalton protein with insulin-like activity [5]. It is clear that acid-ethanol extracts of plasma contain a variety of molecules with insulin-like properties.

The two cell types examined exhibited considerable variation in their ability to respond to eluted fractions. Maximal stimulation was obtained in chick chondrocytes with fractions A, G, and H and in rat chondrocytes with fractions A and D (Fig. 1). Estimations of insulin-like activities of peptide fractions extracted from plasma differed markedly when based on results of bioassays and receptor assays. Thus, among fractions exhibiting major stimulatory effects on lipid synthesis (fractions A, D, H), only fraction D produced significant displacement of ^{125}I-insulin from receptor-rich placental membranes (Table I). Conversely, fractions C and E possessed 40%–50% of the displacement activity of fraction D, while their bioactivities were relatively unimpressive compared with the effects of the more active fractions, especially in chick chondrocytes (Fig. 2). These findings suggest that attachment of plasma peptides to receptors with an affinity for insulin may not stimulate the expected "insulin-like" responses in certain cell types and tissues.

The low-molecular-weight (~800) chick-chondrocyte-stimulating activity isolated from Cohn fraction IV-1 appears to be a peptide and also may be present in other Cohn fractions (III-0 and V). Chromatography of human plasma on molecular sieving gels gives rise to a low-molecular-weight chick-chondrocyte-stimulating activity when processed at low pH (3.3), but not a physiological pH (7.4). This observation parallels the previous finding of Bala et al [4] that low-molecular-weight (500–2,000) chick-chondrocyte-stimulating activities are released from plasma under acidic conditions and suggests that such conditions may dissociate this activity from carrier proteins. It is possible that this factor is a bioactive breakdown product of the somatomedins; however, preliminary experiments have indicated that extension of the time of the acid-ethanol or the formic acid extraction steps does not increase the yield of this activity. This suggests that this factor may be a naturally occurring endogenous peptide which contributes to the ILA of human plasma.

In summary, these experiments give evidence for the existence in human plasma of a low-molecular-weight factor, peptidic in nature, which markedly stimulates the metabolism of cultured chick chondrocytes.

ACKNOWLEDGMENT

This work was supported in part by USPHS grant CA-27129.

REFERENCES

1. Rinderknecht E, Humbel RE: J Biol Chem 253:2769, 1978.
2. Fryklund L, Uthne K, Sievertsson H: Biochem Biophys Res Commun 61:957, 1974.
3. Svoboda ME, Van Wyk JJ, Klapper DG, Fellows RE, Grissom FE, Schlueter RJ: Biochemistry 19:790, 1980.
4. Bala RM, Blakeley ED, Smith GR: J Clin Endocrinol Metab 43:1110, 1976.
5. Plovnick H, Ruderman NB, Aoki T, Chideckel EW, Poffenbarger PL: Am J Med 66:154, 1979.
6. Puck TT, Cieciura SJ, Robinson A: J Exp Med 108:945, 1958.
7. Pickart L, Thaler M: Prep Biochem 5:397, 1975.
8. Van Wyk JJ, Underwood LE, Baseman JB, Hintze RL, Clemmons DR, Marshall RN: Adv Metab Disorders 8:127, 1975.
9. Pickart L, Thaler M, Millard M: J Chromatogr 175:65, 1979.
10. Morgan CR, Lazarow A: Diabetes 12:115, 1963.
11. Garland JT, Lottes ME, Kozok S, Daughaday WE: Endocrinology 90:1086, 1972.
12. Pickart L, Thaler M: Biochem Biophys Res Commun 74:961, 1977.
13. Lactic Acid Determination Procedure, Number 826, Sigma Chemical Co., St. Louis, MO.
14. Cahn RD, Coon HG, Cahn MB: In Wilt FH, Wessels NK (eds): "Methods in Developmental Biology." New York: Academic Press, 1967, pp 517–529.

Purification of Human Interleukin 1

Lawrence B. Lachman, Stella O. Page, and Richard S. Metzgar

Department of Microbiology and Immunology, Duke University Medical Center, Durham, North Carolina 27710

Interleukin I (IL-1) is a lymphocyte stimulant released by human monocytes cultured for 18–24 hours in tissue culture medium containing 5% serum and the non-specific immunostimulant lipopolysaccharide (LPS). Human IL-1 is found in the conditioned medium in a low molecular weight ($\sim$13,000) and a high molecular weight ($\sim$85,000) form. The high MW activity may result from the formation of a complex between IL-1 and serum constituents. During the course of purification, the low MW IL-1 activity is often recovered in a high MW form. Hollow fiber diafiltration and membrane ultrafiltration has been found to rapidly separate low MW IL-1 from all measurable protein with a yield of 4% of the original activity. The IL-1 which converts to the high MW form during the purification is recoverable, 21% of the original activity, but contains small amounts of serum proteins. Isoelectric focusing (IEF) of the low MW IL-1 resulted in a very highly purified sample which was analyzed by polyacrylamide gel electrophoresis (PAGE). Utilizing a new staining procedure which detects less than 1 ng of protein per band, the IEF-purified IL-1 revealed trace quantities ($<$ 1 ng) of a slowly migrating protein similar to immunoglobulin and no other bands. There were no bands which corresponded with the known electrophoretic mobility of IL-1. Since the samples applied to the gel contained significant biological activity, this result implies that human IL-1 is biologically active in picogram quantities.

Key words: lymphocyte activating factor (LAF), Interleukin I, purification of human IL-1, hollow fiber diafiltration, isoelectric focusing, polyacrylamide gel, electrophoresis, human monocytes, endotoxin stimulation, IL-1 release, thymocyte mitogenic activity

Lymphocytes and macrophages in culture release a number of soluble substances (factors) which may affect lymphoid cells undergoing an immune response [1]. The term "lymphokine" has been used to describe lymphocyte-derived factors, and the term "monokine" has been proposed for macrophage- or monocyte-derived factors. Since the description of the first lymphokine, MIF (migration, inhibitory factor for macrophages) [2], the number of reported lymphokines and monokines has nearly defied categorization [3]. Interleukin 1 (IL-1, previously known as Lymphocyte Activating Factor – LAF), is perhaps the most well studied monocyte-derived factor [4–7]. IL-1 is

Received April 7, 1980; accepted July 22, 1980.

0091-7419/80/1304-0457$02.00 © 1980 Alan R. Liss, Inc.

found in the culture medium of human adherent mononuclear cells (monocytes) [8], mouse peritoneal exudate cells [4], mouse macrophage cell lines [9], and human monocytic leukemia cells [10]. Purification of IL-1 has been directed toward human peripheral blood-monocyte derived IL-1, and IL-1 derived from the P388D$_1$ macrophage cell line [9, 11, 12]. Human monocyte IL-1 has been partially purified by sequential gel filtration-ion exchange chromatography [5] and by hollow fiber ultrafiltration and isoelectric focusing [6]. The present report will describe our current procedure for large scale fractionation of IL-1 and evaluation of the purified activity.

MATERIALS AND METHODS

IL-1 Production

IL-1 was prepared on a large scale in a manner similar to the procedure of Gery et al [8]. Heparinized (50 U/ml) whole blood was mixed with one quarter volume of pyrogen-free Plasmagel (HTI Corp., Buffalo, NY), and the red blood cells were allowed to sediment for 30 min in a 37° water bath. The white blood cells were sedimented at 300g for 20 min, and the cells washed twice with Minimum Essential Medium-Earle's salts (MEM) supplemented with 2mM glutamine, 10 mM HEPES, penicillin (100 units/ml), and streptomycin (100 µg/ml) (Grand Island Biochemical Co., NY). Condition medium containing IL-1 was prepared by culturing the white blood cells in MEM at 0.5×10^6 cells/ml with 5% allogeneic (Grand Island Biochemical) or autologous human serum 10^{-5}, M 2-mercaptoethanol (Fisher Scientific Co., Medford, MA), and where indicated 20 µg/ml of E coli lipopolysaccharide W (LPS, 0.55:B5 or 0.27:B8, Difco Laboratories, Detroit, MI). Cell suspensions were cultured in sterile petri dishes (100 ml per 150×25 mm dish, Bioquest, Cockeysville, MD) at 37° in a humidified atmosphere of 5% CO_2 or spinner culture flasks in a 37° warm room for 24 hours. Cultures containing MEM and 5% serum without cells and LPS were incubated as controls. The cells were removed from the conditioned medium by centrifugation at 1000g for 20 min, and the medium could be stored for several months at −20°.

IL-1 Assay

To prepare thymocytes for the IL-1 assay, mouse thymuses (CD-1 mice, female 8–10 weeks old, Charles River Breeding Laboratories, Wilming, MA) were removed aseptically and homogenized in 10 ml MEM using teflon and glass hand homogenizer.

The suspension of cells was allowed to settle for 10 min; the thymocytes were separated from settled debris with a sterile Pasteur pipette; and the cell count adjusted to 10×10^6 cells/ml in MEM containing 5% normal human serum from a single donor and 10^{-5} M 2-mercaptoethanol. One tenth ml of the cell suspension was dispensed into each well of sterile tissue culture plates (IS-MRC-96-TC, Linbro Company, Hamden, CT). IL-1-containing samples were assayed in triplicate by adding 0.01 ml of sample to the wells. The plates were incubated for 72 hours in a humidified atmosphere of 5% CO_2 with the addition of 0.2 µCi of (^{3}H) thymidine (1.9 Ci/mmole, Schwartz/Mann, Orangeburg, NY) in 0.01 ml of 0.85% NaCl after 48 hours. The cells were then collected on glass fiber filters (Reeve Angel, Clifton, NJ) using an automatic cell harvester (Otter Hiller Company, Madison, WI). Liquid scintillation counting of the oven dried (30 min, 120°) filters was performed in a scintillation fluid containing 5 gm PPO and 0.5 gm POPOP/100 ml toluene (New England Nuclear, Boston, MA). All results are expressed as the average of triplicate cultures. A unit of IL-1 activity was determined based on the procedure of Gillis et al [13].

Hollow Fiber Diafiltration and Ultrafiltration

An Amicon hollow fiber filtration device (Model DC2, Amicon Corp., Lexington, MA) was used sequentially for diafiltration of the starting conditioned medium and removal of albumin by ultrafiltration. In the diafiltration step, the conditioned medium was extracted with HEPES (0.05M, pH 7.4)-NaCl (0.85%) buffer using a 50,000 MW cut-off hollow fiber cartridge (H1X50). The diafiltrate (ten times the volume of the original sample) was then ultrafiltered through the same 50,000 MW cartridge (after washing) to remove the small amounts of protein ($\sim$ 0.04 mg/ml) present in the diafiltrate. The ultrafiltrate, containing the low MW IL-1 activity, was then concentrated to $\sim$ 45 ml by ultrafiltration using a stirred cell and a YM 10 (10,000 MW cut-off) membrane (Amicon Corp.).

Sephadex Gel Filtration Chromatography

Sephadex G100 (Pharmacia, Uppsala, Sweden) chromatography of 20-ml samples was carried out at 4° with a water-jacketed column (2.5 × 90 cm) using HEPES-Cl buffer. To concentrate samples of the original conditioned medium, the extracted conditioned medium, and the diafiltrate for Sephadex gel filtration chromatography, a 5,000 MW cut-off hollow fiber cartridge (H1P5) was used. The flow rate was 0.5 ml/min, and fractions of 11.4 ml were automatically collected with a refrigerated fraction collector (LKB Instruments, Rockville, MD).

Isoelectric Focusing

Isoelectric focusing (pH 8–4) was performed on a 40-ml sample using an LKB instrument (model 8100, 110 ml, capacity, Stockholm, Sweden) with a 5–50% sucrose gradient. Voltage was applied for 18–20 hours with an LKB constant power source (Model 2103, Stockholm, Sweden) using the settings of 1600 maximum voltage and 15 watts maximum power. Fractions of 1.5–2.0 ml were collected, measured for pH, and stored at 4°C. To remove Ampholine and sucrose prior to thymocyte assay, IEF fractions were dialyzed at 4° against HEPES-Cl buffer for 20 hours with one change of buffer. The dialyzed fractions received 0.05 ml of fetal calf serum (FCS) per ml of sample before sterile filtering (Disposable filters, 0.22 μm, Millipore, Bedford, MA). Samples could be sterile filtered without the addition of serum by using 0.2 μm Acrodisc membrane filters (Gelman, Ann Arbor, MI) instead of Millex filters.

Polyacrylamide Gel Electrophoresis

Polyacrylamide gel electrophoresis (PAGE) was performed at 4° using 7% slab gels of 1-mm thickness with a pH 8.9 Tris-glycine buffer and a 2.5% stacking gel according to the procedure of Davis [14]. Isoelectric focusing fractions found to contain IL-1 activity (pH 6.8–7.2) were dialyzed against the running buffer for 20 hours at 4°, mixed with one-tenth volume of 0.1% bromophenol blue containing 10% sucrose, and PAGE was performed on 25 μl of each fraction. Slab gels were stained with $AgNO_3$ according to the procedure of Switzer et al [16].

RESULTS

Hollow Fiber Diafiltration and Ultrafiltration of IL-1 Conditioned Medium

IL-1 containing conditioned medium (1000 ml) from spinner culture was extracted with 10 liters of HEPES-Cl buffer (as explained in Materials and Methods). This procedure allows all medium components of MW less than 50,000 daltons to collect with the diafiltrate. For the purpose of Sephadex chromatography, the original conditioned medium, extracted medium, and the diafiltrate were rapidly concentrated to 20 ml with a hollow fiber cartridge of 5,000 MW exclusion (H1P5). The concentrated samples were subjected to Sephadex G100 chromatography and the collected fractions compared with the original conditioned medium for IL-1 activity (Fig. 1). The results clearly indicate (Fig. 1C) that the < 50,000 MW diafiltrate contains only low MW IL-1 activity and is not contaminated with any of the high MW IL-1 activity observed in the unfractionated medium (Fig. 1A) or the medium from which the low MW activity has been partially extracted (Fig. 1B).

The hollow fiber diafiltrate (< 50,000 MW) contained small amounts of protein (~40 µg/ml), of which most was serum albumin. The reason small amounts of albumin and other high MW proteins are found in the < 50,000 MW fraction is that hollow fiber cartridges have an *average* MW cut-off, and thus allow small amounts of components with a MW greater than the listed cut-off to pass through the pores of the fibers. The residual protein was removed by ultrafiltration of the diafiltrate through the hollow fiber cartridge (50,000 MW exclusion) which was used in the original step. This procedure concentrated the protein in the > 50,000 MW retentate (50 ml) but also retained considerable IL-1 activity (Table I). The ultrafiltrate (9,950 ml), when concentrated to 43 ml using a stirred cell and YM 10 membrane, contained only about 20% of the IL-1 activity found in the diafiltrate. The ultrafiltration step of the diafiltrate is the poorest yield step in the purification procedure, but reveals a very interesting property of IL-1. The IL-1 activity was

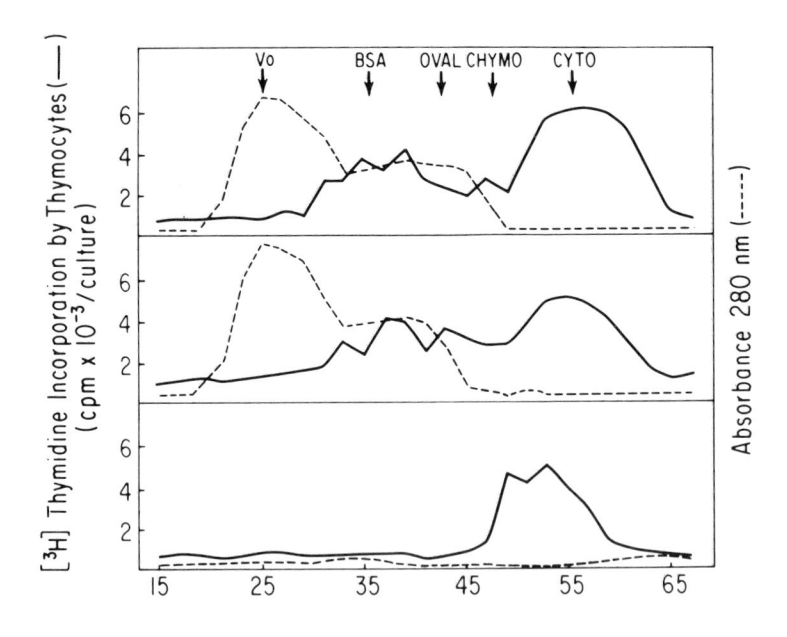

Fig. 1. Sephadex G100 chromatography of IL-1 conditioned medium and hollow fiber-separated, high- and low-MW IL-1 fractions. A) Unfractioned conditioned medium; B) the > 50,000 MW fraction; and C) the < 50,000 MW fractions of conditioned medium were prepared by hollow fiber diafiltration and concentration as described in Materials and Methods. Column fractions were sterile filtered and assayed for IL-1 activity in the thymocyte (^{3}H) thymidine uptake assay.

TABLE I. Separation of Low-Molecular-Weight IL-1 Activity From Conditioned Medium by Hollow Fiber Diafiltration and Ultrafiltration

Purification step[a]	MW range	Volume	Undiluted	(^{3}H) Thymidine incorporation by thymocytes (cpm/culture) 1:2	1:4	1:8	1:6
Conditioned medium	–	1,000	25,219[b]	21,428	15,739	7,565	
Extracted medium	>50,000	1,000	19,010	14,898	9,028		
Diafiltrate	<50,000	10,000	3,821	24			
Concentrated diafiltrate[c]	>50,000	50	25,784	25,283	25,580	19,470	15,485
Ultrafiltrate[c]	<50,000	9,950	8,660	1,102	1,062	372	
Concentrated ultrafiltrate	10–50,000	43	18,200	19,489	13,042	8,688	
Control medium without white blood cells			2,584				
Medium			1,980				

[a]The hollow fiber purification procedure is described in Results.
[b]Background values for control cultures not containing white blood cells have been subtracted.
[c]As explained in the text, the IL-1 activity which was able to pass through the 50,000-MW hollow fiber cartridge during diafiltration was only partially able to pass through the same hollow fiber cartridge during ultrafiltration.

able to pass through the hollow fiber cartridge during diafiltration (Table I and Fig. 1) but when ultrafiltered through the same cartridge, ~80% of the IL-1 activity remained concentrated in the >50,000 MW fractions. This phenomenon has been partially explained by the research of Togwa et al [15], who found that low MW IL-1 would form a high-MW complex when concentrated in the presence of serum proteins. We believe that as the protein concentration increases in the >50,000 MW fraction, the ability of IL-1 to pass through the hollow fiber cartridge is greatly reduced. The hollow fiber ultrafiltration step greatly reduces the yield of IL-1 activity, but substantially improves the purity. The ultrafiltrate, when concentrated to 43 ml in a stirred well with a YM 10 membrane, was found to contain insufficient protein to be measured by any conventional technique.

The H1P5 hollow fiber cartridge (5,000 MW cut-off) used to concentrate the hollow fiber fractions for Sephadex G100 chromatography was found to irreversibly bind the high MW IL-1, and thus explained why the chromatogram of the concentrated diafiltrate (Fig. 1C) revealed only the uncomplexed, low MW IL-1 activity.

Isoelectric Focusing of Hollow-Fiber-Purified, Low-MW IL-1

The concentrated ultrafiltrate was subjected to pH 8–4 sucrose gradient isoelectric focusing, and the resulting fractions were assayed for IL-1 activity (Fig. 2). The center of the peak of IL-1 activity is pH 7.1. In this and other isoelectric focusing experiments, the most active fractions were always found in the pH 6.8–7.2 region. A second much smaller peak of IL-1 activity is occasionally found in the pH 5.3–5.8 region. Isoelectric focusing of the concentrated diafiltrate, however, reveals a large peak of IL-1 activity in the pH 5.3–5.8 region as well as the peak of activity in the pH 7 region. These data indicate that the IL-1 activity found in the concentrated diafiltrate is present as both the pH 7 activity seen in the low-MW concentrated ultrafiltrate and as a pH 5 activity which we believe represents the high-MW complex of IL-1 with a serum protein. The isoelectric point of 5.3–5.8 may indicate that IL-1 has complexed to human serum albumin.

PAGE of Isoelectric Focusing Purified IL-1

The isoelectric focusing fractions from the concentrated ultrafiltrate which contained the IL-1 activity (pH 6.7–7.3 and 5.3–5.5) (Fig. 2A) were dialyzed overnight against the Tris-glycine running buffer and 30-μl samples containing 5 μl of tracking dye was applied to each well of a polyacrylamide slab gel. Staining of the gel with AgNO$_3$ was performed as described [16]. The sensitivity of the staining procedure was confirmed by detection of bovine serum albumin samples containing 1 ng of protein (Fig. 3M). The concentrated ultrafiltrate, prior to isoelectric focusing, contained approximately 2.0 μg of protein per ml based upon the intensity of the visible band of albumin (Fig. 3C). This technique for determining protein concentration is strictly an estimate and is limited by the inability of the staining procedure to detect bands containing less than 1 ng of protein per mm of gel, and the possibility that some proteins may not be detected by this procedure.

The fractions of IEF-purified IL-1 from the pH 6.7–7.3 region (Fig. 3D–H) revealed a very faint band (less than 1 ng of protein) with an R$_f$ of 0.1 in the first four fractions (Fig. 3D–G) and a more definite band in the final fraction (Fig. 3H). The fractions

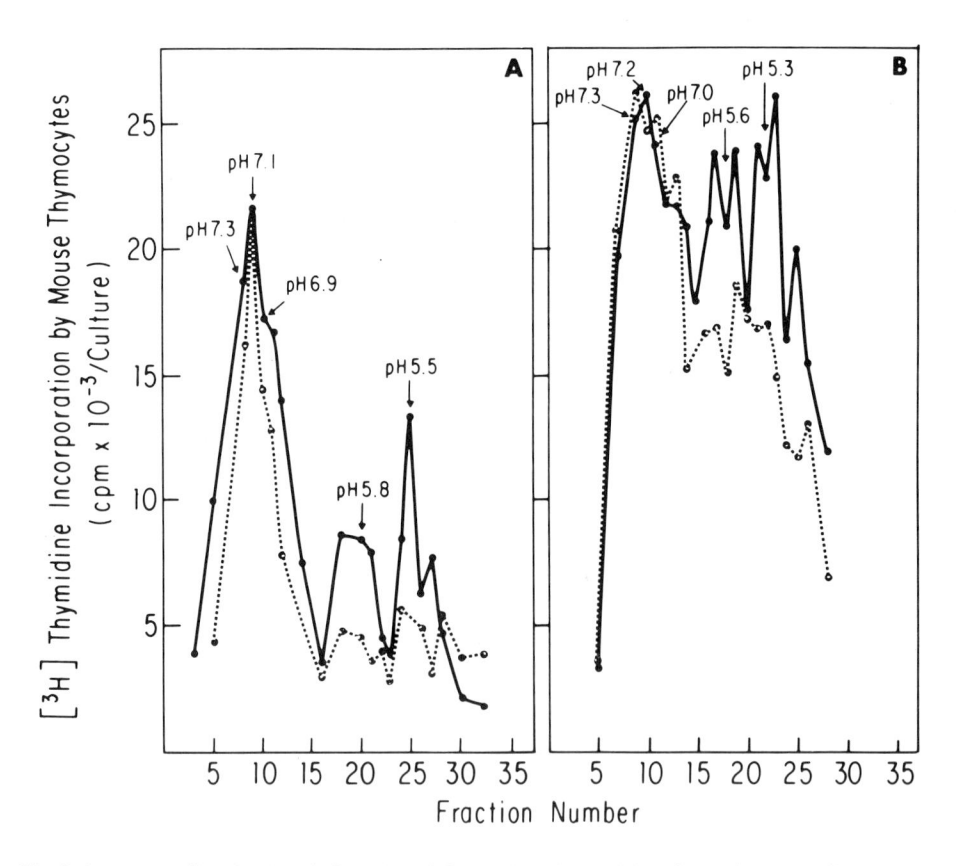

Fig. 2. Sucrose gradient isoelectric focusing of IL-1 activity in a pH 8 to 4 Ampholine gradient. The isoelectric focusing fractions of the concentrated ultrafiltrate (A) and the concentrated diafiltrate (B) from one liter of conditioned medium (Table I) were dialyzed overnight to remove Ampholine and sucrose prior to the thymocyte (^{3}H) thymidine incorporation assay. The dialyzed fractions were assayed undiluted (———) and diluted one to ten (-------). The thymidine incorporation by unstimulated thymocytes was approximately 1,000 cpm/culture.

from the pH 5.5–5.3 region (Fig. 3 I, K) revealed small quantities of albumin and an extremely faint band at R_f 0.1. The R_f and isoelectric point of these faint bands would indicate that they are immunoglobulin. The 0.1 R_f bands do not correspond with the electrophoretic mobility of IL-1 (0.36 vs dye) in this gel system (data not shown) or the R_f or IL-1 relative to albumin (0.65 R_f). In addition, the intensity of the staining does not correspond with the relative biological activity since the samples in wells E, F, and G (Fig. 3) contained the most IL-1 activity. The conclusions reached from this very critical experiment are that 1) the IEF-purified IL-1 activity (pH 6.7–7.3) is very highly purified and contains only trace quantities of protein (<1 ng/25 μl), and 2) no detectable bands of proteins were evident which corresponded with the electrophoretic mobility of IL-1. Selective staining of similarly prepared gels with lipid and carbohydrate specific stains is currently being performed.

DISCUSSION

Separation of IL-1 activity from conditioned medium by hollow fiber ultrafiltration is based on the ability of low-MW IL-1 to pass through the pores of hollow fiber

Fig. 3. Polyacrylamide gel electrophoresis of isoelectric focusing purified IL-1. Electrophoresis of the 7% slab gel with a 2.5% stacking gel in non-denaturing buffer was performed as described in Materials and Methods. The gel was stained for the presence of protein using the Ag staining procedure of Switzer et al [16]. Each well contained 25 μl of sample and 5 μl 0.1% bromophenol blue. The samples were: A) unfractionated IL-1 containing medium diluted 100-fold; B) concentrated diafiltrate (Table I); C) concentrated ultrafiltrate; D–H) isoelectric focusing fractions pH 7.4–6.8 (Fig. 3A); I, J) isoelectric focusing fractions pH 5.8 and pH 5.5 (Fig. 3A); crystalline bovine serum albumin – 100 ng (K), 10 ng (L), and 1 ng (M).

devices while passage of serum proteins is retarded. The hollow fiber procedure originally described [6] has been improved in efficiency and speed of operation by the availability of hollow fiber equipment designed for large scale purification procedures. A major problem with all types of ultrafiltration is that, as the solution becomes concentrated, the ability of the molecule of interest to pass through the pores of the membrane is reduced. We have circumvented this difficulty by the technique of diafiltration. This procedure allows for separation of the <50,000 MW components from the conditioned medium without increasing the protein concentration of the starting sample. IL-1 activity was found to be in both the >50,000 MW fraction (the extracted conditioned medium) and the diafiltrate (<50,000 MW). This result is consistent with the observation of high MW (∼85,000 daltons) peak of IL-1 activity during Sephadex G100 chromatography [5, 15]. In these earlier studies, Sephadex chromatography was performed after the conditioned medium had been concentrated by ultrafiltration: and thus the possibility existed that the high MW IL-1 activity could result from binding of low-MW IL-1 to a carrier protein such as albumin. In the present study, high-MW IL-1 activity is found following diafiltration, when the serum concentration is not increased. If a high-MW complex exists, it appears to be a stable complex, since the original conditioned medium is extracted with ten volumes of buffer during diafiltration. Another possibility is that IL-1 is diafiltered more slowly than would be expected for a ∼14,000 MW molecule, and that the remaining IL-1 activity in the extracted medium simply reflects incomplete diafiltration. Sephadex G100 chromatography of the unfractionated conditioned medium and hollow fiber fractions (Fig. 1) clearly indicates that the <50,000 MW diafiltrate contains low-MW IL-1 activity (Fig. 1C) and that diafiltration with ten volumes of buffer was not sufficient to remove the high-MW IL-1 activity from the conditioned medium (Fig. 1B). In addition, the appearance of low-MW IL-1 during Sephadex chromatography of the >50,000 MW extracted medium could be due to dissociation of the IL-1 protein complex during chromatography.

When the diafiltrate is concentrated using a stirred cell with a YM 10 membrane, a high MW complex of IL-1 is formed. The evidence for this complex formation is that 80% of the IL-1 activity which previously passed through the 50,000 MW hollow fiber cartridge is now retained. Formation of a high-MW complex of IL-1 with serum protein has been previously reported [15]. The IL-1 activity which does not pass through 50,000 MW hollow fiber cartridge, the concentrated diafiltrate (Table I), is not discarded. This activity is recoverable following isoelectric focusing, but purity of this activity is not nearly at the level of purity of the activity which was able to pass through the hollow fiber. We are currently investigating techniques which can reduce complex formation of IL-1 with serum proteins, and thus shift the balance toward recovery of the most highly purified activity. We have tested 10 mM 2-mercaptoethanol and 0.01% Triton-X 100 for this purpose, and found that neither reduced the level of the complex. Another interesting point to remember is that this complex is readily absorbed to PM ultrafiltration membranes manufactured by the Amicon Corporation. We found that the complex is readily absorbed to a 5,000-MW hollow fiber cartridge containing a PM membrane (Results) and also believe the frequently mentioned instability of IL-1 during concentration may be due to absorption of this complex to ultrafiltration membranes.

Isoelectric focusing of the concentrated ultrafiltrate is not only a purification step, but demonstrated a slight charge heterogeneity for low-MW IL-1 (Fig. 2A). IL-1 activity is always found between pH 6.8 and 7.2, and occasionally small peaks of IL-1 activity in the pH 5.5–6.0 region are present. The pH 5.3–5.8 peaks have never been greater

TABLE II. Summary of IL-1 Purification

Purification step	Volume (ml)	Protein (mg)	Specific activity (units/mg)[a]	Purification	Yield (%)
Conditioned medium	1,000	3,800[b]	1.4	1	100
Concentrated ultrafiltrate	43	0.1[c]	2,110	1,465	4
Isoelectric focusing of concentrated ultrafiltrate	9	if <0.0001[d]	>768,000	>500,000	1.5

[a] See text and Table I for a complete explanation of the purification procedure and calculation of units IL-1 activity.
[b] Determined by the microbiuret procedure.
[c] The protein concentrations in these solutions are too low to be estimated by conventional techniques. The values given are estimated from the intensity of stained bands compared with known concentrations of standard proteins.
[d] The sensitivity of the silver-nitrate staining procedure is 1 ng of protein per 2-mm band.

than one third the area of the pH 6.8–7.2 peak, and in several purifications the small peaks were not detected. We believe the pH 5.3–5.8 peaks may be the high-MW IL-1 complex, since they were present in much greater quantity in the fraction of activity which was unable to pass through the 50,000 MW hollow fiber cartridge. The recovery of IL-1 activity following IEF is about 1.5% of the original activity or 3% of the desired low MW IL-1, since the original medium contains about equal amounts of high and low MW IL-1. Isoelectric focusing is not a high yield procedure (Table II), but offers a significant increase in purity (Fig. 3). The final degree of purification of IL-1 can only be estimated, since the IEF fraction contains too small an amount of protein to be measured by conventional techniques (Table II). When IEF-purified human IL-1 was applied to nondenaturing polyacrylamide gels, biological activity could be eluted from the gels as a single peak [6]. An identically run gel did not reveal any protein bands when stained with Coomassie brilliant blue. This result indicated that IL-1 was not a protein or that IL-1 was not present in sufficient quantity to visibly bind the protein stain. Currently, staining or polyacrylamide gels with Ag^+, a technique which detects as little as 1 ng of protein per band, has not revealed any protein bands in the IEF-purified IL-1 samples which correspond to IL-1. The polyacrylamide gels clearly indicated that the purity of the IEF sample is excellent since only trace quantities of proteins are present in the sample, and none of these in a concentration of greater than 1 ng/25 μl of sample. These results also directly indicate that a quantity of IL-1 which is sufficient to stimulate a strong response in mouse thymocyte cultures* does not reveal a distinct band of at least 1 ng of protein which correlates with the known electrophoretic mobility of IL-1 [6, 17, 18]. Possible explanations for this finding are that IL-1 is present in only picogram quantities, that IL-1 may not have stained by this technique, or that IL-1 is only partially composed of protein.

Inhibition of IL-1 release by monocytes when cultured in the presence of cycloheximide is not direct evidence that IL-1 is a protein. Recent experiments have demonstrated that IL-1 is enzymatically inactivated by trypsin and chymotrypsin and chemically

*Each thymocyte culture of 0.1 ml is stimulated with 10 μl of sample. Each well of the gel contained 25 μl of IEF-purified IL-1. The IEF-purified IL-1 applied to the gel (Fig. 3D–H) contained 20 units of activity/ml in fractions E, F, and G and 10 units/ml in fractions D and H.

inactivated by cleavage with cyanogen bromide. Radioiodination of IEF-purified IL-1 has been performed using ^{125}I-labeled Bolton-Hunter reagent [19]. PAGE of the iodinated sample did not reveal a radioactive peak which corresponded with the electrophoretic mobility of IL-1 (data not shown).

IL-1 from the P388D$_1$ and J774.1 mouse macrophage cell lines has been partially purified using a procedure similar to the one described in this paper [12]. The results indicate that culture medium from the LPS-stimulated macrophage cell lines also contain a high ($\sim$ 60,000–80,000) MW and a low ($\sim$ 17,000) MW IL-1 activity. Similar to the human activity, the high MW IL-1 is stable during diafiltration. The major isoelectric point, pH 5.0–5.4, of the low-MW mouse macrophage cell line IL-1 differs significantly from the isoelectric point of pH 6.9–7.1 for human IL-1. In addition the isoelectric point of the human IL-1 is sharp, indicating lack of charge heterogeneity, while the isoelectric point of the mouse macrophage cell line IL-1 is broad [12].

ACKNOWLEDGMENTS

The authors wish to sincerely thank Mrs. Sandra Price for the careful preparation of this manuscript.

This work was supported by grant AM 08054 from the National Institute of Arthritis, Metabolism, and Digestive Diseases to Duke University, and grant CA 08975 from the National Cancer Institute.

REFERENCES

1. Waksman BM, Namba Y: Cell Immunol 21:161, 1976.
2. Cohen S, Pick E, Oppenheim JJ (eds): "Biology of the Lymphokines." New York: Academic Press, 1979.
3. de Weck AL (ed): "Biochemical Characterization of Lymphokines," New York: Academic Press, 1980.
4. Unanue ER, Kiely JM: J Immunol 119:925, 1977.
5. Blyden G, Handschumacher RE: J Immunol 118:1631, 1977.
6. Lachman LB, Hacker MP, Handschumacher RE: J Immunol 119:2019, 1977.
7. Mizel SB, Oppenheim JJ, Rosentriech DL: J Immunol 120:1504, 1978.
8. Gery I, Wasksman BH: J Exp Med 136:143, 1972.
9. Lachman LB, Hacker MP, Blyden GT, Handschumacher RE: Cell Immunol 134:416, 1977.
10. Lachman LB, Moore JO, Metzgar RS: Cell Immunol 41:100, 1978.
11. Mizel SB: J Immunol 122:2167, 1979.
12. Lachman LB, Metzgar RS: J Reticuloendothelial Soc 27:621, 1980.
13. Gillis S, et al: J Immunol 120:2027, 1978.
14. Davis BJ: Ann NY Acad Sci 121:404, 1964.
15. Togwa A, Oppenheim JJ, Mizel SB: J Immunol 122:2112, 1979.
16. Switzer RC III, Merril CR, Shifrin S: Anal Biochem 98:231, 1979.
17. Koopman WJ, et al: J Immunol 119:55, 1977.
18. Koopman WJ, Farrar JJ, Fuller-Bonar J: Cell Immunol 35:29, 1978.
19. Bolton AE, Hunter WM: Biochem J 133:529, 1973.

Visualization of Thrombin Receptors on Mouse Embryo Fibroblasts Using Fluorescein-Amine Conjugated Human α-Thrombin

Darrell H. Carney

Department of Human Biological Chemistry and Genetics, University of Texas Medical Branch, Galveston, Texas 77550

The localization of thrombin receptors on mouse embryo (ME) cells has been examined by direct fluorescence microscopy using a fluorescein amine-labeled thrombin. Two fluorescein amines, 4-(N-6-aminoethyl thioureal)-fluorescein and 4-(N-6-aminohexyl thioureal)-fluorescein, were synthesized and attached to the carbohydrate moiety of highly purified human α-thrombin by periodate oxidation of the carbohydrate and selective reduction of the Schiff's base using sodium cyanoborohydride. Preparations of fluorescent thrombin with from 1 to 4 fluoresceins per molecule of thrombin retained their ability to proteolytically cleave fibrinogin to form fibrin clots, to bind to thrombin receptors on ME cells, and to initiate cell division. After incubating mitogenic concentrations of the fluorescein amine labeled thrombin with ME cells at 4°C, a diffuse fluorescent pattern was observed over the surface of the ME cells. This diffuse pattern was specific: it was not observed on cells from parallel cultures incubated with fluorescent thrombin plus a 20-fold excess of unlabeled thrombin. Thus, thrombin receptors appear to be distributed randomly over the surface of ME cells prior to interaction with thrombin. Increasing the temperature to 37°C following binding at 4°C resulted in a rapid dissociation of the fluorescent pattern from the cells leaving only the autofluorescent vesicles. This result may reflect the unique ability of thrombin to proteolytically cleave its own receptor.

Key words: thrombin, initiation of cell division, receptor visualization, fluorescent labeling, proteolysis of receptors

Highly purified human α-thrombin initiates division of fibroblast-like cells from several different sources [1–3]. Studies to determine the molecular mechanism of this initiation have been recently reviewed [4]. Briefly, these studies have shown that thrombin action at the cell surface is sufficient to initiate division of chick embryo cells [5, 6] and that specific binding of thrombin at the cell surface is necessary for initiation of cell division [7]. The receptors for thrombin have been identified on mouse embryo fibroblasts by photoaffinity labeling experiments as molecules with a Mr = 50,250 ± 5,700 (n = 4) [8]. An additional thrombin binding protein "protease nexin" has been characterized in cultures of

Received April 23, 1980; accepted August 11, 1980.

0091-7419/80/1304-0467$02.30 © 1980 Alan R. Liss, Inc.

human fibroblasts with an apparent molecular weight of 38,000 [9]. The binding of thrombin to this protein, however, does not correlate with initiation of cell division [J. Baker, personal communication].

To determine whether proteolytic cleavage of the thrombin receptor is necessary for thrombin to initiate cell division, we recently prepared several proteolytically inhibited thrombin derivatives and examined their ability to bind to thrombin receptors and to initiate division of chick, mouse, human and Chinese hamster fibroblasts [10]. Two of these derivatives (with active site inhibitors PMSF or DIP) were able to bind to thrombin receptors on mouse and Chinese hamster cells equally as well as unmodified α-thrombin. However, none of the derivatives initiated cell division. These experiments indicate that receptor binding alone is not sufficient for initiation of cell division and suggest that proteolytic cleavage of the thrombin receptor is necessary for initiation.

Further evidence indicating that proteolytic cleavage of the thrombin receptors is necessary for initiation of cell division has come from recent studies with responsive and nonresponsive chick embryo cells. In these cells, thrombin cleavage of an iodinated cell surface protein corresponding in size to the thrombin receptors correlates with initiation of cell division by thrombin [11]. This protein is cleaved and lost from the surface of responsive cells after treatment with mitogenic concentrations of thrombin, but is not lost from cells which are not initiated to divide in response to the same concentration of thrombin [11]. Thus, this evidence as well as that obtained with proteolytically inhibited thrombin indicates that proteolytic cleavage of the thrombin receptor is necessary to initiate cell division.

Proteolytic cleavage of the thrombin receptor could generate a transmembrane signal in at least two ways. First, a fragment of the receptor might be released directly into the cells. A precedent for this type of signal has come from studies with epidermal growth factor (EGF), where a correlation between receptor degradation and initiation of cell division suggests that generation of a receptor fragment could be a part of the mitogenic signal [12]. Second, proteolytic cleavage of the receptor could be necessary for redistribution which might iteself generate a mitogenic transmembrane signal. Indeed, recent studies have indicated that insulin, EGF and many other molecules which bind to receptors aggregate on the cell surface and are then internalized by receptor-mediated endocytosis [13–16]. At present, however, it is unclear whether receptor aggregation and internalization are causally related to generation of a mitogenic signal.

To investigate the possible role of receptor redistribution in the initiation of cell division by thrombin, we have prepared fluorescent thrombin derivatives which retain their biological activity and ability to bind to the thrombin receptor. Using these derivatives we have been able to visualize thrombin receptors on mouse embryo fibroblasts using low mitogenic concentrations of fluorescent thrombin. Incubating the fibroblasts at 37°C following initial binding at 4°C does not lead to aggregation and internalization of the fluorescent thrombin as has been observed for insulin and EGF. Instead, the fluorescence disappears from the cells within minutes. This difference between thrombin and other nonproteolytic growth factors such as insulin and EGF may reflect the unique ability of thrombin to proteolytically cleave its own receptor.

MATERIALS AND METHODS

Cell Cultures

Primary cultures of mouse embryo fibroblasts were prepared from the body walls of 11- to 13-day-old NIH Swiss outbred mouse embryos and cultured in Dulbecco-Vogt-modified Eagles (DME) medium supplemented with 10% calf serum (Irvine Scientific) as previously described [3]. After 3 to 5 days, the primary cultures were subcultured into 35-mm dishes (6.2×10^4 cells cm^{-2}), or 35-mm dishes with 20 mm^2 coverslips (1.5×10^4 cells cm^{-2}). After 16 hr the cells were rinsed and the medium was changed to serum-free DME medium. Binding, growth and fluorescent labeling experiments were performed 2 days after changing cells to serum-free medium. By this time the cells were nonproliferating as judged by cell number and cell cycle analysis with flow microfluorimetry.

Preparation of Fluorescein Amine-Labeled Thrombin

A complete description of this labeling procedure will be published elsewhere [Carney, Jones and Weigel, manuscript in preparation]. Briefly, two different fluorescein amines were synthesized by reacting fluorescein isothiocynate (FITC) with a 10-fold excess of ethylenediame or hexanediamine in absolute ethyl alcohol to give 4-(N-6-aminoethyl thioureal)-fluorescein or 4-(N-6-aminohexyl thioureal)-fluorescein, respectively. The fluorescein amines were isolated from unreacted FITC and ethylene or hexanediamine by preparative thin-layer chromatography using ethylacetate:acetic acid:water (3:2:1) as a developing solvent.

These fluorescein amines were linked to the carbohydrate moiety of highly purified human α-thrombin (generously provided by Dr. John W. Fenton, II, [17]) by a modification of a procedure described by Yang, Fenton, and Feinman [18] for linking dansyl hydrazine to α-thrombin. Briefly, 1 ml of human α-thrombin (1.83 mg/ml, about 4,000 NIH Units/mg) was dialyzed overnight at 4°C against 0.1 M sodium acetate buffer at pH 5.6 with 0.15 M NaCl and reacted at 0°C in the dark for 30 min with sodium meta periodate (Sigma) at a final concentration of 5 μM. The protein was then dialyzed at 4°C against three changes of 50 mM phosphate buffer at pH 7.6 with 150 mM NaCl. Approximately 1 mg of 4-(N-6-aminoethyl thioureal)-fluorescein or 4-(N-6-aminohexyl thioureal)-fluorescein was dissolved in buffer and added to the oxidized protein. The pH was adjusted to 9.0 by addition of NaOH and the solution was stirred gently at 4°C. After 60 min, 1 mg of sodium cyanoborohydride was added to the mixture and stirring at 4°C was continued. After 18 hr, the fluorescent thrombin preparation was removed from the tube and dialyzed for 2 days against five changes of 50 mM phosphate buffer at pH 7.6 with 750 mM NaCl. Protein concentration and average molar ratio of fluorescein: thrombin were determined as described previously [19] using absorption at 495 and 280 nm. Thrombin samples prepared as described above had maximal molar fluorescein thrombin ratios of 1.1:1 when aminoethyl thioureal fluorescein was used and 4:2:1 when aminohexyl thioureal fluorescein was used. Aliquots of these labeled proteins were frozen and stored at −60°C or stored at 4°C. In all cases experiments were performed within 2 weeks since thrombin labeled in this manner was not stable at 4°C or −60°C for prolonged periods.

Measurement of Thrombin Biological Activity: Proteolytic Activity, Receptor Binding and Initiation of Cell Division

To determine proteolytic activity we measured the ability of thrombin and thrombin derivatives to convert fibrinogen to fibrin as described by Lundblad et al [20] except that polyethylene glycol was substituted for acacia [21].

Ability of thrombin derivatives to bind to thrombin receptors on mouse embryo fibroblasts was determined by comparing their ability to compete with ^{125}I-labeled α-thrombin for specific binding sites as described previously [7]. Briefly, competition studies were performed in situ on cultures of nonproliferating mouse embryo cells containing approximately

5×10^5 cells per 35-mm dish. The cultures were rinsed and allowed to equilibrate with binding medium (serum-free medium containing 0.5% albumin, buffered with 15 mM Hepes at pH 7.0) for 30 min at 22°C. The medium was then changed to binding medium containing 10 ng per ml of ^{125}I-thrombin with indicated concentrations of unlabeled α-thrombin or fluorescein amine-labeled thrombin and incubated at 22°C for 120 min. Binding was terminated by removing the binding medium and rinsing the cultures through four beakers of phosphate buffered saline (PBS) at 4°C. The cultures were dissolved in 1 ml of 0.5 M NaOH and ^{125}I radioactivity was determined.

Ability of thrombin derivatives to initiate cell division was assessed by adding indicated concentrations of α-thrombin or fluorescein amine-labeled thrombin to nonproliferating cultures of secondary mouse embryo cells in serum-free medium prepared as described above. After 48 hr, cells were detached from their 35-mm culture dishes using 1 ml of PBS containing 0.02% EDTA and 0.05% trypsin. They were then diluted with PBS and counted in a Coulter electronic particle counter. Each experimental point is the mean of duplicate determinations which usually varied less than 5%.

Visualization of Fluorescent Thrombin by Direct Fluorescence Microscopy

Coverslip cultures of nonproliferating mouse embryo fibroblasts (approximately 1.5×10^4 cells per cm^2) were rinsed and equilibrated with binding medium for 30 min at 4°C. The binding medium was then changed to binding medium containing fluorescein amine-labeled thrombin (250 ng/ml) with or without unlabeled thrombin (5 μg/ml) and the cultures were incubated for 5 hr at 4°C. To terminate binding, cultures were rinsed through four beakers of PBS at 4°C. Some of these cultures were rinsed and immediately fixed with freshly prepared 3% formaldehyde for 20 min, others were increased to 37°C prior to rinsing and fixation.

Coverslip cultures were inverted onto microscope slides and examined by fluorescence microscopy using a Leitz Orthoplan microscope equipped with "ploemopack" illuminator and an I$_2$ fluorescent filter cube. Photographs were all taken through this microscope with 100 X phase objective (numerical apperture of 1.30). All photographs were 30-sec exposures on Kodak Tri-X film pushed to 1600 ASA by developing in Ethol Blue developer, printed on Agfa Rapidoprint contrast 4 single weight paper with identical exposures, and processed with a Kodak Ektamatic rapid processor.

RESULTS

Preparation of Biologically Active Fluorescent Thrombin

Labeling human α-thrombin with fluorescein isothiocyante or other molecules which couple to free-amino groups quantitatively inhibited thrombin activity and ability to bind to thrombin receptors. Therefore, we synthesized two fluorescein amines, 4-(N-6-aminoethyl thioureal)-fluorescein and 4-(N-6-aminohexyl thioureal)-fluorescein. When these fluorescein amines were linked directly to free carboxyl groups on thrombin using carbodiimide condensation there was also a considerable loss of proteolytic activity as judged by the ability of these thrombins to cleave fibrinogen and produce a fibrin clot. As an alternative form of linkage, we attached these same fluorescein amine molecules to the carbohydrate moiety on α-thrombin by periodate oxidation and reduction with sodium cyanoborohydride. As shown below, this procedure yielded thrombin with up to four fluoresceins per thrombin molecule with little, if any, effect on its biological activity.

Proteolytic activity of human α-thrombin did not appear to be affected by linking fluorescein amines to the carbohydrate moiety. As shown in Figure 1, thrombin labeled with 4-(N-6-aminoethyl thioureal)-fluorescein (molar ratio of 1.1 fluoresceins per thrombin) was at least 95% as active as unlabeled thrombin in cleaving fibrinogen to produce a fibrin clot.

To determine whether the fluorescent thrombin could bind to thrombin receptors, we incubated nonproliferating cultures of mouse embryo cells with ^{125}I-thrombin (10 ng/ml) for 2 hr at 22°C in the presence of increasing concentrations of unlabeled and aminoethyl fluorescein-labeled thrombin (molar ratio 1.1:1). Under these steady state conditions approximately 90% of the ^{125}I-thrombin specifically bound is associated with receptors on the cell surface [7]. As shown in Figure 2, the competition curves generated with unlabeled and fluorescein-labeled thrombin were virtually identical. This demonstrates that the fluorescein amine-labeled thrombin retained its full ability to bind specific receptors on the mouse embryo cells.

Fluorescein amine-labeled thrombin was also able to initiate division of mouse embryo cells. Figure 3 compares the mitogenic activity of unlabeled and aminohexyl fluorescein-labeled thrombin (molar ratio 4.2 fluoresceins per thrombin). As shown, at thrombin concentrations of 250 ng/ml the increase in cell number over controls with no addition after 56 hr was approximately 58% for aminohexyl fluorescein thrombin and 61% for unlabeled thrombin. At concentrations above 500 ng/ml the ability of fluorescein amine-labeled thrombin to initiate cell division appeared to decrease while stimulation by unlabeled thrombin continued to increase. This may indicate that high concentrations of carbohydrate modified fluorescein thrombin interfere with the normal initiation process. It should also be noted that there was approximately a twofold difference in concentration of labeled and unlabeled thrombin required for half maximal initiation. A similar difference in specific activity of this particular preparation was observed in proteolytic activity and

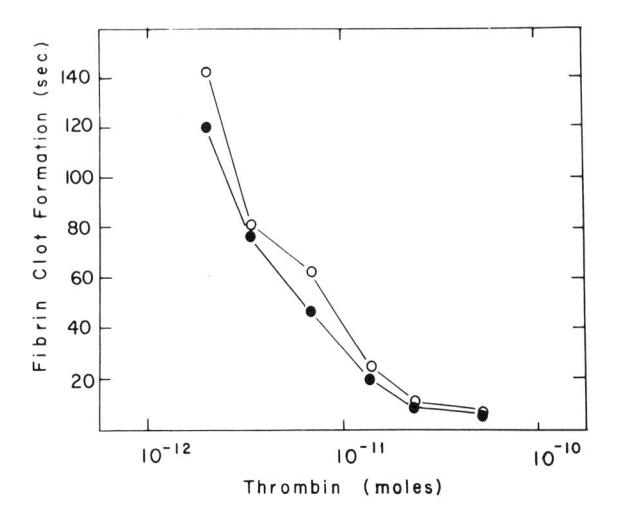

Fig. 1. Fibrinogen clotting activity of fluorescein amine-labeled and unlabeled human α-thrombin. Indicated concentrations of aminoethyl fluorescein-labeled thrombin (○) and unlabeled thrombin (●) were added to a standard fibrinogen solution at 22°C and the time required for fibrin clot formation was determined. Molar ratio of fluorescein to thrombin in this preparation was 1.1.

concentration required to compete for ^{125}I-thrombin binding. The shape of these activity curves, however, were identical to those of unlabeled thrombin (data not presented). This suggests that inactive molecules in this population do not interfere with activity or binding of labeled molecules to the thrombin receptors. Other preparations tested with fluorescein: thrombin ratios of closer to one, retained nearly all of their mitogenic activity consistent with their capacity to bind to thrombin receptors and cleave fibrinogen.

Visualization of Fluorescent Thrombin on ME Cells

In the above results, we have demonstrated that fluorescein amine-labeled thrombin preparations retain their proteolytic activity, can bind to thrombin receptors, and can initiate cell division. These studies, however, relied on average molar ratio of fluorescein to thrombin to determine the extent of labeling, leaving the possibility that the fluorescence was localized on highly fluorescinated, inactive molecules which would only bind nonspecifically to cells or that the fluorescent label might be dissociating from thrombin during incubation. To determine whether the actual molecules which were labeled with fluorescein were binding specifically, we performed competition experiments with fluorescein amine-labeled thrombin with or without a 20-fold excess of unlabeled thrombin and then examined these cultures with fluorescence microscopy. As shown in Figure 4A, when aminohexyl fluorescein-labeled thrombin (250 ng/ml, 4.2 fluoresceins/thrombin) was incubated for 5 hr at 4°C with ME cells, a diffuse pattern of fluorescence could be observed and photographed without the aid of video intensification. In contrast, in parallel cultures incubated with the same fluorescein amine-labeled thrombin plus a 20-

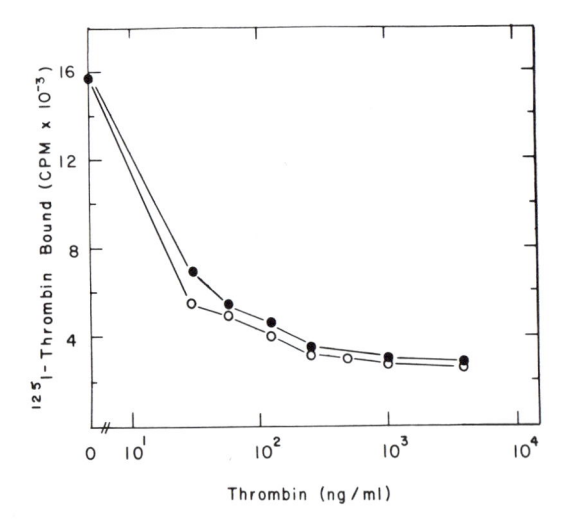

Fig. 2. Ability of fluorescein amine-labeled and unlabeled thrombin to compete for binding of ^{125}I-thrombin to ME cells. Nonproliferating cultures of ME cells (5.5 × 10^5 cells per 35-mm dish) prepared as described in Materials and Methods were incubated at 22°C for 2 hr with ^{125}I-thrombin (10 ng/ml, prepared as described previously [7]) with the indicated concentrations of aminoethyl fluorescein-labeled thrombin (○) or unlabeled thrombin (●). Cells were rinsed and the total ^{125}I-thrombin radioactivity bound to the cells was determined. The molar ratio of fluorescein:thrombin in this preparation was 1.1.

fold excess of unlabeled thrombin (Fig. 4B), the only fluorescence observed was that which could be attributed to autofluorescence (compare Fig. 4B and 4C). This demonstrated that the fluorescein amine-labeled thrombin molecules were themselves binding specifically to receptors on ME cells.

It should be noted that in these experiments virtually all of the cells incubated with fluorescein amine-labeled thrombin alone had about the same level of diffuse fluorescence. Thus, the brightly fluorescent cell depicted in Figure 4A does not merely represent a select population of cells which take up fluorescent dye. There was some variation between cells and between cultures in the number of bright spots which could be observed in the cells. These bright spots, however, do not represent aggregated receptors or aggregates of free fluorescein label because they were also observed in cells incubated with excess unlabeled thrombin and in control cells which were not exposed to any fluorescent label (Note Fig. 4C). Thus, these bright spots appear to be due to aggregates or vesiculated autofluorescent molecules within the cells.

The more diffuse brighter fluorescent pattern in all cases was specific for cells incubated with fluorescent thrombin alone and was easily distinguishable from the nonspecific and autofluorescent patterns. In many cases this diffuse fluorescence appeared to have a mottled or fibrous appearance (see also Fig. 5A). This pattern reflects the normal contours and projections of the cell surface. Thus this somewhat irregular diffuse fluorescent pattern suggests that the fluorescein-labeled thrombin is specifically bound under these conditions to receptors randomly distributed over the entire surface of the cells.

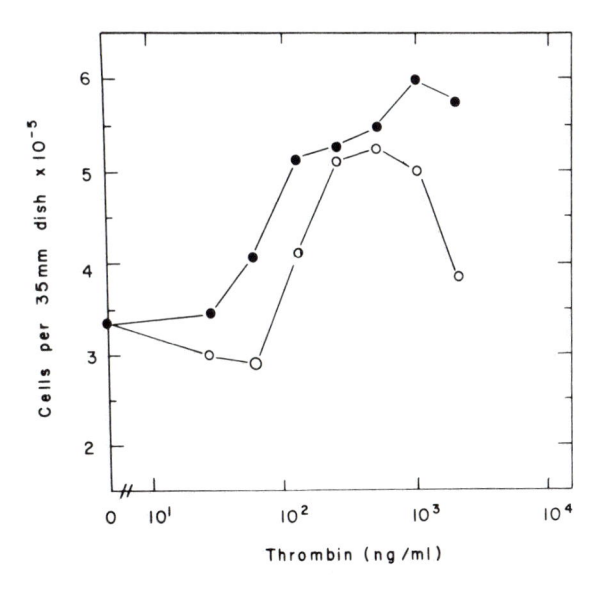

Fig. 3. Effect of various concentrations of fluorescein amine-labeled and unlabeled thrombin on initiation of ME cell division. ME cells were plated in 35-mm dishes at a density of 6.2×10^4 cells per cm^2. After complete cell attachment (16 hr) the cultures were rinsed and the medium changed to serum-free medium. After 2 days the indicated concentrations of aminohexyl fluorescein-labeled thrombin (○) or unlabeled thrombin (●) were added to each culture. Cell number was determined 56 hr later. The molar ratio of fluorescein:thrombin in this preparation was 4.2.

To further insure that the diffuse pattern observed did not reflect fluorescent molecules which were released from thrombin during incubation, cells were incubated with free fluorescein amines under conditions identical to those used in the experiments depicted in Figure 4. In all cases, with free fluorescein concentrations up to 50 ng/ml (about 10 × the amount of fluorescein in 250 ng of fluorescent thrombin), there was no fluorescence observed above that which was due to autofluorescence alone (data not presented). Cells incubated with thrombin plus free fluorescein amines also did not appear to bind or internalize the free fluorescent molecules. Thus from these experiments it has been possible to show that the fluorescein amine-labeled thrombin binds specifically to receptors on the surface of mouse embryo cells at 4°C, that free fluorescein amines cannot duplicate this pattern, and by the diffuse but somewhat complex appearance of the fluorescent pattern, that at 4°C the thrombin receptors to which these molecules bind are randomly distributed over the entire cell surface.

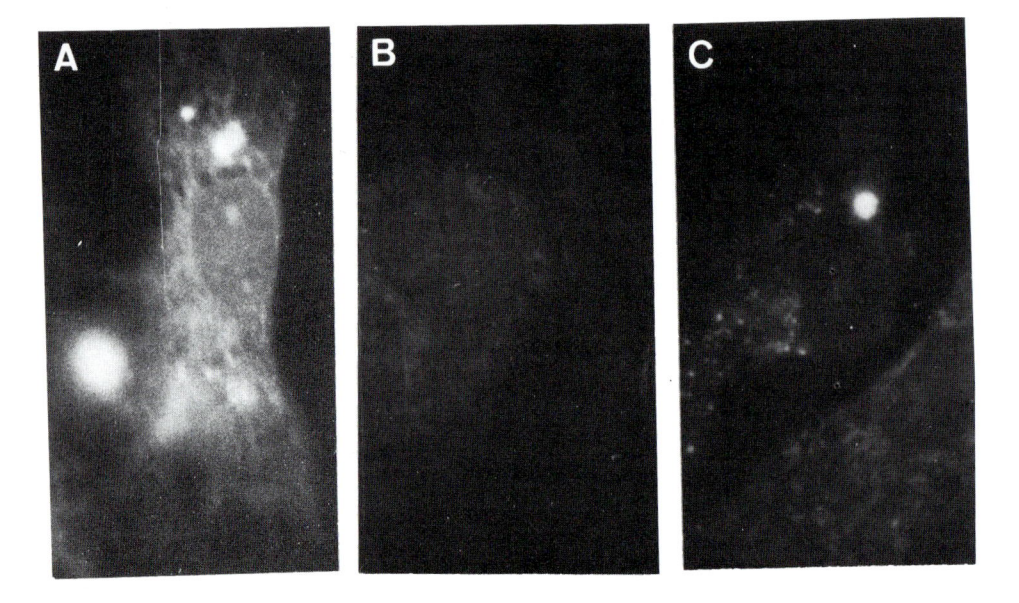

Fig. 4. Specificity and pattern of aminohexyl fluorescein thrombin binding to ME cells at 4°C. Aminohexyl fluorescein-labeled thrombin (250 ng/ml) was incubated with nonproliferating cultures of ME cells plated on glass coverslips at 4°C for 5 hr in the presence or absence of unlabeled thrombin (5 μg/ml). These coverslip cultures and ones incubated without fluorescent thrombin were then rinsed four times with PBS at 4°C, fixed with 3% formaldehyde and photographed using identical exposures through a fluorescent microscope as described in Materials and Methods. (× 4,250) A) Aminohexyl fluorescence thrombin alone. B) Aminohexyl fluorescein plus 5 μg/ml unlabeled thrombin. C) Autofluorescence of cells incubated without fluorescent thrombin.
Note: The molar ratio of fluorescein to thrombin in this preparation was 4.2. A similar diffuse pattern was also observed with aminoethyl fluorescein thrombin preparations (molar ratios of ~ 1.0); however, these patterns were much dimmer and only clearly discernable as specific patterns on cultures with low autofluorescence. It should also be noted that after a 5-hr incubation at 4°C, thrombin binding is approaching steady state with almost all of the cell-associated thrombin molecules on the cell surface [7].

Receptors for other peptide growth factors and hormones have also been reported to be evenly distributed over the surface of cells at 4°C [14–16]. In these cases, warming the cells to 22° or 37°C following 4°C binding resulted in aggregation of the fluorescent hormone-receptor complexes and their internalization into vesicles by receptor mediated endocytosis.

To determine whether a similar process occurred with the thrombin receptor complexes, we incubated cells for 5 hr at 4°C with fluorescein amine-labeled thrombin as described above and then increased the temperature to 37°C. As shown in Figure 5, rather than aggregating or being internalized into vesicles, as the temperature was increased the fluorescent thrombin seemed to disappear from the ME cells. This phenomena was observed in separate experiments with both ethylamine and hexylamine fluorescein-labeled thrombin preparations. In all cases as early as 1 min after increasing the temperature to 37°C there was considerably less diffuse fluorescence and by 5 min the diffuse pattern was gone leaving just the autofluorescent vesicles. These unexpected results suggest that the interaction between thrombin and its receptor might be quite different from that of other nonproteolytic growth factors and hormones.

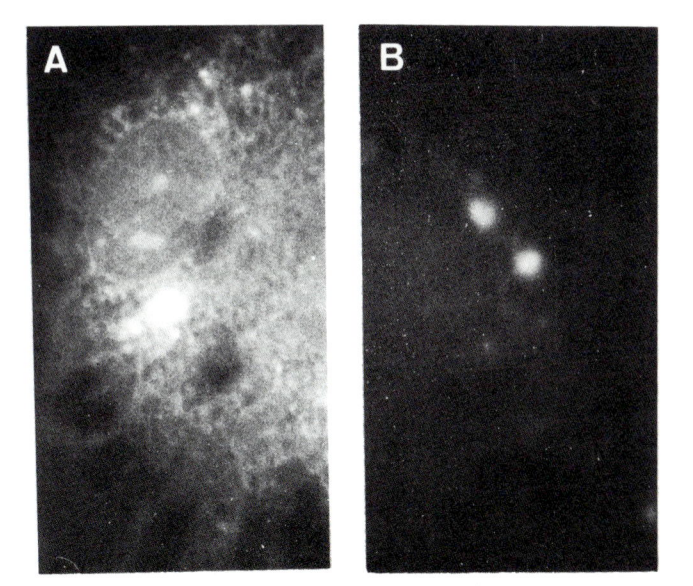

Fig. 5. Effect of 37°C incubation on the cellular pattern of fluorescent thrombin. Coverslip cultures of ME cells were prepared and incubated with aminohexyl fluorescein thrombin (250 ng/ml) as described in Figure 4. After 5 hr of incubation at 4°C, the cultures were either rinsed and fixed for examination (A) or incubated for 5 min at 37°C and then rinsed and fixed for examination (B). The cultures were then photographed as in Figure 4 using identical exposures, processing and printing (see Materials and Methods).

DISCUSSION

As an initial step in determining whether thrombin receptor redistribution on the surface of fibroblasts plays a role in thrombin initiation of cell division, we have synthesized a new fluorescent derivative and coupled this molecule to the carbohydrate moiety on human α-thrombin. Fluorescent thrombin prepared in this manner retains its proteolytic activity, its ability to bind to thrombin receptors on ME cells, and its ability to initiate ME cell division. Most importantly, with this derivative, thrombin receptors on the surface of ME cells can be visualized by direct fluorescence microscopy.

A number of recent studies have utilized fluorescent derivatives of insulin, EGF, and α-2-macroglobulin to visualize the interaction of these molecules with receptors on the surface of cells [13,14]. These molecules appear to bind to receptors which are diffusely distributed and then to rapidly aggregate and be internalized into vesicles [16]. From ultrastructural studies with ferritin derivatives it appears that such molecules are internalized into coated vesicles [15, 16]. These coated vesicles, however, are much smaller than the large fluorescent agggregates observed with video intensified fluorescence microscopy. Therefore, it is not clear exactly what these large patches represent. A further problem in some of these studies has been a lack of specificity of the fluorescently labeled hormone. For example, highly fluorescent insulin has been prepared by conjugating FITC-labeled lactalbumin to insulin. This preparation retained less than 9% of its ability to bind to insulin receptors and less than 2% of its biological activity [22]. In such cases, it is difficult to be convinced that visualized redistributions are truly receptor mediated.

To get around problems of possible loss of specificity in labeled thrombin molecules, we have utilized a new procedure for direct fluorescent labeling of human α-thrombin using 4-(N-6-aminoethyl thioureal)-fluorescein or 4-(N-6-aminohexyl thioureal)-fluorescein [Carney, Jones and Weigel, manuscript in preparation]. This fluorescein amine was linked to the carbohydrate moiety of thrombin using a modification of a procedure described by Yang, Fenton and Feinman [18] for coupling dansyl hydrazine to human α-thrombin. This gentle procedure for fluorescent labeling should have wide applicability for a number of glycoprotein molecules. In the case of thrombin, molar ratios of 1.1 and 4.2 fluoresceins per thrombin with ethyl amine and hexyl amine derivatives were achieved, respectively, suggesting that the length of the spacer arm separating the fluorescein and carbohydrate molecules is important in determining the extent of labeling. These fluorescent thrombin preparations retained virtually all of their ability to cleave fibrinogen and form a fibrin clot and their ability to bind to the thrombin receptors. In addition, they were able to initiate division of nonproliferating ME cells 50–90% as effectively as unlabeled human α-thrombin. Thus, we have prepared fluorescent thrombin with up to four fluoresceins per molecule with little effect upon interactions between thrombin and the cell surface or its substrate molecules.

When fluorescein amine-labeled thrombin (250 ng/ml) was incubated for 5 hr at 4°C with ME cells, a diffuse pattern of fluorescence was observed over the surface of the cells. In parallel cultures, incubated with both fluorescent thrombin and a 20-fold excess of unlabeled α-thrombin, the diffuse pattern was not detectable indicating that this pattern was specific for fluorescent thrombin bound to thrombin receptors. This concentration of fluorescent thrombin produced a 58% increase in number of ME cells relative to controls over a 56-hr incubation. Thus, with this fluorescent thrombin preparation the interactions between thrombin and its receptors were visualized at normal mitogenic concentrations.

In the present study, when fluorescein amine-labeled thrombin was incubated with ME cells at 4°C for 5 hr and then the temperature increased to 37°C, there was no observable receptor redistribution on cell surfaces or internalization into vesicles. Instead, the diffuse pattern disappeared. This was not accompanied by an apparent increase in the number or intensity of fluorescent vesicles, which in these cells appear to be due to autofluorescence.

There are several possible explanations for the disappearance of the fluorescent pattern as the temperature was increased on these cells. For example, it is possible that the carbohydrate label was cleaved from thrombin by cell surface enzymes which were inactive at 4°C. In kinetic experiments with ^{125}I labeled thrombin at the same concentration, we have observed that at 37°C there is a dissociation of labeled thrombin from ME cells at about 1 hr [Crossin and Carney, unpublished observations]. This suggests that a dissociation of thrombin from cells following receptor binding might be a normal step in the sequence of events leading to a mitogenic signal. It is also possible that in the present experiments incubating the ME cells at 4°C altered a cytoskeletal anchorage of the receptors and accelerated the release of thrombin or the thrombin receptor complex as the temperature was increased. Along these lines, we have recently discovered that drugs which interfere with microtubule polymerization affect thrombin binding and its mitogenic acticity on ME cells [Crossin and Carney, manuscript in preparation].

Finally, the rapid dissociation of thrombin at 37°C could represent proteolytic cleavage of thrombin receptors dramatized in these experiments in which binding occurred at 4°C where thrombin is proteolytically inactive. Indeed, several lines of evidence indicate that proteolytic cleavage of the thrombin receptor is necessary for thrombin to initiate cell division [10, 11]. Thus, the dissociation of fluorescent thrombin observed in the present experiments at 37°C may reflect cleavage of the thrombin receptor. This would also explain why this type of dissociaton has not been observed with nonproteolytic mitogens. Further studies utilizing other types of labeled thrombin, indirect immunofluorescence, and a more sensitive video intensified fluorescence microscope are underway to sort out these possibilities.

Establishing that thrombin receptors on ME cells can be visualized using mitogenic concentrations of fluorescent thrombin provides the basis for determining whether or not changes in distribution or release of thrombin from the cell surface correlate with initiation of cell division. The unexpected finding that the fluorescent thrombin pattern was lost after increasing the temperature to 37°C, if substantiated by other techniques, may point to the unique ability of thrombin as a proteolytic mitogen to cleave its own receptor and thus initiate cell division.

ACKNOWLEDGMENTS

We thank Dr. John W. Fenton II, for gifts of highly purified human α-thrombin [17], Peggy Jones for her excellent technical assistance, and Dr. Paul H. Weigel for his help and suggestions in preparation of the fluorescein amine-labeled thrombins. This work was supported by grant AM 25807 from the National Institute of Arthritis, Metabolism and Digestive Diseases and by institutional grants DHEW 5 SO7RR 05427, American Cancer Society IN 112B, and National Cancer Institute Grant CA 17701-04.

REFERENCES

1. Chen LB, Buchanan JM: Proc Natl Acad Sci USA 72:131, 1975.
2. Pohjanpelto P: J Cell Physiol 91:387, 1977.
3. Carney DH, Glenn KC, Cunningham DD: J Cell Physiol 95:13, 1978.

4. Cunningham DD, Glenn KC, Baker JB, Simmer RL, Low DA: J Supramol Struct 11:259, 1979.
5. Carney DH, Cunningham DD: Cell 14:811, 1978.
6. Carney DH, Cunningham DD: J Supramol Struct 9:337, 1978.
7. Carney DH, Cunningham DD: Cell 15:1341, 1978.
8. Carney DH, Glenn KC, Cunningham DD, Das M, Fox CF, Fenton JW II: J Biol Chem 254:6244, 1979.
9. Baker JB, Low DA, Simmer RL, Cunningham DD: J Supramol Struct (Suppl) 4:170, 1980.
10. Glenn KC, Carney DH, Fenton JW II, Cunningham DD: J Biol Chem: 255, 6609, 1980.
11. Glenn KC, Cunningham DD: Nature 278:711, 1979.
12. Das M, Fox CF: Proc Natl Acad Sci USA 75:2644, 1978.
13. Schlessinger J, Shechter Y, Willingham MC, Pastan I: Proc Natl Acad Sci USA 75:2659, 1978.
14. Maxfield FR, Schlessinger J, Shechter Y, Pastan I, Willingham MC: Cell 14:805, 1978.
15. Haigler HT, McKanna JA, Cohen S: J Cell Biol 81:382, 1979.
16. Goldstein JL, Anderson RGW, Brown MS: Nature 279:679, 1979.
17. Fenton JW II, Fasco MJ, Stackrow AB, Aronson DL, Young AM, Finlayson JS: J Biol Chem 252:3587, 1977.
18. Yang CC, Fenton JW II, Feinman RD: (Submitted to Biochemistry).
19. Kawamura A: "Fluorescent Antibody Techniques and Their Application," Tokyo: University Park Press, 1977, p 55.
20. Lundblad RL, Kingdom HS, Mann GK: In Colowich SP, Kaplan NO (eds): "Methods in Enzymology." New York: Academic Press 45: 1976, p 156.
21. Fasco MJ, Fenton JW II: Arch Biochem Biophys 159:802, 1973.
22. Shechter Y, Schlessinger J, Jacobs S, Chang K-J, Cuatrecassas P: Proc Natl Acad Sci USA 75: 2135, 1978.

Activation of T Lymphocytes by the Fc Portion of Immunoglobulin

Marilyn L. Thoman, Edward L. Morgan, and William O. Weigle

Department of Immunopathology, Scripps Clinic and Research Foundation, La Jolla, California 92037

T lymphocytes are stimulated to release T-cell-replacing factors in response to Fc fragments of human IgG. Lyt 1^+23^- T cells are directly triggered to factor production by Fc subfragments, derived from intact Fc fragments by macrophage-dependent enzymatic cleavage. These factor(s) replace T cell function in two Fc-mediated immune responses; induction of polyclonal antibody synthesis, and potentiation of anti-SRBC responses.

Key words: T cell factors, polyclonal antibody formation, Fc fragments, interleukin 2

Antigen-antibody complexes, aggregated human gamma globulin, and Fc fragments derived from mammalian immunoglobulin mediate a variety of immunological functions. They are potent activators of B lymphocytes, stimulating proliferation and polyclonal antibody formation, and modulate specific antibody responses both in vivo and in vitro [1–7]. Both macrophages and T lymphocytes are necessary for the Fc-induced polyclonal antibody response [3, 4]. Macrophages process the Fc fragments through enzymatic cleavage to an active subfragment of 14,000 daltons. The subfragment stimulates both proliferation and polyclonal antibody responses in macrophage-deficient B cell preparations [4, 5]. Fc fragment activation of T lymphocytes is less well characterized, although T cells do not proliferate in response to Fc [1]. Since both the Fc-induced polyclonal response and the adjuvant effects of Fc fragments require T cells, these two systems have been employed to analyze Fc effects on T cells. As will be shown, certain T lymphocyte subsets are stimulated by Fc to produce factors which exert T cell replacing functions. These studies have expanded the understanding of the mechanisms of Fc fragment regulation of immune responses and may provide insight into possible regulatory effects exerted by antigen-antibody complexes. A model for the possible regulation of in vivo antibody responses by Fc fragments derived from immune complexes is discussed.

Received May 2, 1980; accepted July 18, 1980.

0091-7419/80/1304-0479$02.00 © 1980 Alan R. Liss, Inc.

MATERIALS AND METHODS

Animals

Male mice of the inbred C57BL/6J strain were obtained from the Jackson Laboratories (Bar Harbor, Maine). Inbred C57BL/6 nude mice (N_4F_4) were obtained from the Scripps Clinic and Research Foundation breeding colony. All mice were between 8–10 weeks of age.

Preparation of Fc Fragments

A human IgG1 myeloma protein (Fi) was a gift from Dr. Hans L. Spiegelberg, Scripps Clinic and Research Foundation. The IgG1 was purified by ammonium sulfate fractionation followed by DEAE cellulose chromatography with 0.01 M phosphate buffer pH 8 used as the eluent.

Fc fragments were obtained by a 5-hr digestion of IgG1 with papain (Sigma Chemical Co., St. Louis, Missouri) in the presence of L-cysteine (Sigma) and ethylenediaminetetraacetic acid (EDTA) (J.T. Baker Chemical Co., Phillipsburg, New Jersey) for 5 hr [8]. Following digestion the material was chromatographed on Sephadex G-100 (Pharmacia Fine Chemicals, Piscataway, New Jersey) to remove any undigested IgG. The Fc and Fab fragments were then separated from each other by DEAE chromatography [9]. Mitogenic Fc subfragments were prepared as described in detail elsewhere [5].

Preparation of Interleukin 2 (IL-2)

IL-2 was prepared as described by Watson et al [10]. Briefly, the spleen cells of 50 mice were cultured in serum-free RPMI-1640 (Flow Laboratories, Rockville, Maryland) supplemented with 0.1 M HEPES, 2 mM L-glutamine, 100 units penicillin, 100 µg streptomycin, and 5×10^{-5} M 2-mercaptoethanol (2-ME) for 24 hours. The culture supernatants were collected by centrifugation, concentrated by vacuum dialysis, and subjected to chromatography on Sephadex G-100. Chromatography was performed with sterile 0.9% saline containing 100 units penicillin and 100 µg streptomycin/ml. Each column fraction was tested for activity in a thymocyte mitogenic assay described in [10]. The active fractions were pooled, dialyzed against 1% glycine, and subjected to preparative isoelectric focusing [11] in a pH gradient from 3–10. The isoelectric focusing bed was divided into 20 fractions and each eluted with 5 ml sterile 0.9% saline. Each fraction was dialyzed to remove the ampholytes, then tested for the ability to replace T cells in the polyclonal response.

Production of Fc-Fragment-Induced T Cell Factors

Spleen cells were incubated 24 hr in RPMI-1640 supplemented with 2 mM L-glutamine, 100 units penicillin, 100 µg streptomycin, 5×10^{-5} M 2-ME and 5% fetal calf serum (FCS) (Grand Island Biological, New York), with 50–250 µg/ml Fc fragments. Culture supernatants were collected and tested for T-cell-replacing activity in the Fc-induced polyclonal response.

Macrophage Depletion

Macrophages were depleted by Sephadex G-10 (Pharmacia) filtration according to the method of Ly and Mishell [12]. 50×10^6 cells were applied to 9-ml columns of Sephadex G-10 which had been equilibrated in Hanks balanced salt solution (BSS) and 5% FCS at 37°C. Cells were eluted with approximately 5–6 mls of warm BSS and 5% FCS.

Antisera Treatment

Antisera and C' treatment was performed essentially as recommended by Shen and Boyse [13]. 50×10^6 cells/ml were incubated with an appropriate amount of antisera at 4°C for 30 minutes. The cells were washed with phosphate-buffered saline containing 5% FCS, and resuspended in RPMI-1640 containing 5% FCS, 1% sodium azide and 25% C' in which they were incubated 30 min at 37°C. The complement, a mixture of rabbit and guinea pig shown to be non-cytotoxic for murine lymphocytes, was a gift from Dr. Sharyn M. Walker. After complement treatment the cells were washed in PBS with 5% FCS and resuspended in complete media.

αLyt 1.2 and αLyt 2.2 were obtained from Dr. F. W. Shen, of the Memorial Sloan-Kettering Cancer Center, New York. Under the conditions employed, 5–15% of thymocytes and 20–30% of spleen cells treated were recovered following Lyt antisera and C' treatment. The anti-IA antisera, A. TH anti-A. TL, was a gift from Dr. J. Ray of the National Institutes of Health. Thirty to forty percent of spleen cells and 20% of thymocytes treated were recovered following treatment with this antisera and C'. Rabbit anti-mouse thymocyte sera (Lot 15088, Microbiological Associates, Bethesda, Maryland) was absorbed with a BALB/c myeloma, XS63, and was cytotoxic for approximately 50% of spleen cells.

Polyclonal Antibody Response Assay

For the generation of the polyclonal plaque-forming cell (PFC) response, spleen cells were suspended to a concentration of 2×10^6 /ml in RPMI-1640 supplemented with 2 mM L-glutamine, 1% BME vitamins, 100 units penicillin, 100 μg streptomycin, 5×10^{-5} M 2-ME, 0.5% fresh mouse serum, and 7.5% FCS. Duplicate cultures of 6×10^5 cells/0.3 ml were incubated in microtiter plates (3042 Microtest II, Falcon Plastics, Oxnard, California) at 37°C in 5% CO_2. The duplicate cultures were harvested on day 3 and assayed for a response to 2,4,6-trinitrophenyl-sheep red blood cells (TNP-SRBC) by the slide modification of the Jerne and Nordin plaque assay [14]. Heavily conjugated TNP-SRBC were prepared according to the method of Kettman and Dutton [15] and were used as the indicator RBC. Guinea pig serum (Pel-Freez, Rogers, AR) was the source of complement used to develop the IgM plaques. Results of the plaque-forming assay are expressed as mean PFC/10^6 original cells cultured ± standard error. Each experiment was performed several times, and the experiments shown are representative of all the data.

Generation of Immune Responses

Priming for secondary in vitro antibody response. Mice were injected with 0.1 ml of 10% suspension of sheep red blood cells (SRBC) (Colorado Serum Co., Denver, Colorado) intraperitoneally (IP). Six to eight weeks after priming they were boosted IP with the same dose of SRBC and used 7 days later.

In vitro response to SRBC. Spleens were removed from primed and boosted or untreated mice, and a single cell suspension was prepared by teasing the spleens apart with forceps into phosphate-buffered saline (PBS), (0.001 M sodium phosphate, 0.15 M NaCl pH 7.4). A modified Mishell-Dutton culture system was employed for the generation of antibody-producing cells [16]. Cells were suspended to a concentration of 6×10^6/ml RPMI-1640 supplemented with 2 mM L-glutamine, 1% BME vitamins (Grand Island Biological Co., New York), 100 units penicillin, 100 μg streptomycin (Microbiological Associates, Bethesda, Maryland), 5×10^{-5} M 2-ME, 7.5% FCS, and 0.5% fresh normal

TABLE I. T Cell Requirement for Fc Fragment Stimulation of a Polyclonal Antibody Response

Cell source	Fc fragments[a]	Anti-TNP PFC/10^6 ± SE
Normal spleen	−	< 10
Normal spleen	+	117 ± 13
T-cell depleted spleen[b]	+	< 10
B cells and T cells[c]	+	193 ± 12
T-cell depleted spleen and IL-2[d]	−	16 ± 4
T-cell depleted spleen and IL-2[d]	+	121 ± 6

[a]50 μg/culture well.
[b]T cells removed for treatment with anti-thymocyte sera + C' as described in Methods.
[c]2.5 × 10^5 B cells (anti-thymocyte + C' treated spleen cells) were mixed with 2.5 × 10^5 nylon wool purified T cells.
[d]10 μl IL-2.

mouse serum. The spleen cells at a concentration of 6 × 10^5 along with various concentrations of SRBC and Fc were cultured in 0.3 final volume in flat-bottom microtiter plates for 4 days at 37°C in 5% CO_2. At the end of this time duplicate cultures were pooled and assessed for PFC to SRBC, as described above. Indirect plaques were developed with a combination of guinea pig serum and rabbit anti-mouse immunoglobulin.

RESULTS

Fc-Induced Polyclonal Antibody Response

Fc fragments trigger anti-TNP polyclonal antibody responses in murine spleen cell cultures (Table I). The response was found to be T cell dependent as treatment with anti-thymocyte sera and C' prevents the response. Polyclonal antibody formation can be restored by adding back nylon wool-purified T cells. Furthermore, the T cell requirement can be replaced by a soluble T cell replacing factor, IL-2. IL-2 completely restores the Fc-induced polyclonal antibody response in T-deficient cultures. IL-2 is generated by Concanavalin A stimulation of murine spleen cells and purified by Sephadex G-100 chromatography and preparative isoelectric focusing. IL-2 displays considerable charge heterogeneity in preparative IEF in the pH range 4–5. Therefore, each individual fraction of a preparative isoelectric focusing gel run on IL-2 was tested for the ability to replace T cell function in two systems: the Fc-induced polyclonal antibody response and the primary anti-SRBC response. As shown in Figure 1, these activities do not reside in the same fractions. The majority of the activity which restores anti-SRBC antibody formation in T-depleted cultures is present in two peaks with pI values of 4.1 and 4.8. In the same preparation, the polyclonal-restoring activity is found in one peak at pH 4.4

The ability of soluble T cell factors to function in the polyclonal response suggested the possibility that Fc might trigger the production of such material. Table II illustrates that Fc fragments incubated with spleen cells for 24 hr do stimulate the release of T cell replacing factors into the culture supernatant. For the sake of clarity, this Fc fragment-induced T cell replacing factor has been called (Fc) TRF to distinguish it from IL-2 pro-

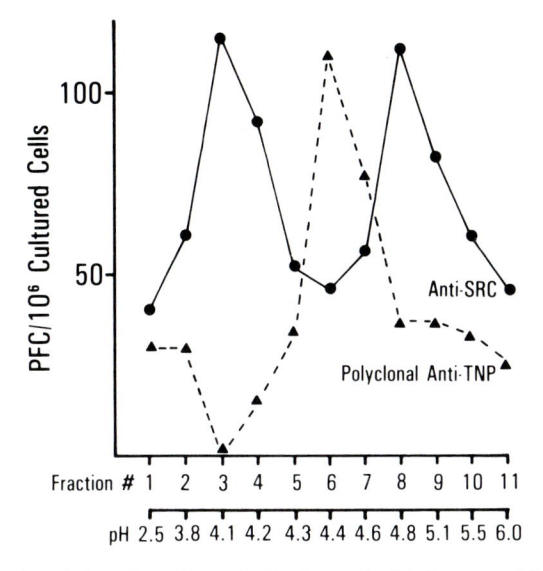

Fig. 1. Preparative isoelectric focusing of interleukin-2. Interleukin-2 prepared by Con A stimulation of murine spleen cells was purified by Sephadex G-100 column chromatography. Fractions containing IL-2 activity were pooled and subjected to preparative isoelectric focusing. The IEF bed was divided into 20 fractions, each fraction eluted with 5 ml sterile saline, dialyzed to remove the ampholytes, and tested for the ability to replace T cell function in primary anti-SRBC responses and Fc-induced polyclonal antibody responses. Only the first ten fractions displayed activity in either of these assays.

TABLE II. Stimulation of TRF Production by Fc Fragments

Stimulating agent[a]	Anti-TNP PFC/10^6 ± SE
–	25 ± 7
50 µg/ml Fc fragments	153 ± 20
100 µg/ml Fc fragments	218 ± 3
150 µg/ml Fc fragments	83 ± 10
2 µg/ml Concanavalin A	160 ± 8

[a]5×10^4 spleen cells were cultured for 24 hr with the stimulating agents. The culture supernatants were then tested for TRF activity in the Fc-induced polyclonal response in T-depleted cultures. Ten µl of culture supernatant/culture well was tested in the standard polyclonal assay containing 50 µg/culture Fc fragments.

duced by Concanavalin A stimulation. The cellular requirements for the production of (Fc) TRF are summarized in Table III. Both macrophages and T lymphocytes are essential for (Fc) TRF synthesis. T-deficient nude spleen cells are incapable of factor generation. Use of anti-Lyt antisera and C′ further defines the essential T cell population. Anti-Lyt 1 and C′ completely abolish the competency of spleen cells to produce (Fc) TRF. Anti-Lyt 2 and C′ treatment has no effect on the subsequent ability to respond to Fc fragments in generation if this material. These results suggest that the Lyt $1^+ 23^-$ T cell subset is

TABLE III. Characteristics of Cells Required for (Fc) TRF Production*

Cell source	Direct anti-TNP PFC/10^6 ± SE
Normal spleen	150 ± 4
Nude spleen	< 10
Sephadex G-10 filtered spleen	14 ± 4
αLy 1.2 + C′ treated spleen	< 10
αLy 2.2 + C′ treated spleen	116 ± 2
α Ia + C′ treated spleen	< 10

*Spleen cells were cultured 24 hr with 50 μg/ml Fc fragments. Culture supernatants (10 μl) were tested for the ability to replace T cell function in Fc (50 μg/well)-induced polyclonal antibody responses.

TABLE IV. Generation of TRF by Fc Subfragment in Macrophage-Depleted Cultures

Cells	Fc[a]	Subfragment[b]	PFC/10^6 [c]
Normal	+	−	146 ± 27
Normal	−	+	144 ± 9
αIa + C′ treated	+	−	< 10
αIa + C′ treated	−	+	154 ± 25

[a] 50 μg Fc/ml in generating cultures.
[b] 2.5 μg Fc subfragment/well in generating cultures.
[c] Direct anti−TNP PFC/10^6 cultured cells + standard error. Assay culture contained 6 × 10^5 cells. 50 μg/well Fc fragments and 10 μl (Fc) TRF containing supernatant.

responsible for factor synthesis. Macrophage depletion by Sephadex G-10 filtration or anti-Ia and C′ prevents factor production. This macrophage requirement can be circumvented by use of macrophage-processed Fc subfragments. Fc subfragments are the biologically active moiety and stimulate B cell proliferation and polyclonal antibody formation in the absence of macrophages [4, 5]. Use of the subfragment allows generation of (Fc) TRF from anti-Ia and C′ treated spleen cells (Table IV). It therefore appears that Fc fragments, following macrophage processing, interact with Lyt 1^+23^- T lymphocytes stimulating the release of (Fc) TRF.

Adjuvant Effect

Fc fragments have adjuvant properties when added with antigen to spleen cell cultures. The observed adjuvant effect is dependent on the dose of antigen employed. An eight-fold enhancement of the secondary IgM anti-SRBC response occurred when Fc fragments were administered with low numbers of SRBC. When the maximum response was obtained, there was no detectable adjuvant effect by Fc (Table V).

The adjuvanticity of Fc appears to be exerted through the T cell population. To determine the role of T cells in the Fc fragment-induced enhancement of the humoral immune response, advantage was taken of the ability of IL-2 to substitute for T cells in an anti-SRBC response [10, 17]. Increasing amounts of IL-2, when added with a constant

TABLE V. Effect of Fc Fragments on Secondary Anti-SRBC Responses* at Several Antigen Concentrations

Culture additions		
Fc[a]	SRBC[b]	PFC/10^6 ± SE
−	10^3	30 ± 2
+	10^3	488 ± 3
−	10^4	57 ± 5
+	10^4	613 ± 45
−	10^5	683 ± 8
+	10^5	790 ± 21

*C57BL/6J were primed and boosted as described in Materials and Methods. Seven days following the boost, spleens were removed and cultured 4 days at 6 × 10^5 cells/well.
[a]100 µg/culture well.
[b]Number of erythrocytes/culture well.

TABLE VI. Inability of Fc Fragments to Enhance Anti-SRBC Responses* When Interleukin-2 Replaces T Cells

Culture additions[a]		IgM Anti-SRBC
IL-2	Fc	PFC/10^6 ± SE
1	−	54 ± 8
1	+	63 ± 3
5	−	128 ± 2
5	+	143 ± 2
10	−	197 ± 23
10	+	155 ± 10
−	−	25 ± 5
−	+	55 ± 10

*Spleen cells from mice primed and boosted as described in Materials and Methods were treated with anti-thymocyte sera and C'. These T-depleted cells were cultured for 4 days with 5 × 10^4 SRBC/well and the other culture additions as described.
[a]Concentrations of IL-2/well in microliters. Fc fragments at 100 µg/well.

low number of SRBC, promote an increasing anti-SRBC PFC response. Adding a constant concentration of Fc fragments does not enhance the response over that obtained with IL-2 alone (Table VI). The enhanced sheep response is not due to the polyclonal response induced by Fc, as the polyclonal response when assayed on day 4 is less than 50 PFC/10^6 cultured cells.

If Fc fragments potentiate humoral immune responses via the release of T cell factors, the Fc-induced TRF might enhance an anti-SRC response in T depleted cultures. As shown in Table VII, such an adjuvant effect is observed when (Fc) TRF is added with

TABLE VII. Enhancement of Anti-SRBC Response* by (Fc) TRF

Culture additions		IgM Anti-SRBC
IL-2[a]	(Fc) TRF[b]	PFC/10⁶ ± SE
5	–	74 ± 1
5	+	129 ± 11
10	–	154 ± 2
10	+	244 ± 6
25	–	145 ± 7
25	+	295 ± 8
–	+	7 ± 1

*Mice were primed and boosted as described in Materials and Methods. Seven days following the boost, spleens were removed and treated with rabbit anti-thymocyte sera and C'. T-depleted cells were cultured 4 days at 6×10^5 cells/well with 5×10^4 SRBC and various culture additives.
[a]Microliters of purified IL-2 added to each culture well.
[b]20 μl culture supernatant containing (Fc) TRF added to each well.

IL-2 to T-deficient cultures. Although the enhancement is not as great as achieved with Fc fragments in intact spleen cultures, the results suggest that the potentiating effect may be mediated by soluble material released by T cells stimulated with Fc.

DISCUSSION

In this report, the role of T cells in the B cell responses induced in murine spleen cell cultures by Fc fragments was examined. Two T-dependent experimental systems were utilized, a polyclonal response triggered by Fc, and the potentiation of an anti-sheep cell response by Fc fragments. In both phenomena, Fc appears to stimulate the generation of helper T cell-replacing factors which exert enhancing effects on proliferating B lymphocytes. Macrophages, which have been shown to be required for the polyclonal antibody response [4], are also necessary for Fc-induced TRF production. Their primary function appears to be to enzymatically cleave the Fc fragments producing biologically active peptides. Purified Fc subfragments bypass the macrophage, triggering both B lymphocyte proliferation [5] and the production of (Fc) TRF by Lyt 1^+23^- T cells. (Fc) TRF provides a differentiative signal to the proliferating B cells, resulting in plasma cell development [18]. The function of this material in mediating the adjuvant effect of Fc fragments is less clear.

Although interleukin 2, prepared by Concanavalin A stimulation of murine spleen cells, contains (Fc) TRF activity, IL-2 and (Fc) TRF may be distinct entities, as both activities are not found in the same fractions of a preparative isoelectric focusing gel. In this case, the interleukin 2 activity was present in two peaks at pH 4.1 and 4.8, while (Fc) TRF was in one peak at pH 4.4. Furthermore, preparations of Fc-induced TRF do not have interleukin 2 activity (data not shown). However, this may be due to differential sensitivities of the assays. The polyclonal response may be sensitive to low concentrations of factor, and suppressed at the higher concentrations which might be necessary to be detected in the thymocyte mitogenic assay and specific antibody response used to measure interleukin 2 activity. Resolution of this question awaits further characterization of (Fc) TRF. Although these two factors may be chemically distinct, the cellular requirements for the production of both are similar [19–23], as both Lyt 1^+23^- T cells and Ia$^+$ macrophages are necessary. The role of the macrophage is defined in the case of (Fc) TRF generation. Macrophages process the Fc fragments to biologically active subfragments. Purified subfragments can bypass the macrophage, directly stimulating T cells to factor production.

Both the requirements for macrophages and T cells in the Fc-induced polyclonal response are somewhat unique to this system. Few, if any, other polyclonal B cell activators have been shown to require macrophages. Lipopolysaccharide [24, 25], dextran sulphate [26], and purified protein derivative [27] are all capable of directly activating B cells. The role of the macrophage in the Fc fragment-induced response is to cleave the Fc fragments, presumably through proteolytic digestion, into mitogenic subfragments that then stimulate B lymphocytes [4, 5].

There have been several reports that T cells modulate the degree of the polyclonal response to bacterial lipopolysaccharide [28, 29] and pokeweed mitogen [30]. The absolute T dependency of the Fc-induced response makes it most similar to the system described by Parker et al [31], in which the polyclonal response triggered by anti-mouse Ig required the addition of culture supernatants from Con A-stimulated murine spleen cells. The capacity to replace both the macrophage and T cell functions with relatively purified soluble factors will allow the analysis of the earliest events in B cell activation, the interaction of the triggers with the target cell membrane, and the biochemical events initiated as a result of such interactions.

The ability of Fc fragments to modulate immune responses is shared by aggregated IgG and immune complexes [1], suggesting complexes might operate by the same mechanism and be involved in regulating ongoing in vivo antibody responses. The following model is proposed for this regulation. Antigen reacts with specific antibody resulting in a conformation change in the antibody which reveals an enzymatic cleavage site within the Fc portion and facilitates binding of the complex to macrophages. Macrophage enzymes cleave the antibody, producing active Fc subfragments. The subfragments interact with B and T lymphocytes, enhancing ongoing antigen-driven responses. In vitro, Fc fragment stimulation results in B cell proliferation and the elaboration of helper T cell-replacing factors from the Lyt 1^+23^- subpopulation of T lymphocytes. These soluble factors potentiate specific immune responses to T-dependent antigens by acting in conjunction with T cell help to enhance the number of specific antibody-secreting cells and by completely replacing T cell action in an Fc fragment-induced polyclonal antibody response. The mechanism of action of this material is not fully understood, or whether the same entity is responsible for both the adjuvant effect, and helper T cell-replacing activity in the polyclonal response.

ACKNOWLEDGMENTS

This is publication no. 2128 from the Department of Immunopathology, Scripps Clinic and Research Foundation, La Jolla, California. These studies were supported in part by United States Public Health Service grants AI-07007 and AI/CA15761, American Cancer Society grant IM-421, and Biomedical Research Support Program grant RRO-5514.

E.L.M. is the recipient of United States Public Health Service Postdoctoral Fellowship AI-05813. We wish to thank Nancy Kantor for excellent technical assistance and Janet Kuhns for secretarial expertise.

REFERENCES

1. Weigle WO, Berman MA: In Pernis B, Vogel HJ (eds): "Cells of Immunoglobulin Synthesis." New York: Academic Press, 1979, p 223.
2. Berman MA, Weigle WO: J Exp Med 146:241, 1977.
3. Morgan EL, Weigle WO: J Exp Med 150:256, 1979.
4. Morgan EL, Weigle WO: J Immunol 124:1330, 1980.
5. Morgan EL, Weigle WO: J Exp Med 151:1, 1980.
6. Morgan EL, Weigle WO: J Immunol 125:226, 1980.
7. Morgan EL, Walker SM, Thoman ML, Weigle WO: J Exp Med 152:113, 1980.
8. Porter R: Biochem J 73:119, 1959.
9. Spiegelberg H: In Miescher P, Muller-Eberhard M (eds): "Textbook of Immunopathology." New York: Grune & Stratton, 1976, p 1101.
10. Watson J, Aarden L, Lfekovits I: J Immunol 122:209, 1979.
11. Schalch W, Braun D: In Lefkovits I, Pernis B (eds): "Research Methods in Immunology." New York: Academic Press, 1978, p 60.
12. Ly I, Mishell R: J Immunol Methods 5:239,,1974.
13. Shen F, Boyse E, Canton H: Immunogen 2:591, 1975.
14. Jerne NK, Nordin AA: Science 140:405, 1963.
15. Kettman J, Dutton RW: J Immunol 104:1558, 1970.
16. Mishell RI, Dutton RW: J Exp Med 126:423, 1967.
17. Schimpl A, Wecker E: Transplant Rev 23:176, 1975.
18. Thoman ML, Morgan EL, Weigle WO: J Immunol 125:1630, 1980.
19. Swain S, Dutton R: J Immunol 124:437, 1980.
20. Pickel K, Hammerling U, Hoffmann M: Nature 264:72, 1976.
21. Amerding D, Eshar Z, Katz D: J Immunol 119:1468, 1976.
22. Larsson EL, Coutinho A: Nature 280:239, 1979.
23. Thoman ML, Weigle WO: J Immunol 124:1093, 1980.
24. Andersson J, Sjöberg O, Möller G: Transplant Rev 11:134, 1972.
25. Yoshinaga M, Yoshinaga A, Waksman BA: J Exp Med 136:956, 1972.
26. Coutinho A, Möller G, Richter W: Scand J Immunol 3:321, 1974.
27. Sulzer BM, Nelsson BS: Nature (New Biol) 240:198, 1972.
28. Shinohara N, Kern M: J Immunol 116:1607, 1976.
29. Goodman M, Weigle WO: J Immunol 122:2548, 1979.
30. Keightley R, Cooper M, Lawton A: J Immunol 117:1538, 1976.
31. Parker D, Fothergill J, Wadsworth D: J Immunol 123:931, 1979.

Regulation of the Balb/c-3T3 Cell Cycle-Effects of Growth Factors

C.D. Stiles, W.J. Pledger, R.W. Tucker, R.G. Martin, and C.D. Scher

Harvard Medical School, Boston, Massachusetts 02115 (C.D.S., C.D.S.), Johns Hopkins University Medical School, Baltimore, Maryland 21218 (R.W.T.), University of North Carolina, Chapel Hill, North Carolina 27514 (W.J.P.), and National Institutes of Health, Bethesda, Maryland 20205 (R.G.M.)

The platelet-derived growth factor (PDGF), which is found in serum but not in plasma, has been purified to homogeneity; it stimulates replication at a concentration of 10^{-10} M. Brief treatment with PDGF causes density-inhibited Balb/c-3T3 cells to become competent to synthesize DNA; pituitary fibroblast growth factor (FGF) or precipitates of calcium phosphate also induce competence. Continuous treatment with plasma allows competent, but not incompetent, cells to synthesize DNA. A critical component of plasma is somatomedin, a group of hormones with insulin-like activity; multiplication-stimulating activity (MSA) or insulin replace plasma somatomedin in promoting DNA synthesis.

We have studied the molecular correlates of competence and the role of SV40 gene A products in regulating DNA synthesis. Treatment of quiescent cells with pure PDGF or FGF causes the preferential synthesis of five cytoplasmic proteins (approximate molecular weight 29,000, 35,000, 45,000, 60,000, and 72,000 detected by SDS-PAGE under reducing conditions). Two of these competence-associated proteins (29,000 and 35,000 daltons) are found within 40 min of PDGF addition; they are not induced by plasma, insulin, or epidermal growth factor (EGF). PDGF, FGF, or calcium phosphate induce an ultrastructure change within the centriole of 3T3 cells; this ultrastructural modification of the centriole is detectable by immunofluorescence within 2 h of PDGF treatment. Plasma, EGF, or MSA do not modify the centriole. SV40 induces replicative DNA synthesis in growth-arrested 3T3 cells but does not cause this alteration in centriole structure.

Gene A variants of SV40, including a mutant with temperature-sensitive (ts) T-antigen (ts A209), a deletion in t-antigen (dl 884), and several ts A209 strains containing t-antigen deletions were used to induce DNA synthesis in Balb/c-3T3 cells. Like wild type SV40, all strains induced DNA synthesis equally well under permissive or nonpermissive conditions. Addition of PDGF or plasma had little effect on SV40-induced DNA synthesis. Thus, the viral function that induces replicative DNA synthesis in Balb/c-3T3 cells is not t and is not temperature sensitive. This SV40 gene function overrides the cellular requirement for hormonal growth factors. It does not induce transient centriole deciliation, a hormonally regulated event.

Key words: PDGF, somatomedin, SV40, cell cycle

Received March 21, 1980; accepted July 25, 1980.

0091-7419/80/1304-0489$02.30 © 1980 Alan R. Liss, Inc.

Circulating blood platelets contain a polypeptide growth factor that stimulates proliferation of cells derived from embryonic mesoderm [1–4]. This platelet-derived frowth factor (PDGF) is sequestered within the α-granules of circulating platelets [5, 6] and is only released into serum when blood clots [2–4]. For this reason, early passage embryo fibroblasts [1], arterial smooth muscle cells [2], normal glial cells [7], and mouse 3T3 cells [8] proliferate much more efficiently in vitro when the cell culture medium is supplemented with clotted blood serum than when the medium is supplemented with platelet-poor plasma. PDGF has recently been purified to homogeneity and characterized as a heat stable, basic (pI 9.8) polypeptide with a native molecular weight of approximately 35,000 daltons by the criterion of SDS-gel electrophoresis under non-reducing conditions [9, 10].

Although PDGF is required for the growth of normal connnective tissue cells, it is not sufficient to promote growth. In 1977, our laboratories discovered that a second set of growth factors, contained in platelet-poor plasma, was required for optimal expression of the mitogenic action of PDGF on Balb/c-3T3 cells [11]. Detailed analysis of the mitogenic response revealed that transient exposure to PDGF rendered quiescent, density-arrested Balb/c-3T3 cells competent to replicate their DNA; however, PDGF-treated, competent 3T3 cells did not progress through the G_0/G_1 phase of the cell cycle into the S phase until they were exposed to a second set of growth factors contained in plasma [12, 13]. Our discovery that the early mitogenic response of density-arrested 3T3 cells to serum growth factors comprises two temporally distinct phases, competence and progression, and has provided a useful conceptual framework in which to analyze the mitogenic action of a wide variety of agents including polypeptide growth factors, tumor promoters, and tumor viruses. The goal of this article is to review and integrate some of these recent data on the relationship between growth factors, tumor promoters, and tumor viruses in regulation of the cell cycle.

THE MITOGENIC RESPONSE OF DENSITY ARRESTED BALB/c-3T3 CELLS TO PDGF AND PLASMA (SERUM)

When grown to the confluent monolayer stage in medium supplemented with 10% calf serum, Balb/c-3T3 cell cultures become growth arrested with a G_1 DNA content. When fresh serum containing medium is added to such cell cultures, replicative DNA synthesis in the cultures is reinitiated. The first cells enter the S phase following a minimum lag of 12 h; thereafter, cells enter the S phase asynchronously in a pseudo-first-order fashion [14]. We have shown that the mitogenic response to serum can be duplicated by the simultaneous addition of PDGF and plasma. Density-dependent growth arrest of Balb/c-3T3 cells reflects starvation for the PDGF component of serum; the addition of PDGF alone to cell-depleted medium will induce mitosis, whereas the addition of plasma to depleted medium will not [11]. The mitogenic response can be plotted on linear graph paper as the percent of cells in S vs time (Fig. 1A) or on semilogarithmic paper as the percent of cells remaining in G_1 vs time, after the fashion of Smith and Martin [14] (Fig. 1B). The latter method has some advantages and will be used in the remainder of this discussion.

PHASE 1 OF THE MITOGENIC RESPONSE: COMPETENCE

Transient exposure to the PDGF component of serum renders quiescent Balb/c-3T3 cells competent to replicate their DNA. However, in the absence of other growth factors contained in the platelet poor plasma component of serum, competent 3T3 cells do not progress through a G_0/G_1 towards S phase. Rather, they remain at growth-arrested at least 12 h before progressing. The cellular response to PDGF is not inhibited by a short treatment with cycloheximide or inhibitors of pinocytosis. As shown in Table 1, several other agents can mimic the action of PDGF by inducing competence in 3T3 cells [13]; these include fibroblast growth factor (FGF) [15], $Ca_3(PO_4)_2$, and wounding [13]. Step 1 of the mitogenic response is summarized schematically in Figure 2.

PHASE 2 OF THE MITOGENIC RESPONSE: PROGRESSION

The addition of platelet-poor plasma to competent Balb/c-3T3 cells initiates progression towards S phase. Under optimal conditions, the first cells enter S phase 12 h after the addition of plasma; thereafter, cells enter S in an apparent first-order fashion. The first-order rate of entry into S is controlled by the concentration of plasma; however, the 12 h lag time is independent of the plasma concentration (Fig. 3). The 12 h lag time is also unaffected by the time of plasma addition. Thus, pretreating the 3T3 cells with plasma

TABLE I. Identification of Progression Factors and Competence Factors*

Factor or treatment	Progression activity		Competence activity	
Somatomedin A (partially pure)	Potent:	1 unit = 20 ng	Weak:	some activity at 330 ng
Somatomedin C	Potent:	1 unit = 2 ng	None:	tested to 200 ng
ILA (partially pure)	Potent:	1 unit = 100 ng	Not tested	
MSA	Potent:	1 unit = 2 ng	None:	tested to 200 ng
BRLC medium	Potent:	1 unit = 6 μl	Weak:	some activity at 50 μl
Insulin	Weak:	1 unit = 160 ng	None:	tested to 120 ng
Hydrocortisone	Weak:	1 unit = 724 ng (10^{-5} M)	None:	tested to 724 ng (10^{-5} M)
Growth hormone (50% pure)	None:	was tested to 200 ng	None:	tested to 200 ng
T-3	None:	tested to 132 ng (10^{-6} M)	Not tested	
NGF	None:	tested to 200 ng	None:	tested to 200 ng
EGF	Partial:	see text	None:	tested to 200 ng
FGF	None:	tested to 200ng	Potent:	2 ng is active
PDGF (partially pure)	None:	tested to 50 μg	Potent:	5 ng is active
$Ca_3(PO_4)_2$	None:	tested to 18 mM	Potent:	3 mM is active
Wounding	None		Potent	
Normal human serum	Potent:	1 unit = 2 μl	Potent:	5% serum is active
Normal human platelet poor plasma	Potent:	1 unit = 2 μl	None:	tested to 100%
SV40	Potent:	1 pfu/cell is active	Potent:	1 pfu/cell is active

*Growth factors were tested for progression factor activity according to the protocol outlined by Stiles et al [13]; a unit of progression factor is defined as the quantity of material that must be added to 3% hypophysectomized-rat plasma to give progression activity comparable to that of 3% normal rat plasma. For competence testing, quiescent density-arrested microtiter cultures of Balb/c-3T3 were exposed for 3 h to various quantitites of growth factor; control cultures were exposed for 3 h to 5% normal human serum. The culture medium was then aspirated. Cell monolayers were washed once with saline containing 28 mM β-mercaptoethanol and once with saline only. The treated cultures were then incubated for 36 h with DME containing 5% normal human plasma and ^{3}H-dThd (5 μCi/ml). After 36 h, the cells were fixed and processed for autoradiography. Under these conditons, less than 5% of cells from control cultures were stimulated to synthesize DNA; those factors with competence activity stimulated 60%–95% of the cells to become labeled. Final concentrations of all growth factors can be derived by multiplying the quantities indicated $\times$ 5ml^{-1}.

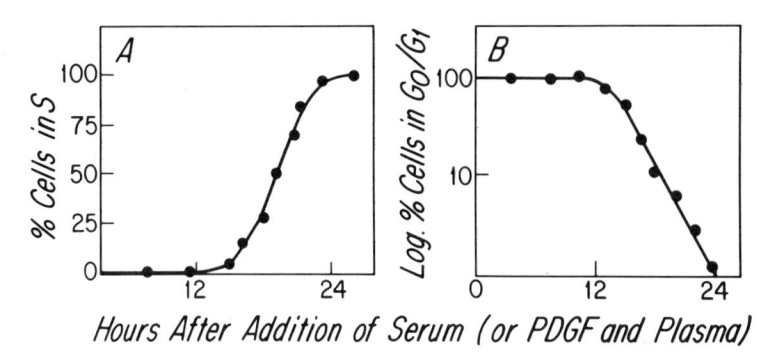

Fig. 1. Schematic illustration of the rate at which density-arrested cultures of Balb/c-3T3 cells enter S phase following the addition of fresh culture medium containing serum (or PDGF and plasma). A) The data are plotted as percent cells in S vs time. B) Same data replotted on a semilogarithmic scale after the fashion of Smith and Martin [14].

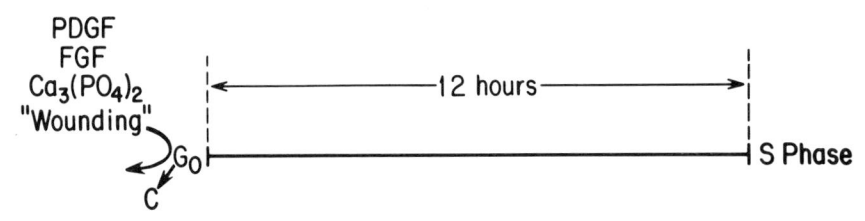

Fig. 2. Schematic illustration of the effects of "competence factors" (PDGF, FGF, $Ca_3(PO_4)_2$, and "wounding") on density-arrested Balb/c-3T3 cells. Transient exposure to any of these agents renders density-arrested Balb/c-3T3 cells in the G_0 phase of the cell cycle competent (C) to replicate their DNA. However, competent cells do not enter S phase until at least 12 h following the addition of platelet-poor plasma to the culture medium. Unlike competence factors that are required only transiently, plasma is required continually for entry into S phase to occur.

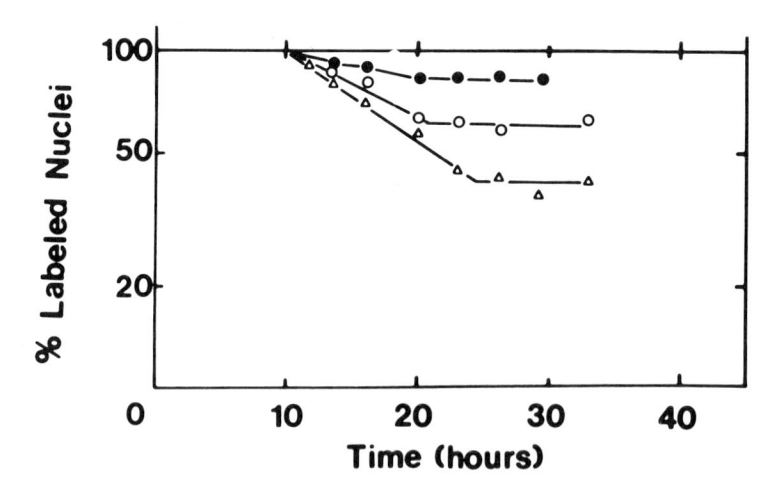

Fig. 3. Effect of concentration of platelet-poor plasma on the rate of cell entry into S phase. Density-arrested Balb/c-3T3 cells were treated with PDGF at 37°C in culture medium supplemented with ^{3}H-thymidine and platelet-poor plasma (●, 0.25%; ○, 2.5%; △, 5%). Cultures were fixed at the times indicated and processed for autoradiography. Adapted from Pledger et al [11].

does not shorten the lag time to S phase. Delaying the addition of plasma to PDGF-treated competent cells delays the onset of DNA synthesis by a corresponding amount of time, so that the minium lag after plasma addition always remains a constant 12 h (Fig. 4). In the presence of a constant concentration of plasma, eg, 5%, the rate that cells enter the S phase and the total number of cells that synthesize DNA is governed by the concentration of PDGF [Pledger et al; in preparation]. In the presence of cycloheximide, progression is totally blocked; cells remain competent, but growth is arrested 12 h prior to S [16].

SOMATOMEDINS ARE REQUIRED FOR PROGRESSION

PDGF-treated, competent Balb/c-3T3 cells were placed in medium containing plasma from surgically hypophysectomized animal donors. Relative to plasma from control animals,

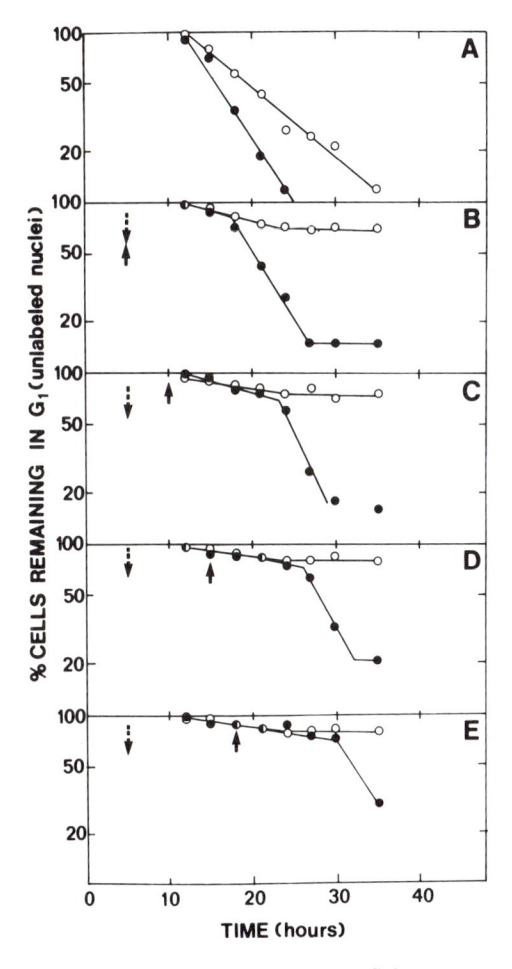

Fig. 4. The stability of the PDGF-induced competent state. A) Cultures were treated with PDGF at 37°C in medium containing 5% platelet-poor plasma and ³H-thymidine. At the indicated times, cultures were fixed and processed for autoradiography. B–E) Cultures were treated with PDGF for 5 h (↓) at 37°C, washed, and returned to medium containing radioactive thymidine but lacking plasma. At the times indicated by (↑), the medium was suplemented with platelet-poor plasma (●, 5%; ○, 0.25%). The cultures were fixed and processed for autoradiography at time intervals. Adapted from Pledger et al [11].

the hypophysectomized animal plasma was deficient in progression activity; only a small fraction of the PDGF-treated cells entered S phase in the medium containing hypophysectomized animal plasma (Fig. 5a). The addition of ng per ml quantities of pure somatomedin C to the hypophysectomized animal plasma totally restored the "progression activity" (Fig. 5b). In a temporal analysis of the cellular response to somatomedin C, density-arrested Balb/c-3T3 cells were exposed briefly to PDGF and then transferred to medium containing plasma from hypophysectomized animals (Fig. 5c) [17]. As noted previously (Fig. 5a), only a small fraction of PDGF-treated cells entered S phase when incubated subsequently in the hypophysectomized plasma medium; however, when pure somatomedin C was added to the PDGF-treated culture that had been incubated in hypophysectomized plasma for 21 h, the cells resumed entry into S phase. The lag time between the addition of somatomedin C and the cellular response was not 12 h; rather, a small but signficiant in-

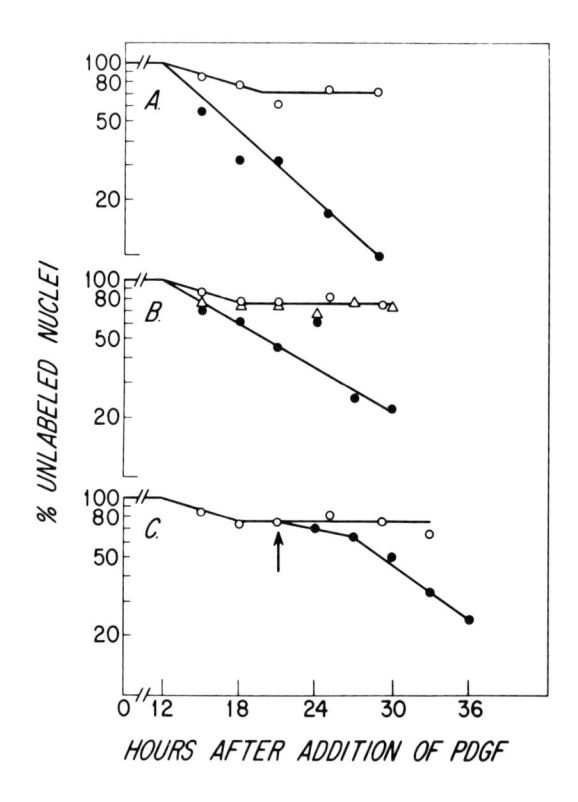

Fig. 5. The addition of pure somatomedin C to plasma from hypophysectomized animals restores "progression" activity. Quiescent density-arrested cultures of Balb/c-3T3 cells were exposed to PDGF for 3 h. The PDGF was removed, and the cells were washed and returned to medium containing ^{3}H-thymidine. A) The medium was supplemented with 3% plasma from either normal (●) or hypophysectomized (○) rats. B) Culture medium was supplemented with 3% plasma from hypophysectomized animals (○), 3% plasma from hypophysectomized animals plus 3 ng/ml pure somatomedin C (●), or 3 ng/ml pure somatomedin C only (△). C) Culture medium was supplemented with 3% plasma from hypophysectomized rats. At the time indicated (↑), somatomedin C (30 ng/ml) was added to some cultures (●), while others (○) received an equivalent volume of saline. At periodic intervals, the cells were fixed and processed for autoradiography. Adapted from Stiles et al [17].

crease in the rate of cellular entry into S phase was noted immediately after the addition of somatomedin C. Then, at 6 h, a major increase in the rate of entry into S phase was noted (Fig. 5c). We interpreted this result in the following way: The "early progress" of PDGF-treated competent cells towards S phase is mediated by an agent contained in platelet-poor plasma. The agent which allows early progress to occur is not under pituitary control since it is unaffected by surgical hypophysectomy. "Late progress" of PDGF-treated competent cells is regulated by somatomedins that, in turn, are regulated by pituitary growth hormone. In medium containing plasma from hypophysectomized animals, most PDGF-treated Balb/c-3T3 cells become growth arrested at a point "V" located 6 h prior to the G_1/S phase boundary; a few of the cells traverse "V" to become growth-arrested at a second point "W" located immediately prior to the G_1/S boundary [12]. The addition of somatomedin to PDGF-treated cells that have been cultured with plasma from a hypophysectomized donor thus causes a biphasic increase in the rate of cellular entry into S: A small fraction of the cell population responds immediately, and the majority responds within 6 h. This interpretation is summarized schematically in Figure 6.

We noted that pure somatomedin C and pure multiplication stimulating activity (MSA) were equipotent in promoting late progress. Impure preparations of somatomedin A and (at high concentration) insulin were also active. Hydrocortisone was also active but only at pharmacologic concentrations. These hormones with progression activity had no competence activity. Conversely, the agents with competence activity (see above) had no progression activity. These results are summarized in Table I.

It should be noted that somatomedins alone will not support the efficient progression of PDGF-treated cells from quiescence into S phase; as indicated in Fig. 5B, another set of components contained in the plasma from hypophysectomized rats is also required. Preliminary studies from our laboratories indicate that the plasma from these rats can be replaced by epidermal growth factor (EGF).

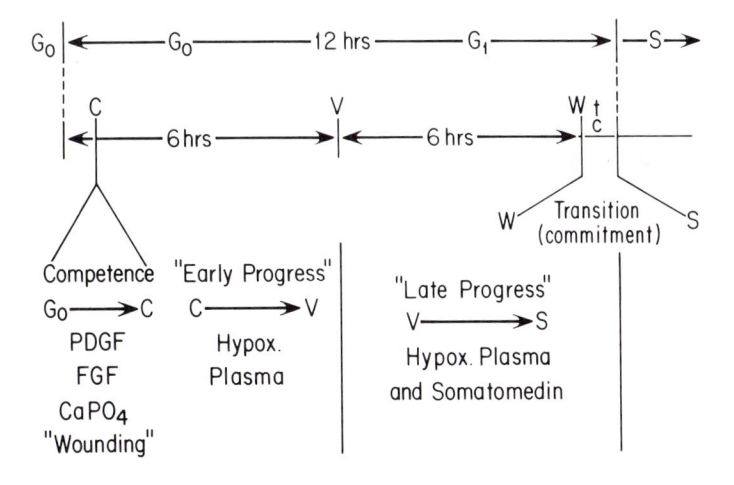

Fig. 6. Hormonal control of early events in the mitogenic response of Balb/c-3T3 cells to serum growth factors. Adapted from Stiles et al [17].

SOMATOMEDIN REGULATORY EVENT IN G_0/G_1 IS RELATED TO NUTRIENTS

In a metabolic analysis of competence and progression, we exposed 3T3 cells to PDGF and then incubated the cells in medium that contained an optimal concentration of plasma but was deficient in essential amino acids [18]. Under these conditions, very few cells entered the S phase. When the missing amino acids were restored to the culture medium, the rate of entry into S increased in a biphasic fashion identical to that noted in our somatomedin experiment. A few of the cell responded immediately to amino acid addition, and after a 6 h lag, a major increase in the rate of cellular entry into S phase was noted. This observation suggests that previous studies on growth-arrest mediated by nutrient starvation were relevant to the phenomena of late progress. In our experiments, amino acid deficiency did not affect the induction of competence by PDGF.

PREPARATION FOR A SECOND ROUND OF CELL DIVISION: PDGF CAN STIMULATE CELLS IN S PHASE

The addition of fresh serum (or of PDGF and plasma) to density-arrested populations of Balb/c-3T3 cells causes them to leave the G_0 state of the cell cycle and replicate their DNA following a minimum lage time of 12 h. Medium that contains serum (or PDGF and plasma) will also sustain the exponential growth of subconfluent cultures of Balb/c-3T3 cells; in exponentially growing 3T3 cell populations, however, the mean lag time between mitosis and the onset of DNA synthesis is 5–6 h. Evidently, serum contains a factor that can prevent cells from becoming growth-arrested when conditions are suitable for sustained exponential growth. We have recently shown that this factor is, in fact, PDGF [19]. In experiments with synchronized cell cultures, we demonstrated that when cells were treated briefly with PDGF during the S, G_2, or very early postmitotic fraction of the cell cycle, a subsequent round of DNA synthesis was initiated following a minimum lag of only a few hours rather than 12 h. Thus, PDGF is a hormone with a dual function. PDGF can stimulate quiescent 3T3 cells to enter the proliferative phase of the cell cycle; PDGF can also stimulate 3T3 cells in the S, G_2, or early postmitotic phase to prevent them from becoming quiescent.

BIOCHEMICAL AND ULTRASTRUCTURAL CORRELATES OF COMPETENCE AND PROGRESSION

Treatment of quiescent Balb/c-3T3 cells with pure PDGF induced the preferential synthesis of five cytoplasmic proteins (approximate molecular weight 29,000, 35,000, 45,000, 60,000, and 72,000 daltons detected by SDS-PAGE under reducing conditions) [Pledger et al, in preparation].Two of these PDGF-induced proteins (29,000 and 35,000 daltons are found within 40 min of PDGF addition; they are not induced by plasma, insulin, or EGF. The PDGF-induced proteins appear to be associated in some way with the cellular state of competence. Fibroblast growth factor from bovine pituitary, which mimics the action of PDGF in the induction of competence, triggers preferential synthesis of the same group of cytoplasmic proteins; by contrast, plasma, insulin, or epidermal growth factor do not affect these cytoplasmic proteins and do not induce competence.

An interesting ultrastructural correlate of competence also exists. As proliferating cells replicate their DNA and proceed through mitosis, the centriole must also duplicate so that chromosomes can be distributed to the daughter cells. Tucker et al [20] demonstrated that density-arrested Balb/c-3T3 cells contain one centriole pair which forms a primary cilium; upon serum stimulation, such cells undergo an early (1–2 h) transient deciliation followed by another deciliation associated with centriole duplication and DNA synthesis. The initial centriole deciliation, observed very rapidly after serum stimulation, is controlled by the PDGF component of serum [21]. In dose-response studies, only doses of PDGF that produced centriole deciliation were capable of inducing competence for DNA synthesis. Plasma alone or somatomedins produced neither centriole deciliation nor competence; however, these agents were required for the optimum progression of competent cells into DNA synthesis and were required for the second centriole deciliation, which was associated with DNA synthesis.

ROLE OF SV40 IN CONTROL OF COMPETENCE AND PROGRESSION

Our cell cycle analysis demonstrated that the mitogenic response of Balb/c-3T3 cells can be resolved into two stages, competence and progression, under the control of separate growth factors. These studies provided a useful new context in which to study the role of tumor viruses and chemical tumor promoters in control of cell growth. SV40 was analyzed in this context. In contrast to growth factors, which provided either competence or progression activity (but never both), SV40 infection provided both competence and progression activity [13]. As shown in Figure 7, induction of host cell DNA synthesis in SV40-infected Balb/c-3T3 cells was strictly a function of virus multiplicty of infection (MOI); neither PDGF nor plasma enhanced the mitogenic response to SV40.

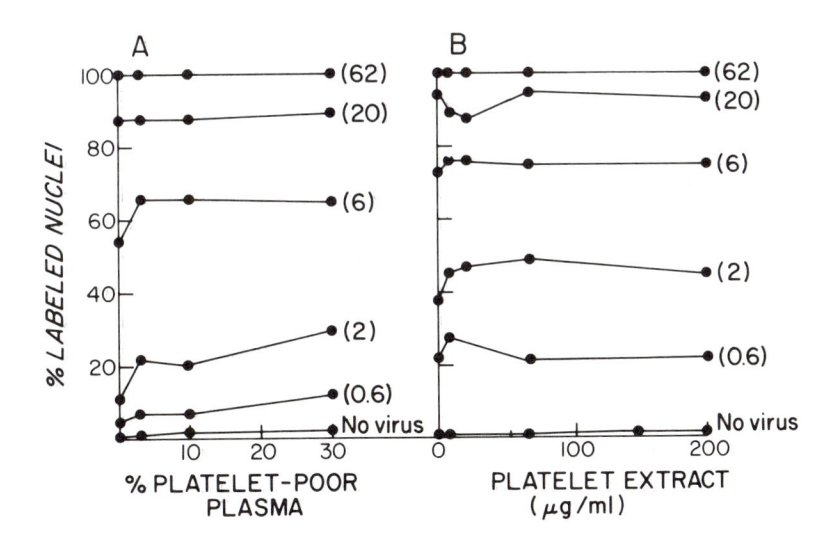

Fig. 7. SV40 induced both competence and progression. A) Gradient-purified SV40 was added to density-arrested Balb/c-3T3 cells at the MOI indicated in parenthesis. After 3 h, the medium was removed, and the cells were washed. Fresh medium containing ^{3}H-thymidine and normal human plasma as indicated (●) was added to the cultures. B) Density-arrested cultures of Balb/c-3T3 cells were treated for 3 h with PDGF as indicated (●). After 3 h, the medium was removed and the cells were washed. Fresh medium containing ^{3}H-thymidine and 0.25% normal human plasma was added together with SV40 at the MOI in parenthesis. After 36 h, all cultures were fixed and processed for autoradiography. Adapted from Stiles et al [13].

The mitogenic response to SV40 differed from the mitogenic response to growth factors in another very signficant way. A transient deciliation of the centriole is associated with the competence stage of the mitogenic response; this transient centriole deciliation was never observed when host cell DNA synthesis was induced by SV40 infection, although the time course of the mitogenic response to serum factors and SV40 was otherwise identical [21].

Our experiments with SV40 suggested that this small DNA tumor virus contained sufficient genetic information to regulate both the competence and progression phases of the mitogenic response. This observation was of special interest since the gene A region of SV40 codes for two transformation specific proteins: The small "t" and large "T" antigens (see [22] for review).

We have begun to use gene A variants of SV40 to explore the separate roles of the "t" and "T"-antigens in control of the mitogenic response. Quiescent Balb/c-3T3 cells were infected with SV40 strains containing a temperature sensitive (ts) T-antigen (ts A209), a deletion in t-antigen (dl 884), and several ts A209 strains containing t-antigen deletions. Surprisingly, all strains of SV40 induced DNA synthesis equally well under permissive or nonpermissive conditions. Treatment with ultraviolet light inactivated the virus preparations ability to induce DNA synthesis. The synthesis of T-antigen by these cells has not been studied. Addition of PDGF or plasma had little effect on DNA synthesis induced by any of the viral strains. Thus, the SV40 viral function(s) that induces replicative DNA synthesis in quiescent Balb/c-3T3 cells does not appear to be little t-antigen and appears to be unaffected by ts lesions in the large T-antigen.

ROLE OF TUMOR PROMOTER IN CONTROL OF HOST CELL DNA SYNTHESIS

The active ingredient of croton oil, 12-0-tetradecanoyl-phorbal-13-acetate (TPA) has been shown to stimulate the proliferation of Balb/c-3T3 cells in culture. TPA is also a widely used model compound in studies on tumor promotion; for this reason, we analyzed the mitogenic action of TPA in the context of competence and progression [23]. Using Balb/c-3T3 cells, we found that TPA functions in a fashion that differs from that of growth factors and also differs from SV40. Growth factors modulate either the competence or the progression events but never both. SV40 provides both competence and progression activity. TPA functions as a competence factor since it enhances the growth stimulatory activity of progression factors. TPA also functions as a progression factor since it enhances the growth stimulatory activity of competence factors. TPA alone, however, does not stimulate cell replication thus distinguishing this compound from SV40.

CONCLUSIONS

Regulation of Cell Cycle Events by Serum Growth Factors

Clotted platelet-rich serum stimulates the growth of normal connective tissue cells. Serum contains PDGF, a polypeptide factor not found in unclotted blood but carried in the α-granules of platelets. Thus, in vivo, PDGF is in a cryptic state and is not available to stimulate the replication of cells.

PDGF initiates the replication of density-arrested Balb/c-3T3 cells by making the cells competent to respond to other growth factors contained in platelet-poor plasma. Several other agents mimic the action of PDGF in inducing competence; these include FGF from bovine pituitary and microprecipitates of calcium-phosphate.

Platelet-poor plasma contains several components required for the growth of PDGF-stimulated Balb/c-3T3 cells; these components include, but are not restricted to, somatomedins; EGF may also play a role. The somatomedins are a family of insulin-like polypeptide hormones whose concentration in blood is controlled to a large extent by a pituitary growth hormone (see [24] for review).

Temporal analysis of the mitogenic response of Balb/c-3T3 cells to serum growth factors indicates that PDGF and somatomedins regulate separate and sequential events in the cell cycle. These data provide a useful context in which to study the molecular action of serum growth factors. Moreover, by examining the cellular growth response to tumor viruses and chemical tumor promoters in the context of competence and progression, it may be possible to establish functional analogies between these agents and specific polypeptide growth factors found in serum.

REFERENCES

1. Balk, SD: Natl Acad Sci USA 68:271, 1971.
2. Ross R, Glomset B, Kariya B, Harker L: Proc Natl Acad Sci USA 71:1207, 1974.
3. Kohler N, Lipton A: Exp Cell Res 87:297, 1974.
4. Scher CD, Shepard RC, Antoniades HN, Stiles CD: Biochim Biophys Acta 560:217, 1979
5. Kaplan DR, Chao FC, Stiles CD, Antoniades HN, Scher CD: Blood 53:1043, 1979.
6. Kaplan KK, Broekman MJ, Chernoff A, Witte L, Linder B: (Abstract). Clin Res 26:349, 1978.
7. Heldin CH, Wasteson A, Westermarke B: Exp Cell Res 109:429, 1977.
8. Scher CD, Pledger WJ, Martin P, Antoniades HN: J Cell Physiol 97:371, 1978.
9. Antoniades HN, Scher CD, Stiles CD: Proc Natl Acad Sci USA 76:1809, 1979.
10. Heldin C, Westmark B, Watson A: Proc Natl Acad Sci USA 76:3722, 1979.
11. Pledger WJ, Stiles CD, Antoniades H, Scher CD: Proc Natl Acad Sci USA 74:4481, 1977.
12. Pledger WJ, Stiles CD, Antoniades H, Scher CD: Proc Natl Acad Sci USA 75:2839, 1978.
13. Stiles CD, Capone GT, Scher CD, Antoniades HN, Van Wyk JJ, Pledger WJ: Proc Natl Acad Sci USA 76:1279, 1978.
14. Smith JA, Martin L: Proc Natl Acad Sci USA 70:1263, 1973.
15. Gospodarowicz D: J Biol Chem 250:2515, 1975.
16. Frantz CH, Stiles CD, Pledger WJ, Scher CD: J Cell Physiol 105:439, 1980.
17. Stiles CD, Pledger WJ, Van Wyk JJ, Antoniades HN, Scher CD: In Sato GH and Ross R (eds): "Hormones and Cell Culture." Cold Spring Harbor Conferences on Cell Proliferation 6:425, 1979.
18. Stiles CD, Isberg R, Pledger WJ, Antoniades HN, Scher CD: J Cell Physiol 99:395, 1979.
19. Scher CD, Stone M, Stiles CD: Nature 281:390, 1979.
20. Tucker RW, Pardee AB, Fujiwara K: Cell 17:527, 1979.
21. Tucker RW, Scher CD, Stiles CD: Cell 18:1065, 1979.
22. Bouck N, Beales N, Shenk T, Berg P, DiMayorca G: Proc Natl Acad Sci USA 75:2743, 19.
23. Frantz CN, Stiles CD, Scher CD: J Cell Physiol 100:413, 1979.
24. Van Wyk JJ, Underwood LE: In Litwach G (ed): "Biochemical Actions of Hormones." 5:101, 1978.

Inhibition of Cell Proliferation and Protease Activity by Cartilage Factors and Heparin

Victor B. Hatcher, Grace Tsien, Martin S. Oberman, and Peter G. Burk

Department of Biochemistry and Medicine, Albert Einstein College of Medicine, Montefiore Hospital and Medical Center, Bronx, New York 10467

Proliferating rat smooth muscle cells and fibroblasts have membrane-associated protease activity. High concentrations of heparin inhibited membrane-associated protease activity and cell proliferation, while low concentration of heparin promoted smooth muscle cell proliferation. The inhibition of protease activity and proliferation was abolished when heparin was treated with protamine sulfate or when acid treated fetal calf serum was used. Heparin required the presence of an acid labile factor(s) in serum for the inhibition of protease activity and proliferation. Heparin and antithrombin III in the presence of acid-treated fetal calf serum did not inhibit cell proliferation or protease activity. Cartilage factors isolated from bovine nasal cartilage containing trypsin inhibitory activity, but not papain inhibitory activity, inhibited rat smooth muscle and fibroblast proliferation and surface associated protease activity. The cartilage factors did not require acid-labile components in the fetal calf serum for the inhibitory activity. The inhibitory activity due to heparin and cartilage factors was not permanent under our experimental condition. Protein synthesis was not inhibited by heparin or the cartilage factors. In rat smooth muscle cells and fibroblasts, the expression of surface-associated protease activity was related to the proliferative state of the cells. Surface protease activity was only present on proliferating cells. When surface protease activity was inhibited by high concentrations of heparin in the presence of an acid-labile serum component(s) or cartilage factors, cell proliferation was also inhibited.

Key words: smooth muscle cell proliferation, fibroblast proliferation, membrane proteases, protease inhibitors, heparin, cartilage factors

The role of proteolytic enzymes in the cellular events accompanying cell proliferation is not clearly understood. Proteases have been implicated in the triggering of lymphocytes [1–3]. A relationship between cell proliferation and surface-associated protease activity has been reported [4, 5]. Proliferating cells have surface protease activity, while a loss of surface protease activity was observed before the termination of cell proliferation [5]. Conversely, protease inhibitors have been shown to inhibit cell proliferation [6–8]. Schnebli and Burger [6] have reported that a number of protease inhibitors including N-α-tosyl-L-lysyl-chloromethyl ketone (TLCK) inhibit the growth of transformed cells. The inhibition occurs either in G_2 phase of the cycle or during mitosis [8]. Several laboratories have demonstrated the inhibitory action of protease inhibitors on lymphocyte activation [9–15].

Received April 9, 1980; accepted July 21, 1980.

0091-7419/80/1401-0033$04.00 © 1980 Alan R. Liss, Inc.

The effect of heparin and various acid mucopolysaccharides on cell growth have been contradictory. Acid mucopolysaccharides have been reported to promote tumor growth [16]. Yang and Jenkin [17] have reported that heparin, at a concentration of 5 µg/ml promotes the growth of prepuce cells, but did not stimulate the growth of BHK-21, MK-2, or Novikoff rat hepatoma cells. Fisher [18] demonstrated that heparin, at concentrations from 20–500 µg/ml inhibited the mitosis of heart fibroblasts, while Lippman [19] reported inhibition of mouse L cells with heparin at 50 µg/ml. Lippman and Mathews [20] have shown that some heparin preparations inhibit mouse L-M cells in suspension culture. The inhibition was due to the attachment of heparin to the cell surface and occurred in late mitosis of the cell cycle. No correlation was observed between the anticoagulant activity of the heparin preparations and the antiproliferative activity. Commerical heparin preparations have also been reported to suppress intimal smooth muscle cell proliferation in a rat model of arterial endothelial injury. Clowes and Karnovsky [21] observed that following arterial de-endothelization in the Sprague-Dawley rat, the intimal smooth muscle cell proliferation could be inhibited by pig mucosa heparin.

Cartilage-derived factors [22–29] have been shown to inhibit tumor-induced vascular proliferation, tumor growth, in vitro bone resorption, and several proteases. The cartilage factors inhibit the growth of endothelial cells [25, 28] and the collagenolytic activity derived from osteosarcoma cells [25] but do not affect the growth of adult fibroblasts [25]. High levels of inhibitor have been found in poorly vascularized tissues such as blood vessel walls, cornea, and dentin [30, 31].

In this investigation we have studied the interaction of commercial heparin on proliferating cultured rat smooth muscle cells. The demonstration that pig mucosa heparin inhibited intimal smooth muscle proliferation in the Sprague-Dawley rat [21] lead to our investigation on the effect of heparin on the proliferation of cultured rat smooth muscle cells. Surface protease activity was measured in order to determine if the expression of protease activity was related to cell proliferation. The effect of high concentration of heparin and antithrombin III on proliferating rat smooth muscle cells was also evaluated. The interaction of cartilage-derived factors on rat smooth muscle cells and fibroblasts was studied in order to determine if cellular protease activity was altered by cartilage-derived factors. In this paper we present evidence that heparin and cartilage-derived factors — under experimental conditions in which cell proliferation is inhibited — also affect the expression of cellular protease activity.

MATERIALS AND METHODS

Cells and Materials

Rat aortic fibroblasts and smooth muscle cells were obtained from Sprague-Dawley rat abdominal aorta. Rat aortic smooth muscle cells (passages 5–12) were dispersed in trypsin (0.25% w/v)–EDTA (0.5 mM) and grown in Dulbecco's modified Eagles medium containing either 10% fetal calf serum (FCS), 5% FCS, 5% pooled human serum, or 5% pooled rat (Sprague-Dawley) serum according to the method described by Ross [32]. Rat smooth muscle cells were also grown in Dulbecco's modified Eagles medium containing 10% acid treated FCS. The FCS was adjusted to pH 3.2 by the addition of 1 N hydrochloric acid incubated at room temperature for 2 hr and then neutralized with 1 N sodium hydroxide. The acid treatment removes some of the endogenous protease inhibitors [33]. Rat aortic fibroblasts were dispersed in trypsin (0.25% w/v)–EDTA (0.5 mM) and grown in Dulbecco's modified Eagles medium containing either 10% FCS or 10% acid-treated FCS.

^{3}H-acetic anhydride (400 mCi/mmole), L-^{3}H-amino acid mixture (1 mCi/ml), and Riafluor scintillation cocktail were purchased from New England Nuclear (NEN). Dulbecco's modifed Eagles medium and bovine fetal calf serum were obtained from Flow Laboratories, Inc. Bovine nasal cartilage was obtained from Max Insel Cohen, Inc., Livingston, NJ. The ovalbumin, myoglobulin, ribonuclease and antithrombin III were from the Sigma Chemical Co. Papain was purchased from Worthington, while trypsin 12X Nat. Form was from I.C.N. Pharmaceuticals. Bovine serum albumin was obtained from Miles (Pentex). Protamine sulfate (1 mg inactivates 90 U.S.P. units of heparin) was obtained from Lilly. Porcine intestinal mucosa heparin (Panheparin, 10,000 U.S.P. units/ml in isotonic sodium chloride, NDC 0074-6738, 02) was purchased from Abbot Laboratories, Chicago. The Sepharose 4B was obtained from Pharmacia Fine Chemicals. The Sprague-Dawley rats were purchased from the Charles River Breeding Laboratories.

Surface Protease Activity

The surface protease determination was based on the measurement of TCA-soluble radioactive peptides released from ^{3}H-labelled casein [4, 5]. The ^{3}H-acetyl-casein assay is capable of detecting proteolysis produced by 50–100 pg of trypsin [34, 35]. The ^{3}H-acetyl-casein, which was prepared by acetylating casein with ^{3}H-acetic anhydride, had a specific activity of 10,099 dpm per pmol, assuming a molecular mass of 121,800 daltons. In order to measure protease activity bound to the cell surface, the cell cultures grown in Falcon petri dishes were washed 6X with Dulbecco's modified Eagles medium without serum. The medium was removed and 300 μl ^{3}H-labelled casein (60 μg casein, 5.3 × 10^6 dpm), which was previously dialyzed against serum-free medium, was added to the plates together with fresh serum-free medium (600 μl). The plates were then incubated for 180 min at 37°C under 7% CO$_2$ in room air. The reaction was terminated by the addition of unlabelled 3% casein (100 μl) dissolved in 1 M KCl followed by chilled 6% TCA (200 μl). The mixture was allowed to remain overnight in an ice slurry after which the samples were centrifuged at 27,000g for 40 min and, 200 μl of supernatant was added to 10 ml of Riafluor (NEN) and then the radioactivity was determined. Control plates without cells were incubated for 24 hr in Dulbecco's modified Eagles medium containing 10% FCS. The plates were washed 6X with Dulbecco's modified Eagles medium without serum. The medium was removed and 300 μl ^{3}H-labeled casein (60 μg casein, 5.3 × 10^6 dpm) was added to the plates together with serum-free medium (600 μl). The plates were incubated and the reaction was terminated as described for the plates containing cells. The TCA soluble counts measured on control plates without cells were subtracted from the TCA soluble counts measured on plates containing cells. Approximately 2.0% of the total radioactivity in the ^{3}H-acetyl casein was TCA soluble following TCA precipitation on the control plates after incubation for 180 minutes. In the plates containing cells from 2.0%–6.3% of the total ^{3}H-acetyl casein, radioactivity was TCA soluble after incubation for 180 minutes. Following each experiment, the cells were trypsinized and counted utilizing a hemocytometer. The viability of cells was examined by Eosin-y exclusion at the end of the experiment. In some experiments the cells were trypsinized, washed 3X, centrifuged, lysed, and the cell button counted in Riafluor. Less than 0.02% of the total amount of ^{3}H-labelled casein was measured in the cell button. Surface protease is only reported under experimental conditions in which no secreted protease was detected.

Secreted Protease Activity

In order to determine if neutral proteolytic activity was secreted during the assay, replicate plates containing serum-free medium (900 μl) were incubated for 180 min at 37°C under 7% CO_2 in room air. The medium was removed, centrifuged at 21,000g for 5 min, and an aliquot (100 μl) incubated for 120 min at 37°C with ^{3}H-labelled casein (40 μl). The reaction was terminated with unlabelled 3% casein (100 μl) and 6% TCA (200 μl) under the conditions described above. An aliquot (100 μl) was counted for radioactivity. The results indicated that under our experimental conditions, no secreted protease activity was detected in the cultured cells. The results do not exclude the possibility that proteases are being secreted, but are being inactivated very rapidly.

Interaction of Heparin With Rat Smooth Muscle Cells

In order to determine the effect of heparin on smooth muscle cell proliferation, heparin (0–333.3 units/ml) was added to the medium containing 5% FCS 24 hr after the cells were subcultured. In experiments in which heparin was inactivated with protamine sulfate, protamine sulfate (3.3 mg/ml) was added to the heparin (33.3 units/ml). The precipitate was removed by centrifugation and the inactivated heparin was added to the medium containing 5% FCS. In experiments in which heparin was added to rat serum or human serum, the cells were subcultured in Dulbecco's modified Eagles medium containing either 5% pooled rat (Sprague-Dawley) serum or 5% pooled human serum. Heparin (333.3 units/ml) was added in either 5% rat serum or 5% human serum 24 hr after subculturing. In one experiment, the heparin was replaced by Dulbecco's modified Eagles medium containing 5% FCS at 48 hours. In the other experiments, no changes in medium or heparin were made during the experiment. At 72 hr and 96 hr, the cells were trypsinized and counted in duplicated plates employing a hemacytometer. Surface protease activity, secreted protease activity, and cell counts were performed 60 min after the heparin was added and at 24-hr intervals.

In one experiment, 72 hr after subculturing, the cells were treated for 30 min with either: 1) 5% FCS; 2) heparin (33.3 units/ml) in serum-free medium; or 3) heparin (33.3 units/ml) in 5% FCS. Surface protease assay was then performed in serum-free medium. Rat smooth muscle cells were also cultured in Dulbecco's modified Eagles medium containing 10% acid-treated FCS. Twenty-four hr after subculturing, either: 1) heparin (333.3 units/ml); 2) antithrombin III (3.0 units/3 ml); 3) antithrombin III (10.0 units/3ml); 4) heparin (333.3 units/ml) + antithrombin III (3.0 units/3 ml); or 5) heparin (333.3 units/ml) + antithrombin III (10.0 units/3 ml) was added to Dulbecco's modified Eagles medium containing 10% acid-treated fetal calf serum (ATFCS). No further changes in medium, ATFCS, antithrombin III, or medium were made during the experiment. Cell counts were performed on duplicate plates at 72 hr, while surface protease activity was performed at 1 hr and 24 hr following subculturing.

Preparation of Cartilage-Derived Factors

The carilage-derived factors were prepared according to the methods described by Keuttner et al [23] and Langer et al [22]. Finely minced bovine nasal cartilage (500 gm) was suspended and extracted in 2500 ml of 1 M guanidine-HCl, pH 6.0, for 24 hr at 4°C. Following removal of the minced tissue by centrifugation, the extracted material was dialyzed exhaustively against distilled water at 4°C. The extract was then centrifuged at 27,000g for 30 min and the supernatant was lyophilized. The guanidine extract (0.5 gm) was dissolved in 40 ml of 0.05 M phosphate buffer, pH 8.1. The cartilage extract was then applied to a trypsin-Sepharose 4B affinity column (2.5 × 8 cm, 50.0 ml/hr) equilibrated with 0.05 M phosphate buffer, pH 8.1. The affinity column was prepared using 0.5 gm of trypsin per 100 ml of packed Sepharose 4B [22]. The column was washed with distilled water and the fraction used in our studies eluted with 0.01 N HCl. Trypsin inhibitory activity was measured on an aliquot of each fraction. The aliquot was preincubated for 30 min at 4°C with trypsin (5 ng) in 0.2 M phosphate buffer, pH 8.1, and ^{3}H-labelled casein was added and proteolysis was measured as previously described [34, 35]. Each fraction with trypsin inhibitory activity was also assayed for papain inhibitory activity utilizing papain (5 ng) and (^{3}H)-labelled casein at pH 6.0 in the presence of cysteine (2 mM). The fraction used in this investigation contained trypsin inhibitory activity, but not papain inhibitory activity.

SDS-polyacrylamide gel electrophoresis on the fraction used in the studies was performed [36] using bovine serum albumin (Mr, 67,000), ovalbumin (Mr, 43,000), myoglobulin (Mr, 16,900), and ribonuclease (Mr, 12,600) as markers. The cartilage fraction used in this investigation contained components of molecular mass of 14,000 daltons, 21,000 daltons, 24,000 daltons, and 32,500 daltons.

Interaction of Cartilage Factors With Rat Smooth Muscle Cells on Rat Fibroblasts

Cartilage-derived factors (5 μg, 30 μg) were added to Dulbecco's modified Eagles medium containing either 5% FCS or 10% ATFCS 24 hr after the rat smooth muscle cells or fibroblasts were subcultured. Surface protease activity, secreted protease activity, and cell number were determined 1 hr after the cartilage factors were added and at 24-hr intervals.

Protein Synthesis

One hr and 24 hr after the heparin (333.3 units/ml) was added to the rat smooth muscle cells, duplicate plates were pulsed for 4.5 hr with L-^{3}H amino acid mixture (1.0 mCi/ml, 10 μl per plate). Rat smooth muscle cells with and without cartilage factor (30 μg/plate) and rat fibroblasts with and without cartilage factors (30 μg/plate) were pulsed for 4.5 hr with L-^{3}H-amino acid mixture (1.0 mCi/ml, 50 μl per plate), 1 hr and 24 hr after the factors were added to the plates. Protein was precipitated with TCA, filtered, washed, and the radioactivity determined by scintillation counting with 10 ml Aquasol (NEN).

RESULTS

The effect of different concentrations of heparin on rat smooth muscle cell (RSMC) proliferation is shown in Table I. In 5% FCS, the cell number increases from an average of 8.2×10^4 cells at 0 hr to 2.2×10^5 cells at 72 hours. At low concentrations, heparin promotes the growth of rat smooth muscle cells. In the presence of low concentrations of heparin (0.3 units/ml, 3.3 units/ml) the cell number increases from an average of 8.2×10^4 cells to 3.7×10^5 cells and 2.9×10^5 cells at 72 hr, respectively. Higher concentrations of heparin (33.3 units/ml, 333.3 units/ml) inhibit the growth of RSMC. Greater inhibition was observed at 72 hr compared to 96 hours. The inhibition was almost completely abolished when the heparin was treated with protamine sulfate. Inhibition was decreased when the plates containing heparin (333.3 units/ml) were changed to 5% FCS at 48 hours. In one experiment (not shown in Table I), in which the RSMC with and

TABLE I. Effect of Heparin on Rat Smooth Muscle Cell Proliferation

	% of Control cell number		
Treatment	1 hr	72 hr	96 hr
Control (5% FCS)	100	100	100
Heparin (0.3 units/ml)	100	168	ND
Heparin (3.3 units/ml)	100	132	ND
Heparin (33.3 units/ml)	100	70	83
Heparin (33.3 units/ml) + protamine sulfate (3.3 mg/ml)	100	94	ND
Heparin (333.3 units/ml)	100	50	70
Heparin (333.3 units/ml) + 5% FCS at 48 hr	100	73	88
Heparin (333.3 units/ml) dissolved in 5% rat serum	100	48	ND
Heparin (333.3 units/ml) dissolved in 5% human serum	100	67	ND

Twenty-four hr after cells were subcultured, heparin was added to the cells. No further changes in medium, serum, or heparin were made, except in the experiment where medium plus 5% FCS was added at 48 hours. Approximately 170 units per mg heparin. Percent of control cell number calculated from the average cell number from two plates counted in duplicate. These data are typical results from a minimum of two independent experiments. ND, not determined.

without heparin (33.3 units/ml) were changed daily, the inhibition of cell growth in the heparin-treated plates was not enhanced at 72 hr or 96 hr compared to the plates in which heparin was added 24 hr after the cells were subcultured. Inhibition of cell growth was also observed when the cells were grown in heparin (333.3 units/ml) in rat serum or human serum.

Surface protease activity on RSMC treated with heparin in serum-free medium and 5% FCS is shown in Table II. Seventy-two hr after the cells were subcultured, either serum-free medium, or heparin in serum-free medium containing 5% FCS was incubated with the cells for 30 minutes. Surface protease activity was then measured for 3 hr in serum-free medium. Surface protease activity was partly inhibited with heparin plus serum-free medium. Complete inhibition of protease activity was observed in the presence of FCS. The effect of heparin on RSMC protease activity is presented in Table III. In the absence of heparin, protease activity was detected at 1 hr, 24 hr, and 48 hours. An increase in the cell number was observed at 24 hr and 48 hours. The presence of protease activity at 48 hr and the increase in cell number at 72 hr (not shown in Table III) was also observed. The presence of protease activity was observed before an increase in cell number was detected. In the presence of heparin, no protease activity was observed at 24 hours. When protease activity was completely inhibited, no increase in cell number was observed. The inhibition of surface protease activity was not observed at 24 hr, and the cells doubled as determined by the increase in the cell number at 48 hours. Under our experimental conditions, the inhibition of protease activity and cell proliferation lasted only for 24 hours. In RSMC, the amount of protease activity did not correlate with the rate of cell proliferation. At 24 hr surface protease activities of $11,309 \pm 2,191$ and $20,514 \pm 3,595$ per 10^5 cells were observed, and in both cases the cells doubled at 48 hours. No significant cell death was observed as examined by Eosin-y exclusion at 24 hr or 48 hours.

TABLE II. Surface-Associated Protease Activity in Heparin-Treated Proliferating Rat Smooth Muscle Cells

Treatment	Surface protease activity (dpm/10^5 cells)
Control (5% FCS)	22,100 ± 1762
Heparin (33.3 units /ml) in serum-free medium	5639 ± 2867
Heparin (33.3 units /ml) in 5% FCS	0

RSMC cells (72 hr after subculturing) were treated for 30 min with: 1) 5% FCS; 2) heparin (33.3 units/ml) in serum-free medium; 3) heparin (33.3 units/ml) in 5% FCS. Surface protease activity was then determined in serum-free medium for 3 hr as described in Materials and Methods. Each value represents mean ± SEM. Approximately 170 units per mg heparin. These data are typical results from a minimum of two independent experiments.

Surface protease activity and cell proliferation of RSMC were also measured in the presence of acid-treated fetal calf serum (ATFCS) (Tables IV, V). In the presence of Dulbecco's modified Eagles medium containing 10% ATFCS, the cells increased from an average of 2.6×10^5 cells at 0 hr to 5.5×10^5 cells at 72 hours. Heparin (333.3 units/ml) in the presence of ATFCS did not inhibit RSMC cell proliferation or protease activity (Tables IV, V). Human antithrombin III (3.0 units/3 ml, 10.0 units/3 ml) did not significantly inhibit cell growth in RSMC (Table IV). Surface-protease activity with antithrombin III (10.0 units/3 ml) was decreased at 1 hr, but not at 24 hr (Table V). No significant inhibition of cell proliferation was observed in the presence of 10% ATFCS, heparin, (333.3 units/ml), and antithrombin III (10.0 units/3 ml) (Table IV). No inhibition of protease activity was observed with heparin (333.3 units/ml) containing antithrombin III (10.0 units/3 ml) (Table V). Thus, ATFCS in the presence of heparin or heparin and antithrombin III did not significantly inhibit cell proliferation or protease activity.

Cartilage factors at a concentration of 5 µg/plate did not inhibit cellular proteases or proliferation. The effect of cartilage factors (30.0 µg/plate) on protease activity on RSMC and rat fibroblasts (RF) is in Table VI. In the RMSC treated with cartilage factors, no surface protease activity was observed at 1 hr and no increase in cell number was observed at 24 hours. RSMC without the cartilage factors contained protease activity at 1 hr and 24 hr, and the cells increased at 24 hr and 48 hours. In the RF without cartilage factors, no surface protease activity was observed at 1 hr, and the cell number only increased from $50 \pm 6 \times 10^4$ cells to $68 \pm 6 \times 10^4$ cells at 24 hours. Protease activity was seen at 24 hr, and a significant increase in cell number was measured at 48 hours. In RF treated with 30 µg of cartilage factors, no protease activity was present at 1 hr or 24 hr, and the cell number did not double at 48 hours. Surface protease activity was present at 48 hours. Thus, cartilage-derived factors also inhibit cellular protease activity as well as inhibiting cell proliferation. No correlation was observed between the amount of cellular protease activity and the rate of cell proliferation. When surface protease was present, the cells proliferated, while very little cell proliferation was observed when the protease activity was inhibited. In order to determine if the inhibition by the cartilage factors occurred in the presence of ATFCS, the cartilage-derived factors were added to RSMC and RF treated with ATFCS. The results are shown in Table VII. Acid treatment

TABLE III. Effect of Heparin on Rat Smooth Muscle Cell Surface Protease Activity and Proliferation

	1 hr		24 hr		48 hr	
	Surface protease activity (dpm/10^5 cells)	Cell number ($\times 10^3$)	Surface protease activity (dpm/10^5 cells)	Cell number ($\times 10^3$)	Surface protease activity (dpm/10^5 cells)	Cell number ($\times 10^3$)
RSMC	11,572 ± 439	150 ± 12	11,309 ± 2,191	290 ± 18	23,670 ± 1,754	620 ± 35
RSMC + heparin (333.3 units/ml)	0	150 ± 12	20,514 ± 3,595	200 ± 24	39,450 ± 2,630	410 ± 29

At 0 time 200,000 RSMC were dispersed. Twenty-four hr later, either 5% fetal calf serum or 5% fetal calf serum and heparin (333.3 units/ml) was added to the cells. No further changes in medium, fetal calf serum, or heparin were made during the experiment. Approximately 170 units per mg heparin. Surface protease activity and cell counts were performed on triplicate plates 1 hr, 24 hr, and 48 hr after the heparin was added. Each value represents mean ± SEM. These data are typical results from a minimum of three independent experiments. RSMC, rat smooth muscle cells.

TABLE IV. Effect of Heparin and Antithrombin III on Rat Smooth Muscle Cell Proliferation

Treatment	% of Control cell number	
	1 hr	72 hr
Control (10% ATFCS)	100	100
+ heparin (333.3 units/ml)	100	96
+ antithrombin III (3.0 units/ml)	100	90
+ antithrombin III (10.0 units/3 ml)	100	98
Antithrombin III (3.0 units/3 ml) + heparin (333.3 units/ml)	100	86
Antithrombin III (10.0 units/3 ml) + heparin (333.3 units/ml)	100	93

Twenty-four hr after RSMC were subcultured in medium containing 10% ATFCS, heparin and/or antithrombin III were added to the cells. No further changes in medium, ATFCS, heparin, or antithrombin III were made. Percent calculated from average cell number from two plates counted in duplicate. Approximately 170 units per mg heparin. These results are typical from a minimum of two independent experiments. ATFCS, acid-treated fetal calf serum.

TABLE V. Surface Protease Activity in Heparin- and Antithrombin III-Treated Rat Smooth Muscle Cells

Treatment	Surface protease activity ($dpm/10^5$ cells)	
	1 hr	24 hr
Control (10% ATFCS)	9016 ± 1949	4626 ± 842
+ heparin (333.3 units/ml)	9707 ± 3340	7369 ± 3472
+ antithrombin III (10.0 units/3ml)	3461 ± 1055	16,905 ± 6441
Antithrombin III (10.0 units/3 ml) + heparin (333.3 units/ml)	9121 ± 2270	13,576 ± 3842

Twenty-four hr after RSMC were subcultured in medium containing 10% ATFCS, heparin and/or antithrombin III were added to the cells. No further changes in medium, ATFCS, heparin, or antithrombin III were made. Surface protease activity was determined 1 hr and 24 hr after heparin and/or antithrombin III was added. Approximately 170 units per mg heparin. These results are typical from a minimum of two independent experiments.

of the FCS did not affect the ability of the cartilage factors to inhibit cell proliferation. The cartilage factors did not require an acid-labile component of FCS to inhibit cellular protease and cell proliferation in RSMC and RF. At the end of the experiments, the cells were viable as examined by Eosin-y exclusion. Protein synthesis in RSMC and RF treated with heparin and cartilage factors are shown in Table VIII. The RSMC and the RSMC + heparin were pulsed for 4.5 hr with 10.0 μl of L-^{3}H amino acid mixture (1.0 mCi/ml), while the RSMC, RSMC and cartilage factors, RF, and RF and cartilage-derived factors were pulsed for 4.5 hr with 50 μl of L-^{3}H amino acid mixture (1.0 mCi/ml). No differences were observed in protein synthesis 1 hr or 24 hr after the heparin was added to the RSMC. Protein synthesis was also similar 1 hr and 24 hr after the cartilage factors were added to

TABLE VI. Effect of Cartilage Factors on Surface Protease Activity and Cell Proliferation

	1 hr		24 hr		48 hr	
	Surface protease activity (dpm/10^5 cells)	Cell number (× 10^3)	Surface protease activity (dpm/10^5 cells)	Cell number (× 10^3)	Surface protease activity (dpm/10^5 cells)	Cell number (× 10^3)
RSMC	12,098 ± 2630	340 ± 17	5260 ± 1578	420 ± 17	ND	770 ± 29
RSMC + cartilage factors (30.0 µg)	0	340 ± 12	2472 ± 350	350 ± 13	ND	540 ± 17
RF	0	50 ± 6	36,820 ± 5699	68 ± 6	25,511 ± 1052	106 ± 6
RF + cartilage factors (30.0 µg)	0	50 ± 6	0	51 ± 6	44,973 ± 3069	70 ± 6

Twenty-four hr after cells were subcultured, either 5% FCS or 5% FCS and cartilage factors was added. No further changes were made. Surface protease activity and cell counts were performed on triplicate plates 1 hr, 24 hr, and 48 hr after the cartilage factors were added. Each value represents mean ± SEM of triplicate determinations on three separate plates. ND, not determined; RSMC, rat smooth muscle cells; RF, rat fibroblasts.

TABLE VII. Inhibition of Cell Proliferation by Cartilage Factors in Acid-Treated Fetal Calf Serum

Treatment	% of Control cell number	
	1 hr	72 hr
RF + 10% ATFCS	100	100
RF + 10% ATFCS + cartilage factors (30.0 μg)	100	50
RSMC + 10% ATFCS	100	100
RSMC + 10% ATFCS + cartilage factors (30.0 μg)	100	53

Twenty-four hr after the cells were subcultured, either 10% ATFCS or 10% ATFCS and cartilage factors were added to the cells. No further changes were made. Percent of control cell number was calculated from the average cell number from two plates counted in duplicate. The results are typical of a minimum of two independent experiments.

TABLE VIII. Protein Synthesis in Cells Treated With Heparin and Cartilage-Derived Factors

Treatment	dpm per 10^5 cells	
	1 hr	24 hr
RSMC + 5% FCS	5394	10,975
RSMC + 5% FCS + heparin (333.3 units/ml)	5497	9994
RSMC + 5% FCS	89,157	51,811
RSMC + 5 FCS + cartilage factors (30.0 μg)	84,949	66,300
RF + 5% FCS	81,530	49,970
RF + 5% FCS + cartilage factors (30.0 μg)	71,799	55,756

The RSMC and RSMC + heparin were pulsed with 10 μl of L-^{3}H-amino acid mixture for 4.5 hr, 1 hr, and 24 hr after the heparin was added. The RSMC, RSMC + cartilage factors, RF, and RF + cartilage factors were pulsed with 50 μl of L-^{3}H-amino acid mixture for 4.5 hr, 1 hr, and 24 hr after the cartilage factors were added. The results are the average of counts on duplicate plates. These data are typical results from a minimum of two independent experiments.

RSMC or RF. Thus, the presence of heparin and cartilage-derived factors for 1 hr and 24 hr did not affect protein synthesis under our experimental conditions.

DISCUSSION

The presence of membrane-associated protease(s) has been confirmed by several investigators [37–41]. Quigley [42] has shown that plasminogen activator was associated with a specific membrane fraction in Rous virus transformed chick embryo fibroblasts. Fulton and Hart [43] have characterized a plasma membrane-associated plasminogen activator on thymocytes. Our studies have shown that proteases other than plasminogen activator are associated with the cell surface of many cell types [3–5]. Under our experimental condition, ^{3}H-acetyl casein is a poor substrate for plasminogen activator from human and rat fibroblasts and RSMC (unpublished results).

The function of membrane-associated proteases in the cell is unknown at the present time. In this study, we have observed that the surface protease is present in proliferating RSMC and RF. Protease activity was followed by an increase in the cell number. The amount of protease activity was not directly related to the rate of cell proliferation. In order to determine if the inhibition of RSMC proliferation due to pig mucosa heparin was similar to the inhibition of intimal smooth muscle cells in the Sprague-Dawley rat, the effect of heparin on cultured Sprague-Dawley smooth muscle cells was evaluated. The addition of heparin at high concentrations to RSMC resulted in the complete inhibition of cell proliferation and surface associated protease activity. The inhibition of cellular protease activity and proliferation was not permanent. Twenty-four hr after the RSMC were incubated with heparin, cellular protease activity reappeared and cell growth resumed. Replacing the heparin each day did not prolong the inhibition of cell proliferation. The inhibition of cell proliferation required an acid-labile component in the fetal calf serum since inhibition was not observed with ATFCS. Acid treatment of FCS is known to inactivate protease inhibitors [33]. Heparin as an anticoagulant interacts with antithrombin III [44, 45]. In order to determine if the interaction of heparin with antithrombin III produced the inhibition of cell proliferation and protease activity, antithrombin III was added to heparin in the presence of RSMC and ATFCS. Our results indicated that heparin and antithrombin III in the presence of ATFCS did not inhibit cell proliferation or cellular protease activity. Either heparin and antithrombin III plus another heat-labile component of FCS was responsible for the inhibition, or heparin and a completely different acid-labile component of FCS were responsible for the inhibition. Heparin and antithrombin III alone did not inhibit cell proliferation or protease activity. The experiments with the ATFCS demonstrated that high concentrations of a glycoconjugate such as heparin were not sufficient to inhibit cell proliferation. Although the commercial heparin which contained multiple components was not further purified, the inhibitory effect of heparin could be removed with protamine sulfate. The heparin was not cytotoxic to the RSMC using methods previously described [35]. Protein synthesis of RSMC was not affected by the presence of heparin in the cultures. The effect of low and high concentrations of heparin also demonstrated that heparin at low concentrations promote cell proliferation, while high concentrations inhibit cell proliferation.

The molecular events involved in the heparin inhibition of RSMC proliferation are not understood. Cell-surface heparan sulfate has been reported to be released from cultured Chinese hamster cells when the cells were incubated with heparin [47]. Heparin has also been shown to interfere with the binding of hormones to specific receptors [48]. Recently, Baker et al [46] have reported that low concentrations of heparin (0.2 μg/ml) prevent the binding of protease nexin, a protease inhibitor complex, to diploid human foreskin fibroblasts. In our experiments, 0.2 μg/ml of heparin promotes RSMC proliferation. At the present time, it is not clear if there is a relationship between the protease-nexin complex to cells and the ability of cells to proliferate. At low concentrations of heparin, a decrease in protease-nexin complexes on the cell may be related to an increase in cell proliferation.

In this investigation, the effect of cartilage factors isolated from bovine nasal septum on RSMC and RF were also evaluated in order to determine if cellular proteases were also inhibited in the presence of cartilage-derived factors. The cartilage factors used in the investigation contained a minimum of 5 components on SDS polyacrylamide gels. Langer et al [28] have shown that the cartilage fraction which contains 4 proteins (Mr, 14,000–28,000) inhibits the vascularization and growth of V_2 carcinoma in the rabbit cornea. Roughley et al [26] have demonstrated that bovine nasal cartilage contains fractions with inhibitory activity against trypsin (Mr, 7,000), papain (Mr, 13,000), and collagenase (Mr, 22,000). The authors suggest that the trypsin inhibitor is Trasylol. Rifkin and Crowe [27] have purified to homogeneity a major trypsin inhibitor from bovine cartilage. The inhibitor was identical to Trasylol, a commercial preparation of Kunitz inhibitor found in bovine tissues and plasma. At the present time, it is not clear if the inhibitory effect of the cartilage factors is due to a single component or many components. It is not known if the effect on cellular protease activity is due to the same component(s) as the effect on cell proliferation. We observed that cartilage factors inhibit cell proliferation and cellular protease activity on RSMC and RF. The effect on RF contradicts the report of Eisenstein et al [25] that the cartilage factors do not affect adult fibroblast proliferation. In comparing our study with the Eisenstein et al [25] report, we utilized higher concentrations of affinity chromatography-purified cartilage components. We did not observe any inhibition of RSMC or RF proliferation or protease activity when 5 μg/plate of cartilage factors were used. The inhibition was not affected by the use of ATFCS, which was not the case when high concentrations of heparin were interacted with ATFCS and RSMC. The cartilage factors do not require an acid labile component in fetal calf serum to inhibit cell proliferation and protease activity. The inhibition was only observed for 24 hr, but the cartilage factors were not added each day. Higher concentrations of factors over several days may inhibit proliferation and protease activity for longer periods. Protein synthesis was not inhibited by cartilage factors.

The functional role of proteases associated with the membrane has not been determined in the present study. Linsley et al [49] have reported that epidermal growth factor (EGF) receptor-complex binding undergoes proteolytic modifications in murine 3T3 cells and human foreskin fibroblasts. Two distinctic proteases, one of unknown specificity and one similar to trypsin, seem to be involved in the cleavages of EGF-receptor complexes [50]. It is not clear if the proteases involved in EGF-receptor complex modification are associated with the plasma membrane. Recently, Bach et al [5] suggested that plasma membrane proteases control the binding of IgE to specific receptors in rat peritoneal mast cells. The present study suggests that protease(s) associated with the plasma membrane are expressed when cells are proliferating and that heparin and cartilage factors, under conditions in which cell proliferation is inhibited, inhibit the cellular protease activity.

ACKNOWLEDGMENTS

The assistance of L. Leibowitz in the preparation of this manuscript is acknowledged. Mel Tiell kindly provided the pooled rat serum. We thank Drs. D. Hamerman, P. Barland, T. Spaet, and A. Grayzel for help and discussion. This research was supported by grants from the U.S. Public Health Service (AG 01732 and HL 16387), New York Heart Association, The Cystic Fibrosis Foundation, and The Arthritis Foundation. Victor B. Hatcher is a Senior Investigator of The New York Heart Association.

REFERENCES

1. Hirschhorn R, Grossman J, Troll W, Weissmann G: J Clin Invest 50:1206, 1971.
2. Kast RE: Oncology 29:249, 1974.
3. Grayzel AI, Hatcher VB, Lazarus G: Cell Immunol 18:151, 1975.
4. Hatcher VB, Wertheim CY, Rhee G, Tsien G, Burk PG: Biochim Biophys Acta 451:499, 1976.
5. Hatcher VB, Oberman MS, Wertheim MS, Rhee CY, Tsien G, Burk PG: Biochem Biophys Res Commun 76:602, 1977.
6. Schnebli NP, Burger MM: Proc Natl Acad Sci USA 69:3825, 1972.
7. Goetz IE, Weinstein C, Roberts E: Cancer Res 32:2469, 1972.
8. Schnebli HP, Haemmerli G: Nature 248:150, 1974.
9. Saito M, Yoshizawa T, Aoyagi T, Nagai Y: Biochim Biophys Res Commun 52:569, 1973.
10. Tchorzewski H, Denys A: Experientia 28:462, 1972.
11. Weissmann G, Zurier RB, Hoffstein S: Am J Pathol 68:539, 1972.
12. Davies R, Krakauer K, Weissman G: Anal Biochem 45:428, 1972.
13. Kaplan JG, Bona C: Exp Cell Res 88:388, 1974.
14. Prokopenko LG, Drobyazg LD: Bull Exp Biol Med 79:558, 1975.
15. Higuchi S, Ohkawara S, Nakamura S, Yoshinaga M: Cell Immunol 34:395, 1977.
16. Ozzello L, Lasfargeus EY, Murray MR: Cancer Res 159:88, 1978.
17. Yang TK, Jenkin HM: Proc Soc Exp Biol Med 159:88, 1978.
18. Fisher A: Protoplasma 26:344, 1936.
19. Lippman M: In Fleischmayer R, Billingham RE (eds): "Epithelial-Mesenchymal Interactions." Baltimore: Williams and Wilkins Co, 1968, p 208.
20. Lippman MM, Mathews MB: Fed Proc 36:55, 1977.
21. Clowes AW, Karnovsky M: Nature 265:625, 1977.
22. Langer R, Brem H, Falterman K, Klein M, Folkman J: Science 193:70, 1976.
23. Kuettner KE, Soble L, Croxen RL, Marczynska B, Hiti J, Harper E: Science 196:653, 1977.
24. Keuttner KE, Hiti J, Eisenstein R, Harper E: Biochem Biophys Res Commun 72:40, 1976.
25. Eisenstein R, Keuttner KE, Neopolitan BS, Soble LW, Sorgente N: Am J Pathol 81:337, 1975.
26. Roughley PJ, Murphy G, Barrett AJ: Biochem J 169:721, 1977.
27. Rifkin DB, Crowe RM: Hoppe-Seyler's Z Physiol Chem 358:1525, 1977.
28. Langer R, Conn H, Vacante J, Klagsbrun M, Folkman J: Fed Proc 37:1338, 1978.
29. Horton JE, Wezeman FH, Keuttner KE: Science 199:1342, 1978.
30. Keuttner KE, Croxen RL, Eisenstein R, Sorgente N: Experientia 30:595, 1974.
31. Sharkey M, Veis A, Keuttner K: J Dent Res 52:134, 1973.
32. Ross RJ: Cell Biol 50:172, 1971.
33. Unkeless JC, Gordon S, Reich E: J Exp Med 139:834, 1974.
34. Kaiser H, Hatcher VB: Conn Tiss Res 5:147, 1977.
35. Hatcher VB, Oberman M, Lazarus GS, Grayzel AI: J Immunol 120:665, 1978.
36. Weber K, Osborn M: J Biol Chem 244:4406, 1969.
37. Tokes ZA, Chambers SM: Biochim Biophys Acta 389:325, 1975.
38. Tokes ZA: J Supramol Struct 4:507, 1976.
39. Tokes ZA, Sorgente N: Biochem Biophys Res Commun 73:965, 1976.
40. Evans I, Bosmann HB: Exp Cell Res 108:151, 1977.
41. Spataro AC, Morgan HR, Bosmann HB: J Cell Sci 21:407, 1976.
42. Quigley JP: J Cell Biol 71:472, 1976.
43. Fulton RJ, Hart DA: J Supramol Struct (Suppl) 4:188, 1980.
44. Damus PS, Hicks MS, Rosenberg RD: Nature 246:355, 1973.
45. Rosenberg RD: Fed Proc 36:10, 1977.
46. Baker JB, Low DA, Simmer RL, Cunningham D: J Supramol Struct (Suppl) 4:170, 1980.
47. Kraemer PM: Biochem Biophys Res Commun 78:1334, 1977.
48. Salomon Y, Amir Y, Azulai R, Amsterdam A: Biochim Biophys Acta 544:262, 1978.
49. Linsley PS, Blifeld C, Wrann M, Fox CF: Nature 278:745, 1979.
50. Linsley PS, Fox CF: J Supramol Struct (Suppl) 4:170, 1980.
51. Bach MK, Bach S, Brashler JR, Ishizaka T, Ishizaka K: Fed Proc 35:808, 1980.

Effects of a Serum Spreading Factor on Growth and Morphology of Cells in Serum-Free Medium

David Barnes, Richard Wolfe, Ginette Serrero, Don McClure, and Gordon Sato

Department of Biological Sciences, University of Pittsburgh, Pittsburgh, Pennsylvania 15260 (D.B.); Roche Institute of Molecular Biology, Nutley, New Jersey 07110 (R.W.); and Department of Biology, University of California at San Diego, La Jolla, California 92093 (G.S., D.M., G.S.)

A heat-sensitive, trypsin-sensitive factor that promoted growth and spreading of cells in serum-free, hormone-supplemented medium was partially purified from human serum. The major portion of the proteins in these preparations migrated upon SDS-polyacrylamide gel electrophoresis with a mobility consistent with molecular weights between 60,000 and 90,000. The spreading activity, which we have termed serum spreading factor, stimulated growth and spreading of a wide variety of cell types. The serum spreading factor was similar to fibronectin in that it showed an affinity for the plastic cell culture substrate but was shown to be distinct from fibronectin by several criteria. This factor may prove useful in studies of cell attachment and spreading and in studies of the relationship of cell shape and cell proliferation.

Key words: serum spreading factor, cell proliferation, cell morphology, cell substratum, serum-free medium

Historically, supplementation of culture medium with serum has been required for maintenance and growth of animal cells in culture. Our laboratory and others have shown recently that for many cell types it is possible to replace the usual serum supplement in culture medium with specific combinations of nutrients, hormones, and purified serum proteins [1–3]. Among the serum proteins stimulatory for growth of some cell types in serum-free medium are factors that mediate the proper attachment and spreading of cells on the plastic or glass culture substrate. In particular, cold-insoluble globulin (CIg), the plasma form of fibronectin, has been useful in this regard [4–6]. In the process of developing serum-free media for various cell lines, we have become involved in the biochemical and biological characterization of another factor present in human serum that appears to be distinct from CIg by several criteria. We have termed this activity serum spreading factor.

Received May 28, 1980; accepted July 17, 1980.

0091-7419/80/1401-0047$05.00 © 1980 Alan R. Liss, Inc.

This factor was first reported to exist in human serum by Holmes in 1967 [7]. Two of us (Barnes and Sato) have previously reported that the serum spreading factor is active on the MCF-7 line of human breast cancer cells [8]. Subsequently, we found that this factor influenced growth and spreading of a wide variety of cell types in serum-free medium. These include cell lines derived from rat glioma (C6), mouse neuroblastoma (N18TG-2), mouse embryonal carcinoma (F9, 1003), rat ovary (RF-1), and mouse embryo (3T3, SV40-transformed 3T3), as well as a mouse preadipocyte line (1246) and human fetal lung fibroblasts (WI38) [9, 10]. In this paper, we report the preliminary biochemical characterization of the partially purified serum spreading factor and describe some of the biological effects of these preparations on the C6, 1246, and SV40-transformed 3T3 cell lines in serum-free medium.

MATERIALS AND METHODS

Materials

Bovine insulin, human transferrin, 4-(2-hydroxyethyl)-1-piperazineethene-sulfonic acid (Hepes), crude pancreatic trypsin, soybean trypsin inhibitor, ovalbumin, bovine serum albumin (BSA), linoleic acid, sodium chloride, calcium chloride, potassium and sodium carbonate, and sodium bicarbonate were obtained from Sigma Chemical Company. Fatty-acid-free BSA was obtained from Miles. Antibiotics and powdered formulations of Ham's F12 and Dulbecco-modified Eagle's medium (DME) were obtained from Grand Island Biological Company. Bovine fibroblast growth factor (FGF), multiplication stimulating activity (MSA), human CIg, and rabbit antiserum to human CIg were obtained from Collaborative Research, Inc. Plastic cell culture labware was obtained from Falcon. Glass beads (Number 1014, Class IV-A) were obtained from Ferro, Cataphote Division, Jackson, Miss. Materials for SDS-polyacrylamide gel electrophoresis (SDS-PAGE) were obtained from Biorad. Plates for double immunodiffusion were obtained from Hyland. Fetal calf serum was obtained from Reheis. Freunds complete adjuvant was obtained from Difco.

Preparation and Characterization of Serum Spreading Factor

Partially purified serum spreading factor was prepared by a modification of the procedure of Holmes [7]. Outdated human plasma was dialysed overnight against 0.8% sodium chloride, and clotted by the addition of 1 mg/ml calcium chloride, and the clot removed by low speed centrifugation (700g, 30 min). The resulting serum was adjusted to pH 8.0 with 1 N sodium hydroxide, and 80 ml was put on a 50 cm by 2.5 cm column previously packed with acid-washed glass beads and equilibrated with 0.6 M sodium bicarbonate (pH 8.0). Bed volume of the column was approximately 250 cm^3. The flow rate of the column was approximately 2 ml/min. The chromatography was carried out at room temperature. The column was eluted sequentially with 250 ml of 0.6 M sodium bicarbonate (pH 8.0), 125 ml 0.6 M sodium bicarbonate-0.2 M sodium carbonate (pH 9.5), 125 ml 0.15 M potassium bicarbonate-0.05 M potassium carbonate (pH 9.5), and 300 ml of 0.6 M potassium bicarbonate-0.2 M potassium carbonate (pH 9.7). Ten milliliter fractions were collected. Spreading activity in fractions eluted from the column was routinely identified by microscopic examination 60–90 min after the addition of 20–50 μl of the fractions to C6 cells seeded in 35 mm diameter plates in 2 ml of serum-free medium containing 2.5 mg/ml BSA. A quantitative spreading assay is described below.

Peak fractions eluted with 0.6 M potassium bicarbonate-0.2 M potassium carbonate were pooled and concentrated with Amicon CF 25 Centriflo membrane cones. The potassium carbonate buffer was exchanged for 10 mM potassium hydroxide by repeated concentration and dilution utilizing the Centriflo membrane cones, and the preparation was sterilized by filtration. All tubes used for collection and storage of the serum spreading factor preparations were polypropylene. Protein assay procedure was the method of Bradford [11]. SDS-PAGE was carried out as described by Weber and Osborn [12]. Polyacrylamide concentration was 10%.

Antiserum to the serum spreading factor preparations was raised in rabbits by injections of the antigen in Freunds complete adjuvant on three occasions at 2 week intervals. At each injection, 400 µg of the serum spreading factor preparation was injected subcutaneously at multiple sites. Antiserum was collected 2 weeks after the third injection. Double immunodiffusion plates were developed in 48 h at room temperature in a humidified atmosphere.

Experiments designed to determine if the serum spreading factor preparations were inactivated by incubation with trypsin were carried out using trypsin covalently bound to polyacrylamide beads (Sigma T8386). Approximately 100 µg of the spreading factor preparation was incubated overnight at room temperature in 1 ml of phosphate buffered saline (PBS) containing 20 units of insoluble enzyme. The spreading activity in aliquots of this incubation mixture was assayed as described, after removal of the beads by centrifugation. Aliquots of controls, in which the trypsin-bead complex was incubated in PBS without the spreading factor and then removed by centrifugation, had no morphological effect on cell spreading induced by active spreading factor preparations. Incubation of spreading factor with a trypsin-bead complex, in which the trypsin had been previously inactivated, had no effect on the spreading activity.

Experiments designed to determine which of the components appearing on stained SDS-polyacrylamide gels was the active protein were carried out by cutting gels of unreduced, unboiled samples of the serum spreading-factor preparations into 25 slices, homogenizing each slice in a glass dounce with 2 ml of serum-free medium, and pretreating plates with this medium as described under Cell Culture Experiments below. Under these conditions, the active factor attached to the plastic culture dish (see Results), while most of the SDS was removed before cells were plated for the standard spreading assay.

Cell Culture Experiments

The C6 rat glioma cell line (C6BU-1) was obtained from Dr. M. Nirenberg, National Institutes of Health. Other C6 sublines from different sources responded in a manner identical to the response of these cells with regard to the properties reported in this paper. Stock cultures of these cells were maintained in DME containing 1.2 g/l sodium bicarbonate, 25 mM Hepes (pH 7.3), 190 IU/ml penicillin, 0.2 mg/ml streptomycin, 25 µg/ml ampicillin, and 10% fetal calf serum. The synthetic nutrient medium used for experiments with C6 was a mixture of 3 parts DME to 1 part Ham's F-12, containing 1.2 g/l sodium bicarbonate, 25 mM Hepes (pH 7.3), 190 IU/ml penicillin, 0.2 mg/ml streptomycin, 25 µg/ml ampicillin, and a mixture of trace elements as previously described [2]. For experiments studying growth and spreading of C6, exponentially growing cells from stock cultures were removed from plates with a solution of 0.1% (W/V) trypsin in a Ca-Mg-free PBS containing 0.9 mM EDTA. The detached cells were suspended in the serum-free, synthetic nutrient medium described above containing 0.2 mg/ml soybean trypsin inhibitor, and centrifuged; the pellet was suspended in fresh serum-free medium without trypsin inhibitor. Cells from the suspension were inoculated into 35 mm culture dishes (2.5 × 10⁴ cells/plate) containing 2 ml of serum-free medium. Supplements as indicated in the figures and tables (eg, insulin, transferrin, FGF) were added to the plates as small volumes of concentrated sterile stocks.

For the C6 growth assay, the number of cells attached to the culture dish was determined 5 days after plating by detaching the cells with the trypsin-EDTA solution and counting the resultant cell suspensions with a Coulter counter, model Z_f. Greater than 99% of the total number of cells in the dish were found to be attached to the dish and counted by this procedure. For the C6 spreading assay, 100 to 250 cells in 5–8 random fields at each concentration of spreading factor were examined and the number of completely rounded cells determined and expressed as a percentage of the total number of cells counted. In experiments in which plates were treated with serum spreading factor preparations, dishes were incubated for 24 h at 37°C with 1 or 2 ml of medium supplemented with the spreading factor at the indicated concentrations and then thoroughly washed with medium before adding the cells. Plates were incubated with the washes for 2 h at 37°C.

The 1246 line is an adipogenic clone derived from a phenotypically unstable myogenic line, T984 [13], isolated from a mouse teratocarcinoma, and was obtained from Dr. M. Darmon, UCSD. Experiments with the 1246 cells were carried out essentially as described for C6 except that the basic nutrient medium for 1246 was a 1:1 mixture of DME and Ham's F12, and cells were detached from flasks or plates by incubating in Ca-Mg-free PBS containing 0.9 mM EDTA, without the use of trypsin or trypsin inhibitor. SV40-transformed Balb/c-3T3 (SV-3T3) cells were obtained from Dr. R. Holley, Salk Institute. Experiments with SV-3T3 cells were carried out essentially as described for C6.

RESULTS

Serum Spreading Factor

Partially purified serum spreading factor was prepared by passing human serum at pH 8.0 over glass beads and eluting with sodium and potassium carbonate buffers (see Materials and Methods). Fractions were collected in 10 ml aliquots. An elution profile of the glass-bead chromatography of human serum (80 ml) under these conditions is shown in Figure 1. The relative sizes of the OD_{280} peaks eluted with the different buffers varied somewhat with different serum batches. This was particularly true with the material eluted with water (fractions 46–50) and the material eluted with 0.15 M potassium bicarbonate-0.05 M potassium carbonate (fractions 59–60). Most of the serum protein passed through the column unretarded (fractions 1–15). Cell spreading activity was found in fractions 1 through 40 and fractions 68 through 73. The latter fractions represent material eluted with 0.6 M potassium bicarbonate-0.2 M potassium carbonate. Fractions 69–71 were pooled, concentrated, desalted, and sterilized. Approximately 3 mg of protein was recovered in fractions 69–71 from approximately 5 g of protein in the starting material. The spreading promoting activity in the material was lost upon incubation at 100°C for 5 min or upon incubation with trypsin.

The SDS-PAGE pattern of the material in fractions 69–71 is shown in Figure 2. Most of the protein in reduced samples of the spreading-factor preparations migrated on SDS-PAGE in a manner consistent with molecular weights between 60,000 and 90,000.

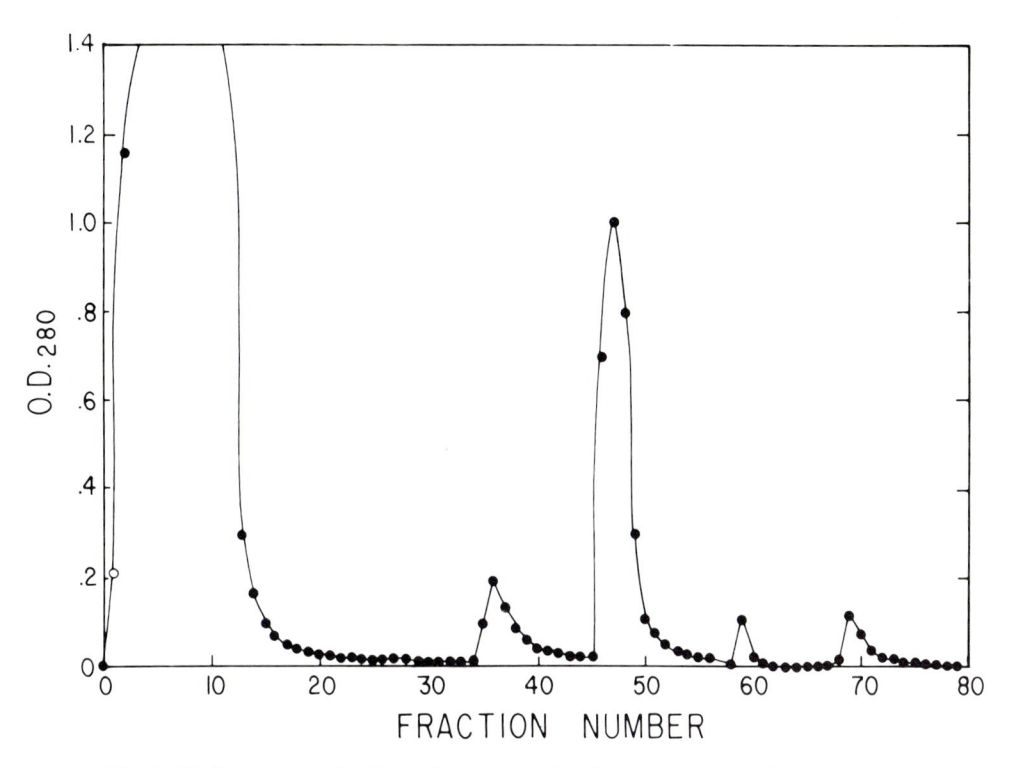

Fig. 1. Elution pattern: glass-bead chromatography of human serum at alkaline pH.

Proteins of higher mobility (15,000–60,000 daltons) were also present. The major band in this area appeared to have a molecular weight of approximately 70,000–80,000, running between BSA (molecular weight 67,000) and human transferrin (molecular weight 86,000). We estimate that this band represents about one third of the total protein in the preparations. No protein migrating in a manner consistent with a molecular weight of 100,000 or higher was detected on the gels. SDS-PAGE of unreduced samples of the serum spreading factor preparations resulted in a pattern similar to that of Figure 2, although the bands were less well resolved.

Fibronectin or CIg is reported to stimulate adhesion or spreading of some types of cells in culture [5, 6, 14–16]. The serum spreading factor described in this paper appeared to be distinct from CIg by several criteria. No 220,000 dalton protein could be detected on SDS-PAGE of spreading factor preparations reduced with mercaptoethanol (Fig. 2). Also, no precipitin band was detected in double immunodiffusion plates in which antiserum to human CIg was tested against the spreading factor preparations (Fig. 3). Likewise, no precipitin band was detected when antiserum raised against the spreading factor preparations was tested against human CIg. Each of these antisera produced precipitin bands against their respective antigens. Finally, active preparations of the spreading factor have been made in our laboratory using as a starting material plasma which had been passed through a gelatin-conjugated sepharose column, which removes CIg [17]. Comparison of theoretical

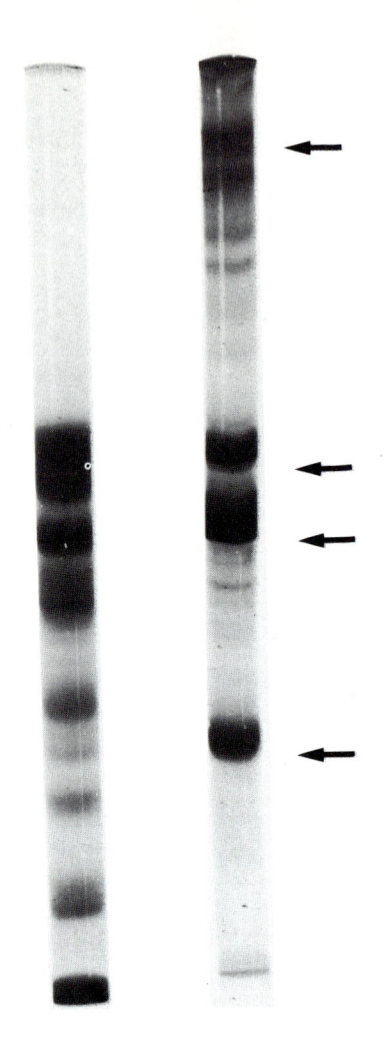

Fig. 2. SDS-polyacrylamide gel electrophoresis of serum spreading factor. Left: 15 µg serum spreading factor, reduced; right: 5 µg molecular weight standards, reduced. Standards, top to bottom (arrows): human CIg; human transferrin; BSA; ovalbumin.

vs actual yield of CIg subsequently eluted from such a column with 8 M urea indicated that all of the CIg had been removed from the plasma which passed through the column under the conditions of the experiment.

Effects of Serum Spreading Factor on C6 Rat Glioma

Figure 4 shows micrographs of C6 cells after 3 days in the presence of serum-free medium containing insulin (25 µg/ml), transferrin (25 µg/ml), FGF (50 ng/ml), and varying concentrations of the partially purified serum spreading factor. Insulin, transferrin, and

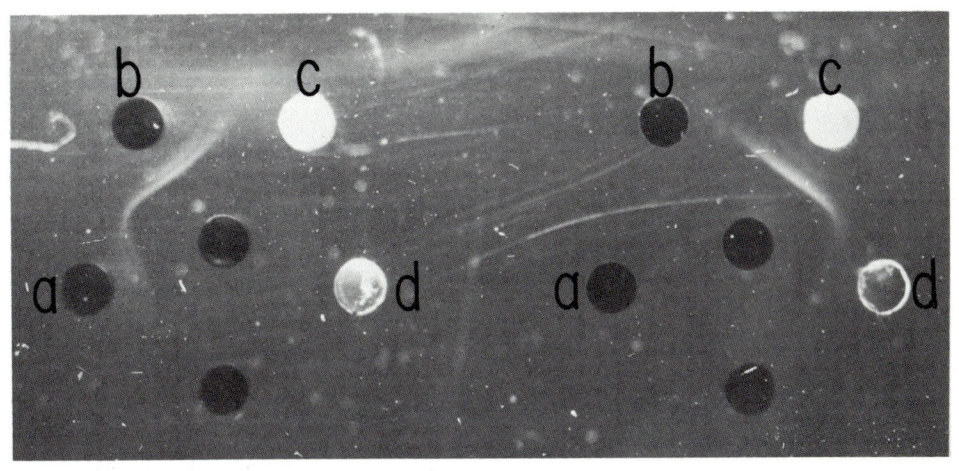

Fig. 3. Double immunodiffusion of serum spreading factor, CIg, and antiserum to each. a: 1 µg of serum spreading factor; b: 2 µg of serum spreading factor; c: 2 µg of CIg; d: 1 µg CIg. Center well, left, 5 µl antiserum to serum spreading factor; center well, right, 5 µl antiserum to CIg.

FGF have been reported previously to be stimulatory for growth of C6 in serum-free medium [18]. Although some effect of the spreading factor on cell morphology could be detected at doses as low as 60 ng/ml, maximum effect was seen at about 4 µg/ml. At this dose, the morphology of the C6 cells was similar to that of cells in medium containing 10% serum. Under the conditions of Figure 4, the spreading factor preparations caused a dose-dependent increase in cell number in addition to influencing cell spreading. Addition of the spreading factor in the absence of insulin, transferrin, and FGF did not cause an increase in cell number.

It is possible to quantitate both the growth-stimulatory effect and the cell-spreading effect. Figure 5 shows the dose-response relationship of spreading-factor preparations for both cell spreading and cell number. Cell spreading was assayed by counting the number of rounded, completely unspread cells in several random fields at each concentration of spreading factor and expressing the number of rounded cells as a percentage of the total number of cells counted. As the concentration of spreading factor was increased from 0 to 5 µg/ml, the number of unspread cells decreased from near 100% to about 5%, and the final cell number increased 3.5-fold. At high concentrations of the spreading factor (10–25 µg/ml), the stimulatory effects were not seen. The data of Figure 5 suggest that there is a reasonably good correlation between the concentration dependency for the spreading activity in the spreading factor preparations and the concentration dependency for the growth promoting activity in these preparations. CIg had no effect on cell number for C6 under these conditions. However, effects of CIg on cell number in the presence of insulin, transferrin, and FGF may be seen if the culture dishes are first coated with polylysine [19].

Serum at concentrations as high as 500 µg of protein/ml added to C6 cultures containing insulin, transferrin, and FGF did not substitute for the spreading factor in regard to cell spreading. This would suggest that the factor responsible for the spreading activity in

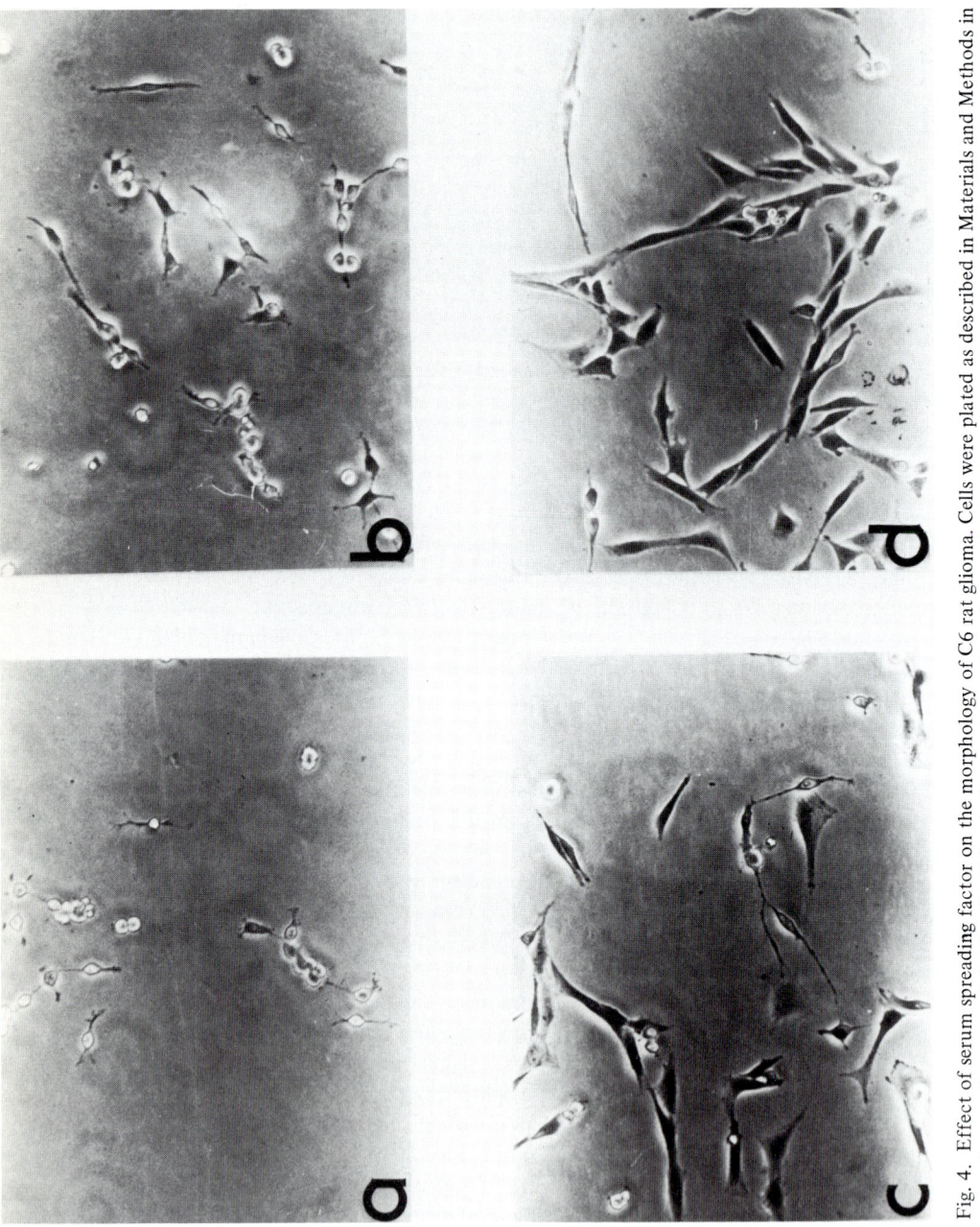

Fig. 4. Effect of serum spreading factor on the morphology of C6 rat glioma. Cells were plated as described in Materials and Methods in serum-free medium containing insulin (25 μg/ml), transferrin (25 μg/ml), and FGF (50 ng/ml) plus no spreading factor (a), or spreading factor at 60 ng/ml (b), 1.3 μg/ml (c), and 3.3 μg/ml (d). Pictures were taken 3 days after plating.

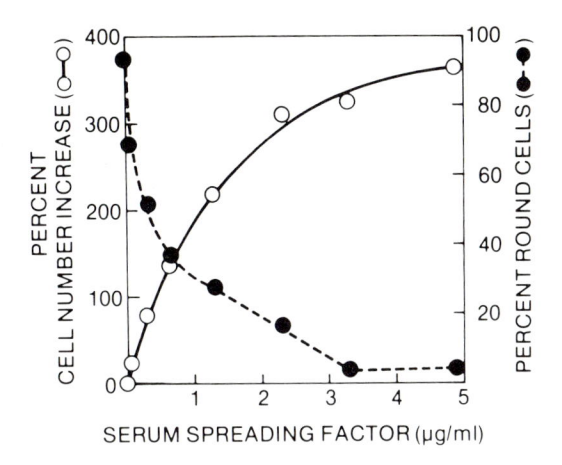

Fig. 5. Effect of serum spreading factor on growth and spreading of C6. Cells were plated as described in Materials and Methods in serum-free medium containing insulin (25 μg/ml), transferrin (25 μg/ml), FGF (50 ng/ml), and the indicated concentrations of the serum spreading factor. Growth and spreading assays were carried out as described in Materials and Methods. Zero percent cell number increase represents cell number in the presence of insulin, transferrin, and FGF in the absence of serum spreading factor. Cell number in this condition was 1.5×10^5 cells/35 mm plate. Plating efficiency under the conditions of this experiment for C6 cells was near 100%.

the preparation was purified 100-fold or more from serum. Such a statement may be misleading, however, since inhibitors of spreading, such as albumin, exist in serum and are separated from the spreading activity by the isolation procedure.

In the experiment described by Figure 5, the spreading factor was added directly to plates containing the serum-free medium in which the cells were incubated. We have found that the factor apparently has an affinity for plastic, and it was possible to pretreat the cell culture dishes with the spreading factor before the cells were added and produce effects identical to those produced if the spreading factor was added directly to the medium. In Figure 6 the increase in cell number due to the spreading factor showed the same concentration dependence whether the spreading factor was added directly to the medium or the dishes were preincubated with medium containing varying concentrations of the spreading factor. The concentration-dependent effect of the spreading factor on cell morphology (as shown in Fig. 4 and Fig. 5) was also maintained on dishes pretreated with the spreading factor. Since some of the inactive proteins in the serum spreading factor preparations probably do not stick to the culture dish, the precoating procedure may be effecting a further purification of the spreading and growth promoting activity.

Effects of Serum Spreading Factor on 1246 Mouse Teratocarcinoma Cells

As with C6, the serum spreading factor caused an increase in cell number for the 1246 cell line in serum-free medium in the presence of insulin, transferrin, and FGF (Table I), and addition of the spreading factor alone did not cause an increase in cell number. Also as with C6, the spreading factor was active on 1246 cells if dishes were pretreated with it rather than adding it to the medium. Unlike C6, the 1246 line showed a similar response to CIg under these conditions, although the response to CIg was somewhat smaller

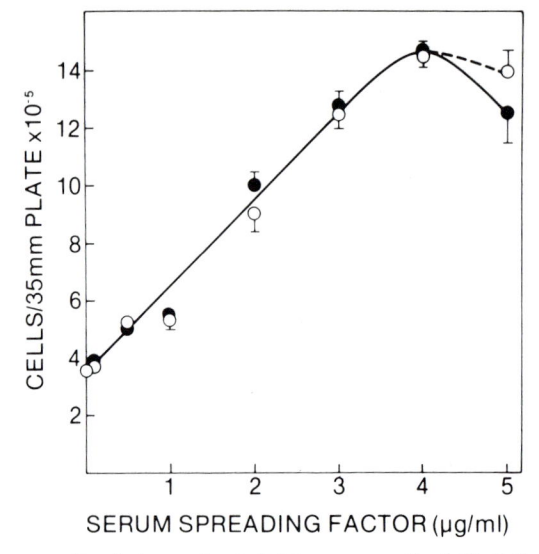

Fig. 6. Effect of serum spreading factor pretreated dishes on growth of C6. Cells were plated, dishes pretreated, and growth assayed as described in Materials and Methods. (●): serum spreading factor added to the medium; (○): serum spreading factor treated dishes.

than the response to the spreading factor. CIg had no effect if added to medium containing insulin, transferrin, FGF, and the spreading factor. The serum spreading factor can also replace the CIg requirement for some other cell lines (eg, RF-1, SV40-transformed 3T3) in serum-free, hormone-supplemented medium [9].

The 1246 cells were capable of attaching and spreading on the culture dish to a greater extent than C6 in the presence of insulin, transferrin, and FGF and the absence of either CIg or the spreading factor. However, addition of the spreading factor did result in a morphological change for these cells from a flattened, epitheloid appearance to a fibroblastic and spindle-shaped morphology, which was more similar to the morphology of these cells in serum-containing medium (Fig. 7). This difference in morphology was maintained in cultures at both low and high cell densities. Under conditions in which cell spreading was inhibited, such as if cells were plated in the presence of 1 mg/ml BSA, spreading of 1246 cells was found to be completely dependent on the presence of the serum spreading factor.

Effects of Serum Spreading Factor on SV40-Transformed 3T3 Cells

A line of SV40-transformed Balb/c-3T3 cells can be grown in serum-free monolayer culture in a medium consisting of a 3:1 mixture of DME and Ham's F12 supplemented with insulin (250 μg/ml) or MSA (100 ng/ml), transferrin (500 ng/ml), fatty-acid-free BSA (1 mg/ml), and linoleic acid (5 μg/ml) [6]. A properly conditioned substratum is also required for the growth of these cells in the absence of serum. This requirement could be met by pretreating plastic tissue culture surfaces with either serum spreading factor or CIg (Fig. 8). As shown, cells plated into either serum spreading factor or CIg-pretreated dishes divided logarithmically with a mean generation time of 20 h and, in each case, reached a final cell density of 2×10^5 cells/cm^2; both of these growth parameters are similar to those

TABLE I. Effect of CIg and Serum Spreading Factor on the Growth of 1246 Cells

Medium supplements[a]		Cells/35 mm plate × 10^{-5}
Insulin + transferrin		0.60
Insulin + transferrin + FGF		1.90
Insulin + transferrin + FGF + CIg,	0.3 μg/ml	2.20
	1.0 μg/ml	2.53
	3.0 μg/ml	3.00
Insulin + transferrin + FGF + serum spreading factor,	0.3 μg/ml	3.10
	1.0 μg/ml	3.70
	3.0 μg/ml	3.80

[a]Concentrations of insulin, transferrin, and FGF were 10 μg/ml, 5 μg/ml, and 100 ng/ml, respectively. Cell number was determined 3 days after plating.

for cells cultured in serum-supplemented medium. There was no significant increase in the number of cells per dish when these cells were seeded directly onto bare plastic surfaces in the absence of either of these proteins.

While the serum spreading factor and CIg acted similarly in that they both were capable of conditioning the tissue culture substratum, making it more amenable for cellular growth, the growth response of SV40 Balb/c-3T3 cells as a function of the concentration used to precoat dishes was quite different for each of these factors (Fig. 9). Under the experimental conditions employed, CIg was maximally active at 25 μg/dish; a half-maximal effect occurred at 3 μg/dish. The highest level of CIg tested (100 μg/dish) was as effective as lower but optimal doses. In contrast, the serum spreading factor was maximally active at 1 μg/dish, while a half-maximal response was seen at 0.25 μg/dish. Increasing the amount of spreading factor in the preincubation medium above 5 μg/dish resulted in a gradual decline in the number of attached cells per dish. At a concentration of 50 μg/dish, the spreading factor preparation was ineffective in supporting cell attachment and growth. Both the spreading factor and CIg, at optimal levels, promoted the attachment and growth of SV-3T3 cells to a similar extent, and a mixture of an optimal amount of each of these factors was no better than when each were added individually.

DISCUSSION

We have described experiments in which a partially purified factor from human serum influenced growth and spreading of several cell types in culture in serum-free medium. Our laboratory has found that the serum spreading factor is also active on a number of other widely divergent cell types in culture [8–10]. Preliminary experiments indicate that the active factor is a protein of molecular weight between 70,000 and 80,000, which composed approximately one third of the total protein in the preparations used in the experiments described in this paper.

It should be emphasized that the glass-bead column procedure described for the preparation of the spreading factor did not remove all of the spreading activity from serum, and we do not wish to suggest that this factor is the only protein in serum capable of influencing cell spreading in vitro. Indeed, CIg, a high molecular weight serum component,

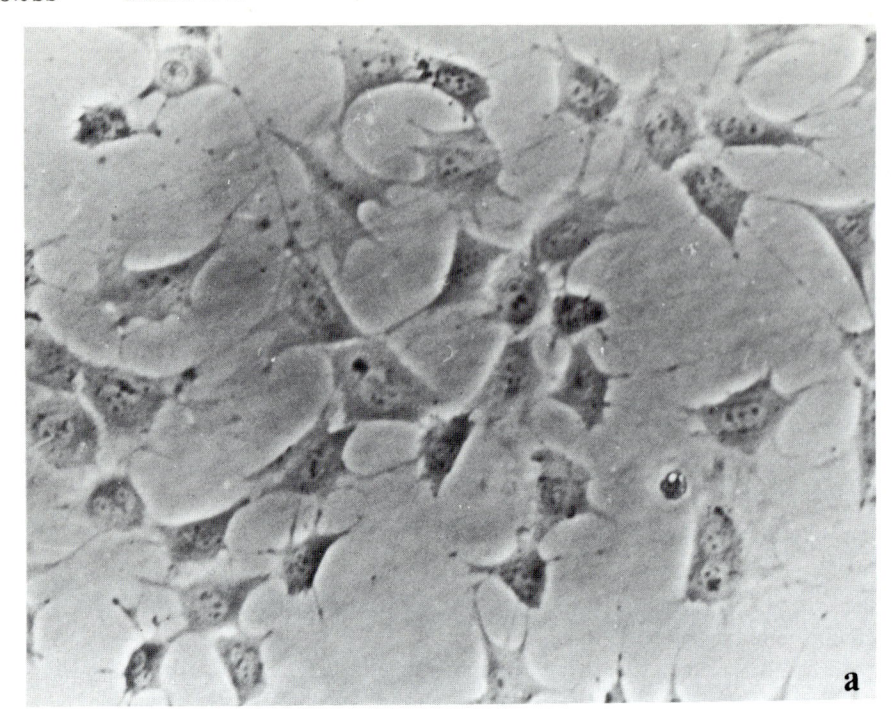

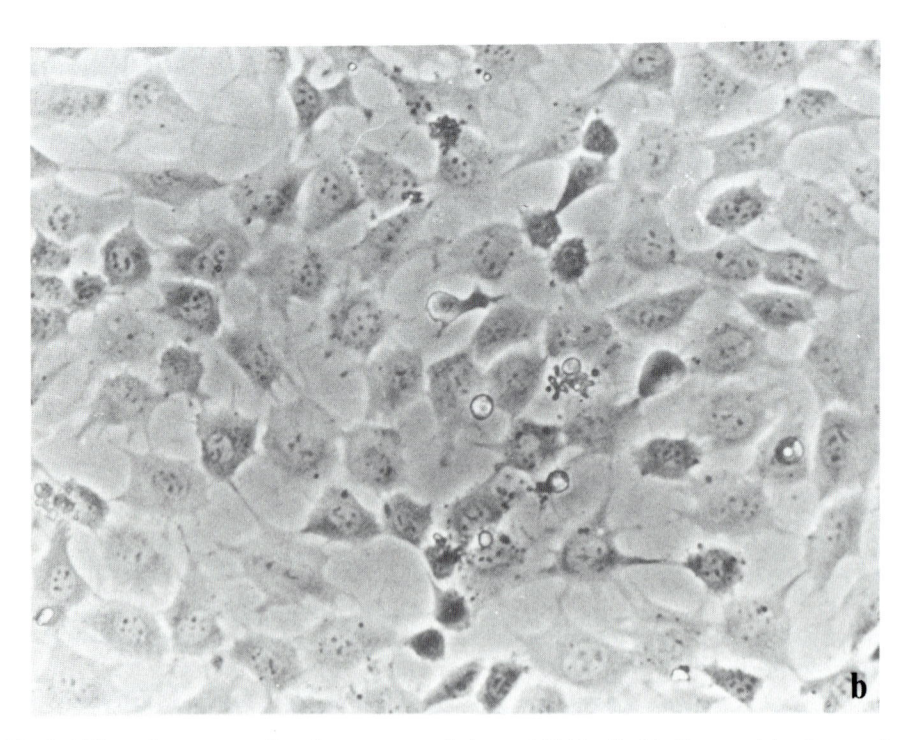

Fig. 7. Effect of serum spreading factor on morphology of 1246 cells. Medium contained as supplements: insulin, transferrin, and FGF (a); insulin, transferrin, FGF, and CIg (b); insulin, transferrin, FGF, and serum spreading factor (c); fetal calf serum (d). Concentrations of the supplements were: insulin, 10 µg/ml; transferrin, 5 µg/ml; FGF, 100 ng/ml; CIg, 3 µg/ml; serum spreading factor, 3 µg/ml; fetal calf serum, 10%.

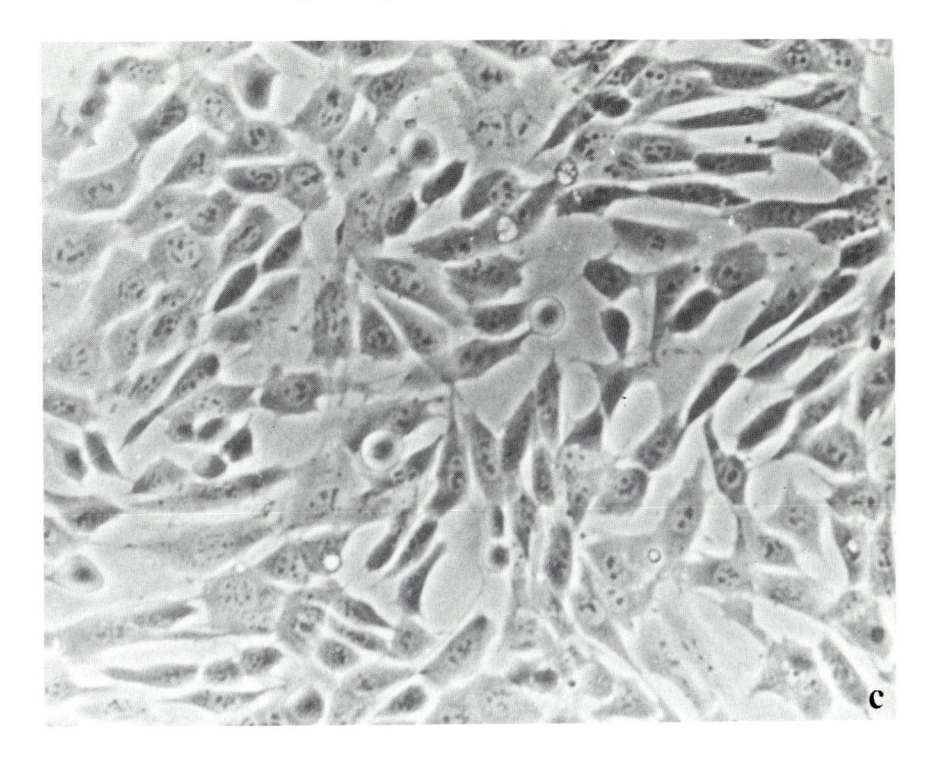

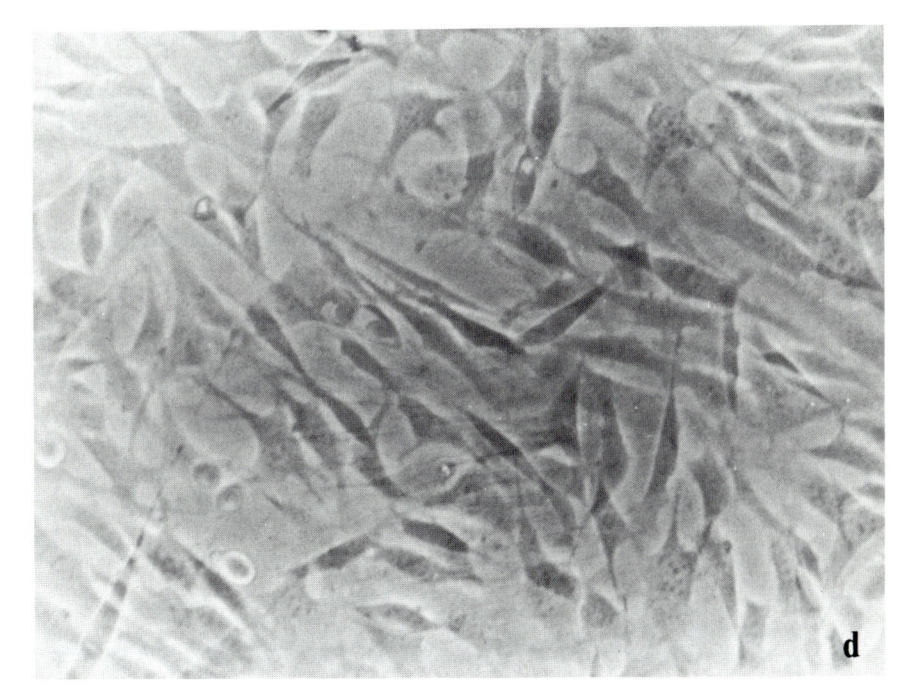

Fig. 7. (continued)

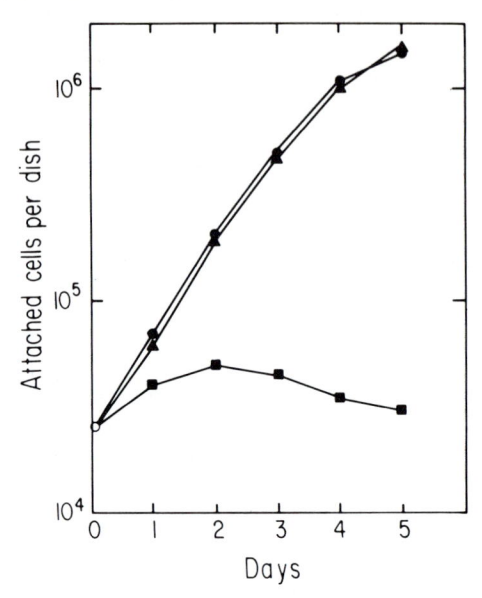

Fig. 8. Effect of serum spreading factor and CIg on growth of SV-3T3 in serum-free medium. Cell culture dishes (35 mm) were pretreated as described in Materials and Methods with 1 ml/plate of serum-free medium (■), serum-free medium supplemented with 5 μg/ml serum spreading factor (●), or 5 μg/ml CIg (▲). SV-3T3 cells (2.5×10^4) were seeded in these plates in serum-free medium and, after the cells attached (3 h), the medium was supplemented with MSA (50 ng/ml), transferrin (5 μg/ml), fatty-acid-free BSA (1 mg/ml), and linoleic acid (5 μg/ml).

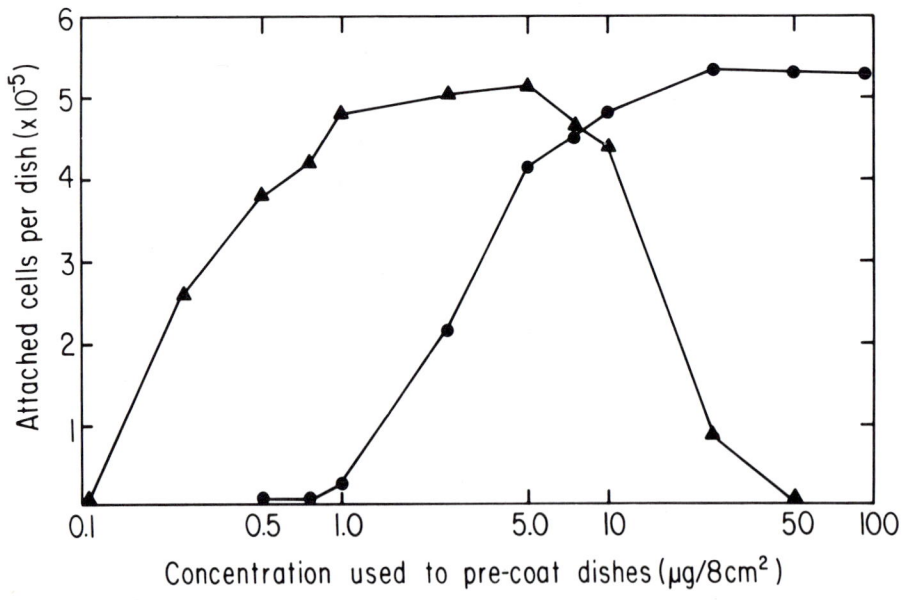

Fig. 9. SV-3T3 growth response as a function of the concentration of serum spreading factor or CIg used to pretreat cell culture dishes. Cell culture dishes were precoated with the indicated concentrations of serum spreading factor (▲) or CIg (●) in 1 ml of serum-free medium as described in Materials and Methods. Two milliliters (2.5×10^4 cells) of a trypsinized cell suspension in medium containing insulin (250 ng/ml), transferrin (5 μg/ml), fatty-acid-free BSA (1 mg/ml) and linoleic acid (5 μg/ml) were plated into each dish. Four days later the number of attached cells in duplicate dishes was determined. Data shown are the average of the duplicate plate cell counts.

is capable of stimulating growth and spreading of several cell types at concentrations comparable to the concentrations that we found to be effective for the spreading factor [4–6, 14–16]. It is likely that several serum factors, some of them as yet unidentified, may play a role in cell adhesion in vitro.

The serum spreading factor shared several properties with CIg. Both exhibit an affinity for plastic and glass surfaces, both are capable of stimulating growth and spreading some cell lines, and both mediate cytokinesis of RF-1 rat ovary cells in serum-free, hormone-supplemented medium [4–6, 14–16]. However, the serum spreading factor seems to be distinct from CIg, since no CIg monomer was detected upon SDS-PAGE of reduced serum spreading factor samples, and no cross reactivity could be demonstrated between antiserum to human CIg and the serum spreading factor preparations or antiserum to the serum spreading factor and human CIg. Knox et al [20] have recently reported that the major portion of the spreading activity in fetal calf serum is separable from CIg by gel filtration and appears in the fractions of serum proteins with molecular weights in the range of 60,000 to 90,000. This factor and the serum spreading factor described in this paper also show some similarity to one of the components of a partially purified serum fraction that promoted cell spreading and has been described by Grinnell [21]. The relationship of the serum spreading factor to these and other serum fractions reported to affect cell spreading [22–24] remains to be determined.

The serum spreading factor preparations used in this study were clearly heterogeneous by examination upon SDS-PAGE, and the possibility exists that the spreading promoting activity and the growth promoting activity reside on different molecules. However, two lines of evidence suggest that the activity may be in the same molecule. The first is the data of Figure 5, in which the concentration dependence of the spreading-factor preparations for both cell spreading and cell number increase were similar. The second is the observation that both the growth-promoting and spreading-promoting activities were absorbed to the cell culture dish if the dishes were incubated with medium containing the spreading factor and these activities could not be removed by several subsequent washes of serum-free medium.

One might speculate that the spreading factor caused an increased cell number for the C6 or 1246 lines 3–5 days after plating simply by improving the initial plating efficiency in cultures to which it had been added. Experiments with both of these lines in which the number of cells attached to the dish 20 h after plating were measured showed that this was not the case and that the same number of cells attached to the dish in the presence or absence of the spreading factor, although gross morphology of the cells under the two conditions was quite different. It is possible that the effect of the spreading factor on cell number was due to a decreased cell generation time in cultures in which it was present. Alternatively, the presence of the spreading factor may allow a higher percentage of the cells plated at the beginning of the experiment to go through the initial rounds of cell division, as is the case with either the spreading factor or CIg for the RF-1 cell line [4, 9]. We have not carried out experiments designed to differentiate between these two possibilities.

Effects of the serum spreading factor and CIg on SV-3T3 cells are shown under two different plating conditions in Figure 8 and Figure 9. In the conditions of Figure 8, cells were first seeded onto pretreated plates in serum-free medium without insulin, transferrin, linoleic acid, or BSA. After the cells had attached to the dish, the above supplements were added. Under these conditions cells attached well, but cell growth was seen only on plates pretreated with CIg or the spreading factor. In the conditions of Figure 9, cells were plated onto pretreated dishes directly in medium containing insulin, transferrin, linoleic acid, and

BSA. The presence of BSA at the time of seeding in this experiment inhibited cell attachment, and this inhibition was reversed by pretreatment of dishes with CIg or the spreading factor. Thus, in Figure 9, part of the increase in cell number due to the spreading factor or CIg was the result of the ability of these factors to promote cell adhesion. However, Figure 8 shows that pretreatment of plates with CIg or the spreading factor caused an increase in cell number even under conditions in which cell adhesion was not a consideration. It may be that these cells, like the RF-1 rat ovary line, require the spreading factor or CIg to carry out cytokinesis.

The striking loss of stimulation at the high concentrations of the spreading factor preparations in Figure 9 may be due to an inhibitory contaminant in these preparations. If this is the case, however, such a contaminant must show an affinity for plastic cell culture surfaces, as do the spreading and growth promoting activities. Alternatively, high concentrations of the spreading factor may act in some way to reverse the stimulatory effect of lower concentrations. Such effects in serum-free medium are well documented in the case of some hormones [29–32]. We have preliminary data indicating that, if cells were pretreated in suspension with concentrations of the spreading-factor preparations that were effective for pretreating dishes and then plated on untreated dishes, cell spreading was extremely poor compared to that seen if untreated cells were plated on pretreated dishes. Cells pretreated in suspension in such a manner and seeded onto pretreated plates spread as well as cells that were not pretreated with the serum spreading factor preparations before plating onto pretreated dishes.

The existence of a purified factor capable of stimulating adhesion and spreading of cells in culture would allow many kinds of interesting experiments, and we are engaged in the further purification of the active 70,000–80,000 dalton protein from the spreading factor preparations used in the experiments described in this paper. Intriguing questions may be asked regarding the role of this protein in vivo. For instance, it is possible that this protein, like fibronectin, is a component of basement membrane [25–28]. The purified factor also should be useful in vitro in studies of the interaction of normal and metastatic and nonmetastatic neoplastic cells with the substratum, studies examining the relationship of cell shape and cell growth, and studies of the signals and processes that control cell shape.

ACKNOWLEDGMENTS

The authors wish to thank Dr. R. Holley, Dr. M. Nirenberg, and Dr. M. Darmon for gifts of the cells used in these studies and Dr. M. Darmon for critical reading of the manuscript. Parts of this work were supported by USPHS GM 17702 and NCI grants 1F32CA06188-01 and CA09290-C2.

REFERENCES

1. Hayashi I, Sato GH: Nature 259:132, 1976.
2. Hutchings SE, Sato GH: Proc Natl Acad Sci USA 75:901, 1978.
3. Barnes D, Sato G: Anal Biochem 102:255, 1980.
4. Orly J, Sato G: Cell 17:295, 1979.
5. Rizzino A, Crowley C: Proc Natl Acad Sci USA 77:457, 1980.
6. Rockwell G, Sato G, McClure D: J Cell Physiol 103:323, 1980.
7. Holmes R: J Cell Biol 32:297, 1967.
8. Barnes D, Sato G: Nature 281:388, 1979.

9. Barnes D, McClure D, Orly J, Wolfe R, Sato G: J Supramol Struct (suppl) 4:180, 1980.
10. Serrero G, Darmon M, Barnes D, Rizzino A, Sato G: In Vitro 16:251, 1980.
11. Bradford M: Anal Biochem 72:248, 1976.
12. Weber K, Osborn M: In Neurath H, Hill R (eds): "The Proteins," New York: Academic, 1975, pp 179–223.
13. Jakob H, Buckingham M, Cohen A, Dupont L, Fiszmann M, Jacob F: Exp Cell Res 114:403, 1978.
14. Hook M, Rubin K, Oldberg A, Obrink B, Vaheri A: Biochem Biophys Res Commun 79:726, 1977.
15. Yamada K, Olden K: Nature 275:179, 1978.
16. Grinnell F, Hays DG: Exp Cell Res 115:221, 1978.
17. Engvall E, Ruoslahti E: Int J Cancer 20:1, 1977.
18. Bottenstein J, Hayashi I, Hutchings S, Masui H, Mather J, McClure D, Ohasa S, Rizzino A, Sato G, Serrero G, Wolfe R, Wu R: In Jakoby WB, Pastan IH (eds): "Methods in Enzymology," vol 45. New York: Academic, 1979, pp 94–109.
19. Wolfe RA, McClure DB, Dibner MD, Sato GH: Fed Proc 39:1926, 1980.
20. Knox P, Griffiths S: Exp Cell Res 123:421, 1979.
21. Grinnell F, Hays D, Minter D: Exp Cell Res 110:175, 1977.
22. Lieberman I, Ove P: J Biol Chem 233:637, 1958.
23. Michl J: Exp Cell Res 23:324, 1961.
24. Fisher HW, Puck TT, Sato G: Proc Natl Acad Sci USA 44:4, 1958.
25. Gospodarowicz D, Greenberg G, Birdwell CR: Cancer Res 38:4155, 1978.
26. Linder E, Vaheri A, Ruoslahti E, Wartiovaara J: J Exp Med 142:41, 1975.
27. Stenman S, Vaheri A: J Exp Med 147:1054, 1978.
28. Birdwell CR, Gospodarowicz D, Nicolson GL: Proc Natl Acad Sci USA 75:3273, 1978.
29. Mather JP, Sato GH: Exp Cell Res 120:191, 1979.
30. Hayashi I, Larner J, Sato G: In Vitro 14:23, 1978.
31. Allegra JL, Lippman ME: Cancer Res 38:3822, 1978.
32. Taub M, Chuman L, Saier M, Sato G: Proc Natl Acad Sci USA 76:3338, 1979.

Polyclonal Activation of Ts Cells With Antiserum Directed Against an IGH-1 Linked Candidate for a T-Cell Receptor Constant Region Marker

Frances L. Owen

Tufts University School of Medicine, Department of Pathology and Cancer Research Center, Department of Pathology and Cancer Research Center, Tufts University School of Medicine, Boston, Massachusetts 02111

An anti-T cell serum raised in allotype congenic mice recognizes the product of a new locus coding for a heavy chain-linked polypeptide found on a subpopulation of T cells. Anti-Ts^d raised in BALB/cAnN mice against selected C.AL-20 T cells reacts with a cell surface antigen in virgin animals that is found on 25% of mature thymocytes and Lyt-bearing T cells, but not on prothymocytes, Lyt1 T cells or B cells. The antigen is restricted to strains bearing the Ig-1^d and Ig-1^e heavy chain allotype haplotypes, and is expressed in the F_1 animal. The antigen is unlinked in expression to the Lyt2, H-2, or kappa light chain loci. The antigen is not detected in the hematopoietic cells in the bone marrow and appears to mark only the mature peripheral pool of T cells. As previously reported, the antiserum blocks the binding of suppressor T cells to the cross-reactive idiotype for arsonate, while reagents specific for Fab, Fc and Ig were ineffective. It seems probable that the marker may represent a T cell constant region marker analogous to the Igh products on immunoglobulin. Antiserum against this marker induces in vivo triggering of Ts cells for a wide variety of T-dependent antigens. All subclasses of anti-hapten antibodies are suppressed; no affinity restrictions or clonotype specificity is observed in suppressed adult mice. Results suggest that precursor T cells regulating major serum idiotypes regulate individual idiotypes.

Key words: T cell, constant region, receptor, suppressor, lymphocyte surface antigen

T-cell receptors for defined antigens have been shown to bear molecules cross-reactive with many of the idiotypes on antibody molecules [1–4]. In addition, framework structures of antibody V_H [5, 6] and V_L [7] chains have been found in close association with T-cell determinants. These pieces of evidence suggest that T cells share overlapping V_H gene repetoires with their B cell counterparts and utilize these structures for antigen recognition. The existence of T-cell constant region determinants which may be analogous to the Igh-1 markers on immunoglobulin subclasses has been hypothesized [8, 9]. Recent studies have shown that Ts^d, a T-cell surface molecule closely associated with the antigen binding site, is a likely candidate for the first constant region marker [9–11]. This marker is expressed on 25–35% of the Lyt2 peripheral cells of mice, restricted to the most mature thymocyte population and present on a high frequency of band I

Received March 6, 1980; accepted June 6, 1980.

0091-7419/80/1402-0175$02.50 © 1980 Alan R. Liss, Inc.

(26% BSA) Con A-induced T-cell blasts. Antisera directed against this determinant inhibits the binding of the Ars-IdX anti-idiotype [12] bearing suppressor T-cell (Ts) to its idiotype bearing target under capping conditions [9], suggesting that there is a physical association between Ts^d and the antigen specific part of the T-cell receptor. This antigen (Ts^d) has been shown to be allotype linked [9] but has no apparent linkage to Lyt2 or to H-2 in contrast to other studies which have shown anti-Lyt2 to block cytotoxic T-cell receptors [13] or H-2 linked determinants in association with allotypic T-cell receptors [14]. Recombinant inbred lines [15] with recombination events between Igh-V and Igh-C have shown this marker to be coded for by a gene(s) located between Igh-5 (δ) and a recombination event between Igh-1 and prealbumin, consistent with the possibility that this marker could be a constant region marker analogous to Igh-1 [11].

If Ts^d is a constant region marker, then all suppressor T-cells with a similar function should express Ts^d as a part of their antigen-binding structures. Since antisera directed against IgM can act as an in vitro polyclonal trigger for cells [16, 17], the possibility that anti-T-cell receptor serum could trigger T-cells was considered. If the antiserum recognizes a conserved determinant, then crosslinking of the membrane and induction of a triggering signal should be possible. If true polyclonal activation were achieved, one would expect to see no antigen restrictions in suppression, and no subclass, idiotype, or affinity alterations, in antibody-suppressed mice. Experiments described here support this hypothesis. However, it was observed that only T-dependent and not type I or II T-independent antibody responses are suppressed, suggesting that the target of that activated suppressor T-cell is preferentially a T-helper cell.

METHODS

Production of Anti-Ts^d

Details of the production and evaluation of this sera have been published elsewhere [9]. Briefly, C.AL-20 mice were treated with anti-Ars IdX serum and KLH-Ars to render them hyperimmune and subsequently suppressed for the major serum idiotype [18]. This protocol results in the generation of large numbers of Ts cells which bear receptors for the idiotype [19]. Mice were sacrificed at the peak of expression of their Ts AID cells, spleen cells treated with 5 μg/ml Con A for 48 hr and blast cells recovered on discontinuous BSA density gradients. After being washed in PBS, cells were injected into Balb/cAnN animals (5X) and serum was pooled from demonstrated positive serum samples.

Suppression with Anti Ts^d

It was previously reported that in vivo injectin of Ts^d (2 λ/mouse) day 4 before antigen lead to induction of T-suppressor cells [10]. That protocol was followed here.

Assay of PFC Response

The method of Cunningham et al [20] was used to evaluate IgM specific PFC against TNP or FITC. TNP was conjugated to sheep red blood cells by the method of Rittenberg and Pratt [21] and FITC was conjugated by the method of Möller [22].

Measurement of Anti-Hapten and Anti-Class Specific Antibodies

A solid-phase radioimmunoassay was used to evaluate the anti-hapten responses of serum antibodies [23]. Briefly, BSA-hapten conjugates were incubated in neutral buffer with polyvinyl chloride trays (Dynatech Laboratories, Alexandria, Va.) for 2 hr at 25°C. Trays were then nonspecifically blocked with 2% horse serum in PBS. Affinity purified anti-hapten antibodies were used to generate a standard curve which was used to quantitate the anti-hapten antibody in limiting dilutions of unknown sera. After a 2-hr room temperature incubation of anti-hapten antibody with the hapten coated plate, anti-hapten antibody was washed away with PBS and ^{125}I-goat anti-isotype serum or ^{125}I-RaMIg serum was added, 150 ng of affinity purified globulin/well, for 12 hr. Plates were washed, individual wells separated using a hot wire cutter, and each well counted in a gamma counter. The assay was sensitive to 5 ng/100 λ of media. Affinity measurements of anti-hapten antibodies were taken by inhibiting the anti-hapten phase of the assay with free hapten in neutral buffer.

RESULTS

Antiserum directed against Ts^d had previously been reported to be effective in inducing Ts cells for the IgM SRBC response when 2 λ was given i.v. 4 days prior to antigen injection [10]. This protocol was used to induce suppression of the IgM anti-TNP response (Table I) or anti-FITC PFC responses. These two haptens were chosen because the carriers, KLH, Ficoll, or LPS, could be varied to induce a response which has been well characterized to be either T-dependent (KLH-TNP [24], KLH-FITC), T-independent type II (Ficoll-TNP [24], Ficoll-FITC), or T-independent type 1 (LPS-TNP [25]). Only the T-dependent anti-hapten response was inhibited; antibody pretreatment reduced the mean PFC response of three mice treated wtih KLH-TNP from 40,000 PFC/10^8 cells on day 3 of the primary response to 17,000 PFC/10^8 cells, a 55% mean inhibition. The anti-KLH FITC response was suppressed 76%, from 5400 PFC to 1300 PFC/10^8 cells. T-independent antigens were not altered with the antiserum as evaluated by PFC formation in the early phases of the response.

TABLE I. Anti-Ts^d Suppression of the T-Dependent and T-Independent IgM PFC Response of A/J Mice*

T-Dependence	Antibody[a]	Antigen[b]	No. of Mice	Mean PFC/10^8 spleen cells[c]	% Suppression
T	0	KLH·TNP	3	40,347 ± 3,000	
T	2λ	KLH·TNP	3	17,450 ± 4,000	58
T	0	KLH·FITC	4	5,436 ± 200	76
T	2λ	KLH·FITC	4	1,332 ± 300	
TI$_1$	0	TNP·Ficoll	3	19,540 ± 4,000	
TI$_1$	2λ	TNP·Ficoll	3	30,378 ± 6,000	Enhancement
TI$_1$	0	FITC·Ficoll	4	12,471 ± 2,000	0
TI$_1$	2λ	FITC·Ficoll	4	13,933 ± 3,000	0
TI$_2$	0	TNP·LPS	3	174,333 ± 20,000	0
TI$_2$	2λ	TNP·LPS	3	123,903 ± 60,000	0

*Adult (5–6 week) mice were used.

[a]Antisera or NMS was diluted in PBS to 200 λ and injected i.v. into the tail vein.

[b]Antigen was diluted to 200 λ in PBS and injected i.v. into the tail vein; KLH·TNP (50 mg), KLH·FITC (50 mg), TNP·Ficoll (10 μg), FITC·Ficoll (10 μg), TNP·LPS (10 μg).

[c]Direct PFC's were evaluated on day 3 of the response.

The serum responses of mice to TNP or FITC was compared with the PFC results. Again only the early primary response is suppressed (Table II), and under repeated antigen pressure, these mice overcame the suppressive signal. Serum from 10 mice in each group was pooled and evaluated for the total anti-hapten antibody using a sensitive solid phase radioimmunoassay. On day 10, mice preinjected with 2 λ of anti-Ts^d serum followed by KLH-TNP made 24 ng/λ of anti-TNP in contrast to 123 ng/λ for the control mice. By day 21 of the secondary response, this difference between groups was no longer observed. In contrast, the TNP-Ficoll response is not altered significantly by pretreatment of mice with antibody. In a parallel experiment (Table II), 10 mice were immunized with KLH-Ars and pretreated with either antibody or NMS. Similar results showed the anti-Ars response was 35 ng/λ for antibody-treated mice and 117 ng/λ for NMS controls (day 17). When mice were treated with anti-Ts^d followed by FITC-KLH or Ficoll-FITC the antibody was effective in suppressing the early primary response from 134 ng/λ to 80 ng/λ for FITC-KLH, but may have somewhat enhanced the T-independent Ficoll-FITC response (day 10). These experiments are consistent with the possibility that anti-Ts^d suppresses the primary response to T-dependent antigens.

A possible isotype restriction in the antibody which is suppressed was evaluated (Table III). With anti-TNP, all isotypes are suppressed, while γ_1 and γ_3 are more completely and persistently altered than μ. In contrast the anti-FITC μ response is more suppressed than γ_1 and γ_3. Quantitatively, the response to TNP is predominately γ_1 and γ_3. In contrast to the FITC response, which is predominately IgM on day 10; the major responding isotype is therefore altered most in each case. There is no evidence

TABLE II. Effect of Anti-Ts^d on a T-Independent Serum Response

		Pool of 10 animals/group				
		1°			2°	
		Day			Day	
Antigen[a]	Antibody[b]	10	15	20	28	31
		mean ng/λ anti-TNP				
KLH·TNP	NMS	123	776	178	302	1409
KLH·TNP	Ab	24	113	74	158	1215
Ficoll·TNP	NMS	51	67	43	61	32
Ficoll·TNP	Ab	62	75	48	63	53
		mean ng/λ anti-ARS				
KLH·Ars	NMS	N.D.	117	N.D.	4374	3947
KLH·Ars	Ab	N.D.	35	N.D.	3645	3773
		mean ng/λ anti-FITC				
KLH·FITC	NMS	134				
KLH·FITC	Ab	89				
Ficoll·FITC	NMS	15				
Ficoll·FITC	Ab	30				

[a]Antigens were administrated i.p. on day 0 in CFA. The following quantities were given: KLH·TNP, 50 μg; TNP·Ficoll, 50 μg; KLH·Ars, 50 μg; KLH·FITC, 50 μg; Ficoll·FITC, 10 μg.

[b]Anti-Ts^d (2 λ/mouse) was administered i.v. 4 days before antigen injection.

for an induction of a shift to enhanced production of another isotype in the presence of antibody [26].

The affinity differences between mice treated with anti-Ts^d and then challenged with antigen on day 15 of a primary response were compared with control mice immunized with NMS and antigen on day 15. Figure 1 shows that there are no gross differences between the affinity of anti-TNP or anti-Ars responses in suppressed or control mice.

DISCUSSION

The network theory of regulation of the immune response [27] states that an extended circuit of antibodies and anti-antibodies interact with one another to control the production of antibody of any given specificity. This original concept has been expanded upon by the recent studies of those working in cellular systems with idiotypic antibodies and anti-idiotypic reagents which recognize major cross-reactive serum idiotypes, related clones which are serologically detectable. It has been shown that heterologous anti-idiotype serum, injected into adult mice can suppress the immune response by generating T-suppressor cells specific for a given idiotype [1, 28] or T-helper cells [1]. In all cases reported, this regulation is clonotype specific. However, the major ques-

TABLE III. Suppression of Class Specific Serum Immunoglobulin Response*

	IgG_1	IgG_3	IgG_{2b}	IgG_{2a}	IgM	Total Ig
			Anti-TNP			
Day 10 – 1° response						
T-cell dependent[a]:						
Ab/NMS ratio	0.14	0.20	0.16	0.21	0.39	0.16
T-cell independent[b]:						
Ab/NMS ratio	1.45	1.89	1.79	3.07	1.15	1.2
Day 15 – 1° response						
T-cell dependent:						
Ab/NMS ratio	0.18	0.19	0.23	0.49	0.50	0.15
T-cell independent:						
Ab/NMS ratio	1.56	1.80	1.56	2.48	1.07	1.10
Day 28 – 2° response						
T-cell dependent:						
Ab/NMS ratio	0.46	0.75	0.64	0.66	1.7	0.54
T-cell independent:						
Ab/NMS ratio	0.71	1.2	0.78	1.5	1.1	0.95
			Anti-FITC			
Day 10 – 1° response						
T-cell dependent[c]:						
Ab/NMS ratio	0.80	0.90	N.D.	N.D.	0.49	0.60

*Isotype specificity was determined by pulsing antigen coated plates with serum antibody and following that 2-hr pulse with ^{125}I-goat anti-isotype specific affinity purified reagents.

[a]KLH·TNP, 50 μg, was injected i.p. in CFA on day 0.

[b]Ficoll·TNP, 50 μg, was injected i.p. in CFA on day 0.

[c]KLH·FITC, 50 μg, was injected i.p. in CFA on day 0.

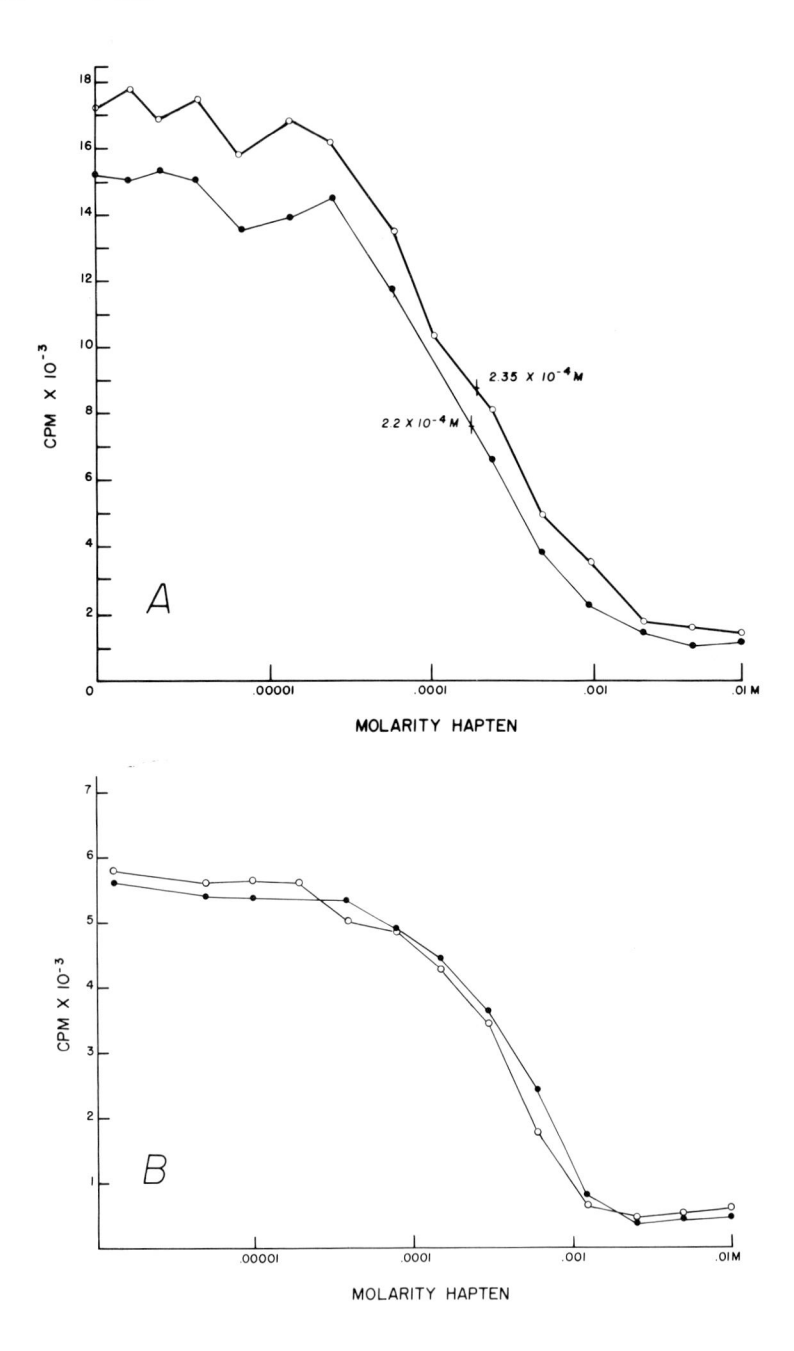

Fig. 1. Affinity measurements of anti-hapten antibodies from mice pretreated with anti-Ts^d or NMS and KLH·TNP or KLH·Ars. Ten mice immunized with TNP (frame A) and aTs^d (0) or NMS (0) showed average affinity on pooled sera at day 10 of 2.35 × 10^{-4}M or 2.2 × 10^{-4}M, respectively, when 10 ng of antibody was blocked with free hapten. The average affinities of 10 anti-Ars producing mice pretreated with Ab (0) or NMS (0) was indistinguishable at 4.5 × 10^{-4}M when 10 ng of day-10 pooled sera was blocked with free hapten.

tion raised by these studies is whether the mechanisms which regulate major dominant idiotypes also apply to disperse clones of cells with idiotypes recurring too infrequently to measure. One might alternatively argue that normal mechanisms which regulate the immune response may be deficient in these mice permitting the observed expression of unusually high numbers of related clones.

This study utilizes antiserum produced against a cell which regulates the anti-Ars IdX, and which appears to be specific for a T-cell receptor bearing anti-idiotype determinants. The antiserum has been shown previously to recognize a conserved determinant outside the antigen binding site which may represent a T-cell constant region marker [9]. We have shown previously that this antiserum, largely λ_1, when injected in vivo can induce T suppressor cells for the primary SRBC response [10]. This study suggests that Ts^d is in effect a polyclonal activator for a functionally restricted subpopulation of cells. The results shown here further confirm that the precursor cells which suppress all T-dependent responses we have measured, independent of clonotype, isotype, or affinity restrictions, share Ts^d as a surface determinant.

It is possible that the cells which control the immune response to T-dependent antigens are regulated more easily than the B cells responding to T-independent antigens. Therefore, the differences seen may be quantitative and not absolute. The results do suggest that the target of suppression is preferentially a T-cell. Whether the cell triggered initially with the antiserum is a suppressor cell or an inducer of that suppressor has not been shown. The NZB animal, which may be defective in the Ly1, 2, 3 antigen specific feedback suppressor, [29] appears to be fully competent in the T-cells expressing Ts^d. Therefore, the T-cell expressing Ts^d may be analogous to the Ts2 cell recently described as an acceptor for the antigen specific signal [30].

It seems probable therefore that similar precursor cells regulate the major dominant idiotypes and are also responsible for regulation of smaller clones, difficult to measure serologically. That is, each individual idiotype may have a mirror image anti-idiotype-bearing Ts cell regardless of its initial clone size. That anti-Ts^d appears to act as a polyclonal activator rather than an idiotype specific regulatory molecule is probably determined by the site of binding to the receptor. Anti-Ts^d may bind to a constant part of the T-cell receptor outside the binding site for idiotype on all Ts cells while anti-idiotype antibodies may trigger idiotype bearing Ts or Th cells which secondarily induce Ts AID bearing cells for only the clone triggered by anti-idiotype.

The availability of monoclonal antibodies directed against a determinant(s) recognized in this strain combination would greatly facilitate our own studies and lead to a reagent of general usefulness in other laboratories. A major effort is now in progress to produce and screen for such a reagent.

ACKNOWLEDGMENT

This work was supported by NIH grant AI-156262.

REFERENCES

1. Eichmann K: Adv Immunol 26:195, 1978.
2. Binz H, Wigzell H: J Exp Med 142:1218, 1975.
3. Weinberger JZ, Germain RN, Ju S-T, Greene MI, Benacerraf B, Dorf ME: J Exp Med 150:161, 1979.

4. Rubin B, Hertel-Wulff B, Kimura A: J Exp Med 150:307, 1979.
5. Lonai P, Ben-Neriah Y, Steinman L, Givol D: Eur J Immunol 8:827, 1978.
6. Finnegan A, Owen FL: Fed Proc 39:4879 (Abst), 1980.
7. Eichmann K, Ben-Neriah Y, Gavish M, Givol D: Eur J Immunol (in press), 1980.
8. Kontianinen S, Feldmann M: Thymus (in press), 1980.
9. Owen FL, Finnegan A, Gates ER, Gotlieb PB: Eur J Immunol 9:948, 1979.
10. Owen FL: J Immunol 124:1411, 1980.
11. Owen FL, Riblet R, Taylor B: (submitted for publication), 1980.
12. Kuettner MG, Wang AL, Nisonoff A: J Exp Med 135:579, 1972.
13. Shinohara N, Sachs DH: J Exp Med 150:432, 1979.
14. Krammer P, Eichmann K: Nature 270:733, 1977.
15. Riblet R, Claflin L, Gibson DM, Matheson BJ, Weigert M: J Immunol 124:787, 1980.
16. Parker DC, Fathergill JJ, Wadsworth DC: J Immunol 123:931, 1979.
17. Mongini P, Friedman S, Wortis H: Nature 276:709, 1978.
18. Pawlak LL, Hart DA, Nisonoff A: J Exp Med 137:1442, 1973.
19. Owen FL, Ju S-T, Nisonoff A: Proc Natl Acad Sci USA 74:2084, 1977.
20. Cunningham AJ, Smith JB, Mercer EH: J Exp Med 124:701, 1966.
21. Rittenberg MB, Pratt, C: Proc Soc Exp Biol 132:575, 1969.
22. Möller G: J Exp Med 139:969, 1974.
23. Klinman NR, Pickard AR, Sigal NH, Gearhart PJ, Metcalf ES, Pierce SK: Ann Immunol (Paris) 127C:489, 1976.
24. Sharon RP, McMaster RB, Kask AM, Owens JD, Paul WE: J Immunol 114:1585, 1975.
25. Mosier DE, Mond JJ, Goldings EA: J Immunol 119:1874, 1977.
26. Woodland R, Cantor H: Eur J Immunol 8:600, 1978.
27. Jerne NK: Annals of Pasteur Instit 125C:373, 1974.
28. Nisonoff A, Ju S-T, Owen FL: Immunol Rev 34:89, 1977.
29. Cantor H, McVay-Boudreau L, Hugenberger J, Neidorf K, Shen FW, Gershon RK: J Exp Med 147:1116, 1978.
30. Sy MS, Dietz MH, Germain RN, Benacerraf B, Greene MI: J Exp Med 151:1183, 1980.

Polypeptide Growth Factors: Some Structural and Mechanistic Considerations

Ralph A. Bradshaw and Jeffrey S. Rubin

Department of Biological Chemistry, Washington University School of Medicine, St. Louis, Missouri 63110

Polypeptide growth factors are substances that stimulate an increase in cell size and/or cell number during embryonic development. In some cases, they have a similar effect on tissues in the mature organism where they function as "maintenance" factors to sustain cell viability. While their profound impact on cell behavior is well recognized, their relationship to other regulators of cell function has remained generally ill-defined. However, the developing appreciation of their hormone-like behavior suggests that they may be conveniently grouped with many other endocrine agents to form a broader group of secondary hormones. The utility of the classification is illustrated by the insulin-related family of molecules. It also serves to emphasize the similarities in function shared by many of these substances including trophic stimulation and modulation of gene expression. Internalization, though, appears to be another common feature. However, whether the uptake of the growth factor mediates an intracellular action or is designed solely to regulate responsiveness at the cell surface and/or degradation remains an important unanswered question. A brief review of two growth factors (nerve growth factor and epidermal growth factor) serves to outline the possible functions that may be served by this endocytotic process.

Key words: primary and secondary hormones, mitogenicity, insulin, insulin-like growth factor, nerve growth factor, relaxin, epidermal growth factor, receptor-mediated endocytosis, lysomes, hormone mechanisms

Growth is defined as an increase in the size of the body or any of its tissues which results from an increase in cell size (hypertrophy) or cell number (hyperplasia). Such changes in cell populations are not only important features of development but also are evident in the regenerative processes of wound healing and in the routine turnover of cells that characterizes most tissues. Regulation of these growth processes is complex, involving hormones, neural elements, and especially proximal contacts by both heterologous and homologous cells [67]. In all instances, there is a transfer of information that is usually mediated by some type of chemical messenger.

Received January 30, 1980; accepted August 12, 1980.

0091-7419/80/1402-0183$05.00 © 1980 Alan R. Liss, Inc.

Among the agents functioning in this capacity are the polypeptide growth factors, hormone-like substances that are released from many different cell types and reach their destination by a variety of routes ranging from local diffusion to systemic transport. In general, they can induce both hypertrophic and hyperplastic responses in their target tissues. The increase in cell size is closely linked to an array of metabolic changes generally defined as a positive pleiotypic response [40], which includes stimulation of metabolite uptake and polysome formation leading to increased protein and nucleic acid synthesis. Hyperplasticity can result from a reduction in the extent of programmed cell death or an increase in the rate of mitosis. Substances acting in the latter capacity are also referred to as mitogens, and many of the growth factors with this property have been grouped into two broad subcategories based on the point in the cell cycle when they appear to act. Agents such as platelet-derived growth factor (PDGF) and fibroblast growth factor (FGF) prime a significant portion of a quiescent cell population (G_o) to enter a state of readiness or "competence," while other agents, notably the somatomedins, stimulate these competent cells to enter the S phase and proceed through mitosis [75]. The perception of distinct roles for these agents emphasizes an important insight into the mechanism of growth control: The stimulation of cell division is a sequential process that is often regulated at different steps and, in many cases, the coordinated action of different factors is required for a maximal response [66].

An important aspect of the action of growth factors may be inferred from the definitions given above: Insofar as growth involves only an increase in size or number of cells, there is no reason to attribute to these agents a role in the differentiation of their target tissues, although in most cases such a function has not been rigorously excluded either. Clearly, they do have an impact on gene expression, which is manifested in the display of specialized traits that characterize maturing, responsive cells. Thus, they can be viewed as modulators of the phenotypic profile, amplifying the distinctive characteristics of a committed cell, as opposed to differentiating agents which switch on (or off) previously unexpressed genes [81]. As modulators, they behave like the classical hormones (vide infra). However, not all of the polypeptide growth factors necessarily fit this description; for example, erythropoietin and the colony-stimulating factors (CSFs) may direct the differentiation as well as the proliferation of their target cells [1, 52]. Thus, although the role of growth factors in differentiation remains largely unresolved, it is not at the present level of understanding a requisite part of their action and will not be further considered in this article.

In addition to the hypertrophic and hyperplastic effects associated with growth and repair processes, polypeptide growth factors have a distinct role as maintenance or survival factors which is not inherent in the definition of a growth-promoting substance. It is clear that most of these substances are normally present in the mature organism, and, based on tissue culture experiments, deprivation in many instances causes responsive cells to be seriously affected or even die. This has also been observed in vivo with nerve growth factor (NGF) where the implementation of a passive- or autoimmune state provides antibodies directed against NGF causing atrophy of neurons in the sympathetic and sensory nervous systems [33, 50]. For some growth factors, this maintenance of cell viability may represent a more fundamental characteristic than their hypertrophic or hyperplastic activities as it would be manifested throughout the lifetime of the target cell, even in situations where growth phenomena had abated.

RELATION TO "CLASSICAL" HORMONES

The first interaction of polypeptide growth factors with their target cells is, at our present level of knowledge, exclusively with the exterior face of the plasma membrane, placing them in a broad category of substances that initiate their biological activity in this manner. For the most part, this contact is a highly specific one because of the presence of membrane-bound receptors that recognize and bind the external ligands both avidly and selectively [12, 47]. Although a myriad of different events is subsequently triggered by the formation of these complexes, these agents share one other common denominator; namely, they transfer information to the cell from its environment. In some cases, such as with toxins of plant or bacterial origin [17], the exchange is of deleterious value, but usually it represents a positive influence, resulting in the stimulation of metabolic machinery with the overall effect of regulating the designated physiological function of that cell. Certain classes of these external regulators, such as hormones and neurotransmitters, are well defined conceptually and are easily recognized by name and, in most cases, function. However, many other such substances are less well categorized, although they have demonstrated physiological significance, often because existing definitions are too rigid to accommodate them in established groupings. Largely because of their diverse nature and limited characterization, the polypeptide growth factors suffer this fate. Although clearly hormone-like in their action, they are, nonetheless, rarely listed as such. In addition, many substances, classically listed as hormones, are now appreciated to act as growth factors as well, which adds further ambiguity.

This problem of hormonal definition and classification is not of recent origin. Many discussions of the subject have been published, one of the most detailed being that of Huxley [43]. He chose to emphasize the transfer of positive information as a basis for definition rather than mode of transport, which would have redefined a wide variety of substances as hormones. Robison et al [64] presented a more limited version of these ideas that subdivided those substances, considered hormones under the classical definition, into two major categories. A logical extension of these ideas provides a convenient means for viewing the polypeptide growth factors in relation to other hormones without the ambiguities engendered by placing them in separate categories.

The principal characteristics of the two classes defined by Robison et al [64] are summarized in Table I. The first group, referred to as messengers, show rapid responses immediately following formation of the hormone-receptor complex. These almost always initially involve an increase in the level of intracellular cyclic AMP produced by the stimulation of membrane (and ultimately receptor) associated adenyl cyclase. In such cases, the cyclic nucleotide becomes a second messenger and is the agent that further dictates the majority of the remaining hormonal responses in that cell. The duration of enhanced cAMP production, and thus the response, are basically proportional to the receptor occupancy, which is typically limited to a short interval because of the relatively rapid turnover of this type of hormone. Members of the second class, which have been variably called maintenance, permissive, or developmental and designated here as secondary hormones,* share some of the features of the primary hormones but are distinguished by

---

*The two classes of hormones are designated primary and secondary in this article, as suggested by Bradshaw and Niall [8], to denote the relative development of their distinguishing characteristic responses (with respect to time). The term "messenger" was found to be less satisfactory because members of both classes can properly be viewed as serving that role.

TABLE I. Classification Scheme for Hormones and Hormone-Like Substances*

Property	Primary (messenger)	Secondary (maintenance, permissive, developmental)
Response	Rapid (sec, min)	Rapid: Slow (hr, day)
Duration of effect	Short	Short: Long
Turnover	Fast	Slow
Principal effect	Production of cAMP	Pleiotypic stimulation: changes in protein synthesis
Mechanism of action	External	External: Internal (?)
Examples	Epinephrine	Thyroid hormones
	Glucagon	Glucocorticoids
	Parathyroid hormone	Insulin
		Growth hormone

*After Robison, Butcher, and Sutherland [64].

several important differences. Principal among these is the extended period of response, often approaching hours or days, that can continue long after hormone-receptor interactions are no longer demonstrable. While the early effects are frequently associated with general anabolic stimulation, the long-term responses, being, for the most part the distinctive features of this group, are clearly related to specific changes in protein synthesis. However, the mechanism by which either the short or long-term responses are produced is not well understood. The rapid effects are certainly initiated by the association with the plasma membrane but have not as yet been shown to involve the generation of a second messenger like the receptor occupancy-dependent production of cyclic nucleotides induced by primary hormones. Nonetheless, a messenger molecule of undetermined chemistry may well be produced that regulates, among other things, the flux of ions and metabolites and, at the same time, initiates the events leading to the modulation of gene expression. However, it is also possible that the two classes of temporal responses are basically separate events that develop independently following formation of the hormone-receptor complex. That is, the "messenger" generated to initiate the general anabolic responses may be unrelated to the stimulus required for the long-term effects. In this regard, the growing evidence that hormones of this class are readily internalized by endocytosis provides a feasible mechanism for the complementation of such a dual mechanism (vide infra). This uptake allows for lysosomal degradation following the appropriate fusion events, but would also permit the hormone or its receptor to act as a new second messenger for events not triggered by the formation of the original hormone-receptor complex. Evidence in support of this mechanism (or variations of it) for one system, NGF, is described below. However, the extent to which internalization is an important feature of the mechanism of other secondary hormones is presently highly speculative.

As noted in Table I, examples of primary hormones are epinephrine and parathyroid hormone, whereas the second class includes such substances as thyroid hormone, glucocorticoids, insulin, and growth hormone. However, Robison et al [64] noted that the distinction between these two groups of hormones "is not always as clear as the human urge to classify things might like it to be." That is, many primary hormones exert maintenance effects despite the fact that they appear, by other criteria, to be classed with the first group, and it is possible that they may, in some circumstances, be internalized as well. Chief among these are the gonadotrophins, thyroid-stimulating hormone and adrenocorticotropin [76]. Nonetheless, this classification remains useful, particularly in dealing with growth factors, because it logically associates them with other substances of similar structural and functional properties.

PROPERTIES OF GROWTH FACTORS

General Considerations

An alphabetical listing of many of the polypeptide growth factors is given in Table II. The entries have been limited to those substances that have been more extensively characterized (see for example, Gospodarowicz and Moran [36]); the existence of many other factors has been inferred from observed biological activities, but has not yet been associated with a unique molecular species. The sources and target cells listed represent the principal tissues used for purification and bioassay, respectively. For many of the substances, a complete knowledge of the physiologically relevant tissues, with respect to either origin or response, is unknown and, in some instances, those reported in Table II may not be significant in the living organism. For example, the site of synthesis of the insulin-like growth factors and somatomedins is uncertain, although liver is strongly suspected [80], and the cells responsive to fibroblast growth factor in vivo have not been firmly established [36].

More extensive data for each of the factors can be found in the references cited in Table II. However, a few general aspects deserve further comment here. First, there exists the distinct possibility that some of the factors listed in Table II represent different names for the same substance. For example, the relationship of the insulin-like growth factors (IGF) I and II, the somatomedins A and C, and multiplication-stimulating activity is presently obscure, but it appears likely that at least some of these will represent the same entity when purification and structural analyses are complete on all of the factors [86].

A second and related problem, in cases where structural analyses are not available, is the association of the biological activity with the correct molecular entity. Recent observations with fibroblast growth factor obtained from bovine brain illustrate this point. Following the initial description by Gospodarowicz et al [34] of the purification and characterization of brain FGF, Westall et al [83] reported that the factor was identical to various fragments derived from the carboxyl terminal region of myelin basic protein (MBP) which had previously been sequenced by Eylar [21]. While preparations of bovine brain FGF made by the procedure of Gospodarowicz et al [34] clearly contain these fragments as their major constituents, Thomas et al [79] have shown that the mitogenic activity of these preparations (the assay originally used to describe FGF) is not associated with the MBP peptides but with an acidic protein present as less than 5% of the sample. This material is clearly distinct from pituitary FGF, which was not reported to be related to MBP [83], as judged by isoelectric focusing criteria.*

Finally, as judged by the molecular weight values shown in Table II, there is no obvious molecular similarity indicative of the group as a whole. However, comparison of amino acid sequences has revealed a relatedness among some of the factors of the kind associated with proteins that have evolved from a common precursor. Such a relationship has been appreciated for some time for the two pituitary hormones, growth hormone and

---

*S.K. Lemmon, M.C. Riley, and R.A. Bradshaw (unpublished observations).

TABLE II. Polypeptide Growth Factors

Factor	Source[a]	Target tissue	Mitogenicity[b]	M_r	Primary sequence	Ref.
Colony-stimulating factor (= macrophage-stimulating factor)	Many human tissues	Immature myeloid precursors, macrophages	Yes	1,300–150,000	No	[52]
Epidermal growth factor	Mouse submaxillary, human urine	Epidermal cells, fibroblasts	Yes	6,045[c]	Yes	[10]
Erythropoietin	Human plasma or urine	Red blood cell precursors	Yes	46,000	No	[1]
Fibroblast growth factor	Bovine pituitary	Mesodermal cells	Yes	13,000[d]	No	[37]
Growth hormone	Pituitary	Liver, muscle, adipose	No	22,005[e]	Yes	[55]
Insulin	Pancreas	Liver, adipose, muscle	Yes	5,796[e]	Yes	[18]
Insulin-like growth factor I	Human plasma	Liver, adipose muscle, cartilage fibroblasts	Yes	7,649	Yes	[86]
Insulin-like growth factor II	Human plasma	Same as IGF-I	Yes	7,471	Yes	[86]
Multiplication stimulating activity	Rat serum, liver	Same as IGF-I	Yes	8,000	No	[58]
Nerve growth factor	Mouse submaxillary, snake venoms	Sympathetic and certain sensory neurons	No	13,259[c]	Yes	[6]
Ovarian cell factor	Bovine pituitary	Ovarian tumor cells	Yes	10,000–13,000	No	[35]
Placental lactogen	Placenta	Mammary gland, liver	ND	22,125[e]	Yes	[55]
Platelet-derived growth factor	Human platelets	Arterial smooth muscle cell, fibroblasts	Yes	13,000[f]–38,000	No	[65]
Prolactin	Pituitary	Mammary gland, ovary, liver	ND	22,500[d]	Yes	[55]
Relaxin	Porcine corpa lutea	Pubic symphysis, cervix, uterus	ND	6,056[g]	Yes	[70]
Somatomedin A	Human plasma	Same as IGF-I	Yes	7,000	No	[80]
Somatomedin C	Human plasma	Same as IGF-I	Yes	7,000	No	[80]
Thrombin	Plasma	Fibroblasts	Yes	30,030[h]	Yes	[59]
Thymosin	Thymus	Lymphoid cells	ND	12,200	No	[84]

[a]From many species unless specified.
[b]ND, not determined.
[c]From mouse.
[d]From bovine pituitary.
[e]From human.
[f]Molecular weight from studies of Antoniades et al [4].
[g]Molecular weight from the sequence of James et al [45].
[h]From cow.

prolactin, and placental lactogen [55]. A second group that has been elucidated more recently is the subset related to insulin [7, 8]. Of the structurally defined members, NGF was the first to be so identified [24], followed by the two closely related insulin-like growth factors I and II [62, 63] and relaxin [45, 69]. Because of the functional similarities and receptor cross-recognition [51], other somatomedins are also expected to resemble insulin. The relationships of this subset are a particularly good illustration of the value of classifying polypeptide growth factors as secondary hormones.

Insulin-Related Subset

The gross biological properties of the insulin-related molecules belie the observed similarities in structure and potentially in mechanism. The prototype of this family, insulin, is a pancreatic hormone more commonly recognized for its profound effects on carbohydrate, lipid, and amino acid metabolism, although it is well known to support growth in vivo and in tissue culture [27] and was selected as the model for defining positive pleiotypic effectors [40]. The IGFs were first defined as the substances displaying insulin-like activity in human serum that were not suppressed by antibodies to insulin and were designated nonsuppressible insulin-like activities (NSILA) [26]. Following fractionation by acid ethanol extraction [44], the soluble component (NSILAs) was found to consist of two closely related polypeptides that were renamed insulin-like growth factors I and II when sequence analysis demonstrated that they were structurally quite similar to the pancreatic hormone [62, 63]. These molecules also act as promoters of sulfate incorporation into the proteoglycans of cartilage (as judged by in vitro assay), and therefore may act in the capacity of somatomedins, the mediators of growth hormone [86].

In contrast to insulin and the IGFs, NGF and relaxin are without effect in the typical target tissues of insulin, ie, liver, adipose tissue, and muscle; rather, they act on the more specialized tissues of the peripheral nervous system [49] and the female reproductive tract [70], respectively. In the embryo, NGF acts as a trophic stimulator of developing sympathetic and selected sensory neurons leading to the proliferation of axonal processes and the ultimate formation of functional synapses. The maintenance of these neurons remains an essential activity of NGF in the adult state [49]. Relaxin appears to exert its effects primarily at parturition when it stimulates the loosening of the pubic symphysis rendering the birth canal more pliable for the passage of the fetus. It is also alleged to soften the cervix, inhibit uterine muscle contraction, and affect mammary gland development [70]. However, unlike the other insulin-related substances, analysis of these biological responses has not been carried out at the molecular or cellular level as yet, and it is therefore uncertain to what extent it exerts these effects through pleiotypic activation.

The structural similarities of the insulin-related molecules underlying these apparently diverse biological activities is summarized in Figure 1. The relatedness, which is manifested in amino acid residues occupying identical positions when the factors are appropriately aligned for comparison, is confined to the A and B chains of insulin that are formed from the biosynthetic single polypeptide chain precursor, proinsulin, by specific proteolytic excision of the intervening C peptide [74]. As shown by the solid boxes, the IGFs exhibit the greatest number of identities with the insulin chains, whereas relaxin and NGF are more distantly related. There are, however, some additional identities found between the factors themselves, not seen in the comparison to insulin, that are denoted by the lined and stippled positions. The boxes containing single bars represent deletions arbitrarily introduced to maximize the identities and should be viewed as nonidentities in the context of this comparison.

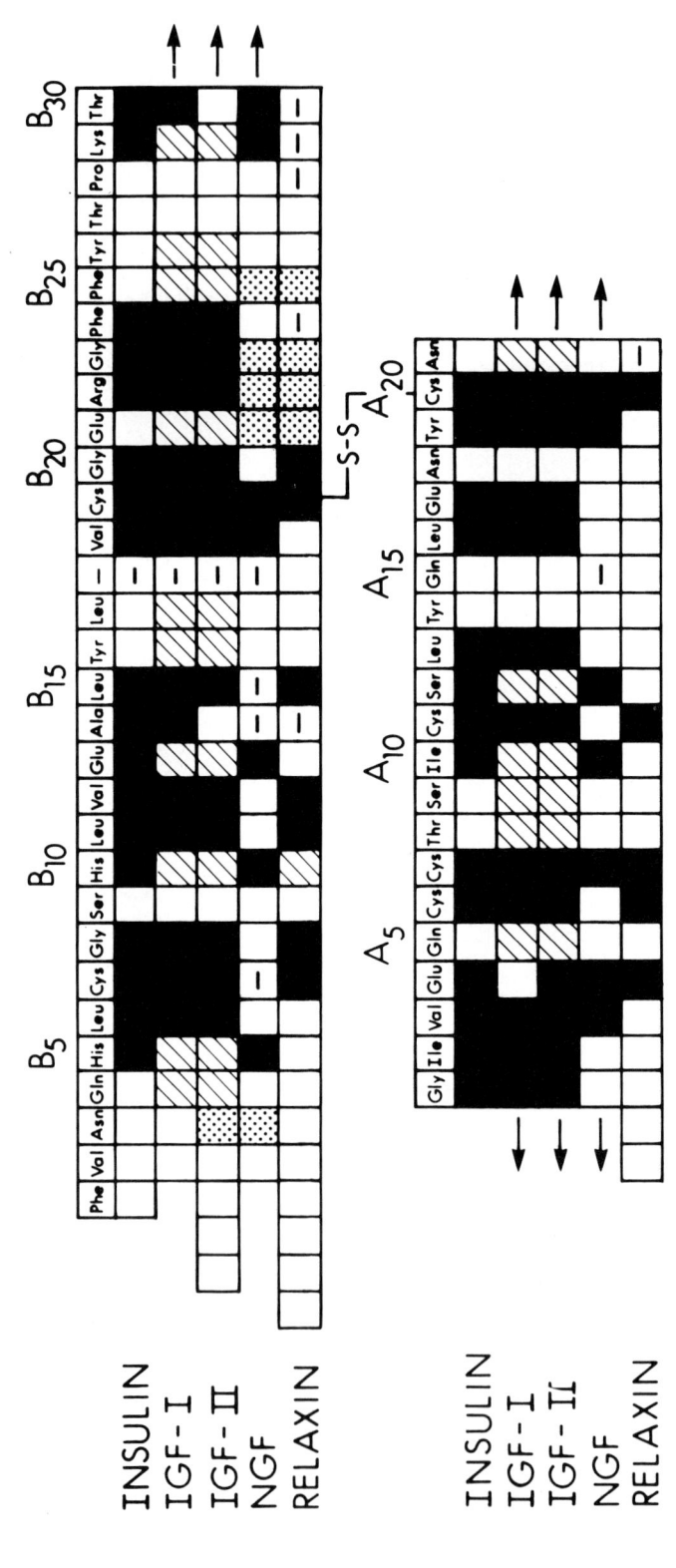

Fig. 1. Diagrammatic comparison of the amino acid sequences of the insulin-related polypeptides in the regions corresponding to the A and B chains of insulin. The actual sequence given in the top line of both segments is that of human insulin [57]. In the block diagram of the insulin structure (second line), solid squares indicate residues in the insulin sequence also found at the same position in one or more of the other factors. Blank squares represent positions in the insulin sequence occupied by different amino acids in the other factors, whereas lined or stippled squares indicate identities in two or more of the growth factors that are not identical to the corresponding residue of insulin. A few deletions, indicated by squares containing dashes, have been introduced arbitrarily to maximize the homology. The single disulfide identically paired in all molecules (B19-A20) is shown. The arrows indicate that the segments listed are part of continuing sequences in the active form of the hormone (see Fig. 2). The polypeptide sequences are human insulin-like growth factor-I (IGF-I) [62], human insulin-like growth factor-II (IGF-II) [63], mouse nerve growth factor (NGF) [6], and porcine relaxin [45].

Further insight into the relationship of these factors as well as the manner in which they have diverged to form unique entities is derived from a consideration of their secondary structures. As shown schematically in Figure 2 (as well as by the arrows in Figure 1), NGF and the IGFs contain the A and B segments shown in Figure 1 as parts of a larger, single polypeptide chain. In this sense they are comparable to proinsulin. In fact, NGF contains a C-peptide region of identical length to the connecting peptide of proinsulin, whereas this segment is foreshortened by about two-thirds in the IGFs. There is no significant homology (identities) with the proinsulin C-peptide in either case. Relaxin, however, is isolated as a two-chain structure comparable to insulin. Preliminary evidence [25] suggests that a prorelaxin molecule analogous to proinsulin is the progenitor of relaxin, but no information about the putative C-peptide is presently available.

A striking additional feature of relaxin and the IGFs is the complete conservation of the disulfide bond pattern of insulin. NGF contains only one of the three disulfide bonds of insulin with the other two being replaced by ones unique to NGF located elsewhere in the molecule (Fig. 2). However, the single conserved disulfide found in all five proteins (emphasized in Fig. 1) is especially important in insulin, as it is located in the re-

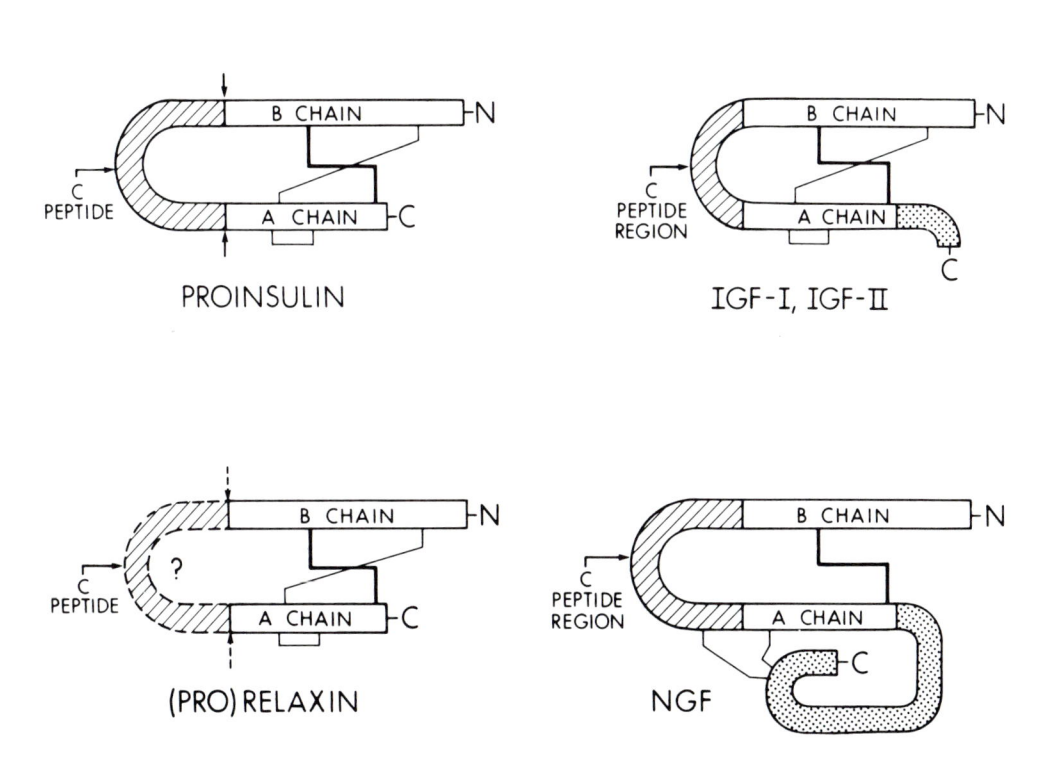

Fig. 2. Schematic comparison of the three-dimensional structure of the five insulin-related growth factors. The diagonally lined segments indicate the connecting "C-bridge" regions (presumed in relaxin) that are not excised in IGF-I, IGF-II, and NGF. The stippled areas represent carboxyl-terminal extensions. The single conserved disulfide is emphasized by the heavy line connecting the A and B chains; other disulfides are shown by lighter lines. For the purposes of these diagrams, the small differences between the IGFs in the length of their C-peptide and the carboxyl-terminal extensions have been ignored [62, 63].

gion of that molecule thought to comprise the receptor-binding site [61]. Both NGF and the IGFs also contain carboxyl terminal extensions that presumably reflect features peculiar to their own unique molecular interactions.

The conversion of insulin and relaxin to two-chain forms and the absence of such processing in NGF and the IGFs may occur as a result of the need to store insulin and relaxin in high concentrations that can be readily released in a large bolus to meet physiological demands [7, 20]. The low steady-state release of NGF and presumably of the IGFs, as judged by the absence of any tissue stores of these molecules, has apparently resulted from evolutionary modifications in the sequence that have eliminated the cleavage points as they became unnecessary. Conversely, this property may have been acquired by insulin and relaxin.

The structural relatedness of the insulin subset is an important illustration of the development of new physiological function through evolutionary change. In addition, it has been instrumental in the emerging ideas concerning the relationship of growth factors and hormones and, most importantly, has provided new views about the mechanism of action of secondary hormones, which are summarized in the ensuing section.

MECHANISM OF ACTION

General Considerations

Although it would be naive to expect that such a diverse group of substances as listed in Table II would share a single mechanism of action, it is not unreasonable to anticipate that some common features will be found. Nonetheless, such generalities have not been easy to identify. On the whole, the polypeptide growth factors stimulate the anabolic metabolism of their target cells following formation of the complex with the cell surface receptor. However, the manner in which this is accomplished is still rather obscure. It does not appear to involve the prolonged production of cyclic nucleotides (although a role for these agents, compatible with transitory changes in concentration, has not been eliminated) and may occur through more than one type of alternative mechanism including the production of other kinds of "second messengers." A more universal feature of these agents appears to be the receptor-mediated endocytosis that ultimately follows the initial interaction with the receptor. However, this is a general phenomenon, found with a large variety of macromolecules [30, 54], and may not be important for function other than as a means for degradation through fusion with lysosomes. On the other hand, it may be of primary importance in the expression of the long-term effects associated with the growth-related processes. The relative merit of this hypothesis is described in the ensuing sections.

Perhaps the most important feature shared by these substances, exemplified by the listing in Table II, is the initiation of both rapid and slow responses. While these are variably expressed in individual cases, they are, nonetheless, a constant feature throughout, and a description of the mechanism of action of any growth factor will necessarily have to account for both phases of activity.

External vs Internal Sites of Action

Although considerable attention is often focused on the long-range effects of growth factors and related substances, particularly those with mitogenic activity, the initial or rapid events that are clearly triggered by the hormone while it is on the cell surface are of equal importance. The relationship of these two phenomena represents the basic enigma that must be resolved before a clear understanding of the principal mechanistic features can be achieved. Several possibilities are currently plausible:

1) The factor interacts with the cell surface to generate a signal that is responsible for both the rapid and slow responses. A close corollary of this hypothesis is that one of the rapid responses in turn produces the signal for the long-term effects, ie, the slow responses could, in theory, be generated without all of the rapid ones if the second signal were introduced separately. In either case, the entire process would not involve internalization (for other than degradative purposes).

2) The factor interacts with the cell surface to generate two independent signals, one producing the rapid effects and the other producing the long-term, growth-related responses. As above, this hypothesis does not require internalization for any part of the activity.

3) The factor interacts with the cell surface to produce only a signal for the rapid events. Internalization is then required to produce the signal for the slow responses.

In any of the models with more than one signal, a concerted mechanism requiring both to act before one or both temporal responses is initiated is also possible. Furthermore, only a single class of receptors is required for any of the models, although two independent receptor types are equally compatible with the last two hypotheses. Also, the models do not specify the nature of the "signal." These could range from second messengers to changes in metabolite or ion concentrations and may well vary from one system to another. Finally, the models do not place emphasis on the relative importance of either phase of activity. In fact, the surface-mediated events have been more extensively studied and in many cases, such as insulin, are probably more important physiologically.

The possibility that internalization by absorptive pinocytosis (or receptor-mediated endocytosis) plays a decisive role in the mechanism of action of any of the secondary hormones is a relatively new idea. Bulk phase pinocytosis, which is simply the engulfment of extracellular fluid by invagination of the plasma membrane and subsequent fusion resulting in the release of a vesicle into the cytoplasm, allows the cell to sample its environment and, therefore, in a general sense, mediates communication between the cell and its surroundings. When this process is modified to include an initial complexation of soluble components to specific receptor molecules on the cell surface, it becomes appreciably more sensitive as the agent can be concentrated as much as 1,000-fold prior to internalization [71]. This feature is certainly of importance for the uptake of some metabolites that enter the cell via this pathway, and may well be of significance in endocrine systems too.

The diversity of substances that are internalized by absorptive pinocytosis underscores the importance of this phenomenon in the overall interaction between cells and their environment. In addition to secondary hormones, toxins, carrier proteins — such as low-density lipoproteins and transferrin — various glycoproteins, lysosomal enzymes, antibodies, and viruses are, at least in part, transported across the plasma membrane by this process [30, 54]. In fact, the detailed studies [see for example, 31, 32] with low-density lipoprotein have been among the most revealing in establishing the essential features of receptor-mediated endocytosis, which appear to have broad applicability to other ligands. Of particular note is the participation of specialized areas in the membrane called "coated pits" because of the bristle coat composed primarily of the protein clathrin underlying these regions, the formation of "coated vesicles" from invagination and fusion of the membrane, and their subsequent association with intracellular organelles, in particular, lysosomes [31].

A fundamental element in this scheme is the concomitant internalization of the receptor along with the ligand. At least in experimental situations, this can lead to a major reduction in the number of receptor molecules on the cell surface. The extent and duration of this effect seem to depend largely on whether or not the internalized receptors are recycled back to the plasma membrane. Receptors for endocrine substances do not appear to be reutilized and must be replaced by de novo protein synthesis, whereas receptors for carrier molecules such as LDL are extensively reused. As a result of the greater time required for the new synthesis of the former type, the deficit is more readily demonstrable. In such cases, the process has been designated receptor *down regulation* [13, 48, 72].

The two categories of receptor treatment following endocytosis may signify an important physiological distinction. Those systems in which receptors are recycled generally involve the transfer of the ligand to the cell for further processing and utilization. In contrast, irreversible receptor consumption is found in those instances where information transfer may be viewed as the overall purpose of the original interaction. This suggests the possibility that the receptor itself might play a major role in any signal resulting from the internalization event. Such a hypothesis has been suggested by Fox and Das [23] in their "Endocytotic Activation Model" for EGF. They proposed that after the agent binds to its surface receptor and enters the cell by endocytosis, it is transported to the lysosome where the contents of the endocytotic vesicle come in contact with proteolytic enzymes. The digestion of the receptor would result in the generation of a new "messenger" or perhaps an enzyme which, in turn, would catalyze the production of an agent responsible for the subsequent long-term effects.

The principal roles that internalization might play in hormone action are summarized in Table III. The first three entries list the basic possibilities that might apply to systems in which internalization is a fundamental part of the mechanism (see model 3 above). In the first two possibilities, lysosomal fusion, and subsequent proteolysis, to produce an active peptide to act as the second signal is a major alternative. However, the hormone or acceptor can be envisioned to act without such modification, too. The third possibility suggesting that the receptor can act in some fashion after translocation to an intracellular structure is consistent with the hypothesis that it is the receptor that must be internalized for long-term effects to develop and that the hormone serves only to initiate this event. The final two entries of Table III, degradation and desensitization, have been shown to occur in experimental systems but still must be established to be of importance physiologically.

Evidence in support of either roles 1 or 3 (Table III) is provided by several reports identifying intracellular receptors for a number of secondary hormones (Table IV). These entities, which have been identified by binding assays, have been found to be associated with elements of the nucleus, Golgi membranes, and other organelles. However, in no case has any functionality been associated with them, and it has been suggested by one group [5] that they represent solely biosynthetic precursors of the plasma membrane receptors. However, it is difficult to conceive how nuclear receptors are related to biosynthetic precursors. At present, no relationship between a cell surface and intracellular receptor has been demonstrated and, in the case of the insulin receptor, naturally occurring antibodies directed against it, which are characteristic of type B syndrome insulin-resistant diabetes, do not cross-react with nuclear receptors [29]. However, either the translocation process or the new intracellular environment could materially alter or screen the antigenic determinants of the nuclear entity. More definitive characterization of both plasma membrane and nuclear receptors will be required to determine if any relationships exist.

TABLE III. Potential Roles of Internalization in Hormone Action

1. Interaction of hormone or receptor or fragment thereof with intracellular receptors.
2. Processing of hormone or receptor to generate "second messenger."
3. Direct action of receptor after translocation to intracellular organelle.
4. Degradation.
5. Regulation of cell sensitivity to hormone by decreasing number of surface receptor.

TABLE IV. Intracellular Receptors for Polypeptide Hormones/Factors

Hormone/factor	Tissue	Location	Ref.
Insulin	Liver	Golgi	[60]
	Heart	Mitochondria	[22]
	Liver	Microsomes	[42]
	Liver	Nucleus	[28, 41]
Nerve growth factor	Dorsal root ganglia	Nucleus	[3]
	Pheochromocytoma (PC12)	Nucleus	[85]
Growth hormone	Liver	Golgi	[5]
	Liver	Microsomes	[56]
Prolactin	Liver	Golgi	[82]
Epidermal growth factor	Liver	Nucleus	[53]

Specific Examples

Nerve growth factor. NGF is one of the most extensively characterized polypeptide growth factors and represents the single case in which there exists considerable data to support a mechanism of action involving the internalization of a peptide hormone [6]. In fact, the idea that polypeptide growth factors or related hormones are taken up by their target cells first received experimental support in studies on NGF [39] and was only subsequently established for other related substances [9, 77]. It should be noted that the demonstration of internalization was, as with other polypeptide hormones, preceded by a period characterized by the prevailing view that hormones acted solely at the external surface of their target cells, a theory engendered in large measure by experiments utilizing insolubilized hormones [6, 16]. While this methodology clearly was not without value, it obscured the discovery of internalization which, at least in the case of NGF, appears to be of fundamental importance.

To gain a proper appreciation of the proposed mechanism of NGF, one must view it in the context of its overall physiological role. It is particularly important in this regard to recognize that the target tissues for this substance, ie, sympathetic and selected sensory neurons, have a unique morphology that changes dramatically during development and have certain specific requirements such as the establishment of appropriate contacts allowing constant communication with their periphery. These unique problems have been effectively managed by modifying the classical endocrine scheme. Rather than having a single site of synthesis, storage, and release, the hormone is elaborated in multiple sites throughout the organism. These tissues ultimately form the end organs that are innervated by the responsive neurons. Systemic transport is replaced by interstitial diffusion, which has the added advantage of providing tropic stimulation that aids in the development of the neuronal network [14] as the neurites presumably grow along gradients of NGF. This delivery system is also highly flexible with the length of the diffusion pathway depending on the extent of axonal maturation; the final limiting distance is the transsynaptic passage.

The mechanism of trophic stimulation initiated when NGF reaches the neuron and binds to its plasma membrane receptor also remains unaltered throughout development, despite the changes that occur in cell shape. The principal destination of NGF is the cell body which it reaches after endocytotic uptake at any stage in the development of the cell. However, as the growth cone of the axon approaches its end organ and the extracellular transit of NGF from its source thereby decreases, the length of the intracellular flow of NGF, in terms of both time and distance, increases. As with other features of the system, this process subserves an additional function for the neurons; namely, it allows the NGF to act as a messenger between the synapse and the perikaryon. It appears that neurons that have formed the correct synaptic junctions are assured a continuous supply of NGF, whereas those that do not are somehow deprived and subsequently expire. Agents acting in this capacity are collectively referred to as chromatolytic messengers, because the interruption of their flow results in the state of neuronal chromatolysis which, if not corrected, will lead to cell death [15]. It seems appropriate to view this chromatolytic function as being synonymous with the survival or maintenance activity of NGF.

The message that NGF brings to the cell body is manifested by changes in specific transcriptional events among which are the induction of tyrosine hydroxylase and dopamine β-hydroxylase [78]. While the manner in which this is accomplished is unknown, it appears probable that the nuclear receptors identified in responsive neurons are somehow involved [3]. These entities display binding and solubility characteristics that clearly distinguish them from plasma membrane receptors of the same tissue, and they are found to be highly concentrated in responsive neurons. Subcellular fractionation experiments suggest that neurons maximally loaded with ^{125}I-NGF introduced by retrograde axonal transport contain about 15–30% the labeled hormone in the nucleus [46]. As judged by EM autoradiography, much of the tracer appears in lysosomal structures [68].

Recently, nuclear receptors for NGF have also been identified in the PC-12 cell line, derived from a pheochromocytoma [85], which adopt a differentiated phenotype in response to NGF [38]. The properties of their receptors are similar to the characteristics of the nuclear receptors identified in the normal target tissues of NGF but were found to be localized in the nuclear envelope. This result is intriguing in view of the finding that initiation of microtubule orientation preceding neurite outgrowth in neuroblastoma cells occurs at or in close proximity to the nuclear envelope [73]. As the stimulation of neurite outgrowth is one of the major effects of NGF which has a delayed onset [49], the interaction of NGF with these receptors may be necessary for this biological effect.

EGF. Unlike NGF, epidermal growth factor (EGF) has mitogenic activity, both in vivo and in vitro [10]. Given the availability of ample quantities, it has been used extensively as a prototype for mitogenic agents in studies on the control of cell division. In addition, biological activities such as precocious eyelid opening and incisor eruption, and inhibition of gastric acid secretion, have been associated with the peptide. However, a complete spectrum of its physiological role encompassing its site of synthesis, means of transport, and target tissues remains substantially undefined. Nonetheless, data collected from experiments in tissue culture have provided considerable information about the mode of action of EGF. As with other polypeptide growth factors, it complexes with high affinity to cell surface receptors, a process which is followed by endocytosis [9]. However, there appear to be significant differences from the internalization of NGF: The process is much more rapid with respect to both uptake and proteolytic degradation [10]. Virtually all of the internalized EGF is extensively degraded in a few hours in contrast to retrogradely transported NGF, which remains in the cell soma for well over a day. Despite the efficacy of internalization, there appears to be compelling support for an "external" mechanism. Indeed, Aharonov et al [2] have shown that the continued occupation of at least a small portion of cell surface receptors during the period of incubation necessary to produce cell division (~18 hrs) is a prerequisite for mitogenicity, and Das [19] has provided evidence for the production of a second messenger, which stimulates DNA synthesis when transferred to nuclei not exposed to EGF. However, it was not established in this study whether the putative messenger was generated from an external or internal location of the EGF. Alternatively, Carpenter et al [11] have demonstrated a specific phosphorylation of the receptor by a kinase that is either closely associated with the receptor or is an integral part of it. Although no functional significance has yet been attached to this finding, it might represent an event triggering the uptake of the complex, which would suggest a mechanistic role for it. It might also be noted that the observations of Aharonov et al [2] are equally compatible with a mechanism that requires a controlled uptake of EGF or its receptor, as opposed to the controlled production of a second messenger at the plasma membrane. The recent observations of Moriarty and Savage [53] of the presence of apparent EGF receptors in the nuclei of hepatocytes would be consistent with such a role for internalization.

CONCLUDING REMARKS

In the last several years polypeptide growth factors have been transformed from biological curiosities to agents of prime importance recognized for their in vivo activities as well as their usefulness in tissue culture and in the study of cell division. Central to this rise in stature has been the progressive realization of their hormonal character. As expounded in this article, it now seems fitting to group these substances in a broad category of secondary hormones organized according to principles that transcend the classical definitions. This has not only served to develop further our understanding of their mode of action but has also expanded our concepts about the action of classical hormones. Among other things, it has brought considerable attention to the phenomenon of receptor-mediated endocytosis, particularly as a potential mechanistic feature. This exciting concept, which still requires substantial clarification and development, has been a primary focus of this article. As with all discussions that are heavily dependent on hypotheses and speculations, it may be anticipated that substantial revision will be necessary when the results of future experimentation become available. In the interim, it may be hoped that these ideas will stimulate that research.

ACKNOWLEDGMENTS

This work was supported in part by National Institutes of Health Research Service Award GM-07200, Medical Scientist, from the National Institute of General Medical Sciences, National Institutes of Health research grant NS-10229, and by American Cancer Society research grant BC-273.

REFERENCES

1. Adamson JW, Brown JE: Symp Soc Dev Biol 35:161, 1978.
2. Aharonov A, Pruss RM, Herschman HR: J Biol Chem 253:3970, 1978.
3. Andres RY, Jeng I, Bradshaw RA: Proc Natl Acad Sci USA 74:2785, 1977.
4. Antoniades HN, Scher CD, Stiles CD: Proc Natl Acad Sci USA 76:1809, 1979.
5. Bergeron JJM, Posner BI, Josefberg Z, Sikstrom R: J Biol Chem 253:4058, 1978.
6. Bradshaw RA: Ann Rev Biochem 47:191, 1978.
7. Bradshaw RA, Niall HD: In Seng CJ, Pin L, Tambyah JA, Shanmugaratnam S, Boon PYP, Ark CNS, Foo KNK, Tau NT (eds): "Endocrinology" (Proc 6th Asia and Oceania Cong of Endocrinology, vol 1). Singapore: Stamford College Press (Pte.) Ltd., pp 358–364, 1978.
8. Bradshaw RA, Niall HD: TIBS 3:274, 1978.
9. Carpenter G, Cohen S: J Cell Biol 71:159, 1976.
10. Carpenter G, Cohen S: Ann Rev Biochem 48:193, 1979.
11. Carpenter G, King L Jr, Cohen S: Nature 276:409, 1978.
12. Catt KJ, Dufau E: Ann Rev Physiol 39:529, 1977.
13. Catt KJ, Harwood JP, Aguilera G, Dufau ML: Nature 280:109, 1979.
14. Chen MM, Chen J, Levi-Montalcini R: Arch Ital Biol 116:53, 1978.
15. Cragg BG: Brain Res 23:1, 1970.
16. Cuatrecasas P: Ann Rev Biochem 43:169, 1974.
17. Cuatrecasas P (ed): "The Specificity of Animal, Bacterial and Plant Toxins" (Receptors and Recognition Series B11, Vol 1). London: Chapman and Hall, 1976.
18. Czech M: Ann Rev Biochem 46:359, 1977.
19. Das M: Proc Natl Acad Sci USA 77:112, 1980.
20. Dodson GG, Isaacs N, Bradshaw RA, Niall HD: In Brandenburg D, Wollmer A(eds): "Insulin: Chemistry, Structure and Function of Insulin and Related Hormones." Berlin: de Gruyter, pp 695–701, 1980.
21. Eylar EH: Proc Natl Acad Sci USA 67:1425, 1970.
22. Forgue ME, Freychet P: Diabetes 24:715, 1975.
23. Fox CF, Das M: J Supramol Struc 10:199, 1979.
24. Frazier WA, Angeletti RH, Bradshaw RA: Science 176:482, 1972.
25. Frieden EH, Yeh L: Fed Proc 35:1628, 1976.
26. Froesch ER, Burgi H, Ramseier EB, Bally P, Labhart A: J Clin Invest 42:1816, 1963.
27. Gey GO, Thalhimer W: JAMA 82:1609, 1924.
28. Goldfine ID, Smith GJ: Proc Natl Acad Sci USA 73:1427, 1976.
29. Goldfine ID, Vigneri R, Cohen D, Pliam NB, Kahn CR: Nature 269:698, 1977.
30. Goldstein JL, Anderson RGW, Brown MS: Nature 279:679, 1979.
31. Goldstein JL, Brown MS: Curr Top Cell Regul 11:147, 1976.
32. Goldstein JL, Brown MS: Annu Rev Biochem 46:897, 1977.
33. Gorin PD, Johnson EM: Proc Natl Acad Sci USA 76:5382, 1979.
34. Gospodarowicz D, Bialecki H, Greenburg G: J Biol Chem 253:3736, 1978.
35. Gospodarowicz D, Jones KL, Sato G: Proc Natl Acad Sci USA 71:2295, 1974.
36. Gospodarowicz D, Moran JS: Ann Rev Biochem 45:531, 1976.
37. Gospodarowicz D, Rudland P, Lindstrom J, Benirschke K: Adv Metab Disord 8:302, 1975.
38. Greene LA, Tischler AS: Proc Natl Acad Sci USA 73:2424, 1976.
39. Hendry IA, Stockel K, Thoenen H, Iversen LL: Brain Res 68:103, 1974.
40. Hershko A, Mamont P, Shields R, Tompkins GM: Nature New Biol 232:206, 1971.
41. Horvat A: J Cell Physiol 97:37, 1978.
42. Horvat A, Li E, Katsoyannis PG: Biochim Biophys Acta 382:609, 1975.
43. Huxley JS: Biol Revs 10:427, 1935.
44. Jakob A, Hauri Ch, Froesch ER: J Clin Invest 47:2678, 1968.
45. James R, Niall H, Kwok S, Bryant-Greenwood G: Nature 267:544, 1977.
46. Johnson EM, Andres RY, Bradshaw RA: Brain Res 150:319, 1978.
47. Kahn CR: In Korn ED (ed): "Methods in Membrane Biology," vol 3. New York: Plenum Press, pp 81–146, 1975.
48. Lesniak MA, Roth J: J Biol Chem 251:3720, 1976.
49. Levi-Montalcini R, Angeletti PU: Physiol Rev 48:534, 1968.
50. Levi-Montalcini R, Booker B: Proc Natl Acad Sci USA 46:384, 1960.

51. Megyesi K, Kahn CR, Roth J, Neville DM Jr, Nissley SP, Humbel RE, Froesch ER: J Biol Chem 250:8990, 1975.
52. Moore MAS: Symp Soc Dev Biol 35:181, 1978.
53. Moriarity BM, Savage CR Jr: Abstrs 6th Intl Cong Endocrinol:175, 1980.
54. Neville DM, Chang TM: In Kleinzeller A, Bronner F (eds): "Current Topics in Membranes and Transport," vol 10. New York: Academic Press, pp 65–150, 1978.
55. Niall HD, Hogan ML, Tregear GW, Segre GV, Hwang P, Friesen H: Rec Prog Horm Res 29:387, 1973.
56. Niall M, Wilmot H, Bradshaw RA: Proc Endo Soc Australia 22:62, 1979.
57. Nicol DSHW, Smith LF: Nature 187:483, 1960.
58. Nissley SP, Rechler MM: Natl Cancer Inst Monogr 48:167, 1978.
59. Noonan KD: In Kleinzeller A, Bronner F (eds): "Current Topics in Membranes and Transport," vol 11. New York: Academic Press, pp 397–461, 1978.
60. Posner BI, Josefsberg Z, Bergeron JJM: J Biol Chem 253:4067, 1978.
61. Pullen RA, Lindsay GD, Wood SP, Tickle IJ, Blundell TL, Wollmer A, Krail G, Brandenburg D, Zahn H, Gliemann J, Gammeltoft S: Nature 259:369, 1976.
62. Rinderknecht E, Humbel RE: J Biol Chem 253:2769, 1978.
63. Rinderknecht E, Humbel RE: FEBS Lett 89:283, 1978.
64. Robison GA, Butcher RW, Sutherland EW: "Cyclic AMP." New York: Academic Press, pp 19–22, 1971.
65. Ross R, Vogel A: Cell 14:203, 1978.
66. Rudland PS, Jimenez de Asua L: Biochim Biophys Acta 560:91, 1979.
67. Rutter WJ: Symp Soc Dev Biol 35:3, 1978.
68. Schwab M, Thoenen H: Brain Res 122:459, 1977.
69. Schwabe C, McDonald JK, Steinetz BG: Biochem Biophys Res Commun 75:503, 1977.
70. Schwabe C, Steinetz B, Weiss G, Segaloff A, McDonald JK, O'Byrne E, Hochman J, Carriere B, Goldsmith L: Rec Prog Horm Res 34:123, 1978.
71. Silverstein SC, Steinman RM, Cohn ZA: Ann Rev Biochem 46:669, 1977.
72. Soll AH, Kahn CR, Neville DM Jr, Roth J: J Clin Invest 56:769, 1975.
73. Spiegelman BM, Lopata MA, Kirschner MW: Cell 16:253, 1979.
74. Steiner DF, Oyer PE: Proc Natl Acad Sci USA 57:473, 1967.
75. Stiles CD, Capone GT, Scher CD, Antoniades HN, Van Wyk JJ, Pledger WJ: Proc Natl Acad Sci USA 76:1279, 1979.
76. Tata JR: Ciba Found Symp 41:297, 1976.
77. Terris S, Steiner DF: J Clin Invest 57:885, 1976.
78. Thoenen H, Angeletti PU, Levi-Montalcini R, Kettler R: Proc Natl Acad Sci USA 68:1598, 1971.
79. Thomas KA, Riley MC, Lemmon SK, Baglan NC, Bradshaw RA: J Biol Chem 255:5517, 1980.
80. Van Wyk JJ, Underwood LE, Hintz RL, Clemmons DR, Voina SJ, Weaver RP: Rec Prog Horm Res 30:259, 1974.
81. Varon SS, Bunge RP: Ann Rev Neurosci 1:327, 1978.
82. Verma AK, Posner BI: Abstrs 61st Mtg Endocrine Soc:80, 1979.
83. Westall FC, Lennon VA, Gospodarowicz D: Proc Natl Acad Sci USA 75:4675, 1978.
84. White A, Goldstein AL: Adv Metab Disord 8:359, 1975.
85. Yankner VA, Shooter EM: Proc Natl Acad Sci USA 76:1269, 1979.
86. Zapf J, Rinderknecht E, Humbel RE, Froesch ER: Metabolism 27:1803, 1978.

Regulation of B Lymphocyte Activation by the Fc Portion of Immunoglobulin

Edward L. Morgan and William O. Weigle

Department of Immunopathology, Scripps Clinic and Research Foundation, La Jolla, California 92037

Murine splenic B lymphocytes are induced to proliferate and undergo polyclonal activation in the presence of Fc fragments, AHGG, antigen-antibody complexes, and CH_3 fragments derived from plasmin digestion of human Ig. The unifying feature of the polyclonal antibody response induced by these agents is that in all cases a portion of the constant region of the Ig molecule (ie, Fc region) is present. Fragments of Ig lacking the Fc piece, such as Fab and $F(ab')_2$ were found not to be stimulatory. In addition, a model is proposed to account for the regulatory effects of antigen-antibody complexes on an ongoing humoral immune response.

Key words: Fc fragments, immune complexes, macrophages, polyclonal antibody response

Antigen-antibody complexes [1–11], aggregated gamma globulin [12], and Fc fragments [11, 13–18] are known to be important in both in vivo and in vitro activation and/or regulation of humoral immune responses. These phenomena require the Fc piece, since they are not manifested by Fab or $F(ab')_2$ fragments.

The mechanism(s) involved in activation of bone marrow-derived (B) lymphocytes by Fc fragments, aggregated gamma globulin, and immune complexes may be the same. Murine B lymphocytes were shown to undergo a proliferative response in the presence of Fc fragments derived from mammalian immunoglobulin (Ig) [11, 13–15], and this observation was recently extended to aggregated gamma globulin [12] and immune complexes [11]. Immune complexes formed after the separate addition of antigen and antibody to in vitro cultures resulted in a pronounced proliferative response. Moreover, immune complexes have been shown to be capable of both suppressing [4] and enhancing [3] immune responses, and recently Fc fragments have been shown to enhance weak antibody responses [17, 18], whereas the addition of Fc fragments to strongly responding cultures results in a suppressed response.

Evidence is presented in this report that activation of mouse B lymphocytes by Ig requires the presence of a portion of the constant region of the Ig heavy chain. In addition a model is proposed that describes a role for the Fc region of Ig in regulation of in vivo immune responses.

Received April 1, 1980; accepted August 21, 1980.

0091-7419/80/1402-0201$02.50 © 1980 Alan R. Liss, Inc.

MATERIALS AND METHODS

Animals

Male mice of the inbred C57BL/6 strain were obtained from L.C. Strong Laboratories (Del Mar, CA) and Jackson Laboratories (Bar Harbor, ME). All mice were between 8 and 10 weeks of age.

Immunizations

Mice were immunized with 200 µg ovalbumin (OVA) (Miles Laboratories) suspended in Freunds complete adjuvant intraperitoneally (ip). The mice were boosted 30 and 60 days postimmunization with 200 µg OVA in Freunds incomplete adjuvant and were bled 10 days following the last injection. The pooled mouse serum (anti-OVA) was used at a concentration of 0.5% in the polyclonal antibody response.

Preparation of Fc Fragments

A human IgG1 myeloma protein (Fi) was a gift from Dr. Hans L. Spiegelberg, Scripps Clinic and Research Foundation. The IgG1 was purified by ammonium sulfate fractionation, followed by DEAE cellulose chromotography, with 0.01 M phosphate buffer pH 8 used as the eluent.

Fc fragments were obtained by digestion of IgG1 with papain (Sigma Chemical Co.) in the presence of L-cysteine (Sigma) and ethylenediaminetetraacetic acid (EDTA) (J.T. Baker Chemical Co.) for 5 h [19]. Following digestion the material was chromatographed on Sephadex G-100 (Pharmacia Fine Chemicals) to remove any undigested IgG. The Fc and Fab fragments were then separated from each other by DEAE chromatography [20].

Preparation of Heat-Aggregated Human Gamma Globulin

Pooled IgG was obtained as Cohn Fraction II through the courtesy of the American Red Cross National Fractionation Center, with the partial support of National Institutes of Health grant HE-138801, and was purified by DEAE cellulose chromatography, with 0.01 M phosphate buffer pH 8.0 used as the eluent. Human gamma globulin (HGG) was heat aggregated as described previously [21]. The HGG was adjusted to a concentration of 20 mg/ml in 0.01 M phosphate buffer pH 8.0 and maintained at a temperature of 63°C for 25 min. The heated material was then left at 4°C for 24 h prior to collection of the aggregated HGG. The aggregates were precipitated twice with 0.62 M sodium sulfate and dialyzed extensively against phosphate-buffered saline (PBS), 0.001 M phosphate pH 7.2, 0.15 M NaCl before use.

Preparation of F(ab')$_2$ Fragments

F(ab')$_2$ fragments were prepared by digestion of pooled HGG with pepsin (Sigma Chemical Co.) for 18 h [22]. Following digestion, the material was chromatographed on Sephadex G-100 (Pharmacia) to remove any undigested HGG. The F(ab')$_2$ fragments were then heat aggregated as described above.

Preparation of Plasmin-Digested IgG1

Plasmin digests were prepared by incubating the IgG1 myeloma protein for 24 h at 37°C [23]. The digested IgG1 was resolved into two peaks by Sephadex G-150 chromotography. Polyclonal activity was found only in the small molecular weight (CH$_3$ domain) peak. The optimal concentration was found to be 25 µg/culture.

Polyclonal Antibody Response Assay

For the generation of the polyclonal plaque-forming cell (PFC) response, spleen cells were suspended to a concentration of 6×10^6/ml in RPMI 1640 supplemented as described previously [16]. Duplicate cultures of 6×10^5 cells/0.3 ml were incubated in microtiter plates (3040 Microtest II, Falcon Plastics) at 37°C in 5% CO_2. For the response to immune complexes, 0.5% mouse anti-OVA was substituted for the 0.5% NMS.

Plaque-Forming Cell Response

The response to TNP was assayed by the slide modification of the Jerne plaque assay [24]. Heavily conjugated TNP-SRBC were prepared by the method of Kettman and Dutton [25] and were used as indicator cells to measure the response to TNP. Data are recorded as PFC/10^6 spleen cells ± standard error. Each experiment was performed a minimum of 3 times, and the experiments shown are representative of all the data.

RESULTS

Generation of the Polyclonal Antibody Response

To determine the role of the Fc portion of antibody in the generation of an in vitro polyclonal antibody response, various reagents were assessed for their polyclonal inducing ability. Fc fragments derived from papain digestion of human IgG-induced murine spleen cells to produce a significant polyclonal antibody response (Table I). The polyclonal antibody response produced by Fc fragments was approximately 50% that observed with LPS. Fab fragments and intact IgG were unable to induce a polyclonal antibody response [11, 13, 16].

The fact that intact Ig does not induce a polyclonal response does not preclude the possibility that modification of intact Ig could render it stimulatory. HGG was heat aggregated (AHGG) and assessed for its ability to stimulate murine spleen cells. A significant polyclonal antibody response was produced with AHGG compared to native HGG (157 PFC vs 4 PFC) (Table II). That the presence of the Fc portion of the AHGG was critical for the preparation to be stimulatory was concluded from the observation that aggregated F(ab')$_2$ was unable to activate spleen cells.

Since immune complexes had previously been shown to be able to induce murine splenic B cells to proliferate, it was important to ascertain whether they could function as polyclonal activators as well. Immune complexes were prepared by substituting 0.5% mouse anti-OVA for the 0.5% NMS in culture and adding increased amounts of OVA to the cultures. The results indicate that the maximum polyclonal response was achieved

TABLE I. Fc Fragment Induced Polyclonal Antibody Response

Stimulator	Direct Anti-TNP PFC/10^6 ± SE[a]
–	6 ± 1
Fc fragment[b]	162 ± 25
LPS[c]	345 ± 55

[a] The response was measured on day 3 of culture.
[b] 100 μg Fc/culture.
[c] 20 μg (E coli 055:B5) LPS/culture.

TABLE II. Aggregated Human Gamma Globulin Induced Polyclonal Antibody Response

Stimulatory	Direct Anti-TNP PFC/10^6 ± SE[a]
–	5 ± 5
HGG[b]	4 ± 3
AHGG[b]	157 ± 13
A(Fab')$_2$[b]	6 ± 1
Fc fragment[b]	171 ± 9

[a]The response was measured on day 3 of culture.
[b]100 μg/culture.

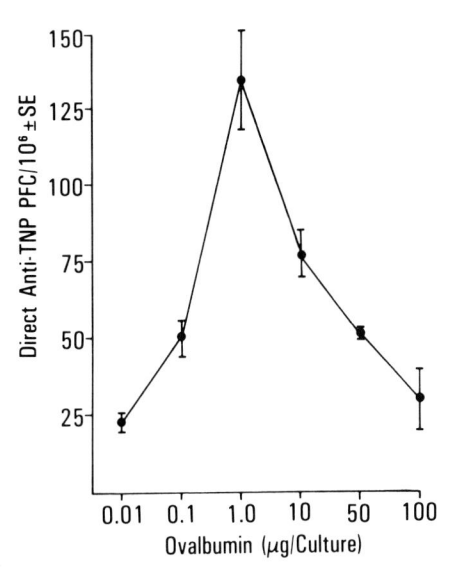

Fig. 1. Increasing amounts of ovalbumin were added to a constant amount of mouse anti-ovalbumin (0.5%). The polyclonal antibody response was measured against TNP-SRBC on day 3 of culture. The direct or IgM plaque-forming response was measured.

when 0.1 μg OVA was added to culture (Fig. 1). It is important to note that as the amount of antibody in mouse serum changes, the peak amount of OVA needed to produce the optimal response also changes (data not shown).

To determine which part of the Fc fragment was needed for the polyclonal response, advantage was taken of the enzyme plasmin, which cleaves between the CH_2 and CH_3 domain, producing Fabc and CH_3 fragments [26]. The small molecular weight plasmin-digested material (ie, CH_3) produced dramatic polyclonal antibody response (19-fold) compared to control cultures (Table III). The polyclonal antibody response generated by plasmin-digested material was always higher than that induced by Fc fragments. Polyclonal

TABLE III. The Ability of Plasmin Digested HGG to Induce a Polyclonal Antibody Response

Stimulator	Direct Anti-TNP PFC/10^6 ± SE[a]
–	36 ± 2
Plasmin digest[b]	678 ± 47
Fc fragment[c]	154 ± 8

[a]The response was measured on day 3 of culture.
[b]25 μg/culture. Sephadex G-150 purified material from plasmin digested HGG.
[c]100 μg/culture.

activiation was never greater than background controls when the large molecular weight plasmin digest was used (data not shown).

DISCUSSION

Murine splenic lymphocytes undergo polyclonal activation in the presence of Fc fragments, AHGG, immune complex, and CH_3 fragments derived from plasmin digestion of human Ig. The unifying feature of the polyclonal antibody response induced by these agents is that in all cases a portion of the constant region of the Ig heavy chain (ie, Fc region) is present. Attempts to induce a polyclonal antibody response with fragments lacking the Fc piece, such as Fab and $F(ab')_2$ resulted in no activation [11, 13].

Fc fragments and AHGG are unique polyclonal activators in that they required the presence of both macrophages and T lymphocytes [12, 16]. The macrophage is responsible for the generation of a mitogenic 14,000 molecular weight Fc subfragment through enzymatic cleavage of the intact Fc fragment or AHGG [12, 15, 16]. These Fc subfragments are responsible for inducing B cells to proliferate, and in the presence of a signal provided by T lymphocytes [16] or soluble factors derived from these cells [27] the B cells differentiate to polyclonal antibody synthesis and secretion. Polyclonal activators such as LPS, dextran sulfate, and purified protein derivative activate B lymphocytes to proliferate and synthesize antibody in the absence of macrophages [8–30]. These results differ from ours in that Fc fragment-induced stimulation has a mandatory requirement for both macrophages and T cells.

The requirement for T lymphocytes has been demonstrated for both the Fc fragment [16] and AHGG [12] induced polyclonal antibody responses. The polyclonal antibody response to immune complexes presumably has parallel cellular requirements, although formal proof is unavailable. That T lymphocytes provide a nonspecific second signal has also been suggested by Parker et al [21]. These authors found that activation of murine B lymphocytes by anti-Ig-coated polyacrylamide beads resulted in proliferation and, in the presence of concanavalin A derived supernate, differentiation to polyclonal antibody production [31].

Other investigators have employed anti-Ig as a probe for studying B lymphocyte activation [31–36]. Parker [32] first reported that purified rabbit anti-mouse Ig covalently linked to polyacrylamide beads could stimulate mouse B lymphocytes to proliferate. Weiner et al [33] observed that the inability of soluble anti-Ig to stimulate mouse B lymphocytes was an age-related phenomenon. The proliferative response to anti-Ig appeared when the mice reached 5 to 7 months of age. More recently, Sieckmann et al [34] have shown that a soluble purified anti-mouse μ chain had the capacity to activate mouse B lymphocytes to proliferate. Moreover, the proliferative response was independent of macrophages and T lymphocytes [35]. Stimulation by anti-Ig does not appear to be related to the Fc-mediated stimulation described here, because in all reports in the literature where anti-mouse Ig was used it was also demonstrated that F(ab')$_2$ fragments were as effective as the intact molecule [33–35].

It was observed by Sidman and Unanue [36–38] that in the absence of the proper accessory cofactor(s) derived from serum, anti-mouse IgM antibodies were non-mitogenic and in fact were potent suppressors of other mitogenic responses. The cofactor required for the anti-IgM-induced mitogenesis was generated from serum by 2-ME. These authors found that F(ab')$_2$ fragments of anti-IgM were as stimulatory as the intact antibodies [36]. Our results are distinguished from these reports, since AHGG and immune complex induced stimulation are dependent on the Fc piece of Ig.

The cellular events in both the activation of B cells by Fc fragments and the regulatory effect of these fragments on the immune response suggest a model for B cell participation in immunoregulation by antigen-antibody complexes. We have recently observed that Fc fragments have potent adjuvant properties [17, 18]. Fc fragments enhance both the in vivo and in vitro antibody response to SRBC. In addition, the in vitro antibody response to the hapten-protein conjugate TNP-KLH was enhanced. The Fc fragment adjuvant effect was found to be mediated through T cells [18].

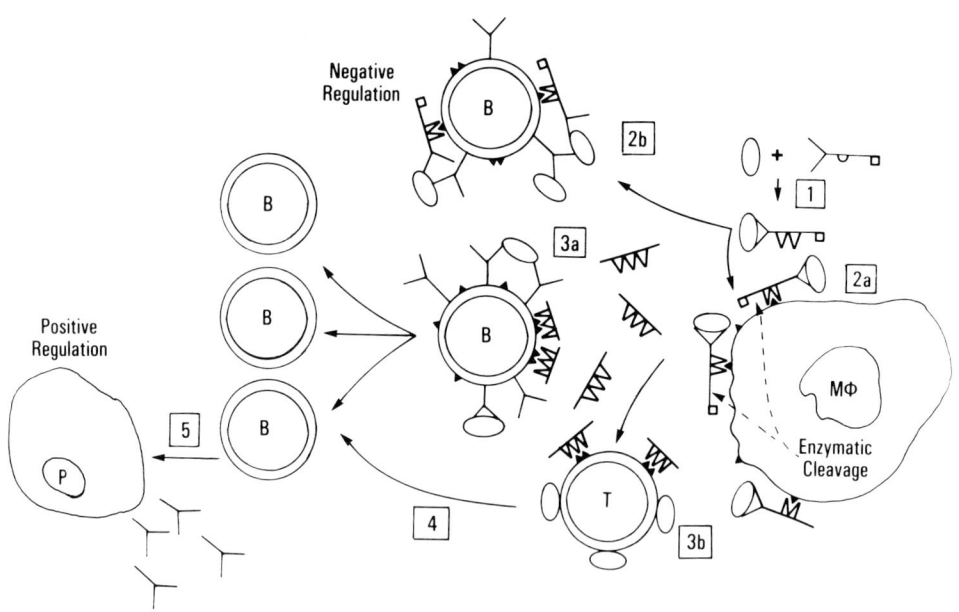

Fig. 2. Proposed immune complex mediated regulation pathway. Antigen-antibody complexes form exposing an Fc attachment site (1); the complex binds to the macrophage Fc receptor, and macrophage enzymes cleave the Fc into an Fc subfragment (2a); the cleavage of the Fc subfragment exposes another binding site, which is recognized by B lymphocyte and T lymphocyte (3b) receptors (3a, b); the binding of the Fc subfragment enhances the antigen-driven response (4,5); when the ratio of antigen to antibody is such that the Fc receptors on the macrophage become saturated, the complexes attach to the B lymphocytes through their Fc receptor and the antigen in the complex crosslinks the Fc receptor and Ig receptor, resulting in a suppression of the response (2b).

Since the cellular events in B cell activation by Fc fragments and antigen-antibody complexes appear to be similar, perhaps the mechanism(s) involved in regulation of B lymphocyte function by Fc fragments may be the same as that involved in regulation by immune complexes. Furthermore, both agents provide positive [3, 11–18] and negative [4, 9–11] signals, which regulate the immune response. The diagram in Figure 2 represents a model for antibody mediated regulation of an immune response. In this model antigen interacts with specific antibody (1), forming an immune complex that results in a conformation change in the antibody molecule potentiating a reactive site in the Fc region. The complex binds either to the surface of macrophages (2a) or B lymphocytes (2b) through an Fc receptor. Since macrophages are responsible for the enzymatic cleavage of Fc and AHGG to mitogenic Fc subfragments [12, 15, 16], binding of the complex to the macrophage surface would facilitate the production of the mitogenic subfragments from these complexes. Cleavage of the subfragment would reveal another binding site, distinct from the site described above, which binds to a receptor an B (3a) and T (3b) lymphocytes. Binding of the subfragment to B and T lymphocytes (3a and 3b) would enhance an ongoing antigen-driven response. If the ratio of antigen to antibody is such that the macrophage receptors become saturated, then the complexes would bind to the B lymphocytes (2b) through an Fc receptor and the antigen in the complex would cross-link the Fc receptor to the Ig receptor, resulting in suppression.

ACKNOWLEDGMENTS

This is publication number 2093 from the Department of Immunopathology, Scripps Clinic and Research Foundation, La Jolla, California. These studies were supported in part by United States Public Health Service grants AI-07007 and AI/CA-15761, American Cancer Society grant IM-421, and Biomedical Research Support Program grant RRO-5514. E.L.M. is the recipient of Damon Runyon-Walter Winchell Cancer Fund Postdoctoral Fellowship grant DRG-239-F and United States Public Health Service Postdoctoral Fellowship AI-05813. We wish to thank Nancy Kantor for excellent technical assistance and Janet Kuhns for secretarial expertise.

REFERENCES

1. Eisen HN, Karush F: Nature 202:677, 1964.
2. Bystryn JC, Schenkein J, Uhr, JW: In Amos B (ed): "Progress in Immunology. First Int Congr Immunol, New York: Academic Press, 1971, p 627.
3. Morrison SL, Terres G: J Immunol 96:901, 1966.
4. Rowley DA, Fitch FW, Axelrod MA, Pierce CW: Immunology 16:549, 1969.
5. Diener E, Feldmann M: Transplant Rev 8:76, 1972.
6. Sinclair NR St C, Lees RK, Abrahams S, Chan PF, Fagan G, Stiller CR: J Immunol 113:1493, 1974.
7. Kontiainen S, Mitchison NA: Immunology 28:523, 1975.
8. Kontiainen S: Immunology 28:535, 1975.
9. Morgan EL, Tempelis CH: J Immunol 119:1293, 1977.
10. Morgan EL, Tempelis CH: J Immunol 120:1669, 1978.
11. Weigle WO, Berman MA: In Pernis B, Vogel HJ (eds): "Cells of Immunoglobulin Synthesis." New York: Academic Press, 1979, p 223.
12. Morgan EL, Weigle WO: J Immunol 125:226, 1980.
13. Berman, MA, Weigle WO: J Exp Med 246:241, 1977.
14. Morgan EL, Weigle WO: J Exp Med 150:256, 1979.
15. Morgan EL, Weigle WO: J Exp Med 151:1, 1980.
16. Morgan EL, Weigle WO: J Immunol 124:1330, 1980.

17. Morgan EL, Walker SM, Thoman ML, Weigle WO: J Exp Med 152:113, 1980.
18. Morgan, EL, Thoman M, Walker SM, Weigle WO: J Immunol 125:1275, 1980.
19. Porter RR: Biochem J 73:119, 1959.
20. Spiegelberg HL: In Miescher PA, Müller-Eberhard HJ (eds): "Textbook of Immunopathology". New York: Grune & Stratton, 1976, p 1101.
21. Chiller JM, Weigle WO: J Immunol 106:1647, 1971.
22. Nisonoff A, Wissler FC, Lipman LN, Woernly DL: Arch Biochem Biophys 89:230, 1960.
23. Connell GE, Painter RH: Can J Biochem 44:371, 1966.
24. Jerne NK, Nordin AA: Science 140:405, 1963.
25. Kettman J, Dutton RW: J Immunol 104:1558, 1970.
26. Winkelhake JL: Immunochemistry 15:695, 1978.
27. Thoman ML, Morgan EL, Weigle WO:(in press).
28. Andersson J, Sjöberg O, Möller G: Transplant Rev 11:134, 1972.
29. Coutinho A, Möller G, Richter W: Scand J Immunol 3:321, 1974.
30. Sulzer BM, Nelsson BS: Nature [New Biol] 240:198, 1972.
31. Parker DC, Fothergill JJ, Wadsworth DC: J Immunol 123:931, 1979.
32. Parker DC: Nature 258:361, 1975.
33. Weiner HL, Moorhead JW, Claman HN: J Immunol 116:1656, 1976.
34. Sieckmann DG, Asofsky R, Mosier DE, Zitron IM, Paul WE: J Exp Med 147:814, 1978.
35. Sieckmann DG, Scher I, Asofsky R, Mosier DE, Paul WE: J Exp Med 148:1678, 1978.
36. Sidman CL, Unanue ER: Proc Natl Acad Sci USA 75:2401, 1978.
37. Sidman CL, Unanue ER: J Immunol 121:2129, 1978.
38. Sidman CL, Unanue ER: J Immunol 122:406, 1979.

Modulation of EGF Binding and Action by Succinylated Concanavalin A in Fibroblast Cell Cultures

Kurt Ballmer and Max M. Burger

Department of Biochemistry, Biocenter of the University of Basel, CH-4056 Basel, Switzerland

The role of the binding of succinylated concanavalin A to tissue culture cells in influencing epidermal growth factor (EGF)-mediated cell proliferation has been studied. Succinylated concanavalin A dramatically reduces the stimulation of 3T6 cells by EGF in Dulbecco's modified Eagle's medium (DME) containing insulin and vitamin B_{12} as additional growth factors, but no serum. Furthermore, binding studies using ^{125}I-labeled EGF have shown that the binding of EGF to the cell surface is reduced upon addition of succinylated concanavalin A.

Key words: growth regulation, epidermal growth factor, density inhibition of growth, extracellular matrix

Many of the approaches to elucidate growth control in tissue culture in the last few years can be grouped into two categories. The first one considers the availability and binding of various nutrient and/or regulatory factors to the cell surface or the cell in general. The second one is concerned with the role of cell—cell interactions in the regulation of cell growth. The two concepts do not necessarily have to be mutually exclusive. Modifications or perturbations of the cell surface which lead to changes in the growth control of the cells might give some clues as to how membranes can regulate cell growth.

It has been shown earlier that a nontoxic lectin-derivative of concanavalin A (succinyl-Con A) is able to inhibit cell growth reversibly in a density-dependent manner in various cell lines [1–3]. Since epidermal growth factor (EGF) acts by first binding to a specific receptor on the cell surface, we decided to investigate the interaction of succinyl-Con A with binding and growth promotion by EGF. For these studies 3T6 cells were used since they can be stimulated to reassume DNA synthesis after serum starvation by the addition of pure mitogenic factors.

Received March 5, 1980; accepted October 10, 1980.

0091-7419/80/1402-0209$02.00 © 1980 Alan R. Liss, Inc.

MATERIAL AND METHODS

Cell Cultures and Materials

3T6 cells were cultured in Falcon tissue culture plates in Dulbecco's modified Eagle's Medium (Gibco) supplemented with 10% calf serum (Gibco). 3T6 cells were a gift from Dr. Rudland at the Imperial Cancer Research Fund in London. Cultures were unfrozen from stocks every 2–3 months and routinely checked for PPLO. Radioactive isotopes were from Amersham or Institut für Reaktorforschung, Würenlingen, Switzerland.

Con A and succinylated Con A were isolated and prepared as described previously [1]. Insulin, bovine serum albumin (BSA), and vitamin B_{12} were obtained from Sigma. Penicillin-streptomycin (1,000 units/ml) was from Gibco. EGF was isolated [4] and labeled [5].

Stimulation of Cells in Presence of Growth Factors

The 3T6 cells (2×10^5) were seeded in 2 ml DME plus 0.5% serum into 35 mm plates. After 3 days the cells were washed twice with serum-free medium, and 2 ml of DME plus 1% penicillin-streptomycin were added. EGF, insulin, and vitamin B_{12} were added 2 days later. The factors were added in phosphate-buffered saline (PBS) containing 0.1% BSA. At the same time the cultures obtained 2 μCi ^{3}H-thymidine (20 Ci/mmole). Succinyl-Con A was added to a final concentration of 200 μg/ml a few hours prior to factor addition.

The cultures were incubated for up to 48 h, fixed in PBS plus 3.5% formaldehyde, and prepared for autoradiography. After 10 days they were developed and stained with Giemsa. Five hundred to 1,000 cells were counted in duplicate samples; the standard deviation was always less than 10%. In 4 different experiments similar results were obtained.

Binding of ^{125}I-EGF

The 3T6 cells (2.5×10^5) were plated in 5 ml DME plus 10% serum into 6 cm plates. Three days later the medium was removed and changed to 2 ml serum-free DME. Succinyl-Con A was added (200 μg/ml) and ^{125}I-labeled EGF (specific activity 2×10^5 cpm/ng) was added in PBS containing 0.1% BSA 2–4 hours later. The total volume in the binding assays was 1 ml. Maximal binding occurred after 30–60 min at 37°C. The cells were washed 4 times with 5 ml ice-cold PBS and dissolved in 0.5 N NaOH. They were counted in a Packard Model 5385 γ-counter. Each point was determined in triplicate; the standard deviation was always below 10%. Similar results were obtained if the cells were kept in serum-free DME for 1 day prior to the addition of EGF.

RESULTS

In order to investigate the question whether succinyl-Con A affects cell stimulation in a defined tissue culture system, we used 3T6 cells, which can be arrested in G_0/G_1 by withdrawing serum for several days. Upon addition of EGF, insulin, and vitamin B_{12} they reenter the cell cycle within about 12 h. Insulin and vitamin B_{12} alone show no stimulatory activity if added in the absence of EGF. Without insulin and vitamin B_{12}, no growth stimulation could be observed after addition of EGF. Figure 1 shows the increase in the percentage of cells showing DNA synthesis after addition of different concentrations of EGF. Insulin and vitamin B_{12} were always added at concentrations of 100 ng/ml and 400 ng/ml, respectively. Above 2–5 ng/ml EGF the cells are not stimulated any further. Maximal stimulation using 3T6 cells achieves in the order of 80% labeled cell nuclei and occurs within 30 h. Longer incubation does not lead to a higher proportion of cells synthesizing DNA.

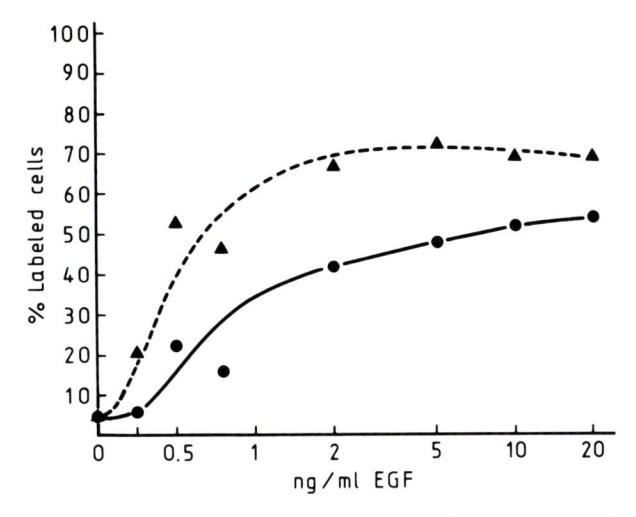

Fig. 1. 3T6 cells were plated and EGF, insulin, and vitamin B$_{12}$ added as described in Material and Methods; (▲ - - ▲) shows stimulation of cells after 30 h in the absence of succinyl-Con A, (●—●) shows cell stimulation in presence of 200 µg/ml succinyl-Con A. If no EGF was added, the labeling index was 4–5%, independent of whether insulin and vitamin B$_{12}$ were present. This background stimulation was also independent of whether succinyl-Con A was present.

If resting 3T6 cells are treated with 200 µg/ml succinyl-Con A 2–4 h prior to addition of growth factors, they re-enter the cell cycle to a considerably lower extent. Figure 1 shows that the highest inhibition occurs between 0.2 and 2 ng/ml EGF added to resting cells. This is the concentration range shown to be below saturation of the biological effect of EGF on resting cells in the absence of succinyl-Con A. The maximal inhibition of EGF stimulation by succinyl-Con A is about 70% (at 0.25 ng/ml EGF). At very high EGF concentrations (above 20 ng/ml) the succinyl-Con A inhibition can be overcome, as is expected for inhibition of the competitive type.

When α-methyl-D-mannoside is added together with succinyl-Con A to the cells, it inhibits the effect of the lectin-derivative completely. The hapten sugar also abolishes the effect of succinyl-Con A if added after the cells are cultured in the presence of succinyl-Con A for several days [1–3]. We conclude therefore that succinyl-Con A does not inactivate EGF in the medium.

In order to investigate the interaction between EGF and the cell membrane, in the presence and absence of succinyl-Con A, we studied the binding of this growth factor to 3T6 cells. Succinyl-Con A clearly reduces the binding of ^{125}I-labeled EGF to its receptors. Figure 2 shows a plot of a representative binding experiment done with 3T6 cells. When 200 µg/ml succinyl-Con A was added to the cells 2–4 hours prior to EGF, it clearly reduced the amount of EGF bound to the cells. Succinyl-Con A has to be present in the binding medium during the whole incubation period with EGF. When the cells were treated with succinyl-Con A for longer periods prior to EGF addition no further inhibition of binding occurred. When α-methyl-D-mannoside was added to the binding medium together with succinyl-Con A, it prevented the lectin-induced decrease in EGF binding (data not shown). The binding data represented in Figure 2 show that the binding of EGR to 3T6 cells in the presence of succinyl-Con A is not saturated at concentrations up to 10 ng/ml. This may in

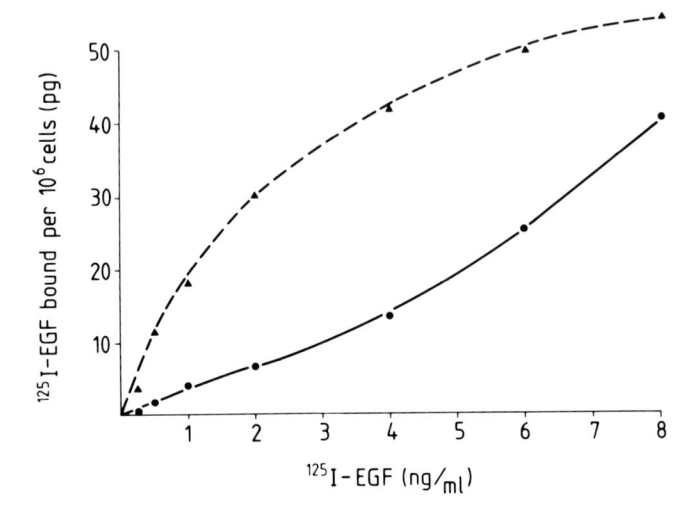

Fig. 2. Binding curve of ^{125}I-labeled EGF to 3T6 cells. Only the specific binding – ie, the amount of EGF bound after subtracting the percentage of cell-associated material in presence of a 100–500-fold excess of unlabeled EGF – is plotted; (▲---▲) shows binding to control cells; (●—●) shows binding to cells pretreated with 200 μg/ml succinyl-Con A.

part be related to the finding that succinyl-Con A increases the unspecific binding of the hormone to 3T6 cells. In our experiments unspecific binding was the amount of ^{125}I-EGF associated with the cells in presence of a 100–500-fold excess of cold material. Up to 70% of the cell-associated material was unspecifically bound in presence of succinyl-Con A (at 8 ng/ml EGF), whereas in control experiments it was always below 10%.

DISCUSSION

Succinyl-Con A is known to reversibly inhibit cell growth in vitro in tissue culture media containing calf serum, a complex mixture of exogenous growth factors [1–3]. Under these conditions succinyl-Con A mimics some aspects of density-dependent growth control in untransformed and virus-transformed cells. Addition of a Con A-specific hapten sugar leads to release of most of the bound lectin, followed by reinitiation of cell growth [3]. This suggests that succinyl-Con A acts at the cell periphery. Several other authors have reported that lectins interfere with the binding of hormones to membrane receptors [6–8].

We investigated the influence of succinyl-Con A on the binding of a specific growth factor to the cell surface. To circumvent the problems arising from the use of complete (and therefore undefined) sera in growing cells in culture, we used a more defined system containing only 3 factors that can stimulate resting cells to enter S-phase [9]. Although EGF, insulin, and vitamin B_{12} stimulate a high proportion of G_0/G_1 cells to enter S-phase without serum addition, we were not able to monitor a fast increase in the mitotic rate of our cultures. Neither is it possible to maintain these cells for more than 10 days under serum-free conditions without a reduction in cell viability. We assume therefore that other factors are necessary to induce the whole cell cycle. Nevertheless it was possible to study the effect of succinyl-Con A with binding and stimulation of DNA synthesis after addition of EGF, insulin, and vitamin B_{12} to resting 3T6 cells. The dose-response curve for EGF is clearly shifted to higher EGF concentrations (Fig. 1). Figure 2 shows that succinyl-Con A also influences the binding of EGF to the cell membrane and/or the subsequent degradation or uptake of the hormone-receptor complex.

To clarify further the role of succinyl-Con A in reducing the EGF-induced cell stimulation, several aspects have to be considered: 1) succinyl-Con A may reduce the binding of EGF to the cell membrane by an allosteric hindrance of the receptor; 2) succinyl-Con A may down regulate the number of EGF receptors — eg, via co-internalization of EGF receptors into lectin-induced endocytotic vesicles; 3) succinyl-Con A may interfere with the degradation of the EGF–receptor complex. All three points need further investigation.

Other authors have described the effect of tumor promoters on EGF binding or action in different tissue culture systems. These data suggest that phorbol esters interact indirectly with EGF receptors and do not bind to the receptor itself. In our system there is good correlation between succinyl-Con A–induced inhibition of cell stimulation and EGF binding. However, the possibility that other mechanisms are involved in growth regulation have also to be considered. Therefore no causal relationship between inhibition of EGF binding and the observed decrease in cell stimulation can be evoked.

The number of binding sites for lectins at the cell surface is much higher than for hormones like EGF [14, 15]. It is possible therefore that several other growth factor binding sites and/or other functions of the cell membrane are impaired by succinyl-Con A. The reversible growth regulation in serum containing tissue cultures observed earlier could be the reflection of many different molecular changes induced by succinyl-Con A on the cell surface. The data presented here do not allow a definite interpretation of the earlier finding that succinyl-Con A–regulated cell growth in a density-dependent way [1]. Other workers found a density dependence of the effect of EGF on cell proliferation [12, 13]. However, nothing is known so far about the mechanism by which cell density affects hormone receptors.

Other effects of succinyl-Con A have been found; among them is a reduction of cell mobility in dense but not in sparse culture (manuscript in preparation). This could proceed through an increase in cell-to-cell adhesion mediated by succinyl-Con A.

Other authors have discussed a possible role of the extracellular matrix or, more generally, of the microenvironment of the cell in growth regulation [16–20]. Since the extracellular matrix is rich in carbohydrate to which lectins can bind, we speculate that a change in the extracellular matrix may be related to the growth inhibition induced by succinyl-Con A. A not yet determined decrease in the affinity of the EGF receptors upon succinyl-Con A treatment may be the reflection of a general change in the structure of the extracellular matrix. The alternative possibility is a direct interaction of succinyl-Con A with the receptor molecules. No data are available so far that make it possible to distinguish between these two hypotheses.

ACKNOWLEDGMENTS

These studies were supported by the Swiss National Foundation Grant, number 3.513-0.79. We thank R. Imhof for her expert technical assistance.

REFERENCES

1. Mannino RJ, Burger MM: Nature 256:19, 1975.
2. Mannino RJ, Ballmer K, Burger MM: Science 201:824, 1978.

3. Ballmer K, Mannino RJ, Burger MM: Exp Cell Res (in press).
4. Savage CR, Cohen S: J Biol Chem 247:7609, 1972.
5. Das M, Miyakawa T, Fox CF, Pruss RM, Aharonov A, Herschman H: Proc Natl Acad Sci USA 74:2790, 1977.
6. Cuatrecasas P, Tell GP: Proc Natl Acad Sci USA 70:485, 1973.
7. Costlow ME, Gallagher PE: Biochem Biophys Res Commun 77:905, 1977.
8. Carpenter G, Cohen S: Biochem Biophys Res Commun 79:545, 1977.
9. Mierzejewski K, Rozengurt E: Biochem Biophys Res Commun 73:271, 1976.
10. Brown KD, Dicker P, Rozengurt E: Biochem Biophys Res Commun 86:1037, 1979.
11. Shoyab M, De Larco JE, Todaro GJ: Nature 279:387, 1979.
12. Pratt RM, Pastan I: Nature 272:68, 1978.
13. Brown KD, Yeh YC, Holley RW: J Cell Physiol 100:227, 1979.
14. Phillips PG, Furmanski P, Lubin M: Exp Cell Res 86:301, 1974.
15. Carpenter G, Lembach KJ, Morrison MM, Cohen S: J Biol Chem 250:4297, 1975.
16. Culp LA: J Supramol Struct 5:239, 1976.
17. Cohn RH, Cassiman JJ, Bernfield MR: J Cell Biol 71:280, 1976.
18. Underhill CB, Keller JN: J Cell Physiol 89:53, 1976.
19. Stoker MGP: Nature 246:200, 1973.
20. Stoker M, Piggott D: Cell 3:207, 1974.

The Cell Substratum Modulates Skeletal Muscle Differentiation

Hannah Friedman Elson and Joanne S. Ingwall

Departments of Biology (H.F.E.) and Medicine (J.S.I.), University of California, San Diego, La Jolla, California 92093

During chick embryogenesis, massive alterations occur in the migrating cell's substratum, or extracellular matrix. The possibility that some of the components of this milieu play a regulatory role in cell differentiation was explored in a cell-culture system derived from embryonic chick skeletal muscle tissue. In particular, the effects of collagen and the glycosaminoglycans were studied. Collagen is required for muscle cell attachment and spreading onto plastic and glass tissue-culture dishes. A major constituent of the early embryonic extracellular space, hyaluronate (HA), while having no significant effect on collagen-stimulated cell attachment and spreading, was found to inhibit myogenesis. The muscle-specific M subunit of creatine kinase was preferentially inhibited. Control experiments indicated that the inhibition was specifically caused by HA and not by other glycosaminoglycans. A general metabolic inhibition of the cultures was not observed. Muscle cells could bind to HA-coated beads at all stages of differentiation but were inhibited only when HA was added within the first 24 h of culture. Endogenous GAG in the culture is normally degraded during the first 24 h after plating as well; this may parallel the massive degradation of HA that occurs in the early embryo in vivo. These findings suggest a regulatory role for HA in modulating skeletal muscle differentiation, with degradation of an inhibitory component of the cell substratum a requirement for myogenesis.

Key words: skeletal muscle, myogenesis, chick embryo, hyaluronic acid, glycosaminoglycan, extracellular matrix

During early embryogenesis, cells of predetermined tissue types often undergo extensive migrations before they express their full developmental potential. Often, it is not until they arrive at their final destination that they turn on the synthesis of tissue-specific

---

Dr. Ingwall is now at Department of Medicine, Harvard Medical School and Peter Bent Brigham Hospital, Boston, MA 02115.

Received April 7, 1980; accepted August 22, 1980.

enzymes and other macromolecules. Since cells migrate through extracellular matrix, it seemed possible that some of the components of the matrix could play a role in modulating the developmental programs [1].

Of particular interest is hyaluronate, a glycosaminoglycan that is a major component in the large interstitial spaces of the early embryo [2–4]. As differentiation proceeds, the hyaluronate is degraded [5–8]. We explore here the possibility that when cells are migrating through hyaluronate, their tissue-specific differentiation is repressed, and when the hyaluronate is degraded, differentiation can proceed.

Cartilage tissue and the chondrogenic cell-culture system provided examples of this response when it was shown that cartilage formation by chondrocytes was inhibited by hyaluronate and stimulated by chondroitin sulfates [9–14]. Other developing systems have also been shown to be responsive to glycosaminoglycans. Work by Lash and his coworkers [15–18] indicated that glycosaminoglycans promoted chick somite morphogenesis and cell differentiation. Similarly, the development of the chick cornea can be affected by glycosaminoglycans [19] and collagen [20].

During the development of skeletal muscle, mononuclear myogenic precursor cells migrate and form aggregates, which then fuse to form muscle fibers. Many aspects of the differentiation process were clarified in studies in cell culture [21–23]. One of the extracellular matrix materials, collagen, has been shown to be required for myogenesis [24–26]. Nameroff and Holtzer showed that myogenesis was inhibited when muscle cells were plated on killed confluent cultures of chondrocytes and liver cells [27]. They concluded that a nondiffusible polysaccharide was the inhibitor. Postmitotic myoblasts were not inhibited, and the inhibition could be reversed with conditioned medium.

This research has indicated that the extracellular matrix may regulate processes in differentiating cells. There is still a need to collect biochemical data on a specific extracellular component's effect on well-defined differentiating systems. This paper describes the effect of one glycosaminoglycan, hyaluronate, on muscle differentiation, using as quantitative indicators of myogenesis, creatine kinase and fusion index.

MATERIALS AND METHODS

Cultures

Primary cultures of trypsin-dissociated skeletal muscle cells were prepared from 10-day chick embryo breast tissue by a modification of the method of Bischoff and Holtzer [28]. Breast muscle was freed of bone, minced, and incubated in 0.25% trypsin in Earle's balanced salt solution, Ca^{++} and Mg^{++}-free, at 37°C for 40 min. The tissue was washed 3 X with growth medium by pelleting, triturated in a Pasteur pipet, and filtered through two pieces of washed lens paper. The cells were plated in 0.5 or 1.0 ml medium 8:1:0.25 at an initial density of 1×10^5 cells on 3 cm diameter Petri dishes (Nunclon) coated with calf skin collagen (Sigma). The growth medium contained 84% modified Eagle's medium (MEM) with Earle's salts, 10% horse serum, 2.5% embryo extract, 1 mM glutamine, and penicillin/streptomycin (125 units/ml each). The medium was changed daily. All medium components were purchased from Gibco. Embryo extract was prepared by extruding whole 11- or 12-day chick embryos through a 50 cc syringe, diluting with an equal volume of balanced salt solution (BSS), and freezing. Upon thawing, the slurry was spun 40 min at 1,500g and the supernatant aliquoted and frozen. Each aliquot was respun upon thawing before use. Under the conditions employed here, mononuclear myoblasts underwent one or two rounds of cell division and then fused to form multinucleate myotubes. Fusion began 1 1/2 days after plating and was complete by 2 1/2 days.

The percent fusion is the percent of the total nuclei present in myotubes. Several random fields in a dish were scored. All nuclei in mononuclear cells, including those in fibroblasts, were counted in the unfused population. The total number of nuclei assayed per datum always exceeded 500. The microscope resolution was sufficiently good to distinguish two overlapping myoblasts from a myotube with two nuclei, and "dimers" were consequently included in the estimate of nuclei in myotubes. A slight underestimate of fused cells may have occurred at very early fusion stages [29, 30].

Assays

Hyaluronic acid (HA), sodium salt from hog skin, chondroitin sulfates (ChS; chondroitin-4-sulfate, super special grade, sodium salt from whale cartilage, and chondroitin-6-sulfate, super special grade, sodium salt from shark cartilage) were purchased from Miles Laboratories, Inc. Several preparations of HA and ChS were used during these experiments. The HA ran as a single band on strip paper electrophoresis [31], and this band was completely destroyed by testicular hyaluronidase.

Creatine kinase (CK) (EC 2.7.3.2, ATP: creatine N-phosphotransferase) was measured by the coupled enzyme reaction as described by Rosalki [32]. Reagents were purchased from Calbiochem. To prepare lysates, cells were washed twice with Dulbecco's phosphate-buffered saline (PBS), scraped into 2 ml of 0.01 M Tris-HCl, pH 7.4, 1 mM EDTA and 1 mM 2-mercaptoethanol, and homogenized for 15 strokes in a ground-glass tissue grinder. Data presented are the averages of 4–6 determinations. Experiments were repeated 3 to 5 times. CK activity is expressed as units per milligram total lysate protein.

Control experiments to rule out an inadvertent effect of HA on the CK assay were done. When various concentrations of HA were added directly to aliquots of a cell lysate and the CK activity was then measured, little inhibition was seen. If the final lysate had a concentration as high as 0.2 mg/ml HA, the enzyme was inhibited by only 13% (Table I). It is not possible for the cells to have accumulated more than one-half of this amount, even assuming they could accumulate all of the HA ever provided in the medium: a 3-day culture was routinely given a total of 3 changes of 0.5 ml of medium containing 0.1 mg/ml HA, or 0.2 mg total, which was then diluted to 2 ml of lysate. If the cells had indeed taken up all the externally supplied HA, the lysate would contain 0.1 mg/ml HA, which would inhibit CK activity in the assay by only 10%. It is unlikely that the cells accumulated this amount of HA.

CK isoenzymes were determined by a method adapted from Klein et al [33]. Protein was determined by a modification of the method of Lowry et al [34]. Cells were generally harvested several hours after a feeding to avoid serum-starved stationary cells. Assays on each dish were performed in duplicate.

The amount of glycosaminoglycan (GAG) present in the culture was assayed by a modified procedure of Hamerman et al [35]. Cultures were grown in medium without phenol red, to avoid interference with the assay for hexuronic acid [36]; 2.7×10^6 cells were plated on 9 cm diameter Petri dishes in 5 ml 8:1:0.25. The medium was in contact with the cells for 24 h prior to collection for the assay, unless otherwise indicated, and the cells were given fresh medium daily. Twenty-five milliliters of combined media from 5

TABLE I. Effect of HA on CK Assay*

	HA (mg/ml)	CK (units/mg)	% Activity
Experiment 1	0	4.35	100 ± 5
	0.002	4.46	102 ± 2
	0.02	3.92	90 ± 2
	0.2	3.80	87 ± 0
Experiment 2	0	6.97	100 ± 1
	0.01	6.69	98 ± 3
	0.1	6.26	90 ± 4

*Lysates were prepared from 3-day (Experiment 1) or 6-day (Experiment 2) muscle cultures. HA was added to them at the final concentrations indicated, and CK was assayed.

dishes were dialysed against 3 liters of 5 mM sodium phosphate buffer, pH 7.5, containing 0.1 mM phenylmethylsulfonyl fluoride (PMSF), at 4°C for 2 or more days in Spectropor 3 tubing. The buffer was changed 6 to 7 times. Samples were lyophilized and redissolved in 1.5 to 3.0 ml 0.1 N sodium chloride and 0.05 M potassium acetate, pH 4.8. Testicular hyaluronidase (Calbiochem, B grade, 702 units/mg) was added to give a final concentration of 250 units/ml, and the solution was incubated for 12 h at 37°C. Trichloroacetic acid, 50%, was added to each sample to give a final concentration of 5%. After incubating in ice for 20 min, the precipitate was spun for 15 min at 1,500g at 4°C, and the supernatant was assayed for uronic acid. Recoveries of HA added to the medium at a final concentration of 25 µg/ml are over 90%. A final concentration of PMSF of 1 mM or less did not interfere with the uronic acid assay. PMSF, at a concentration of 0.01 M, will also not interfere with hyaluronidase activity, as determined by electrophoresis of HA on cellulose acetate strips, nor will it interfere with the strip assay itself. When GAG in the cell layer was determined, the cells were scraped up in a total of 5 ml PBS, dialyzed, and treated in the same manner as the medium above. The basal level of GAG in the growth medium varied from 2.8 to 3.6 µg/ml. The same preparation of medium was used throughout an experiment.

HA was covalently bound to polyacrylamide gel beads by a modification of the methods of Hoare and Koshland [37] and Cuatrecasas [38]. All glassware was siliconized. Eighty-three milligrams of Aminoethyl Bio-Gel P-150 beads (Biorad, 1.2 mEq/g) were hydrated at room temperature for 4 h in 5 ml water with stirring and allowed to settle. The supernatant was decanted, leaving 2 ml of packed beads. Two milliliters of HA, containing 50 mg and adjusted to pH 4 with HCl, were added. One milliliter of a solution of 200 mg/ml water-soluble carbodiimide, 1-ethyl-3 (3-dimethylaminopropyl) carbodiimide (EDAC, Biorad), adjusted to pH 4.5, was added dropwise with stirring over 5 min. The pH was maintained with HCl at pH 4–5. The solution was stirred for 4 h at room temperature and an additional 14 h at 5°C. Unreacted groups on the beads were blocked by adding 1 mmole acetic acid and another 200 mg EDAC dropwise, incubating for 2 h at room temperature and another 2 h at 5°C. The beads were washed with 0.005 M sodium phosphate buffer, pH 7.5, containing 0.1 mM PMSF and stored at 4°C. Before binding cells, they were washed extensively in the binding medium. Assaying the HA bound to beads gave a yield of 24 µg HA bound per milligram of beads.

TABLE II. Time Course for Accumulation of CK Activity in the Presence and Absence of HA*

Days after plating	Creatine kinase (units/mg)	
	−HA	+HA
1	0.66 ± 0.26	0.49 ± 0.14
2	0.88 ± 0.23	0.76 ± 0.25
3	4.54 ± 0.87	2.27 ± 0.45
4	4.73 ± 0.41	3.93 ± 0.35

*Cells were plated in medium 8:1:0.25 with or without 0.1 mg/ml HA and fed daily. Duplicate cultures were harvested at 24 h intervals by rinsing the dishes in PBS and storing them at −70°C. All the lysates were prepared and assayed simultaneously for CK and protein. Values given are means and one standard deviation.

To prepare untrypsinized cells for binding to beads coated with HA, minced muscle tissue was incubated in collagenase (Sigma, Fraction A, from Clostridium histolyticum, 200–500 units per mg), 0.2 mg/ml for 30 min at 37°C in BSS.

Incorporation of thymidine was assayed with thymidine (methyl-^{3}H-) (New England Nuclear, 6.7 Ci/mmole) at a final concentration of 5 μCi/ml. Included in the incubation was 20 μM unradioactive thymidine. Incorporation of amino acids was measured with a tritiated L-amino acid mixture (New England Nuclear, 1.0 mCi/ml) also at 5 μCi/ml. Cells were maintained in growth medium. Incorporations were assayed by rinsing cells 3 times in cold PBS, scraping them up twice into a total of 1.0 ml 0.01 N NaOH per dish, and incubating at 37°C for 15 min. One milliliter of 10% trichloroacetic acid containing 1 mg/ml bovine serum albumin (BSA) was added, and macromolecules were precipitated after 30 min at 0°C. Precipitates were collected on glass fiber filters (Whatman GF/C), dried and counted in 3 ml of scintillation fluid containing 2 parts toluene, one part Triton X-100, and 5 g 2,5-diphenyloxazole (PPO) per liter. The counting efficiency was 28%.

RESULTS

HA Modulates Differentiation Indicators of Muscle Cell Cultures

Time course of CK activity. One of the enzymes that increases following fusion of myoblasts is creatine kinase (CK) [39, 40]. The increase in culture has been shown to be due primarily to the muscle-specific isoenzyme, the MM form; thus, the increase in total CK is a valid indicator of myogenesis after 2 days in culture [41–44].

A time course of CK activity is shown in Table II. The activity increases 10-fold from day 2 to day 3. The addition of 0.1 mg/ml HA to the cultures diminishes the rate of the increase about twofold (Table II). In the presence of HA, the CK activity on day 3 is reduced to 50% of the control activity. By day 4, however, the activity is reduced by only 22%. Eventually, full activity is reached. It appears that HA delays, but does not permanently repress, the appearance of CK under these culture conditions.

Dose response curve and specificity. When cells were given various concentrations of HA at the beginning of the culture period, there was a substantial depression in the CK specific activity on the third day (72 h) of culture (Fig. 1). At 0.001 mg/ml HA there was

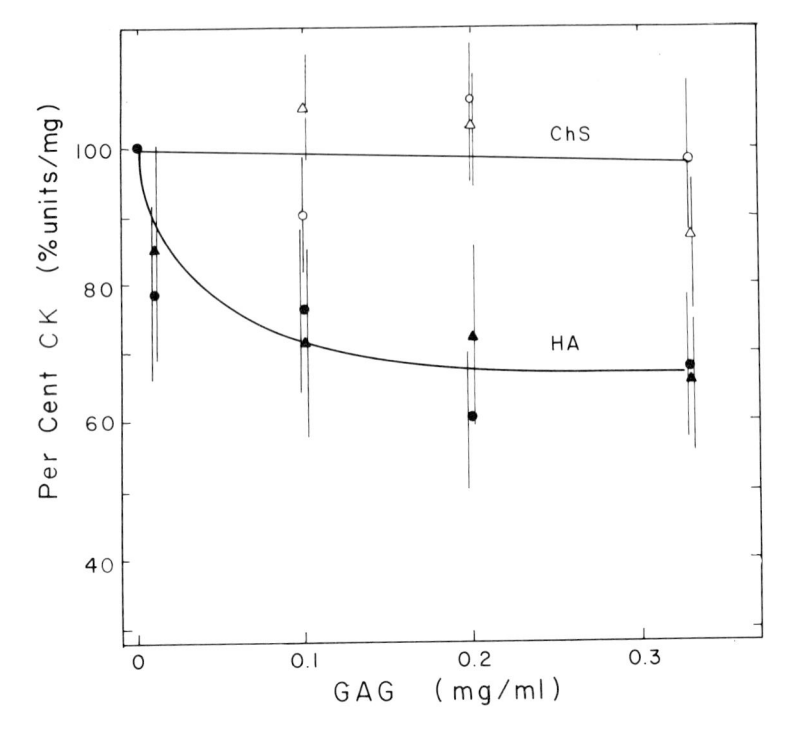

Fig. 1. Percent CK specific activity as a function of HA and ChS concentrations in medium containing 2.5% (8:1:0.25) or 10% (8:1:1) embryo extract. Muscle cells from 10-day-old chick embryos were plated in the indicated medium and incubated for 4 h to allow complete attachment. The medium was replaced with similar medium containing the indicated concentrations of HA or an equimolar mixture of chondroitin-4-sulfate and chondroitin-6-sulfate. The cells were harvested after 3 days and assayed in duplicate for CK and protein. Average results are reported in terms of percent of activity of cells incubated in the absence of glycosaminoglycans. Error bars indicate one standard deviation. (—o—), ChS in 8:1:1; (—△—), ChS in 8:1:0.25; (—●—), HA in 8:1:1; (—▲—, HA in 8:1:0.25.

little loss in CK activity, whereas significant loss occurs at 0.01 mg/ml HA. The maximum inhibition at 72 h varied between 40% and 50% at 0.1 mg/ml HA, with no further inhibition even at 0.33 mg/ml.

HA specificity was tested by assaying CK activity in the presence of other glycosaminoglycans. Since HA is a polyanion, it was possible that its effect was an electrostatic one and could be mimicked by other polyelectrolytes. The chondroitin sulfates are also polyanions found in the extracellular matrix. However, an equimolar mixture of chondroitin-4-sulfate and chondroitin-6-sulfate had no effect on the specific activity of CK after 3 days of culture (Fig. 1). Therefore, the inhibition by HA was not likely to have been an electrostatic effect alone. Nor is HA, as a polyanion, sequestering divalent cations necessary for myogenesis [56, 57]. At 0.1 mg/ml HA, there are approximately 0.2 mEq/liter negative charges. MEM contains 1.8 mM Ca^{++} or 3.6 mEq/liter positive charges. At most, only 5% of the Ca^{++} could be sequestered by the HA.

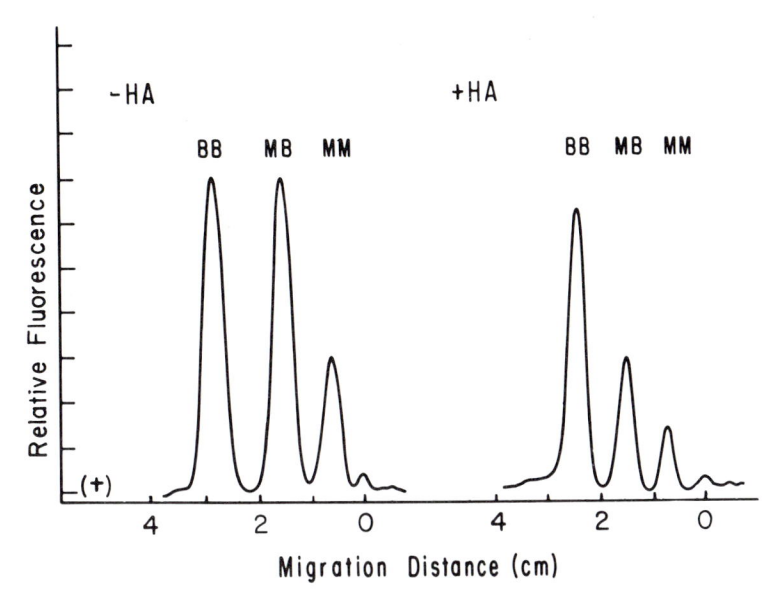

Fig. 2. A comparison of isoenzymes of CK from cultures treated with hyaluronate to those from untreated cultures. Peaks representing the 3 dimeric forms of CK are labeled, from left to right, BB, MB, and MM. The scan on the left shows the distribution in the control culture (−HA), and the scan on the right is the distribution from the treated culture (+HA).

Also indicated in Figure 1 are the dose-response curves for HA and ChS at two different embryo extract concentrations. Medium 8:1:1 contained 10% embryo extract, and medium 8:1:0.25 contained 2.5%. The CK specific activity was about 30% higher in 8:1:1. In both, the same relative dose-response curves for HA-induced inhibition were obtained. In both, ChS gave no inhibition. Thus, it is not likely that the HA effect is mediated or prevented by a component of embryo extract.

CK isoenzymes. Creatine kinase is a dimeric protein containing two subunits, B or M, in 3 isoenzymes, BB, MB, and MM. During early myogenesis, the B-subunit predominates; it is later supplanted by an M-subunit [39–44]. An isoenzyme analysis of CK is shown in Figure 2. In 3-day cultures, the predominant isoenzymes are the BB and MB dimers. HA reduced the MB and MM isoenzymes. In several experiments, the M-subunit accounted for 33–36% of the total control activity. In the presence of hyaluronate, this M-subunit activity decreased to 23–30% of its control level. The muscle-specific subunit thus appears to be more inhibited than total CK activity.

Time course of fusion. The effect of HA on the muscle cells' fusion index was also tested. The rate of fusion of cultures treated with HA parallels that of untreated cultures, although fusion was delayed by 3 h by HA (Fig. 3). A 3 h lag was found each time this experiment was performed. The slope of the fusion curve was 3.02% per hour in the control and 3.21% per hour in the HA-treated culture (as determined by a least-squares analysis).

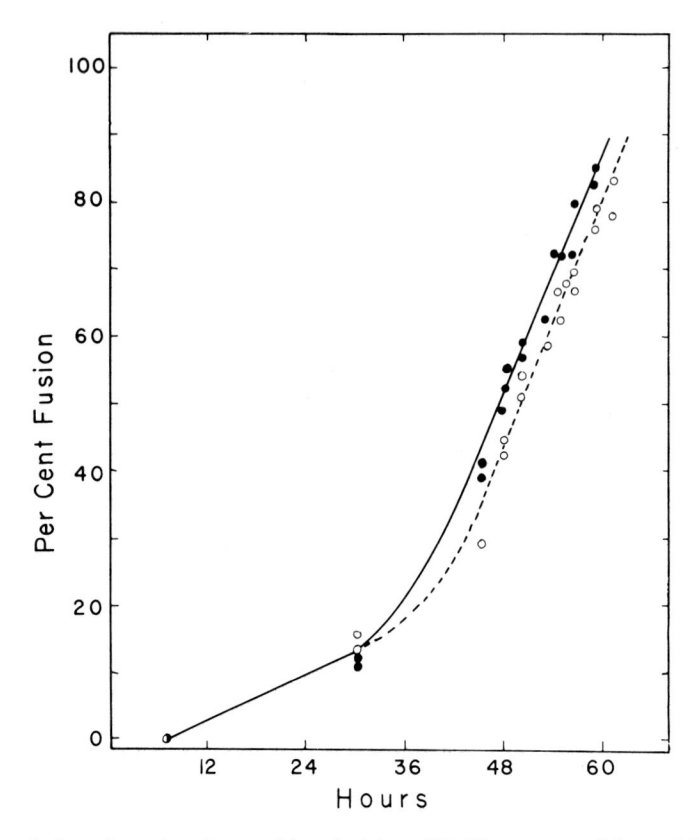

Fig. 3. Percent fusion of muscle cultures with and without HA. The percent of the nuclei in myotubes was determined as described in Methods. Fourteen dishes in all were followed. Two or 3 dishes were averaged per datum. (—●—), control; (---o---), 0.1 mg/ml HA.

Cells could be secreting factors that might be interacting with HA to either produce or diminish the inhibition by HA. Diluting these factors should diminish their effect. An experiment in which the cells' growth medium was changed every 12 h was performed to see what these semi-perfusion conditions would do to the HA effect. Control cells that were fed frequently showed delayed fusion and accumulation of CK activity (not shown), which is consistent with the observations of Konigsberg [45], who grew cells in perfused medium and found that fusion was delayed. HA under these frequent feeding conditions further delayed fusion by about 11 h and similarly delayed the appearance of CK activity (not shown). Changing the growth medium frequently, which would dilute conditioning factors, thus increased the length of the delay in CK activity and fusion caused by HA, suggesting that something in the medium may ordinarily overcome the HA-inhibition.

In these and other experiments, the maximum level of CK specific activity and fusion is eventually obtained in the presence of HA. It is therefore unlikely that HA has an effect on cell viability or on relative populations of myoblasts and fibroblasts. Furthermore, the overall appearance of the cells grown in HA is normal. It would thus appear that HA is affecting just the rate of onset of myogenesis.

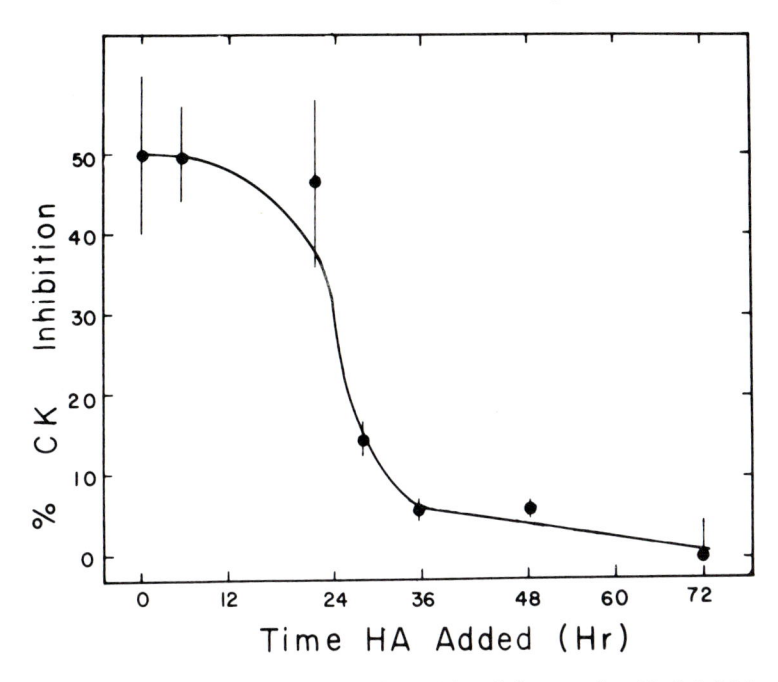

Fig. 4. Effect of time of addition of HA on later CK activity. Cells were plated in 8:1:0.25, and medium was changed daily. HA, 0.1 mg/ml, was added at the various times (hours) indicated on the abscissa. CK specific activity was assayed at 72 h after plating. The data indicate the means and one standard deviation.

Mechanism of the HA Effect

Critical time. The transient nature of the HA effect suggested that something was occurring in the culture to overcome it. This conclusion was supported by the experiment indicated in Figure 4. Exogenous HA had to be added within the first 20 h after plating in order to produce a delay in CK activity. Thereafter, it had no effect.

Effect on attachment. The inhibition of HA on attachment and spreading of cells on their collagen substratum was ruled out in the experiment illustrated in Figure 5. Freshly plated cells on the bottom of a Petri dish were randomly photographed, and the fraction of cells that were non-spherical was counted as spread. While collagen is required for efficient spreading of cells to Petri dishes, HA has no significant effect on collagen-mediated attachment and spreading in full growth medium by 1 h after plating.

Metabolic effects. Since the first day in culture is one during which there is cell growth and mitosis [46], the effect of HA on protein and DNA synthesis was examined. The incorporation of ^{3}H-amino acids or ^{3}H-thymidine was, however, not inhibited by HA (not shown). There is a negligible difference in the rate of uptake of these precursors in 1- or 4-day cultures. The total increase in cell culture protein over a week in culture is identical with and without HA, suggesting that there is no general metabolic inhibition.

GAG is degraded in early myogenic cell cultures. Freshly prepared trypsinized cells were found to contain a small amount of glycosaminoglycan (GAG) bound to them. During cell culture, the amount of hexuronic acid-containing GAG found in the cell layer declined

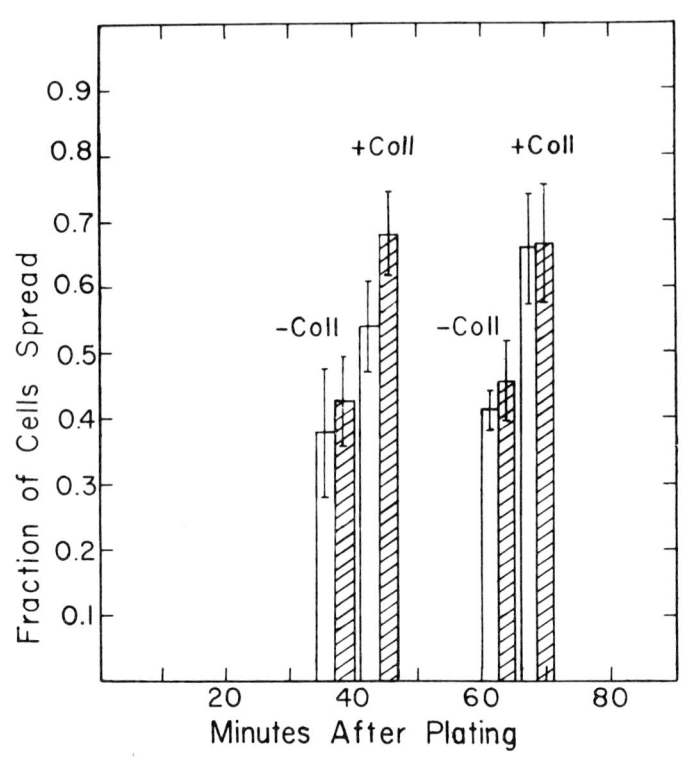

Fig. 5. Effect of HA on attachment and spreading. Trypinized cells were plated at a density of 2.5×10^5 cells per 35 mm dish, which contained 8:1:0.25 prewarmed to 37°C. The dishes were either coated with collagen or not, as indicated. At about one-half hour after plating, 4–5 random fields in 3 dishes were photographed. This was repeated 1 h after plating. The fraction of cells spread is the ratio of spread cells (nonspherical or cells with processes coming out) to the total cells counted on the bottom of the dish in the field. The 4–5 fields were averaged, and one standard deviation is shown. The total number of cells counted per field ranged from 19 to 129 at 1/2 h after plating, with an average of 67, and 45–109, with an average of 73, at 1 h after plating. Open bars, minus HA; hatched bars, plus HA.

for the first 12 h in full growth medium and then slowly increased (Fig. 6a). In particular, the specific activity of the GAG declined dramatically during that time (Fig. 6b). GAG in the medium increased rapidly for the first 4 h, accounting for the loss from cells in the first 4 h, and then also declined during the next 20 h in culture. Subsequently, the amount of GAG found on both the cell layer and in the medium slowly increased. In other experiments, it was found that the specific activity of GAG in the cell layer continued to decline slowly, but the specific activity of the amount found in the medium over cells, produced within 24 h, remained constant for up to 5 days in culture.

It appears from these results that some of the GAG bound in vivo to cells survives trypsinization and is either given up or secreted into the culture medium over the first 4 h. In the first 24 h after plating, however, there is a decline in the total amont of these present. Thereafter, they accumulate steadily.

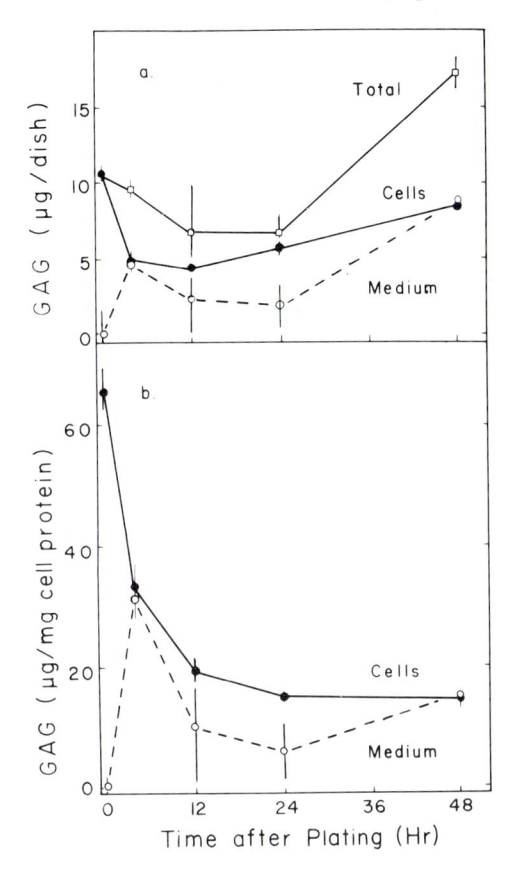

Fig. 6. GAG content of muscle cell cultures. The ordinate represents the uronic acid-containing GAGs and was obtained by multiplying the uronic acid content by 2.5. The abscissa is the hours after plating. (a) GAG associated with the cells and in the medium was assayed as described in Methods. In this experiment, the medium was not changed over the 48 h indicated. The curve of total GAG is the sum of the other two curves. The background in the medium was 2.80 µg/ml and was subtracted from the GAG found in the medium. Error bars are one standard deviation. (b) Specific activity of GAG is µg GAG/dish divided by the amount of cell-layer associated protein per dish. The protein content stayed constant for the first 4 h and then increased linearly for the next 44 h. Error bars indicate one standard deviation.

Cells bind to HA at all stages. Freshly prepared muscle cells, isolated with either trypsin or protease-free collagenase, will bind to beads that contain covalently bound HA (Fig. 7). Binding occurred in PBS, MEM, BSS, or 8:1:0.25, and within 15 min at 22°C or 5 min at 37°C. The binding can be partially competed with free HA (2 mg/ml) or 2–4 mM each of glucuronic acid and N-acetyl-glucosamine, the alternating subunits of HA. Binding is not competed by these concentrations of sialic acid and galacturonic acid. Increasing amounts of EDTA (0.1–4 mM) caused aggregation of collagenized cells and did not inhibit their binding to HA beads.

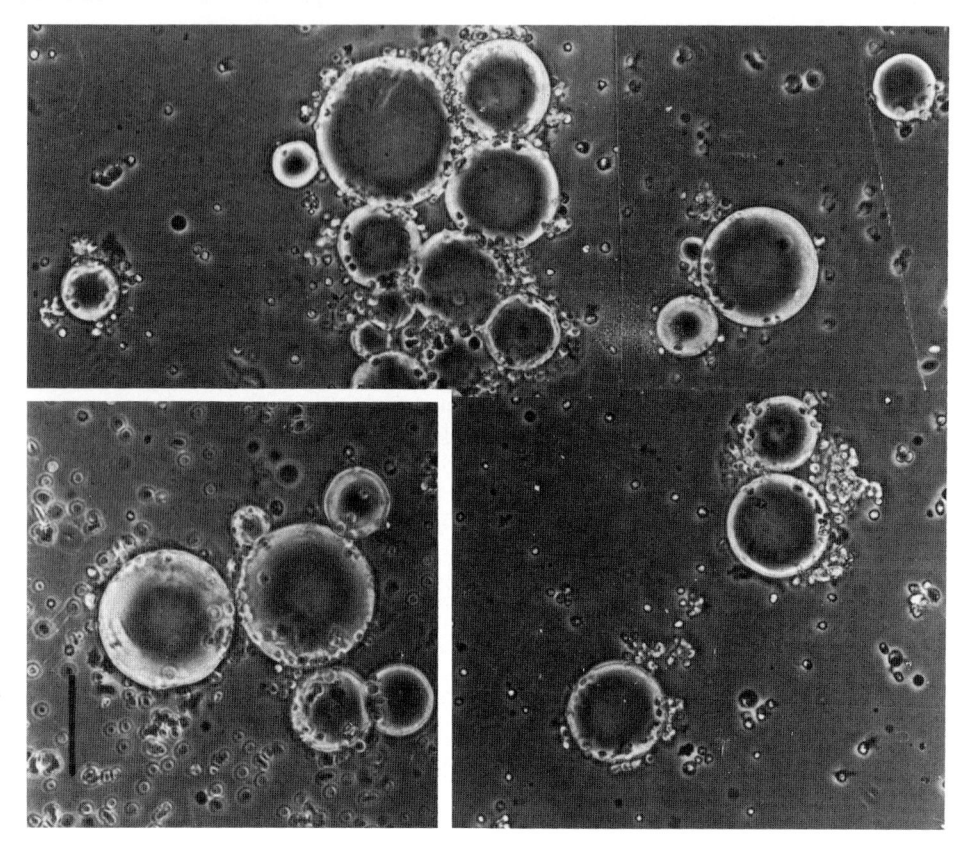

Fig. 7. Binding of fresh skeletal muscle cells to HA beads. Trypsinized cells were incubated with HA beads in 8:1:0.25 or (insert) PBS + 0.2 mg/ml BSA for 10 min at room temperature. Bar = 100 μm.

Chick erythrocytes do not bind to the beads at all. A population of white blood cells did bind, as well as what appeared to be platelets. Colchicine (5 μM) or PMSF (10 μM) do not prevent binding of muscle cells to HA beads.

P-200 beads to which no HA has been attached did not bind cells (Fig. 8). Two days after plating cells with P-200 beads, the beads remained in the medium over the differentiated cells. When HA beads were present, however, the cells attached to the beads and pulled them down within the muscle network, with myotubes extending up over the beads (Fig. 9). There were also cells on the free HA beads. Close inspection of the cells found on the floating beads showed a predominance of myotubes and cells with a myoblast morphology. Only one fibroblast-like cell was observed on the many beads examined. These observations indicate that muscle cells are capable of HA recognition and binding and that the ability to bind to HA is not lost following differentiation to myotubes.

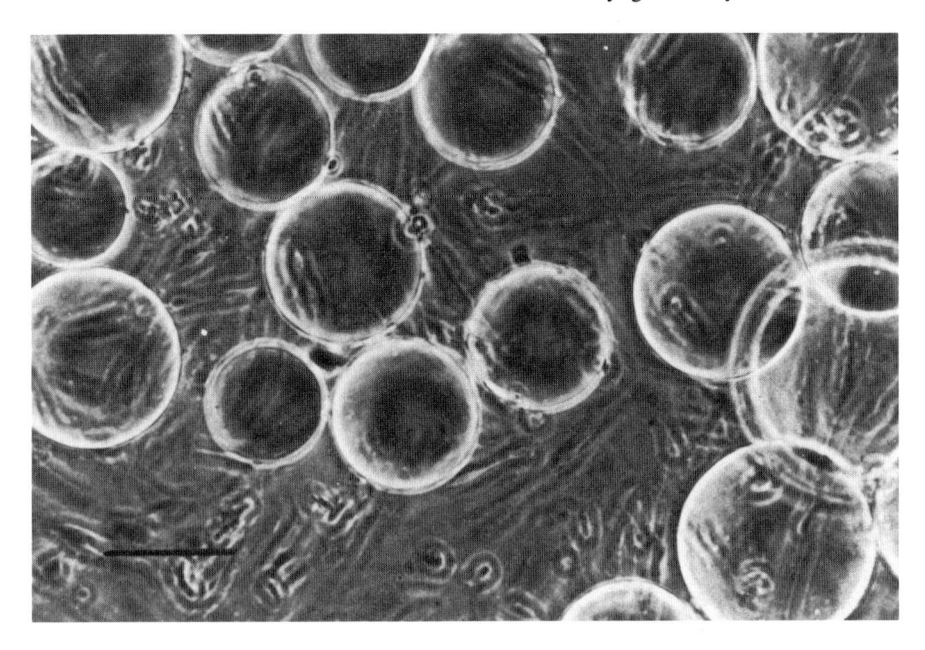

Fig. 8. Cells cultured with control beads. Cells dissociated with collagenase were mixed with P200 polyacrylamide beads (Biorad) that had not been coated with HA. The cells were plated and photographed after 2 days in culture. Bar = 100 μm.

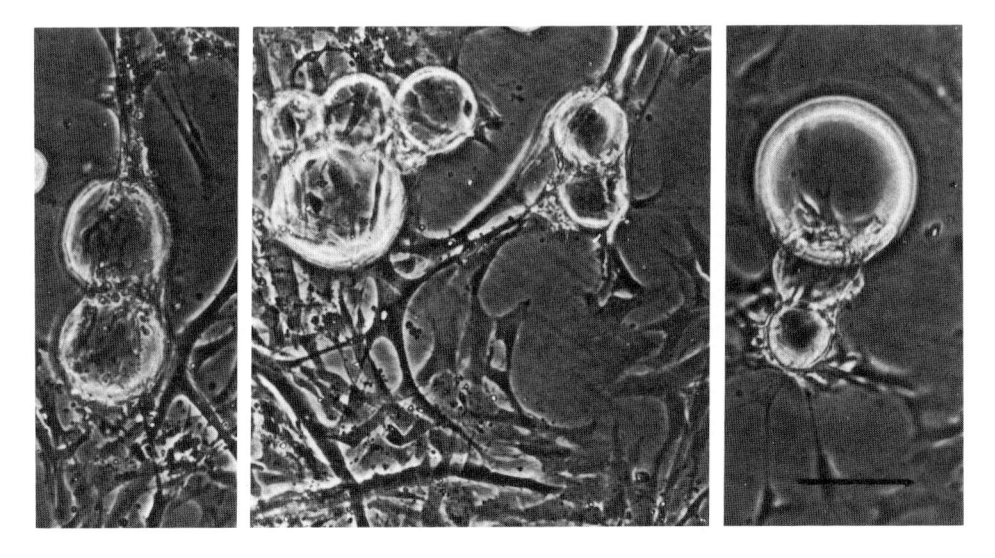

Fig. 9. Cells cultured with HA beads. Cells prepared as in Figure 8 were mixed with HA beads, plated, and photographed after 3 days in culture. Bar = 100 μm.

DISCUSSION

We explore here the possibility that a component of the extracellular matrix can modulate the expression of a differentiating cell. The term "modulation" is used here in the specific sense defined by Weiss [47], in which a cell's environment can produce a reversible response in the cell's behavior or program. In the case presented here, modulation by hyaluronate may serve the purpose of giving cells time to finish morphogenetic movements and other early events before differentiating to terminal forms. Adding HA in large amounts to skeletal muscle cell cultures delays the onset of two myogenic indicators, fusion and accumulation of CK activity [48, 49]. The delay in the appearance of CK activity is striking and represents a good biochemical parameter for further molecular studies into the mechanism of the delay. Other glycosaminoglycans tested, chondroitin sulfates, do not inhibit CK activity, which suggests that the effect is specific for HA and not for just polyanions. The muscle-specific isoenzyme of CK (MM) is preferentially reduced compared to the embryonic CK isoenzyme (BB).

Ahrens et al analyzed the glycosaminoglycans accumulated in primary chick muscle cultures over a 24 h period 72 h after plating [50]. They found that muscle cultures synthesized hyaluronate, chondroitin, chondroitin sulfates, and heparin sulfates. Most of the material was found in the culture medium, and specifically, very little HA remained in the cell layer. Rubin [51] showed that secondary chick fibroblasts secrete HA, and it is possible that the fibroblasts are producing HA in primary skeletal muscle cell cultures as well. Other workers have also noted GAG synthesis by primary muscle cultures [52–54]. Analysis of endogenous hexuronic acid levels in these cultures confirmed that the cultured cells secrete glycosaminoglycans. The very early culture results, however, are quite startling. The basal level of HA and chondroitins in the growth medium, contributed by embryo extract and horse serum, is 2.80 μg/ml. This rapidly rises, but the increase can be accounted for by the loss from the cell layer. Over the first 12 h in culture, however, the total hexuronic acid-containing GAG content of the cell layer, as well as in the medium, declines. Thereafter, it increases concomitantly with cell protein accumulation. The simplest explanation for these observations is that there may be hyaluronidase activity in the culture whose effects are apparent within the first 24 h after plating the cells. The source of the endogenous hyaluronidase activity is unknown, although there is some evidence for hyaluronidase production by cultured cells. Chick fibroblasts obtained from skin have been shown to secrete hyaluronidase in culture [55]. The low pH optimum of this enzyme makes it unlikely that it could function in a standard growth medium unless it does so in association with other factors. It is also possible that GAG in the medium could be taken up by the cells to be degraded within lysosomes. If such hyaluronidase activity as seen in culture is related to the known degradation of HA that occurs in vivo, then these results suggest that part of the early developmental program in these cells may involve a degradation of the extracellular GAG in the matrix around myoblasts.

The fact that high concentrations of exogenous HA can only inhibit when added at early times suggests that HA is removed during an early step in myogenesis and that once muscle cells have gone beyond this step they can not be turned off by HA.

Feeding the cells frequently delayed overall myogenesis, possibly by diluting conditioning factors or inducers of differentiation. HA added in excess under these semi-perfusion conditions was found to exaggerate further the differentiation delay. Konigsberg [45] showed that continuous perfusion of myogenic cultures delayed fusion. He postulated that myogenesis required depletion of medium components or accumulation of cell products in the medium, or a combination of the two. It is possible that the HA/hyaluronidase system postulated here was involved.

How HA, a major component of the early embryo extracellular matrix, can cause a delay in differentiation is unknown. It was shown not to affect significantly attachment and spreading, cell viability, or metabolism. A direct interaction between HA and the cells is indicated by the ability of cells to bind to HA-coated beads in serum-free buffers. The cells may have membrane receptors for HA, or the binding could be mediated by tightly bound trypsin- and collagenase-resistant peripheral molecules. One mechanism that might explain these results is that HA bound to the cell surface alters the cell's microenvironment. Local changes in ion distributions or charged lipid distributions could result in changes in membrane properties. For instance, myoblast membranes undergo large changes in membrane fluidity before they fuse [58–60]; if bound HA interfered with these membrane alterations, it could inhibit differentiation.

The inhibitory effect is not permanent; the cells' differentiation, as assayed by fusion and CK, has been delayed. The fact that the inhibition is reversible is not surprising. In vivo, the HA inhibition would eventually have to be overcome for myogenesis to proceed. The model of Toole [6] depicts HA as an inhibitor of differentiation during the cells' migratory phase, with a reversal of this inhibition when cells reach their final destination. If this model is correct, then it is possible that the cell culture system mimics this in vivo situation.

By extending the cell-culture observations presented here to the in vivo case, one could postulate that differentiation of muscle precursor cells, or myoblasts, is regulated by components of the cells' microenvironment, including the extracellular matrix. Hyaluronate can be added to the growing list of regulating components, which already contains collagen and fibronectin. These may modulate the behavior of cells until they are in their final positions in space and time. The cells' specific gene products can subsequently be generated by the removal of inhibitors or the appearance of inducers, or both, so that development can proceed.

ACKNOWLEDGMENTS

This work was supported by a research grant from the Muscular Dystrophy Association (HFE) and the Cancer Research Coordinating Committee of the University of California (HFE) and by grant HL 17682 from the NHLBI (to J.S.I.). The assistance of Raymer Seavers, Michael Woo, Glenn Greene, and Norman Hall is gratefully acknowledged.

REFERENCES

1. Hay ED: In Brinkley BR, Porter KR (eds): "International Cell Biology, 1976–1977." New York: Rockefeller University Press, 1977, pp 50–57.
2. Manasek FJ: Current Topics Dev Biol 10:34, 1975.
3. Solursh M: Dev Biol 50:525, 1976.
4. Solursh M, Fisher M, Singley CT: Differentiation 14:77, 1979.
5. Toole BP, Gross J: Dev Biol 25:57, 1971.
6. Toole BP: Dev Biol 29:321, 1972.
7. Toole BP, Trelstad RL: Dev Biol 26:28, 1971.
8. Iwata H, Urist MR: Clin Orthop 90:236, 1973.
9. Toole BP: Am Zool 13:1061, 1973.

10. Wiebkin OW, Muir H: FEBS Lett 37:42, 1973.
11. Solursh M, Vaerewyck SA, Reiter RS: Dev Biol 41:233, 1974.
12. Handley CJ, Lowther DA: Biochim Biophys Acta 444:69, 1976.
13. Nevo Z, Dorfman A: Proc Natl Acad Sci USA 69:2069, 1972.
14. Solursh M, Hardingham TE, Hascall VC, Kimura JH: Dev Biol 75:121, 1980.
15. Lash J, Kosher RA: In Slavkin HC, Greulich, RC (eds): "Extracellular Matrix Influences on Gene Expression." New York: Academic Press, 1975, pp. 671–676.
16. Kosher RA, Lash JW: Dev Biol 42:362, 1975.
17. Kosher RA, Lash JW, Minor RR: Dev Biol 35:210, 1973.
18. Lash JW, Vasan NS: Dev Biol 66:151, 1978.
19. Meier S, Hay ED: Proc Natl Acad Sci USA 71:2310, 1974.
20. Meier S, Hay ED: Dev Biol 38:249, 1974.
21. Konigsberg IR: Exp Cell Res 21:414, 1960.
22. Stockdale FE, Holtzer H: Exp Cell Res 24:508, 1961.
23. Yaffe D, Feldman M: Dev Biol 11:300, 1965.
24. Hauschka SD, Konigsberg IR: Proc Natl Acad Sci USA 55:119, 1966.
25. Hauschka SD, White NK: In Banker B, Pryzbylski R, van der Meulen J, Victor M (eds): "Research Concepts in Muscle Development and the Muscle Spindle." Amsterdam: Excerpta Medica, 1972, pp 53–71.
26. de la Haba G, Kamali HM, Tiede DM: Proc Natl Acad Sci USA 72:2729, 1975.
27. Nameroff M, Holtzer H: Dev Biol 19:380, 1969.
28. Bischoff R, Holtzer H: J Cell Biol 36:111, 1968.
29. Lipton BH, Konigsberg IR: J Cell Biol 53:348, 1972.
30. Shimada Y: J Cell Biol 48:128, 1971.
31. Seno N, Anno K, Kondo K, Nagase S, Saito S: Anal Biochem 37:197, 1970.
32. Rosalki SB: J Lab Clin Med 69:696, 1967.
33. Klein MS, Shell WE, Sobel BE: Cardiovasc Res 7:412, 1973.
34. Lowry OH, Rosebrough NJ, Farr AL, Randall RJ: J Biol Chem 193:265, 1951.
35. Hamerman D, Todaro GJ, Green H: Biochim Biophys Acta 101:343, 1965.
36. Bitter T, Muir HM: Anal Biochem 4:330, 1962.
37. Hoare DG, Koshland DE: J Biol Chem 242:2447, 1967.
38. Cuatrecasas P: J Biol Chem 245:3059, 1970.
39. Reporter MC, Konigsberg IR, Strehler BL: Exp Cell Res 30:410, 1963.
40. Coleman JR, Coleman AW: J Cell Physiol 72 (suppl 1):19, 1968.
41. Eppenberger HM, Eppenberger M, Richterich R, Aebi H: Dev Biol 10:1, 1964.
42. Lough J, Bischoff R: Dev Biol 57:330, 1977.
43. Turner DC, Gmur R, Siegrist M, Burckhardt E, Eppenberger HM: Dev Biol 48:258, 1976.
44. Morris GE: Biochem Soc Trans 6:509, 1978.
45. Konigsberg IR: Dev Biol 26:133, 1971.
46. Stockdale F, Okazaki K, Nameroff M, Holtzer H: Science 146:533, 1964.
47. Weiss P: In Parpart AK (ed): "Chemistry and Physiology of Growth." Princeton: Princeton University Press, 1949, pp 135–186.
48. Elson HF: J Cell Biol 79:156a, 1978.
49. Elson HF, Ingwall JS: J Supramol Struct Suppl 4:180, 1980.
50. Ahrens PB, Solursh M, Meier S: J Exp Zool 202:375, 1977.
51. Rubin H: Nature 254:65, 1975.
52. Holtzer H, Rubinstein N, Fellini S, Yeoh G, Chi J, Birnbaum J, Okayama M: Q Rev Biophys 8:523, 1975.
53. Hermann H, Havaranis AS, Doetschman TC, J Cell Physiol 85:557, 1975.
54. Angello JC, Hauschka SD: Dev Biol 73:322, 1979.
55. Orkin RW, Jackson G, Toole BP: Biochem Biophys Res Commun 77:132, 1977.
56. Shainberg A, Yagil G, Yaffe D: Dev Biol 25:1, 1971.
57. Cox PG, Gunter M: Exp Cell Res 79:169, 1973.
58. Elson HF, Yguerabide J: J Supramol Struct 12:47, 1979.
59. Prives J, Shinitzky M: Nature 268:761, 1977.
60. Herman BA, Fernandez SM: J Cell Physiol 94:253, 1978.

ND# Mapping the Mitotic Clock by Phase Perturbation

R. R. Klevecz, G. A. King, and R. M. Shymko

Division of Biology, City of Hope Research Institute, and Division of Radiation Oncology, City of Hope Medical Center, Duarte, California 91010

In synchronized V79 cells perturbed by serum, heat shock, or ionizing radiation at half-hour intervals through a modal 8.5-hour cell cycle, phase-response curves show a characteristic biphasic pattern of advances and delays in subsequent cell divisions. These observations, together with previous observations of quantizement of generation times in this and other cell lines have led us to consider a model incorporating, in the simplest case, a two-component oscillator with two threshold crossings required per cell cycle. By assuming that oscillator variables respond in a simple way to the experimental perturbations, for example, by first order destruction due to heat shock, a map of the qualitative features of the oscillator can be obtained by matching simulated with experimental phase response curves. Random fluctuations in oscillator variables about a fixed trajectory lead to subthreshold oscillations and result in a distribution of generation times which is roughly a negative exponential, but quantized within this exponential envelope. The extent of the random fluctuations can be determined from comparison with data on desynchronization of a cell population after mitotic selection. The same parameters which correctly simulate phase response and the desynchronization data also give good agreement with generation time distribution data.

Key words: cell cycle, transition probability, limit cycle oscillator, generation time, phase response, division delay, cellular clock

The response of cells to external influences is often strongly cell cycle phase dependent. This fact has encouraged many workers to attempt to obtain information about the cell cycle timekeeping mechanism by observing the phase change in some marker event (usually mitosis) after administration of a perturbing agent. The rationale behind this approach lies in the hope that the chosen perturbation affects the timekeeping mechanism directly, so that the phase response reflects the properties of the underlying clock. However, since the biochemical nature of the clock, and consequently the effects on this clock of any given perturbation, are largely unknown, it is difficult to confidently identify cellular phase responses with clock responses. Some or all of the observed phase shifts, for example, after any given perturbation may be due to defects in cellular functions normally under the control of an underlying clock, and not due to effects on the clock itself.

Received May 5, 1980; accepted August 6, 1980.

0091-7419/80/1403-0329$04.00 © 1980 Alan R. Liss, Inc.

More critically, the very notion of a cellular timekeeper has come into question in recent years as a result of observations made on unperturbed cell populations. Individual cell intermitotic times have been cataloged for many different cell types and have been found to be distributed not in a normal fashion about some mean but in a skewed, roughly negative exponential distribution [1–3]. Such observations have been taken as evidence for a stage in the cell cycle from which cells leave in a perfectly random fashion. The introduction of such a purely stochastic element into the cell cycle has led to an appealingly simple model of the cell cycle first proposed by Burns and Tannock [1] and later revived by Smith and Martin [2] in which the cell cycle is divided into two parts, one which cells enter after mitosis and leave randomly at a constant rate per unit time, and a second state through which cells move uniformly, traversing this state in a constant time T_B. The random exit from the "A-state" [2] generates the exponential tail in generation time distribution, and the uniform traversal through the "B-phase" produces a shoulder at time T_B in the graph of fraction of cells undivided vs. age (the "α-curve").

To counter the objections that cells do appear to be influenced in subsequent generations by their prior history and that the α-curves for most cells are not, strictly speaking, exponential, Brooks et al [4] have convoluted the original model by the addition of a second random transition and a second fixed interval state "L."

We have assembled evidence to argue that the cell cycle is timed by a macromolecular oscillator with limit cycle properties [5, 6]. Such a model is supported by a formal mathematical structure and is based empirically on numerous observations of oscillations in enzyme levels and in the rates of RNA and DNA synthesis [7–9]. Stable limit cycle oscillators have been shown to provide a useful representation of metabolic oscillations in glycolytic intermediates [10], of the mitotic clock of Physarum [11], and of circadian rhythms of eclosion in Drosophila [12, 13]. Even so, in attempting to describe a timekeeping oscillator in a biological system the situation is quite unlike that found in engineering or in the physical sciences, where the state variables are usually well defined and measurable. In biological systems the structure of the oscillator must be imputed by perturbing the organism and then observing some manifestation of the underlying oscillator [14]. By constructing a phase response curve — in this instance, the change in time of mitosis between perturbed and control cultures — an assessment of the relative value of the state variables of the underlying oscillator can be performed. Although our model was developed to explain the phase response curves of cells following perturbation, we recognized that it might also be possible to generate an exponential distribution of generation times using a limit cycle oscillator. Our purpose in this paper is to show that such accurate timekeeping can be reconciled with the apparent stochastic properties of the cell cycle, and to deduce some of the qualitative and quantitative properties of the timekeeping mechanisms from the response of cells to phase perturbations.

MATERIALS AND METHODS

Cell Culture, Synchronization and Time-Lapse Video Tape Analysis

V79 cells were synchronized by mitotic selection from roller bottles using the automated synchrony system described in detail elsewhere [6]. Cell divisions were scored by continuously monitoring a field (100–200 cells/field) of each of the synchronous cultures using time lapse video tape microscopy at 50- to 100-fold time compression as described previously [15]. Cell divisions were registered by writing the time of occurrence of each anaphase directly onto the video screen over the dividing cells. For graphical presentation these times were subsequently tallied and grouped into half-hour classes. Modal class values were determined by inspection. Median values were determined by interpolation within class intervals. Intermitotic times were determined in much the same way by following individual cells and their daughters through several generations.

Phase Response Curves

Phase response curves were generated by determining the midpoints of the first mitotic wave following selection synchrony. Midpoints were compared for each pair of perturbed and control cultures as a function of time in the cycle at which the perturbation was begun. The change in phase, $\Delta\phi$, was calculated as the algebraic sum of the median of the generation time of the control cultures minus the median of the paired perturbed culture, $\Delta\phi = T_{gc} - T_{gp}$. Positive values of $\Delta\phi$ indicate that shocked cultures divided sooner than the controls, while negative values indicate that division occurred later than the controls.

RESULTS

Dynamics of the Oscillator

Several models of oscillating chemical reactions which generate limit cycles in the X,Y phase plane have been described. In our stimulations we have tried different reaction schemes and have found that the qualitative behavior of our model system is not strongly dependent on the particular scheme used. Results presented here use the trimolecular model of Prigogine and Lefevre [16], which has since been called the Brusselator [17]. It can be represented by the following reaction scheme:

$$A \underset{k_{-1}}{\overset{k_1}{\rightleftharpoons}} X$$

$$B + X \underset{k_{-2}}{\overset{k_2}{\rightleftharpoons}} Y + D$$

$$2X + Y \underset{k_{-3}}{\overset{k_3}{\rightleftharpoons}} 3X$$

$$X \underset{k_{-4}}{\overset{k_4}{\rightleftharpoons}}$$

In this system the "precursors" A are assumed to be in vast excess, and therefore $k_{-1} \simeq 0$. It is also assumed that the final "products" D and E are instantly removed from the vicinity and therefore that effectively $k_{-2} = k_{-4} = 0$, and that $k_{-3} = 0$. If for further simplicity the remaining forward rate constants are set equal to 1 then the dynamical properties of the oscillator are given by the set of differential equations

$$dX/dt = A - BX + X^2 Y - X$$
$$dY/dt = BX - X^2 Y$$

The steady state of this system is given by the equations

$$0 = A - (B+1)X + X^2 Y$$
$$0 = BX - X^2$$

from which we obtain the solutions $X = A$ and $Y = B/A$. Assuming A and B are both positive as required by the chemical equations, the term under the radical is always less in magnitude than $B - A^2 - 1$ or is negative. Therefore, if $B - A^2 - 1$ is greater than zero the steady state is unstable. For $B > A^2 + 1$ stable oscillations of the variables X and Y are obtained. A detailed analysis of this system will be presented elsewhere.

General Behavior of the Oscillator

Two extreme forms of the Brusselator are shown in Figures 1A and B. As the value of A increases such that $A^2 \to B-1$, oscillations become more sinusoidal and the time required for initial values of X and Y to approach the limiting values likewise increases. Such an oscillator is described as "soft" and perturbations to the system may be righted only after multiple loops or oscillator periods are accomplished. This is in contrast to the situation where A decreases such that $A^2 \ll B-1$, causing the oscillator to assume properties much like a relaxation oscillator. Such a "hard" oscillator is righted almost immediately following peturbation. If cellular timekeeping were accomplished by means of a soft oscillation, then long, multiple cycle delays in cell division would be expected for appropriate choice of time and intensity of the perturbing stimulus. Alternatively, if a hard oscillation is involved, delays longer than one cycle should be difficult to achieve. Work to date suggests that if there is an oscillator it is more hard than soft. We assume that cells travel around a limit cycle shown in Figure 1B with a critical event in the cycle being triggered when Y exceeds a threshold value θ. We assume further that cells are not confined strictly to the locus of the cycle; rather they travel around the cycle in a cloud. If the uppermost part of the cloud is only slightly above the triggering threshold, then the cells in the lower part of the cloud will fail to reach threshold and will not proceed through the cycle. If these cells continue to be distributed in a random cloud, some fraction will have the possibility of skipping yet another threshold event in the next cycle. A phase plane trajectory of the oscillator used in this study is shown in Figure 1C. For high threshold values of θ, and sufficiently large random walk parameters, D_x and D_y, the cloud may be broadened enough so that only a small fraction of cells will cross threshold. In Figure 1C the path of a single cell and one of its daughters following each division is tracked through a total of five cell divisions. A cell may pass below threshold in a given cycle and may therefore require two or more circuits before exceeding threshold, as shown. The cell cycle generation time distribution will be polymodal, with peaks at multiples of the limit cycle time, but the envelope enclosing these peaks will be exponential or nearly so. Further, for cells with long generation times the cloud of cells may be very broad and can yield a distribution which is very nearly a simple exponential.

The Two-Loop Model

The above discussion has been limited to the case of one limit cycle period per cell cycle. However, our observations of subcyclic phenomena suggests that more than one loop is required per cycle. We envision that X and Y are macromolecules, most probably proteins, with constants for synthesis and degradation of a value to give oscillations with a nominal 4-hr period. The values of such constants and the half times required for synthesis and degradation to give such a periodicity are in agreement with values generally seen in animal

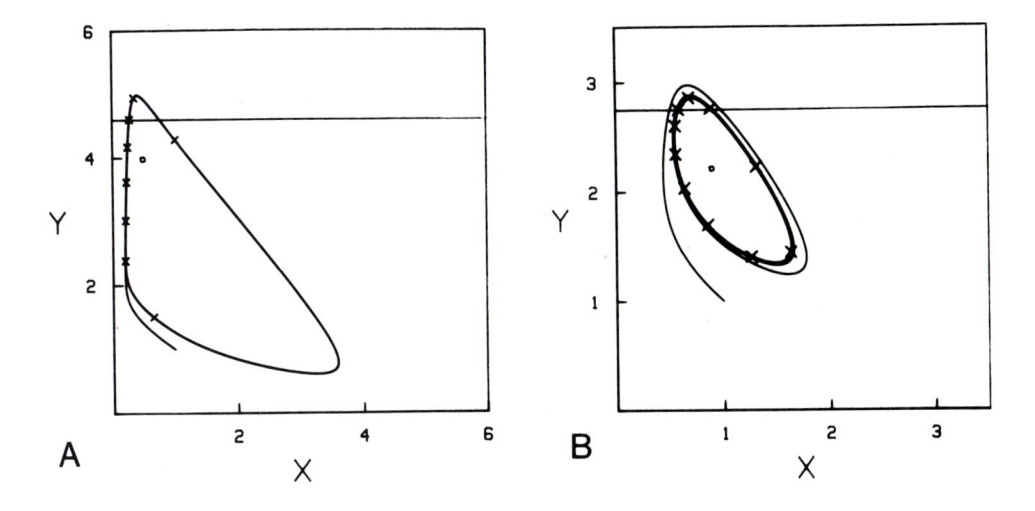

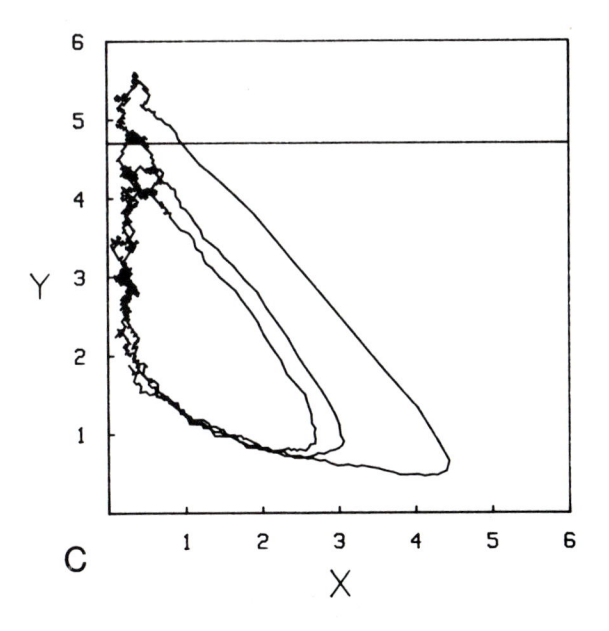

Fig. 1. Phase plane trajectories of the Brusselator. Stable oscillations in the variables X and Y occur for values of parameters $B > A^2 + 1$. In the oscillatory region, the system moves in a clockwise direction about the steady state ($X_{ss} = A$, $Y_{ss} = B/A$). For different values of A and B, the behavior can approximate that of a relaxation oscillator [(1A) A = 0.5, B = 2.0] or can be more "sinusoidal" [1B) A = 0.9, B = 2.0]. Phase plane portrait of the oscillator used to simulate results in this study is shown in 1C (A = 0.5, B = 2.0, $X_0 = 1$, $Y_0 = 1$ Random walk; $D_X = D_Y = 0.024$). Threshold θ is indicated by the horizontal line. Crosses indicate intervals of 1/10 cycle. All simulations were done using the system in Figure 1C.

cell systems. We consider that in the shortest mammalian cell cycle (8 hr), two crossings of threshold must be accomplished, the first possibly associated with the initiation of a cycle of DNA replication and the second associated with the triggering of mitosis. As we will show later, phase perturbation data also suggests that there are two rather than one "clock" periods per mitotic cycle. Lengthy prereplication stages seen in cells at confluence or in serum or amino acid depleted media occur in this model as a consequence either of altered levels of the "precursors" A, or increased threshold θ, resulting in repeated sub-threshold oscillations. Notice that if the threshold value in the first loop is higher than that in the second, then cells will show greater asynchrony in the execution of early versus later cell cycle events. The results presented here will be qualitatively similar if the two thresholds are either sufficiently different or are equal; for the sake of simplicity we have performed all simulations with the assumption of equal thresholds.

There is evidence for approximately 4-hr subcycling behavior in cell lines with generation times considerably longer than 8 hr [9, 18]. It may be, therefore, that these cells

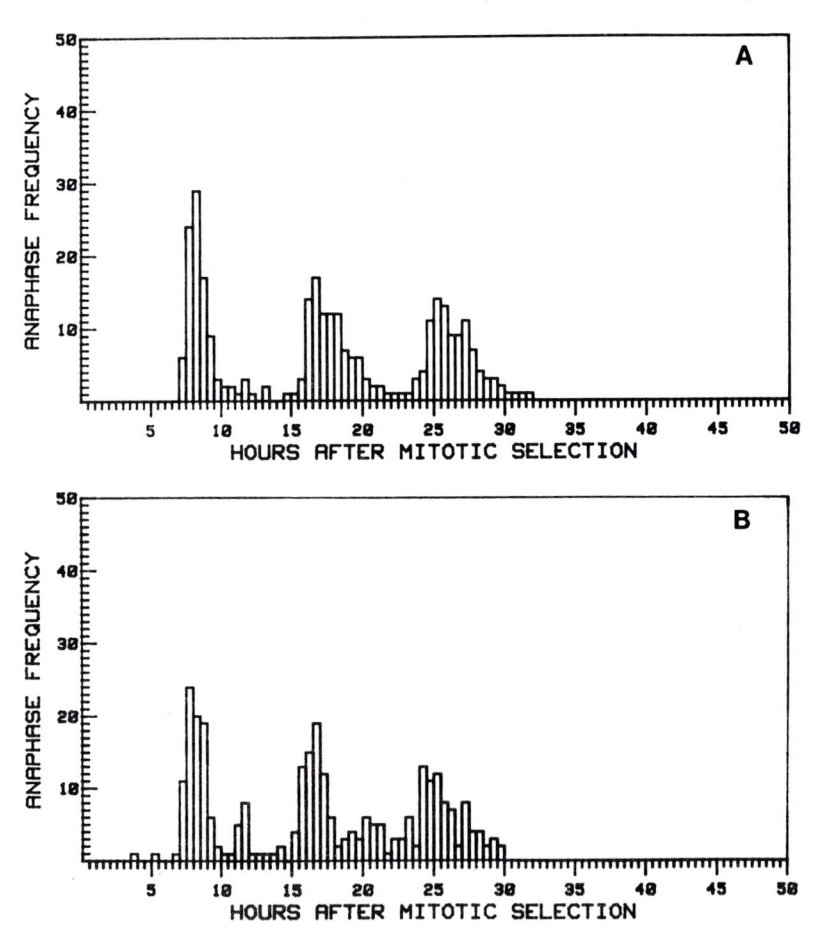

Fig. 2. Desynchronization histograms. Histograms of anaphase frequency for several generations of synchronous V79 cells are shown under conditions of slow (2A) and rapid (2C) decay of synchrony. 100–200 cells were followed for two or three modal generation times after mitotic selection. Figures 2B and D are paired simulations of 2A and C which vary only in the choice of threshold θ. In B, $\theta = 4.5$, in D, $\theta = 4.7$.

require more than two threshold crossings per cycle (see Fig. 3C), each crossing triggering successive cycle events. Alternatively, execution of events normally triggered by a threshold crossing may be delayed until some independent cell parameter, perhaps related to cell size, reaches a critical level. In this way, additional clock periods are inserted into the cell cycle.

Desynchronization of Mitotic Waves

We have simulated the behavior of a population of 100 cells following a limit cycle trajectory by supposing that two thresholds must be crossed per cell cycle. Simulated cell populations were initiated synchronously just after mitosis, and allowed a random walk with equal magnitude steps in both the X and Y direction in each of 200 time intervals per loop of the limit cycle. After each step in the random walk the dynamical motion in that time interval was calculated. Preliminary calculations and simulations were done to determine a threshold value such that an appropriate fraction of cells failed to reach threshold on any given pass. Each individual cycle time was monitored and a histogram of cycle times was plotted.

In Figures 2A and B, the decay of synchrony in synchronous populations of V79 cells and simulations of the decay of synchrony are shown. The loss of synchrony in a population is accelerated by a number of things including temperature shifts or cold collection,

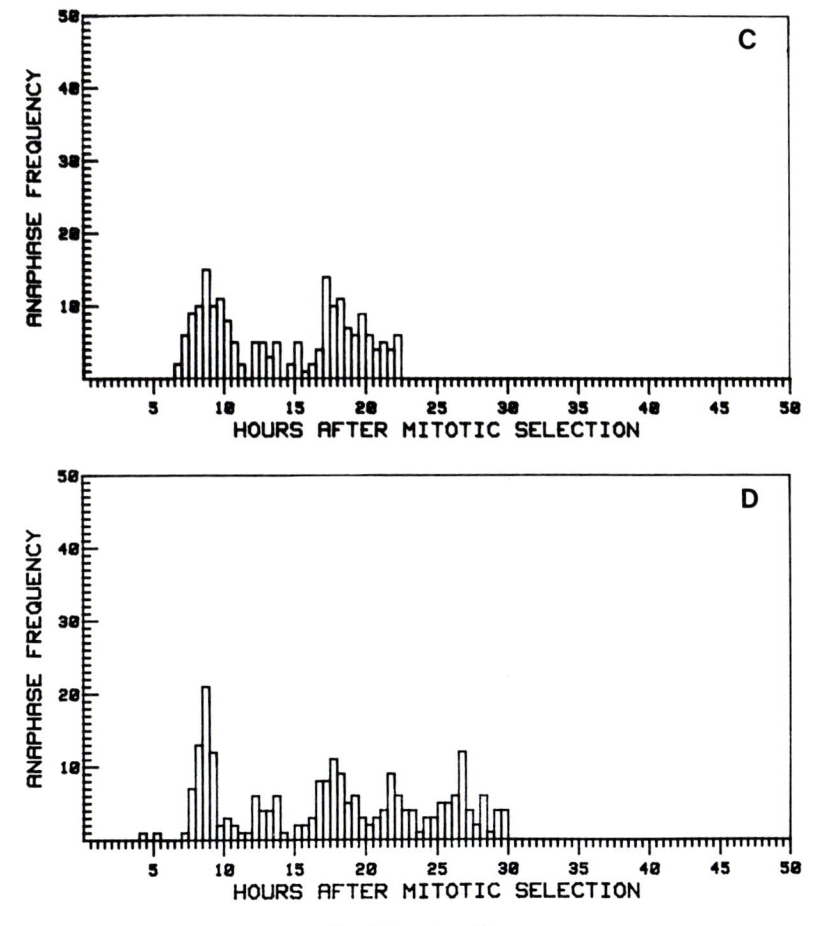

Fig. 2 (continued).

changes in serum concentration and the method of achieving synchrony. In this study a synchronous culture obtained under optimized conditions (Fig. 2A and B) is compared with a culture that experienced a serum concentration shift prior to mitotic selection (Fig. 2C and D). Desynchronization can occur in the model as a result of either an increase in random step size, or an increase in the threshold value θ. In the latter, the loss of synchrony occurs due to subthreshold oscillations and consequent skipping of threshold crossings. The change in θ from 4.5 to 4.7 was sufficient to generate the increase in desynchronization.

Generation Time Distributions — α and β Curves

Generation time distributions of V79 cells were obtained by following individual cells and their progeny in video tape recordings. These intermitotic times were then tallied. In the case of WI-38 the pedigree data of Absher et al [19] were replotted as intermitotic times. In Figure 3A, B, and C, the distribution of generation times of individual cells in populations of V79 and WI-38 are plotted as the fraction of the population that has not yet divided vs. the generation time (the α curve [2]). Simulated distributions generated by the oscillator model are shown for comparison. Simulations using the 2-loop oscillator were performed using the parameter values obtained from desynchronization experiments shown in Figure 2. For θ = 4.5, nearly all cells cross both thresholds on the first attempt and generation times are tightly distributed about the modal value. For slightly higher values of the threshold, significant numbers of cells fail to cross one or both of the thresholds in the first pass and hence the distribution shows considerable skewing and some quantizement.

Additional simulations were done to obtain the distribution of differences in generation times between sister cells (the β-curve [4]). As expected, in this model, sister cell generation times are uncorrelated, and the β-curves parallel the tails of the α-curves (data not shown). In order that cell populations show correlated sister cell generation times as well as exponential β-curves, there must be an aspect of the cellular timekeeping which is variable in the population but constant between sister cells, and another aspect which is purely random. We speculate that here again cell size may play a role in that sister cells will both tend to be smaller or larger than normal according to whether the mother is smaller or larger than normal. In the picture we present here, a sub- or supra-normal size at birth may lead to a greater or lesser number of oscillator loops, respectively, before first threshold crossing is reached. This will tend to correlate sister generation times, while subsequent random subthreshold passages will generate nearly exponential β-curves.

Phase Response to Perturbation

Figure 4 shows a series of synchronized V79 cells given single high serum pulses, heat shocks, or ionizing radiation treatment at half-hour intervals through a modal 8.5-hr cell cycle. Each data point represents the difference between paired treated and control cultures of 100–200 cells/field.

The phase response curves display a biphasic pattern of advances and/or delays in subsequent cell divisions. A characteristic phase response curve with a repeating 4-hr periodicity was generated. In Figure 4a the times to the first and second synchronous waves following synchronization were compared for each pair of serum pulsed and control cultures as a function of time from the beginning of the serum pulse. Some differences were noted in the phase shift accomplished by different serum lots. This may be reflected in the fact that the phase response curve shows a relatively broad band of responses. Beginning 0.5 hr after mitotic selection, pulses with serum produce delays in the midpoint

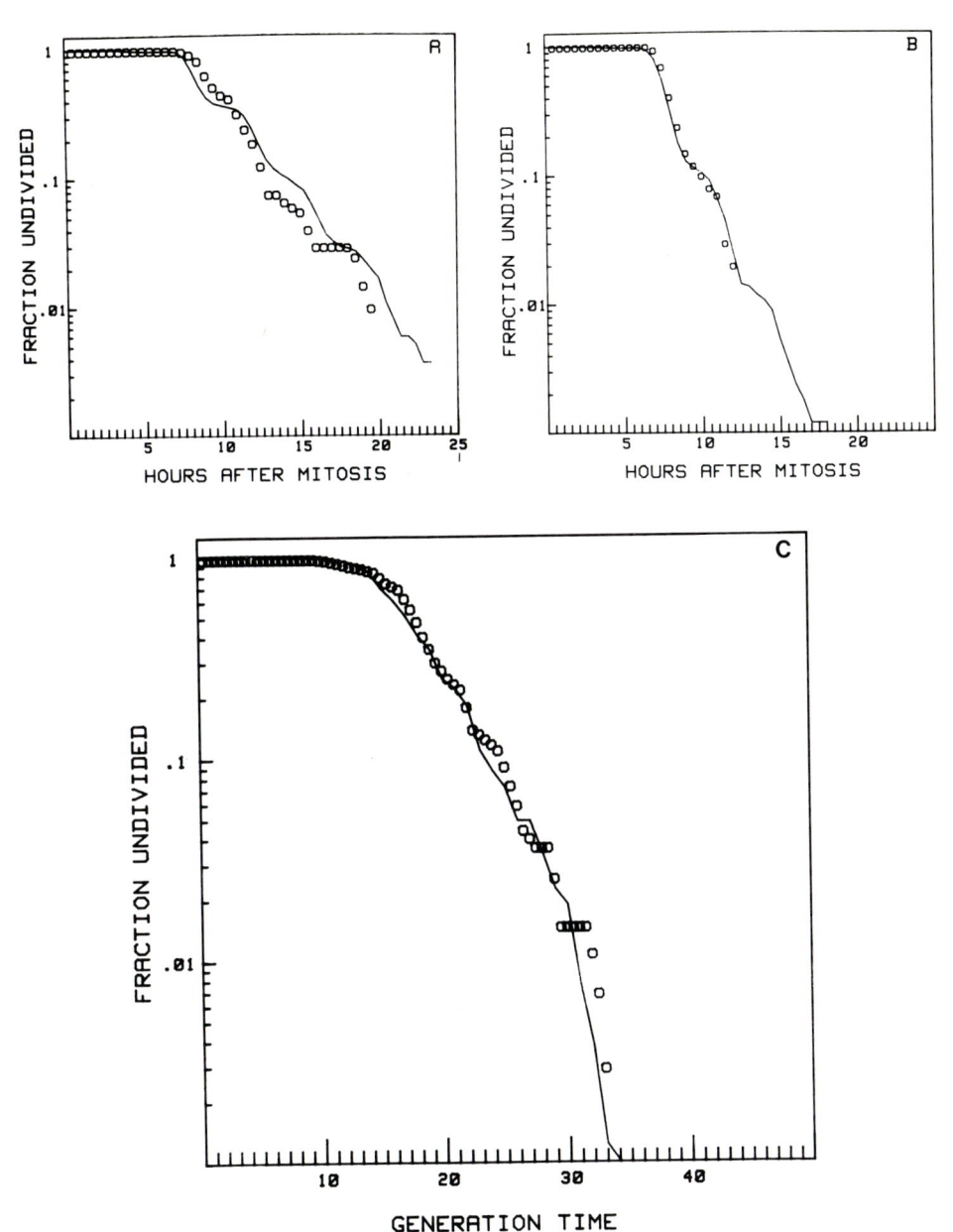

Fig. 3. Alpha curves of V79 and WI-38 cells and their simulation by the oscillator model. Alpha curves describing the undivided fraction of cells in generation time distribution curves are shown in Figures 3A, B and C. 200 ± 50 generation times are represented in each curve. In A and B, circles indicate distributions of generation times in V79 cells growing under suboptimal (A) and optimal (B) conditions. Simulation of these curves (solid lines) shows that the distribution of generation times is exponential but quantized within the exponential envelope. In A threshold $\theta = 4.7$, in B $\theta = 4.5$. All other parameters are unchanged from those in Figure 1C. In Figure C, circles indicate generation time distributions of WI-38 cells and the lines give the simulated distributions. Here parameters are the same except that cycle time is increased by requiring that the number of oscillator loops in one cycle is increased from 2 to 4. Note that with the long cell cycle generated by this model, the resulting distribution of generation times is smoother and approaches a straight line at long generation times.

of the subsequent mitotic waves (delay is maximum at 1.5 hr). Delays give way abruptly to advances at 2.5 hr and the amount of advance then decreases as pulses are given between 3 and 5 hr into the cycle. At 5 hr decreasing advances become delays which increase for serum pulses occurring between 5 and 6 hr. Delays again give way abruptly to advances at 6 hr and again the amount of advance decreases through the late portion of the cycle. Pulses very late in the cycle appear to generate phase delays.

Similarly in Figure 4B the results of 10-min 45°C heat shocks given at 0.5-hr intervals through the cycle are shown. Cells pulsed soon after mitosis are slightly delayed in the subsequent mitoses relative to the paired unshocked control. Minimum delay, and in some

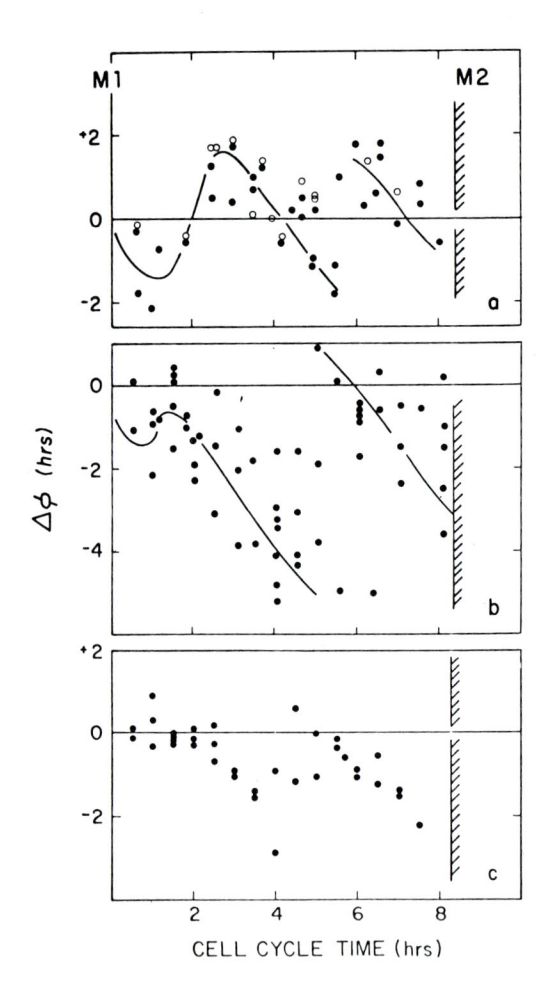

Fig. 4. Phase response of synchronous V79 cells to perturbation by serum, heat shock and ionizing radiation. (A) Serum pulses. At intervals following mitotic selection, serum concentration in the medium was increased from 5 to 20%. Midpoints of first (●) and second (○) mitotic waves. (B) Heat shock. Midpoints of the first mitotic wave following synchronization and a 10-min 45°C heat shock are compared for each pair of heat shocked and control cultures as described in (a). (C) Ionizing radiation. Synchronous V79 cells were exposed to 150 rads from a Cobalt-60 source at 30-min intervals through the first synchronous cell cycle. Analyses of division advance or delay were determined as described in (A) and (B).

instances a slight advance, occurs when shocks are given 1.5 hr after mitosis. There follows a pattern of increasing delays up to 5 hr, when an abrupt shift in response occurs giving a second minimum in delay at 5.5–6 hr of the cell cycle. Pulses given later than 6 hr in the cycle give a pattern of increasing delays up to the subsequent mitosis. The response curve appears as two parallel lines sloping downward to the right, with a small cluster of values between 4 and 6 hr of the cycle showing a constant 2-hr delay. In some experiments heat shocks given after 4 hr showed a splitting of the anaphase frequency histogram, suggesting that at these values of time and perturbing stimulus member cells may be either slightly advanced or delayed, or quantally delayed by $\Delta\phi + 4$ hr. It is often the case in calculating the mean or midpoint of the population that the value will be found to lie between the two discrete peaks [6]. This may serve to explain why in earlier division delay studies the results were described as showing a transition point with constant delays in response to perturbations late in the cycle. This capacity to phase jump is further shown when on occasion the midpoint of a population shocked at 7 hr is delayed, not by 0.5–1.0 hr, but by 5 hr, or a full subcycle.

Figure 4C shows the phase response to 150 rads of ionizing radiation applied at one-half-hour intervals through the V79 cell cycle. Cells pulsed up to 2 hr after mitotic selection are slightly delayed or advanced in the subsequent mitosis relative to the paired untreated control. There follows a pattern of increasing delays up to 4 hr when an abrupt shift in response occurs, giving a second minimum delay, and in some cases slight advances at 4.5 and 5 hr. Pulses given after 6 hr in the cycle show increasing delays up to the initiation of the first postselection mitotic wave.

Simulation of Phase Response to Heat Shock

To simulate the phase response behavior of V79 cells to heat shock, oscillator variables X and Y were reduced by 90% at various intervals and the system was then allowed to evolve from this new state. In each case the system relaxed toward the stable trajectory, and eventually crossed threshold. Phase shifts in the time of mitotic threshold crossings were computed and compared with the heat shock data of Figure 4. The results are shown in Figure 5. Note that any extensive first order destruction of oscillator variables would place the system close to the origin and yield qualitatively similar phase response curves.

DISCUSSION

It does not appear necessary to resort to a strictly stochastic model in order to generate the commonly observed distribution of cell generation times. The limit cycle model simulates an exponential or pseudo-exponential distribution by permitting random walk of individual cells below threshold. The oscillator model predicts (and fits) a biphasic response to heat shock and other perturbations in V79 cells with 8-hr generation times. The model suggests that prolonged failure to execute early cycle processes will occur when the value of the parameter A, the precursor material in this reaction scheme, changes, moving the steady state and trajectory and causing repeated subthreshold oscillations.

The tendency of many cellular processes to oscillate with a common period that is shorter than that of the cell cycle [7–9, 20–23] and the consequent quantizement of generation times [23] suggests that cellular timekeeping involves an oscillatory mechanism having much in common with other biological rhythms. That such periodicities are not routinely observed in heterogeneous tumor cell populations or in cells with relatively long generation times, may be more a reflection of the choice of experimental system and synchrony technique than an intrinsic difference between cell types. Setting aside the existence of oscillations of periods less than one-cell generation time, it seems appropriate to require the crossing of two thresholds for cell division to take place since there are by most accounts a minimum of two dependent events in the cell cycle. DNA synthesis must occur before cell division and cell division must occur between rounds of DNA synthesis, although exceptions to those two conditions can be found. It should also be mentioned that pseudo-exponential generation time distributions (but not the correct phase response curves) also were achieved in a series of simulations in which the limit cycle period equaled the cell cycle period.

The model presented here does not quantitatively describe the common observation that big mother cells give rise to relatively larger daughter cells which therefore divide faster than their more ordinary sized counterparts, nor does it account for the need for a cell to achieve some minimum size before DNA synthesis can occur. Elsewhere we have extended the model by adding cell size along a third (Z) axis [6]. In that model a threshold crossing will only initiate an event such as DNA synthesis if such minimum size has already been obtained. This more general model predicts a correlation of sister generation times and may possibly explain the negative correlation of mother/daughter cell generation times. In addition exponential β-curves are generated.

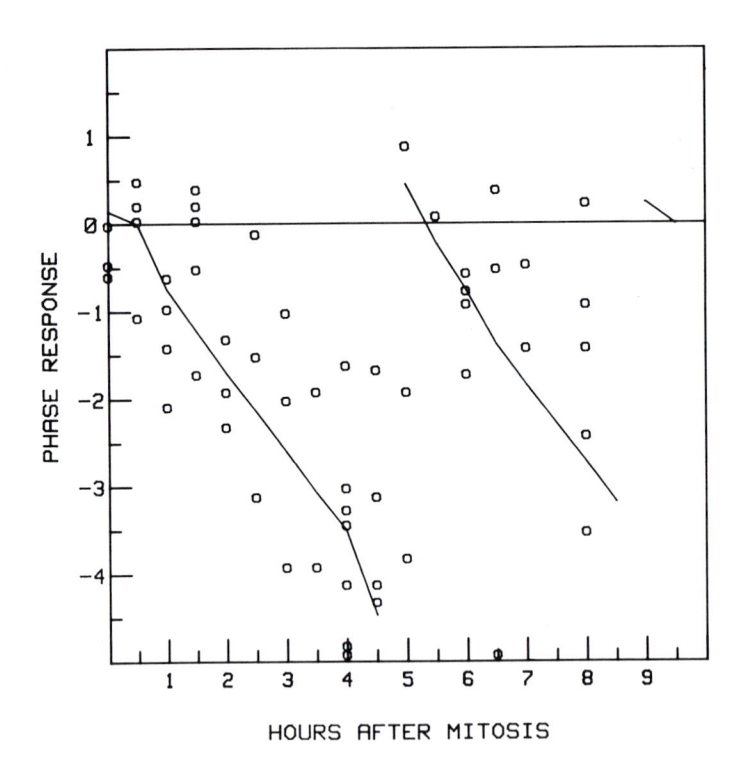

Fig. 5. Simulation of phase response of V79 cells to heat shock. The heat shock phase response curve generated in Figure 4 by treatment of synchronous cell cultures for 10 min at 45°C through the cycle is shown as open circles (○). Solid lines indicate a simulation performed using the oscillator model with parameter values of A = 0.5, B = 2, θ = 4.7. Perturbation at each half-hour point in the cycle was assumed to destroy 90% of X and Y by first order kinetics.

Limit cycle oscillators are attractive as models for timekeeping because they can combine both stochastic and deterministic elements. In unperturbed conditions, in the absence of any noise or random walk, the trajectory followed by a cell would be fixed and the time to pass an arbitrary threshold would be constant in each loop. Either normal variation or a perturbation to the state variables, or a change in the system parameters will alter the course of a cell on the trajectory in ways that often seem very "real" from the biological perspective. This property of phase responsiveness led to the initial experiments on phase perturbation and the resulting 2-loop model for the minimum cell cycle.

The set of strong perturbations described above provide information about the period and about the asymmetry and continuity properties of the clock. All the curves are consistent with two repeated response curves, each with period roughly equal to half of the 8-hr cell generation time. This suggests that there is a 4-hr timekeeper, responsible for the quantizement of generation times, which is being affected by the perturbations. For heat shock or serum perturbations, each of the two branches of the phase response curves have slope of approximately -1, suggesting that the perturbations drive the timekeeping system to a constant phase in the cycle, after which all perturbed cells take an equal time to pass through mitosis. This provides evidence that it is the clock which is being affected by the perturbation, since a response not based on clock perturbation would be highly unlikely to yield a delay exactly proportional to cycle phase.

The phase response data in Figure 4 provide information about the shape of the oscillator cycle. If, for example, heat destroys each of the oscillator variables with first-order kinetics, the system will be driven toward the origin as a result of the application of heat. After heat shock all cells are found in a small area in X-Y space, and have only a small range of phases. Consequently, when cells resume their cycling they take nearly the same time to reach threshold. Since cells which the have just passed threshold show small advances and those which are just reaching threshold show large delays, the origin $(X = Y = 0)$ must be at a phase representing a very early part of the cycle. Therefore, as shown here, the part of the cycle just after threshold, during which the system moves through decreasing values of X and Y, must be traversed very rapidly. Conversely, the part of the cycle during which the system rises toward threshold must be traversed relatively slowly (Fig. 1A).

Evidence for a continuous oscillator, rather than a strict relaxation oscillator comes from the capacity of cells to skip an event such as mitosis and display quantized generation times [6]. Furthermore, at the junction between the two repeated branches of the response curves for heat shock and serum pulses, the phase difference rapidly changes by approximately 4 hr, equal to the period of the assumed underlying timekeeper. This magnitude of change is exactly what is expected for a clock with a threshold triggering of some cell cycle event (ie, a "point of no return"). To visualize this, consider two cells, one with phase just prior to the threshold crossing and one just after. If the clock is continuous and the perturbation affects clock variables only, both cells should be driven to very nearly the same phase by identical perturbations. Since one cell is at the end of its cycle and the other is at the beginning, the difference in phase change due to the perturbation is one period of the timekeeper, ie, different by $\simeq 4$ hr. This 4-hr discontinuity should be seen for all "large" perturbations of a continuous oscillator, but will occur in a relaxation oscillator only if a particular perturbation resets the clock variable to exactly the same level starting from either its high prerelaxation or low postrelaxation value.

The above discussion points out the fact that phase response data as commonly presented are often not sufficiently stringent in their assessment of oscillator dynamics. Perturbations are characteristically chosen because they give maximum phase response consistent with viability. Under such conditions variables are reset to nearly identical phase and the response curve so generated shows an increasing delay with increasing progress through the cycle up to some execution point. In our experiments, two such points appeared — one 4.5 hr into the cycle, after the initiation of DNA synthesis and coincident with the replication of the bulk of the DNA, and the other at or near mitosis. To properly assess the "interior" of the oscillator the cells should be subject to gentler perturbations where, as Winfree [13] has done for Drosophila eclosion rhythms, for appropriate choice of phase and perturbing stimulus an apparent equilibrium point or singular state is approached, with the consequence that arrhythmic emergences are observed. It is important to state explicitly that the apparent discontinuity observed here in the phase response curves does not imply that timekeeping is accomplished by a relaxation oscillator. Rather it is more consistent with a continuous oscillator which gates certain mutually dependent, largely irreversible and therefore discontinuous cellular events such as DNA synthesis and mitosis.

REFERENCES

1. Burns FJ, Tannock IF: Cell Tissue Kinet 3:321, 1970.
2. Smith JA, Martin L: Proc Natl Acad Sci 70:1263, 1973.
3. Brooks RF: Cell 12:311, 1977.
4. Brooks RF, Bennett DC, Smith JA: Cell 19:493, 1980.
5. Klevecz RR: Cell Repr 12:139, 1978.
6. Klevecz RR, Kros J, King G: Cytogenet Cell Genet 26:236, 1980.
7. Klevecz RR, Forrest GL: In Cristofalo VJ and Rothblat G (eds): "Growth, Nutrition, and Metabolism of Cells in Culture." New York: Academic Press, 1977, pp 149–196.
8. Klevecz RR, Keniston BA, Deaven LD: Cell 5:195, 1975.
9. Kapp LN, Painter RB: Exp Cell Res 107:429, 1977.
10. Winfree AT: Arch Biochem Biophys 149:388, 1972.
11. Kauffman SA, Wille JJ: J Theor Biol 55:47, 1975.
12. Pavlides T, Zimmerman WF, Osborn J: J Theor Biol 18:210, 1968.
13. Winfree AT: Science 183:970, 1974.
14. Pittendrigh CS: In Hastings JW and Schweigert E (eds): "Dahlem Workshop on the Molecular Basis of Biological Clocks." Berlin: Dahlem Press, 1977, p 1.
15. Klevecz RR, Kros J, Gross SD: Exp Cell Res 116:285, 1978.
16. Prigogine I, Lefevre R: J Chem Phys 48:1695, 1968.
17. Tyson J, Light J: J Chem Phys 59:4164, 1973.
18. Klevecz RR, Kapp LN: J Cell Biol 58:564, 1973.
19. Absher PM, Absher RG, Barnes WD: Exp Cell Res 88:95, 1974.
20. Klevecz RR: Science 166:1536, 1969.
21. Klevecz RR: In "The Cell Cycle in Malignancy and Immunity." 13th Annual Hanford Biology Symposium, p 1, 1975.
22. Forrest GL, Klevecz RR: J Cell Biol 78:441, 1978.
23. Klevecz RR: Proc Natl Acad Sci 73:4012, 1976.

Cell Surface Receptors for Endogenous Mouse Type C Viral Glycoproteins and Epidermal Growth Factor: Tissue Distribution In Vivo and Possible Participation in Specific Cell-Cell Interaction

U.R. Rapp and Thomas H. Marshall

Laboratory of Viral Carcinogenesis, National Cancer Institute, National Institutes of Health, Bethesda, Maryland 20205

We have described previously the detection and tissue distribution of free cell surface receptors for ecotropic R-MuLV envelope glycoprotein and the growth factor EGF in vivo [1]. More recently, we have reported the chromosomal map position of the ecotropic viral receptor and its conservation between subspecies of the genus Mus [2]. This work has shown, for the first time, the presence of multiple, independently segregating cell surface receptor genes specific for different classes of ecotropic type C viral envelope glycoprotein. In this report we extend these findings and identify chromosome 2 as coding for the receptor used by M813, an ecotropic MuLV from a feral Asian mouse. This new receptor is probably also used by oncogenic, recombinant (MCF class) MuLV of C3H origin.

Key words: cell surface receptors, type C viral glycoproteins, growth factors

Organization of cells into tissues and the induction of specific patterns of differentiation in these tissues presumably requires cell surface components that can mediate specific intercellular adhesion [3] as well as growth stimulation and morphogenesis [4]. Endogenous and transforming type C viruses code for or induce gene products that can act as ligands for both classes of cell surface receptors; those involved in cell adhesion and others participating in mitogenesis. Thus, the viral envelope glycoprotein gp70, which can be expressed independent of complete virus production [5], specifically binds to cell surface receptors [6] and may thereby affect the social behavior of cells. There are multiple receptors for the different classes of endogenous mouse type C viruses [2]. At least one of them shows a tissue-specific distribution in vivo [1, 7], and the linkage group to which its gene has been assigned is conserved between different species of the genus Mus [2]. Sorting out of cells into tissue-specific combinations from mixtures that were derived from different species also shows that organ specificity of cell adhesion is maintained across species barriers [8]. These findings are consistent with the hypothesis that endogenous viral envelope gene products or their cellular homologues (progenitors), together with the corresponding receptors, may mediate tissue-specific cell-cell adhesion [2].

Received May 29, 1980; accepted September 19, 1980.

Published 1980 by Alan R. Liss, Inc.

Mammalian sarcoma viruses, on the other hand, code for or induce polypeptides, which interact with cell surface receptors that are responsible for mitogenesis and, possibly, morphogenesis. A specific example of this is sarcoma growth factor (SGF) that is produced by MSV- or FeSV-transformed cells and acts through the EGF receptor on cells [9].

During chronic infection of mouse cells with endogenous mousetropic MuLV, variants emerge that have gained the ability to induce infected cells to growth in soft agar and/or to show an altered tissue tropism for oncogenesis [10, 11]. At least some of these variants have a substitution in their envelope gene [12] resulting in the recognition of a new cell surface receptor for infection, as is shown in this report. Thus, it appears that variants of endogenous mouse type C viruses that differ in their tissue tropism for infection and others that induce infected cells to autonomous growth can be used as tools for the identification of ligands, as well as receptors, that may be important in the establishment and maintenance of differentiated cells.

MATERIALS AND METHODS

Cells

All cell lines were maintained in Eagle's minimal essential medium (MEM) with 10% heat-inactivated fetal calf serum (FCS). The origin and characteristics of the cell lines used for virus growth and virus assays have been described [13]. The C3H/10T1/2 C18 line was developed from a C3H/Heston mouse embryo. This strain has a low "spontaneous" leukemia incidence. The 10T1/2 line and chemically transformed derivative lines have previously been shown to have endogenous type C viruses that can be activated [14].

Viruses

The ecotropic viruses M813 and Moloney MuLV (from M cervicolor popaeus) were maintained in NIH3T3 cells [15]. MCF class recombinant MuLV clones Z6 and Z9 were obtained from endogenous C3H MuLV as described previously [10, 12].

Preparation of Tissues for Binding Assays With gp70 and EGF

Cells from all major mouse tissues were obtained as previously described [1]. Cell or tissue fragments were suspended at 1×10^6 cells per ml, or approximately 0.2 mg protein per ml, and washed twice with 2 ml portions of Dulbecco's modification of Eagle's medium containing 1 mg/ml of bovine serum albumin and 50 mM N,N-bis-(2-hydroxyethyl-2-aminoethanesulfonic acid (BES), pH 6.8 (binding buffer). Binding assays were performed by incubating 10 ng of radiolabeled gp71 with cells in 1 ml of binding buffer for 2 h at 22°C. At the end of this incubation, the cells were centrifuged for 10 min at 500g, the pellets were washed 3 times with binding buffer and lysed, and the bound radioactivity was determined by liquid scintillation counting as previously described [1]. The same procedure was followed when the binding of iodinated mouse EGF was determined.

Radioimmunoassay for gp71

The 71,000 mol wt glycoprotein purified from R-MuLV was iodinated to high specific activity [1]. Competition radioimmunoassays were performed by incubating goat ant-Gross MuLV antiserum (Viral Oncology Program Resources, National Cancer

Institute, Bethesda, MD) and competing antigen at 37°C for 1 h in 0.2 ml reaction mixtures containing 10 mM Tris HCl, pH 7.8, 1 mM EDTA, 0.4% Triton X-100, 0.1 M NaCl. The ^{125}I-labeled antigen (12,000 cpm) was then added and the mixture incubated for an additional hour at 37°C and 18 h at 4°C. This was followed by the addition of 50 µl of rabbit anti-goat immunoglobulin G to each reaction mixture. Incubation proceeded for 1 h at 37°C and 3 h at 4°C, after which time precipitates were pelleted by centrifugation at 1,300 g for 15 min. The supernatants were aspirated, the pellets washed twice with the above described buffer, and radioactivity in the precipitate was measured in a gamma counter.

Construction of Somatic Cell Hybrids and Isozyme Analysis

Hybrids between mouse and Chinese hamster fibroblasts were obtained as described previously [2]. Three sets of hybrids were used in this study [1] between M cervicolor and Chinese hamster E36 cells [2] and between M musculus strain C57BL/6 G IX$^+$ and E36 cells [17]. For isozyme analysis soluble extracts were prepared as described previously [2]. The following isozymes were determined: dipeptidase 1 (DIP-1,3,4,11), phosphoglucomutase-2 (PGM-2,2.7.5.1), phosphoglucomutase-1 (PGM 1,2.7.5.1), 6-phosphogluconate dehydrogenase (6 PGD, 1,1.1.44), glucosephosphate isomerase (GPI,5,3.1.9), mannosephosphate isomerase (MPI, 5.3.1.8), nucleoside phosphorylase (NP, 2.4.2.1), tripeptidase-1 (TRIP-1, 3.4.11), dipeptidase-2 (DIP-2,3,4,11), adenine phosphoribosyl-transferase (APRT, 2.4.2.7), acid phospatase (ACP-1,3.1.3.2), adenylate kinase (AK-1, 2.7.4.3), dipeptidase-D (DIP-D, 3,4.11.9), and hypoxanthine phosphoribosyl transferase (HPRT, 2.4.2.8), Glyoxylase-1 (GLO-1,4.4.1.5), triosephosphate isomerase (TPI, 5.3.1.1.), galactokinase (GALK, 2.7.1.6), glutathione reductase (GR, 1.6.4.2), malic enzyme (MOD-1,1.1.1.40).

RESULTS

Tissue Distribution of Free Receptors for RLV gp70 and EGF

Table I summarizes the data obtained from the binding of ^{125}I-labeled RLV gp70 and EGF to cells freshly prepared from CBA mice at 4 weeks of age. The highest binding activity per mg protein was found in the bone marrow, followed by cells from the other major lymphoid tissues. Among the nonlymphoid tissues, brain showed the highest level of binding that cannot be explained by the presence of contaminating lymphocytes, since this organ is known to be void of lymphatic structures. Free receptors for ^{125}I-labeled EGF showed a more limited distribution. Liver was the only major organ with high levels of free receptors. Lung, kidney, and brain also bound radiolabeled EGF to a limited extent (between 0.3 and 0.5%) relative to liver.

Tissues that were negative for binding of either RLV gp70 or EGF were also low or negative in expression of endogenous viral gp70 or EGF [1]. Thus, the absence of binding activity was not due to the presence of blocked receptors.

Evidence for Multiple Receptors Specific for Different Classes of Endogenous MuLV Envelope Glycoproteins

Somatic hybrids between hamster and mouse cells were used to determine which mouse chromosomes were essential for infection of cells by different classes of MuLV. The Chinese hamster fibroblasts E36 are negative for the salvage pathway enzyme hypox-

TABLE I. Results of Binding of ^{125}I-Labeled RLV gp70 and EGF to Cells Prepared from CBA Mice at 4 Weeks*

Assay	Heart	Brain	Kidney	Lung	Liver	Bone marrow	Thymus	Spleen	Small intestine
cpm of ^{125}I-EGF bound/mg protein	40	138	175	225	47,700	0	0	0	0
cpm of ^{125}I-gp71 bound/mg protein	3,100	28,900	5,000	7,750	12,500	176,000	105,000	120,700	24,400
ng EGF per mg protein	<1	<1	4.6	<1	<1	<1	<1	<1	<1
ng gp71 per mg protein	14	14	16	44	4	376	79	64	72

*Binding reactions were carried out in 1 ml of binding buffer containing either 10 ng of [^{125}I] gp 71 (49,000 cpm) or 3 ng of [^{125}I] EGF (210,000 cpm). Each assay contained approximately 0.20 mg of the tissue examined. The cellular concentrations of gp71 and EGF were determined on aliquots of the samples using a radioimmunoassay as described in Methods. The data shown are for specific binding, which is obtained by subtracting the counts per minute bound to an identical reaction mixture containing 5 μg of unlabeled gp71 (180–420 cpm). All major tissues were freed of lymphatic structures such as lymph nodes or Peyers patches before preparation for binding assays.

anthin-guanin phosphoribosyltransferase (HGPRT-) and therefore do not grow in medium containing hypoxanthin, aminopterin, and thymidine (HAT). These cells were fused with mouse spleen and thymus cells and colonies of hybrid cells selected in HAT medium. The retention of mouse chromosomes was analyzed by electrophoresis for 19 isozyme markers specific for individual chromosomes. All hybrid cells had in common the retention of the mouse X chromosome (due to HAT drug selection) and loss of chromosome 11. The other mouse chromosomes were present in variable numbers and combinations.

Ecotropic MuLV From Mus musculus Uses a Chromosome 5 Coded Receptor

Various sets of hybrids between hamster and Mus cervicolor or hamster and Mus musculus cells were thus analyzed for mouse isozymes and tested for their ability to replicate Moloney MuLV and to bind RLV-gp70. The results from such an experiment with hamster × Mus cervicolor hybrids is shown in Table II, which shows synteny between Moloney replication, RLV-gp70 binding, and retention of PGM-1. PGM-1 is a marker enzyme for chromosome 5 of the mouse [18]. Marker enzymes for other chromosomes did not show synteny [2].

When hybrid cells between hamster and Mus musculus cells were tested in a similar fashion, Moloney MuLV replication and RLV gp70 binding were again syntenic with retention of chromosome 5 and asyntenic with all other isozymes tested [2].

Ecotropic MuLV From Mus cervicolor Uses a Chromosome 2 Coded Receptor

The presence of multiple receptors for different classes of ecotropic as well as for xenotropic MuLV has been reported previously by our laboratory [2]. Here we extend these findings and show synteny between M813 replication, an ecotropic MuLV from Mus cervicolor, and adenylase kinase AK-1, a marker enzyme for chromosome 2 of Mus

TABLE II. Moloney MuLV Replication, gp70 Binding, and Chromosome 5 Marker in M cervicolor × Hamster Hybrid Cells

Hybrid clone	Replication[a]	gp70 Binding[b]	PGM-1
C36 1	−	1.89	−
2	−	1.53	−
3	−	0.95	−
4	−	1.30	−
5	+++	17.4	+
6	+++	32.0	+
7	NT	1.02	−
8	+++	24.1	+
9	+++	16.2	+
11	+++	25.3	+
13	−	0.26	−
14	−	1.46	−
15	+++	47.4	+
16	++	54.4	+
19	+++	31.9	+
20	−	2.54	−
21	+++	49.3	+

[a]NT, not tested
[b]Femtomoles bound per 10^6 cells. E36 and NIH cells bound 2.20 and 34.5 fmoles respectively.

musculus [19]. These data are summarized in Table III. RLV-gp70 binds to mouse × hamster hybrid (or mouse) cells that replicate M813 provided they retain mouse chromosome 5. Hybrids that have lost chromosome 5 still replicate M813. However, all the clones that replicate M813 retain chromosome 2. Using additional hybrid cells resulting from an independent fusion, we observed concordance between M813 replication and mouse adenylate kinase-1 (AK-1) in 59 of 62 hybrids.

Oncogenic Recombinant Viruses (MCF-class) Derived From C3H MuLV May Also Use the Chromosome 2 Coded Receptor

M813 MuLV induces lymphomas in inbred NFS/N mice after adaptation to high titered growth in culture [Rapp and Callahan, unpublished data]. Therefore, we decided to test whether oncogenic recombinant MCF-class MuLV derived from endogenous C3HMuLV as described previously [10–12] might also use a new cell surface receptor, different from that used by its ecotropic parent MuLV and possibly identical to the one used by M813. Table IV shows that the two MCF class MuLV from C3H, Z6, and Z9 do indeed use a new cell surface receptor as evidenced by the fact that RLV gp70 still binds to productively infected cells. M813 infection also does not intefere with this binding, whereas this is completely abolished upon infection with Rauscher MuLV. We therefore used Z6 and Z9 MuLV for infection of hybrid cells that had lost either chromosome 2 or chromosome 5 (Table V). Both MCF-class MuLV from C3H replicate in chromosome 2 positive, chromosome 5 negative hybrid cells but they do not infect a hybrid clone that is positive for chromosome 5 and negative for chromosome 2. M813 shows a pattern of replication that is identical to that observed with Z6 and Z9.

DISCUSSION

Tissue Distribution of Receptors for RLV gp70 and Epidermal Growth Factor, EGF

We have described here that free receptors for gp70 prepared from R-MuLV are present on cells from all major lymphoid organs of young CBA mice. Certain nonlymphoid tissues, most notably brain, also have large numbers of free gp70 cell surface receptors. The absence of detectable receptors does not result from the expression of excess endogenously produced gp71.

Cell surface receptors for EGF appear to be highly restricted in vivo. The only tissue from CBA mice with high binding activity was liver. Lung, kidney, and brain showed approximately 0.5% activity/mg protein relative to liver, and most other tissues did not bind EGF at all. Again, lack of binding was not due to expression of endogenous EGF.

It might be argued that our method for the preparation of cells for the in vitro binding experiments destroyed active receptors in some of the tissues. We attempted to control for this possibility by mixing experiments between cells from receptor-positive and receptor-negative organs. None of these experiments provided any evidence for the presence in non-binder tissues of trans-acting components with the ability to block specific binding of either gp70 or EGF (data not shown).

The high binding activity of liver cells for EGF might indicate that this tissue is the main site of action in adult mice. Our experiments did not screen all mouse tissues for binding, however, and thus it is still possible that cell types present, for example, in skin, might also have high concentrations of free receptors. The use of primary cell

TABLE III. Rauscher gp70 Binding to Cells Previously Infected With M813 Ecotropic MuLV

Mouse × hamster hybrid clone	gp70[a] binding	Chromosome 5	M813 replication	Chromosome 2
FV 1	21.0	+	+	+
NF 1	14.7	+	+	+
FV 5	13.6	+	+	+
NF 4	14.4	+	+	+
NF 5	1.1	−	+	+
NF 7	1.0	−	+	+
NF 10	0.90	−	+	+
NF 12	1.1	−	+	+
FV 11	1.3	−	−	−
FV 13	0.70	−	−	−
Control cells				
Mouse (NIH)	31.4			
Cat (FEC)	1.6			
Hamster (E36)	1.3			

[a]fmoles gp70 bound per 10^6 cells.

TABLE IV. Rauscher gp70 Binding

Culture cells	gp 70 binding
Uninfected	
NIH (mouse)	+
E36 (hamster)	−
NIH infected with MuLV	
RLV	−
Z6	+
Z9	+
M813	+

TABLE V. Replication of MuLV on Hybrid Cells Retaining Mouse Chromosomes 2 or 5

Hybrid cell	M813/NIH[a]	M813/M813[b]	M813/A10924[c]	Z6	Z9	Moloney
Retains 2 but not 5	+	+	+	+	+	−
Retains 5 but not 2	−	−	−	−	−	+

[a]M813 virus after long-term passage on NIH cells.
[b]Early freeze-down of primary culture from which M813 was originally isolated.
[c]M813 virus isolated from co-cultivation of tumor produced by injecting an NIH mouse with M813/NIH virus.

culture systems [20] for cell types that were not readily accessible to in vitro testing by the methods employed in this study should permit us to test this possibility. Our experiments also did not definitively establish whether binding of EGF to liver cells was mediated by a cell surface receptor such as has been purified from established cell lines [21] or by the presence of an EGF transport protein that might be produced in this tissue. Further characterization of the ligand—receptor complex from liver will be needed to settle this question.

Genes Controlling Receptors for Different Classes of Mouse Type C Viruses

Although the physiological significance of growth factor receptors is readily apparent, this is not so for receptors which bind type C viral glycoproteins. It may be worthwhile at this point to digress for a moment and consider some basic features of endogenous type C viruses that have a bearing on the interpretations of our findings reported here. The available data on the distribution of type C viruses in the genomes of vertebrate species allow two alternative views of their origin. Either they were deposited there as a consequence of horizontal spread of infectious viruses, or they evolved from within the vertebrate genome, only occasionally giving rise to highly infectious forms that could spread between members of a species as well as possibly infect different species. In the former case the presence in the hosts genome of multiple divergent copies would be a consequence of the eventual decay of the original infecting viral genome. If they evolved from within the genome to highly infectious forms, their progenitors might have been either sets of genes with a genetic structure analogous to that of type C viruses, such that only small changes, if any, had to occur before they could be expressed as an infectious virus. Or the precursors of type C viruses were genes that are normally not physically linked and jointly expressed. The formation of an infectious viral genome would then be a rare evolutionary artifact resulting from the assembly of genes that otherwise are used individually as building blocks for normal cells. A mechanism by which this might occur has been suggested by Temin [22].

In the first case, endogenous type C viral functions, as a rule, would not be expected to participate in normal cellular processes, except in rare instances where the host has learned to make use of them for its own ends. In the second case, component parts of the virus would have cellular homologues. But even if one were to take the view that all endogenous type C viruses are stranded, exogenous viruses, those of their gene products which had to interact specifically with cellular structures for the virus to be a successful parasite presumably had to mimic a normal host function. Thus, type C viral gene products may be used to probe into vital cell processes.

What are some of the interactions between virus and cell that would lend themselves to the identification of important cellular functions? Little is known in this regard, but 4 (poorly defined) targets seem apparent. One is the cell surface receptor, to which the viral envelope glycoprotein has to bind before entry into the cell is gained. The envelope gene also encodes a polypeptide, p15E, which appears to act as a cellular receptor for Clg, a component of complement [23]. A third receptor or receptor-binding site appears to be located within the viral gag (internal structural protein) gene [24]. In this case interaction is with an intracellular target that can restrict the course of virus infection. The host gene(s) controlling this latter function, Fv-1, has been mapped. No evidence of a possible normal function of this gene has so far been obtained. A fourth class of cellular target molecules is affected by the products of transforming genes carried by oncogenic type C viruses. In this case it is immediately obvious that the cellular target(s) is vital for normal function, since expression of a transforming gene in the appropriate cell drastically alters its phenotype.

In this report we have described in some detail host genes controlling cell surface receptors for type C viruses. We, and others, have shown previously that a prototype ecotropic MuLV from inbred mice, RLV, uses a chromosome 5 coded receptor for infection of cells [2, 25–27].

The linkage between a gene controlling the RLV gp70 receptors and PGM-1 was retained in hybrids between Chinese hamster × Mus cervicolor cells. Thus, this receptor appears conserved between different species of the genus Mus. Here we show that replication of another ecotropic MuLV, M813, which was isolated from Mus cervicolor popaeus spleen, is associated with retention of mouse adenylate kinase-1 (AK-1). This enzyme is a marker for chromosome 2 [19]. The data specifically exclude any association between M813 replication and chromosome 5. The level at which M813 virus replication is controlled by a gene on chromosome 2 is probably the M813-specific cell surface receptor. This appears most likely, since absence of chromosome 2 does not affect replication of ecotropic MuLVs other than M813.

The specific synteny between chromosome 2 and the M813 virus receptor excludes a relationship between receptor and the major histocompatability complex. (Two hybrid clones that are negative for chromosome 17 but retain chromosome 2 are permissive for M813 replication). However, chromosome 2 does contain a minor histocompatability locus, as is also true of chromosome 5.

We have examined the question of whether use of a certain cell surface receptor for infection would correlate with the pathogenicity in vivo of a given MuLV. M813 virus does induce lymphomas upon inoculation into newborn NFS/N mice (data not shown); so does RLV and Mol MuLV, which infect cells via a chromosome 5 coded receptor. However, the latter viruses invariably generate envelope gene recombinants prior to or during induction of disease, and we therefore cannot conclude that oncogenicity is independent of receptor specificity. To test this possibility further, we determined the receptor used for infection by two such recombinant MCF class MuLV, which we had previously isolated and characterized [10, 12]. As we have shown in this report, both of these oncogenic viruses do not use the chromosome 5 coded receptor; rather, they replicate in hybrid cells that have lost this chromosome as long as they retain chromosome 2. In the course of disease development induced by these latter viruses, no variants that reverted to the use of the chromosome 5 coded receptor emerged [U. R. Rapp, unpublished data]. Thus, we can conclude that interaction with the chromosome 5 coded receptor is not necessary for malignant transformation by MuLV. Interaction between viral envelope glycoprotein and a chromosome 2 coded cell surface receptor, however, may be a critical factor in the transformation of sensitive target cells by MuLV, either because it mediates infection of such cells without being involved in the subsequent transformation process, or because binding of viral gp70 to this receptor directly effects transformation in chronic producer cells. In vitro infection of appropriate target cells with M813 or the MCF class recombinant viruses Z6 and Z9 versus treatment of such cells with viral envelope glycoprotein may allow us to distinguish between these possibilities.

We have hypothesized in a previous report [2] that binding of viral envelope glycoprotein in one cell to the corresponding receptor on the surface of the other cell could provide a basis for modulating cellular recognition and organization in normal and pathological processes [2] The present results extend this suggestion concerning the role of endogenous viral gp70 and its corresponding receptor in at least two ways. By defining a new gene controlling a new ecotropic receptor, it supports the concept of receptor diversity in mouse cells. The hypothesis that certain interactions between cells are mediated by viral related genes would seem to require multiple receptors, each with its own specificity. In addition, it suggests that genetic recombination, which is a mechanism of generating diversity among type C viruses, may lead to altered receptor specificity. Similar processes may generate diverse viral glycoproteins, which may accumulate in the cell membrane and thus generate new intercellular interaction mediated by specific receptors.

We have shown that mouse cells have the genetic capacity to express distinct receptors for closely related ecotropic type C viruses. Recently, it has been shown that differential expression of specific receptors for similar type C viruses can be related to leukemic transformation in certain mouse cells [28]. Preferential binding of leukemogenic viruses was demonstrated using the fluorescence-activated cell sorter [28]. Our results showing separate genetic control of receptors for two ecotropic type C viruses provide a genetic basis for such processes.

REFERENCES

1. De Larco JE, Rapp UR, Todaro GJ: Int J Cancer 21:356–360, 1978.
2. Marshall TM, Rapp UR: J Virol 29:501–506, 1979.
3. Holtfreter T: Arch Exp Zellforsch 23:169–209, 1939.
4. Tiedemann H: Naturwissenschaften 46:17–26, 1959.
5. Lerner RA, Lentis BW, Del Villano BC, McConahey PJ, Dixon FJ: J Exp Med 143:151–165, 1976.
6. De Larco JE, Todaro GJ: Cell 8:365–371, 1976.
7. Fowler AK, Twardzik DR, Reed CD, Winslow OS, Hellman H: J Virol 24:729–73, 1977.
8. Moscona AA: Int Rev Exp Pathol 1:371–529, 1962.
9. Todaro GJ, De Larco JE, Cohen S: Nature 264:26–31, 1976.
10. Rapp UR, Todaro GJ: Proc Natl Acad Sci USA 75:2468, 1978.
11. Rapp UR, Todaro GJ: Proc Natl Acad Sci USA 77:1–5, 1980.
12. Devare SG, Rapp UR, Todaro GJ, Stephenson JR: Virology 93:582–588, 1979.
13. Rapp UR, Nowinski RC: J Virol 10:411–417. 1976.
14. Rapp UR, Nowinski RC, Reznikoff CA, Heidelberger C: Virology 65:392–409, 1975.
15. Jainchill TL, Aaronson SA, Todaro GJ: J Virol 4:549, 1969.
16. De Larco JE, Todaro GJ: Cell 8:365–371, 1976.
17. Stockert E, Boyse EA, Obata Y, Ikeda H, Sarkar H, Hoffman H: J Exp Med 142:512–517, 1975.
18. Hutton JJ, Roderick TH: Biochim Genet 4:339–350, 1970.
19. Francke U, Lalley PA, Moss W, Ivy J, Minna JD: Cytogenet Genet 19:57–84, 1977.
20. Rheinwald JG, Green H: Cell 6:331–343, 1975.
21. Wrann MM, Fox CF: JBC 254:8083–8086, 1979. See also Fox F this volume, and Cohen S, this volume.
22. Temin HM: J Nat Cancer Inst 46:III–VII, 1971.
23. Bartholomew RM, Esser AF, Meuller-Eberhardt HJ: J Exp Med 147:844–853, 1978.
24. Schindler T, Hynes R, Hopkins N: J Virol 23:700, 1977.
25. Ruddle NH, Conta BS, Linwand L, Kozak C, Ruddle F, Besmer P, Baltimore D: J Exp Med 148: 451–465, 1978.
26. Oie HK, Gazdar AK, Lalley PA, Russel EK, Minna JD, De Larco JE, Todaro GJ, Francke U: Nature 274:60–62, 1978.
27. Hilkens JA, Colombatti M, Strand M, Hilgers J: Cold Spring Harbor Meeting on RNA Tumor Viruses, 41, 1978.
28. McGrath MS, Decleve A, Lieberman M, Kaplan HS, Weissman IL: J Virol 28:819–827, 1978.

The Role of Colony-Stimulating Factor in Granulopoiesis

Richard K. Shadduck, Giuseppe Pigoli, Abdul Waheed, and Florence Boegel

Department of Medicine, Montefiore Hospital, University of Pittsburgh School of Medicine, Pittsburgh, Pennsylvania 15213

The proliferation and maturation of granulocytic-monocytic stem cells appears to be controlled by a series of closely related glycoproteins termed "colony-stimulating factors" (CSFs). Recently, we devised a 6-step scheme for the purification of murine fibroblast (L-cell)-derived CSF. Ten liter pools of conditioned media were concentrated by ultrafiltration, precipitated by ethanol, and separated on DEAE cellulose, Con-A Sepharose, and Sephadex G150. The CSF was separated from trace contaminants, including endotoxin, by density gradient centrifugation. The purified material was radioiodinated and used to define the serum half-life and in vivo distribution. Following IV injection there was a biphasic serum clearance with a $t_{1/2}$ of 24–40 min and 2–2½ hours in the first and second phases. Approximately 25% of the tracer was excreted in the urine at 6 h; however, urinary radioactivity was due to low molecular weight peptides. Simultaneous studies by radioimmunoassay showed a similar rapid serum clearance of unlabeled CSF but virtually no urinary CSF activity. Thus, assays for urinary CSF may not provide useful measures of in vivo CSF activity. Further in vitro studies have defined the interaction of CSF with responsive cells in the marrow. Varying doses of CSF were incubated with 10^7 marrow cells for intervals of 24–48 h. The major increment in cell-associated radioactivity occurred between 6 and 16 h. The reaction was saturable with 1–2 ng/ml CSF. Binding was prevented by cold CSF, but not by other proteins. Irradiation yielded only a minimal reduction in CSF binding. The interaction of CSF with marrow cells appeared to require new protein synthesis, as binding was completely inhibited by cycloheximide and puromycin. Irradiated mice injected with antibodies to CSF showed an inhibition of granulopoiesis by marrow cells in peritoneal diffusion chambers; however, granulopoiesis in the intact bone marrow was unaffected. Granulpoiesis in long-term marrow cultures was also unaffected by anti-CSF. These different responses may be due to accelerated clearance of injected CSF in nonirradiated mice or to extensive stromal interactions that modulate and perhaps control granulocytic differentiation in the intact bone marrow microenvironment.

Key words: granulopoiesis, colony stimulating factor, diffusion chamber granulopoiesis, radioimmunoassay for colony stimulating factor, long-term marrow cultures, purification of colony stimulating factor, binding of colony stimulating factor

Received May 16, 1980; accepted August 18, 1980.

0091-7419/80/1404-0423$05.00 © 1980 Alan R. Liss, Inc.

During the past decade there has been a progressive accumulation of data suggesting that granulopoiesis is controlled by a humoral mechanism. In the preceding 10 years granulocyte differentiation was extensively characterized with definition of pool sizes, proliferative capacity, and transit times [1, 2]. The advent of techniques for the clonal growth of granulocytes and macrophages in vitro provided a further dimension for exploration of potential regulatory factors [3, 4].

Incubation of bone marrow cells in semi-solid medium in the presence of a stimulatory material termed colony-stimulating factor (CSF) leads to the development of discrete colonies of granulocytes and macrophages. Studies with the cell cycle active agent hydroxyurea showed that the colony-forming cells had a high rate of replication [5], which served to distinguish these progenitor cells from the spleen colony-forming units or pluripotential stem cells; the latter generally manifested a slow rate of cellular turnover. Based on the high rate of DNA synthesis and the apparent restriction to a single common (granulocyte and macrophage) line of differentiation, most investigators believe that the agar gel assay measures granulocyte, and perhaps monocyte, committed stem cells. Since colony formation results from the growth of individual granulocytic stem cells in response to CSF in vitro, it seemed reasonable to postulate that similar events might be occurring in vivo.

A series of correlative studies have shown an inverse relationship between the number of circulating neutrophils and serum levels of CSF [6–9]. Sera or urine from animals rendered neutropenic by X-irradiation, administration of an antineutrophil antibody, or injection of cytotoxic drugs show substantial increments in CSF activity. Moreover, serum samples from animals with cyclic neutropenia [10, 11] or from patients [12] with similar oscillations in their neutrophil counts also show periodic fluctuations in CSF activity. Higher levels of CSF are found during the neutropenic phase, with depression of CSF in conjunction with peak neutrophil values.

The use of a further experimental model — namely, the growth of granulocytic cells in peritoneal diffusion chambers [13] — has also provided data suggestive of a CSF effect in vivo. Cell growth is modest in control animals, whereas pretreatment with radiation [14] or cyclophosphamide [15] markedly enhances cellular proliferation. Since these treatments increase serum CSF activity, it is tempting to speculate that the accelerated granulopoiesis is due to this factor. Indeed, treatment of animals with an antibody to CSF markedly restricts the degree of diffusion chamber granulopoiesis [16].

Further studies with bacterial vaccines or lipopolysaccharides are also indicative of a CSF effect in vivo. For instance, a single injection of pertussis vaccine leads to a 48–72 h increment in circulating CSF, which in turn is followed by a heightened wave of granulocytic differentiation [17]. Various endotoxin preparations also stimulate an acute increase in CSF activity [18]. Following repetitive injection there is rapid tolerance to the CSF-inducing effect [19]. Nonetheless, animals treated in this fashion develop striking granulocytic hyperplasia of the bone marrow [19].

The rapid turnover rate and the marked lability of the granulocyte system have thus far impeded efforts to document the effect of CSF in vivo. Endotoxins and various foreign proteins, in addition to increasing CSF activity, cause an immediate release of mature neutrophils from the marrow [2]. This effect that is mediated by a neutrophil-releasing factor [20], might also be expected to stimulate cell production by reduction in a possible cell-contact negative feedback mechanism. Investigators have described numerous inhibitors of in vitro colony formation; however, it remains to be shown whether these materials are active in vivo. High molecular weight inhibitors are readily demonstrated in the serum [21], yet the levels of these materials do not change in neutropenic or neutrophilic states [22]. Low molecular weight substances have been isolated from mature neutrophils [23]; however, such inhibitors are also derived from lymphocytes and various somatic tissues, which suggests a lack of specificity [24]. Recent data do indicate that neutrophil derived lactoferrin may serve a regulatory function by limiting the response of granulocytic progenitor cells both in vitro and in vivo [25].

Although the injection of crude sources of CSF produces a modest neutrophilia in experimental animals, the aforementioned concerns about endotoxin and foreign proteins clearly indicate the need for pure material. Recently, 3 groups have succeeded in purifying several sources of murine CSF [26–28]. We have used purified L-cell CSF for production of antibodies [29], for development of a radioimmunoassay [30], and for in vivo studies of plasma clearance and tissue distribution [31]. Both purified CSF and anti-CSF have been used for studies of granulopoietic control mechanisms using both diffusion chamber [16] and long-term marrow culture techniques [32].

METHODS AND RESULTS

Purification of CSF

CSF was produced by growth of L-cells in serum-free CMRL 1066 medium [28]. Each 10 l pool of conditioned medium was concentrated 100–250-fold by ultrafiltration and subjected to precipitation in 40% and 50% cold ethanol. The supernatant, which contained all of the active CSF, was applied to DEAE cellulose and eluted with a linear sodium chloride gradient (Fig. 1A). The active fractions were concentrated and applied to concanavalin-A Sepharose, wherein approximately 1/3 of the CSF was nonadherent; the remainder was bound and specifically eluted with alpha-methyl glucoside (Fig. 1B). Each active peak was separately concentrated and applied to coupled columns of Sephadex G-150. The CSF eluted in a volume 1.2 to 1.3 times the void volume of the column. With a series of protein markers, each peak eluted with an apparent but anomalous molecular weight of approximately 190,000 (Fig. 2). In the final purification step, CSF was separated from trace contaminants by sucrose density gradient centrifugation (Fig. 3). This proved successful as the high molecular weight contaminants, which eluted with CSF on Sephadex gels, were effectively sedimented to the bottom of the gradient, whereas CSF was retained in the upper portion. After ultracentrifugation, the resultant materials from both the Con-A adherent and nonadherent fractions appeared homogeneous after electrophoresis in varying concentrations of acrylamide, in SDS acrylamide, and by Ouchterlony gel diffusion [28]. The final materials were purified greater than 1,000-fold and showed a specific activity of $2-5 \times 10^7$ units/mg of protein. Maximum colony formation was detected using approximately 3 ng/ml in agar gel cultures.

Based on the migration in SDS gels with marker proteins (Fig. 4, top panel) and calculations using the sedimentation coefficient (3.0S) and Stoke's radius (50 Å), both CSF fractions had a molecular weight of 60,000–70,000. Both peaks contain substantial quantities of carbohydrates as judged by staining with PAS, by inactivation with periodate, and by altered electrophoretic behavior after incubation with neuraminidase. Reduction with 2-mercaptoethanol in the presence of SDS yielded two subunits of approximately 35,000 daltons (Fig. 4, bottom panel), however, these products were devoid of biologic activity.

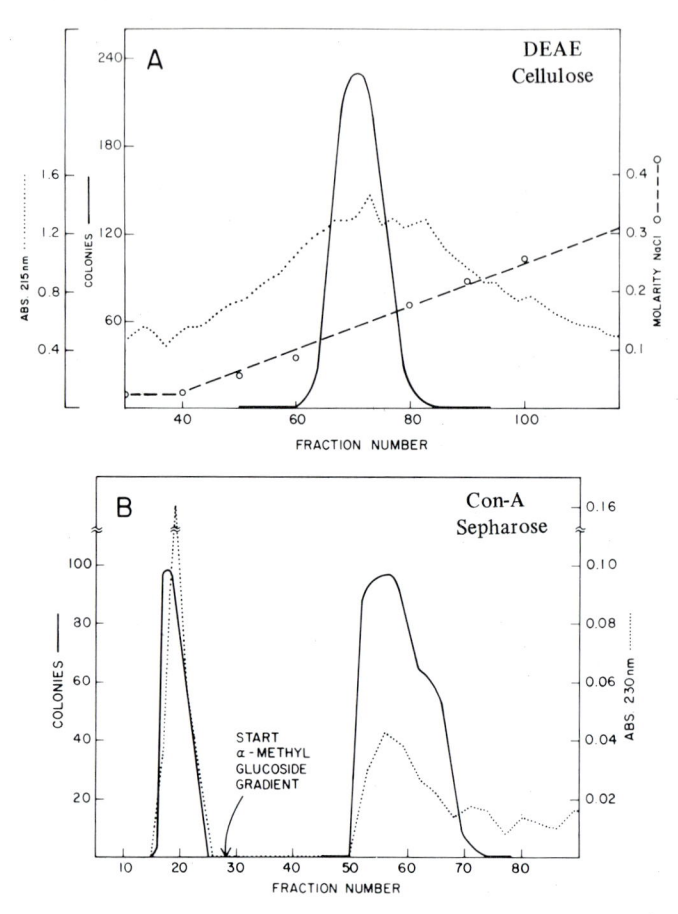

Fig. 1. A. Chromatographic separation of L-cell CSF with DEAE-cellulose. Biologic activity was detected in the 0.10–0.18 M NaCl fractions. B. Affinity chromatography of concentrated CSF from the DEAE-cellulose fraction on Con-A-Sepharose. Two peaks of activity were detected; nonadherent CSF (peak 1) and adherent (peak 2). Reproduced from Waheed and Shadduck: Purification and properties of L-cell derived colony-stimulating factor. J Lab Clin Med 94:180–194, 1979, with permission.

Clearance and Distribution Studies

Purified CSF has been radioiodinated using both the lactoperoxidase [30] and modified chloramine-T techniques [31]. With the inclusion of 0.06 M DMSO it has been possible to preserve 75–100% of the biologic activity with stability of the tracer for a minimum of 3 weeks after iodination. These various tracers have been used to define the clearance of CSF in vivo. After IV injection there was a rapid initial plasma clearance with a $t_{1/2}$ of 24–40 min through the first hour (Fig. 5). Thereafter, the disappearance rate was essentially linear, with a $t_{1/2}$ of 2–2½ hours. Plasma levels of radioactivity were lower after IP administration, with peak values at 1 h and linear clearance thereafter, with a half-life of 3½ hours.

By use of a radioimmunoassay [30] it was also possible to monitor the serum clearance of unlabeled CSF. In these studies CSF disappeared from the plasma in a near linear fashion, with a $t_{1/2}$ of 1½ hours. The differences in behavior between radiolabeled

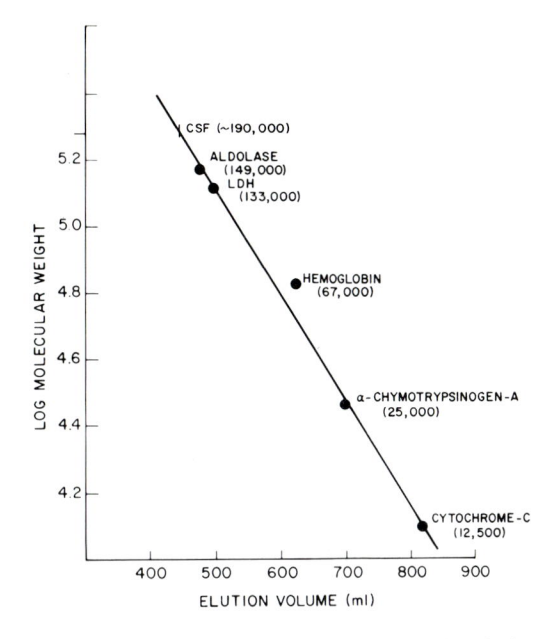

Fig. 2. The apparent molecular weight of CSF on Sephadex G150. The elution of CSF was compared to a series of marker proteins and plotted as a log of the molecular weight vs elution volume.

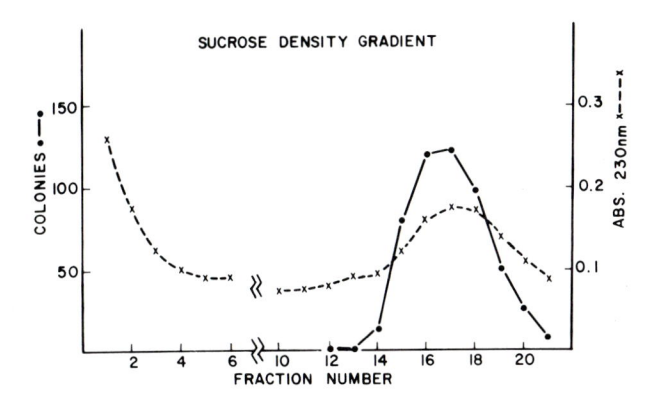

Fig. 3. Final separation of a peak 2 CSF fraction by sucrose density gradient centrifugation. High molecular weight contaminants were removed in fractions 1–4, which were obtained from the bottom of the centrifuge tube. CSF was detected in fractions 14–20, which represented the upper portion of the gradient. Reproduced from Waheed and Shadduck: Purification and properties of L-cell derived colony-stimulating factor. J Lab Clin Med 94:180–194, 1979, with permission.

and unlabeled material may be more apparent than real, as tracer studies only utilized 10–20 units of CSF whereas 125,000 units of cold material were employed.

Although in these experiments 15–30% of the radioactivity was excreted in the urine over 6 h, only 0.3–0.6% of intact CSF could be detected by immunoassay (Table I). Separation of the urinary radioactivity on Sephadex gel showed that virtually all of the tracer was degraded into low molecular weight peptides.

CCDD:A:269

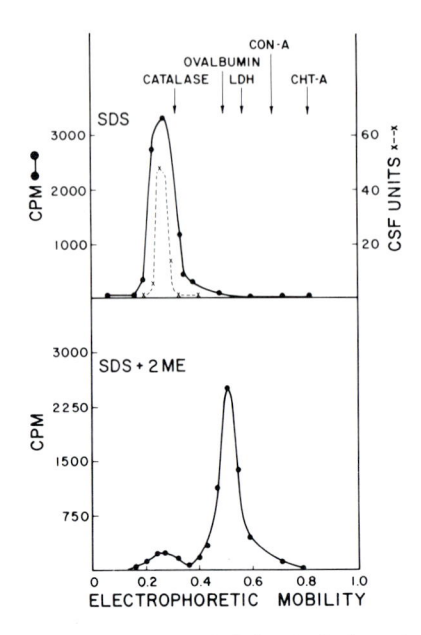

Fig. 4. A. Electrophoretic mobility of step VI, peak 2 CSF in SDS-acrylamide. B. Mobility of step VI, peak 2 CSF after reduction with 2-mercaptoethanol. Abbreviations refer to lactate dehydrogenase (LDH), concanavalin A (Con-A), and alpha chymotrypsinogen A (CHT-A). Reproduced from Waheed and Shadduck: Purification and properties of L-cell derived colony-stimulating factor. J Lab Clin Med 94:180–194, 1979, with permission.

TABLE I. Serum and Urinary CSF Values 6 Hours After Injection*

Injection	Measurement	Serum	Urine
^{125}I-CSF + 31,000 U cold CSF	Radioactivity	8.6% ± 0.4%	28.4% ± 4.2%
^{125}I-CSF + 125,000 U cold CSF	Radioactivity	6.4% ± 0.3%	14.3% ± 2.4%
31,000 units cold CSF	RIA	4.6% ± 0.5%	0.6% ± 0.05%
125,000 units cold CSF	RIA	2.9% ± 0.2%	0.3% ± 0.03%

*Groups of 5 mice each were injected with unlabeled cold CSF alone or mixed with ^{125}I-labeled CSF. Six-hour serum and urinary CSF was measured by either radioactive counting or radioimmunoassay (RIA). Values are means ± 1SE. Reproduced from Shadduck, Waheed, Porcellini, Rizzoli and Pigoli: Physiologic distribution of colony-stimulating factor in vivo. Blood 54:894–905, 1979, with permission.

In these studies the great majority of CSF was rapidly removed from the circulation, and within 5 min approximately 40% of the injected dose was detected in the liver (Fig. 6). Lesser quantities were found in the kidneys, lungs, and spleen, with clearance patterns that approximated those of the plasma. It is intriguing to note that only extremely small quantities of the isotope accumulated in the marrow. This could be due to the relatively low quantity of tracer injected, to a delay in cell binding, or perhaps to sequestration of the responsive cells in ecologic niches created by marrow stromal cells.

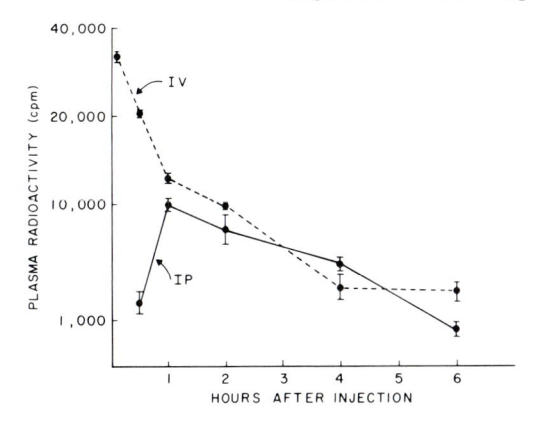

Fig. 5. Plasma clearance of radiolabeled CSF after IV or IP injection of the tracer. All mice received 100,000 cpm of tracer and were bled at intervals of 5 min to 6 h. Values represent total plasma radioactivity (means ± 1 SE). Reproduced from Shadduck, Waheed, Porcellini, Rizzoli and Pigoli: Physiologic distribution of colony-stimulating factor in vivo. Blood 54:894–905, 1979, with permission.

Binding of CSF to Bone Marrow Cells

In further in vitro experiments radioiodinated CSF was directly incubated with murine bone marrow cells for periods up to 48 h [31]. Control cultures contained an 80-fold excess of cold CSF to provide a measure of nonspecific binding. With 10^7 marrow cells, only minimal cell-associated radioactivity was detected over 1–3 h (Table II). The major increment in cell binding occurred between 6 and 16 h, with a near plateau thereafter. Saturation was achieved with concentrations of approximately 1 ng/ml, wherein 8% of the tracer was bound.

Cellular binding appeared specific as competition studies showed a prevention of binding with excess cold CSF but no inhibition with human serum, bovine albumin, ovalbumin, chorionic gonadotropin, porcine insulin, thyroxine, or dexamethasone. Furthermore, ^{125}I CSF did not bind to lymph node, thymic, hepatic, or renal cells, and showed markedly reduced binding with splenic cells, which are known to contain only small numbers of granulocyte-macrophage colony-forming cells. L-cell CSF is specific to murine progenitor cells; incubation of crude or purified material with human marrow does not generate colony formation. As a further measure of specificity ^{125}I CSF was incubated with 2 samples of normal human marrow. With 4×10^{-11} M CSF, no binding was observed at 24 h with the human samples, whereas approximately 9% of the tracer was bound to murine cells.

Although colony-forming cells show a high rate of DNA synthesis [5], only a modest decrease in cellular binding of radiolabeled CSF was seen with X-irradiation. As shown in Figure 7, binding was not inhibited by treatment of the cultures with 500 rads; a 24% reduction was observed with 1,000 rads. Similar findings were noted with cycle, active, phase-specific agents such as cytosine arabinoside or hydroxyurea. Preliminary autoradiographic studies done in collaboration with Dr. Lewis Schiffer of Allegheny General Hospital, Pittsburgh, PA, are in accord with these findings. As shown in Figure 8, large mononuclear or blast cells were labeled after 1–6 h exposure to the tracer. With increasing incubation time, virtually all cells of the granulocytic series, including mature neutrophils, were labeled.

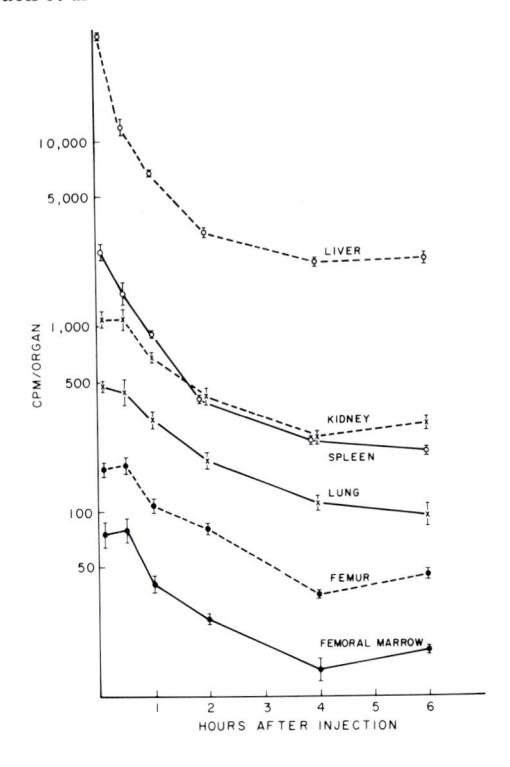

Fig. 6. Patterns of radioactivity in the tissues after IV injection of radiolabeled CSF. Values represent total tissue radioactivity and are means ± 1 SE for 5 animals/point. Reproduced from Shadduck, Waheed, Porcellini, Rizzoli and Pigoli: Physiologic distribution of colony-stimulating factor in vivo. Blood 54:894–895, 1979, with permission.

TABLE II. Bone Marrow Cell Uptake of ^{125}I-CSF*

	cpm/10^7 Marrow cells		
Time (h)	^{125}I-CSF	^{125}I-CSF plus cold CSF	Specific binding
1	1,882	642	1,240
3	2,532	728	1,804
6	5,663	1,009	4,654
10	13,934	1,619	12,315
16	31,501	2,840	28,661
24	40,003	7,550	32,453
48	46,480	10,510	35,970

*Marrow cells (10^7) were incubated for 1–48 h in 1 ml of supplemented McCoy's medium at 37°C, with 400,000 cpm of CSF. Control tubes for nonspecific binding contained an 80-fold excess of unlabeled CSF. Reproduced from Shadduck, Waheed, Porcellini, Rizzoli and Pigoli: Physiologic distribution of colony-stimulating factor in vivo. Blood 54:894–905, 1979, with permission.

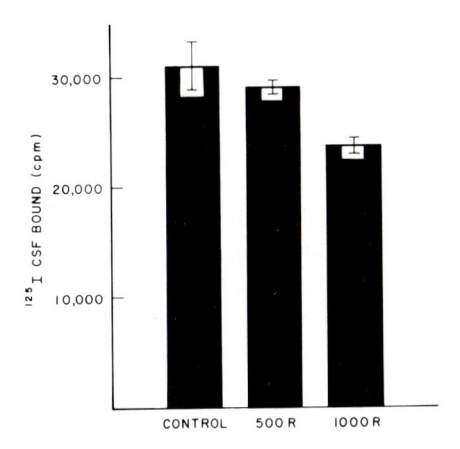

Fig. 7. The effect of irradiation on binding of radiolabeled CSF to murine bone marrow cells. Each culture contained 10^7 marrow cells and 400,000 cpm of tracer. Cultures were irradiated immediately after preparation; values represent specific cell-associated radioactivity following 24 h incubation.

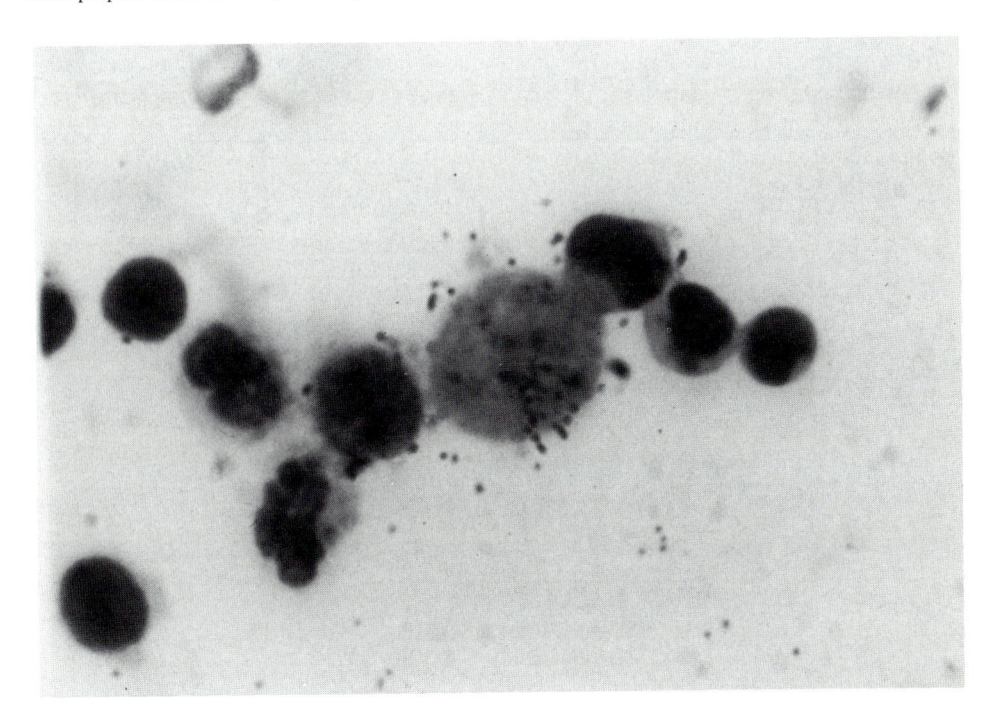

Fig. 8. Autoradiograph obtained after 6 h exposure of the radioiodinated CSF to murine bone marrow cells. A large mononuclear or blast cell is labeled, whereas lymphocytes and erythroid cells show no cell-associated radioactivity.

The delay in cellular binding of CSF appears to require synthesis of new protein as both cycloheximide (1 µg/ml) and puromycin (10 µg/ml) inhibited cellular uptake of radioactivity.

In Vivo Studies With Anti-CSF

Although foreign proteins may stimulate granulopoiesis by accelerating release of neutrophils from the marrow or by other nonspecific mechanisms, such materials are not known to depress granulocyte production. It was therefore reasonable to postulate that neutralization of all circulating CSF by antiserum should diminish or perhaps abolish cellular production. Initially, it was necessary to document the ability of rabbit antiserum directed against L-cell CSF to neutralize completely serum colony-stimulating activity. Colony formation by both L-cell and murine serum CSF was completely inhibited with a 1:512 dilution of antiserum. The in vivo serum disappearance of anti-CSF was studied using a highly purified antibody fraction, which had been obtained by an immunoadsorbent technique [33]. In addition this material was repeatedly adsorbed with limulus lysate by Dr. Jack Levin of Johns Hopkins University School of Medicine to remove detectable levels of endotoxin. A total of 25,000 units of this purified anti-CSF was injected IP into each of 9 mice, and serum was obtained 2, 7, and 12 h thereafter. Circulating antibody activity was measured by determining the ability of the various serum samples to precipitate ^{125}I-labeled CSF. When compared to a standard curve, the 2 h serum value reflected a 10.4% recovery of the injected dose (Fig. 9). Clearance was essentially linear through 12 h, with a $t_{1/2}$ of approximately 2 h. This compares quite favorably with the clearance of CSF, thereby suggesting that antibody clearance is due to interaction with newly synthesized CSF. Bioassay of serum samples from 2 and 7 h showed no CSF; only 80 units/ml were detected at 12 h. This study and the in vitro incubation of anti-CSF with mouse serum clearly indicate that this antibody completely neutralizes circulating CSF activity.

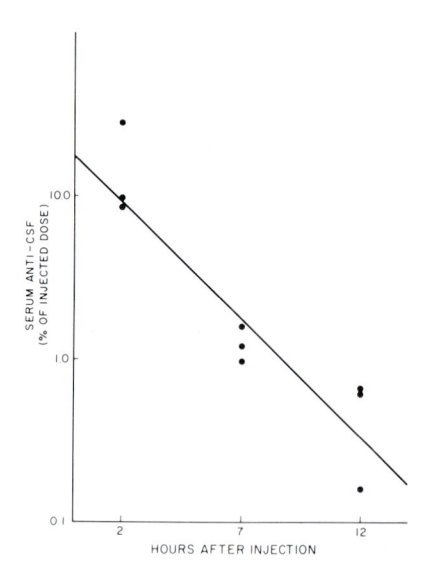

Fig. 9. Serum half-life of purified anti-CSF. Animals were injected with 25,000 units of purified anti-CSF and circulating antibody determined at intervals thereafter. Values represent the % of injected antibody activity recovered from serum at the indicated time intervals.

Following these experiments, groups of mice were injected with 35,000 units of anti-CSF IP every 12 h for 7 days. Peripheral blood granulocytes were reduced from 815 to 510/μl on day 7; however, bone marrow cellularity was essentially unchanged. Further studies using 175,000 units of purified anti-CSF showed essentially no decrease in circulating neutrophils or marrow cellularity. However, a modest decrease in bone marrow granulocyte-macrophage colony-forming cells from 18,000 to 8,000 per femur was observed.

One explanation for this relatively limited inhibition of granulopoiesis may relate to the rapid development of antibodies against the injected anti-CSF. Although high levels of circulating anti-CSF were detected on days 1 and 3 of this study, virtually none could be detected on days 5 and 7, despite continued administration every 12 h. Modified radioimmunoassays showed that the recipient mice had developed precipitating antibody against the anti-CSF.

If further experiments, mice were treated with 2.5 mg of cortisone acetate on day −1, day 2, and day 5 in an attempt to limit their development of antibodies directed against anti-CSF. Groups of mice received high titer anti-CSF serum (230,000 units of anti-CSF) every 12 h and were evaluated on days 1−7 of treatment. Although circulating anti-CSF could be detected for up to 5 days, no decrease in circulating granulocytes or in the proliferative or nonproliferative compartments in the marrow was noted (Fig. 10). Thus, these studies show that the in vivo administration of anti-CSF has only minimal effects in intact mice.

Diffusion Chamber Studies With Anti-CSF

As noted above, increased proliferation of granulocytic cells occurs in diffusion chambers implanted in neutropenic as compared to normal hosts. Studies were undertaken in collaboration with investigators at Brookhaven National Laboratories to determine whether CSF might be responsible for this effect. Groups of mice received 700 R total body irradiation and were each implanted with 2 intraperitoneal diffusion chambers containing 5×10^5 nonadherent marrow cells. Each group was injected with 0.5 ml of saline, control serum, or CSF antiserum (titer, 1:128) every 12 h for 4 days. Diffusion chambers were harvested at 2, 4, and 7 days for evaluation of cell counts, differentials, and colony-forming assays [16]. The results showed marked inhibition of granulocytic differentiation in chambers from antibody recipients (Fig. 11). Both control serum and saline groups had a 12- to 18-fold increase in granulocyte production. In addition, chamber CFU-C content was markedly depressed in antibody recipients (Fig. 12). These findings, as well as the continued requirement for CSF in agar cultures, suggest that CSF is necessary both for the growth and maturation of differentiated cells and for proliferation of the granulocytic stem cell compartment. Thus, these observations are strongly suggestive that CSF may function as an in vivo as well as an in vitro granulopoietin.

Effect of CSF and Anti-CSF on Long-Term Bone Marrow Cultures

As noted above the stromal microenvironment may represent an important determinant of stem cell proliferation and differentiation. Recently, a system was devised in which a bone marrow-derived adherent cell layer provides the necessary environment for sustained hemopoiesis [34]. Murine bone marrow cells are suspended in culture medium supplemented

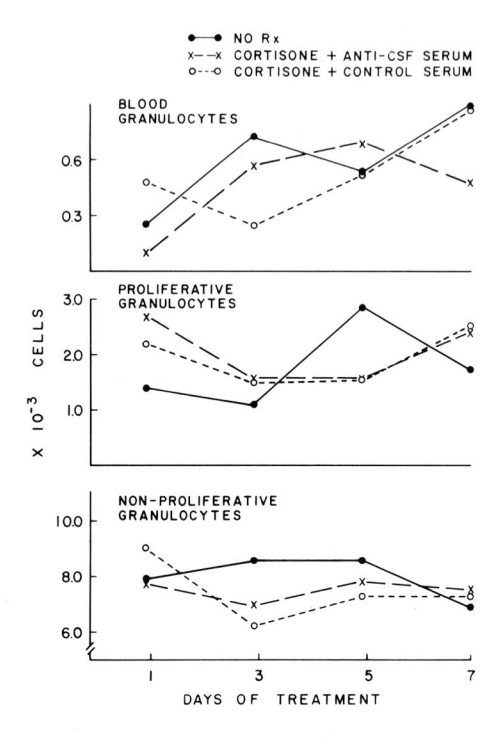

Fig. 10. Effect of anti-CSF on in vivo granulopoiesis. Values represent absolute peripheral blood neutrophil counts or quantitative determinations of bone marrow granulocytic cells at the times indicated. Test animals received 0.5 ml of antiserum or control serum every 12 h throughout the experiment. Values are means from 5 animals/point.

with 20% horse serum and incubated at 33°C. With half depopulation and refeeding on a weekly basis, the nonadherent cells show continual production of granulocytes, macrophages, and maintenance of granulocytic and pluripotential colony-forming cell populations.

Studies were undertaken in collaboration with Dr. Michael Dexter of Manchester, England, to determine whether exogenous CSF or anti-CSF influenced these long-term cultures. Previous experiments had shown virtually no CSF in culture supernatants [34, 35] and a remarkable decrease in cellular differentiation following addition of crude sources of CSF. In our recent studies impure CSF derived from heart conditioned medium or standard L-cell CSF caused a progressive decline in culture cellularity and granulocytic differentiation [32]. In contrast, purified L-cell CSF had no such effect. This suggests that the decline in cellular differentiation was probably due to exogenous materials in the impure CSF preparations. In the opposite type of experiment, CSF antiserum had essentially no effect on granulocyte differentiation [32]. Despite addition of 10% antiserum to these cultures, there was no decline in cellularity, granulocytic growth, or proliferation of CFU-C or CFU-S compartments.

These findings as well, as the previous inability to detect CSF in the long-term bone marrow cultures, call into question the involvement of CSF in this type of granulocytic differentiation. To examine this issue further, supernatant material from long-term cultures

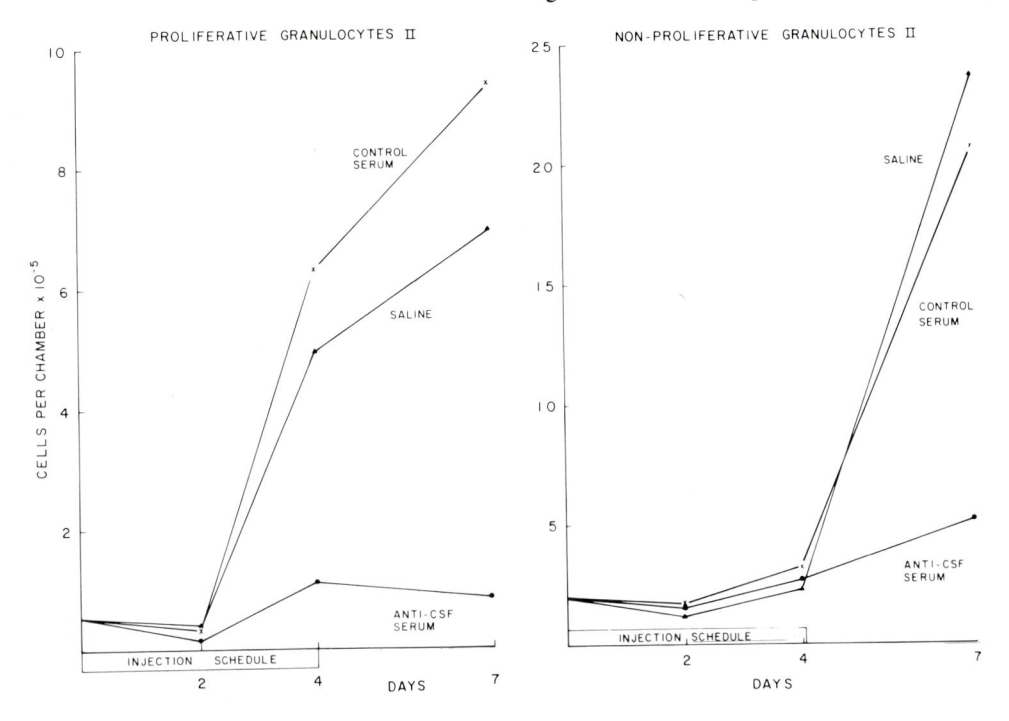

Fig. 11. Effect of anti-CSF serum on diffusion chamber granulopoiesis. After 700 R irradiation, groups of mice were injected with 0.5 ml of saline, control serum, or antiserum every 12 h through 4 days. Shown are the total number of proliferative and nonproliferative granulocytes in each treatment group on the 2nd, 4th, and 7th days of study. Reproduced from Shadduck, Carsten, Chikkappa, Cronkite and Gerard: Inhibition of diffusion chamber (DC) granulopoiesis by anti-CSF serum. Proc Soc Exp Biol Med 158:542–549, 1978, with permission.

was kindly provided by Dr. Michael Dexter and Dr. Joel Greenberger (Harvard Medical School). The results of CSF assays are shown in Table III. Although no CSF activity could be determined by agar gel bioassay, substantial quantities were found in all samples by a sensitive radioimmunoassay. Preliminary experiments suggest that the inability to detect CSF in these cultures may result from high levels of inhibitory substances that prevent colony formation. Clearly, the further delineation of granulopoietic control mechanisms in these long-term cultures may have important implications concerning regulatory mechanisms that may be operative in vivo.

DISCUSSION

As shown in these and other recent studies [26, 27], murine fibroblast or L-cell CSF has been purified to homogeneity. This factor has a molecular weight of 60,000–70,000 daltons and contains substantial quantities of carbohydrate as evidenced by PAS staining, inactivation by periodate, and altered electrophoretic mobility following incubation with neuraminidase. Based on the variable binding to concanavalin-A, the L-cell CSF appears to have some heterogeneity in carbohydrate residues; however, this difference

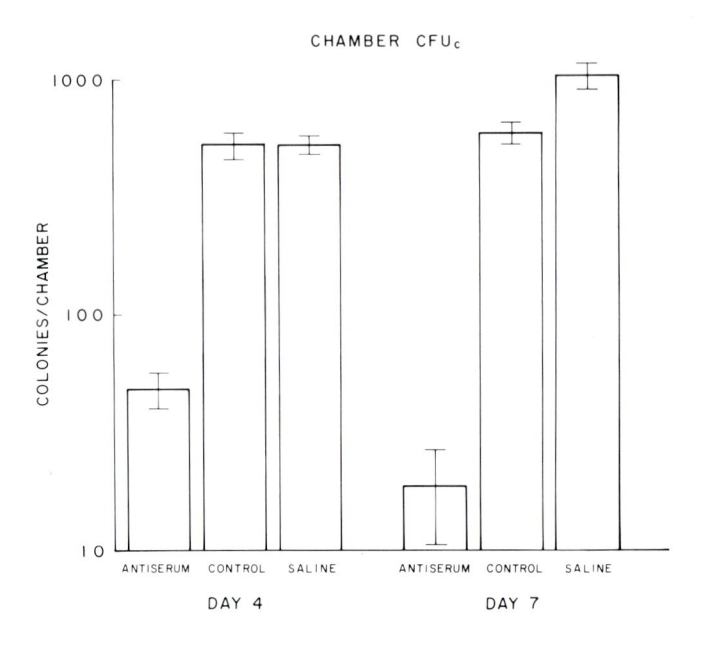

Fig. 12. Recovery of granulocyte-macrophage colony-forming cells or CFU-C from diffusion chambers following treatment with CSF antiserum. Hosts were treated as in Figure 11. Values are means ± 1 SE. Modified from Shadduck, Carsten, Chikkappa, Cronkite and Gerard: Inhibition of diffusion chamber (DC) granulopoiesis by anti-CSF serum. Proc Soc Exp Biol Med 158:542–549, 1978, with permission.

does not influence specific activity [28] or colony morphology [33]. The purified CSF is active in concentrations of 5×10^{-12} M and yields maximal colony formation with 5×10^{-11} M, or approximately 3 ng/ml.

Although in our studies [28] 50–70% of the biologic activity is retained after purification, the final yields of material only range from 200 to 300 µg/10 l pool. This problem and the marked time delay inherent in the separative procedures have substantially hampered large scale production and purification of this factor. Recently, using "monospecific" antibodies to CSF [33], we have found that the crude L-cell material can be purified by a single-step immunoadsorbent technique [36]. Yields have ranged from 65% to 100%, and specific activities are essentially identical to those obtained by the 6-step procedure. This improved technique may now provide the relatively large quantities of material required for in vivo studies and for further characterization of subunit structure.

One of the major problems in the study of in vivo granulopoiesis has been the invariable contamination of test solutions with endotoxin. Since such lipopolysaccharides cause both the production of CSF in vivo [18] and stimulation of granulopoiesis [19], purified CSF must be rendered virtually free of biologically active levels of such contaminants. By virtue of the widely differing molecular weights, it has now been possible to markedly reduce the level of endotoxin in purified CSF preparations [37]. Density gradient centrifugation sediments greater than 95% of endotoxin activity to the bottom of the tube, whereas CSF is retained in the upper portion of the gradient. Levels of endotoxin are generally reduced to < 0.01 µg/ml – a concentration that is essentially devoid of effects on hemopoiesis [37].

TABLE III. CSF Activity in Long-Term Marrow Cultures*

Duration of culture	Bioassay (units/ml)	Radioimmunoassay (units/ml)
1–3 weeks (Dexter)	4.4 (2–6)	79.2 (44–145)
4–13 weeks (Greenberger)	3.0 (0–8)	69.6 (47–135)

*Values represent units/ml of CSF in supernatants harvested from long-term marrow cultures at the indicated times. One unit of CSF is defined as that activity which stimulates 1 colony in the agar-gel bioassay. Units were calculated, when feasible, from the linear portion of the dose-response curve.

With radiolabeled CSF, it has been possible to define the plasma disappearance rate and partially characterize the degradative pathways. The major plasma clearance has a $t_{1/2}$ of approximately 2 h. Only minimal activity is detected in target marrow cells, whereas > 40% of the tracer rapidly accumulates in the liver. Within 6 h as much as 30% of the tracer is excreted in the urine. However, this is composed entirely of low molecular weight peptides. Parallel studies with unlabeled CSF show a similar plasma clearance; however, less than 1% of the intact molecule is recovered in the urine. It appears clear from these studies that urinary CSF levels may not reflect serum levels of this factor. Despite the fact that human urine is a good source of CSF for murine cultures, these findings, as well as the inability of human urinary CSF to stimulate human cultures, cast serious doubt on the value of such determinations.

In vitro incubation of iodinated CSF with murine marrow cells leads to delayed but extensive binding. The kinetics differ markedly from most other hormone–receptor interactions by virtue of a 6–16 h time lag. The binding is specific and saturable, and it is not influenced by addition of various exogenous proteins. Preliminary results indicate that once CSF is bound to the cells it cannot be displaced by a 100- to 1,000-fold excess of cold material. Perhaps the delay in binding results from irreversible receptor occupancy in vivo such that in vitro binding requires genesis of new cellular receptors.

Recently, increasing attention has been directed at colony morphology. Various sources of CSF yield colonies of granulocytes (G), macrophages (M), or a mixture thereof (GM) when scored after 7 days of incubation. Based on this terminology, L-cell CSF is believed to represent a pure M type CSF [38]. However, when colonies are examined on the 3rd or 4th day of growth in agar gels [33] or in plasma clots [39], substantial numbers of granulocytic cells are observed.

Our preliminary autoradiographic studies also indicate that CSF binds to cells of the granulocytic series. Between 35% and 60% of large mononuclear cells are labeled after 6–24 h incubation. Fifty percent of myeloblasts and promyelocytes are labeled at 6 h, with an increase to 95% at 24 h. Between 30% and 60% of later cells in the myeloid series are labeled after 24 h incubation [40]. Thus, it would seem clear that this variety of CSF binds to and causes limited proliferation of cells in the granulocytic as well as monocyte-macrophage lineages.

Initially, it was believed that if CSF is a biologically important regulator of granulopoiesis, simple in vivo neutralization by antibodies should lead to granulocytic aplasia. Although anti-CSF completely neutralizes circulating CSF, studies with crude antiserum indicated no depression of in vivo granulopoiesis [16]. As shown in this report, highly purified anti-CSF is also devoid of activity. In part this may result from second antibody formation, with rapid clearance of anti-CSF within 5 days of repetitive injection. Animals treated with anti-CSF and cortisone to prevent second antibody formation still manifest active granulocytic differentiation. Thus, it must be concluded that CSF in the serum may not be a controlling factor in granulopoiesis.

Although anti-CSF has not suppressed bone marrow granulopoiesis, such treatment is markedly inhibitory to diffusion chamber granulopoiesis. Either crude antiserum [16] or, in recent experiments with Drs. Cronkite and Carsten, the highly purified antibody reduced both the number and size of diffusion chamber granulocytic colonies. This striking difference from the effect in intact animals may be due to altered clearance of antibody in irradiated mice — a step necessary to induce heightened diffusion chamber granulopoiesis. Alternatively, there may be extensive local stromal production of CSF in the intact marrow, which is difficult to neutralize by systemic administration of antibody. Diffusion chamber studies obviate this problem by selectively removing marrow cells from their intact microenvironment.

To examine further the question of stromal interaction, the effect of anti-CSF was determined in long-term marrow cultures. In this system adherent "stromal" layers are established after 3–4 weeks of marrow cell culture [34]. With refeeding on a weekly basis, granulopoiesis continues for several months without addition of CSF. Early studies showed no CSF production in these cultures [34, 35]. However, the results reported herein indicate this is an artifact caused by coexistent inhibitors. Virtually no CSF is detected in supernatants by agar gel bioassay, whereas substantial quantities of CSF are found by radioimmunoassay.

Although CSF is present in these cultures, addition of anti-CSF does not prevent continuing granulocytic differentiation [32]. Thus, it appears clear both from the in vivo studies and from the long-term "Dexter-type" cultures that granulopoiesis cannot be inhibited by anti-CSF when the responsive colony-forming cells are in contact with stromal elements of the marrow. When removed from the microenvironment either in the agar gel [29] or in the diffusion chamber system [16], marrow cells are markedly suppressed by this antiserum. This suggests that granulopoiesis is normally CSF mediated by a short-range cell–cell contact within the intact marrow microenvironment.

ACKNOWLEDGMENT

These studies were supported, in part, by a grant from the National Institutes of Health, R01 CA15237-05.

REFERENCES

1. Cronkite EP: Natl Cancer Inst Monogr 30:51, 1969.
2. Boggs DR: Semin Hematol 4:359, 1967.
3. Pluznik DH, Sachs L: J Cell Physiol 66:319, 1965.
4. Bradley TR, Metcalf D: Aust J Exp Biol Med Sci 44:287, 1966.
5. Rickard KA, Shadduck RK, Stohlman F Jr: Proc Soc Exp Biol Med 134:152, 1970.
6. Morley A, Rickard KA, Howard D, Stohlman F Jr: Blood 37:14, 1971.
7. Shadduck RK, Nagabhushanam G: Blood 38:559, 1971.
8. Shadduck RK, Nunna NG: Proc Soc Exp Biol Med 137:1479, 1971.

9. Vogler WR, Mingioli ES, Garwood FA, Smith BA: J Lab Clin Med 79:379, 1972.
10. Yang TJ, Jones JB, Jones ES, Lange RD: Blood 44:41, 1974.
11. Hammond WP, Engelking ER, Dale DC: J Clin Invest 63:785, 1979.
12. Moore MAS, Spitzer D, Metcalf D, Pennington DG: Br J Haematol 27:47, 1974.
13. Boyum A, Borgstrom R: Scand J Haematol 7:294, 1970.
14. Boyum A, Carsten AL, Laerum OD, Cronkite EP: Blood 40:174, 1972.
15. Tyler WS, Niskanen E, Stolhman F Jr, Keane J, Howard D: Blood 40:634, 1972.
16. Shadduck RK, Carsten AL, Chikkappa G, Cronkite EP, Gerard E: Proc Soc Exp Biol Med 158:542, 1978.
17. Shadduck RK, Nunna NG, Krebs J: J Lab Clin Med 78:53, 1971.
18. Shadduck RK: Exp Hematol 2:147, 1974.
19. Quesenberry P, Halperin J, Ryan M, Stohlman F Jr: Blood 45:789, 1975.
20. Gordon AS, Handler ES, Siegel CD, Dornfest BS, Lobue J: Ann NY Acad Sci 113:766, 1964.
21. Chan SH, Metcalf D, Stanley ER: Br J Haematol 20:329, 1971.
22. Zidar BL, Shadduck RK: J Lab Clin Med 91:584, 1978.
23. Rytomaa T, Kiviniemi K: Cell Tissue Kinet 1:329, 1968.
24. Shadduck RK: J Lab Clin Med 87:1041, 1976.
25. Broxmeyer HE, Smithyman A, Eger RR, Meyers PA, de Sousa M: J Exp Med 148:1052, 1978.
26. Burgess AW, Camakaris J, Metcalf D: J Biol Chem 252:1998, 1977.
27. Stanley ER, Heard PM: J Biol Chem 252:4305, 1977.
28. Waheed A, Shadduck RK: J Lab Clin Med 94:180, 1979.
29. Shadduck RK, Metcalf D: J Cell Physiol 86:247, 1975.
30. Shadduck RK, Waheed A: Blood Cells 5:421, 1979.
31. Shadduck RK, Waheed A, Porcellini A, Rizzoli V, Pigoli G: Blood 54:894, 1979.
32. Dexter TM, Shadduck RK: J Cell Physiol 102:279, 1980.
33. Shadduck RK, Waheed A, Pigoli G, Boegel F, Higgins L: Blood 53:1182, 1979.
34. Dexter TM, Allen TD, Lajtha LG: J Cell Physiol 91:335, 1977.
35. Williams N, Eger RR, Moore MAS, Mendelsohn N: Differentiation 11:59, 1978.
36. Shadduck RK, Waheed A: Clin Res 28:324A, 1980.
37. Shadduck RK, Waheed A, Porcellini A, Rizzoli V, Levin J: Proc Soc Exp Biol Med 164:40, 1980.
38. Stanley ER: Proc Natl Acad Sci USA 76:2969, 1979.
39. Cronkite EP, Carsten AL, Cohen R, Miller ME, Moccia G: Blood Cells 5:331, 1979.
40. Pigoli G, Shadduck RK, Schiffer LM: Clin Res 28:320A, 1979.

Direct Linkage of EGF to Its Receptor: Characterization and Biological Relevance

Peter S. Linsley and C. Fred Fox

Department of Microbiology and Parvin Cancer Research Laboratories, Molecular Biology Institute, University of California, Los Angeles, California 90024

A small portion of the ^{125}I-EGF that binds specifically to intact cells or isolated membranes from a variety of sources becomes directly and irreversibly linked to EGF receptors. This provides a simple technique for affinity labeling the EGF receptor. Membranes isolated from the human epidermoid carcinoma cell line A431, which posesses extraordinarily high numbers of EGF receptors, gave rise to three major direct linkage complexes of MW = 160,000, 145,000, and 115,000. The time course for formation of each is similar, showing that ^{125}I-EGF can form direct linkage complexes with several preexisting forms of the EGF receptor. The direct linkage of EGF to receptor is slow in comparison to ^{125}I-EGF binding, but both processes have similar susceptibilities to competition by unlabeled EGF.

EGF was modified chemically with the amino site-specific reagent, N-hydroxysuccinimidyl biotin. The biotinyl-EGF had a reduced capacity to engage in direct linkage complex formation with no concomitant reduction in its ability to bind to EGF receptors. Since native and biotinyl EGF have identical abilities to stimulate the uptake of ^{3}H-thymidine into DNA when incubated with cultured murine 3T3 cells, the direct linkage of EGF to its receptor does not appear to play an important role in EGF-stimulated mitogenesis.

Key words: EGF receptors, biotinyl EGF, covalent EGF-receptor complexes — and 3T3 cell growth regulation, on human placental membranes, on cultured cells

Epidermal growth factor (EGF) is a potent polypeptide mitogen that initiates its biological activity by interacting with the cell surface [1]. Radioiodinated derivities of EGF bind to specific high-affinity receptors on responsive cells [2–6]. Rapid internalization and lysosomal degradation of EGF follow binding and have been demonstrated by a

---

Abbreviations: EGF, epidermal growth factor; DMEM, Dulbecco's modified Eagle medium; ST buffer, 0.25 M sucrose in 10 mM Tris·HCl at pH 7.4; PBS, 0.15 M NaCl in 0.1 M sodium phosphate at pH 7.4; DBH, DMEM minus NaHCO$_3$, but containing 0.1% bovine serum albumin in 10 mM HEPES buffer at pH 7.4; BSA, bovine serum albumin; SDS, sodium dodecyl sulfate; ESB, electrophoresis sample buffer (63 mM Tris·HCl at pH 6.8, 3% SDS, and 10% glycerol); NHS Biotin, N-hydroxysuccinimidyl biotin; MW, molecular weight (M$_r$).

Received August 25, 1980; accepted October 3, 1980.

0091-7419/80/1404-0441$05.50 © 1980 Alan R. Liss, Inc.

variety of biochemical and optical techniques [5, 7–11]. Because the process of down regulation (ie, EGF-mediated loss of its own binding capacity) occurs concomitantly with the internalization and degradation of receptor-associated EGF in the lysosomes, the EGF receptor was presumed to be processed similarly [7]. In this laboratory, we have investigated directly the fate of receptor after EGF addition. Photoaffinity probes, which crosslinked radioiodinated EGF to its receptor, were used initially for receptor identification [12]. The photoaffinity crosslinked EGF–receptor complex, like EGF, was internalized and degraded in the lysosomes [13]. Other strategies for identifying EGF receptors by photoaffinity labeling have been reported [14].

Recently we [15] and Baker et al [16] independantly observed that a small portion of the radiolabeled EGF that binds specifically to murine 3T3 cells or human foreskin fibroblasts becomes directly and irreversibly linked to EGF receptors. The EGF–EGF receptor direct linkage complex has properties identical to those of the EGF–EGF receptor complex formed by photoaffinity labeling. Thrombin may also undergo direct linkage to its receptor during mitogenic stimulation [16].

Direct linkage affords an obvious technical advantage since photoaffinity probes, which are difficult to synthesize and use, need not be employed to specifically label EGF receptors. While the biological role of direct linkage was not obvious, we were intrigued by the possibility that it might be physiologically important, perhaps representing the small fraction of "very high affinity binding sites," which Schechter et al [17] proposed to mediate the biological effects of EGF.

Because of its technical utility and potential biological importance, we have investigated the direct linkage process in detail, but we find no evidence for its involvement in EGF-induced mitogenesis. In a companion paper [18], we use this technique to document the striking protease sensitivity of EGF receptors in isolated membranes and the specific EGF-stimulated phosphorylation [19] of two fragments of the receptor.

EXPERIMENTAL PROCEDURES

Materials

Male mouse submaxillary glands were from Pel-Freeze. Swiss 3T3 cells (clone 42) were obtained initially from G. Todaro, National Cancer Institute; HF-15 cells (human foreskin fibroblasts), from D. Cunningham, University of California, Irvine; and A431 cells, from G. Todaro and J. De Larco. Iodoacetic acid, biotin and p-nitrophenyl-N-acetyl-β-D glucosaminide were from Sigma. TPCK-treated trypsin (254 units/mg) was obtained from Worthington. Ficoll and Sephadex were obtained from Pharmacia; Iodine-125, from Amersham; and ^{3}H-thymidine, from New England Nuclear.

Cell Culture

Cultures were maintained in DMEM plus 10% fetal calf serum. Cells were grown on plastic dishes and plastic or glass roller bottles in an atmosphere of 10–15% CO_2.

Membrane Preparations

Three procedures were used; all result in substantial enrichment of ^{125}I-EGF binding over that present in crude homogenates.

Procedure I. For preliminary experiments, a crude particulate fraction was prepared after cells were homogenized in hypotonic solution known to disrupt osmotically fragile lysosomes [20]. Cells grown on plastic dishes were washed free of serum and harvested by scraping into DMEM. The cell pellet was collected by sedimentation at 750g for 5 min, suspended in 10 volumes of 10 mM Tris·HCl 7.4, containing 2 mM phenylmethyl sulfonyl fluoride; the suspension was incubated at 20°C for 10 min. Cells were disrupted by vigorous mechanical homogenization (35 strokes in a hand-driven Dounce B homogenizer), and unbroken cells and nuclei were collected by sedimentation at 750g for 5 min. The supernatant fraction was layered over a cushion of ST buffer. After sedimentation at 24,000g for 30 min, the crude membrane pellet was suspended in ST buffer. These membranes were stored at −20°C for several days with no change in EGF binding activity.

Procedure II. This membrane preparation was utilized in all experiments not specifying procedures I or III. Cells were grown to confluence in roller bottles, growth medium was removed, and the monolayers were washed with 10 mM Tris·HCl, pH 7.4, containing 0.15 M NaCl, and scraped into ST buffer. Cell pellets were collected by centrifugation at 2,000g for 5 min and suspended in 9 volumes of ST buffer. The suspension was homogenized with a tight-fitting, motor-driven Teflon pestle (20 strokes). Nuclei and unbroken cells were removed by centrifugation at 2,600g for 5 min. The particulate matter in the postnuclear supernatant was sedimented at 35,000g for 30 min, and the pellet was suspended in 4 times the original packed cell volume in 35% Ficoll in ST buffer. Ten milliliters of this suspension were overlaid with 10 ml each of 25% and 12% Ficoll in ST buffer. A final overlay of 6 ml of ST buffer was added, and the tubes were sealed and centrifuged for 1.5 h at 167,000g in a Beckman VTi 50 verticle rotor. The material at the 0/12%, 12/25%, and 25/35% Ficoll interfaces was individually isolated, diluted with 10 mM Tris·HCl, pH 7.4, and pelleted at 35,000g for 30 min. The pellets were suspended in a small volume of ST buffer and diluted prior to use. For convenience, membranes were routinely stored at −20°C for periods of several months with little effect on ^{125}I-EGF binding. In Table A-I (see Appendix) the characterizations of two representative preparations are described with respect to ^{125}I-EGF binding, a plasmalemmal marker, and the activity of a lysosomal marker, N-acetyl-β-D-glucosominidase [21]. Material isolated at the 0/12% Ficoll interface contained the highest ^{125}I-EGF binding activity relative to the lysosomal marker enzyme activity; this fraction was the source of membranes used in all experiments unless another procedure is specifically cited.

Procedure III. In some experiments, membranes were prepared by a published procedure [22] in which cells were lysed by dilution into hypotonic buffer prior to a series of differential centrifugations. Membranes prepared by this procedure had a ^{125}I-EGF specific binding activity of 41 pmoles/mg and a specific activity of 0.96 A_{400} units (Table A-1), for N-acetyl-β-D-glucosaminidase.

Preparation and Iodination of EGF

EGF was purified by the method of Savage and Cohen [23] and radioiodinated using chloromine T [12]. The ^{125}I-EGF preparations used in these experiments had specific activities ranging from $6-35 \times 10^8$ cpm/nmole.

EGF Binding and Direct Linkage Complex Formation

The procedures used for measuring ^{125}I-EGF binding have been described [12]. Unless indicated otherwise, direct linkage complexes were formed with either intact cells or isolated membranes during a 1 h binding assay at 20°C. In studies with intact cells, the cultures were washed thoroughly with DBH and incubated with the desired concentration of ^{125}I-EGF in DBH. Unbound ^{125}I-EGF was removed by washing the cell monolayers with DBH, and cell bound radioactivity was determined after dispersion of the cells in 0.5 N NaOH. Total EGF binding was corrected for nonspecific binding measured under identical conditions, but in the presence of a large excess of unlabeled EGF. When direct linkage was assessed, the final washes were done with protein-free medium to remove BSA (which interfered with gel electrophoresis), and the cell monolayers were solubilized in ESB. In studies with isolated membranes, the reactants were mixed in a total reaction volume of 0.1 ml; the final concentration of BSA in the assay was at least 0.01%. When binding alone was measured, membrane-bound ^{125}I-EGF was separated from the mixture by filtration on Whatman GFC filters, and the filters were washed thoroughly with DBH. For direct linkage measurements, membranes were sedimented in a Brinkman microfuge through a cushion of 10% sucrose in DBH (15 min at 12,800g) to remove unbound ^{125}I-EGF. The membrane pellet was given a final wash with BSA-free medium prior to addition of ESB. Nonspecific binding was measured in the presence of 2.4 μM unlabeled EGF. Roughly equivalent binding affinities were detected on intact cells and on membranes washed either by filtration or sedimentation.

Gel Electrophoresis and Autoradiography

The formation of the direct linkage complex was examined using the discontinuous pH system of Laemmli [25]. Best results were obtained with slab gels having a 5–12% gradient acrylamide resolving gel, measuring 1.5 mm thick × 150 mm wide × 180 mm long, topped with a 20 mm high stacking gel. Intact cells and membranes were solubilized in ESB, and boiled for 5 min prior to their application to a gel. When intact cells were used, the monolayers were solubilized by the addition of 0.1 ml of ESB per 10 cm^2 of growth area, and the resulting highly viscous solution was subjected to vigorous shearing through a 20–22-gauge hypodermic needle prior to boiling. Gels were stained with Coomassie blue, destained, dried on filter paper, and autoradiographed on Kodak x-ray film. Only the resolving gels, which contained the bulk of the applied radioactivity in all cases, have been photographically reproduced in this paper. Relative molecular weights were determined by comparison with the mobilities of red blood cell ghost proteins, which were visualized by staining, and by using the published MW values for bands 1, 2, 3, 4.1, 4.2, and 5 [26]. BSA was present in all binding assays and was visible as a contaminant in most samples as a consequence of its binding to membranes; it served as a convenient marker of MW = 68,000. A plot of log MW versus mobility was linear in the MW = 70,000–240,000 region. Identical results were obtained with a commercially available (Bio Rad) kit with molecular weight standards ranging from 43,000 to 200,000. The molecular weights reported here are strictly relative, and as such, differ slightly from those previously reported for EGF–EGF receptor direct linkage complexes [15, 27]. The slight deviation from agreement with previous results is reasonable considering the differences in gel systems and molecular weight standards used.

Determination and Quantitation of Radioactivity in the Directly Linked Complex

The regions of the dried gel which contained direct linkage complexes were localized by autoradiography and excised, and the radioactivity was measured by gamma scintillation spectrometry. The radioactivity present in direct linkage complexes was corrected for background radioactivity in adjacent regions of the same lane or corresponding regions of a lane containing an identical sample, where excess unlabeled EGF was included during the initial incubation.

Chemical Modification of EGF

Synthesis of biotinyl EGF. N-hydroysuccinimidyl biotin (NHS Biotin) was synthesized by dicyclohexyl carbodiimide-mediated condensation of biotin and N-hydroxysuccinimide [28]. Acylation of EGF was accomplished at alkaline pH, a condition that predominantly modifies free amino groups [29], of which EGF has only one, the amino terminus [30]. One hundred micrograms (16.7 nmoles) of EGF in 0.147 ml of H_2O were mixed with 0.153 ml of 0.5 M Na_2CO_3 at pH 9.0, containing 0.3% NaN_3 and 0.1% Triton X-100. Modification was accomplished by the addition of 8 sequential aliquots of NHS biotin (0.286 ng or 0.84 μmoles in 5 μl of dry dimethyl formamide) during a 2 h incubation at 20°C. The reaction was terminated by the addition of 20 μmoles of ammonium acetate in 10 μl of H_2O, and the solution was stored at −20°C. The resulting biotinyl EGF was iodinated using chloromine T to specific activities comparable to those achieved with native EGF. Chromatography on Sephadex G-10, which was routinely used to separate ^{125}I-EGF from free ^{125}I, also served to purify ^{125}I-biotinyl EGF (as well as iodinated forms of the other derivatives) to an extent that permitted their use in the described experiments. The extent of acylation of EGF was estimated by mixing ^{125}I-biotinyl EGF with an excess of avidin and subjecting the mixture to column chromatography on Sephadex G-50. ^{125}I-biotinyl EGF was included in the column, but when avidin was added, more than 80% of the applied radioactivity was eluted in the void volume. When avidin was mixed with excess free biotin and then with ^{125}I-biotinyl EGF, the radioactivity was eluted from Sephadex G-50 in the position of free ^{125}I-biotinyl EGF. We conclude that the preparation of biotinyl EGF used in these experiments was at least 80% modified with NHS biotin, presumably at the amino terminus. In order to determine the amount of unmodified EGF present in this preparation, we subjected biotinyl EGF to DEAE chromatography at pH 7.1 (Fig. A-3, Appendix). This procedure separates native EGF from amino modified EGF [10]. Two EGF-containing peaks were detected. One, which represented less than 5% of the total EGF in the preparation, had a reduced ability to bind avidin relative to the material in the major peak and probably contained unmodified EGF. Because of its high avidin binding capacity, the major peak contained biotinyl EGF. Fractions taken from across this peak displayed a constant amount of non-avidin-binding material (approximately 20%). Since the non-avidin-binding material in the major peak is not readily separable from that which does bind avidin, it most likely does not represent unmodified EGF. The existence of non-avidin-binding material may result from the modification of EGF with a form of biotin incapable of binding to avidin. This derivative could have been present originally as an impurity in the commercial biotin preparations, or it could have been formed by chloramine T oxidation during the iodination procedure. The amount of unmodified EGF in this biotinyl EGF preparation (less than 5%) was not sufficient to explain the residual direct linkage complex forming ability of biotinyl EGF (see Table IV).

Trypsin treatment of biotinyl EGF. Biotinyl EGF was subjected to limited trypsin treatment to prepare an EGF derivitive lacking the carboxy terminal 5 amino acids [30]. Twenty five micrograms of EGF (4.2 nmoles) were treated with TPCK-treated trypsin at an EGF:trypsin ratio of 300:1 (w:w). The mixture was incubated for 2 h at 37°C. The product was stored without purification at −20°C.

Iodoacetate modification of EGF. EGF was treated with iodoacetate under conditions reported to lead to modification of histidine and, to a lesser extent, methionine residues [31]. Twenty-five micrograms of EGF (4.2 nmoles EGF, containing 4.2 nmoles histidine) in 37 μl of H_2O were mixed with 63 μl of 0.2 M sodium acetate at pH 5.6, containing 3.2 M urea. An excess of iodoacetic acid (107 nmoles in 10 μl of H_2O) was added, and the mixture was incubated under an atmosphere of toluene at 30°C for 48 h. The product was stored at −20°C.

Reduction and carboxymethylation of EGF. EGF was reduced essentially by the procedure of Savage et al [30]. Twenty-five micrograms of EGF (4.2 nmoles) in 37 μl of H_2O were mixed with 63 μl of 0.1 M sodium phosphate at pH 8.5, containing 8 M urea, and 2-mercaptoethanol (1μl) was added. The mixture was maintained under an atmosphere of toluene at 30°C for 48 h. The reduced EGF was alkylated with iodoacetic acid (15 μmoles added in 10 μl of 0.1 M sodium phosphate, at pH 8.5) for 5 h in the dark and stored at −20°C. The specific activity of this derivitive after chloromine T-mediated iodination was considerably less than that of iodinated native EGF.

Determination of EGF-Stimulated ^{3}H-Thymidine Incorporation Into DNA

3T3 cells were plated in DMEM plus 10% fetal calf serum. After 48 h this medium was replaced with starvation medium (DMEM plus 0.8% fetal calf serum); and the cells were starved for 3 days. The efficacy of this procedure in producing quiescent cultures was experimentally confirmed by in situ autoradiography. To initiate the G–S transition, the medium was replaced with DMEM + 0.8% fetal calf serum containing EGF. At the time of maximal DNA synthesis during S phase, a time determined in preliminary experiments to by 18 h, ^{3}H-thymidine was added for 2 h at 37°C, and the cells were processed for determination of ^{3}H-thymidine uptake into DNA [13].

Determination of Protein Concentration

Protein was determined by either a conventional [32] or modified [33] Lowry procedure, using BSA as the standard. In one instance (see Table II) the total cellular protein of the intact A431 cells used to determine ^{125}I-EGF binding was measured using a dye binding assay [34], again using BSA as the standard.

RESULTS

Selection of a Source of Material for Characterization of the Direct Linkage Complex

Direct linkage complex formation was observed initially on intact 3T3 cells [15]. A small portion of the ^{125}I-EGF that was bound specifically to these cells migrated as an MW = 160,000 protein having the characteristics of an EGF–EGF receptor complex (Fig. 1, lane A). A small amount of an additional complex of MW = 145,000 was sometimes observed; the formation of both was inhibited when binding proceeded in the presence of unlabeled EGF (lane B). Direct linkage complexes of the same MW are formed with other cell lines (data not shown). Rapid lysosomal processing of the complexes [15] limits the usefulness of intact cells for studying the direct linkage process. We therefore used isolated membranes, which lack the normal cellular degradative pathway.

Membranes from a variety of sources were surveyed for direct linkage complex forming ability. Membranes from 3T3 cells, HF-15 cells (normal human foreskin fibroblasts), and A431 cells (a human tumor line that hyperproduces EGF receptors [4]) give rise to several direct linkage complexes; the formation of all is inhibited by excess unlabeled EGF (Fig. 1, lanes C–H). The lower molecular weight direct linkage complexes arise as the result of endogenous proteolytic modification of EGF receptors during membrane preparation [18]; these modifications do not detectably alter the affinity of EGF receptors for

TABLE I. Direct Linkage Complex Formation in Membrane Preparations From Different Cell Types*

	Source of membranes	Specific activity ^{125}I-EGF binding (pmoles/mg)	Bound EGF in direct linkage complexes (%)
Experiment 1[a]	3T3	0.50	1.45
	HF-15	0.59	0.72
	A431	8.0	0.78
Experiment 2[b]	Human placenta	0.53	0.78

*Experimental details are described in Figure 1 and Experimental Procedures. Crude membranes were prepared from cultured cells by Procedure I (Experimental Procedures). Human placental membranes were prepared according to O'Keefe et al [6]. After electrophoresis, the radioactivity present in all forms of the direct linkage complex was measured, corrected for background radioactivity and expressed as a percentage of the total radioactivity applied to the gel.
[a]Assays contained 8.3 nM ^{125}I-EGF (2.6 × 10^9 cpm/nmole) and 64 µg (3T3), 54 µg (HF-15), or 80 µg (A431) of membrane protein. The incubation was for 1 h at 20°C.
[b]The assay contained 8.3 nM ^{125}I-EGF (1.6 × 10^9 cpm/nmole) and 160 µg of human placental membrane protein. The incubation was for 4 h at 20°C.

^{125}I-EGF. The percentage of specifically bound ^{125}I-EGF in all direct linkage complex components varied in different experiments and was usually less than 2% of the total radioactivity bound to membranes from all sources (Table I). Although the ratio of direct linkage complex formed to EGF bound was similar for all membranes studied, those derived from A431 cells bound extraordinarily high amounts of EGF and gave rise to correspondingly high amounts of direct linkage complexes. Therefore A431 membrane preparations were used in most experiments.

Effects of Time and Temperature on Direct Linkage Complex Formation in A431 Membranes

Binding of ^{125}I-EGF to its receptor on A431 membranes reaches a maximum within 15 min at temperatures ranging from 0° to 37°C, but direct linkage occurred more slowly (Fig. 2) [15]. The rate of direct linkage complex formation was decreased more by low temperature than was binding (data not shown and [15]). The rate of formation of all major direct linkage complex forms is similar (Fig. A-1, Appendix), showing that the lower molecular weight components do not arise by proteolytic cleavage of 160,000 dalton EGF-linked progenitor.

Dependence of Direct Linkage on EGF Concentration

The susceptibilities of ^{125}I-EGF binding and direct linkage complex formation to competition by unlabeled EGF were similar (Fig. 3; Fig. A-2, Appendix). When ^{125}I-EGF was present at half-saturating concentration, half-maximal inhibition of both binding and direct linkage complex formation was achieved at an unlabeled EGF concentration of 87 nM. Saturation of direct linkage complex forming activity with ^{125}I-EGF was similar to that of the binding reaction (data not shown) [15]. The percentage of bound EGF involved in complex formation with A431 receptors varied somewhat with increasing EGF concentration, ranging from 0.33% at 14 nM, to 0.21% at 112 nM. The K_F of direct linkage com-

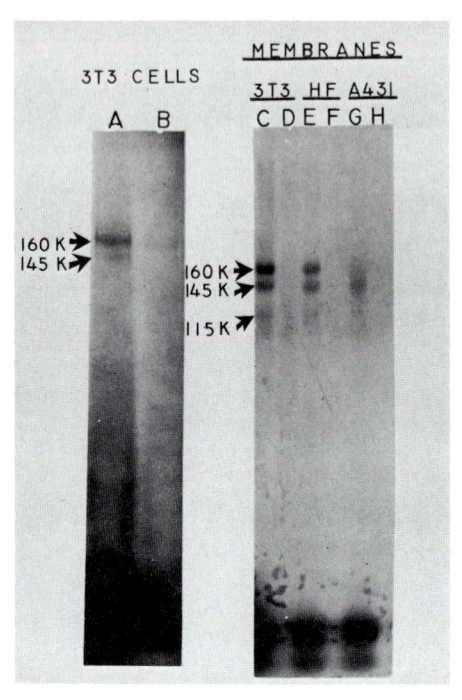

Fig. 1. Direct linkage complex formation in intact 3T3 cells and isolated membranes from different cell types. Cells: confluent 3T3 cell monolayers were incubated with 16.7 nM ^{125}I-EGF (1.2×10^9 cpm/nmole) in DBH. One set of cultures contained unlabeled EGF at 158 nM. After the standard binding reaction (Experimental Procedures), the cells were washed and solubilized in ESB, and an aliquot of 200 μl was subjected to electrophoresis. Membranes: Membranes from 3T3, HF-15, and A431 cells were isolated by Procedure I (Experimental Procedures). The binding assays contained 8.3 nM ^{125}I-EGF (2.6×10^9 cpm/nmole); 2.4 μM unlabeled EGF (where indicated) and membrane protein at 400 μg/ml (3T3); 340 μg/ml (HF) or 500 μg/ml (A431) in DBH. Binding was arrested by sedimentation through a sucrose shelf, the membrane pellet was washed with DMEM and then solubilized with ESB, and the solubilized proteins were subjected to electrophoresis (Experimental Procedures). Arrows indicate the molecular weights of the direct linkage products relative to the mobilities of the standards visible on the stained, dried gel. The formation and significance of each product is discussed in the text.
Lanes: A, direct linkage complexes formed with intact 3T3 cells; B, as in A, plus unlabeled EGF; C, direct linkage complexes formed with 3T3 membranes; D, as in C, plus unlabeled EGF; E, direct linkage complexes formed with HF-15 membranes; F, as in E, plus unlabeled EGF; G, direct linkage complexes formed with A431 membranes; H, as in G, plus unlabeled EGF.

plex formation (a constant analagous to the K_A of ^{125}I-EGF binding) can be utilized to describe the concentration dependence of direct linkage. In A431 membranes, the K_F values for all forms of the direct linkage complex are similar, although not identical, to the K_A of ^{125}I-EGF binding (data not shown). A similar correspondence of K_F and K_A was observed with 3T3 cells, which bind EGF with an affinity at least 10 times greater than A431 [15]. The data are consistent with the conclusion that binding of ^{125}I-EGF to its receptor is obligatory for direct linkage complex formation.

Chemical Stability of the Direct Linkage Complexes

Direct linkage complex formation was influenced by pH. The percentage of bound radioactivity present in direct linkage complexes was constant at pH values less than neutrality but was increased at higher pH values. Once formed, direct linkage complexes

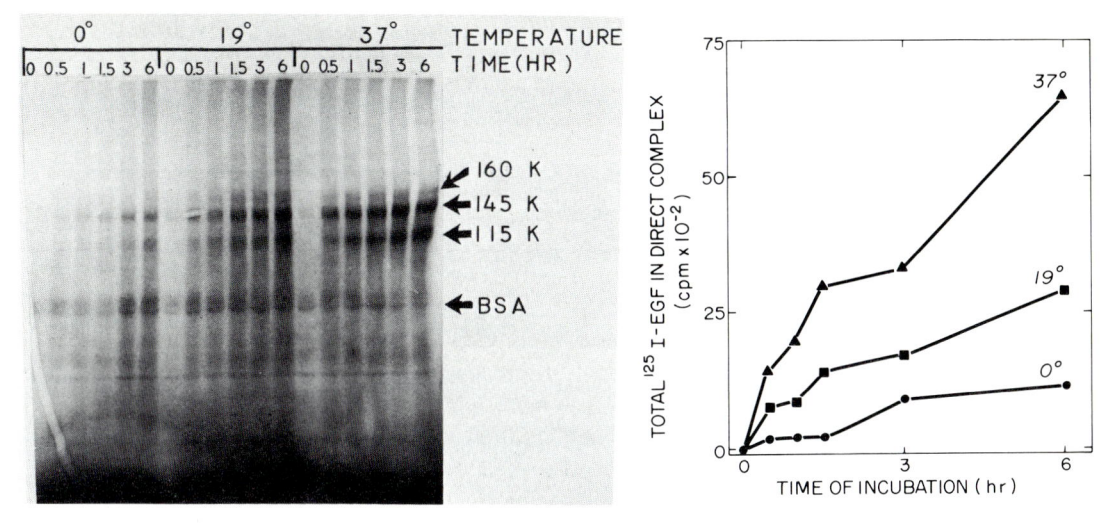

Fig. 2. Time and temperature dependance of direct linkage in A431 membrane. Samples containing 190 μg of A431 membrane protein in 0.59 ml of DBH (plus 0.01% BSA) were equilibrated at the indicated temperatures for 5 min. Binding was initiated by the addition of 0.11 ml of ^{125}I-EGF (1.7 × 10^9 cpm/nmole) to give a final concentration of 88 nM. At the indicated times 0.1 ml aliquots were removed, and the membranes were collected and washed by sedimentation, frozen, and solubilized in ESB immediately prior to electrophoresis. Left panel: autoradiogram showing the time and temperature dependence of direct linkage complex formation. Arrows indicate the relative molecular weights of each direct linkage component. The position of the complex migrating with BSA is also indicated. Right panel: quantitation of radioactivity in direct linkage complexes as a function of time and temperature. The radioactivity in all direct linkage complexes was determined and corrected for the amount of radioactivity present at time zero. The time and temperature dependences of the formation of each individual component are presented in Figure A-2, Appendix.

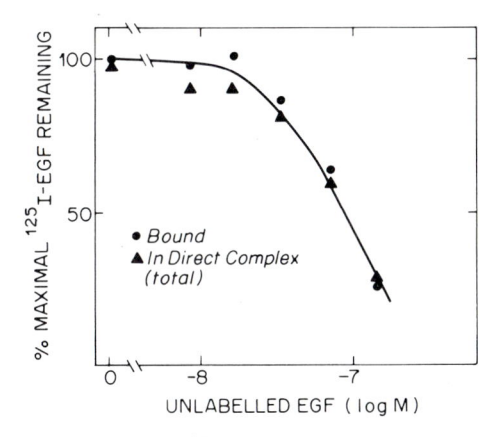

Fig. 3. Competition of unlabeled EGF for ^{125}I-EGF binding and direct linkage complex formation with A431 membranes. Experimental conditions were identical to those given in Figure 5, except that the ^{125}I-EGF concentration was held constant at 28 nM and the concentration of unlabeled EGF was varied from 0 to 143 nM. The K_d of this membrane preparation for ^{125}I-EGF was 29 nM. ^{125}I-EGF was thus present at a concentration sufficient for half-maximal receptor occupancy. ●—● ^{125}I-EGF binding in the presence of increasing amounts of unlabeled EGF. Each point is the average of duplicate determinations. Maximal binding was 0.14 pmoles, which was approximately 90% specific. ▲—▲ Total radioactivity in direct linkage complexes. Maximal radioactivity present in the major forms of the direct linkage complex was 0.78 fmoles. The competition for the formation by each individual direct linkage complex was virtually identical and is presented in Figure A-1 (Appendix).

were stable in solution at pH values ranging from 3.5 to 11.5 (Table II). Direct linkage complexes were also insensitive to hydroxylamine, another condition that disrupts ester linkages [31, 35]. The direct linkage complexes are thus markedly stable and probably are covalent products.

Effects of Chemical Modification of EGF on Direct Linkage

EGF was subjected to various chemical modifications, and the derivitives were assayed for both EGF binding and direct linkage complex formation (Table III). Of the various derivatives tested, only biotinyl EGF bound to 3T3 cells equally as well as native EGF and had a reduced ability to engage in direct linkage complex formation. Removal of the 5-COOH terminal amino acid residues of biotinyl EGF by limited trypsin treatment [30] reduced its binding ability but did not further lower the extent to which the bound derivitive formed direct linkage complexes. A more detailed experiment confirmed the initial observation that biotinyl EGF was reduced in its ability to form direct linkage complexes. Native and biotinyl EGF exhibited almost identical K_d values for binding to 3T3 cells (Fig. 4), but the number of direct linkage complexes formed by biotinyl EGF in the same experiment was reduced by 63 ± 7% (Table IV). The reduction in direct linkage was less than the extent of modification of the EGF in this preparation by biotin (at least 80%, as discussed in Experimental Procedures). This disparity suggests that the observed inhibition is not due simply to blocking of a nucleophilic site — presumably the amino terminus of EGF — involved in covalent bond formation. Instead, acylation may result in the formation of a conformationally altered EGF molecule, in which the site(s) involved in complex formation is sterically shifted to a position unfavorable for direct linkage complex formation. Alternatively, the amino terminus may be one, but not the only, site on the EGF molecule involved in covalent bond formation giving rise to direct linkage complexes. The observed inhibition may reflect only the extent of the amino terminus participation in direct linkage.

TABLE II. Chemical Stability of Direct Linkage Complexes*

Treatment	Radioactivity in direct linkage complexes (%)
None	2.1
0.1 N acetic acid	1.9
0.1 N NaOH	2.0
0.45 M hydroxylamine	1.8

*Membrane protein (36 μg) was mixed with 67 nM ^{125}I-EGF (1.1×10^9 cpm/nmole) in a volume of 1 ml of 10 mM sodium phosphate at pH 8.5 containing 0.3 M sucrose, and the samples were incubated for 1.5 h at 37°C. The membranes were collected by sedimentation and suspended in 1 ml of the same solution at pH 7.4. Duplicate 90 μl aliquots of this suspension were mixed with 10 μl of H_2O (as a control), 1 N acetic acid (to give a final pH of 3.5), 1 N NaOH (final pH was 11.5), or 4.5 M hydroxylamine at pH 7.0. After a 5 h incubation at 20°C, all samples were neutralized, mixed with twice-concentrated ESB, and subjected to electrophoresis. The amount of radioactivity present in direct linkage complexes was determined and is expressed as percentage of the total radioactivity applied to the gel.

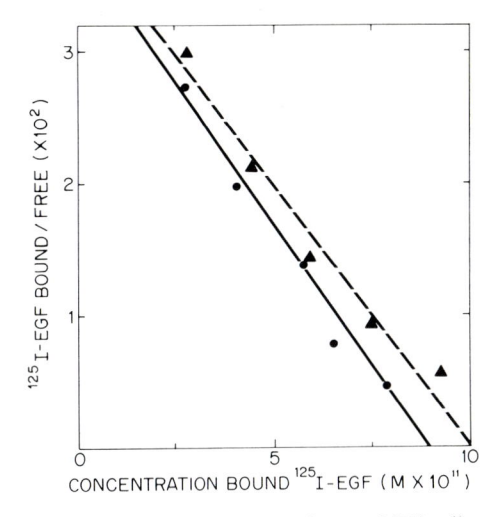

Fig. 4. Binding of native and biotinyl EGF. Duplicate cultures of 3T3 cells grown on 28 cm² plastic dishes were treated with ^{125}I-native or biotinyl EGF at 1.0, 2.1, 4.2, 8.3, and 16.7 nM, and nonspecific binding was determined in samples also incubated with 118 nM unlabeled EGF. The specific activity of native EGF was 1.68×10^9 cpm/nmole; biotinyl EGF, 1.43×10^9 cpm/nmole. After thorough washing, monolayers were solubilized in ESB, and cell-bound radioactivity was determined. Specific binding is plotted according to Scatchard [24]. ▲--▲ Specific binding of ^{125}I-native EGF. ●—● Specific binding of ^{125}I-biotinyl EGF.

TABLE III. Effects of Chemical Modification of EGF on EGF Binding and Direct Linkage Complex Formation*

	Specific binding (% of untreated EGF)	EGF in direct linkage complexes (% of specific binding)
None	100	1.33
Iodoacetate	127	1.08
N-hydroxysuccinimidyl biotin	109	0.53
N-hydroxysuccinimidyl biotin, followed by trypsin treatment	48	0.60
2-Mercaptoethanol	0	0

*EGF was subjected to chemical modification with several site-specific reagents (Experimental Procedures). The derivitives were radioiodinated by the chloramine T procedure to the following specific activities: iodoacetate-treated EGF, 2.4×10^9 cpm/nmole; 2-mercaptoethanol-treated EGF, 0.23×10^9 cpm/nmole; biotinyl EGF, 2.4×10^9 cpm/nmole; trypsin-treated, biotinyl EGF, 3.7×10^9 cpm/nmole. Native EGF was radioiodinated to a specific activity of 3.2×10^9 cpm/nmole. After radioiodination, each derivitive was incubated at 5.3 nM in DBH with 3T3 cells for 1 h at 37°C. The binding of each derivitive was measured on duplicate monolayers and corrected for the nonspecific binding measured in the presence of 167 nM unlabeled EGF. Binding is expressed as a percentage of the amount of iodinated native EGF bound per monolayer (27 fmoles). The amount of bound EGF present in the direct linkage complexes was then determined as described in Experimental Procedures.

Biological Properties of Native and Biotinyl EGF

The reduced ability of biotinyl EGF to form direct linkage complexes presented an opportunity to test the role of direct linkage in EGF-induced DNA synthesis. The same preparations of native and biotinyl EGF assayed for their abilities to form direct linkage complexes (Table IV) were compared for their abilities to stimulate the uptake of ^{3}H-thymidine into DNA (Fig. 5). In two separate experiments, the concentration dependence and both the half-maximal (Fig. 5, right) and maximal extent of stimulation (Fig. 5, left) were identical for native and biotinyl EGF. Since direct linkage was inhibited by over 60% without affecting the dose-response curve for EGF-induced stimulation of ^{3}H-thymidine into DNA, it is unlikely that it serves in a role causal to the biological activity of EGF. In another experiment, native and biotinyl EGF were tested for their abilities to stimulate EGF receptor down regulation at low physiological EGF concentrations. No differences were observed, indicating no involvement of direct linkage in triggering of EGF receptor down regulation (data not shown).

Does a Restricted Subclass of EGF Receptors Engage in Direct Linkage Complex Formation?

The low yield of direct linkage complex formation relative to EGF binding raised concern that a small subclass of EGF receptors, possibly one not representative of the total, participated in complex formation. This could severely limit the usefulness of direct linkage complex formation as an affinity labeling technique. The test for this possibility is described in Table V. If membranes are first incubated with unlabeled EGF or EGF derivatized with nonradioactive iodine, a small reactive subclass of EGF receptors might engage

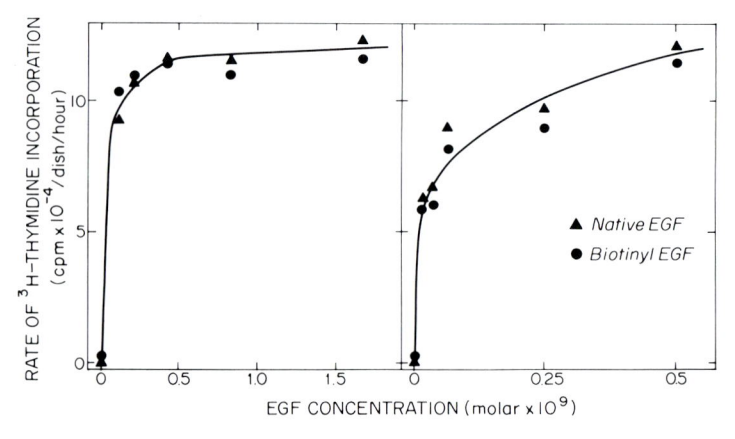

Fig. 5. Biological activity of native and biotinyl EGF. Quiescent cultures of 3T3 cells (prepared on 2 cm^2 plastic wells as described in Experimental Procedures) were incubated with 1 ml of DMEM plus 0.8% fetal calf serum containing native or biotinyl EGF at the indicated concentrations. After 18 h at 37°C, 10 μl of ^{3}H-thymidine (50 Ci/mmole, diluted to 0.1 mCi/ml with DMEM), was added to each culture. After an additional 2 h at 37°C, acid-precipitable radioactivity was measured by liquid scintillation counting as described in Experimental Procedures. Each point is the average of duplicate determinations corrected for the radioactivity incorporated in the absence of EGF (36,200 cpm/dish/h for the left panel and 108,300 cpm/dish/h for the right panel. In the experiment in the left panel, each well contained an average of 17 μg of total protein; protein in the experiment in the right panel was not determined. Left panel: rate of thymidine incorporation by native (▲) and biotinyl EGF (●) over a wide range of EGF concentrations. Right panel: rate of thymidine incorporation at low EGF concentrations.

in the formation of unlabeled direct linkage complexes, thereby reducing the amounts of labeled direct linkage complexes formed during a second incubation with ^{125}I-EGF. Alternatively, if the EGF receptors that engage in direct linkage complex formation are members of a larger class that is more representative of the total population, then the small amount of unlabeled complex formed after even an extended incubation period would not noticeably alter the number of direct linkage complexes formed later with ^{125}I-EGF. The data presented in Table V support the second alternative. Prior incubation of membranes with noniodinated EGF or with ^{127}I-EGF was ineffective in reducing subsequent direct linkage complex formation with ^{125}I-EGF.

TABLE IV. Direct Linkage Complex Formation by Native and Biotinyl EGF*

^{125}I-EGF (nM)	^{125}I-EGF in direct linkage complexes (moles × 10^{16}) Native	Biotinyl	% Reduction
1.04	4.88	1.78	63.5
2.08	12.3	3.56	71.1
4.17	12.5	5.88	53.0
8.3	12.1	4.87	59.8
16.7	17.9	5.77	67.8
		Mean	63.0
		SD	7.06

*Aliquots (100 µl) of each of the samples used to determine specific binding in Figure 10 were subjected to electrophoresis for the determination of the amount of bound EGF in direct linkage complexes.

TABLE V. Effects of Prior Incubation of A431 Membranes With Unlabeled EGF or ^{127}I-EGF on Direct Linkage Complex Formation*

Addition, first incubation (0–24 h)	Second incubation (24–48 h) ^{125}I-EGF bound (fmoles)	^{125}I-EGF in Direct linkage complexes (fmoles)
None	161	3.8
Unlabeled EGF	151	3.1
^{127}I-EGF	182	3.3

*Duplicate aliquots of membrane protein (72 µg, Procedure III) in PBS plus 0.5% NaN$_3$ were mixed with no EGF, 67 nM unlabeled EGF, or 67 nM ^{127}I-EGF (prepared as described [12], but using nonradioactive Na^{127}I instead of Na^{125}I), and incubated for the first 24 h period at 37°C. Membranes were collected by sedimentation, washed and suspended in PBS plus 0.5% NaN$_3$ containing 67 nM ^{125}I-EGF (3.2 × 10^9 c pm/nmole). Samples were then incubated for a second 24 h period at 37°C, washed again by sedimentation, and dissolved in ESB. Membrane-bound radioactivity was determined and corrected for the radioactivity bound in the presence of 5 µM unlabeled EGF (25% of the total). After electrophoresis, the radioactivity present as direct linkage complexes was determined as described in Experimental Procedures. During the first 24 h incubation, identical prepared samples specifically bound 223 fmoles of ^{125}I-EGF, of which 6.57 fmoles was in direct linkage complexes.

DISCUSSION

We have extended our previous observation that specific binding of radioiodinated EGF is accompanied by direct linkage of EGF to its receptors on intact cells and isolated membranes from a variety of sources [15]. Evidence that direct linkage complexes represent ^{125}I-EGF attached to its receptors can be summarized as follows: 1) The direct linkage complexes and the EGF–EGF receptor complex produced by photoaffinity labeling have identical electrophoretic behavior [12, 15]. 2) The complexes have biological and chemical properties expected of EGF–EGF receptor complexes [15] — eg, production of the same lysosomal processing products noted previously with photoaffinity labeled EGF receptors [13]. 3) Cell surface proteins having properties identical to those of the direct linkage complexes can be identified by surface-specific iodination of EGF-responsive cells [36]. 4) After A431 membranes have been solubilized with solutions containing Triton X-100, direct linkage complex forming activity cofractionates with ^{125}I-EGF binding activity during gel chromatography and velocity sedimentation [42].

The formation of direct linkage complexes in A431 membranes requires EGF binding; its dependence on ^{125}I-EGF concentration and susceptibility to inhibition by unlabeled EGF are similar to those of the binding reaction. Complex formation is markedly slower than binding. The direct linkage complex formed with intact 3T3 cells is precipitated from detergent solution by an antiserum to EGF [15]. These observations indicate that direct linkage involves attachment of a relatively intact EGF molecule to its receptor(s) during a reaction that occurs after binding.

The direct linkage complexes once formed are stable, and no test performed to date provides insight into the mechanism of formation. The complexes are not disrupted by boiling in SDS solutions containing 2-mercaptoethanol. Nor are they disrupted by extremes of pH or treatment with high concentrations of a strong nucleophile, hydroxylamine. Direct linkage complex formation is not affected by inhibitors of transglutaminase reactions [15] — eg, EDTA or exogenously added amines [37]. Specific interference with direct linkage was achieved by acylation of EGF with a derivative of biotin. Since the extent of inhibition of direct linkage (63%) is substantially less than the extent of EGF acylation (at least 80%), the observed inhibition may not be due to the modification of a specific amino acid residue involved in covalent bond formation. Since EGF and biotinyl EGF have identical dose-response curves for stimulation of DNA synthesis in 3T3 cells, direct linkage complex formation is not essential for this cellular response to EGF.

We have not completely excluded the possibility that direct linkage results from a chemically reactive EGF derivitive formed during the harsh oxidative conditions necessary for iodination. This possibility is not supported by the following evidence. First, ^{125}I-EGF forms the direct linkage complexes after several weeks of storage in aqueous solution. Second, direct linkage complexes were formed to a similar extent when the ^{125}I-EGF was prepared by a milder procedure than that used here [36]. Third, prior incubation of membranes with ^{127}I-EGF does not inhibit subsequent direct linkage complex formation with ^{125}I-EGF to a greater extent than does unlabeled EGF (Table V). Fourth, the ability of ^{125}I-EGF to form direct linkage complexes was not decreased by its prior incubation with membranes for periods of up to 24 h at 37°C (data not shown). We have made several attempts to use antiserum to EGF to precipitate direct linkage complexes from nonionic detergent extracts of metobolically labeled cells incubated with unlabeled EGF. In preliminary studies [15], a protein similar in size to the direct linkage complex was immune-precipitated. We subsequently became aware that EGF binding to the EGF receptor persists in detergent solution [42], even though at that time there were reports in the literature that EGF binding activity was inactivated by detergent treatment [14, 38].

There is precedent in the literature for the formation of a covalent complex between two reversibly bound proteins. Chymotrypsin forms significant amounts of covalent intermediates with its substrates, including the insulin B chain, during the initial burst of its catalytic activity [39]. Kunitz soybean trypsin inhibitor and trypsin also form a covalent adduct [40]. The latter finding is particularly noteworthy, as EGF has been shown to possess a distinct structural homology with another trypsin inhibitor known as pancreatic secretory trypsin inhibitor [41]. These complexes are not identical with the direct linkage complexes. They can be isolated only under extremely acidic conditions, and once isolated they are disrupted under alkaline conditions, to which the direct linkage complexes are impervious (Table IV) [40]. The low yield of direct linkage complex formation could indicate a vestigial function of the EGF molecule, or at least one unrelated to its mitogenic activity. Interestingly, the covalent complex formed between thrombin and its receptor on human fibroblasts [16] closely resembles a reported linkage between thrombin and antithrombin III, a prominent inhibitor of thrombin in serum [43].

Regardless of the mechanism of direct linkage complex formation or its biological significance, this process remains an exquisitely simple and powerful technique for EGF receptor affinity labeling. This has enabled us to make several observations concerning the transmembrane distribution of functional sites on EGF receptors [18].

ACKNOWLEDGMENTS

This work was supported by grant BC-79 from the American Cancer Society and in part by USPHS grant AM 25826-01 during the terminal stages. P.L. was recipient of a predoctoral USPHS-National Research Service Award in Tumor Cell Biology (CA 09056) and also received support from an American Cancer Society Institutional grant (IW-131) to the UCLA Jonsson Comprehensive Cancer Center.

We wish to thank Cindy Blifeld for invaluable assistance at the initial stage of this work, Pamela Billings for assistance in the development of a membrane preparative procedure, and Betty Handy for typing the manuscript.

REFERENCES

1. Carpenter G, Cohen S: Annu Rev Biochem 48:193–216.
2. Hollenberg MD, Cuatrecasas P: Proc Natl Acad Sci USA 70:2964–2968, 1973.
3. Carpenter G, Lembach KJ, Morrison MM, Cohen S: J Biol Chem 250:4297–4304, 1975.
4. Fabricant RM, DeLarco JE, Todaro GJ: Proc Natl Acad Sci USA 74:565–569, 1977.
5. Aharonov A, Pruss RM, Herschman HR: J Biol Chem 253:3970–3977, 1978.
6. O'Keefe E, Hollenberg MD, Cuatrecasas P: Arch Biochem Biophys 164:518–526, 1974.
7. Carpenter G, Cohen S: J Cell Biol 71:159–171, 1976.
8. Schlessinger J, Schechter Y, Willingham MC, Pastan I: Proc Natl Acad Sci USA 75:2659–2663, 1978.
9. Schechter Y, Schlessinger J, Jacobs S, Chang K, Cuatrecasas P: Proc Natl Acad Sci USA 75:2135–2139, 1978.
10. Haigler H, Ash JF, Singer SJ, Cohen S: Proc Natl Acad Sci USA 75:3317–3321, 1978.
11. Haigler HT, McKanna JA, Cohen S: J Cell Biol 81:382–395, 1979.

12. Das M, Miyakawa T, Fox CF, Pruss RM, Aharonov A, Herschman H: Proc Natl Acad Sci USA 74:2790–2794, 1977.
13. Das M, Fox CF: Proc Natl Acad Sci USA 75:2644–2648, 1978.
14. Hock RA, Nexo E, Hollenberg MD: Nature 277:403–405, 1979.
15. Linsley PS, Blifeld C, Wrann M, Fox CF: Nature 278:745–748, 1979.
16. Baker JB, Simmer RL, Glenn KC, Cunningham DD: Nature 278:743–745, 1979.
17. Schechter Y, Hernaez L, Cuatrecasas P: Proc Natl Acad Sci USA 75:5788–5791, 1978.
18. Linsley PS, Fox CF: J Supramol Struct 14:461–471, 1980.
19. Carpenter G, King L Jr, Cohen S: Nature 276:409–410, 1978.
20. Steck TL: In Fox CF (ed): "Membrane Molecular Biology." Stamford, Connecticut: Sinaeur Associates Inc, 1972, p 87.
21. Touster O, Aronson NN Jr, Dulaney JT, Hendrickson H: J Cell Biol 47:604–618, 1970.
22. Thom O, Powell AJ, Lloyd CW, Rees DA: Biochem J 168:187–194, 1977.
23. Savage CR Jr, Cohen S: J Biol Chem 247:7609–7611, 1972.
24. Scatchard G: Ann NY Acad Sci 51:660–672, 1949.
25. Laemmli VK: Nature 277:680–685, 1970.
26. Steck TL: J Cell Biol 62:1–19, 1974.
27. Wrann M, Linsley PS, Fox CF: FEBS Lett 104:415–419, 1979.
28. Heitzmann H, Richard FM: Proc Natl Acad Sci USA 71:3537–3541, 1974.
29. Anderson GW, Zimmerman JE, Callahan F: J Am Chem Soc 86:1839–1842, 1964.
30. Savage RC Jr, Inagami T, Cohen S: J Biol Chem 247:7612–7621, 1972.
31. Glazer AN, Delange RJ, Sigman DS: Modification of protein side chains: Group specific reagents. In Work TS, Work E (eds): "Chemical Modification of Proteins." New York: American Elsevier Publishing Co, Inc, 1975, pp 68–120.
32. Lowry OH, Rosebrough NJ, Farr AL, Randall RJ: J Biol Chem 193:265–275, 1951.
33. Markwell MAK, Haas SM, Bieber LL, Tolbert NE: Anal Biochem 87:206–210, 1978.
34. Bradford MM: Anal Biochem 72:248–254, 1976.
35. Morrison RT, Boyd RN: "Organic Chemistry." Boston: Allyn and Bacon, 1966, pp 675–678.
36. Wrann M, Fox CF: J Biol Chem 254:8083–8086, 1979.
37. Siefring GE, Apostol AB, Velasco PT, Lorand L: Biochemistry 17:2598–2604, 1978.
38. Sahyoun N, Hock RA, Hollenberg MD: Proc Natl Acad Sci USA 75:1675–1679, 1978.
39. Dahlqvist U, Wahlby S: Biochem Biophys Acta 391:410–414, 1975.
40. Huang J-S, Liener IE: Biochemistry 16:2474–2478, 1977.
41. Hunt LT, Barker WC, Dayhoff MO: Biochem Biophys Res Commun 60:1020–1028, 1974.
42. Linsley PS, Fox CF: J Supramol Struct 14:511–525, 1980.
43. Baker JB, Low DA, Simmer RL, Cunningham DD: Cell 21:36–45, 1980.

APPENDIX

TABLE A-I. Distribution of Membrane Markers Upon Subcellular Fractionation*

Fraction	Protein (mg)	^{125}I-EGF binding activity Total (pmoles)	Specific activity (pmoles/mg)	N-Acetyl-β-D-glucosaminidase activity Total (A_{400}/h)	Specific activity (A_{400}/h/mg)
Experiment I					
Whole cell	825	7331 (100)[a]	8.89	–	–
2,600g supernatant	224	3573 (48.7)	16.0	–	–
35,000g pellet	–	2321 (31.7)	–	–	–
Band					
0/12	7.2	513 (7.0)	71.2	–	–
12/25	6.0	267 (3.6)	44.5	–	–
25/35	3.2	107 (1.5)	33.4	–	–
Experiment II					
Whole cell	1040	9460	9.10	1040	1.0
Band					
0/12	3.6	297	82.5	1.14	0.32
12/25	3.2	184	57.5	2.08	0.65
25/35	3.2	143	44.7	1.28	0.40

*Membranes were prepared by Procedure II (Experimental Procedures) from A431 cells grown in roller culture on approximately 10,000 cm^2 surface area. The indicated fractions were assayed for membrane marker activities in order to assess purification. EGF binding was measured at an ^{125}I-EGF concentration of 25 nM (1.7 × 10^9 cpm/nmole). The lysosomal marker, N-acetyl-β-D-glucosaminidase, was assayed by a modification of a published procedure [21], using p-nitrophenyl N-acetyl-β-D-glucosaminide as a substrate. Triton X-100 was added to a final concentration of 0.5% (v:v) prior to assay to improve substrate accessibility. Color was developed by the addition of 1 ml of a 0.1 M Na$_2$CO$_3$ solution at pH 10.7, containing 0.1% SDS, and the increase in absorbance at 400 nm was recorded.

CCDD:A:299

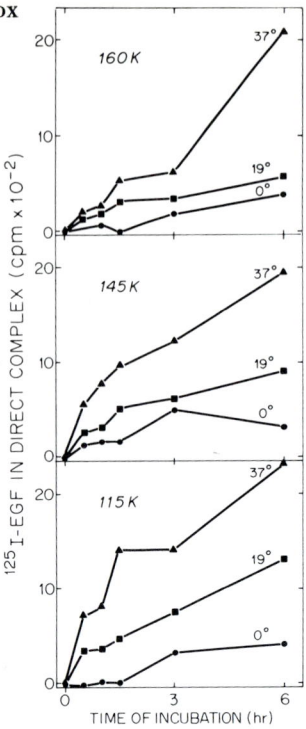

Fig. A-1. Time and temperature dependence of the formation of the individual direct linkage complexes. The time and temperature dependence of the formation of each individual direct linkage complex was determined in the experiment presented in Figure 2 of the text. Approximately 21% of the radioactivity in direct linkage complexes was present as the MW = 160,000 component, 36% in the MW = 145,000 component, and 43% in a MW = 115,000 component. (●——●) Direct linkage complexes were formed at 0°; (■——■), 19°; (▲——▲), or 37°. Top panel: Formation of the MW = 160,000 component of the direct linkage complex. Middle panel: Formation of the MW = 145,000 component of the direct linkage complex. Bottom panel: Formation of the MW = 115,000 component of the direct linkage complex.

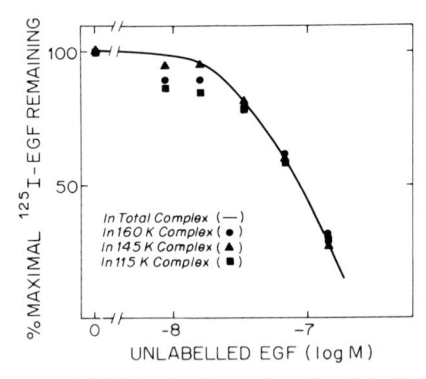

Fig. A-2. Competition by unlabeled EGF for the formation of each individual direct linkage complex. The ability of unlabeled EGF to compete for the direct linkage of ^{125}I-EGF to each individual proteolytic product of the EGF receptor was determined for the experiment presented in Figure 3 of the text. ^{125}I-EGF was present at a half-saturating concentration. Of the total radioactivity present as direct linkage complexes, 25% was in the form of a MW = 160,000 component, 35% was present as a MW = 145,000 component, and 39% was present as a MW = 115,000 component. —— Competition for total ^{125}I-EGF in all forms of the direct linkage complex (0.785 fmoles total). ●——● Competition for ^{125}I-EGF in the MW = 160,000 direct linkage complex. ▲——▲ Competition for ^{125}I-EGF in the MW = 145,000 direct linkage complex. ■——■ Competition for ^{125}I-EGF in the MW = 115,000 direct linkage complex.

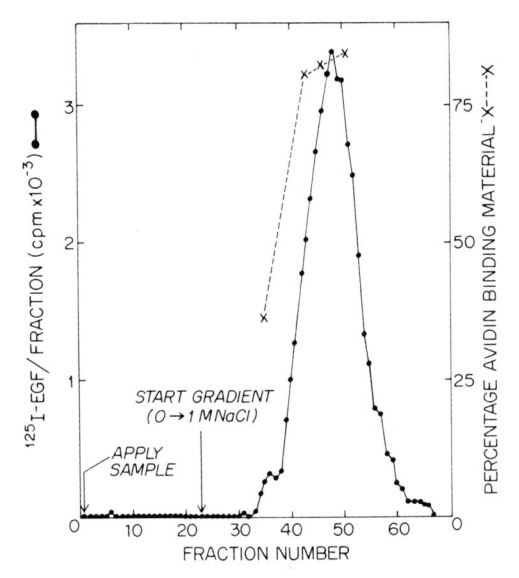

Fig. A-3. Fractionation of biotinyl EGF by analytical ion-exchange chromatography. Unlabeled EGF (85 nmoles in 1.2 ml of an aqueous 0.5 M Na_2CO_3 at pH 9.0, containing 0.1% Triton X-100 and 0.3% NaN_3) was mixed with a trace quanity of ^{125}I-EGF to achieve a final specific activity of 9.0×10^5 CPM/μmole. EGF was modified with NHS biotin as described in Experimental Procedures. Unreacted NHS biotin was first removed by chromatography on Sephadex G-25 equilibrated with 0.02 M Tris·HCl at pH 7.1, and the EGF was then applied to DEAE cellulose for separation of unreacted EGF from acylated EGF [10]. One milliliter fractions were collected, and EGF-containing fractions were detected by monitoring the effluent for radioactivity. Total recovery of EGF from the column was determined to be 86%. Individually iodinated aliquots of the indicated fractions were assayed for avidin binding as described in Experimental Procedures. Only a small amount of material was eluted in the position characteristic of native EGF (fractions 33–40).

Controlled Proteolysis of EGF Receptors: Evidence for Transmembrane Distribution of the EGF Binding and Phosphate Acceptor Sites

Peter S. Linsley and C. Fred Fox

Department of Microbiology and Parvin Cancer Research Laboratories, Molecular Biology Institute, University of California, Los Angeles, California 90024

A small quantity of the ^{125}I-EGF (epidermal growth factor) bound specifically to EGF receptors on the human epidermoid carcinoma cell line A431 associates covalently. The direct linkage complex formed migrates during gel electrophoresis as a single diffuse band of MW = 160,000–170,000. In contrast, direct linkage complexes of 160,000, 145,000, and 115,000 daltons are formed when EGF is incubated with membranes isolated from these cells; these arise from EGF receptor modification during membrane isolation. None of these modifications affected the affinity of the EGF binding site for ^{125}I-EGF.

The electrophoretic mobilities of the MW = 160,000 and 145,000 direct linkage complexes were similar to those of the major ^{32}Pi-labeled products of the EGF-stimulated phosphorylation reaction described by Carpenter et al [Nature 276:409–410, 1978], indicating that proteolytic fragments of EGF receptors are the major phosphate acceptors in this reaction. EGF receptors on intact A431 cells accepted phosphate effectively from γ-^{32}Pi-ATP only when the cells were permeabilized with lysolecithin. This shows that the EGF binding and phosphate acceptor sites lie on opposing faces of the membrane. When the 145,000 dalton form of receptor is labeled with EGF or ^{32}Pi and the labeled peptides subjected to tryptic hydrolysis under identical conditions, all phosphate is lost from high molecular weight products under conditions where the EGF-receptor covalent complex is converted largely to a 115,000 dalton form. This suggests that the phosphate acceptor site lies on the cytoplasmic side of the membrane on a region of receptor extending 30,000 daltons from the 115,000 dalton fragment containing the EGF binding site.

Key words: protein phosphorylation, permeabilized cells, EGF receptors – transmembrane distribution, fragmentation by trypsin, phosphate acceptor site

Epidermal growth factor (EGF) initiates its action by binding to specific cell surface receptors on responsive cells [1]. Cell-bound EGF is internalized in endocytic vesicles, which ultimately fuse with lysosomes [2], where it is proteolytically degraded into its component

---

Abbreviations: EGF, epidermal growth factor; DMEM, Dulbecco's modified Eagle medium; DBH, DMEM minus NaHCO$_3$, containing 10 mM HEPES at pH 7.4 and 0.1% bovine serum albumin; ST buffer, 0.25 M sucrose in 10 mM Tris · HCl at pH 7.4; BSA, bovine serum albumin; ESB, electrophoresis sample buffer (3% sodium dodecyl sulfate and 10% glycerol in 63 mM Tris · HCl at pH 6.8); MW, molecular weight (M$_r$); buffer A, 0.15 M sucrose, 0.08 M KCl, 5 mM potassium phosphate, 5 mM MgCl$_2$, and 0.5 mM CaCl$_2$ in 35 mM HEPES at pH 7.4.

Received August 25, 1980, accepted October 3, 1980.

0091-7419/80/1404-0461$03.50 © 1980 Alan R. Liss, Inc.

amino acids [3]. Since a loss of cellular binding capacity for EGF (known as EGF-induced receptor down regulation) accompanies the internalization and degradation of EGF, cell surface receptors for EGF were presumed to be processed similarly [3]. Studies performed in this laboratory to determine the metabolic fate of a photoaffinity crosslinked ^{125}I-EGF–EGF receptor complex demonstrated that EGF receptor down regulation proceeds by receptor internalization and degradation by lysosomal proteases [4, 5]. The identical EGF dose responses for EGF receptor down regulation and induction of DNA synthesis led to the hypothesis that proteolytic processing of EGF or its receptor is an obligatory step in the mechanism of hormone action [6].

Recently, we [7, 8] and Baker et al [9] observed that a small portion of the ^{125}I-EGF specifically bound to its receptor on a variety of cell types undergoes a spontaneous and irreversible "direct linkage" to its receptors. Because of the potential utility of direct linkage as an affinity labeling technique for EGF receptors, we conducted a detailed investigation of the process. The complexes formed with isolated membranes were generally of lower molecular weight and more heterogeneous than the complexes formed with intact cells [10]. This showed that EGF receptors are proteolytically processed during membrane isolation. Here we have exploited these EGF–EGF receptor complexes to demonstrate that a phosphate acceptor site on a segment of the EGF receptor has a transmembrane distribution from the EGF binding site.

EXPERIMENTAL PROCEDURES

Materials

Trypsin (type XII: 2X crystallized) and soybean trypsin inhibitor (type II S: 1 mg inhibits 0.9 mg trypsin) were from Sigma. TPCK-treated trypsin (254 units/mg) was obtained from Worthington. γ-^{32}Pi-ATP was obtained from Amersham. Sources of other materials are indicated elsewhere [10].

Methods

Procedures for cell culture, membrane isolation, EGF preparation and iodination, gel electrophoresis, EGF binding, and direct linkage complex formation are described elsewhere [10]. EGF-stimulated phosphorylation of membrane proteins was accomplished by the procedure of Carpenter et al [11, 12].

RESULTS

Comparison of the EGF–EGF Receptor Direct Linkage Complexes Formed With Intact A431 Cells and Isolated Membranes

The direct linkage complex formed with intact A431 cells migrated as a diffuse band of MW = 160,000–170,000 (Fig. 1, lane A), and this complex was not readily degraded to specific products by trypsin added either before (lane B) or after (lane C) direct linkage complex formation. When A431 cells were first scraped from their substratum, and then incubated with ^{125}I-EGF, they did not give rise to the diffuse MW = 160,000–170,000 band in gel electrophoresis. Instead, they yielded direct linkage complexes of MW = 160,000, 145,000, and 115,000 (Figs. 1 and 2).

The MW = 145,000 complex formed with isolated membranes (lane D) was also the major radiolabeled product visible on gels when cells labeled by direct linkage were scraped from their substratum in ST buffer at 4°C 2 min prior to their solubilization (data not

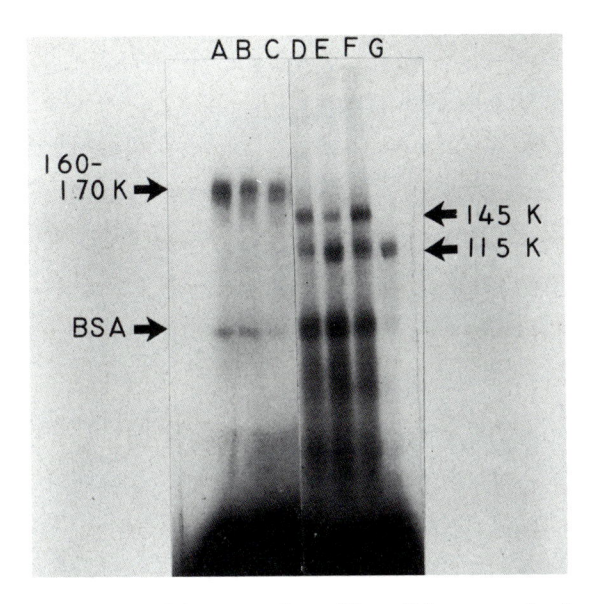

Fig. 1. Characterization of the direct linkage complexes. Direct linkage complex formation with intact cells. Intact monolayers of A431 cells in 28 cm^2 plastic dishes were incubated at 20°C for 1 h with 33.3 nM ^{125}I-EGF (2.6 × 10^9 cpm/nmole) to form direct linkage complexes between EGF and receptor. Where indicated, cultures were treated before or after ^{125}I-EGF binding for 30 min at 4°C with 5 ml of a solution containing 0.25 mg/ml of trypsin in DMEM. The cultures were washed thoroughly after trypsin treatment. The culture that was treated with trypsin before labeling was incubated for 15 min at 4°C with 2.5 mg/ml of soybean trypsin inhibitor in DMEM and washed again prior to the initiation of the binding reaction. Each culture was solubilized in ESB as described in Experimental Procedures. An 80 μl aliquot was subjected to electrophoresis.

Direct linkage complex formation with membranes. Membranes (6 μg of protein, prepared according to Procedure II [10]) were suspended in 0.1 ml of DMEM containing, where indicated, 25 μg of trypsin and 250 μg of soybean trypsin inhibitor. The mixtures were incubated at 4°C for 30 min and sedimented for 15 min at 12,800g. The pellets were washed with DMEM. Each membrane pellet was suspended in 0.1 ml of DBH containing 2.25 mg/ml of soybean trypsin inhibitor and 130 nM ^{125}I-EGF. After a 1 h incubation at 20°C, the membranes were collected by sedimentation and the pellets were suspended in 0.1 ml of DMEM with or without 0.25 mg/ml of trypsin. After a 30 min incubation at 4°C, the membranes were collected by sedimentation and dissolved in ESB, and the proteins were subjected to electrophoresis. All samples were resolved on the same acrylamide slab. Intervening lanes were cut from the photograph of the autoradiogram. The radioactivity in the 68,000 dalton region is a complex between ^{125}I-EGF and BSA; its formation was not inhibited by unlabeled EGF and was not investigated further.

Lane A: direct linkage complex formed in intact A431 cells; lane B: as in A, except that the culture was treated with trypsin prior to ^{125}I-EGF binding; lane C: as in A, except that culture was treated with trypsin after ^{125}I-EGF binding.

Lane D: direct linkage complexes formed with isolated membranes (experimental conditions were analogous to those in lane A); lane E: as in D, with trypsin treatment prior to ^{125}I-EGF binding; lane F: as in E, except that trypsin was mixed with soybean trypsin inhibitor prior to its addition to membranes; lane G: as in D, except that trypsin treatment took place after ^{125}I-EGF binding.

shown). This finding shows that the EGF receptors were degraded during or shortly after the scraping procedure. The activity(ies) responsible for the production of a lower molecular weight, MW = 145,000 form of the direct linkage complex was not inhibited by inclusion of 10% fetal calf serum in the isotonic buffer solution into which the cells were scraped from the tissue culture dish. The physiological role of this uncharacterized proteolytic

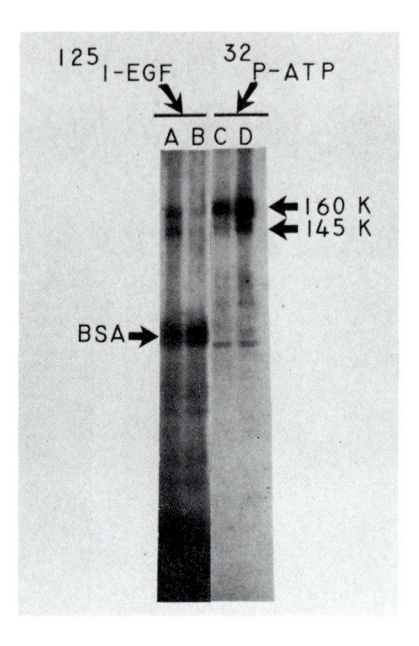

Fig. 2. Comparison of phosphorylated membrane proteins with the direct linkage complexes. Direct linkage complex formation was assayed using 0.1 ml reaction mixtures containing 8.75 μg of membrane protein, 139 nM ^{125}I-EGF (1.6 × 10^9 cpm/nmole), and where indicated, 2.38 μM unlabeled EGF. After a 90 min incubation at 37°C, the membranes were collected by sedimentation and washed once. The final membrane pellets were dissolved in 0.1 ml of ESB, and 40 μl of each sample was subjected to electrophoresis.

The membrane proteins labeled by the rapid, EGF stimulated phosphorylation described by Carpenter et al [11] were compared with those that link directly to ^{125}I-EGF. Membrane protein (Preparation III [10], 28 μg) was suspended in 38 μl of a solution containing 2.8 μmoles of HEPES, pH 7.4, 0.28 μmoles of MnCl, 19 μg of BSA, and 8.3 pmoles of unlabeled EGF, where indicated. The samples were incubated for 10 min at 0°C prior to the addition of 12 μCi of γ-^{32}Pi-ATP (1 mCi/ml, 3,000 Ci/mmole). After a 2.25 min incubation at 0°C, the reactions were terminated by the addition of an equal volume of twice concentrated ESB. Forty microliters of each sample was subjected to electrophoresis.

Since the phosphorylated and iodinated samples required different exposure times to achieve comparable grain densities on x-ray film, the dried gel was cut in half; each isotope was exposed separately and a final composite figure was derived from photographs of each half. The amount of radioactivity present in the MW = 145,000 and 160,000 regions of each lane was determined as described in Experimental Procedures for ^{125}I or by Cerenkov counting for ^{32}Pi. Of the ^{125}I-EGF present as direct linkage complexes, 63% was present as the MW = 160,000 dalton product. Of the total ^{32}Pi in the MW = 160,000 and 145,000 regions, 75% was present as the MW = 160,000 product. EGF stimulated by 2.6-fold the ^{32}Pi incorporated in the MW = 160,000 region and by 2.2-fold that incorporated in the MW = 145,000 region.

Direct linkage, lane A: formation of direct linkage complexes (MW = 160,000 and 145,000). – A complex between ^{125}I-EGF and BSA was formed in this experiment; unlabeled EGF did not block formation of this complex; lane B: as in A, plus unlabeled EGF.

Phosphorylation, lane C: phosphorylation of membrane proteins in the absence of EGF; lane D: EGF-stimulated phosphorylation of membrane proteins; as in C, plus EGF.

activity is unclear, although the product of its action, the MW = 145,000 component, is sometimes detected in small amounts in populations of mitogenically competent murine cells [10]. For the purpose of this discussion we refer to this activity as "scraping protease."

The MW = 115,000 direct linkage complex is formed by further modification of higher molecular weight forms of EGF receptors or EGF–EGF receptor direct linkage complexes. The receptor portion of the 145,000 dalton complex is sensitive to trypsin added either before (lane E) or after (lane G) direct linkage complex formation; the direct linkage complex product is a radiolabeled MW = 115,000 fragment that may be identical to that which accumulates physiologically on cells incubated in the presence of chloroquine [5, 7]. The MW = 145,000 complex formed with isolated membranes was rapidly degraded at one-eighth the trypsin-receptor ratio that produced much less hydrolysis of the MW = 160,000–170,000 complex formed on intact cells (lanes B and C). This finding suggests that the protease-sensitive site that gives rise to the 115,000 dalton form of the EGF receptor direct linkage complex becomes exposed during membrane isolation. Protease treatment is widely used for determining the vectorial orientation of membrane proteins [13]. The insensitivity of EGF receptors to trypsin in intact cells relative to membranes provides evidence that a trypsin-sensitive portion of the receptor extends into the cytoplasm. The cleavage site for the scraping protease is also likely to be located on the cytoplasmic portion of the receptor, since broken cell preparations do not contain activities that degrade EGF receptors on intact cells (data not shown).

The endogenous proteases responsible for the lower molecular weight direct linkage complexes act prior to rather than after direct linkage complex formation. The time courses for formation of the MW = 160,000, 145,000, and 115,000 direct linkage complexes in membranes show that the corresponding EGF receptor precursors bind EGF and form direct linkage complexes independently [10]. The data in Figure 1 provide additional documentation for the retention of EGF binding activity by fragments of EGF receptors [5, 10]. The MW = 115,000 fragment is formed when membranes are treated with trypsin before EGF addition (lane E). Since a concentration of soybean trypsin inhibitor sufficient to inhibit the action of any residual trypsin (lane F) was present at the time of EGF addition, the component giving rise to the MW = 115,000 form of the direct linkage complex existed prior to EGF addition.

The K_d for ^{125}I-EGF binding was similar when measured with intact cells or with isolated membranes displaying varying amounts of the MW = 160,000, 145,000, and 115,000 forms of the direct linkage complex (Table I). The similar K_d values for receptors on intact cells or isolated membranes show that no major changes in the affinity for ^{125}I-EGF are produced by the action(s) of the endogenous protease activity that produces the MW = 145,000 component or that which produces the MW = 115,000 component.

Correlations in the Electrophoretic Behavior of Direct Linkage Complexes and Proteins Labeled by EGF Stimulated Phosphorylation

The electrophoretic behavior of direct linkage complexes formed with A431 membranes was compared with that of the proteins phosphorylated in the EGF-stimulated reaction described by Carpenter et al [11]. The proteins phosphorylated specifically in response to EGF addition migrate during gel electrophoresis with mobilities similar to those of the MW = 160,000 and 145,000 fragments of the direct linkage complex (Fig. 2). The slight differences in mobility between the fragments of the direct linkage complex and the phosphorylated proteins (Fig. 3) can be accounted for by the 6,000 daltons contributed by EGF. While co-migration is not a rigorous criterion for protein identity, the fact that two related fragments of the EGF–EGF receptor direct linkage complex each co-migrate with a protein that is phosphorylated in the EGF-stimulated reaction provides strong evidence that the EGF receptor is itself the major protein phosphorylated in this reaction.

Differential Trypsin Sensitivity of Radiophosphate-Labeled and Direct Labeled EGF Receptors

Two identical samples of membranes were first radiolabeled — one sample by direct linkage of ^{125}I-EGF to receptor, the second, by EGF-induced phosphorylation — and then

TABLE I. Effects of Endogenous EGF Receptor Proteolysis on ^{125}I-EGF Binding*

System	Distribution of radioactivity in direct linkage complexes products[a]	K_d
Intact cells		
Experiment 1	160 K (61%)	23 nM
	145 K (39%)	
Experiment 2	160 K (45%)	23 nM
	145 K (55%)	
Membranes		
Experiment 1	145 K[b]	28 nM
Experiment 2	160 K (23%)	29 nM
	145 K (41%)	
	115 K (37%)	

*^{125}I-EGF binding was determined at several EGF concentrations with intact cells and isolated membranes giving rise to differing amounts of the main direct linkage complexes. Binding to intact cells was compared using two separate populations of A431 cells obtained from G. Todaro. ^{125}I-EGF (3.5 × 10^9 cpm/nmole) was added at concentrations ranging from 0.8 nM–100 nM. Specific binding was determined as described in Experimental Procedures and plotted according to Scatchard [17]. The K_d of binding was determined from the slopes of the line best fitting the data. Identical cultures were incubated with 16.7 nM ^{125}I-EGF in order to form the direct linkage complexes. After electrophoresis, the relative amounts of radioactivity in the indicated regions of the gel were determined as described elsewhere [10].
[a]Measurements of binding and direct linkage complex formation with isolated membranes were performed in an analagous fashion, as described elsewhere [10].
[b]No data are available for the relative distribution of other direct linkage products in this experiment; the 145 K fragment was predominant.

treated with varying amounts of trypsin (Fig. 3). The radiophosphorylated protein migrating in the band corresponding to the 145,000 dalton form of the direct linkage complex was far more sensitive to tryptic digestion than were the direct linkage labeled receptors (lanes B and G). This could result from structural modifications of receptor caused by covalent coupling with phosphate or EGF. This difference also could be related to the finding that there are forms of the receptor having differential phosphate acceptor activities [14]; these forms could have inherently different susceptibilities to tryptic digestion.

The direct linkage labeled receptor yielded tryptic digestion products not observed with radiophosphate-labeled receptor. Digestion of the direct linkage complex first gives rise to the MW = 115,000 component, and then to a broad band of radiolabeled material centering on MW = 85,000 (lanes B and E). No products in either of these molecular weight ranges were observed following tryptic digestion of the radiophosphate-labeled protein corresponding to the 145,000 dalton direct linkage complex (lanes G–J). Several explanations can be advanced to account for this behavior. First, the phosphate may reside on a low molecular weight fragment of the protein produced by trypsin treatment. A phosphate-labeled band migrating at approximately 30,000 daltons was detected after trypsin treatment (lane G), but this could have arisen from the digestion of other phosphate-labeled, higher molecular weight components. Second, tryptic digestion of membranes might accelerate cleavage of the bond linking phosphate to receptor.

Inaccessability of EGF-Stimulated Protein Kinase in Intact A431 Cells to Externally Added γ-^{32}Pi-ATP

EGF receptors in intact A431 cells were tested for their ability to accept ^{32}Pi from γ-^{32}Pi-labeled ATP (Fig. 4, lanes A and B). Negligible radioactivity was incorporated migrating with bands at the electrophoretic mobility of EGF receptors. When the permeability of

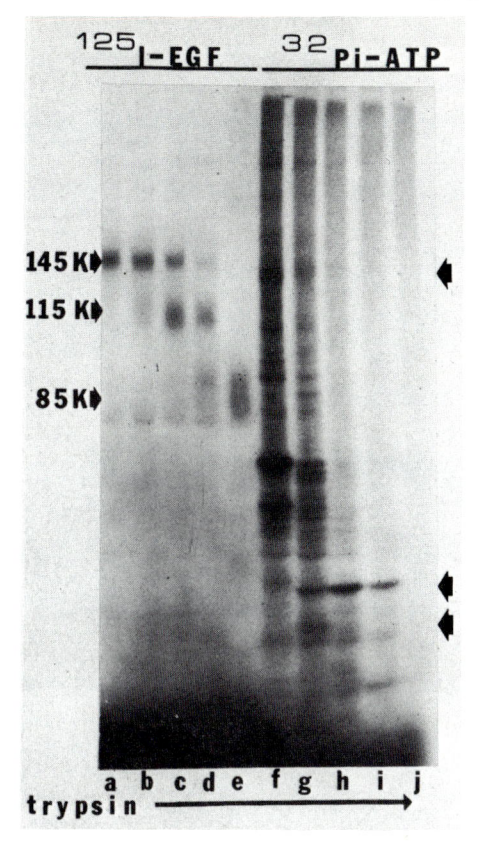

Fig. 3. Trypsin treatment of direct linkage complexes and phosphorylated membrane proteins. Two identical samples of A431 membranes (360 µg protein) were incubated for 1 h at 37°C with 104 nM labeled (1.2 × 10^9 cpm/nmole) or unlabeled EGF in a final volume of 0.1 ml of ST buffer. Both mixtures were then diluted with an equal volume of a solution of 0.1 M HEPES at pH 7.4, containing 0.4 M NaCl, 10 mM MnCl, and 0.05% BSA. The sample containing unlabeled EGF was mixed with 100 µCi of γ-^{32}Pi-labeled ATP (1 mCi/ml, 3,000 Ci/mmole); the sample containing ^{125}I-EGF was treated with H$_2$O as a control. The phosphorylation reaction proceeded for 1.5 min on ice. EDTA was then added to both samples to a final concentration of 13 nM. The membranes were washed by sedimentation through a sucrose shelf as described elsewhere [10], washed with DMEM, and suspended in 110 µl of DMEM. Identical 20 µl aliquots of both the direct linkage labeled (lanes A–E) and radiophosphate-labeled (lanes F–J) membranes were then mixed with 5 µl of TPCK treated trypsin in DMEM. After an additional 30 min at 0°C, each sample was mixed with an equal volume of twice-concentrated ESB. Aliquots (20 µl) of each sample were then subjected to electrophoresis. The arrows on the left-hand side of the figure denote the positions of the MW = 145,000 direct linkage complex and its major tryptic products. The uppermost arrow on the right-hand side denotes the position of the major protein phosphorylated in an EGF-induced reaction (MW = 145,000). The bottom two arrows denote low molecular weight labeled proteins which accumulate during trypsin treatment (MW = 30,000–40,000); the appearance of the lower of these follows most closely the loss of radioactivity from the MW = 145,000 ^{32}Pi-labeled species. Lanes A, F: no trypsin; lanes B, G: 0.0125 µg trypsin; lanes C, H: 0.125 µg trypsin; lanes D, I: 1.25 µg trypsin; lanes E, J: 12.5 µg trypsin.

cells to radiolabeled nucleotide was enhanced by mild lysolecithin treatment [15], EGF-stimulated incorporation of ^{32}Pi into components in the molecular weight range of EGF receptors was increased dramatically (lanes C and D). The ability of these components to ac-

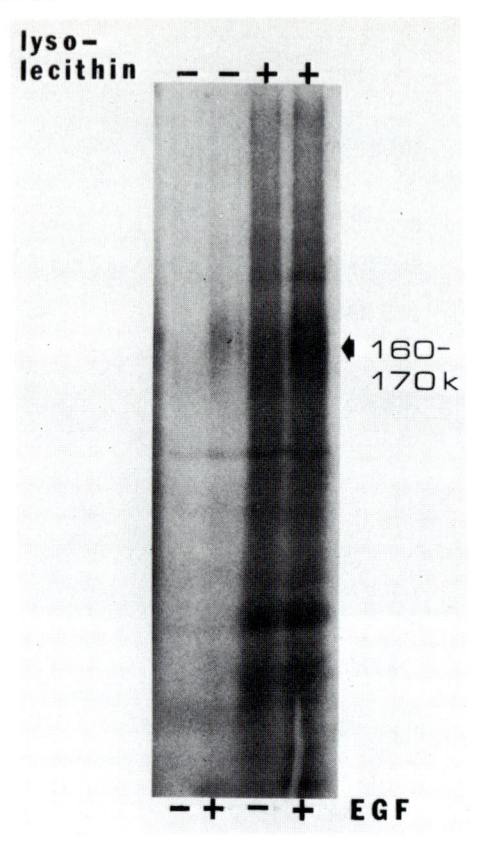

Fig. 4. Phosphorylation of EGF receptors in intact A431 cells. Identical 28 cm² monolayer cultures of A431 cells were treated with (lanes B and D) or without (lanes A and C) 67 nM unlabeled EGF in DBH. After incubation for 1 h at 23°C, the cells were washed twice with ice-cold buffer A [15], and two cultures (lanes C and D) were permeabilized as described by Miller et al [15]. After the addition of 0.1 mCi/well of γ-^{32}Pi-ATP (1 mCi/ml, 3,000 Ci/mmole), incubation was continued at 4°C for an additional 15 min. All cultures were then solubilized in ESB, and 20 μl aliquots were subjected to electrophoresis [10].

cept phosphate from ATP after disruption of the cellular permeability barrier is consistent with the localization of the phosphate acceptor site on a region of receptor not exposed to the extracellular space.

DISCUSSION

Direct linkage complex formation was exploited to make two observations on the structural properties of EGF receptors: 1) The EGF receptor is highly susceptible to proteolytic digestion during membrane isolation. Some proteolytic fragments observed with isolated membranes are also present in populations of mitogenically stimulated cells [16]; the ability of defined fragments of the EGF receptor to interact with EGF should be considered in the assessment of the proposed role of receptor processing during mitogenic activation [6]. 2) Two molecular weight forms of the EGF receptor comigrate during gel electrophoresis with the major proteins phosphorylated in the rapid, EGF-stimulated reaction described by Carpenter et al [11]. This indicates that the EGF receptor is itself the major phosphate acceptor in this reaction.

TREATMENT		DIRECT LINKAGE COMPLEX
NONE	Inside / Outside, Pi, (A431 only), EGF	160K (160-170K)
ENDOGENOUS "SCRAPING" PROTEASE	10K, Pi, 15K + membrane, EGF	145K
ENDOGENOUS PROTEASE OR TRYPSIN	Pi, 30K + membrane, EGF	115K

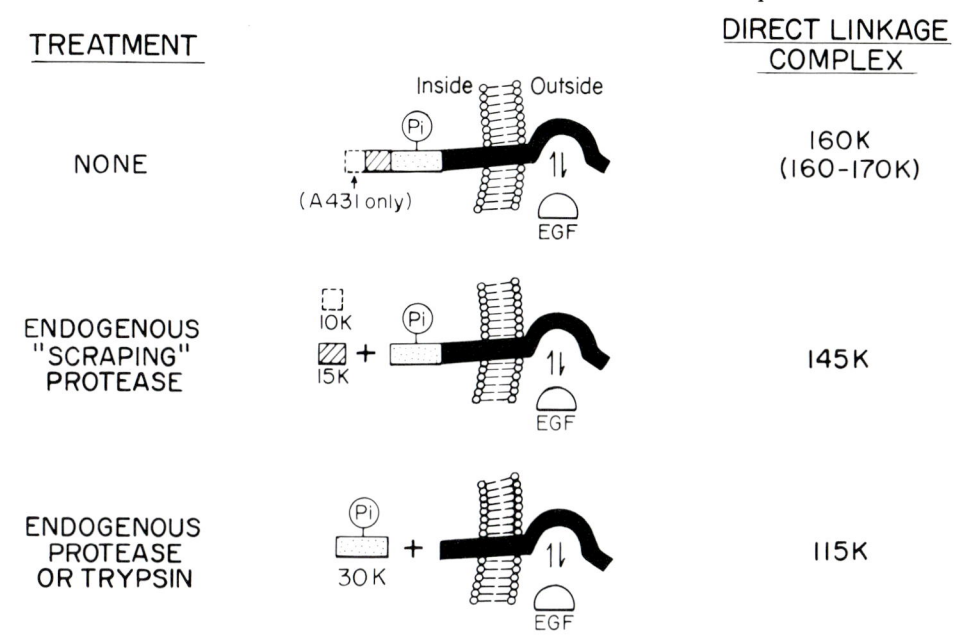

Fig. 5. A structural model of the EGF receptor. The data presented in Figure 1 and Table I demonstrate the sequential, endogenous proteolytic modifications of EGF receptors which do not appreciably affect their ability to bind ^{125}I-EGF. The EGF receptors in intact A431 cells form a direct linkage complex which migrates in gel electrophoresis as a diffuse band of MW = 160,000–170,000. This diffuse band migrates slightly behind the much sharper band of the MW = 160,000 complex formed with intact HF-15 [9] or 3T3 cells [P.S. Linsley and C.F. Fox, unpublished data] or that formed with isolated membranes from A431 cells (Fig. 2). A MW = 145,000 component of the direct linkage complex is formed by the action of an endogenous "scraping protease" activated when cells are scraped from their substratum. This is the prominent form of receptor in membranes isolated by the procedures employed. In some membrane preparations a direct linkage complex of MW = 115,000 is formed. Its electrophoretic behavior is identical to that of the complex generated by treatment of isolated membranes (but not intact cells) with trypsin. Both the MW = 145,000 and 115,000 direct linkage complexes remain membrane bound.

Since the MW = 145,000 component of the direct linkage complex comigrates during gel electrophoresis with a form of the EGF receptor that is phosphorylated in response to EGF addition (Fig. 2), the site of phosphorylation is not removed by the scraping protease. The ^{32}Pi incorporated in both the MW = 160,000 and 145,000 forms of the receptor in response to EGF addition is completely removed when membranes are treated with trypsin under conditions that lead to the formation of the MW = 115,000 form of the direct linkage complex (Fig. 3). The phosphorylated site on the receptor is therefore likely to be located on a MW = 30,000 region released by trypsin treatment.

The EGF receptor in intact A431 cells is relatively insensitive to exogenously added trypsin. Broken cell preparations also are ineffective in cleaving the MW = 160,000–170,000 direct linkage complex formed on intact cell monolayers (data not shown). The MW = 145,000 component of the direct linkage complex present in isolated membranes is converted totally to the MW = 115,000 component by concentrations of trypsin that do not degrade the MW = 160,000–170,000 complex present on intact cells to discrete products (Fig. 1 and data not shown). The cleavage sites for both the scraping protease and trypsin are therefore most likely displayed on the cytoplasmic side of the membrane.

A hypothetical structural model of EGF receptors based on our observations is presented in Figure 5. This model identifies three distinct sites exposed as the permeability barrier of the cell is disrupted. Two of these are protease-sensitive sites (see Fig. 1), and the third site accepts phosphate from ATP in a reaction stimulated by EGF addition (see Figs 2 and 4).

One of the protease-sensitive sites gives rise to a 145,000 dalton direct linkage complex of EGF and receptor; the second, to a 115,000 dalton form. The text of the legend of Figure 5 presents the rationale for the presence of the phosphate acceptor site on the region of receptor that is lost when the MW = 145,000 ^{125}I-EGF–receptor complex is cleaved to yield the 115,000 dalton complex.

Cohen et al have observed a gel electrophoretic doublet in the 150,000–170,000 MW range as the major products of EGF-induced phosphorylation in detergent extracts of A431 cells; they concluded that this is likely to be the EGF receptor [18]. These bands probably are analogous to the 145,000 and 160,000 dalton direct linkage ^{125}I-EGF–receptor complexes and the corresponding products that we observe after EGF-induced radiophosphorylation (Fig. 4). Additionally, the major product of EGF-induced radiophosphorylation in lysolecithin "permeabilized" cells is in the 160,000–170,000 dalton range (Fig. 4). We conclude that the higher molecular weight form of the direct linkage EGF receptor complex and its radiophosphate-labeled counterpart is the intact receptor, and that the lower molecular weight forms are artifacts produced by endogenous protease activity. The endogenous activities released during membrane isolation degrade receptor to fragments that accept ^{125}I-EGF to produce 145,000 and 115,000 dalton labeled complexes. Treatment of the 145,000 dalton ^{125}I-EGF–receptor complex with trypsin also gives rise to a 115,000 dalton fragment, indicating a receptor region that contains a site or sites sensitive to hydrolysis by either trypsin or endogenous protease activity.

Proteolytic digestion of EGF receptors labeled by direct linkage with ^{125}I-EGF or by phosphorylation with γ-^{32}Pi-ATP proceeds at different rates. These different rates of digestion could result from alterations in receptor conformation induced by covalent addition of EGF, phosphate, or both. Alternatively, the root of these differences in rate of digestion by protease may lie in the observation that the nonionic detergent-soluble and -insoluble forms of receptor can be labeled by direct linkage with ^{125}I-EGF, whereas the detergent-solubilized receptors are readily labeled by EGF-induced phosphorylation and the detergent-insoluble receptors are not [14]. The direct linkage labeled receptors may therefore consist of a mixed population, some of which are readily sensitive to proteolytic digestion and some of which are not.

When A431 membranes are first radiophosphate labeled in the presence of EGF and then treated with trypsin, products of lower molecular weight arise as the phosphate-labeled receptor bands disappear. One of these products has a molecular weight of approximately 30,000. This might be the fragment released during the stepwise degradation of the 145,000 dalton direct linkage labeled receptor to the 115,000 dalton receptor fragment. It is important to establish with certainty whether the fragments of phosphorylated receptor released by proteolysis retain phosphate following their formation. If so, phosphorylated fragments of receptor are attractive candidates for carriers of second messenger activity in EGF action [6].

ACKNOWLEDGMENTS

This work was supported by grant BC-79 from the American Cancer Society and in part by USPHS grant AM 25826-01 during the terminal stages. P. L. was recipient of a predoctoral USPHS-National Research Service Award in Tumor Cell Biology (CA 09056) and also received support from an American Cancer Society Institutional Grant (IW-131) to the UCLA Jonsson Comprehensive Cancer Center.

We wish to thank Terry Lipari and Steve Ellis for invaluable technical assistance, and Betty Handy for typing the manuscript.

REFERENCES

1. Carpenter G, Cohen S: Annu Rev Biochem 48:193–216, 1979.
2. Haigler HT, McKanna JA, Cohen S: J Cell Biol 81:382–395, 1979.
3. Carpenter G, Cohen S: J Cell Biol 71:159–171, 1976.
4. Das M, Miyakawa T, Fox CF, Pruss RM, Aharonov A, Herschman H: Proc Natl Acad Sci USA 74: 2790–2794, 1977.
5. Das M, Fox CF: Proc Natl Acad Sci USA 75:2644–2648, 1978.
6. Fox CF, Das M: J Supramol Struct 10:199–214, 1979.
7. Linsley PS, Blifeld C, Wrann M, Fox CF: Nature 278:745–748, 1979.
8. Linsley PS, Das M, Fox CF: Affinity labeling hormone receptors and other ligand binding proteins. In Greaves M, Cuatrecasas P (eds): "Receptors and Recognition." London: Chapman and Hall (in press).
9. Baker JB, Simmer RL, Glenn KC, Cunningham DD: Nature 278:743–745, 1979.
10. Linsley PS, Fox CF: J Supramol Struct 14: 511–525, 1980.
11. Carpenter G, King L Jr, Cohen S: Nature 276:409–410, 1978.
12. Carpenter G, King L Jr, Cohen S: J Biol Chem 254:4884–4891, 1979.
13. Steck TL: J Cell Biol 62:1–19, 1974.
14. Linsley PS, Fox CF: In Middlebrook J, Kohn L (eds): "Receptor Mediated Internalization of Hormones and Toxins." New York: Random House, 1980.
15. Miller MR, Castellot JL, Pardee AB: Biochemistry 17:1073–1080, 1978.
16. Linsley PS, Fox CF: J Supramol Struct 14: 441–459, 1980.
17. Scatchard G: Ann NY Acad Sci 51:660–672, 1949.
18. Cohen S, Carpenter G, King L Jr: J Biol Chem 255:4835–4842, 1980.

Selective Protein Transport: Identity of the Solubilized Phosvitin Receptor From Chicken Oocytes

John W. Woods and Thomas F. Roth

Department of Biological Sciences, University of Maryland Baltimore County (UMBC), Catonsville, Maryland 21228

By two independent methods, the solubilized receptor for phosvitin (PV) has a subunit MW of 116K. Affinity chromatography, showed that only 2 of the more than 25 proteins present in the total detergent solubilized oocyte membrane extract were retained on a PV–agarose column. These proteins of MW of 116K and 100K could be eluted from PV–agarose with free PV. By gel exclusion chromatography, the receptor-^{125}I-PV complexes elute in the void volume of a Biogel A-1.5 column. When these void fractions were assayed by SDS-PAGE only a single protein of MW of 116K was observed in addition to ^{125}I-PV.

Key words: protein transport, phosvitin, receptor, coated vesicles

Selective protein transport, mediated by specific receptors in association with coated pits and coated vesicles is a fundamental cellular process. The vital importance of this transport process is particularly manifest during reproduction. The selective transport of maternal immunoglobulins into the offspring provides the newborn with a passive maternally derived immunity until it becomes immunocompetent. Well studied examples include the illial cells of the rat that sequester maternal IgG from the mother's milk and release it into the neonatal circulation [1, 2]. A different tissue mediates a similar function in the rabbit. In the rabbit, IgG is transported into the developing fetus via the cells of the yolk sac splanchnopleure [3, 4]. In the chicken, IgG crosses into the oocyte via a receptor-mediated mechanism [5, 6].

In addition to IgG, the specific transport of other maternal proteins is also essential for successful reproduction. A particularly well studied example is the selective uptake of vitellogenin via coated vesicles into the developing oocytes of all oviparous animals. In oviparous animals vitellogenin is stored in the oocyte until it is degraded during embryogenesis to provide nutrients for the developing embryo. Well documented studies on the uptake of vitellogenin have been carried out in the mosquito [5, 7], saturnid moths [8, 9], the amphibian Xenopus laevis [10, 11], and the domestic chicken [12, 13].

Receptor-mediated, protein transport has also been well documented in systems unrelated to species reproduction. Low density lipoprotein, a serum cholesterol carrier, appears

Received May 30, 1980; accepted October 16, 1980.

0091-7419/80/1404-0473$03.00 © 1980 Alan R. Liss, Inc.

to enter fibroblasts exclusively via a receptor-coated vesicle associated system [14, 15]. Epidermal growth factor [16] and α_2-macroglobulin [17] also appear to be sequestered by a similar mechanism.

The developing chicken oocyte provides an excellent model system in which to study specific protein transport. During the final stages of maturation up to 1 gm of protein per day is transported into the developing oocyte. Kinetic studies in our laboratory have demonstrated the existence of specific receptors for vitellogenin [13], PV [18, 19], IgG [6], LDL and VLDL [20], in association with the oocyte membrane. Morphological studies indicate that virtually the entire oocyte surface is coated, thus implicating coated pits and coated vesicles in the transport process [5, 21, 22].

We have previously characterized the kinetic binding parameters of PV, a 30K dalton subunit of vitellogenin, to isolated oocyte membranes [18]. These studies indicated that PV bound to a specific receptor with a K_D of 3.3×10^{-6} M. We also described experiments that suggested that the membrane-associated receptor could be obtained in a soluble form by extracting the membranes with Triton X-100. In the present report we present evidence tentatively identifying the solubilized receptor as a single polypeptide of MW 116K.

MATERIALS

Biogel A-1.5m, Affi-gel 10, acrylamide, bisacylamide, and molecular weight standards for SDS-PAGE (sodium dodecyl sulfate-polyacrylamide gel electrophoresis) were obtained from BIO-RAD Laboratories. Carrier-free Na^{125}I was obtained from the Amersham Radiochemical Centre. Phosvitin and Triton X-100 were from Sigma Chemical Co. All other chemicals were of reagent grade and were obtained from commercial sources. Developing chicken oocytes were obtained from a local slaughter house.

All experimental procedures were carried out in an incubation buffer (IB) containing 0.01 M 2-N-morpholinoethane sulfonic acid (pH 6.0), 0.14 M NaCl, 5 mM KCl, 0.83 mM MgSO$_4$, 0.13 mM CaCl$_2$ plus 0.02% sodium azide. Solubilization and chromatography experiments were carried out in IB containing 0.1% (w/v) Triton X-100 (IB-TX).

METHODS

Preparation of Solubilized PV Receptors

Developing oocytes from freshly killed white leghorn laying hens were placed in IB at 0°C. Oocytes approximately 1.5–2 cm in diameter, which are rapidly sequestering vitellogenin, were slit, drained of yolk and returned to ice cold IB. Adherent yolk was removed by gentle shaking in the IB solution. The membrane complex consisting of the oocyte plasma membrane, a fibrous perivitelline layer, a monolayer of follicular epithelial cells, and an acellular basement lamella was dissected free of the overlaying connective tissue and placed in fresh IB. Dissected membranes from 10 oocytes were homogenized in 5 ml of IB containing 1% (w/v) Triton X-100 in a Teflon glass homogenizer. The homogenate was then centrifuged for 60 min at 100,000g, after which the supernatant fraction (soluble extract) was used immediately. All isolation and solubilization procedures were carried out at 4°C.

Affinity Chromatography Procedure

PV–agarose was prepared using Affi-gel 10, a commercially prepared N-hydroxysuccinimide ester of a succinylated aminoalkyl derivative of agarose available from BIO-RAD Laboratories. Affi-gel 10 forms covalent crosslinks to proteins via their free amino groups with the concomitant release of N-hydroxysuccinimide. PV—agarose used in our experiments was prepared by incubating 200 mg of PV in 10 ml of 0.1 M phosphate buffer (pH 8.0) with 10 ml of Affi-gel 10 resin for 16 h at 4°C. One ml of 1.0 M ethanolamine (pH 8.0) was added to quench any remaining reactive groups for a further 2 h at 4°C. The final product, which contained 9.5 mg of PV bound per ml of packed gel, was placed in a small 15 mm diameter column and washed extensively with IB-TX.

For affinity chromatography experiments, 3 ml of the Triton soluble extract was applied to the column at 4°C. The resin was then washed with 70 ml of IB-TX, and 2 ml eluate fractions were collected. The volume of the washing solution was empirically determined by assaying the protein composition of the eluate fractions by SDS-PAGE. If any proteins were detected by SDS-PAGE in the last fraction of the wash, the wash volume was increased in subsequent experiments. In the experiments described herein no proteins were detectable in the final wash fraction by SDS-PAGE. In our initial experiments we routinely assayed the protein concentration of the eluate fractions by both their optical density at 280 nm and the BIORAD Protein Assay. However, we observed that the protein concentration of the eluate fractions determined by either of these methods would appear to be zero when proteins could be detected by SDS-PAGE. Thus, it was apparent that neither of the assay methods was as sensitive as SDS-PAGE for detecting small quantities of protein. We therefore routinely used SDS-PAGE to assay for the presence of protein components in eluate fractions. Following the 70 ml wash with IB-TX, 40 mg of PV, chicken IgG or chicken serum albumin in 3 ml of IB-TX was applied to the column followed by a further 70 ml of IB-TX. The composition of the eluate fractions was then assayed for protein components by SDS-PAGE. Except for the application of 40 mg of free PV and extensive washing with IB-TX, our PV—agarose columns were not regenerated by the application of low pH or high salt solutions between experiments. Our columns yielded similar results in at least 5 separate experiments using the same PV—agarose.

Gel Exclusion Chromatography Procedures

Glass columns (0.4 × 53 cm) packed with Biogel A-1.5 m and equilibrated with IB-TX at 4°C were used for all gel-exclusion chromatography experiments. One ml samples were applied to the columns and eluted with IB-TX. One ml fractions were collected at a flow rate of 4 ml/h. Vitamin B-12 and blue dextran were added to the samples as internal markers. The protein composition of the eluate fractions was assayed by SDS-PAGE. In order to demonstrate soluble PV binding activity, one ml aliquots of soluble extract were incubated with 10^{-6} M ^{125}I-PV plus or minus a 100-fold molar excess of unlabeled PV for 60 min at 4°C prior to chromatography. ^{125}I-PV was prepared as described previously [18]. ^{125}I activity in the eluate fractions was determined in a well-type gamma counter.

SDS-PAGE Procedure

SDS-PAGE was carried out using the discontinuous buffer system of Laemmli [23]. Fifty μl samples were reduced by incubation with 10% β-mercaptoethanol and then applied to slab gels (0.1 × 16 × 17 cm) and electrophoresed for 16 h at 60 V. Ovalbumin, MW 43K; bovine serum albumin, MW 68K; phosphorylase B, MW 94K; β-galactosidase, MW 116K; and myosin, MW 200K were used as MW standards. Gels were fixed and stained in 500 ml of 0.5% Coomassie blue R in 30% isopropanol—20% acetic acid for 4 h and then destained in 10% isopropanol—10% acetic acid until no further background color could be removed. After destaining, gels were dried and photographed using a yellow filter.

RESULTS

Affinity Purification

We wished to determine if phosvitin (PV) covalently coupled to agarose beads could be used to affinity purify PV receptors from detergent solubilized oocyte membranes. In these experiments, a Triton X-100 extract was applied to a PV–agarose column as described in Methods. After an initial wash with IB-TX, excess free PV was applied to the column in order to competitively displace any PV binding proteins, including the solubilized PV receptor, bound to the PV–agarose. Using SDS-PAGE, the protein composition of the soluble extract, the material that initially washed through the column, and the material that eluted after the addition of free PV were analyzed. The gels (Fig. 1) demonstrate that at least 25 different proteins can be resolved in the soluble extract. Comparison of the material that eluted from the column in the initial wash and the soluble extract indicated that two proteins of MW 116K and 100K are retained on the column. Significantly, these same two proteins elute from the column after the addition of excess free PV. We obtained similar results in five separate experiments. Other proteins were observed to elute after the addition of free PV. However, these appeared to be very minor components and did not appear to be depleted from the material that initially washed through the column. It should be noted that PV is not observed in the gels because it binds Coomassie blue poorly.

Additional experiments were carried out to demonstrate that the elution of the 116K and 100K proteins by PV was the result of a specific interaction between PV and these proteins. When chicken IgG or chicken serum albumin were applied to the PV–agarose column after the application of soluble extract, neither chicken IgG or chicken serum albumin resulted in the elution of the 116K and 100K proteins from the column. Furthermore, the subsequent application of PV to the column resulted in the elution of both the 116K and 100K proteins (data not shown). For these elution experiments, IgG, albumin, and PV were used at the same mg/ml concentrations. In order to verify that the elution of the 116K and 100K proteins from the PV–agarose column was not a simple ionic effect, we attempted to elute these proteins with high concentrations of salt. Washing the PV–agarose column with 1.0 M NaCl did not result in the elution of any detectable protein from the column (data not shown).

In a separate experiment, 40 mg of PV were applied to a PV–agarose column that had not been exposed to the oocyte detergent extract. When the eluent fractions were analyzed by SDS-PAGE, no protein bands were detected by Coomassie blue staining (data not shown).

Gel Exclusion Chromatography

We have previously used gel exclusion chromatography to demonstrate the existence of a soluble PV binding component in Triton X-100 extracts of oocyte membranes [18]. The rationale behind the use of gel exclusion chromatography for these experiments is that soluble receptor–PV complexes must necessarily be larger than free PV itself. The results of a typical gel exclusion chromatography experiment (Fig. 2) show that when detergent extracts, preincubated with ^{125}I-PV, are applied to a Biogel A-1.5m column, three peaks of ^{125}I activity are obtained. The first peak, which eluted in the void volume of this column, was tentatively identified as containing soluble receptor-^{125}I-PV complexes. This peak contained 12% of the eluted ^{125}I activity. The second peak contained 69% of the eluted activity and appears to be free PV because it elutes in the same position and has the same distinctive profile of free ^{125}I-PV. The third peak, which contained 19% of the eluted ^{125}I activity,

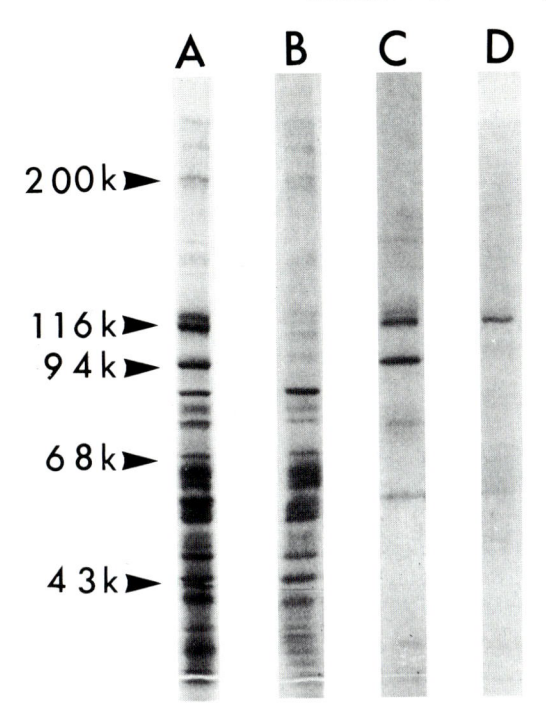

Fig. 1. SDS-PAGE analysis of Triton solubilized extract (A). Solubilized extract that did not adhere to a PV–agarose affinity column (B). Material that was eluted from PV–agarose by the addition of 40 mg of free PV (C). Solubilized extract that eluted in the void volume of a Biogel A-1.5m column (D). SDS-PAGE was carried out on 8–12% gradient slab gels using the discontinuous buffer system of Laemmli [23]. MW standards indicated by arrows are myosin, 200K; β-galactosidase, 116K; phosphorylase B, 94K; bovine serum albumin, 68K; and ovalbumin, 43K.

appears to be a small proteolytic breakdown product of ^{125}I-PV. This material is not observed when ^{125}I-PV is chromatographed separately and does not contain any detectable protein components when assayed by SDS-PAGE. Interestingly, the formation of this peak is not inhibited by the addition of both aprotinin at 100 units/ml and PMSF at 10 μg/ml. In order to demonstrate that the formation of ^{125}I-PV-receptor complexes was specific, a parallel experiment was carried out. In this experiment an aliquot of soluble extract was incubated with 10^{-6} M ^{125}I-PV plus a 100-fold excess unlabeled PV. When this sample was chromatographed over Biogel A-1.5m, only two peaks of ^{125}I activity were observed. Under these conditions no activity was observed to elute in the void volume of the column. The two peaks that were observed in the presence of unlabeled PV corresponded to the free PV peak and the proteolytic fragment peak obtained in the previous experiment. This supports our tentative identification of the void material in the preceding experiment as containing the receptor-^{125}I-PV complexes.

After solubilized extract was incubated with PV and chromatographed over Biogel A-1.5m, the protein composition of the eluate fractions was determined by SDS-PAGE (Fig. 3). The material eluting in the void volume contained one major protein of MW 116K as well as trace amounts of several other proteins. This result, in conjunction with the pre-

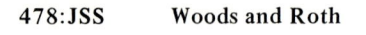

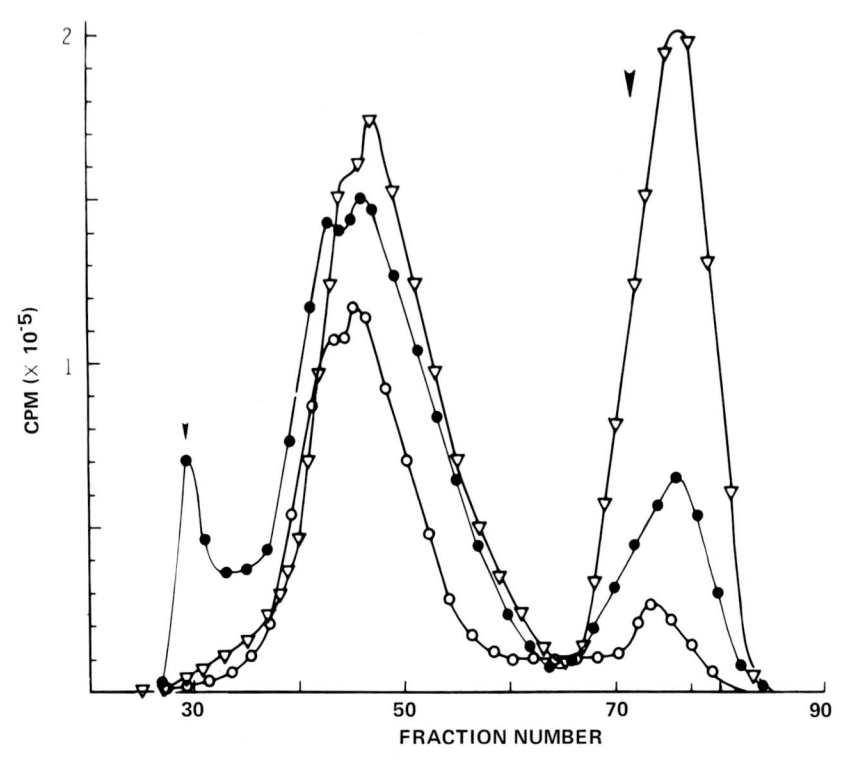

Fig. 2. Assay of ^{125}I-PV binding to soluble extract. One ml aliquots of soluble extract were incubated with 10^{-6} M ^{125}I-PV (2.5×10^6 cpm) in the presence ($\triangledown$) or absence ($\bullet$) of 10^{-4} M unlabeled PV for 60 min at 4°C. Following incubation the samples were immediately applied to two identical Biogel A-1.5m columns and eluted with IB-TX as described in Methods. Elution profile of free ^{125}I-PV (1.2×10^6 cpm) ($\circ$). The small arrow indicates the elution position of blue dextran, the large arrow indicates the elution position of vitamin B-12. Both blue dextran and vitamin B-12 were routinely included in our samples as internal standards.

vious experiment (Fig. 2) that demonstrated that receptor-PV complexes elute in the void volume, suggests that this 116K protein is in fact the receptor. In addition, this 116K protein corresponds to one of the proteins that was retained by the PV–agarose column and released by the addition of free PV. When the void volume fraction (Fraction 30, Fig. 3) was electrophoresed on the same slab as the affinity purified material, the 116K components obtained by either of these methods appeared to comigrate (Fig. 1).

In 4 separate experiments, we compared the protein composition of the eluate fractions obtained from two identical Biogel A-1.5m columns. One column was used to chromatograph a soluble extract preincubated with PV whereas the other was used to chromatograph a soluble extract preincubated in the absence of PV. In four separate experiments the 116K protein was always observed to elute in the void volume regardless of the presence or absence of PV in the preincubation. In addition, the protein composition of the fractions eluting after the void were found to be the same in both columns, regardless of the presence or absence of PV in the preincubation. Because the exclusion limit of Biogel A-1.5m is approximately 1.5 million daltons for globular proteins we can assume that the 116K protein

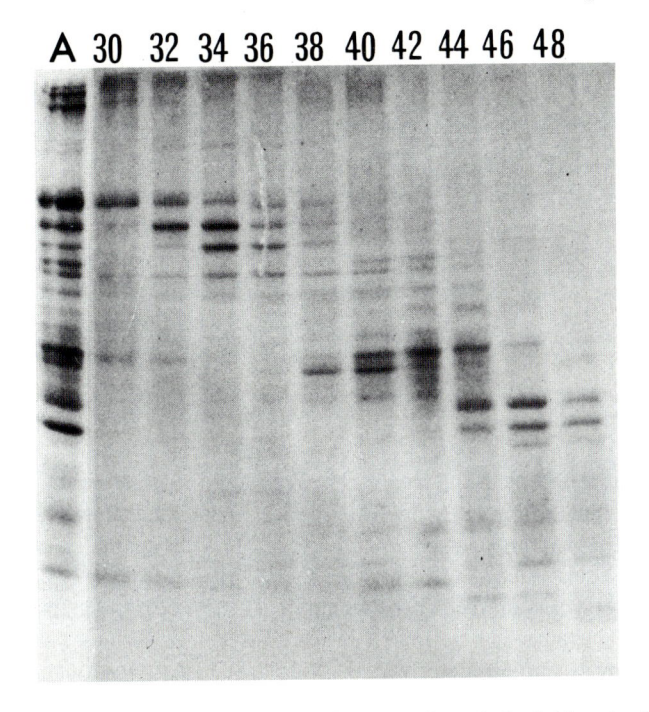

Fig. 3. SDS-PAGE analysis of Biogel A-1.5m eluate fractions. One ml of soluble extract was incubated with 10^{-5} M PV for 60 min at 4°C. Following incubation the sample was immediately applied to a Biogel A-1.5m column and eluted with IB-TX as described in Methods. The eluate fractions were then analyzed by SDS-PAGE. Fraction numbers indicated over appropriate lanes correspond to fraction numbers from related chromatography experiments shown in Figure 2 (void volume is Fraction 30). Lane A contains unchromatographed soluble extract. SDS-PAGE was carried out on a 10% slab gel using the discontinuous buffer system of Laemmli [23].

that elutes in the void volume must be a subunit of a larger complex. Because PV was not required in order for the 116K protein to elute in the void, and no other proteins were present, it may well be that the 116K protein self-associates to form a multimeric receptor complex.

DISCUSSION

Previous reports from our laboratory have shown that both vitellogenin and PV, a proteolytic fragment of vitellogenin, bind to specific receptors associated with isolated oocyte membranes [19]. In addition, we have evidence that both vitellogenin and PV bind to the same receptor [19]. These results lead us to suggest that phosvitin is the component of vitellogenin recognized by the membrane-associated receptor and that this binding is the initial event in the subsequent transport of vitellogenin into the developing oocyte. We have therefore utilized PV as a probe in our continuing studies on the mechanism of vitellogenin transport into developing chicken oocytes. In addition, it should be noted that PV is more readily obtained than vitellogenin and is stable in solution thus making it the ligand of choice.

Recently, we have shown that the nonionic detergent, Triton X-100, can be used to extract the receptor in an active form from isolated oocyte membranes [18]. The soluble form of the receptor was assayed by gel-exclusion chromatography. ^{125}I-PV, preincubated in the presence of Triton X-100 extracts of oocyte membrane, eluted as a higher MW species than free ^{125}I-PV. This effect was not inhibited by the presence of high concentrations of chicken IgG or chicken serum albumin, but was completely inhibited by the presence of high concentrations of unlabeled PV.

In the present report, we have used two independent methods to identify the protein component of the solubilized receptor. One method, affinity chromatography, showed that two proteins with MW of 116K and 100K were capable of binding to PV—agarose. Furthermore, both of these proteins could be competitively displaced from the PV—agarose by the addition of free PV, but not by the addition of IgG or albumin. The second method, gel-exclusion chromatography, took advantage of the observation that ^{125}I-PV-receptor complexes eluted in the void volume of a Biogel A-1.5m column. When the protein composition of the material eluting in the void was assayed by SDS-PAGE, a protein of MW 116K was observed. We believe these results suggest that the 116K protein represents the solubilized form of the membrane-associated PV receptor. Although unlikely, it is also possible that the 116K protein is not the receptor, but rather the receptor may be one of the minor components observed on our gels. For example, if the receptor is a glycoprotein it may stain poorly with Coomassie blue and therefore appear to be a minor component on our gels. Such a possibility will be investigated. Direct evidence supporting our proposition will await the completion of experiments designed to determine whether or not the purified 116K protein will specifically bind PV.

The coelution of the 116K and 100K proteins from the PV—agarose column by PV suggests that both proteins may bind PV or may both be part of a multisubunit complex. However, the results of our gel-exclusion chromatography experiments indicate that only the 116K protein actually binds PV. This suggests that the 116K and 100K proteins may associate with one another, but that the long time and infinite partitioning afforded by gel-exclusion chromatography allowed these two proteins to be separated from one another. Experiments to test this possibility are currently in progress.

ACKNOWLEDGMENTS

This investigation was supported in part by NIH grants HD 09549 and HD 11519 from the National Institute of Child Health and Human Development.

REFERENCES

1. Rodewald R: J Cell Biol 58:189, 1973.
2. Rodewald R: J Cell Biol 71:666, 1976.
3. Sonoda S, Schlamowitz M: J Immunol 108:1345, 1972.
4. Tsay DD, Schlamowitz M: J Immunol 115:939, 1975.
5. Roth TF, Cutting JA, Atlas SD: J Supramol Struct 4:527, 1976.
6. Jackson J, Roth TF: (in preparation).
7. Roth TF, Porter KR: J Cell Biol 20:313, 1964.
8. Telfer WH: J Biophys Biochem Cyto 9:747, 1961.
9. Melius ME, Telfer WH: J Morph 129:1, 1969.
10. Wallace RA, Dumont JN: J Cell Physiol 27(suppl):73, 1968.
11. Wallace RA, Jarad DW: J Cell Biol 69:345, 1976.

12. Schjeide OA, Wilkens M, McCandless RG, Munn R, Peterson M, Carlsen E: Am Zool 3:167–184, 1963.
13. Yusko SC, Roth TF: J Supramol Struct 4:89, 1976.
14. Anderson RGW, Goldstein JL, Brown MS: Proc Natl Acad Sci USA 73:2434, 1976.
15. Anderson RGW, Brown MS, Goldstein JL: Cell 10:351, 1977.
16. Gordon P, Carpenter J, Cohen S, Orci L: Proc Natl Acad Sci USA 75:5025, 1978.
17. Willingham MC, Maxfield FR, Pastan IH: J Cell Biol 82:614, 1979.
18. Woods JW, Roth TF: J Supramol Struct 12:491, 1979.
19. Yusko SC, Roth TF, Smith T: (submitted).
20. Krummins S, Roth TF: J Cell Sci (in press).
21. Perry MM, Gilbert AB, Evans AJ: J Anat 125:481, 1978.
22. Heuser J, Roth TF: (unpublished).
23. Laemmli UK: Nature 227:680, 1970.

Photoaffinity Labeling of the Insulin Receptor in H4 Hepatoma Cells: Lack of Cellular Receptor Processing

Cecilia Hofmann, Tae H. Ji, Bonnie Miller, and Donald F. Steiner

Department of Biochemistry, The University of Chicago, Chicago, Illinois 60637 (C.H., B.M., D.F.S.), and Department of Biochemistry, University of Wyoming, Laramie, Wyoming 82071 (T.H.J.)

Photoaffinity labeling techniques were used to identify insulin-binding components of the plasma membrane in insulin-responsive, monolayer-cultured hepatoma cells. The activated, photosensitive reagent, an n-hydroxysuccinimide ester of 4-azidobenzoic acid, was coupled with highly purifed insulin, and the hormone derivative was subsequently iodinated, bound to cell surface receptors of intact H4 cells, and photoactivated. After dissolution of the cells, labeled proteins were analyzed by SDS/polyacrylamide gel electrophoresis under reducing conditions. The main labeled band exhibited an apparent molecular weight of 130,000. Two minor components of apparent mol wt 95,000 and 40,000 were also identified. Specific labeling of all 3 bands was inhibited by simultaneous incubation of the cells with native insulin, but not by the heterologous hormone, glucagon, prior to photoactivation. Binding of azidobenzoyl-insulin to H4 cells was time-dependent, as was the correlated labeling of receptor components. Band-labeling by the photosensitive insulin derivative was totally light-dependent; spontaneous covalent linking of insulin and receptor was not observed. The labeled receptor-related proteins were not degraded by the cells under our experimental conditions.

Key words: insulin receptors, photoaffinity labeling, electrophoresis

The initial event leading to the biological effects of insulin in target cells is generally considered to be hormone binding to plasma membrane receptors [1–3]. Following hormone binding, aggregation of receptors occurs [4–6], succeeded by endocytic uptake of the hormone-receptor complex leading to lysosome-mediated insulin degradation [7–9] and extracellular release of hormonal degradation products [7, 8, 10]. Although these events are probably important in the metabolism of insulin in vivo [8], their role(s) in mediating cellular responses to insulin remains unknown. To identify receptor components directly and to examine their intracellular metabolic fate following hormone binding in intact cells, we have synthesized an activated heterobifunctional cross-linking reagent, the n-hydroxy-succinimide ester of 4-azidobenzoic acid, which can be coupled to insulin

Received September 3, 1980; accepted September 12, 1980.

0275-3723/81/1501-0001$04.00 © 1981 Alan R. Liss, Inc.

through its free amino groups. The radiolabeled hormone derivative was then covalently linked to receptor components of cultured hepatoma cells via light-induced conversion of the aryl azide moiety to a highly reactive aryl nitrene.

In this report, we describe the specific labeling of 3 high-molecular weight receptor-related proteins in intact, biologically responsive, hepatoma cells using the radioactive, photoreactive insulin derivative. Following solubilization of labeled cells, SDS polyacrylamide gel electrophoresis under reducing conditions showed labeling of a major band having a mol wt of about 130,000 and 2 minor bands of 95,000 and 40,000. These receptor-related proteins were not degraded in the cells during postincubation conditions.

MATERIALS AND METHODS

Cell Cultures

The H4-II-E-C$_3$ hepatoma cells [11] used for this study were supplied by Dr. Alan Horwitz, Department of Pediatrics, The University of Chicago, and were continuously cultured as monolayers as described previously [12].

Reagents

Monocomponent porcine insulin was a generous gift of Novo Research Laboratories (Copenhagen), and glucagon was kindly provided by Dr. Howard S. Tager, Department of Biochemistry, The University of Chicago.

Cross-Linking Reagent Preparation

The N-hydroxysuccinimide ester of 4-azidobenzoic acid (NHS-ABA) was prepared by conversion of 4-aminobenzoic acid to 4-azidobenzoic acid [13], which in turn was treated with N-hydroxysuccinimide and dicyclohexylcarbodiimide. N-hydroxysuccinimide (41 mmoles in 15 ml of dry dioxane) and dicyclohexylcarbodiimide (42 mmoles in 10 ml of dry dioxane) were introduced sequentially into 40 mmoles of 4-azidobenzoic acid dissolved in 40 ml of dry dioxane. This mixture was stirred for 5 h and the precipitate recovered by filtration. The filtrate was evaporated to dryness and the resulting powder washed with boiling petroleum ether to remove unreacted N-hydroxysuccinimide.

ABA Derivatization of Insulin

NHS-ABA was freshly dissolved in dimethylformamide to a concentration of 200 mM and was added to a 1-mg/ml solution of insulin (in phosphate-buffered saline (pH 7.4)–50% dimethylformamide) to a final concentration of 1, 2, 5, or 10 mM reagent. The mixture was incubated in the dark for 60 min at 23°C with occasional mixing and then stored at −20°C.

Iodination of ABA Insulin

Derivatized insulin was labeled with Na^{125}I to a specific activity of 50–100 μCi/μg by a modification [14] of the method of Freychet et al [15].

ABA-^{125}I-Insulin Binding and Cross-Linking to H4 Cells

Confluent monolayer cultures of H4 cells (1–4 × 10^6 cells/60-mm plate) were washed with Hank's buffer and incubated with 1.5 ml. Swim's S-77 medium containing 20 mM HEPES, 5 mM sodium bicarbonate, and 1% bovine serum albumin, pH 7.4. ABA-^{125}I-insulin was added to a final concentration of about 3×10^{-9} M under dark conditions. Unless specified otherwise, binding was carried out at 15°C for 2 h in the dark. Cells were then rinsed 8 times with 2 ml ice-cold Hank's buffer, and photoactivated cross-linking of bound ABA-^{125}I-insulin was induced by a 5-min irradiation of the cell monolayer (in 2 ml Hank's buffer) with a UVS-11 Mineralight (Ultraviolet Products, Inc., San Gabriel, CA) at a distance of 1.5 cm. The cell monolayer was then rinsed, solubilized in 1.5 ml of 88% formic acid, and transferred to counting tubes to determine the amount of radioactivity bound. Samples were then dried under vacuum and stored at −20°C. Prior to electrophoresis, samples were solubilized in appropriate buffers as described below.

Electrophoresis and Autoradiography

Samples of ABA-^{125}I-insulin were solubilized in 1 M acetic acid containing 8 M urea and were run on pH 4.5 polyacrylamide gels in 8 M urea [16] to determine the extent of derivatization by the reagent. Migration of derivatives was evaluated by counting 1.5-mm gel slices in a Gamma scintillation spectrometer (Packard Instrument Company, Downers Grove, IL).

Evaluation of the photolinked insulin-receptor complex was made by running cell samples prepared as described previously on discontinuous slab gels according to the technique of Laemmli et al [17]. Samples were solubilized in 0.0625 M Tris (pH 6.8), 2% SDS, 0.001% bromophenol blue, 8 M urea, 5% mercaptoethanol and boiled for 3 min prior to electrophoresis at 30 mamp for 4–5 h on 1-mm-thick gels containing 5.0%, 7.5%, 10.0%, or 15.0% acrylamide with 3.0% or 5.0% stacking gels.

Gels were stained overnight in 0.25% Coomassie Blue R (Sigma) dissolved in 45% methanol, 9% acetic acid, and destained in 45% methanol, 9% acetic acid. Following drying, −70°C autoradiographic exposures of the gels were made on Kodak XR-5 film using the DuPont Cronex Lightning-Plus Intensifying Screen for enhanced sensitivity. Standards used and their molecular weights were as follows: ovalbumin, 45,000 (Sigma), bovine serum albumin, 64,000 (Sigma), phosphorylase A, 92,500 (Sigma), β-galactosidase, 116,000 (P-L Biochemicals, Milwaukee), xanthine oxidase, 137,000 (Dr. Gene Nathans, University of Chicago), and RNA polymerase, 39,000, 155,000, 165,000 (Boehringer Mannheim Biochemicals).

RESULTS

ABA Derivatization of Insulin

The extent of derivatization of insulin by NHS-ABA was determined by electrophoresis of iodinated, derivatized insulin samples on pH 4.5 tube gels [16] containing 8 M urea. Derivatization was dependent upon the concentration of NHS-ABA present in the reaction mixture as shown in Figure 1. Table I provides a quantitative summary of these data and shows that with 1-mM reagent present, the predominant reaction product was monoazidobenzoyl insulin, while approximately 1/3 of the insulin remained unaltered. With 2-mM reagent, mono-derivatized insulin predominated, but about 1/4 of the insulin was di-derivatized. At 5-mM NHS-ABA, the resultant insulin mixture was composed primarily of di- and triazidobenzoyl products with about 1/4 of the insulin remaining as mono-derivative. Using 10-mM reagent, more than 90% of the reaction products were di- or triazidobenzoyl derivatives of insulin, the remainder being mono-derivatized insulin.

Monoazidobenzoyl-insulin was further analyzed to determine the preferential posi-

ACID GEL ELECTROPHORESIS OF INSULIN DERIVATIVES

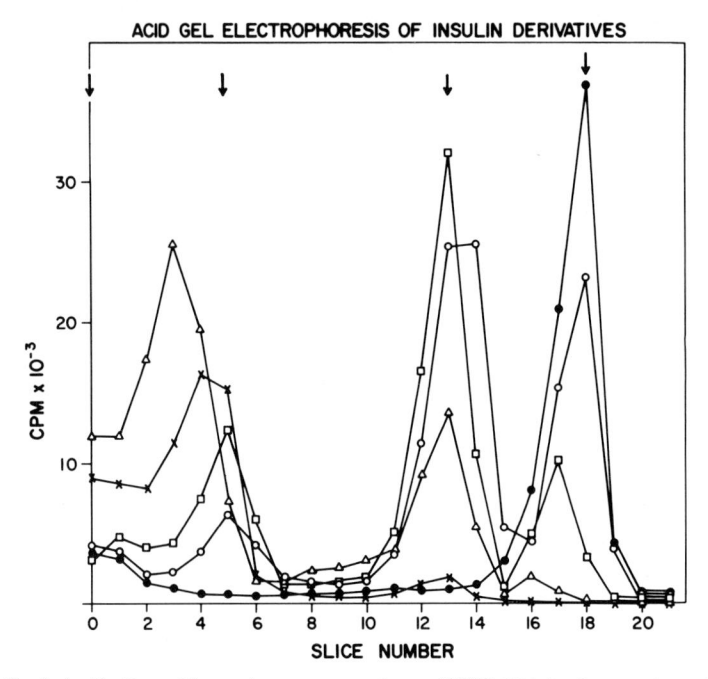

Fig. 1. Insulin derivatization with varying concentrations of NHS-ABA in the reaction mixture. The extent of derivatization was determined by electrophoresis of iodinated, azidobenzoyl-insulin samples on pH 4.5 polyacrylamide gels [16] containing 8 M urea. From left to right, arrows indicate migration position of tri-, di-, mono-, and underivatized insulin (●—● = 0 mM NHS-ABA in reaction mixture; ○—○ = 1 mM; □—□ = 2 mM; △—△ = 5 mM; ×—× = 10 mM). Gels were calibrated with desamidoinsulin and with a monoderivatized insulin kindly provided by Richard Assoyan, Department of Biochemistry, University of Chicago.

TABLE I. Insulin Derivatization by Photoactivatable Reagent NHS-ABA Expressed as % of Total Products*

Reagent concentration	Un-	Mono-	Di-	Tri-
1 mM	34.0	51.7	14.3	0
2 mM	15.6	52.6	24.0	7.8
5 mM	2.5	26.3	71.1	
10 mM	0	6.7	93.3	

*Insulin was derivatized using 1, 2, 5, or 10 mM NHS-ABA as described in Experimental Procedures. The extent of insulin derivatization was evaluated by running samples on pH 4.5 polyacrylamide gels [16] containing 8 M urea. Gels were sliced and counted, and quantitation of the results was done by weighing cutout peak areas on an analytical balance.

tion of derivatization. Following oxidative sulfitolysis (3 h at 37°C in 0.2 M phosphate buffer at pH 7.5, containing 75 mM $Na_2S_4O_6$ and 270 mM Na_2SO_3 with 8 M urea) the resulting peptides were examined by paper electrophoresis in 30% formic acid. Under these conditions, only a small fraction of the A-chain exhibited altered mobility, while a major portion of the B-chain migrated more slowly. Thus, the photoactivatable reagent reacted mainly with amino groups at position B1 or B29. Numerous previous studies have indicated that modifications of insulin in these positions does not drastically alter its biological activity [18].

Photoactivated Cross-Linking of ABA-^{125}I-Insulin to Albumin and Monolayer-Cultured H4 Cells

When 50 μl of 0.15 M phosphate buffer (pH 7.0) containing 12.5 mg of normal human albumin and 7 μCi of azidobenzoyl-^{125}I-insulin prepared with 1 or 10 mM NHS-ABA were exposed to ultraviolet irradiation with a UVS-11 mineralight for 30 min at room temperature, significant cross-linking was observed. Analysis of the samples in comparison with non-irradiated samples by gel chromatography on Biogel P-30 in 3 M acetic acid is shown in Table II. Clearly, photoactivation of ABA-^{125}I-insulin resulted in an increase of radiolabeled high-molecular-weight products concomitant with a decrease in the insulin peak. These results are consistent with a light-dependent covalent linking of insulin to albumin. While the proportion of high-molecular-weight light-independent aggregates was slightly increased with the more highly derivatized insulin preparation (10 mM NHS-ABA), covalent multimers of insulin were never observed.

In similar studies, direct photoaffinity labeling of H4 hepatoma cell surface receptors using ABA-^{125}I-insulin was carried out as described in Methods. Following insulin binding, the monolayers were rinsed, photoactivated, dissolved in 88% formic acid, and analyzed directly by chromatography on Bio-Gel P-30 in 3 M acetic acid. As in the previous experiment, increased formation of ABA-^{125}I-insulin-labeled high-molecular-weight complexes was observed (Fig. 2). Again, complex formation was light-dependent and presumably represented ^{125}I-insulin covalently linked to receptors.

The cross-linking efficiency of ABA-^{125}I-insulin was dependent upon the degree of azidobenzoyl derivatization of the insulin as demonstrated by the experiments summarized in Table III. H4 cells were initially labeled by a 2-h, 15°C incubation with ABA-^{125}I-insulin (1–2 μCi). Excess insulin was rinsed from the monolayer cultures, and cells were either 1) solubilized for determination of initial counts bound, 2) photoactivated and postincubated at 37°C for 10 min in fresh medium, or 3) postincubated at 37°C without photoactivation. Following postincubation, the difference between cell-bound counts in photoactivated and unactivated samples was apparently due to covalent insulin-receptor coupling. Although highly derivatized insulin preparations (5, 10 mM NHS-ABA in coupling mixture) showed remarkable cross-linking efficiency, the efficacy of the modified hormone in initial

TABLE II. P-30 Column Elution Profile for Insulin–Normal Human Albumin Mixture*

Derivatizing concentration of NHS-ABA	UVS-11 photoactivation	%Total counts Void peak	Insulin peak	Column peak
1 mM	+	10.8	88.4	0.8
1 mM	−	1.3	97.7	1.0
10 mM	+	13.4	84.5	2.1
10 mM	−	3.0	94.9	2.1

*P-30 column elution profile for insulin–normal human albumin mixtures. Fifty μl of 0.15 M phosphate buffer (pH 7.0) containing 12.5 mg of normal human albumin and 7 μCi of ABA-^{125}I-insulin prepared with 1 or 10 mM reagent were exposed to ultraviolet irradiation for 30 min at room temperature and chromatographed on Biogel P-30 in 3 M acetic acid. Nonirradiated samples were run as controls. The void volume peak represented labeled high-molecular-weight products, and the column peak represented small degradation products of the labeled insulin.

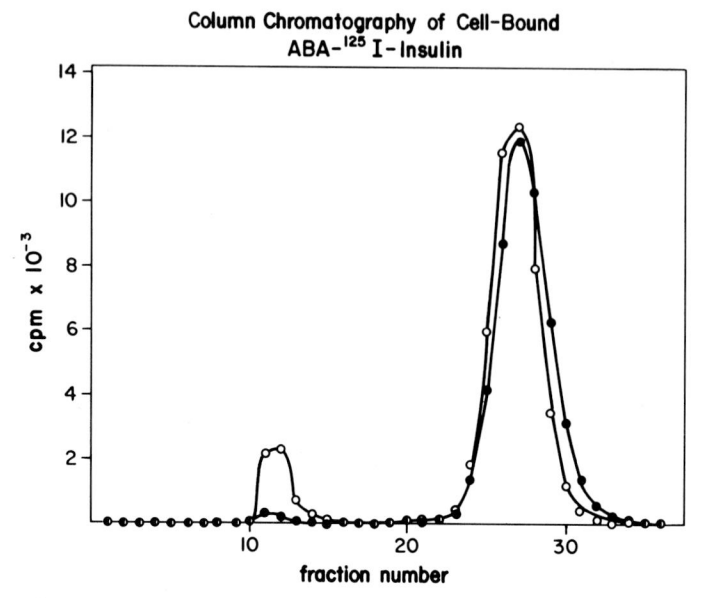

Fig. 2. Column chromatography of cell-bound ABA-^{125}I-insulin. Following insulin binding in the dark, the monolayer-cultured H4 cells were rinsed, photoactivated (●–●) or maintained dark (○–○), dissolved in 88% formic acid, and chromatographed on Bio-Gel P-30 in 3 M acetic acid. The peaks on the right represent ^{125}I-insulin, and the smaller peaks on the left represent ^{125}I-insulin associated with high-molecular-weight products, presumably receptor proteins. Clearly, photoactivation of the receptor-bound ABA-^{125}I-insulin is essential for formation of the high-molecular-weight products.

binding was markedly reduced. In contrast, minimally derivatized insulin (1 mM NHS-ABA) bound to cells more readily, but had decreased crosslinking capabilities. Combining both the binding and cross-linking properties of the various batches of azidobenzoyl-insulin by determining the percent of counts initially added that ultimately become cross-linked, the preparation giving the most efficient labeling was that which had been prepared with 1 mM NHS-ABA in the reaction mixture and consisted mainly of the monoderivative. The overall efficiency of this preparation was approximately 2, 3, and 5 times greater than that of the 2 mM, 5 mM, and 10 mM preparations, respectively. For this reason, insulin derivatized with 1 mM NHS-ABA was used for the remainder of these experiments.

Electrophoretic Analysis of the Insulin-Receptor Complex

To investigate further the nature of the products generated by photoactivation of receptor-bound ABA-^{125}I-insulin, cell samples were prepared as described in Methods and electrophoresed under reducing conditions on 7.5% polyacrylamide-SDS gels with 5.0% stacker gels according to the discontinuous system of Laemmli [17]. As shown in Figure 3A, autoradiography of dried gels indicates the light-dependent generation of 3 high-molecular-weight bands; a major band of approximate mol wt 130,000 was observed along with minor bands at 95,000, and at 40,000. A diffuse band of material at the top of the gel slab representing material of mol wt greater than 250,000 was also observed. Under non-reducing conditions, the proportion of this higher molecular weight material was markedly increased.

TABLE III. Cross-Linking Efficiency of ABA-^{125}I-Insulin to H4 Cells*

Derivatizing concentration of NHA-ABA	Photoactivation	Postincubation	Counts bound as (% of counts added)	% Bound counts remaining cell-associated	Apparent % cross-linking	% Counts added that become cross-linking
1 mM						
a	−	−	3.7%			
b	+	+		24.7%		
c	−	+		6.7%	18.0%	0.67%
2 mM						
a	−	−	2.1%			
b	+	+		23.1%		
c	−	+		9.8%	13.3%	0.28%
5 mM						
a	−	−	0.7%			
b	+	+		37.7%		
c	−	+		11.1%	26.6%	0.19%
10 mM						
a	−	−	0.5%			
b	+	+		41.8%		
c	−	+		14.8%	27.0%	0.14%

*Cross-linking efficiency of ABA-^{125}I-insulin to H4 cells. H4 cells were labeled by a 2-h, 15°C incubation with 1–2 μCi of ABA-^{125}I-insulin. Unbound insulin was rinsed from the cultures, and cells were either 1) solubilized for determination of initial counts bound, 2) photoactivated and postincubated at 37°C for 10 min in fresh medium, or 3) postincubated at 37°C without photoactivation. Following postincubation, the difference between cell-bound counts in photoactivated and unactivated samples was apparently due to covalent insulin-receptor coupling.

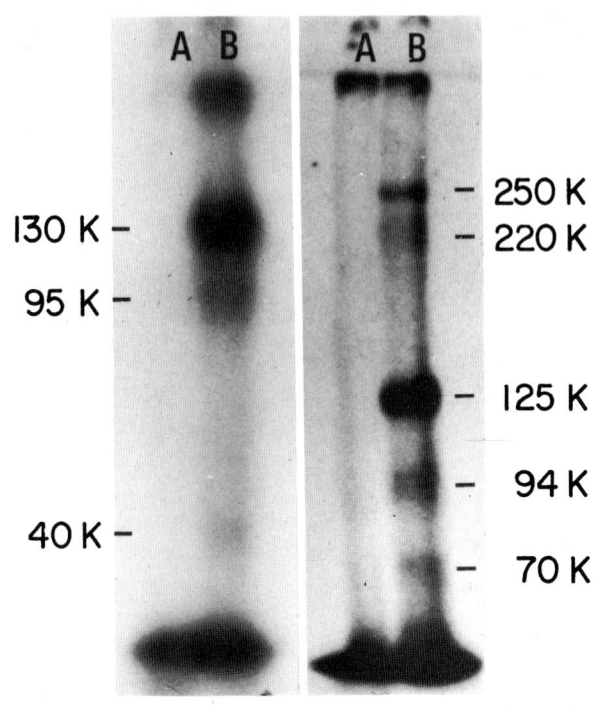

Fig. 3. SDS-polyacrylamide gel electrophoresis of labeled receptor proteins under reducing conditions. Left. Autoradiography of a dried 7.5% gel with a 5.0% stacker indicates light-dependent generation of 3 high-molecular-weight bands — a major band of apparent mol wt 130,000 and minor bands at 95,000 and 40,000. A, without photoactivation; B, with photoactivation. Right. Autoradiography of a dried 5.0% gel with a 3.0% stacker shows, from top to bottom, labeled bands at apparent molecular weights 250,000, 220,000, 125,000 predominating bands 94,000, and 72,000. A, without photoactivation; B, with photoactivation. The material at the bottom of the gel represents free insulin and/or its chains. Estimated molecular weights are indicated along the margins.

When identically labeled cell samples were run on 5.0% polyacrylamide-SDS gels with 3.0% stacker gels (Fig. 3B), the diffuse band at the top of 7.5% gels was resolved as a nonspecific component remaining at the gel top, a specific component with a mol wt of 250,000, and a diffuse minor band with an approximate mol wt of 220,000. The predominant band at approximately 125,000 and the minor band at 94,000 are both correspondent with bands previously identified on the 7.5% gels. Although the smallest receptor component (40,000 on the 7.5% gel) migrated with the tracking dye on the 5.0% gel, an additional minor component of about 70,000 was also resolved. When labeled components were further analyzed on gels of varying acrylamide concentrations in the range of 5–10%, aberrant migration of the 3 predominent labeled bands at 130k, 95k, and 40k was not observed. While altered migration due to the presence of carbohydrate might be anticipated [19], these results suggest that the proportion of receptor-associated carbohydrate is probably relatively small.

The 3 main bands (130k, 95k, 40k) become labeled in a time-dependent and parallel fashion, as shown in the autoradiogram in Figure 4. In this experiment, cells were incubated for 10, 20, 30, 60, or 120 min at 15°C with ABA-^{125}I-insulin and were rinsed, photoactivated, and solubilized and run on 7.5% Laemmli gels with 5.0% stackers [17]. Time-de-

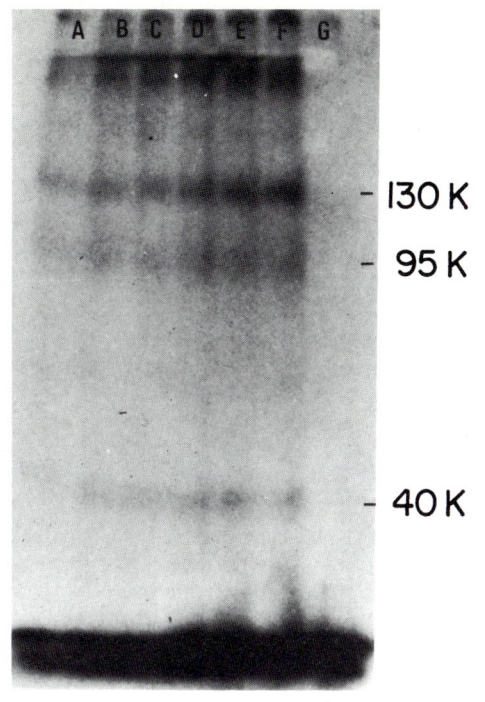

Fig. 4. Time dependency for receptor complex formation. Cells were incubated at 15°C with ABA-^{125}I-insulin for 10 (A), 20 (B), 30 (C), 60 (D), 90 (E), or 120 (F) min, and subsequently rinsed, photoactivated, solubilized, and run on a 7.5% gel with a 5.0% stacker. G represents the control condition of 120 min incubation with ABA-^{125}I-insulin, but without photoactivation. Estimated molecular weights are indicated on the right margin.

pendent formation of the labeled receptor component bands was further substantiated by direct counting of bands cut from the dried slab gels. When binding of ABA-^{125}I-insulin was carried out at 37°C, the same bands were labeled, although binding occurred more rapidly, as expected.

Furthermore, the binding of ABA-^{125}I-insulin to H4 hepatoma cells was specific, as was the light-dependent generation of the radioactively labeled 130k, 95k, and 40k bands. Figure 5 shows the progressive displacement of ABA-^{125}I-insulin binding by increasing concentrations of native insulin and demonstrates that glucagon did not inhibit binding. Autoradiograms prepared from slab gel analysis of similar experimental samples demonstrated that all of the photolabeled bands were displaced equally well by native hormone (data not shown). The slight decrease in K_D for this preparation (Fig. 5) in comparison to the value previously observed for native insulin [12] indicates that the monoderivatized material has about 1/3–1/2 the affinity of native insulin for the receptor. For this reason it was not considered necessary to purify this preparation further.

To determine whether insulin receptors were processed by the cells as previously reported for epidermal growth factor [20], receptors were initially covalently labeled as described in Methods and subsequently postincubated at 37°C in fresh medium for periods up to 3 h. Under these conditions, we failed to observe the generation of any radiolabeled receptor breakdown products as shown in Figure 6, even though control experiments (not

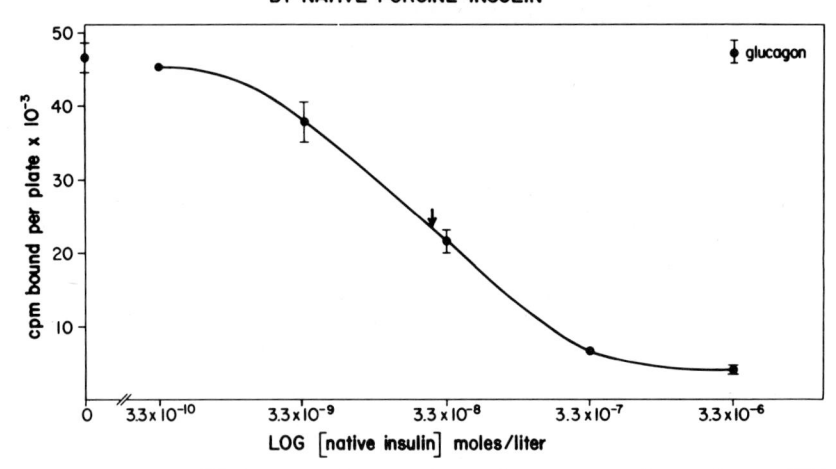

Fig. 5. Specificity of ABA-^{125}I-insulin binding by H4 cells. Confluent monolayer cultures of H4 cells were incubated for 30 min at 30°C with ABA-^{125}I-insulin in the presence of varying concentrations of native insulin, as indicated, or with the heterologous hormone, glucagon. The K_D value of 26 nM for ABA-^{125}I-insulin (arrow) was similar to, though slightly higher than, the value of 14 nM obtained from studies with ^{125}I-insulin [23].

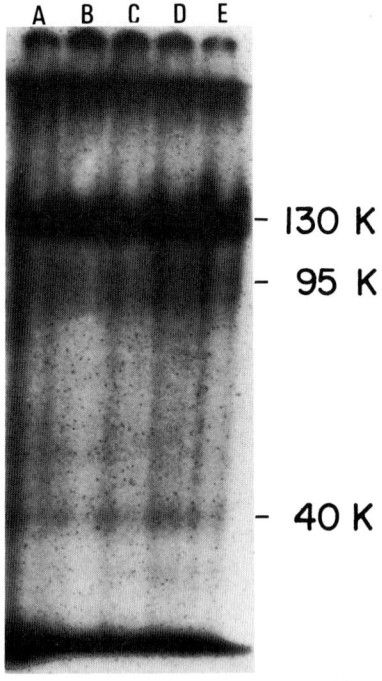

Fig. 6. Cellular receptor processing experiment. Cell-bound ABA-^{125}I-insulin (120 min, 15°C incubation) was covalently linked to receptor components by photoactivation, and cells were subsequently postincubated at 37°C for various intervals: A = 0 min; B = 5 min; C = 10 min; D = 20 min; E = 60 min. Following rinsing and solubilization, labeled components were analyzed on 7.5% gels with 5.0% stackers. Bars indicate the 3 receptor-associated protein bands (130k, 95k, 40k), which do not appear to change in intensity with time.

shown) showed nearly normal processing of insulin by cells exposed to ultraviolet irradiation. When labeled receptor component bands were analyzed in 3 separate experiments, there did not appear to be any substantial loss of radioactivity during the course of post-incubation. Also, under our experimental conditions, we did not detect any covalent cross-linking of native hormone and receptor as reported for thrombin [21] and epidermal growth factor [22], or as suggested recently for α_2-macroglobulin [23].

DISCUSSION

Using a photoreactive, iodinated derivative of insulin (ABA-^{125}I-insulin) to specifically label cell surface components of the insulin receptor complex in insulin-responsive hepatoma cells [24], a predominant band of apparent* molecular weight 130,000 was observed by SDS polyacrylamide gel electrophoresis. This major receptor component is likely to be the same component as was identified in previous investigations using rat liver and adipocyte plasma membranes [25–31], as well as human placental membranes [26]. Variability in the molecular weight reported for this major receptor component from 125,000 to 135,000 presumably has resulted from the use of different protein standards or systems. Minor labeled components of the H4 cell insulin-receptor complex with molecular weights of 95,000 and 40,000 were also observed. Yip et al [30] have previously reported the labeling of a 90,000 mol wt band in liver plasma membranes by azidobenzoyl-insulin derivatives, while Jacobs et al [26] have observed the presence of a similar band as a minor component in the preparation of purified rat liver insulin receptors, along with other minor bands at 75,000 and 45,000. Iodination of these purified insulin receptors resulted in the labeling of major bands of apparent mol wt 135,000 and 45,000 on SDS gels [26]. These bands probably correspond to the photolabeled bands at 130,000 and 40,000 described here. Thus, while similar receptor subunit bands have been observed previously [26–30], they have only been simultaneously present in early stages of preparative receptor isolation [26]. The fact that all these components become labeled when intact, insulin-responsive cells are studied implies that they are all contiguous and externally oriented in the unperturbed membrane. Indeed, previous studies on insulin receptors in isolated plasma membranes have shown that the molecular weight of the intact receptor complex is about 300,000 [26, 27, 32].

The photolabeled bands observed here (130k, 95k, and 40k) were insulin-specific, since their labeling could be prevented by simultaneous incubation with native insulin. Also, association of the radioactive, photoactivatable insulin derivative with these receptor-related bands was both time- and light-dependent. We can therefore tentatively conclude that the major protein band at 130,000 may represent the main insulin-binding subunit of the receptor and that the minor protein bands at 95,000 and 40,000 represent components of the receptor complex or other receptor-associated proteins. The smaller components may indeed play important structural roles as, for example, in orientating the receptor complex within the membrane, or alternatively, they may be important in mediating, either directly or indirectly, the biological effects of insulin [1–3].

It seems highly unlikely that the 40,000 and 95,000 receptor components simply represent degradative fragments of the main 130,000 component, since they were observed in experiments where binding occurred at 15°C, a temperature which prevents uptake and

---

*The actual molecular weight of these components would presumably be roughly 3,000 less than observed due to the weight of the associated photoligand (derivatized insulin B chain).

degradation of insulin and, presumably, of the receptor as well [7, 8]. Also, when the receptor was insulin-labeled at 15°C, photoactivated for hormone-receptor cross-linking, and postincubated at 37°C for intervals up to 3 h, the smaller subunits (95k, 40k) failed to appear in increased proportions. Furthermore, we did not detect the appearance of any labeled breakdown products of the covalently linked insulin-receptor complex as previously observed for epidermal growth factor [20], although it remains possible that the modified receptor does not behave normally in terms of its uptake and intracellular distribution. However, no substantial loss of radioactivity from the labeled bands could be observed over periods up to 3 h. Our failure to observe receptor breakdown is consistent with observations in our laboratory that H4 cells fail to show significant receptor down regulation even in the presence of 1 μg/ml insulin for 24 h [Miller, Hofmann, and Steiner; unpublished results]. The results are thus consistent with proposals that intact receptors may be recycled to the plasma membrane [33, 34] or that modification of insulin receptors or of essential cellular components by cross-linking and/or irradiation may alter their behavior.

ACKNOWLEDGMENTS

We wish to thank Ms. Myrella Smith for her assistance in preparing this manuscript. Additionally, we thank Dr. Howard Tager for critically reading the manuscript and for many helpful discussions.

This work has been supported by The Lolly Coustan Memorial Fund and by grants AM-05361, AM-13914, AM-20595, CA-19265 from the United States Public Health Service and by grant 78-09815 from the National Science Foundation.

REFERENCES

1. Czech MP: Annu Rev Biochem 46:359–384, 1977.
2. Kahn CR: J Cell Biol 70:261–286, 1976.
3. Pilkis SJ, Park CR: Annu Rev Pharmacol 14:365–388, 1974.
4. Maxfield FR, Schlessinger J, Schechter Y, Pastan I, Willingham MC: Cell 14:805–810, 1978.
5. Schlessinger J, Schechter Y, Willingham MC, Pastan I: Proc Natl Acad Sci USA 75:2659–2663, 1978.
6. Schechter Y, Schlessinger J, Jacobs S, Chang K, Cuatrecasas P: Proc Natl Acad Sci USA 75:2135–2139, 1978.
7. Terris S, Steiner DF: J Biol Chem 250:8389–8398, 1975.
8. Terris S, Hofman C, Steiner DF: Can J Biochem 57:549–568, 1979.
9. Gorden P, Carpentier JL, Freychet P, LeCam A, Orci L: Science 200:782–785, 1978.
10. Gliemann J, Sonne O: J Biol Chem 253:7857–7863, 1978.
11. Pitot HC, Peraino C, Morse PA, Potter VR: Natl Cancer Inst Monogr 13:229–245, 1964.
12. Hofmann C, Marsh JW, Miller B, Steiner DF: Diabetes 29:865–874, 1980.
13. Ji TH: J Biol Chem 252:1566–1570, 1977.
14. Starr JI, Rubenstein AH: In Jaffe B (ed): New York: Academic Press, Ch 31.
15. Freychet P, Roth J, Neville DM: Biochem Biophys Res Commun 43:400–408, 1971.
16. Reisfeld RA, Lewis VJ, Williams DE: Nature 195:281–283, 1962.
17. Laemmli UK: Nature 227:680–685, 1970.
18. Pullen RA, Lindsay DG, Wood SP, Tickle IJ, Blundell TL, Wollmer A, Krail G, Brandenburg D, Zahn H, Gliemann J, Gammeltoft S: Nature 259:369–373, 1976.
19. Banker GA, Cotman CW: J Biol Chem 247:5856–5861, 1972.
20. Das M, Fox CF: Proc Natl Acad Sci USA 75:2644–2648, 1978.
21. Baker JB, Simmer RL, Glenn KC, Cunningham DD: Nature 278:743–745, 1979.
22. Linsley PS, Blifeld C, Wrann M, Fox CF: Nature 278:745–748, 1979.
23. Davies PJA, Davies DR, Levitzki A, Maxfield PM, Willingham MC, Pastan IH: Nature 283:162–167, 1980.

24. Hofmann C, Marsh JW, Steiner DF: XIth International Congress of Biochemistry, Abstracts, p 599.
25. Jacobs S, Schechter Y, Bissell K, Cuatracasas P: Biochem Biophys Res Commun 77:981–988, 1977.
26. Jacobs S, Hazum E, Schechter Y, Cuatracasas P: Proc Natl Acad Sci USA 76:4918–4921, 1979.
27. Pilch PF, Czech MP: J Biol Chem 254:3378–3381, 1979.
28. Pilch PF, Czech MP: J Biol Chem 255:1722–1732, 1980.
29. Yip CC, Yeung CWT, Moule M: J Biol Chem 253:1743–1745, 1978.
30. Yip CC, Yeung CW-T, Moule ML: In Brandenberg D, Wollmer A (eds): "Insulin. Chemistry, Structure and Function of Insulin and Related Hormones." Berlin: Walter de Gruyter, 1980, pp 337–344.
31. Wisher MH, Thamm P, Saunders D, Sönksen, Brandenberg D: In Brandenberg D, Wollmer A (eds): "Insulin. Chemistry, Structure and Function of Insulin and Related Hormones." Berlin: Walter de Gruyter, 1980, pp 345–351.
32. Harrison LC, Billington T, East I, Nichols RJ, Clark S: Endocrinology 102:1485–1495, 1978.
33. Goldstein JL, Anderson RGW, Brown MS: Nature 279:679–685, 1979.
34. Terris S, Steiner DF: In Brandenberg D, Wollmer A (eds): "Insulin. Chemistry, Structure and Function of Insulin and Related Hormones." Berlin: Walter de Gruyter, 1980, pp 277–284.

Proteolytic Domains of the Epidermal Growth Factor Receptor of Human Placenta

Edward J. O'Keefe, Teresa K. Battin, and Vann Bennett

Department of Dermatology, University of North Carolina, Chapel Hill, North Carolina 27514 (E.J.O'K., T.K.B.) and Department of Cell Biology and Anatomy, The Johns Hopkins School of Medicine, Baltimore, Maryland 21205 (V.B.)

Microsomal membranes from human placenta, which bind 5–20 pmol of ^{125}I-epidermal growth factor (EGF) per mg protein, have been affinity-labeled with ^{125}I-EGF either spontaneously or with dimethylsuberimidate. Coomassie blue staining patterns on SDS polyacrylamide gels are minimally altered, and the EGF-receptor complex appears as a specifically labeled band of 180,000 daltons which is not removed by urea, neutral buffers, or chaotropic salts but is partially extracted by mild detergents. Limited proteolysis by alpha chymotrypsin and several other serine proteases yields labeled fragments of 170,000, 130,000, 85,000, and 48,000 daltons. More facile cleavage by papain or bromelain rapidly degrades the hormone-receptor complex to smaller labeled fragments of about 35,000 and 25,000 daltons. These fragments retain the binding site for EGF, are capable of binding EGF, and remain associated with the membrane. Alpha chymotryptic digestion of receptor solubilized by detergents yields the same fragments obtained with intact vesicles, suggesting that the fragments may represent intrinsic proteolytic domains of the receptor.

Key words: epidermal growth factor, epidermal growth factor receptor, integral membrane proteins, hormone receptor, limited proteolysis

The membrane receptor for epidermal growth factor (EGF) has been identified in microsomal preparations from human placenta [1] and cultured cells [2, 3] by affinity labeling using photoactivated bifunctional cross-linking reagents. Using these methods and labeling the receptor with ^{125}I-labeled EGF, a single band of M_r 190,000 was observed in autoradiograms by Das and Fox [2] and by Baker et al [3] in fibroblasts, while two closely spaced bands of approximately 180,000 and 160,000 daltons were described by Hock et al in human placenta [1]. A single band of 150,000–170,000 daltons was described in A-431 epidermoid carcinoma cells by Wrann et al [4]. Reducing agents did not affect the migration of the receptor on SDS gels, and the labeled band was immunoprecipitable by anti-EGF IgG, indicating that labeled EGF remains associated with this 180,000-dalton protein [5].

Received May 8, 1980; accepted September 17, 1980.

0275-3723/81/1501-0015$04.00 © 1981 Alan R. Liss, Inc.

Incubation of cells containing receptor previously labeled with ^{125}I-labeled EGF results in the production of three fragments of 59,000–70,000 daltons, 38,000–50,000 daltons, and 30,000–37,000 daltons in Swiss 3T3 cells and human fibroblasts, but not in A-431 cells, which have 10- to 20-fold more receptors than do fibroblasts [2–5]. Inhibitors of lysosomal enzymatic activity result in the production of a 130,000-dalton protein from the originally labeled 190,000-dalton band and prevent formation of the three lower molecular weight bands; this 130,000-dalton band is seen following tryptic digestion of labeled cells [2]. In other experiments using cells treated with proteases after brief exposure to label, ^{125}I-labeled EGF was found to be cleaved from the receptor [3].

These experiments indicate that the EGF receptor is probably composed of a single polypeptide that is cleaved in intact fibroblasts by lysosomal hydrolases to yield discrete fragments. It has been suggested that one or more of these fragments is mitogenic and mediates the action of EGF in cells [2]. It is not known whether similar fragments can be generated by controlled proteolysis of the receptor in membrane preparations or whether such fragments would be water-soluble or would remain associated with or buried in the membrane. In order to study such fragments generated by controlled proteolysis, we have used human placental microsomal membranes affinity labeled with ^{125}I-labeled EGF and then exposed to proteases. These studies demonstrate that similar discrete fragments are produced by enzymes of differing specificities although with different facility, and that the EGF binding site on the receptor is extremely resistant to proteolytic digestion. The smallest fragment visualized, of approximately 25,000 daltons, remains associated with the membrane and contains the EGF binding site. The fragments generated may depend on the intrinsic tertiary structure of the receptor, since identical fragments are produced after solubilization in detergents. The relationship of the fragments generated in intact cells to those obtained by limited proteolysis is unknown.

METHODS

Placenta Membranes

Placentas were obtained fresh after delivery, separated from the umbilical cord, and manually cut into 5–10 cubic centimeter pieces in cold Dulbecco's phosphate buffered saline to remove contained blood. The washed pieces were then transferred to 0.25 M sucrose to remove salt and then to 1 volume of 0.25 M sucrose for homogenization in a Waring blender at 4°. The suspension was made 2 mM in iodoacetate and 2 mM in EDTA, and 200 µg per ml phenylmethylsulfonyl fluoride was added immediately prior to homogenization. The material was then further homogenized for 60 seconds at full speed with a Polytron homogenizer and centrifuged at 4° in a Sorvall GSA rotor at 2,400 rpm for 10 minutes. The sedimented material was again homogenized with the Polytron and centrifuged as before. The sedimented material was again homogenized with the Polytron, sedimented, and the supernatants were pooled and centrifuged at 10,000g for 30 minutes and the pellet was discarded. The supernatant was then centrifuged at 40,000g for 4 hours to obtain the microsomal fraction. To purify this material further, the pellet was resuspended in 35% (w/v) sucrose and centrifuged at 40,000g for 1 hour, and the supernatant diluted with 1 mM EDTA to a final concentration of 0.25 M sucrose and centrifuged at 40,000g for 4 hours. The pellet was resuspended in 10 mM sodium phosphate buffer, pH 7.4, containing 1 mM EDTA, or in 25 mM Tris buffer, pH 7.4, at a concentration of 20 mg per ml and frozen at −70° and is referred to hereafter as "membranes."

^{125}I-Labeled EGF

EGF was purified to homogeneity by the method of Savage and Cohen [6] and labeled by the method of Hunter and Greenwood [7] to a specific activity of 50–175 uCi per μg. The labeled material was purified on a Sephadex G-25 column equilibrated with 0.1 M sodium phosphate buffer containing 0.1% bovine serum albumin. Following reduction of Chloramine T with sodium metabisulfate, 5 mM tyrosine was added 60 seconds prior to chromatography to prevent iodination of albumin.

SDS Gel Electrophoresis

Electrophoresis was performed according to the method of Fairbanks [8]. Gels contained 6% acrylamide and 0.45% bis acrylamide and were polymerized with 0.03% TEMED and 0.2% ammonium persulfate. Samples were added to electrophoresis buffer to achieve a final concentration of 1% SDS, 10 mM Tris-HCl, 10 mM EDTA, pH 8.0, 5% sucrose, 40 mM dithiothreitol, and 0.1% bromophenol blue. Samples containing 40–70 μg protein were added to 5-mm wells and run on 3-mm thick gels in a Hoefer slab gel apparatus at 20 volts constant voltage. Running buffer contained 40 mM Tris, 20 mM sodium acetate, 2 mM EDTA, 0.2% SDS, and was adjusted to pH 7.4 with acetic acid. Standards used were human erythrocyte membranes, phosphorylase, beta galactosidase, and bovine serum albumin. Autoradiograms were produced by exposing the dried gels to Kodak X-O-Mat XR-1 film at −70° in the presence of Dupont Cronex lightning-plus intensifying screens.

Affinity Labeling of the EGF Receptor

^{125}I-labeled urogastrone or murine EGF (5×10^{-8}M) was incubated for 60 minutes at 4° with placenta membranes (10 mg protein per ml) in 10 mM sodium phosphate buffer, pH 7.4, containing 1 mM EDTA. The spontaneously labeled membranes [3, 5] were diluted in 5 volume 1.0 M Tris-acetate buffer, pH 7.4, sedimented at 160,000g for 20 minutes, and resuspended in various buffers, usually 10 mM sodium phosphate, pH 7.4. For covalent labeling with dimethylsuberimidate [9], the pellet was resuspended in 0.1 M sodium borate buffer, pH 8.0, containing 0.5 mg per ml dimethylsuberimidate and incubated at 24° for 20 minutes, diluted with 5 volumes 1.0 M tris-acetate buffer, and centrifuged and resuspended as above. From 5% to 20% of radioactivity in such washed membranes is associated with the receptor on SDS polyacrylamide gels.

Binding Assays

Binding of ^{125}I-labeled EGF to placenta membranes was measured as previously described [10].

Materials

SDS gel electrophoresis reagents were from BioRad. Triton X-100 and dimethylsuberimidate were from Sigma, and Ammonyx-LO was a gift from the Onyx Chemical Company, Jersey City, New Jersey. Papain, bromelain, elastase, alpha chymotrypsin, trypsin, phosphorylase, and beta galactosidase were from Worthington; Carrier-free ^{125}I NaI was from Union Carbide. Chemicals were from Fisher or Sigma and were reagent grade or better. Urogastrone was a kind gift of Dr. H. Gregory.

Enzymatic Digestions

After labeling of the receptor in intact membranes and resuspension in appropriate buffers, various enzymes were added to the membranes and incubated as described in the figure legends. Digestions were terminated by addition of electrophoresis buffer, heating to

70° for 2 minutes, and incubation at 37° for 20 minutes. In some cases, inhibitors such as phenylmethylsulfonyl fluoride or iodoacetate were added before electrophoresis buffer to inhibit serine or sulfhydryl proteases.

RESULTS AND DISCUSSION

In binding studies with ^{125}I-labeled EGF the placenta membrane preparation yields a curvilinear Scatchard plot indicating the presence of high-affinity binding sites (K_D = 1 × 10^{-9}M) and a capacity of approximately 6 pmols of ^{125}I-labeled EGF bound per mg protein (Fig. 1a) [10]. For a protein of M_r 180,000 this indicates that the receptor may comprise 0.2% of the protein of the microsomal membrane preparation. Transmission electron microscopy of the preparation reveals vesicles and sheets with the appearance of membranes. Gaps are visible in the vesicles indicating that they are probably not sealed and suggesting that enzymatic digestion probably affects both cytoplasmic and external sides of the plasma membrane. Membranes incubated with ^{125}I-labeled EGF in saturating concentrations and then washed and electrophoresed in 6% polyacrymamide gels in SDS show either a single band on autoradiograms or two closely spaced bands of about 180,000 and 170,000 daltons. If the membranes are homogenized in the presence of inhibitors of proteolysis (see Methods), a single band is labeled reproducibly, whereas the label on membranes prepared without inhibitors or from frozen placenta shows two bands and occasionally a third band of 130,000 daltons. The same lower molecular weight bands are produced in membranes previously incubated at 24° or 37° for 15–60 minutes. These bands are labeled specifically, since saturating concentrations of native EGF prevent labeling of all bands equally (Fig. 1b). The label seen on autoradiograms coincides with a visible band on Coomassie blue-stained gels, and the label as well as this band, which comigrates with band 2.3 of erythrocyte membranes, is especially heavy in membranes prepared in the presence of inhibitors with attention to maintenance of low temperatures.

The labeled material is not extractable from membranes by neutral buffers such as 10 mM sodium phosphate/1 mM EDTA, 50 mM EDTA, 1 M KCl, or by chaotropic salts such as 1 M KBr, KSCN, or Tris-acetate, and it resists extraction by 8 M urea or 0.2 M acetic acid. Although some proteins are removed by this washing and extraction, Scatchard plots indicate that substantial enhancement of binding capacity per mg protein is not achieved by these procedures, and some degradation of the 180,000-dalton band to lower molecular weight bands is apparent after exposure to urea. Binding affinity is altered by 0.2 M acetic acid, which reduces high-affinity binding without substantially altering binding capacity (data not shown).

The affinity label produced by incubating ^{125}I-labeled EGF alone [3, 5] with placenta membranes is identical on SDS gels to that produced in the presence of low concentrations of the bifunctional crosslinker, dimethylsuberimidate, but is less heavily labeled. A time- and concentration-dependent enhancement of the labeled 180,000-dalton band is seen using 0.1–2.0 mg per ml of cross-linker for 10–90 minutes at 24°, although at higher concentrations and longer incubation times extensive cross-linking of many membrane proteins is seen and labeled material fails to enter 6% acrylamide gels. At 0.5 mg dimethylsuberimidate per ml for 15–30 minutes at 24°, there is minimal or undetectable alteration of staining patterns and a two- to tenfold enhancement of labeling, which is confined to the bands labeled in the absence of dimethylsuberimidate. Satisfactory labeling is seen with either murine or human EGF (urogastrone) using dimethylsuberimidate, although murine EGF has only one available amino group [11], whereas urogastrone has in addition two

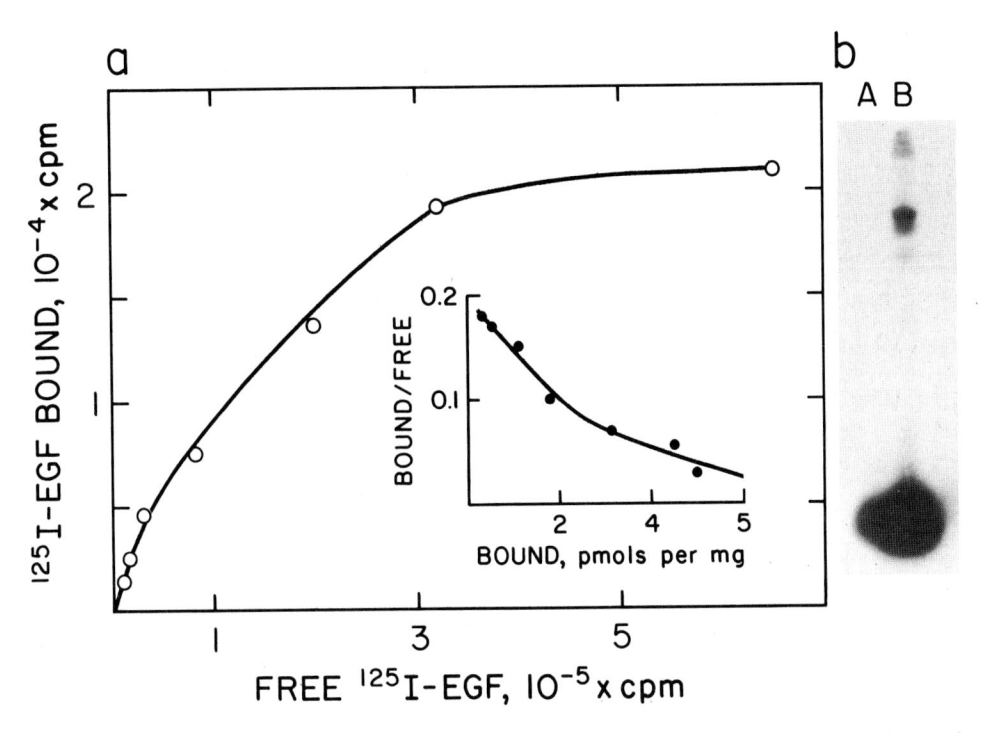

Fig. 1. Binding of ^{125}I-labeled EGF and affinity labeling of the EGF receptor in placenta membrane. a) Various concentrations of ^{125}I-labeled EGF (0.4 Ci/μmol) were incubated (30 min, 24°) with 9.2 μg of placenta membranes in 0.2 ml buffer and filtered as described [10]. Only specific binding is shown. Inset shows Scatchard plot of binding data. b) Dimethylsuberimidate cross-linking of ^{125}I-labeled EGF with (lane A) and without (lane B) 10^{-6} M native EGF added.

lysine groups [12]. ^{125}I-labeled EGF is retained in labeled membranes, since the 180,000-dalton band can be immunoprecipitated by anti-EGF antiserum and visualized on autoradiograms of SDS gels (data not shown).

Although solubilization of the labeled 180,000-dalton band is readily achieved by 0.5% Ammonyx-LO or other mild detergents with parallel partial solubilization of the Coomassie blue staining band identified with the radioactive band on autoradiograms, it is notable that complete solubilization of the labeled material is not achieved even with concentrations of detergent (eg, 10%) far above the critical micellar concentration (Fig. 2) both in placenta membranes and in KB cells extracted with Triton X-100 or Ammonyx-LO. Extraction of radioactivity reaches a plateau as determined by residual radioactivity pelleted at 160,000g for 1 hour, approximately 20% of radioactivity not removed by washing membranes in 1 M Tris-acetate remaining sedimentable after treatment at 4° with 5% (v/v) Ammonyx-LO. The Coomassie blue-stained 180,000-dalton band is apparently heterogeneous, since low ionic strength buffers may remove small amounts of material from this band that reveal no radioactive label on autoradiograms, and detergent is required to release labeled material (Fig. 2).

Proteolytic digestion of affinity-labeled membranes with chymotrypsin or trypsin results in sequential degradation of the receptor and production of a labeled 170,000-dalton band followed by 130,000 and 80,000–85,000-dalton bands on autoradiograms (Fig. 3).

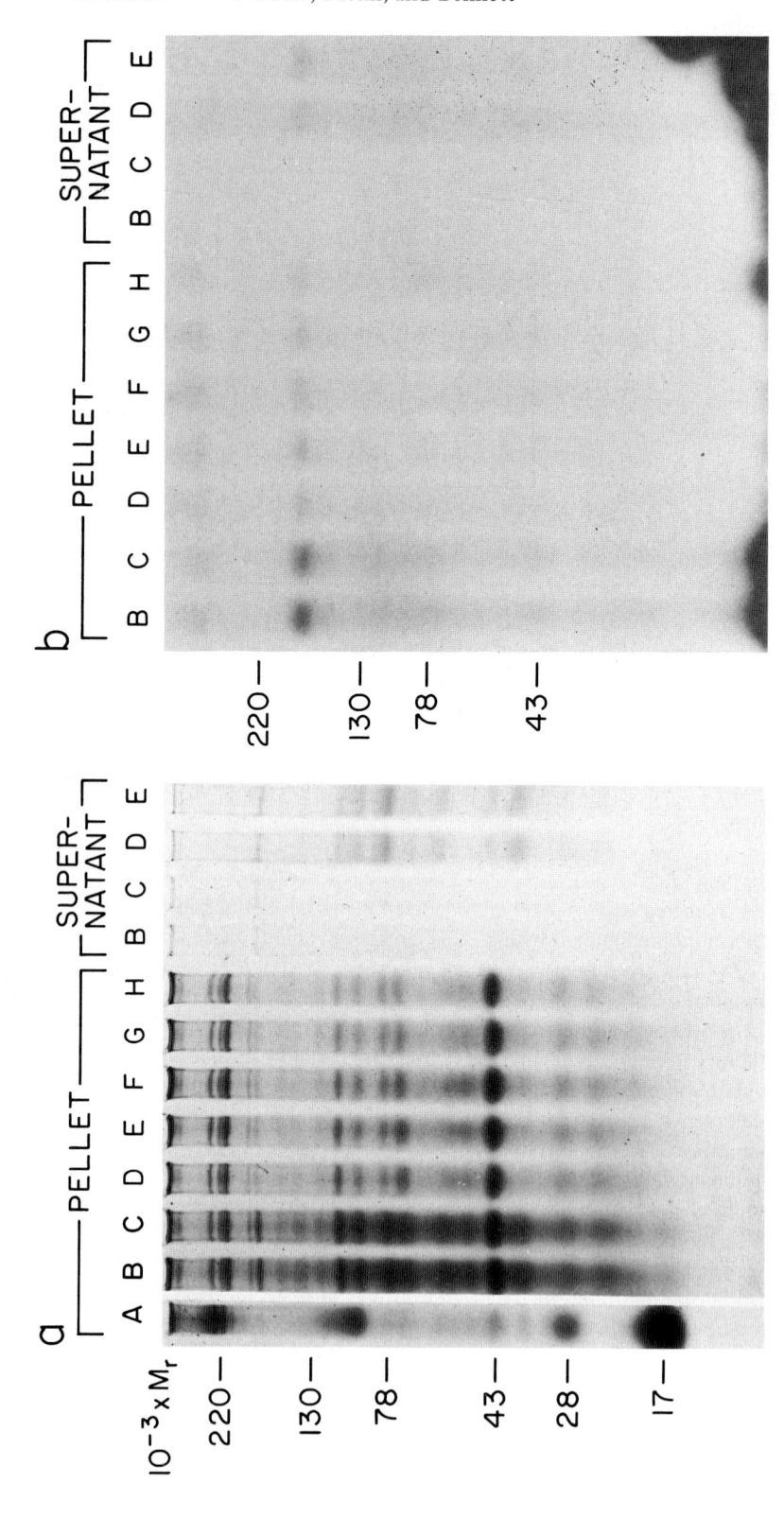

Fig. 2. Solubilization of the labeled EGF receptor with Ammonyx-LO. Membrane suspensions prepared from human placenta were incubated with 10 volumes of detergent diluted in 10 mM sodium phosphate buffer, pH 7.4, at 4° for 15 minutes and centrifuged at 190,000g$_{av}$ for 40 minutes. Pellets were resuspended in 10 mM sodium phosphate buffer, pH 7.4, and 40 μg electrophoresed. Supernatants were added directly to electrophoresis buffer and processed as described in Methods. Solubilization reached 80–90% at 2% (v/v) Ammonyx-LO under these conditions when pellets were measured for protein by the method of Lowry [15]. a) Coomassie blue-stained gel; b) autoradiogram. Ammonyx-LO concentration, percent (v/v): B) 0, C) 0.1, D) 0.5, E) 1.0, F) 2.0, G) 5.0, H) 10.0. A) erythrocyte standard.

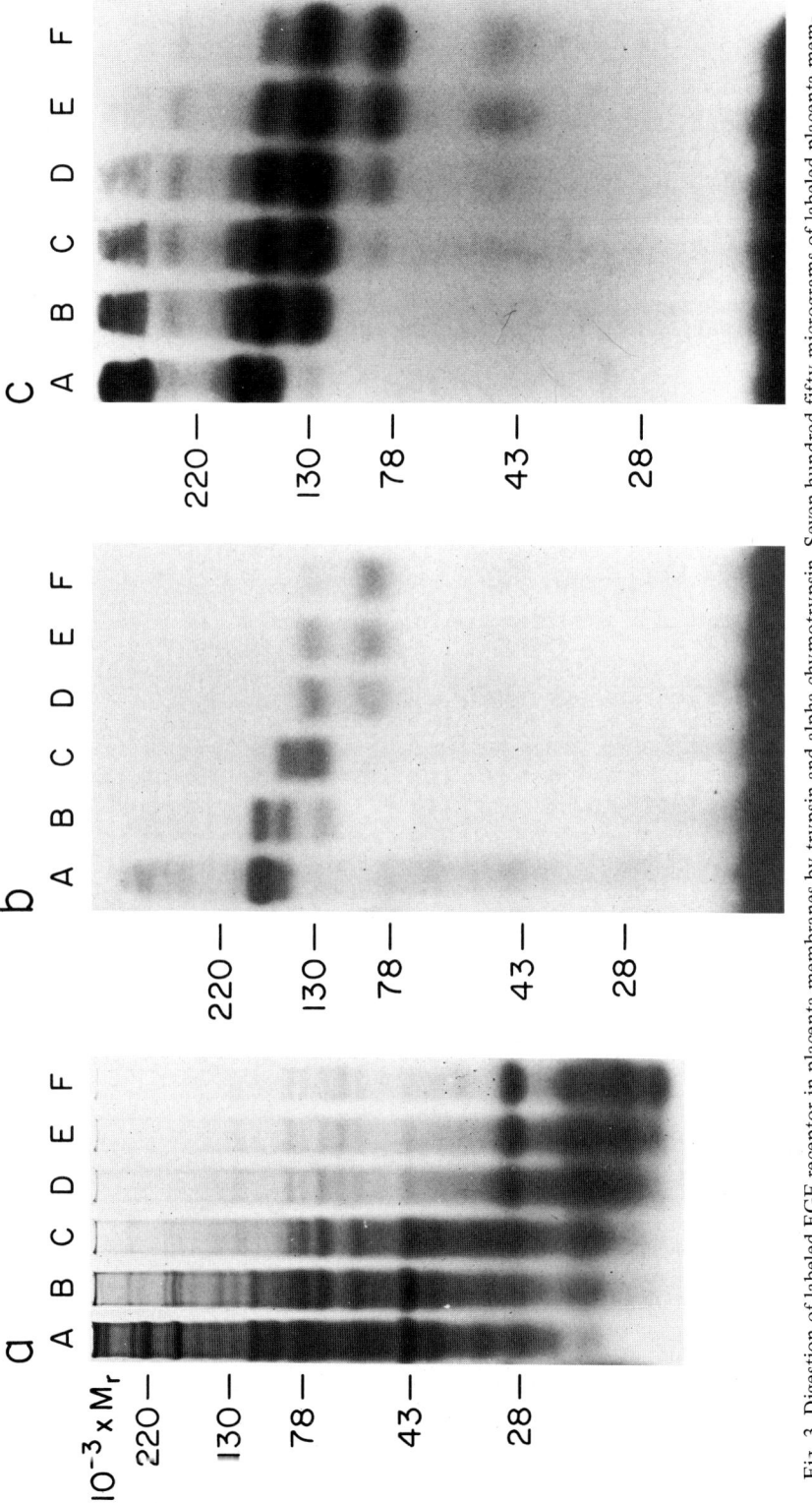

Fig. 3. Digestion of labeled EGF receptor in placenta membranes by trypsin and alpha chymotrypsin. Seven hundred fifty micrograms of labeled placenta membranes were incubated with enzyme for 15 minutes at 24° in 10 mM sodium phosphate buffer, pH 7.4, containing 2 mM $CaCl_2$. The reaction was terminated with 200 μg per ml phenylmethylsulfonyl fluoride and processed as described in Methods using 75 μg protein per lane. a) and b) trypsin; c) alpha chymotrypsin. a) Coomassie blue-stained gel; b) and c) autoradiograms. Concentration of enzyme, μg per ml: A) 0, B) 20, C) 50, D) 100, E) 250, F) 500.

Further degradation is difficult even in the presence of 10 mM–100 mM CaCl$_2$, and controlled proteolysis is hampered by variable degrees of degradation in different experiments. Added proteases fail to degrade further the membranes after as little as 15–30 minutes under these conditions, and repeated additions produce little additional degradation, suggesting that remaining proteins may be shielded in vesicles from further enzymatic attack.

As shown in Figure 3 the EGF receptor is among the more resistant of membrane proteins to proteolysis. Similar degradation is seen with pepsin (Fig. 4) with production of identical fragments on autoradiograms, but we have not observed degradation beyond the 130,000-dalton fragment with elastase (Fig. 4). Thrombin has minimal effects on the membranes. Digestion of membranes with sulfhydryl enzymes papain and bromelain produces fragments identical to those seen with the serine proteases and pepsin but at much lower concentrations, and the degradation proceeds readily beyond 130,000- and 80,000-dalton

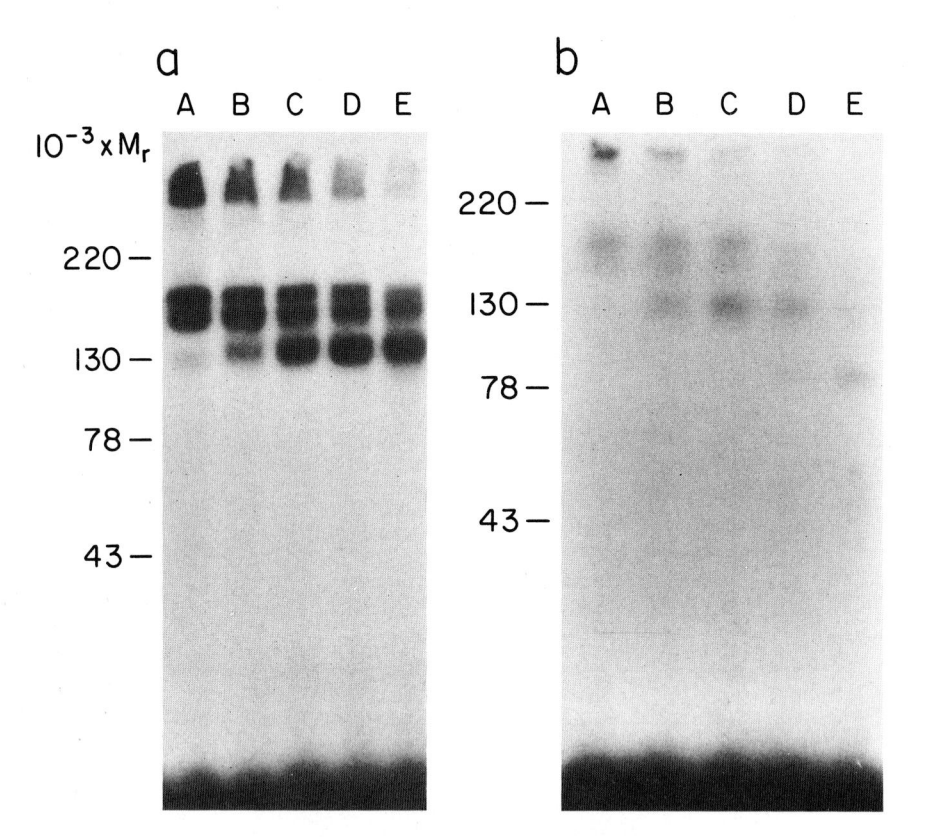

Fig. 4. Digestion of labeled EGF receptor in placenta membranes by elastase and pepsin. a) Elastase. Five hundred micrograms of placenta membranes labeled with ^{125}I-EGF were treated with elastase in 50 mM sodium phosphate buffer, pH 7.4, for 15 minutes at 24° in 0.125 ml reaction volume. The sample was dissolved in electrophoresis buffer and heated, 60 µg of protein electrophoresed, and autoradiograms made as described in Methods. Elastase concentrations, µg per ml: A) 0, B) 10, C) 50, D) 200, E) 500. b) Pepsin. Labeled membranes were suspended in 1 volume of 0.5 M glycine buffer pH 2.5 and exposed to pepsin for 20 minutes at 24°. Samples were added to electrophoresis buffer, heated, and 50 µg electrophoresed. Pepsin, µg per ml: A) 0, B) 10, C) 50, D) 250, E) 500.

fragments to produce fragments of 48,000, 35,000, and 25,000 daltons (Fig. 5). These fragments are not artifacts of the labeling procedure using dimethylsuberimidate since identical results are obtained in spontaneously labeled preparations, although less radioactivity is incorporated, and identical proteolytic fragments are seen whether or not high molecular weight aggregates are produced by dimethylsuberimidate. The ability of proteases of varying specificities to produce identical sequential cleavage fragments indicates that production of these fragments is dependent on the structure of the substrate rather than on the specificity of the protease.

Prior addition of phenylmethylsulfonyl fluoride in 2:1 molar ratio to trypsin and chymotrypsin prevented digestion completely, but addition to membrane suspensions inhibited less effectively. Although this demonstrates that proteolytic action is responsible for the described degradation, digestions were usually terminated by inhibitors and prompt solubilization of membranes in SDS (electrophoresis buffer) with heating as described in the figure legends. Inhibition of pepsin with pepstatin and of sulfhydryl enzymes with iodoacetate was studied with similar results.

Although cleavage of EGF from the receptor cannot be measured accurately because of the high background of labeled, noncovalently bound, ^{125}I-labeled EGF, which migrates in the 6,000-dalton range, sufficient label remains with the fragments to permit their visualization. Bound ^{125}I-labeled EGF therefore appears to be relatively inaccessible to proteases in comparison with the major part of the receptor. Affinity labeling of membranes previously degraded by proteases results in labeling of all fragments seen with proteolysis of the undegraded receptor (data not shown); the resistance of the EGF binding site to proteolysis therefore does not depend on protection by previously bound hormone. For the same reason, the pattern of fragments produced does not depend on the presence of bound EGF but results from intrinsic proteolytic susceptibility of the receptor. The smaller fragments generated by sulfhydryl proteases are occasionally produced by trypsin and chymotrypsin, and a somewhat diffuse band is commonly seen in the 45,000–50,000-dalton range. Such heterogeneity may arise from variations in carbohydrate components of cleaved fragments.

The characteristic degradative fragments of the labeled receptor produced by various proteases could be determined in part by the disposition of various segments of the receptor in the membrane, since portions of the receptor that are not physically buried within the membrane could be more susceptible to enzymatic attack in membrane vesicles. Susceptibility to proteolysis of the solubilized labeled receptor was therefore studied. If labeled membranes are solubilized in 1% (v/v) Ammonyx-LO and centrifuged at 290,000g_{av} for 60 minutes, proteolytic treatment of the supernatant with alpha chymotrypsin yields fragments identical to those produced in intact vesicles (Fig. 6). Cleavage of the 180,000-dalton EGF-receptor complex is seen, however, at 50- to 100-fold lower enzyme concentrations. Transition to the 130,000-dalton fragment is complete with 2 μg per ml in vesicles. Disappearance of label from the region of 130,000 daltons or higher requires 100–200 μg per ml in solubilized preparations under these conditions but requires 200–500 μg per ml in vesicles. The receptor retains some tertiary structure following solubilization in 0.2% SDS, since limited proteolysis produces identical fragments under these conditions as well. Degradation by alpha chymotrypsin of the SDS-solubilized receptor which has been heated at 80° for 2 minutes does not yield the characteristic fragments, but completely degrades all labeled material, indicating that this treatment probably extensively denatures the receptor [13].

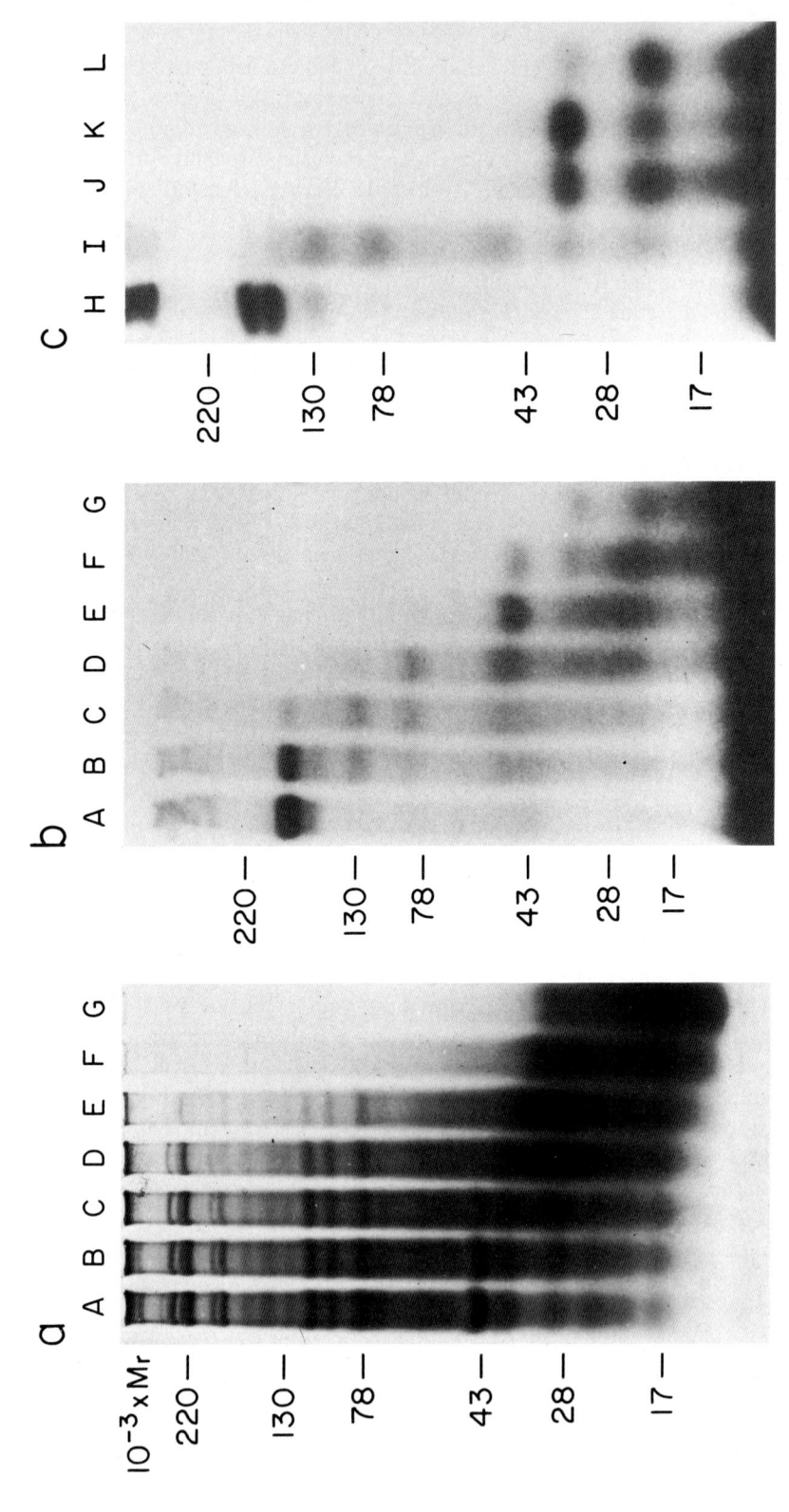

Fig. 5. Digestion of labeled EGF receptor in placenta membranes by papain and bromelain. Four hundred micrograms of labeled placenta membranes were incubated for 15 minutes at 24° in 0.75 ml buffer with enzyme which had been preincubated for 10 minutes with 1 mM EDTA and 5 mM cysteine in 0.1 M sodium phosphate buffer, pH 7.4. The reaction was terminated with iodoacetimide (2 mM final concentration) and the sample was dissolved in electrophoresis buffer and processed as described in Methods using 80 μg of protein per lane. Digestion with papain (a) and (b) or bromelain (c). a) Coomassie blue-stained gel; b) and c) autoradiograms. Concentration of enzyme, μg per ml: Papain, A) 0, B) 0.01, C) 0.05, D) 0.1, E) 0.2, F) 1, G) 5. Bromelain, H) 0.2, I) 2, J) 20, K) 100, L) 500.

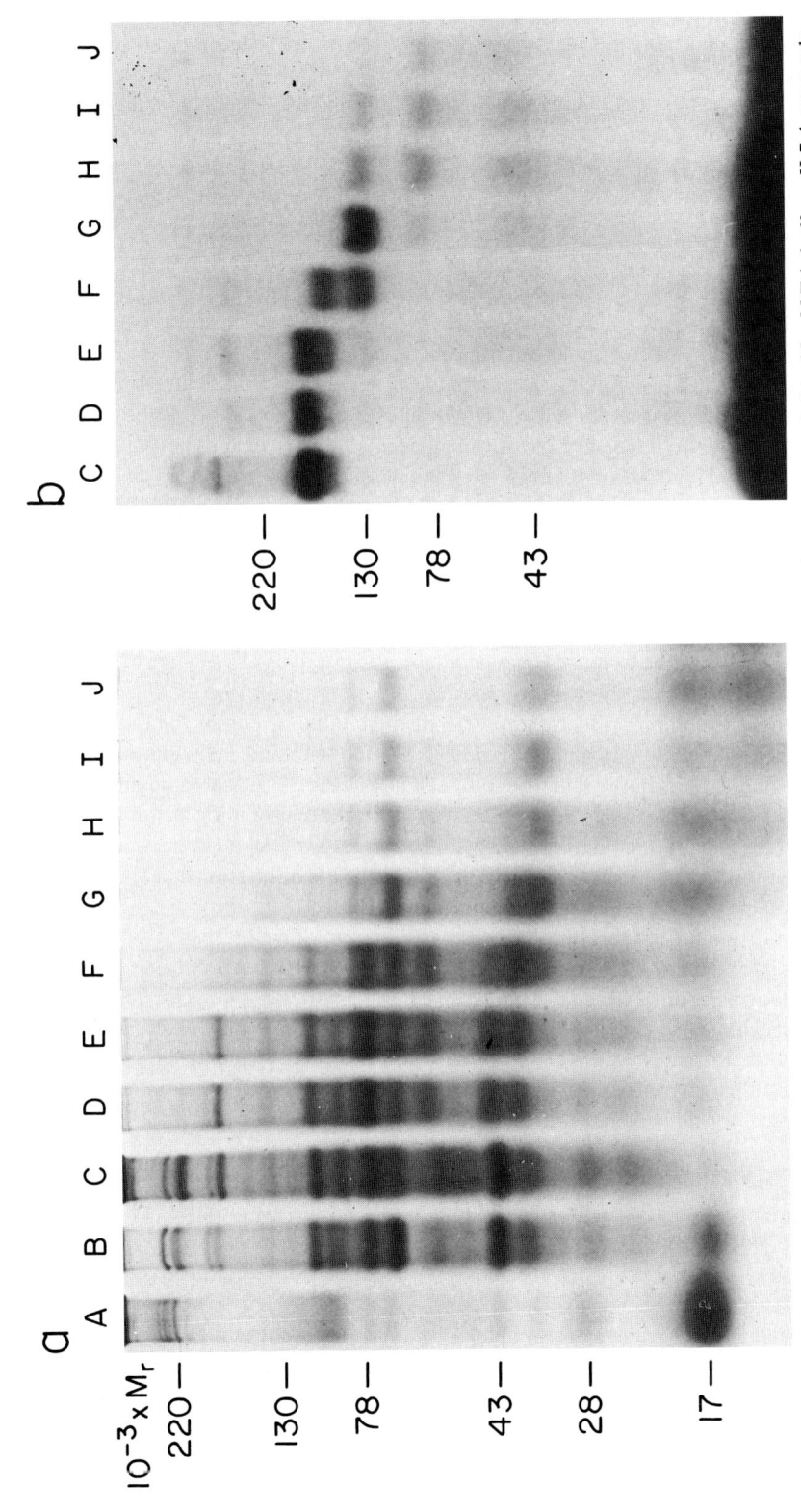

Fig. 6. Digestion of solubilized labeled EGF receptor by alpha chymotrypsin. Labeled placenta membranes resuspended in 25 mM Tris buffer, pH 7.4, were solubilized in 5 volumes of 2% Ammonyx-LO for 20 minutes at 4° and centrifuged at 180,000g$_{av}$ for 60 minutes. Samples of supernatant were digested with enzyme for 20 minutes at 24°, added to electrophoresis buffer, and heated and electrophoresed as described in Methods. a) Coomassie blue-stained gel, b) autoradiogram. A) erythrocyte standard, B) unlabeled membrane standard, C) control labeled membranes, D–J) supernatant from solubilized labeled membranes. Alpha chymotrypsin concentration, µg per ml: D) 0, E) 0.1, F) 0.5, G) 2, H) 10, I) 50, J) 100.

These studies utilizing a cross-linking reagent to affinity-label the EGF receptor confirm previous demonstrations that the receptor is probably a single polypeptide chain of approximately 180,000 daltons that behaves as an integral membrane protein. Reducing agents do not alter its migration, and lower molecular weight fragments, including the component of the doublet described by Hock et al in placenta [1], are proteolytic fragments. Although the approximate molecular weights of the receptor reported in placenta [1], A-431 epidermoid carcinoma cells [4], and human [3] and murine [2] fibroblasts differ, we find that the high molecular weight labeled bands from KB cells, human placenta, and human fibroblasts coelectrophorese (data not shown).

The relationship of endogenously produced fragments in fibroblasts to those seen following protease treatment of labeled placenta membranes is not known. It is of interest that the KB cell, a human epidermoid carcinoma cell, resembles the EGF receptor "hyper-producing" A-431 cell [4] in binding at least 10^6 EGF molecules per cell and in failing to degrade the labeled receptor as do fibroblasts (data not shown). Following solubilization of the labeled receptor with detergents, part of the labeled material remains particulate even in 10% Ammonyx-LO, suggesting that the receptor may be in part associated with structural or other proteins not released with the dissolution of the membrane by mild detergents. Similar findings are also obtained using either Triton X-100 or Ammonyx-LO with KB cells, in which physiologic significance is more likely than in membrane preparations.

Proteases cleave the receptor, producing several unlabeled fragments, without removing bound ^{125}I-labeled EGF. Since EGF alone is readily degraded by papain and other proteases at low concentrations to non-TCA precipitable radioactivity, EGF appears to be protected when bound to the receptor. Furthermore, the failure of cleavage of EGF from the receptor does not depend on protection by EGF of the binding region of the receptor, since fragments produced by prior proteolytic degradation still bind EGF and can be visualized on autoradiograms in the same pattern seen in Figures 3–6. Scatchard analysis of the binding data from protease-treated membranes indicates that binding affinity is unchanged following papain or alpha-chymotrypsin treatment even with extensive degradation, but receptor number is decreased (data not shown). It is of interest that fragments generated in intact fibroblasts also retain EGF binding ability [2].

The smallest fragment observed in these studies, of approximately 25,000 daltons, may approach in size the residual hydrophobic core of the receptor, since it could not be further degraded and is not extracted from the membrane without detergent. The EGF binding site, which may be preserved on this fragment, may be sterically inaccessible to these larger proteases. The bulk of the receptor, which is cleaved sequentially by proteases, may be at least in part hydrophilic, although the possibility that additional unlabeled segments of the receptor remain buried in phospholipid cannot be evaluated from these studies.

The receptor does not appear to undergo any dramatic conformational change in this preparation as a result of EGF binding, since proteolytic fragments produced from labeled membranes (ie, previously exposed to EGF) are identical with those produced by proteolysis of unlabeled membranes and then specifically labeled. Sensitivity to proteases and extraction by detergents are also unchanged.

Limited proteolysis provides a useful index of loss of tertiary structure of the EGF receptor [13] and may be helpful in defining conditions during solubilization and purification, which are not denaturing. Binding of labeled EGF can also be used for this purpose, but binding affinity is substantially reduced in detergent [14] and requires more rigorous conditions for measurement, such as decreased detergent and salt concentrations, than does limited proteolysis.

ACKNOWLEDGMENTS

These studies were supported by U.S. Public Health Service grant AM24871 to E. J. O'Keefe and by grants from the Dermatology Foundation, the Burroughs Wellcome Co., and the Medical Faculty Grants Committee and University Research Council of the University of North Carolina.

REFERENCES

1. Hock RA, Nexø E, Hollenberg MD: Nature 277:403, 1979.
2. Das M, Fox CF: Proc Natl Acad Sci USA 75:2644, 1978.
3. Baker J, Simmer RL, Glenn KC, Cunningham DC: Nature 278:743, 1979.
4. Wrann MM, Fox CF: J Biol Chem 254:8083, 1979.
5. Linsley PS, Blifeld C, Wrann M, Fox CF: Nature 278:745, 1979.
6. Savage CR Jr, Cohen S: J Biol Chem 247:7609, 1972.
7. Hunter WM, Greenwood FC: Nature 194:495, 1962.
8. Fairbanks G, Steck TL, Wallach DFH: Biochemistry 10:2606, 1971.
9. Davies GE, Stark GR: Proc Natl Acad Sci USA 66:651, 1970.
10. O'Keefe E, Hollenberg MD, Cuatrecasas P: Arch Biochem Biophys 164:518, 1974.
11. Savage CR Jr, Ingami T, Cohen S: J Biol Chem 247:7612, 1972.
12. Gregory H, Preston BM: Int J Peptide Protein Res 9:107, 1977.
13. Tanford C, Reynolds JA: Biochim Biophys Acta 457:133, 1976.
14. Carpenter G: Life Sci 24:1691, 1979.
15. Lowry OH, Rosebrough NJ, Farr AL, Randall RJ: J Biol Chem 193:265, 1951.

Interaction of Serum and Cell Spreading Affects the Growth of Neoplastic and Non-Neoplastic Fibroblasts

R. W. Tucker, C. E. Butterfield, and J. Folkman

Cell Proliferation Laboratory, The Johns Hopkins Oncology Center, Baltimore, Maryland 21205 (R.W.T.) and The Children's Hospital Medical Center, Harvard Medical School, Boston, Massachusetts 02114 (C.E.B., J.F.)

Both growth factor availability and cell-to-cell contact have been mechanisms used to explain cell growth regulation at high cell density. Recently Folkman and colleagues have shown that changes in cell shape, rather than cell-to-cell contact, can regulate the growth of fibroblasts. However, in those studies the relation between serum and shape regulation of growth was not studied, nor were neoplastic and non-neoplastic cells compared. In this report we have studied these aspects by varying cell spreading and serum concentration independently for 2 non-neoplastic and 3 neoplastic cell lines. Cell spreading (projected cell area) was controlled by decreasing the adhesiveness of tissue culture plastic plates with poly (hydroxyethyl methacrylate) [poly (HEMA)]. Cell growth was measured as the increase in cell number/day. We have found that more spreading increased net growth of both neoplastic and non-neoplastic cells, while less spreading (toward rounded configuration) depressed growth. There were also quantitative differences between neoplastic and non-neoplastic cells. Neoplastic cells continued to grow under conditions of cell rounding, which completely prevented the growth of their non-neoplastic counterparts. Some neoplastic cells also tended to show little or no increase in net cell number for serum concentrations above 10% as cells became more spread; in contrast, all non-neoplastic cells grew more with increasing concentrations of serum as they became well spread. Thus, in normal cells, it appears that the sensitivity of cells to humoral factors is governed by cell spreading. This interaction between serum and cell shape is less prominent in some neoplastic cells.

Key words: serum, neoplastic, non-neoplastic, growth, cell area, cell spreading

It is well established that non-neoplastic fibroblasts stop growing at high cell density. Both humoral and cell contact mechanisms have been suggested to explain this phenomenon. For example, it has been proposed that cell crowding decreased the amount of serum available in the microenvironment of each cell [1–5]. Other studies suggested that contact between cells was itself sufficient to slow cell growth [6–9]. The fact that this effect of cell contact could be overcome by adding either serum [9] or pure mitogenic factors [10] suggested to some investigators that humoral mechanisms were the dominant factors.

Received June 2, 1980; accepted October 8, 1980.

0275-3723/81/1501-0029$03.50 © 1981 Alan R. Liss, Inc.

More recently, the implication that humoral and cell contact are competing factors has been questioned, as has the nature of the signals in cell contact. In fact, recent work [11] has shown that cell contact prevents growth simply by preventing cell spreading. Thus, sparsely plated cells held in the same rounded shape as crowded cells showed the same decreased growth rate. Similarly, Westermark [6] observed that restricted cell spreading on haptotatic palladium islands inhibited cell growth in the absence of cell-to-cell contact. Other workers have concentrated on the interaction between growth factors and cell shape. Using a combination of in vivo and in vitro studies on corneal cells, Gospodarowicz et al [12] showed that 2 different cell shapes had a profound influence on the cell's sensitivity to 2 different mitogens. O'Neill et al [1] recently demonstrated the importance of cell shape or cell spreading in the growth of hamster fibroblasts in vitro under conditions in which serum but not substratum conditions were varied. Thus, there are a number of studies demonstrating the importance of both cell shape and growth factors, but the description of the interaction of cell spreading and humoral factors is incomplete. No one study has examined the growth effect of serum and cell shape when they were varied independently.

In the present study, we have changed cell spreading and serum *independently* for both neoplastic and non-neoplastic cells. Cell spreading was varied by reducing the adhesiveness of tissue culture plastic with coatings of different concentrations of poly (hydroxyethyl methacrylate) [poly (HEMA)]. Cells were kept sparse, so that cell-to-cell contact was not a significant factor contributing to growth. We have found that over a wide range of different amounts of cell spreading (projected cell area) and serum concentrations, the growth of 3 non-neoplastic rodent fibroblasts was more sensitive to serum when the cells were extensively spread. Unspread (rounded) non-neoplastic cells were realtively resistant to growth stimulation by serum. In contrast, some neoplastic cells showed little change to serum above 10% as cells became more spread. Neoplastic cells also continued to grow under conditions of cell rounding and low serum concentration, which prevented the growth of non-neoplastic cells.

MATERIALS AND METHODS

Cell Cultures

3T3 and SV40-transformed 3T3 cells were both cultured in Dulbecco's modified Eagle's medium (DME, GIBCO Laboratories) supplemented with 10% calf serum (CS; GIBCO Laboratories). 3T3 cells, clone A-31, were obtained from C. Stiles, and SV40-transformed 3T3 cells were obtained from C. Scher. New ampules of 3T3 cells were thawed every 4 to 6 weeks to ensure that cells were growth-inhibited at high cell density.

The non-neoplastic cell line, CHEF 18-1, and the neoplastic line, CHEF 16-2, were established from the same Chinese hamster embryo [13]. Cells were grown in Alpha-MEM (Kansas City Biological) supplemented with 10% fetal calf serum (Microbiological Associates). New ampules of line 18-1 were thawed every 4 to 6 weeks to ensure that the cells remained density inhibited and nontumorigenic. Both cells lines were obtained from R. Sager and P. Novac.

HT1080, a human fibrosarcoma established from a tumor specimen [14], was grown in DME and 10% CS.

Poly (HEMA) Coating of Plastic

Plastic plates were coated with films of poly (2-hydroxyethyl methacrylate) [poly (HEMA)] to change the spreading of cells as previously described [11]. Poly (HEMA), obtained as a purified powder from Hydron Laboratories (New Brunswick, NJ), was dissolved in 95% ethanol to make a 12% stock solution (w/v). The stock solution was diluted from 10^{-1} to 10^{-4} with 95% ethanol. Various dilutions of the alcoholic solution of poly (HEMA) were distributed into Falcon 24 well multiwell culture plates with 200 μl in each well. The plates were allowed to dry in a 37°C warm room for 48 h with the covers in place. A hard, optically clear film remained bonded to the surface of the plate. Two to five $\times 10^4$ cells were then added to each well. Plates coated with the highest concentrations of poly (HEMA) (10^{-1} dilution) were the least adhesive, and cells remained spheroidal and weakly adherent. Plates coated with lower concentrations (10^{-2} to 10^{-4} dilutions) were more adhesive, and cell spreading increased as the poly (HEMA) dilutions increased.

Measurement of Cell Growth

Two to four $\times 10^4$ cells were plated initially into each well so that 3 days growth on plastic substratum was not quite confluent. Adherent cells were removed from the plates with trypsin (GIBCO Laboratories) 1, 2, and 3 days following plating of the cells. The total number of adherent and suspended cells was then counted in a Coulter Counter, and cell growth was expressed as the total number of cells accumulated over a 2- or 3-day period. Cell counts on the previous days were checked to be sure that an increase in the number of cells over 3 days represented progressive cell growth. When cells were clumped in suspension [eg, on 10^{-1} poly (HEMA)], 0.1% collagenase (grade II, Worthington Biochemical Inc.) was added to assist in breaking up the clumps. The incubation with collagenase was continued until gentle mechanical agitation produced a single cell suspension. Cell counts were done in triplicate and expressed as the mean ± SD.

Growth rates (ie, population doubling) from day 1 to day 3 were measured by dividing final cell number on day 3 by the initial cell number on day 1 after plating cells. Ratio of >1 indicates net growth.

Measurement of Cell Spreading

The projected cell area for each individual cell was measured from the cell outline in 35 mm pictures. At 24 h after plating cells were photographed through an inverted Nikon phase microscope, and the developed film was projected onto a digital image analyzer (ZEISS MOP-3) on which the cell area was directly measured. The average cell area ± SEM was obtained for 40 cells for each combination of serum and substrata.

RESULTS

Well Spread Non-Neoplastic Cells on More Adhesive Substrata Were More Sensitive to Serum

As Figures 1A and 2A show, non-neoplastic cell lines 3T3 and 18-1 grew more (larger cell number on day 3) as serum was increased from 0.5% to 30%. Moreover, on the more adhesive surfaces [plastic, 10^{-3} poly (HEMA)], growth rate (fold increase) was increased compared to that on the less adhesive surfaces [10^{-2} and 10^{-1} poly (HEMA)]

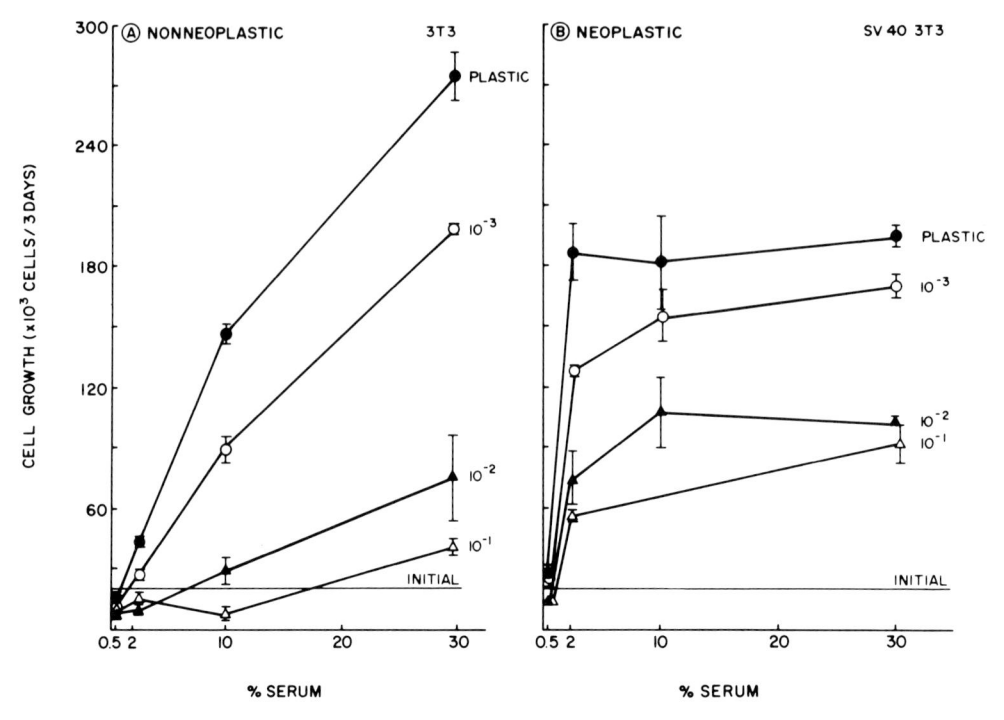

Fig. 1. Adhesiveness of the substratum changes of cells. 3T3 (A) and SV40-transformed 3T3 (B) cells were plated at same initial density (2×10^4/well) then counted daily. Total number of cells/well after 3 days of growth is plotted vs percent serum concentration in medium for different poly (HEMA) concentration. Note increasing slope of growth vs serum lines for 3T3 cells (A) as substrata were made more adhesive. In contrast, SV40-transformed 3T3 cells (B) showed little change in slope above 2% serum supplementation. Data plotted are mean ± SD. Symbols ● = for plastic; ○⁻ = 10^{-3} poly (HEMA); ▲ = 10^{-2} poly (HEMA); △ = 10^{-1} poly (HEMA).

(see Table I). The increased cell number on day 3, therefore, primarily reflects changes in growth rate and not plating efficiency. We conclude that on the more adhesive surfaces, non-neoplastic cells were more sensitive to growth stimulation by serum, grew at a higher rate, and reached a higher final cell number.

Rounded Non-Neoplastic Cells on Poorly Adhesive Substrata Did Not Grow

Figures 1A and 2A also demonstrates that non-neoplastic cells on poorly adhesive substrata [10^{-1} and 10^{-2} poly (HEMA)] did not show net growth in 10% serum or less, and only marginal growth in 30% serum. In these conditions, the number of cells at day 3 was not above the initial number plated, and monitoring of cell counts on days 1 and 2 revealed no transient increase in growth (see Table I). Thus, these non-neoplastic fibroblasts did not grow when prevented from spreading on poorly adhesive substrata, although high serum concentration (30%) could partially overcome this growth inhibition.

Well-Spread Neoplastic Cells on Adhesive Substrata Were More Sensitive to Serum

As Table I shows, neoplastic lines (mouse SV40 3T3, hamster 16-2, and human HT1080) grew faster at any given serum concentration as substratum was made more adhesive. However, unlike non-neoplastic cells, on a given substratum net cell number did

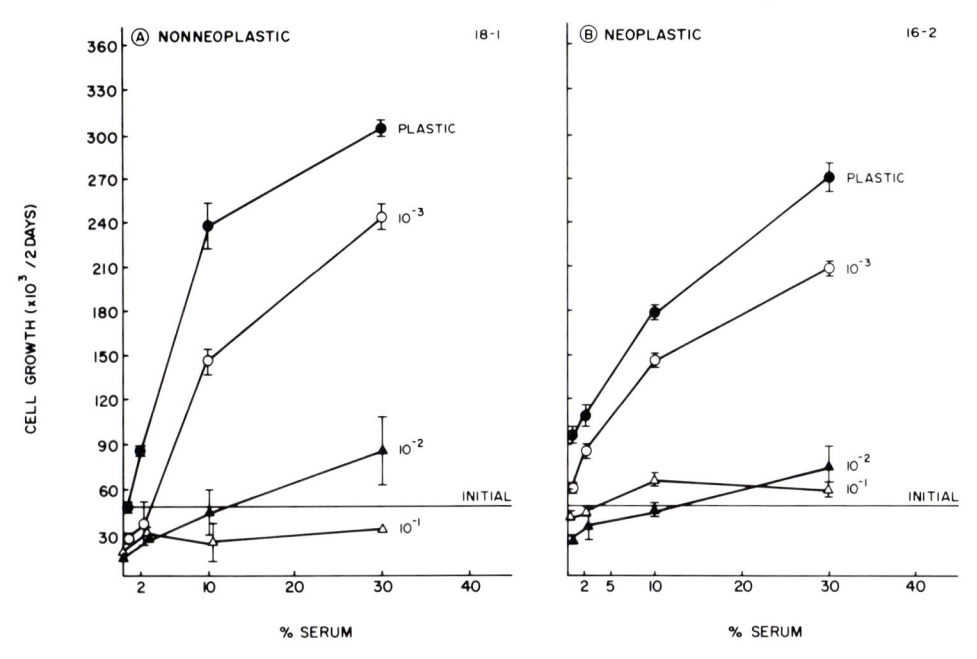

Fig. 2. Experimental details same as Figure 1. Cell lines 18-1, 16-2 were plated at same initial density (5×10^4/well). Note that slopes are similar, but that growth of neoplastic 16-2 is above initial cell number on poorly adhesive substrata (10^{-1}), whereas, that for 18-2 cells is not.

not increase above that stimulated by 2% serum for SV40 3T3 and by 10% for HT1080 (Figs. 1B, 2B, 3). Neoplastic line 16-2 showed an increase in net cell number similar to that of non-neoplastic line 18-1. Thus, neoplastic cell lines showed increased sensitivity to growth stimulation by serum as the adhesiveness of the substratum was increased, but growth for SV40 3T3 was maximal at 2%, and for HT1080 growth was maximal at 10% serum. This increase in sensitivity to growth stimulation by serum is also reflected in the net cell growth of all 3 neoplastic cells in 0.5% serum.

Rounded Neoplastic Cells on Poorly Adhesive Substrata Grew

As Figures 1B, 2B, and 3 show, neoplastic lines all grew above the initial number of cells plated on poorly adhesive substrata [10^{-1} and 10^{-2} poly (HEMA)]. This growth occurred even in the presence of low amounts of serum (2%) for cell lines SV40 3T3 and HT1080. This is in contrast to the absence of growth of non-neoplastic cells on 10^{-1} and 10^{-2} poly (HEMA).

Non-Neoplastic and Neoplastic Cell Pairs in Identical Culture Conditions Were Spread to the Same Extent

Despite the marked differences in the effect of substrata on growth, neoplastic and non-neoplastic cells from common origin changed their cell spreading to the same extent as substrata and serum concentrations were changed. As Figures 4 and 5 show, the amount of cell spreading on the different dilutions of poly (HEMA) was remarkably similar for both cell pairs — 3T3 vs SV40-transformed 3T3 cells, and 18-1 vs 16-2. All cell lines showed little change in the amount of spread cell area as serum concentrations changed

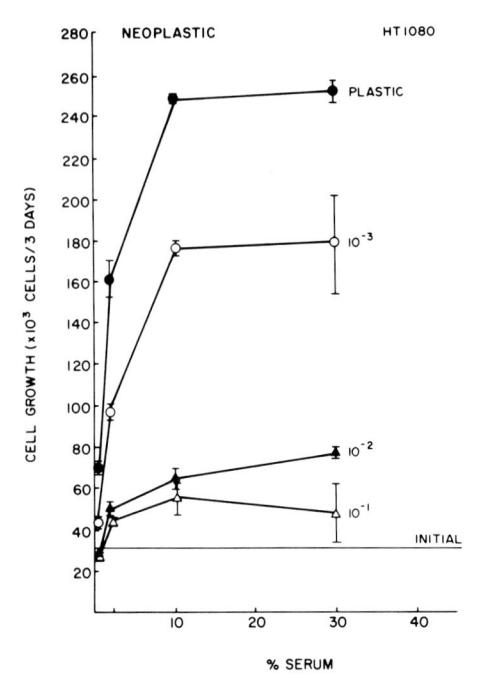

Fig. 3. Experimental details same as Figure 1. Cell line HT1080 was plated at initial density of 3×10^4 cells/well. Note flat curves (no change in cell number) above 10% serum supplementation.

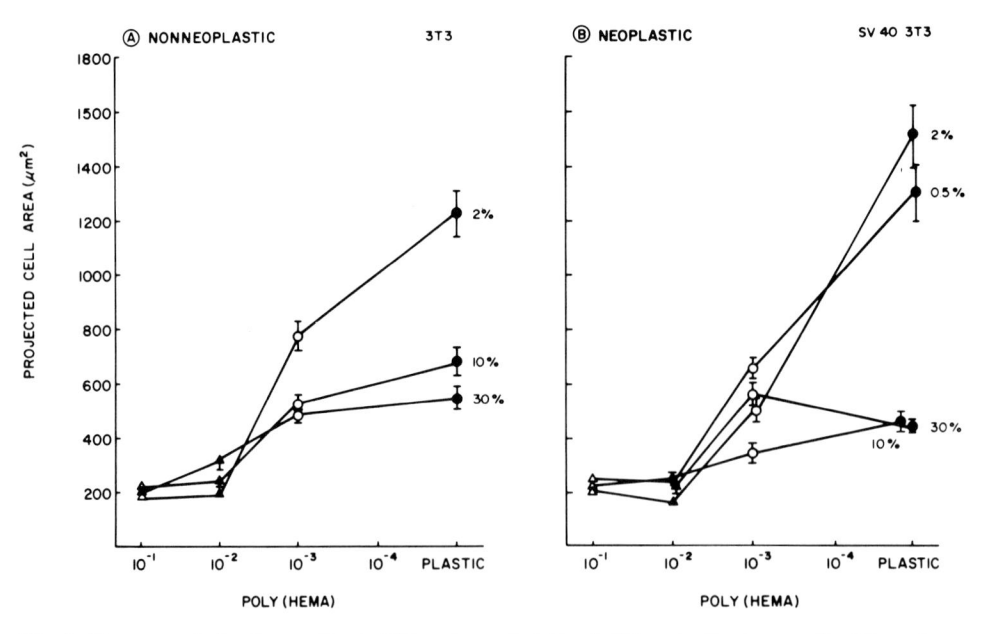

Fig. 4. The amount of cell spreading is different on substrata of different adhesiveness. See Figure 1 for experimental details. Cell area measured of at least 40 cells 24 h after plating. Note both 3T3 (A) and SV40-transformed 3T3 cells (B) showed increased cell spreading (projected cell area) on 10^{-3} poly (HEMA) and plastic with 2% serum supplementation. Data plated are mean ± SEM.

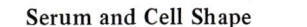

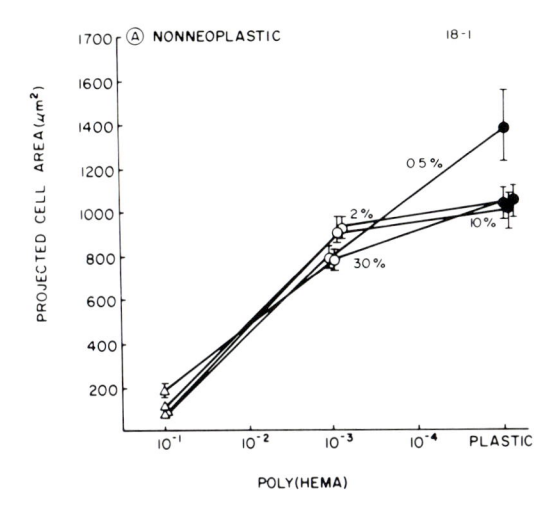

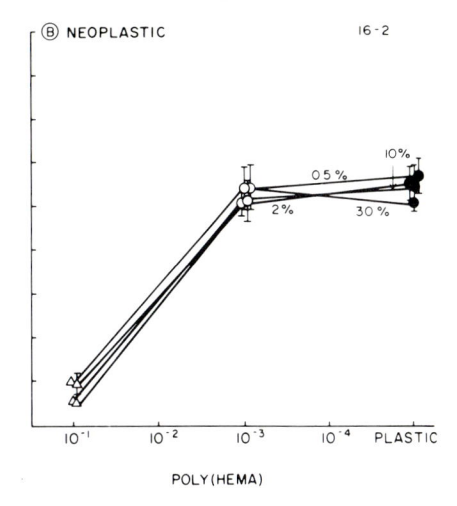

Fig. 5. Experimental details same as Figures 2 and 4. Note very little effect of serum on cell spreading except for 0.5% serum supplementation of 18-1 cells on plastic.

TABLE I. Effect of Serum and Substratum on Growth Rates

Non-neoplastic cell line	Serum (%)	Fold increase on[a] PL, 10^{-3}, 10^{-2}, 10^{-1}	Neoplastic cell line	Fold increase on[a] PL, 10^{-3}, 10^{-2}, 10^{-1}
3T3	30	3.6, 3.6, 1.4, 1.2	SV3T3	2.3, 1.7, 1.4, 1.2
	10	2.5, 1.5, 0.6, 1.0		1.6, 1.2, 1.0, –
	2	1.5, 0.6, 0.3, 0.7		6.0, 4.8, 3.8, 2.7[b]
	0.05	0.4, 0.2, 0.4, 0.5		2.0, 1.4, 1.4, 1.6
18-1	30	4.3, 4.7, 1.1, 0.3	16-2	1.8, 1.6, 0.9, 1.0
	10	5.1, 3.8, 0.8, 0.2		1.0, 1.1, 0.5, 0.7
	2	2.2, 2.9, 0.5, 0.2		1.1, 1.1, 0.8, 0.6
	0.05	1.8, 1.4, 0.6, 0.4		1.1, 0.8, 0.6, 0.8
	30		HT1080	2.8, 2.7, 2.9, 1.8
	10			5.6, 3.8, 1.5, 1.0
	2			2.2, 1.8, 1.1, 0.9
	0.5			1.5, 1.2, 1.2, 0.8

[a]Fold increase is net number of cells on day 3 divided by net number of cells on day 1 of growth.
[b]High values of fold increases in 2% serum reflect the high sensitivity of these cells to serum, so that growth was maximal in 2%, and in 10% and 30%, much of final growth had already occurred by day 1 after plating.

until poly (HEMA) was 10^{-3} diluted or 0 (plastic). Moreover, the amount of cell spreading was similar for cell line pairs at each combination of poly (HEMA) and serum; Figures 4A and 5A are virtually superimposable with 4B and 5B, respectively. Figures 4 and 5 also demonstrate that on an adhesive substratum like plastic, both cell lines were more spread in the medium with lower serum concentrations. Thus, both serum concentration and substratum adhesiveness interacted to affect the extent of cell spreading, and this amount of cell spreading was remarkably similar for paired non-neoplastic and neoplastic cell lines. Figure 6 demonstrates the very small changes in spreading by HT1080 in the different serum concentrations.

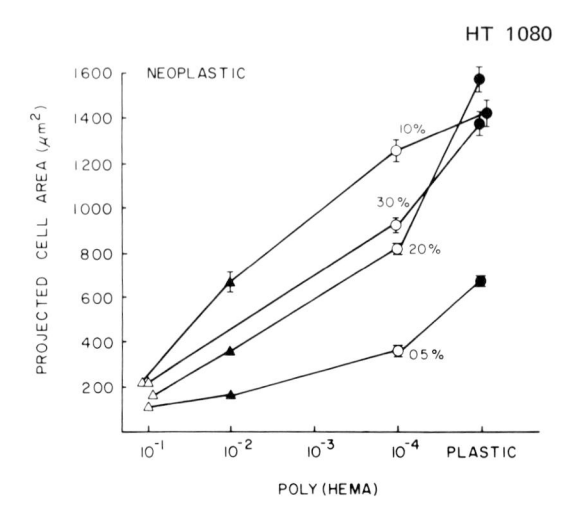

Fig. 6. Experimental details same as Figures 3 and 4. Note that lowered serum did not cause increased cell spreading for this neoplastic cell line, HT1080.

Serum and Cell Spreading Interacted in the Growth Stimulation of Non-Neoplastic Cells

Figures 7A and 8A show that both 3T3 and 18-1 cells were more sensitive to serum as they became more spread (larger projected area). For a given increment in cell area, there was a larger increment in net cell growth at the higher serum concentrations. As we have seen before, these changes in net cell growth for a given serum concentration reflect increased growth rates (see Table I). Thus, increased spreading of non-neoplastic cells increased their sensitivity to growth factors in serum.

Serum and Cell Spreading Were Not Coupled in the Growth Stimulation of Some Neoplastic Cells

Figures 7B and 9 show that neoplastic cells SV403T3 and HT1080 showed little change in the increment of net cell growth for a given change in projected cell area as serum concentration was changed from 10% to 30%. Cell line 16-2 (Fig. 8B) showed some change in the net cell growth as cell area was changed at different serum concentrations, but quantitatively less than its non-neoplastic counterpart (line 18-1 in Fig. 8A). Thus, the relationship between net cell growth and projected cell area for some neoplastic lines showed little or no change as serum concentration was varied. In other words, growth of these neoplastic cells was already maximal at low serum concentrations, so that changes in projected cell area on different substrata affected growth, but changes in serum concentration did not. In this sense, exogenous serum and cell spreading effects on cell growth were not coupled for neoplastic cells as they were for non-neoplastic cells (see Figs. 7A and 8A).

DISCUSSION

For the first time it has been possible to vary serum and cell shape independently and to study their combined effects on cell growth. This study shows that cell spreading and serum effects on growth are much more tightly coupled in 2 non-neoplastic cell lines

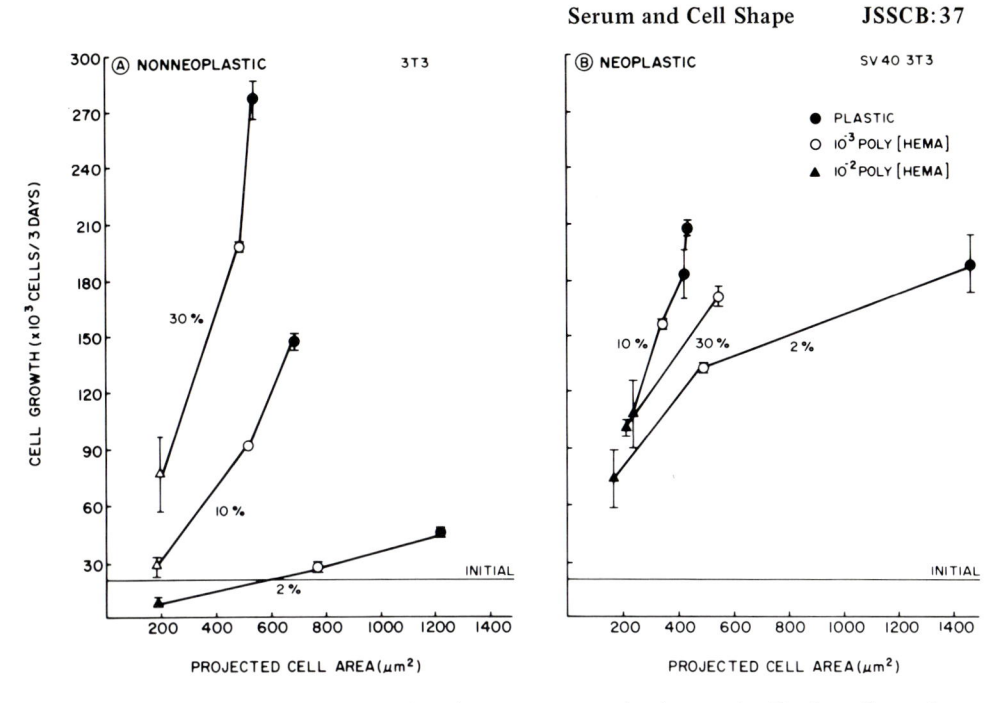

Fig. 7. Cell spreading (projected cell area) and serum concentration interact in affecting cell growth. See Figures 1 and 4 for experimental details. Note the marked differences in graphs of cell growth vs projected cell area of 3T3 at different serum concentrations (A). In contrast, the corresponding curves for SV40-transformed 3T3 cell (B) are more similar.

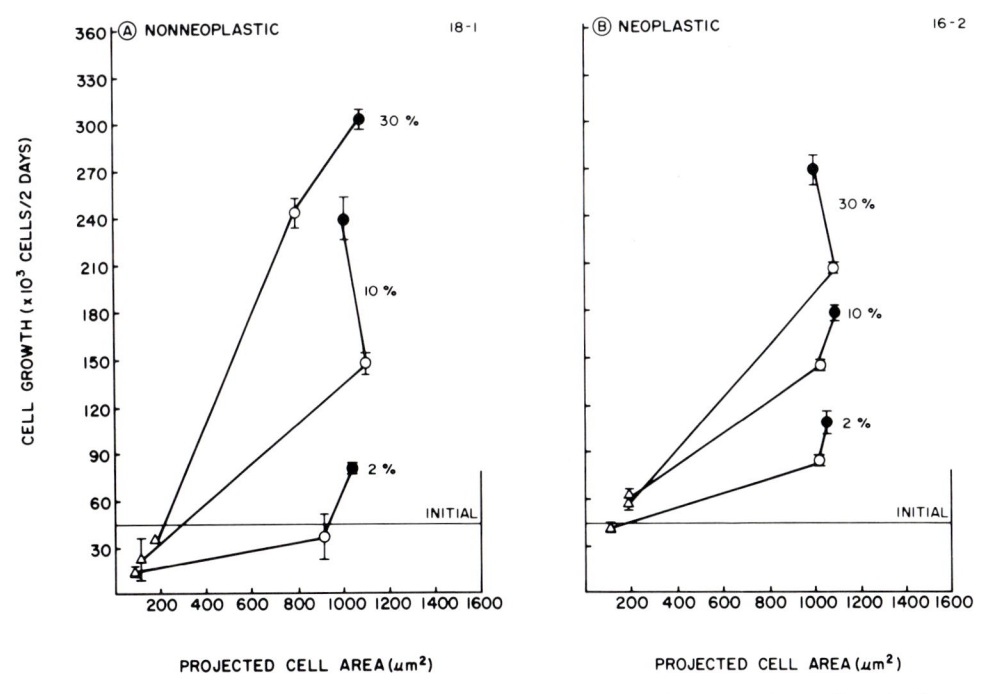

Fig. 8. Experimental details same as Figure 7. Note that the graphs of cell growth vs projected cell area at different serum concentrations for cell line 16-2 are closer together than the same graphs for line 18-1.

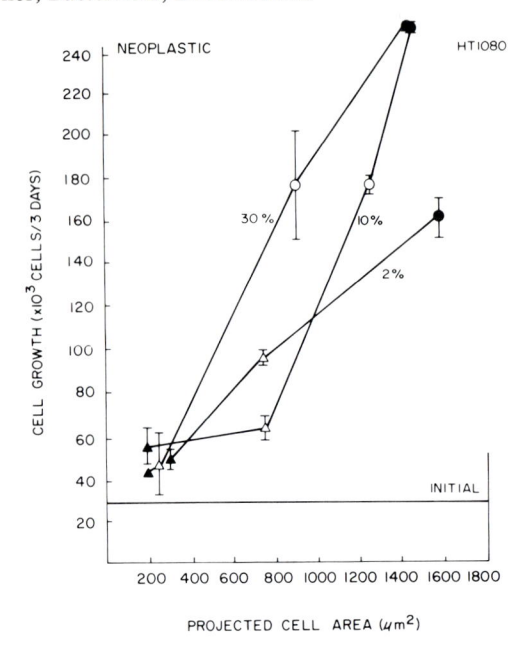

Fig. 9. Experimental detail same as Figure 7. Note the marked similarity in the slope of the graphs of cell growth vs projected cell area in 10% and 3% serum concentrations.

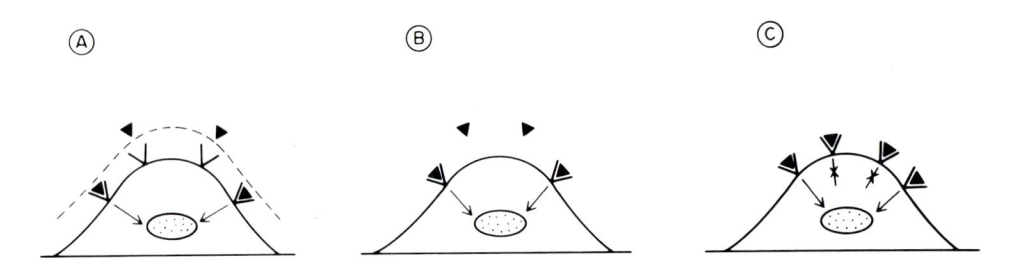

Fig. 10. Different theoretical reasons for decreased growth in rounded non-neoplastic cells secondary to decreased effective growth factor effect. A. Diffusion limited hypothesis. Growth factors (▲) are prevented from reaching receptors (V) on cell membrane. B. Receptor hypothesis. Receptors (V) for growth factors decrease in number so that fewer growth factors (▲) are bound. C. Cytoskeletal hypothesis. Growth factors bind to receptors, but the cytoskeletal link between receptor binding and nuclear DNA synthesis is ineffective.

than in 3 neoplastic cell lines we studied. In the 2 non-neoplastic cell lines increased cell spreading was associated with an increased sensitivity to serum concentration; neoplastic cells also increased net cell growth with similar changes in cell spreading, but growth was already maximal at low serum concentrations. At the same time, rounded cells of all 3 neoplastic lines grew on poorly adhesive substrata even in low (2%) serum. Thus, these neoplastic cells required little serum and no attachment to grow, but still increased their growth rate if allowed to spread. Neoplastic cells, which depended the least on increases to serum (SV403T3), were either producing their own growth factors [15] whose function was also sensitive to cell shape or were so sensitive to exogenous serum that concentrations above 2% or 10% did not stimulate additional growth. The important point is that cell spreading and exogenous serum do not always cooperate in producing neoplastic cell growth to the same extent that they do in non-neoplastic cells. The most fundamental control appears to be the effect of cell spreading with which exogenous serum and other growth factors interact in various ways.

The interdependence of serum (growth factor) and cell spreading in producing growth also provides new interpretations of previously reported phenomena. The requirement for increased serum [9] or increased growth factor [10, 16] to overcome density inhibition of growth can now be viewed as resulting from the resistance of rounded non-neoplastic cells to growth stimulation by serum. Similarly, the increased growth of neoplastic cells in suspension when attachment is provided by floating beads [17] shows the sensitivity of even neoplastic cells to changes in cell spreading. The growth of some non-neoplastic cells in suspension when very high serum concentrations are used [1] emphasizes that the effect of cell shape on serum sensitivity is a quantitative, not an all-or-none, change.

The mechanism of the coupling of cell spreading and serum effects in non-neoplastic cells is unknown. Possible explanations of the inhibiton of growth of rounded non-neoplastic fibroblasts include either increased inhibitors or decreased stimulators. So far, increased inhibitors have been found only for epithelial cells [18], so we will concentrate here on stimulators in discussing fibroblasts. Decreased actions of growth-stimulatory factors could result from a diminished availability of diffusion-limited growth factors associated with the postulated decreased surface area in rounded cells [1]. The result would be fewer growth factor receptors occupied (Fig. 10A). Fewer bound receptors could also result from changes in number or distribution of receptors (Fig. 10B), although Westermark [19] has shown that changes in number of receptors are not important in EGF stimulation of glial cells. Another possibility is that receptors are occupied but unable to stimulate DNA synthesis in the rounded cell (Fig. 10C). How cell shape itself might affect response to growth factors is unclear, but Benecke et al [20] have recently shown that cytoskeletal events associated with cell spreading affect protein synthesis and mRNA metabolism. Whatever the mechanism, the coupling of humoral information and conformational (spatial) information is important in the regulation of cell growth. Aspects of this coupling appear quantitatively different in neoplastic cells. Our work here suggests that changes in a cell that occur with increased spreading (increased projected area) are associated with growth regulatory changes.

ACKNOWLEDGMENTS

Supported by NCI grant CA14019-04 (to Dr. Folkman) and a grant to Harvard University from the Monsanto Company. Dr. Tucker was supported by a grant from the American Cancer Society, Massachusetts Division, and a fellowship from the Medical Foundation, Incorporated, Boston, Mass. The generous support and advice of Dr. A. B. Pardee is also gratefully acknowledged.

REFERENCES

1. O'Neill CH, Riddle PN, Jordan PW: Cell 16:909, 1979.
2. Stoker MGP: Nature 246:200, 1973.

3. Holley RW: Nature 258:487, 1975.
4. Rozengurt E: In Dumont, Brown BL, Marshall NJ (eds): "Eukaryotic Cell Function and Growth." New York: Plenum Press, 1976, p 71.
5. Noonan KD, Burger MM: Prog Surface Membr Sci 8:245, 1974.
6. Westermark B: Biochem Biophys Res Commun 69:304, 1976.
7. Dulbecco R: In Wolstenholm GE, Knight J (eds): "Growth Control in Cell Cultures." Edinburgh and London: Churchill Livingstone, 1971, p 71.
8. Stoker MGP: Virology 24:164, 1964.
9. Dulbecco R: Nature 227:802, 1970.
10. Rozengurt E: Nature 269:155, 1977.
11. Folkman J, Moscoma A: Nature 273:345, 1978.
12. Gospodarowicz D, Greenburg G, Birdwell CR: Cancer Res 38:4155, 1978.
13. Sager R, Novac PE: Somatic Cell Genet 4:375, 1978.
14. Rasheed S: Cancer 33:1027, 1974.
15. Delarco JE, Todaro GJ: Proc Natl Acad Sci USA 75:4001, 1978.
16. Vogel A, Ross R, Raines E: Personal communication, 1980.
17. Paul D, Henaham M, Walter S: J Natl Cancer Inst 53:1499, 1974.
18. Holley RW, Armour R, Baldwin JH: Proc Natl Acad Sci USA 75:1864, 1978.
19. Westermark B: Proc Natl Acad Sci USA 74:1619, 1977.
20. Benecke BJ, Ben-Ze'en A, Penman S: Cell 16:931, 1978.

Stimulatory Activity of PHA-LCM for Normal Human Hemopoietic Progenitors and Leukemic Blast Cell Precursors: Separation by Isoelectric Focusing

A. A. Fauser and H. A. Messner

The Ontario Cancer Institute, Department of Medicine, and Institute of Medical Science, University of Toronto, Toronto, Ontario, Canada M4X 1K9

Medium conditioned by leukocytes in the presence of phytohemagglutinin (PHA-LCM) promotes growth of human hemopoietic progenitors (CFU-GEMM, BFU-E, CFU-C) and precursors of leukemic blast cells. PHA-LCM was separated by isoelectric focusing and each fraction tested with nonadherent cells of normal individuals as well as blast cells from two patients with acute myelogenous leukemia. Activity profiles for CFU-GEMM, BFU-E and CFU-C ranged from pH 5.0—6.5. The profile for activity stimulatory for leukemic blast cells was broader and ranged from pH 5.5—7.5. Although some overlap was observed, the main peaks of stimulatory activity for normally differentiating progenitors and precursors of leukemic blast cells were separable with respect to their isoelectric point.

Key words: hemopoiesis, leukemia, hemopoietic progenitors, cell culture, stimulatory molecules, isoelectric focusing

The addition of plant lectins [1—7] such as phytohemagglutinin (PHA), pokeweed mitogen (PWM), or concanavalin A (Con A) to short-term cultures of hemopoietic subpopulations facilitates the release of growth-promoting stimulators for early hemopoietic progenitors. These molecules were instrumental in the development of culture assays for murine [8, 9] and human pluripotent hemopoietic progenitors (CFU-GEMM) [10, 11], and for precursors of human leukemic blast cells [12]. Murine pluripotent hemopoietic progenitors form large mixed colonies of granulocytes, erythroblasts, megakaryocytes, and macrophages when cultured with media conditioned by spleen cells in the presence of PWM [8, 9]. A similar type of mixed colony can be observed in cultures of human bone marrow or peripheral blood when media conditioned by peripheral leukocytes under the influence of phytohemagglutinin (PHA-LCM) are added [10, 11]. This material usually also enhances growth of committed erythroid (BFU-E) [13] and granulocytic precursors (CFU-C). Furthermore,

Received April 21, 1980; accepted October 20, 1980.

0275-3723/81/1501-0041$02.50 © 1981 Alan R. Liss, Inc.

it was observed that the same crude preparation of PHA-LCM stimulates the formation of colonies with blast cell properties in cultures of peripheral blood specimens derived from patients with acute myelogenous leukemia (AML) [12]. Similar growth requirements for leukemic blast cells have been reported by other authors [14, 15].

We have attempted to separate activities in PHA-LCM that stimulate CFU-GEMM, BFU-E, CFU-C, and cells that form leukemic blast colonies. A crude PHA-LCM preparation was fractionated by isoelectric focusing, and the resulting fractions were tested on bone marrow cells of normal individuals and leukemic blast cell populations from peripheral blood samples of patients with AML.

MATERIALS AND METHODS

Patient Material

Bone marrow samples were obtained from 2 normal bone marrow transplant donors, 1 bone marrow transplant recipient after stable engraftment, and 1 patient after successful treatment for pure red cell aplasia. All patients had normal hemopoietic parameters at the time of study. Specimens were aspirated into heparinized syringes, and the buffy coat was prepared by centrifugation at 150g for 10–15 min. Subsequently, adherent cells were removed as previously described [16] to minimize the endogenous production of stimulatory activities [16, 17]. The nonadherent fraction was further depleted of red blood cells and mature granulocytes by centrifugation in LSM (Litton Bionetics) at a density of 1.077 gm/ml. The resulting nonadherent, mononuclear cells were used to test the stimulatory activity of crude and fractionated PHA-LCM.

Peripheral blood specimens from two patients with acute myeloid leukemia in relapse were drawn into heparinized syringes, and mononuclear cells of density less than 1.077 gm/ml were prepared as described above. The cell population was further depleted from E-rosette-forming cells by incubation with sheep red blood cells (SRBC) at 4°C and subsequent centrifugation in LSM [18, 19]. Cells were stored at −70°C (Kelvinator) in 10% DMSO, 20% fetal calf serum (FCS), and thawed immediately prior to plating [20].

Preparation of PHA-LCM

Active PHA-LCM is released by peripheral blood cells of normal individuals. In addition, two patients with hemochromatosis who regularly undergo therapeutic phlebotomies have provided cells for the production of PHA-LCM. Material for this study was exclusively prepared using peripheral blood cells from 1 of these 2 patients. Cells were incubated in modified Dulbecco's minimum essential medium (DMEM) [21] with 10% FCS and 1% PHA [5]. The supernatant was harvested after 7 days of culture.

Separation of PHA-LCM

PHA-LCM was separated by column isoelectric focusing in a density gradient [22]. Briefly, the material was admixed with 1% LKB Ampholine carrier of pH 5–8 and loaded into the isoelectric focusing column (LKB 8101, volume 110 ml) incorporated into a sucrose density gradient (5–50% w/v). Current was applied for 16 h at 4°C. Four milliliter fractions were collected and dialysed over 48 h with 2 exchanges of modified DMEM. After dialysis, the pH of each fraction was adjusted to 7.4.

Each fraction was tested for its growth-promoting activity for CFU-GEMM, BFU-E, CFU-C, and leukemic blast cell progenitors.

Colony Assays for Hemopoietic Progenitors

Mixed hemopoietic colonies, erythroid bursts, and granulocytic colonies were grown as previously described [10, 11]. Experiments were usually performed with 2×10^5 nonadherent mononuclear target cells per culture plate. In some control experiments, unseparated buffy coat cells were used. Cells were admixed with modified DMEM, 30% FCS, 0.9 methylcellulose, and PHA-LCM preparations as indicated below. Aliquots of 0.9 ml were placed in 35 mm Petri dishes and incubated at 37°C in a humidified atmosphere supplemented with 5% CO_2. One unit of erythropoietin (EPO) (Step III, Connaught Laboratories, Willowdale, Ontario) was added on day 4 of culture. Each dish was examined after a total incubation of 14 days for the presence of mixed colonies, erythroid bursts, and granulocytic colonies.

Crude PHA-LCM was added to control plates at a previously established optimal concentration of 5% [23]. In experimental groups, the crude preparation of PHA-LCM was replaced by material of individual fractions obtained by isoelectric focusing. Each fraction was routinely tested at a concentration of 0.5%. This concentration was established for the peak fraction (pH 5.9) as being equivalent to the stimulatory activity in 5% crude PHA-LCM.

Leukemic blast cell colonies were grown from E-rosete-forming cell-depleted preparations [20] immobilized in 0.9% methylcellulose with modified DMEM, 30% fetal calf serum, and 5% crude PHA-LCM or 0.5% of material derived from each fraction. EPO was not added. Colonies were counted after 5 to 7 days of culture. Their blast-like phenotype was verified by subjecting randomly removed colonies to further analysis by Wright stain, peroxidase reaction, and rosette formation with SRBC. Blast colonies contained E-rosette-negative cells with blast morphology and negative or slightly positive peroxidase reaction.

RESULTS

Growth Requirements of Normal and Leukemic Target Cell Populations

The growth of human hemopoietic colonies is dependent upon appropriate culture conditions. Some of the required stimulatory activities are released into the culture by adherent cells. We attempted to eliminate the endogenous production of stimulators for CFU-GEMM, BFU-E, and CFU-C by removing adherent cells. The quality of the depletion was tested in 3 experiments by examining the influence of PHA-LCM and EPO on colony growth. Colony formation was found to be dependent upon the addition of exogenous stimulators. While granulocytic colonies required only PHA-LCM, it was necessary to add PHA-LCM and EPO to promote growth of mixed hemopoietic colonies and erythroid bursts (Table I). It is of note that erythroid bursts developed only occasionally in cultures that contained EPO but no PHA-LCM. Thus, nonadherent, mononuclear bone marrow cells are appropriate target cells to analyze the stimulatory effects of PHA-LCM and its fractions in cultures that also contain EPO. No colonies were observed without PHA-LCM.

Frozen cells from both patients with AML formed E-rosette-negative, peroxidase-negative, or weakly positive colonies of blast cell morphology in the presence of 5% PHA-LCM.

Stimulatory Activity Profiles for Normal Hemopoietic Progenitors

The fractions of PHA-LCM obtained by isoelectric focusing were analyzed for their ability to promote growth of CFU-GEMM, BFU-E, and CFU-C by performing 3 types of experiments: first, to establish the activity profiles using a constant concentration of each

TABLE I. Effect of EPO and PHA-LCM on NA Cells in Culture

Patient	Culture conditions	CFU-C[a]	BFU-E[a]	CFU-GEMM[a]
M.R.	—	0	0	0
	EPO	0	1	0
	PHA-LCM	27	0	0
	PHA-LCM + EPO	29	32	2
C.C.	—	0	0	0
	EPO	0	1	0
	PHA-LCM	81	0	0
	PHA-LCM + EPO	74	57	2
G.H.	—	0	0	0
	EPO	1	0	0
	PHA-LCM	28	0	0
	PHA-LCM + EPO	24	34	1

[a]Number of colonies by 2×10^5 NA-cells/plate.

fraction; second, to examine the presence of inhibitors; third, to study the influence of increasing concentrations of the fraction with highest activity and a fraction that did not promote colony growth at 0.5%.

Nonadherent, mononuclear bone marrow cells were plated with one unit of EPO and material from each fraction at a concentration of 0.5%. Representative profiles from one of four experiments are depicted in the top panel of Figure 1. Stimulatory activities for CFU-GEMM, BFU-E, and CFU-C were present in fractions of pH 5–6.5. The plating efficiency observed for the peak fraction was comparable to that of 5% crude PHA-LCM. Although some differences became apparent in the shape of the activity profiles for different hemopoietic progenitors, it was not feasible to separate molecules with restricted specificity.

In order to examine the fractions for putative inhibitors, unseparated buffy coat cells were cultured with material of each fraction and one unit of EPO to assess the plating efficiency of BFU-E. Buffy coat cells were utilized since the addition of EPO permitted the formation of some erythroid bursts without PHA-LCM. As indicated in Figure 2, burst formation in the presence of material from each fraction was at least comparable to that of the control group (49 erythroid bursts per 2×10^5 target cells). None of the fractions led to reduction of the plating efficiency below the control value. A consistent increase was documented in 3 experiments for fractions with an isoelectric point of pH 5.9. The peak was identical to that identified with nonadherent, mononuclear target cells. In addition, a second peak was demonstrated in fractions with material of acid isoelectric point. This peak was regularly absent in cultures of nonadherent, mononuclear target cells.

Material of the peak fraction (pH 5.9) and of one of the inactive fractions (pH 7.6) was added in concentrations increasing from 0.125% to 5% (Fig. 3). Nonadherent, mononuclear cells and unseparated buffy coat cells served as target cells and were assessed for their ability to form erythroid bursts with 1 unit of EPO. The plating efficiency increased from 1 to 34 bursts for 2×10^5 nonadherent cells and from a background level of 18 to 49 bursts for 2×10^5 buffy coat cells with increasing concentrations of material with an isoelectric point at pH 5.9. The highest observed frequency was considerably above the control values of 18 and 29 bursts, respectively. The fraction with isoelectric point of pH 7.6 did not contain any stimulators when tested up to concentrations of 5%. No indication of inhibitory activities was obtained for unseparated buffy coat cell preparations.

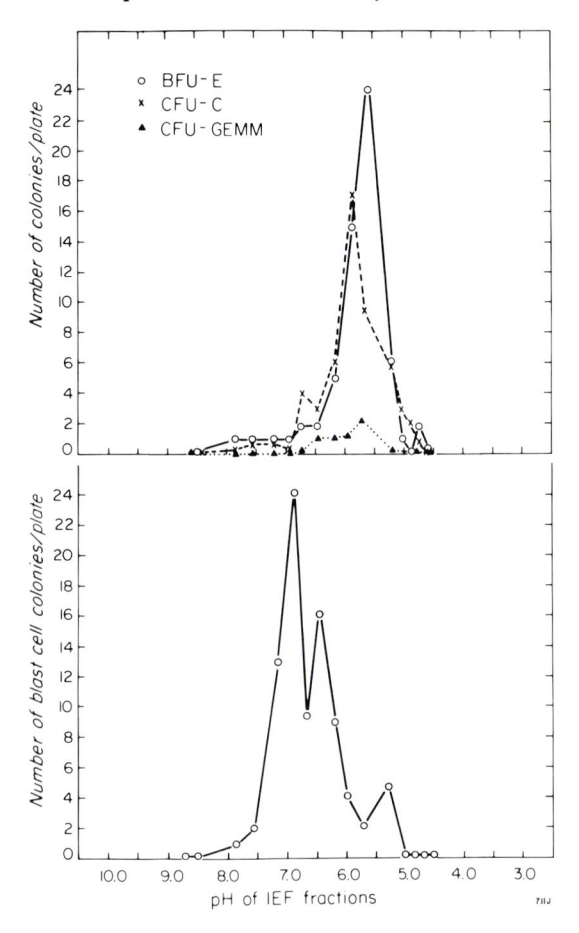

Fig. 1. Top panel. Fractions of PHA-LCM separated by isoelectric focusing were assessed for activities that promote growth of CFU-GEMM, BFU-E, and CFU-C by culturing nonadherent, mononuclear cells in the presence of 1 unit of EPO. No colonies were observed in the absence of crude or fractionated PHA-LCM. Addition of 5% crude PHA-LCM to controls yielded 2 mixed colonies, 27 erythroid bursts, and 20 granulocytic colonies. Bottom panel. A cryopreserved, E-rosette forming, cell depleted blast cell population of a patient with AML in relapse was used to identify activities that stimulate blast colony formation. Controls without crude or fractionated PHA-LCM did not yield any colonies.

Stimulatory Activity Profiles for Leukemic Blast Colony-Forming Cells

Peripheral blood cells from 2 patients with AML in relapse were depleted of E-rosette-forming cells. The resulting mononuclear, E-rosette-depleted population was used to test the fractions for activities that promote growth of leukemic blast cell progenitors. Each fraction was examined in cultures without EPO at a concentration of 0.5%. Growth of leukemic blast cell colonies was supported by material derived from fractions of pH 5.5 to 7.5 as depicted for one patient in the bottom panel of Figure 1.

The profile differed from those observed for normal progenitors and included mainly material with a more alkaline isoelectric point. It overlapped with activities for normal hemopoietic progenitors and appeared more heterogeneous, with 3 subcomponents. A bio-

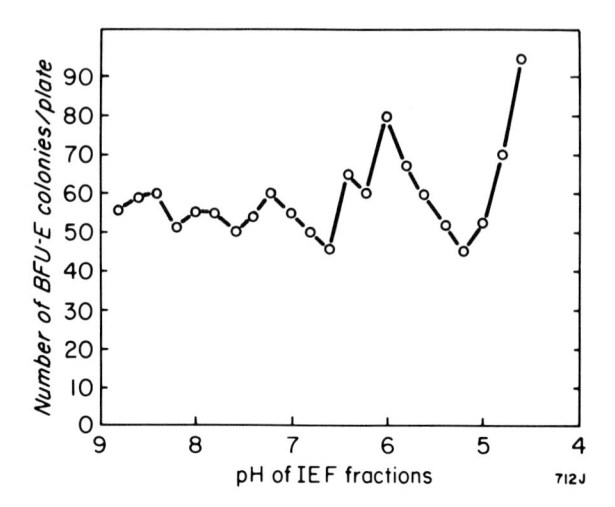

Fig. 2. Influence of PHA-LCM fractions obtained by isoelectric focusing on erythroid burst formation using 2×10^5 buffy coat cells as targets. All cultures contained 1 unit of EPO. Controls without crude or fractionated PHA-LCM gave rise to 49 colonies in the presence of 1 unit of EPO.

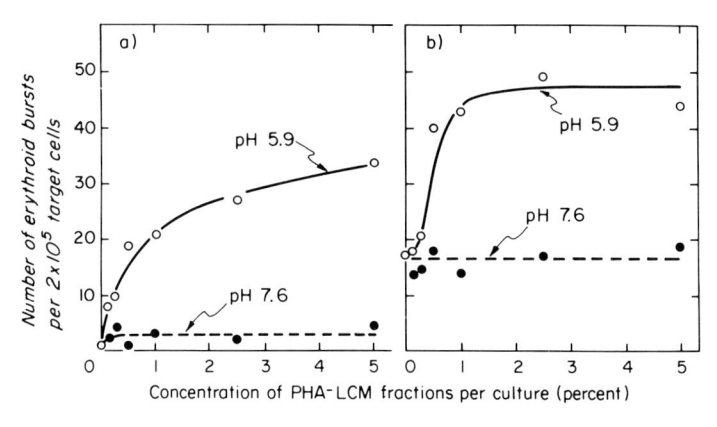

Fig. 3. Influence of increasing concentrations of 2 fractions of PHA-LCM obtained by isoelectric focusing. Both fractions were tested on nonadherent, mononuclear cells and unseparated buffy coat cells. Controls with 5% crude PHA-LCM yielded 18 bursts for nonadherent cells and 44 bursts for buffy coat cells.

logical difference was not observed since colonies grown in the presence of material with pH 5.9 and 7.2 were composed of cells with similar primitive blast-like appearance.

DISCUSSION

Stimulatory activities in PHA-LCM for CFU-GEMM, BFU-E, and CFU-C can be separated partially by isoelectric focusing from activities that promote growth of leukemic blast cell precursors. Nonadherent, mononuclear bone marrow populations of normal individuals and mononuclear, E-rosette-depleted peripheral blood cells from patients with AML were found to be appropriate target cells for the different stimulatory activities since colony formation of these populations was dependent upon the addition of PHA-LCM.

The activity profiles for CFU-GEMM, BFU-E, and CFU-C are very similar with respect to their isoelectric point, indicating that this method does not permit separation of molecules with specificity for cells of a restricted hemopoietic lineage. These data are consistent with observations by Metcalf [personal communication] for growth-promoting molecules in murine hemopoiesis. Stimulators that were recovered in the peak fraction yielded, at a much lower concentration (0.5%), a plating efficiency of CFU-GEMM, BFU-E, and CFU-C similar to that obtained with optimal concentrations of 5% crude PHA-LCM. As demonstrated for BFU-E, the plating efficiency increased above control values with addition of higher concentrations of the peak material. This observation could be explained by a number of alternatives. The removal of inhibitory activities by isoelectric focusing was not substantiated since the formation of endogenous bursts by unseparated buffy coat cells was not reduced by any of the fractions. The fractionation procedure may lead to enrichment of stimulators and then to general improvement of the culture system. Alternatively, additional subpopulations of BFU-E may be recruited. Supporting evidence for the latter may be gathered by examining the resulting bursts for such markers as fetal hemoglobin [13].

The additional peak of stimulatory activity consistently observed with unseparated buffy coat cells suggests that colony formation may be controlled by more than 1 type of stimulator in PHA-LCM. Some stimulators may interact directly with appropriate target cells; others may influence adherent cells directly or indirectly and induce the production and release of a second mediator. PHA could be considered as a candidate. Further depletion and reconstitution experiments are required to assess these postulated interacting populations and their mediator molecules more completely.

The most promising prospect of our data is related to the possibility of separating stimulatory activities for normal hemopoietic progenitors from those of leukemic blast precursors. The availability of material with specificity for leukemic blast cells would facilitate selective cloning of leukemic populations and may help to explore the relationship between normal hemopoietic progenitors and leukemic blast-forming cells.

ACKNOWLEDGMENTS

This work was supported by MRC, University of Toronto, Leukemia Research Fund, Toronto. A. A. Fauser is a Fellow of the Medical Research Council of Canada.

REFERENCES

1. Parker JW, Metcalf D: J Immunol 112:502, 1974.
2. Cerny J: Nature 249:63, 1974.
3. Ruscetti FW, Chervenick PA: J Immunol 114:1513, 1975.
4. Cline MJ, Golde DW: Nature 248:703, 1974.
5. Aye MT, Niho Y, Till JE, McCulloch EA: Blood 44:205, 1974.
6. Prival JT, Paran M, Gallo RC, Wu AM: J Natl Cancer Inst 53:1583, 1974.
7. Shah RG, Caporale LH, Moore MAS: Blood 50:811, 1977.
8. Johnson GR, Metcalf D: Proc Natl Acad Sci USA 74:3879, 1977.
9. Metcalf D, Russell S, Burgess AW: Transplant Proc 10:91, 1978.
10. Fauser AA, Messner HA: Blood 52:1243, 1978.
11. Fauser AA, Messner HA: Blood 53:1023, 1979.
12. Buick RN, Till JE, McCulloch EA: Lancet 1:862, 1977.
13. Fauser AA, Messner HA: Blood 54:1384, 1979.
14. Dicke KA, Spitzer G, Ahearn MJ: Nature 259:129, 1976.

15. Park C, Savin MA, Hoogstraten B, Amare M, Hathaway P: Cancer Res 37:4595, 1977.
16. Messner HA, Till JE, McCulloch EA: Blood 42:701, 1973.
17. Eaves CJ, Eaves AC: Blood 52:1196, 1978.
18. Minden MD, Buick RN, McCulloch EA: Blood 54:186, 1979.
19. Wybran J, Levin AS, Spitler LE, Fudenberg HH: N Engl J Med 288:710, 1973.
20. Minden MD, Till JE, McCulloch EA: Blood 52:592, 1978.
21. Aye MT, Seguin JA, McBurney JP: J Cell Physiol 99:233, 1979.
22. Scandurra R, Cannella C, Elli R: Sci Tools 16:17, 1969.
23. Messner HA, Fauser AA, Lepine J, Martin M: Blood Cells (in press).

Cleavage of Cell Surface Proteins by Thrombin

Martin Moss and Dennis D. Cunningham

Department of Microbiology, College of Medicine, University of California, Irvine, California 92717

This study was based on our previous findings that the mitogenic action of thrombin on cultured fibroblasts can result from interaction of thrombin with the cell surface in the absence of internalization, and that the proteolytic activity of thrombin is required for stimulation of cell division. This prompted us to look for thrombin-mediated cleavages using 2-dimensional gel electrophoresis of labeled cell surface proteins. Surface membrane components were labeled by 3 procedures: 1) proteins were labeled by lactoperoxidase-catalyzed iodination using $^{125}I^-$; 2) galactose and galactosamine residues of glycoproteins were oxidized with galactose oxidase and reduced with ^3H-NaBH$_4$; and 3) glycoproteins were metabolically labeled by incubating cells with ^3H-fucose. Labeling with the first 2 procedures was carried out after thrombin treatment; in contrast, cells metabolically labeled with ^3H-fucose were subsequently treated with thrombin to look for proteolytic cleavages. Collectively, these studies indicated that only about 5 cell surface proteins were thrombin-sensitive, consistent with the high specificity of this protease. Each of the labeling procedures revealed a thrombin-sensitive cell surface glycoprotein which was identified as fibronectin by immunoprecipitation experiments. In addition, cell surface proteins of about 140K and 55K daltons were thrombin-sensitive. However, cell surface proteins of about 45K daltons and 130K to 150K daltons were increased after thrombin treatment. These experiments were conducted on an established line of Chinese hamster lung cells with the eventual goal of studying thrombin-mediated cleavages of cell surface proteins in a large number of cloned populations derived from this line that are either responsive or unresponsive to the mitogenic action of thrombin. This approach should permit identification of proteolytic cleavages that are necessary for thrombin-stimulated cell division.

Key words: labeling of cell surface proteins, two-dimensional gel electrophoresis, fibronectin

Addition of thrombin to cultured nonproliferating fibroblastic cells leads to about one round of DNA synthesis and cell division. This stimulation has been observed with several kinds of early passage cells in culture, including strains derived from chick embryo, mouse embryo, and human neonatal foreskins [1–4]. In contrast, many established cell lines appear to be refractory to the stimulation by thrombin [4], although 3 different

Received August 18, 1980; accepted October 21, 1980.

0275-3723/81/1501-0049$04.00 © 1981 Alan R. Liss, Inc.

Chinese hamster cell lines derived from ovary, lung, and whole embryo readily divide after thrombin treatment [5, 6]. With the above fibroblastic cells, stimulation of cell division can be brought about by adding highly purified thrombin to serum-free cultures in the absence of other mitogens. In addition, thrombin can potentiate the action of certain growth factors on fibroblastic cells [7, 8]. With human vascular endothelial cells, thrombin is not mitogenic by itself but will potentiate the action of epidermal growth factor, fibroblast growth factor, and platelet-derived growth factor [9, 10]. Although the biological role of thrombin-stimulated cell division has not yet been directly examined, it is reasonable to suggest that it might be involved in tissue repair following injury [1]. Large quantities of thrombin are produced at very localized regions of tissue damage and could stimulate cell division either by itself or by augmenting the action of other growth-promoting agents.

Studies on the mechanism by which thrombin stimulates fibroblast proliferation have been greatly aided by its ability to produce cell division in chemically defined serum-free medium. Early experiments were directed at identifying the cellular site of thrombin action. By covalently linking thrombin to carboxylate-modified polystyrene beads, it was possible to show that cell surface action of thrombin was sufficient to bring about cell division [11, 12]. Although these studies did not rule out an additional intracellular action of thrombin, they demonstrated that internalization was not *necessary* for stimulation. Subsequent experiments on the binding of ^{125}I-thrombin revealed cell surface binding sites or receptors that were highly specific for thrombin [13, 14]. Studies on the mechanism by which cells bind thrombin have shown that several cellular components are involved. One of these has some novel properties that distinguish it from most cellular receptors. This component (protease-nexin) is released from the cells and forms a covalent linkage with thrombin; these complexes then bind back to cells via the protease-nexin portion of the complex [15]. Thus, the binding of thrombin to fibroblastic cells is complex; studies are in progress to evaluate the role of each of the thrombin-binding components in thrombin-stimulated cell division. A fundamental question regarding the mechanism by which thrombin initiates cell division is whether the stimulation requires its proteolytic activity. This issue was examined with 2 preparations of thrombin that had been derivatized at the catalytic-site serine. These enzymatically inactive thrombins bound to mouse embryo and Chinese hamster lung cells as effectively as active thrombin but did not cause the cells to divide, thus demonstrating a requirement for proteolysis [16].

The above results prompted a search for thrombin-mediated cell surface cleavages that are required for stimulation of cell division. The first experiments were conducted with secondary chick embryo cells, which are responsive to the mitogenic action of thrombin, and chick embryo cells at their 25th population doubling, which are responsive to serum stimulation but which do not divide after thrombin treatment [17]. These studies revealed a 43K dalton cell surface component that was thrombin-sensitive on the responsive cells. An apparently identical component was present on the cells that were unresponsive to thrombin, but on these cells this component was not cleaved by thrombin. Thus, cleavage of the 43K dalton component on responsive chick cells appears necessary for thrombin-stimulated cell division [17].

In the present studies we extended the search for thrombin-mediated cell surface cleavages by employing 2-dimensional gel electrophoresis to better resolve individual cell surface proteins. In addition, 3 different approaches were used to label cell surface proteins to maximize the likelihood of detecting cleavages by thrombin. The first involved labeling tyrosine residues of cell surface proteins using ^{125}I$^-$ and lactoperoxidase [18]. The second involved labeling galactose and galactosamine residues of cell surface glycoproteins with ^{3}H-NaBH$_4$ following oxidation with galactose oxidase [19]. In the third approach, cell surface glycoproteins were metabolically labeled with ^{3}H-fucose [20]. These experiments were conducted with a line of Chinese hamster lung (CHL) cells, which are very sensitive to the mitogenic action of thrombin. Our eventual goal is to determine which cleavages are necessary for the stimulation of cell division by examining thrombin-mediated cell surface alterations in a series of cloned responsive and unresponsive cells derived from this line.

MATERIALS AND METHODS

Cells and Cell Culture

CHL fibroblasts from the V79 strain were obtained from Dr. John J. Wasmuth, University of California, Irvine. Cell stocks were grown in Dulbecco-Vogt-modified Eagle's medium (DV medium) supplemented with 5% calf serum (Irvine Scientific), designated DV-5.

Radioactive Labeling of Cell Surface Proteins

Lactoperoxidase-catalyzed iodination. Replicate cultures of CHL cells (1.8 × 10^5 to 3.6 × 10^5 cells/cm^2) were used for thrombin treatment and subsequent radioactive labeling. Serum-containing growth medium was aspirated from culture dishes and replaced with serum-free medium (DV-0) after one rinse with DV-0. Thrombin-treated cultures received highly purified human thrombin [21] (4,300 NIH units/mg, kindly provided by Dr. John W. Fenton, II) at a final concentration of 20 μg/ml in phosphate-buffered saline (PBS), while control cultures received PBS alone. The cultures were then incubated at 37°C for 30 min. After the incubation, each culture dish was rinsed 4 times with Dulbecco's PBS (0.137 M NaCl, 2.7 mM KCl, 8.0 mM Na$_2$HPO$_4$, 1.47 mM KH$_2$PO$_4$, 0.87 mM CaCl$_2$·2H$_2$O, 0.49 mM MgCl$_2$·6H$_2$O, pH 7.2). Na ^{125}I (Amersham Radiochemical Centre, 15–17 mCi/μg iodine) in Dulbecco's PBS (400 μCi/ml) was then added to the cultures, followed by 85 μl/ml Enzymobeads (Biorad) as a source of lactoperoxidase and glucose oxidase. Cell surface labeling was initiated by addition of glucose to a final concentration of 10 mM. After 20 min at 23°C, radioactive labeling was terminated by aspirating the reaction mixture and rinsing the cultures 4 times with phosphate-buffered iodide (10 mM Na$_2$HPO$_4$–NaH$_2$PO$_4$, 150 mM NaI pH 7.2). Preliminary experiments showed that labeling with Enzymobeads led to specific activities that were 4- to 8-fold greater than those obtained with soluble lactoperoxidase plus soluble glucose oxidase, immobilized lactoperoxidase plus soluble glucose oxidase, or immobilized lactoperoxidase plus H$_2$O$_2$. ^{125}I-labeled cell lysates were prepared by scraping cells from the culture dishes in cell lysis buffer (0.5% NP40, 0.1 mM MgCl$_2$, 50 mM Tris HCl, pH 7.0). The lysates were then treated with nucleases for 10 min at 0°C (Ribonuclease A, Miles Laboratories, 0.1 mg/ml; Deoxyribonuclease I, Worthington Biochemicals 0.1, mg/ml). Following nuclease digestion, sodium dodecyl sulfate (SDS) and 2-mercaptoethanol were each added to a final concentration of 2% and the ^{125}I-labeled lysates were heated for 5 min at 100°C. The lysates were stored at −20°C.

Labeling of cell surface glycoproteins with galactose oxidase and ^{3}H-NaBH$_4$. This procedure was performed essentially as described by Baumann and Doyle [19]. As for iodination, replicate cultures of CHL cells were treated with thrombin or PBS for 30 min at 37°C. The cultures were then rinsed 4 times with Dulbecco's PBS containing 2 mM phenylmethyl sulfonylfluoride (PMSF) and then incubated for 5 min at 23°C with nonlabeled 1 mM NaBH₄ in Dulbecco's PBS. Terminal sialic acid residues were removed by treatment with neuraminidase (Calbiochem) at a final concentration of 5 units/ml in Dulbecco's PBS containing 2 mM PMSF for 15 min at 37°C. After rinsing the neuraminidase-treated cultures 3 times with Dulbecco's PBS containing 2 mM PMSF, galactose oxidase (Sigma) was added to a final concentration of 6 units/ml in Dulbecco's PBS containing 2 mM PMSF. The cultures were then incubated at 37°C for 15 min. Tritiation of the oxidized sugar residues was accomplished by incubating the cultures in Dulbecco's PBS containing 2 mCi/ml ^{3}H-NaBH₄ (Amersham, 7.26 Ci/mM) for 5 min at 23°C. ^{3}H-labeled cells were then lysed as described for iodination.

Metabolic labeling with ^{3}H-fucose. Cultures of log-phase growing CHL cells in DV-5 were incubated with 10 μCi/ml L-[6^{3}H] fucose (Amersham, 26 Ci/mmole) for 36 h. The cultures were then rinsed with DV-0 and treated with thrombin or PBS as described above for iodination. Cell lysates were prepared as described above.

Two-Dimensional Gel Electrophoresis

Cell lysates were analyzed by 2-dimensional gel electrophoresis with modifications of the techniques of O'Farrell [22] and Ames and Nikaido [23]. Equal amounts of TCA-precipitable radioactivity from lysates of control and thrombin-treated CHL cells were applied to isoelectric focusing (IEF) tube gels (3 mm × 10 cm) prepared as described by O'Farrell. Before IEF each SDS-denatured sample was adjusted to 8.0 M urea and 17% NP40 using O'Farrell solution A, 50% NP40, and ultrapure urea. After IEF for a total of 7,200 volt-hours [22], each IEF gel for 2-dimensional analysis was equilibrated for 2 h in O'Farrell solution O [22] containing 9 M urea. The urea was necessary to electrophorese some iodinated cell surface proteins out of the IEF gel. Duplicate IEF gels which were not equilibrated were cut into 5 mm slices; each slice was incubated in 10 mM KCl for 2 h and then analyzed for pH. The SDS-equilibrated IEF gels were loaded on top of a 7½–15% polyacrylamide gradient slab gel (1 mm × 10 cm × 15 cm), which had a 1 cm stacking gel made of 5% acrylamide without any sample wells. The use of gradient polyacrylamide slab gels instead of constant concentration gels for the second dimensions greatly increased the resolution. The IEF gel was then overlaid with O'Farrell solution P [22] (without 2-mercaptoethanol). Molecular weight markers were applied to a single sample well formed on one side of the IEF gel-agarose overlay. The molecular weight markers were as follows: human plasma fibronectin (FN) (240K), β-galactosidase (120K), phosphorylase A (90K), bovine serum albumin (68K), pyruvate kinase subunit (57K), muscle actin (43K), carboxypeptidase B (35K), and soybean trypsin inhibitor (21K). SDS gel electrophoresis was carried out at 23 mA/gel for 4 h. The gels were then fixed and stained overnight in 0.25% Coomassie brilliant blue, 50% ethanol, 7% acetic acid, and subsequently destained in 5% ethanol 7% acetic acid. Two-dimensional gels of tritium-labeled samples were infiltrated with 2,5-diphenyloxazole as described by Bonner and Laskey [24] and then fluorographed at −80°C with Kodak X-Omat R film. Gels containing ^{125}I-labeled samples were autoradiographed at −80°C using Dupont Cronex "Lighting-Plus" intensifying screens. Average exposure times were 10–20 days for ^{3}H-fucose gels, 2–4 days for ^{3}H-NaBH₄ gels, and 24 h for ^{125}I gels.

Immunoprecipitation of ^{125}I-Cell Surface Proteins With Anti-Fibronectin IgG

Confluent cultures of CHL cells were labeled with ^{125}I⁻ as described above. They were then lysed in PBS, pH 7.2 containing 1.0% NP40 and 2 mM PMSF. The lysates were nuclease-treated as above and then incubated with rabbit anti-human-FN IgG (kindly supplied by Mr. Channing Der, University of California, Irvine), for 30 min at 23°C and 6 h at 4°C. ^{125}I-labeled-immunoprecipitates were isolated with protein-A Sepharose beads (Pharmacia). Briefly, 50 μl of a 50% slurry of protein-A Sepharose beads in PBS were combined with 150 μl of the lysate-anti-FN IgG mixture. The samples were occasionally swirled during a 30 min incubation at 23°C, after which the beads were removed by centrifugation. The beads were then washed extensively in 0.5% NP40, 0.05 M NaCl, 0.01 M Na_2HPO_4-NaH_2PO_4, pH 7.5. Immunoprecipitated ^{125}I-proteins were then released from the beads by treatment with 2% SDS, 2% 2-mercaptoethanol, 50 mM Tris-HCl, pH 7.0 for 3 min at 75°C.

RESULTS

The required thrombin treatment for producing a maximal cell number increase (about 60%) with nonproliferating CHL cells in serum-free medium is 500 ng/ml (14 nM) for 48 h (data not presented). However, during this extended incubation in the absence of serum, a small fraction of the cells round up from the culture dish and eventually lose viability as judged by the loss of their ability to exclude trypan blue. Because these cells might be permeable to thrombin as well as the enzymes for radioactively labeling cell surface proteins, we chose a thrombin treatment of 20 μg/ml for 30 min to examine thrombin-mediated cleavages of cell surface components. This short incubation in the absence of serum did not cause the cells to round up or increase their permeability to trypan blue. Although this thrombin concentration was much higher than those used in the long-term treatments to stimulate cell division, it was chosen to maximize the opportunity for detecting thrombin-sensitive cell surface proteins. If a given protein must be cleaved by thrombin for cell division to occur, cleavage of only a fraction of the total cell surface population of that protein might be required, just as occupancy of only a fraction of the total cell surface receptors for a given hormone is frequently sufficient for production of the hormone response [25, 26]. Thus, thrombin-mediated cleavages that are necessary for cell division might be difficult to detect using mitogenic concentrations of thrombin. With this issue in mind, the present studies were initiated to identify thrombin-sensitive cell surface proteins with the realization that it would next be important to analyze whether a given cleavage was necessary for the biological response.

An early step in these experiments involved optimizing the 2-dimensional gel electrophoresis procedures of O'Farrell [22] and Ames and Nikaido [23] as described in Materials and Methods to obtain maximal resolution of individual membrane proteins from Chinese hamster lung cells. Autoradiograms of 2-dimensional gels containing membrane proteins, which had been radioactively labeled by several different procedures (Figs. 2–5), revealed that there was some streaking or "stuttering." In contrast, Figure 1 shows the Coomassie-stained total cellular proteins of one such gel. As can be seen, the procedures employed yielded a high degree of resolution of total cellular proteins. Thus, the apparent streaking of the radioactively labeled membrane proteins was attributable to the properties of these polypeptides and was probably a result of carbohydrate modifications, which produced microheterogeneity in membrane-abundant glycoproteins.

It should be emphasized that the thrombin-mediated changes in cell surface components described below are ones that were *consistently* observed in a large number of experiments. Although close inspection of the autoradiograms in Figures 2, 4, and 5 suggest some additional changes brought about by thrombin, some of these were not reproducible.

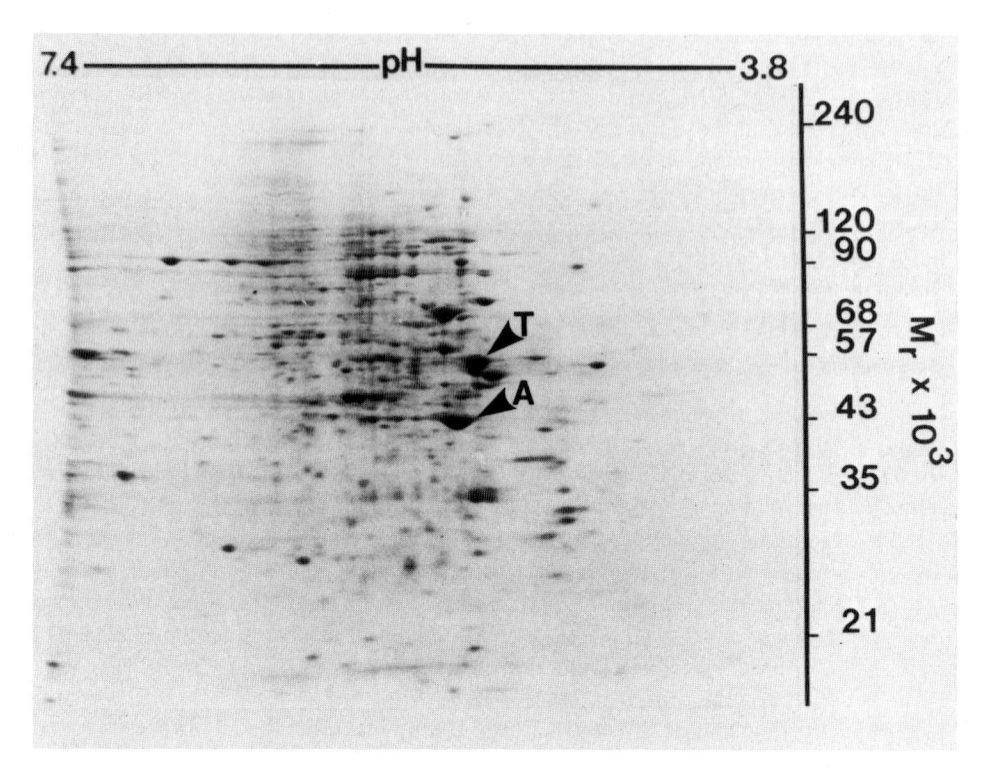

Fig. 1. Coomassie-stained total CHL cell proteins separated by 2-dimensional gel electrophoresis. CHL cells were lysed as described in Materials and Methods. Approximately 80 µg of total cellular protein was separated by IEF in the first dimension and SDS-polyacrylamide gel electrophoresis (7½–15% gradient) in the second dimension. The gels were fixed and stained with Coomassie brilliant blue as described in Materials and Methods. The arrows indicate migration of actin (A) and tubulin (T). Migrations of molecular weight standards are shown at right.

Figure 2 shows autoradiograms of ^{125}I-labeled cell surface proteins of control (left panel) and thrombin-treated (right panel) CHL cells. Control experiments indicated that the proteins labeled by this procedure that were thrombin-sensitive were cell surface rather than intracellular components: 1) there was no detectable cell lysis judged by trypan blue exclusion; 2) these proteins were not labeled when lactoperoxidase was omitted; and 3) *these* proteins were not similarly labeled after iodination of lysed cells. Also, the thrombin-sensitive proteins were cellular rather than serum proteins since they were not detected after iodination of serum. In Figure 2, the lactoperoxidase-catalyzed iodinations were carried out *after* the thrombin treatment; thus, the "absence" of a component from the thrombin-treated autoradiogram could be a result of thrombin-mediated internalization of the intact component in addition to cleavage or release of it from the cells.

The arrows in Figure 2 denote 4 changes that were consistently observed in the autoradiograms prepared from thrombin-treated cells. The component that becomes visible at about 35K daltons is apparently thrombin that was bound to the cell surface and subsequently iodinated. The arrow at about 240K daltons shows a heterogeneous component

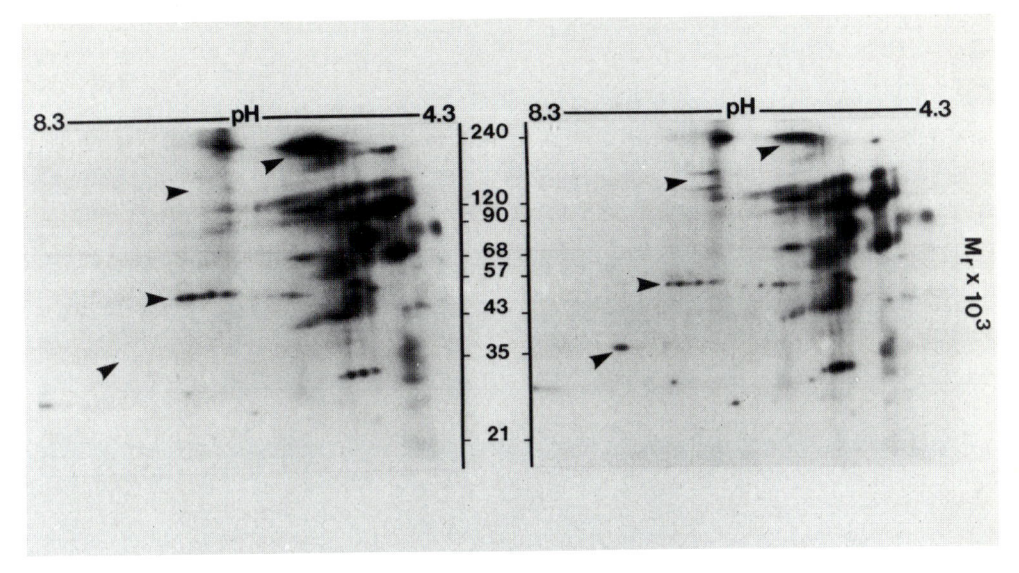

Fig. 2. Two-dimensional gel electrophoresis of ^{125}I-labeled cell surface proteins from control (left panel) and thrombin-treated (right panel) cells. The procedures are described in Materials and Methods. The arrows denote thrombin (35K daltons) and proteins affected by thrombin treatment, which migrated with apparent molecular weights of 240K, 130K to 150K, and 55K daltons.

that migrated to a pH of about 6.0 and was decreased by thrombin. The arrow at about 130K to 150K daltons shows two components that migrated to a pH of about 6.5 that were increased after thrombin treatment. Whether these are fragments resulting from thrombin-mediated cleavages or whether they are components whose appearance in the cell surface labeling pattern is stimulated by thrombin has not yet been evaluated. The arrow at about 55K daltons denotes a "series" of components migrating to a pH of about 6.5 that were deminished by thrombin treatment. In several but not all experiments we also observed a thrombin-sensitive component of approximately 45K daltons that migrated to a pH of 6.1. It should be emphasized that the apparent "stutter" patters of some iodinated components was not due to multiple iodinations; similar patterns were observed when cell surface glycoproteins were labeled with ^{3}H-NaBH$_4$ after oxidation with galactose oxidase (Fig. 4) or metabolically labeled with ^{3}H-fucose (Fig. 5).

Because of its size and relative abundance in the gel profile of ^{125}I-labeled cell surface proteins, the 240K thrombin-sensitive cell surface protein appeared similar to fibronectin (FN). To determine its relatedness to FN, immunoprecipitation of cell lysates containing ^{125}I-labeled cell surface proteins was carried out using rabbit anti-FN IgG and protein-A Sepharose beads. Figure 3 shows the 2-dimensional gel profiles of ^{125}I-labeled cell surface proteins (upper left gel), ^{125}I-labeled cell surface proteins that were not immunoprecipitated (upper right gel), and ^{125}I-labeled cell surface proteins that were immunoprecipitated with anti-FN IgG and protein-A Sepharose beads (lower gel). Three major ^{125}I-labeled cell surface proteins were apparent in the immunoprecipitated samples: one at 240K, labeled F (pH migration of 6.0 to 6.5), one at 85–90K, labeled X (pH migration of 5.0), and a series at 55K labeled Y (pH migration centering around 6.5). Comparison of the migration in 2-dimensional gels of purified human plasma FN and the immunoprecipitated ^{125}I-labeled

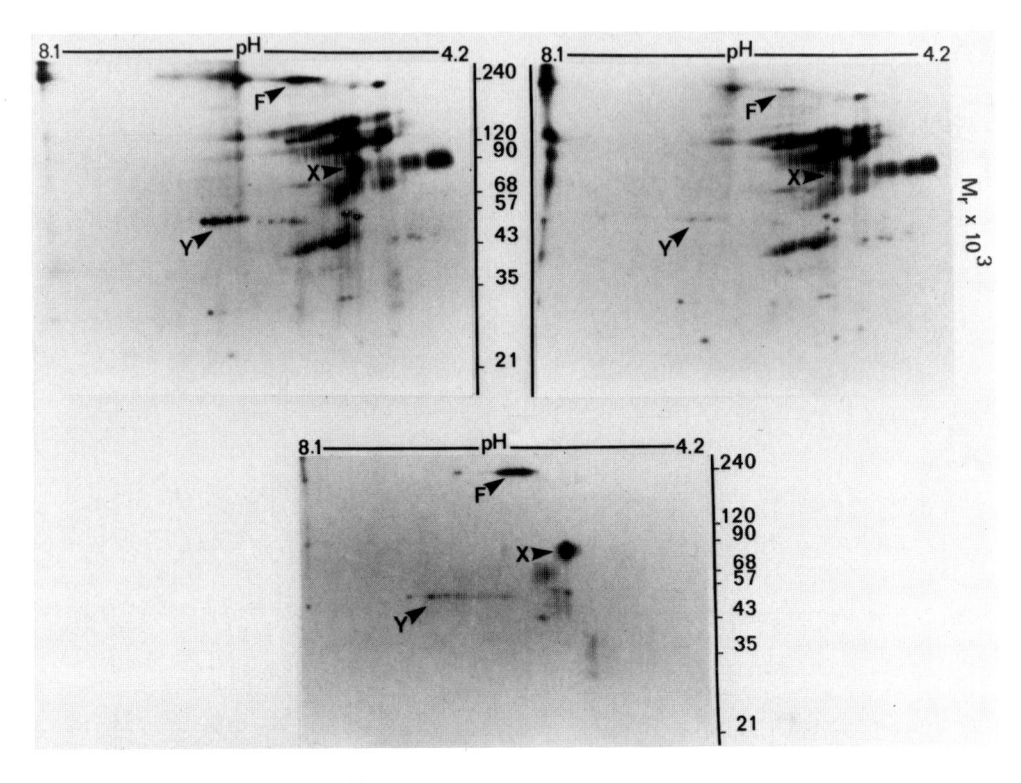

Fig. 3. Immunoprecipitation of ^{125}I-labeled cell surface proteins with rabbit anti-human fibronectin IgG. Cell surface proteins of CHL cells were labeled with ^{125}I, solubilized in NP40, immunoprecipitated with rabbit anti-FN IgG, and separated by 2-dimensional gel electrophoresis as described in Materials and Methods. The upper left panel shows the profile of labeled cell surface proteins. The upper right panel shows ^{125}I-labeled cell surface proteins that were not immunoprecipitated. The lower panel shows ^{125}I-labeled cell surface proteins that were immunoprecipitated. The arrows denote 3 major immunoprecipitated polypeptides with apparent molecular weights of 240K (F), 90K (X), and 55K (Y).

240K component showed them to have the same apparent molecular weight and very similar pH migrations. However, the ^{125}I-labeled cell surface component migrated over a broader pH range than the human plasma FN, indicating a greater charge variance among individual molecules of the cell-associated 240K component. This was possibly due to additional carbohydrate heterogeneity. In fact, the data presented in Figures 4 and 5 indicate that the cell-associated 240K component is a heterogeneous glycoprotein since it was labeled by ^{3}H-NaBH$_4$ after galactose oxidase treatment and after incubating cells with ^{3}H-fucose. Further relatedness between the ^{125}I-labeled 240K cell surface protein and authentic FN was demonstrated by displacement of immunoprecipitated ^{125}I-labeled 240K with excess nonlabeled human plasma FN (data not shown). The immunoprecipitated 55K dalton ^{125}I-labeled component (Fig. 3, component Y) appears to be the same series of 55K dalton polypeptides that were found to be thrombin-sensitive (Fig. 2). The coprecipitation of the 55K dalton component and FN initially suggested a possible interaction between it and FN. This possibly appears unlikely, however, since we observed that immunoprecipitation

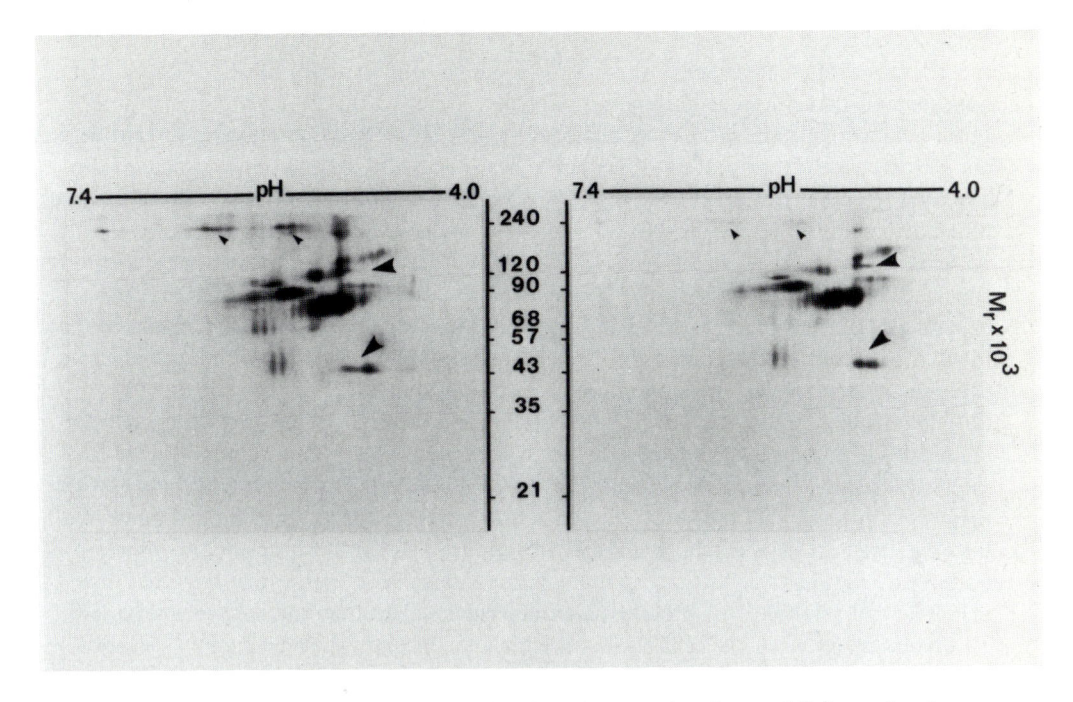

Fig. 4. Two-dimensional gel electrophoresis of surface glycoproteins of control (left panel) and thrombin-treated (right panel) cells labeled by ^{3}H-NaBH$_4$ after galactose oxidase treatment. The procedures are described in Materials and Methods. The arrows denote glycoproteins migrating with apparent molecular weights of 240K, 140K, and 45K daltons that were affected by thrombin treatment.

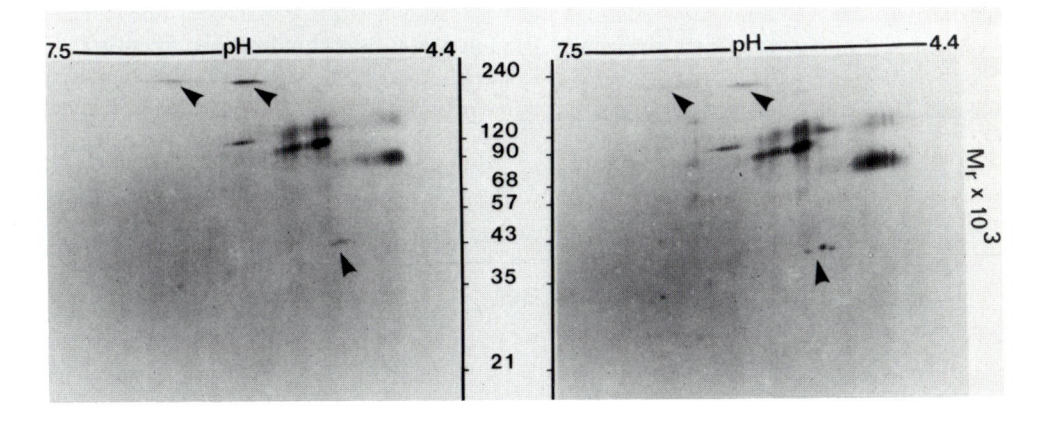

Fig. 5. Two-dimensional gel electrophoresis of surface glycoproteins of control (left panel) and thrombin-treated (right panel) cells metabolically labeled with ^{3}H-fucose. The procedures are described in Materials and Methods. The arrows denote glycoproteins of 240K and 45K that were affected by thrombin treatment.

of the X and the Y components was not appreciably altered by preadsorption of the rabbit anti-FN IgG with unlabeled FN, whereas the F component was significantly reduced (data not shown). Furthermore, additional experiments suggested that several ^{125}I-labeled cell surface components nonspecifically adsorbed to protein-A Sepharose beads. Thus, while our washing protocol of protein-A Sepharose beads was sufficient to remove some of the adsorbed ^{125}I-labeled proteins, the X and Y components appeared to represent cell surface proteins that adsorb very tightly to the protein-A Sepharose beads.

To evaluate the thrombin-sensitivity of cell surface glycoproteins, control and thrombin-treated cells were exposed to galactose oxidase and subsequently to ^{3}H-NaBH$_4$. Control experiments like the ones described for lactoperoxidase-catalyzed iodination indicated that the thrombin-sensitive components were cell surface glycoproteins. As with the iodination experiments, the absence of a component after thrombin treatment could be due to thrombin-mediated internalization of that component rather than cleavage of it, since the radioactive labeling was conducted after the cells were treated with thrombin. Figure 4 shows that several cell surface glycoproteins were altered after thrombin treatment. Two of these were markedly reduced by thrombin and migrated within the size and pH range that the experiment in Figure 3 showed for FN. In addition, a slight but repeatedly observed decrease occurred in a cell surface component of about 140K daltons that migrated to a pH of 4.7 and a doublet at a slightly lower molecular weight that migrated to a pH of 5.0. Thrombin treatment of the cells also resulted in an increase in the labeling by ^{3}H-NaBH$_4$ after galactose oxidase treatment of a component of about 45K daltons that migrated to a pH of 4.7.

The last procedure that was used to examine the thrombin sensitivity of cell surface components involved metabolic labeling with ^{3}H-fucose, a glycoprotein precursor that is incorporated mainly into cell surface glycoproteins [20]. In these experiments, the cells were labeled with ^{3}H-fucose and subsequently treated with thrombin. Since thrombin treatment followed the labeling of cellular proteins, the absence of a component in thrombin-treated cells would indicate cleavage of the component of removal of it from the cells, and not simply internalization without cleavage. Figure 5 shows that this procedure revealed a thrombin-mediated decrease of 2 components that migrated to an apparent molecular weight of about 240K daltons and a pH range of 6.0 to 6.5, as does FN. It is noteworthy that some of each component remained after thrombin treatment, consistent with a small intracellular pool of ^{3}H-FN that would be expected with metabolic labeling. A small but reproducible increase was also noted after thrombin treatment in a component of about 45K daltons which migrated to a pH of about 4.7 to 5.0. This might be the same component shown in Figure 4 that was more effectively labeled by galactose oxidase and ^{3}H-NaBH$_4$ following treatment of cells with thrombin. There is not yet enough information to decide whether such increased labeling is a result of a rapid stimulation of appearance of the component at the cell surface or whether the component represents a glycoprotein fragment resulting from proteolysis by thrombin.

DISCUSSION

The present studies were based on our previous findings that cell surface action of thrombin is sufficient to stimulate cell division [11, 12] and that the proteolytic activity of thrombin is required for the stimulation [16]. Thus, cleavage of one or more cell surface proteins is a critical step for thrombin-stimulated cell division; this limited proteolysis may trigger the series of events that finally leads to DNA synthesis and cell division. At this point in our limited understanding of molecular events that can lead to cell proliferation, it is important to identify cell surface proteins that are susceptible to cleavage by thrombin.

To maximize our ability to detect proteolytic cleavages, the present experiments were conducted on labeled surface membrane components that were separated by 2-dimensional gel electrophoresis [22, 23]. The methods to label the membrane components included one that labeled tyrosines of surface proteins (^{125}I$^-$ and lactoperoxidase), a procedure that labeled galactose and galactosamine residues of surface glycoproteins ^{3}H-NaBH$_4$ and galactose oxidase) and a procedure that metabolically labeled glycoproteins (^{3}H-fucose). With the first 2 procedures we employed the standard controls to show that cell surface rather than intracellular proteins were labeled and that the thrombin-sensitive components were not serum proteins [27]. It has previously been shown that glycoproteins metabolically labeled with ^{3}H-fucose occur mostly at the cell surface [20]. In the experiments employing ^{3}H-fucose, cells were treated with thrombin *after* metabolic labeling. Thus, a component that was judged thrombin-sensitive on the autoradiograms was either removed from the cell (perhaps as a result of thrombin cleaving an adjacent membrane protein) or directly cleaved by thrombin. However, with the 2 other labeling procedures, cells were labeled *after* thrombin treatment. In these cases, thrombin sensitivity of a cell surface component could be a result of thrombin-mediated internalization without cleavage, as well as removal from cells or direct cleavage by thrombin. In addition, thrombin could produce cell surface alterations that might change the accessibility of a given protein to the labeling reagents. Collectively these procedures indicated that only about 5 cell surface proteins were thrombin-sensitive, consistent with the high specificity of this serine protease [28].

With each of the 3 labeling procedures, a glycoprotein that was identified as FN was found to be thrombin-sensitive. Previous studies have shown that thrombin stimulates the release of apparently intact FN from cultured human fibroblasts [29]. However, experiments with chick embryo fibroblasts did not reveal a reduction of cell surface FN after thrombin treatment [17, 30, 31]. Thermolysin, papain, and elastase, however, readily removed cell surface FN from chick embryo fibroblasts [31]. It is noteworthy that thrombin is very mitogenic for chick embryo fibroblasts but that these cells do not divide after treatment with thermolysin or elastase [31]. Thus, it appears that, with chick cells, removal of cell surface FN is neither necessary nor sufficient for protease-stimulated cell division [31]. However, since cell surface FN is apparently involved in the attachment of many cells to their extracellular matrix [32], thrombin-mediated alterations in cell surface FN could be involved in perturbations of this attachment and perhaps changes in cell migration or other events that might be a part of the total wound healing process in vivo.

A thrombin-sensitive cell surface component of 205K daltons was revealed in earlier studies on chick embryo fibroblasts [33]. However, this protein was also removed by non-mitogenic proteases including α-protease, thermolysin, and papain [31]. Thus, the 205K polypeptide is not simply a negative effector molecule whose removal from the cell surface is sufficient to stimulate cell division. These results are consistent, however, with the possibility that proteolysis of 205K by thrombin produces a specific peptide that is a positive signal in or on the cell. In the present experiments we did not detect on CHL cells a 205K dalton component that was thrombin-sensitive. This could be a result of a difference in the cells that were used or the procedures to analyze cell surface proteins.

Previous studies have shown that a cell surface component of about 43K daltons that is labeled by ^{125}I$^-$ and lactoperoxidase on chick embryo fibroblasts appears to be involved in thrombin-stimulated cell division [17]. This component was thrombin-sensitive on chick embryo cells that were responsive to the mitogenic action of thrombin. On 4 separately isolated populations of chick embryo cells that divided after serum treatment but not after thrombin treatment, there was an apparently identical component, but it was not thrombin-sensitive.

In view of these results, we looked for a similar component in the present studies on CHL cells. In several experiments we found a cell surface component of about 45K daltons that migrated to a pH of 6.1, which was iodinated with $^{125}I^-$ and lactoperoxidase in control but not in thrombin-treated cells. However, in subsequent experiments we could not detect this component; the reason for this apparent discrepancy is not yet clear. It is noteworthy that the present experiments, in which CHL cell surface glycoproteins were labeled with ^{3}H-NaBH$_4$ after galactose oxidase treatment or with ^{3}H-fucose, did not reveal a 45K dalton component that was diminished or absent after thrombin treatment.

A cell surface component with an apparent molecular weight of about 45K daltons, which migrated to a pH of about 4.7 to 5.0 upon isoelectric focusing, was found in the present studies to be increased after thrombin treatment. The increase was not large but was reproducible and detected both by ^{3}H-NaBH$_4$ and galactose oxidase labeling after thrombin treatment and by thrombin treatment after metabolic labeling with ^{3}H-fucose. It was not detected by $^{125}I^-$ and lactoperoxidase labeling. Thrombin treatment could bring about an increased amount of a given cell surface protein either by stimulating its appearance in the membrane via a synthetic route or by direct proteloysis of a cell surface protein, which would yield the given protein fragment. At this time we do not have the data that would permit a choice between these possibilities.

Cell surface proteins of about 55K and 140K daltons were found to be diminished by thrombin in the present experiments. Although it is not possible to describe the relationship, if any, of these changes to thrombin-stimulated cell division, we are in the process of developing approaches to examine the biological significance of cell surface changes produced by thrombin. The key element of this approach is the isolation of cloned CHL cell lines that are either responsive or unresponsive to the mitogenic action of thrombin. By examining thrombin-mediated cell surface changes in a large number of responsive and unresponsive clones, it should be possible to identify which of the surface changes are *necessary* for thrombin-stimulated cell division. A cell surface cleavage brought about by thrombin that occurs in all of the responsive clones but that does not take place or is altered in some of the unresponsive clones is likely to be necessary for the biological response.

ACKNOWLEDGMENTS

This work was supported by a research grant (CA-12306) from the National Institutes of Health. D.D.C. is a recipient of a Research Career Development Award (CA-00171) from the National Institutes of Health. M.M. received support from a National Institutes of Health Postdoctoral Fellowship (GM-07465). We thank Cindy Rofer for technical assistance and Dr. John W. Fenton II for generous gifts of human thrombin [21].

REFERENCES

1. Chen LB, Buchanan JM: Proc Natl Acad Sci USA 72:131, 1975.
2. Pohjanpelto P: J Cell Physiol 91:387, 1977.
3. Carney DH, Glenn KC, Cunningham DD: J Cell Physiol 95:13, 1978.
4. Buchanan JM, Chen LB, Zetter BR: In Schultz J, Ahmad F (eds): "Cancer Enzymology." New York: Academic Press, 1976, p 1.
5. Simmer RL, Baker JB, Cunningham DD: J Supramol Struct 12:245, 1979.
6. Moss M, Wiley HS, Low DA, Cunningham DD: unpublished results.
7. Zetter BR, Sun TT, Chen LB, Buchanan JM: J Cell Physiol 92:233, 1977.
8. Cherington PV, Pardee AB: J Cell Physiol 105:25, 1980.

9. Gospodarowicz D, Brown KD, Birdwell CR, Zetter BR: J Cell Biol 77:774, 1978.
10. Zetter BR, Antoniades HN: J Supramol Struct 11:361, 1979.
11. Carney DH, Cunningham DD: Cell 14:811, 1978.
12. Carney DH, Cunningham DD: J Supramol Struct 9:337, 1978.
13. Carney DH, Cunningham DD: Cell 15:1341, 1978.
14. Perdue J, Kivitz E, Lubenskyi W: J Cell Biol 83:255a, 1979.
15. Baker JB, Low DA, Simmer RL, Cunningham DD: Cell 21:37, 1980.
16. Glenn KC, Carney DH, Fenton JW II, Cunningham DD: J Biol Chem 255:6609, 1980.
17. Glenn KC, Cunningham DD: Nature 278:711, 1979.
18. Hubbard AL, Cohn ZA: J Cell Biol 64:438, 1975.
19. Baumann H, Doyle D: J Biol Chem 253:4408, 1978.
20. Atkinson PH, Summers DF: J Biol Chem 246:5162, 1971.
21. Fenton JW II, Fasco MJ, Stackrow AB, Aronson DL, Young AM, Finlayson JS: J Biol Chem 252:3587, 1977.
22. O'Farrell PH: J Biol Chem 250:4007, 1975.
23. Ames GF-L, Nikaido K: Biochemistry 15:616, 1976.
24. Bonner WM, Laskey RA: Eur J Biochem 46:83, 1974.
25. Hollenberg MD, Cuatrecasas P: Proc Natl Acad Sci USA 70:2964, 1973.
26. Kahn CR: J Cell Biol 70:261, 1976.
27. Hynes RO: In Pain RH, Smith BJ (eds): "New Techniques in Biophysics and Cell Biology." London: John Wiley and Sons, 1976, vol 3, p 147.
28. Blomback B, Hessel B, Hogg D, Cloesson G: In "Chemistry and Biology of Thrombin." Ann Arbor: Ann Arbor Science, 1977, p 275.
29. Mosher DF, Vaheri A: Exp Cell Res 112:323, 1978.
30. Blumberg PM, Robbins PW: Cell 6:137, 1975.
31. Zetter BR, Chen LB, Buchanan JM: Cell 7:407, 1976.
32. Vaheri A, Mosher DF: Biochim Biophys Acta 516:1, 1978.
33. Teng NNH, Chen LB: Nature 259:578, 1976.

Release of Erythropoietin From Macrophages by Treatment With Silica

I.N. Rich, V. Anselstetter, W. Heit, E. Zanjani, and B. Kubanek

Department of Transfusion Medicine (I.N.R., B.K.) and Department of Inner Medicine III (V.A., W.H.), University of Ulm, Federal Republic of Germany and Department of Medicine, Veterans Administration Medical Center, University of Minnesota, Minneapolis, Minnesota 55417 (E.Z.)

An erythropoietic stimulating factor (ESF) can be shown to be released from preincubated macrophage-containing cell suspensions from mice by the macrophage-specific, cytotoxic agent, silica. A concentrated silica-treated spleen cell supernatant containing ESF is shown to cause a dose-dependent increase in ^{59}Fe incorporation into red blood cells using the in vivo polycythemic mouse bioassay. The ESF from the same supernatant can also be neutralized by anti-erythropoietin. A second concentrated supernatant fractionated using wheat germ lectin-Sepharose 6MB and compared to either unfractionated or fractionated step III erythropoietin (Ep), tested in vitro using the erythroid colony-forming technique and 12-day fetal liver as target cells, indicates parallelism of all linear dose-response lines. This, together with the in vivo data, strongly suggests that the ESF released from macrophages treated with silica is, in fact, Ep. Substituting Ca^{2+} ions for fetal calf serum in the preincubation procedure results in the same activity being released compared to the presence of 1% or 20% fetal calf serum.

Key words: erythropoietin, macrophages, silica, erythrocytic colony-forming units, polycythemic mouse bioassay, anti-erythropoietin

Crystalline silica (silicon dioxide, SiO_2) has been shown to be specifically cytotoxic for macrophages [1, 6]. We [12, 13] have recently demonstrated that an erythropoietic stimulating factor (ESF) can be released when cell suspensions from 14-day fetal liver and adult bone marrow and spleen are preincubated with silica. In addition, it was shown that the activity of the ESF released was dependent on the erythropoietic status of the animal [13]. The ESF was detected using the erythroid colony-forming technique [17] and 12-day fetal liver as a source of erythrocytic colony-forming units (CFU-E). It has been previously shown that 12-day fetal liver CFU-E are extremely sensitive to erythropoietin (Ep), thus providing an in vitro assay for this hormone [10, 11].

In this communication we extend our previous results by attempting to show that the ESF released from macrophages after treatment with silica is Ep, as demonstrated by its effect in vitro and in vivo, and by its neutralization by anti-Ep. In addition, evidence is presented showing that release of Ep can be obtained under essentially serum-free conditions.

Dr. I. N. Rich is now at the Department of Biochemistry, University of Chicago, Chicago, IL 60637.

Received April 29, 1980; accepted October 20, 1980.

0275-3723/81/1502-0169$02.50 © 1981 Alan R. Liss, Inc.

METHODS

Female CBA/Ca and C57/65J mice were used. Preparation of single cell spleen suspensions has been described elsewhere [10]. However, for clarity, spleens were first homogenized in a loosely fitting glass homogenizer using Hank's balanced salt solution (BSS) containing 3% fetal calf serum (FCS). After the large debris sank to the bottom the suspension was decanted into a plastic tube (Falcon Plastics, Oxnard, CA) and again left on ice for about 5 min. Thereafter, the suspension was withdrawn using a 25-gauge needle and the volume measured. Nucleated cells were counted (Coulter Counter Model ZF).

Treatment of cells with silica has been detailed previously [13]. In summary, 8×10^6 spleen cells/ml were preincubated with 1×10^{-4} gm/ml silica (Min-U-Sil; particle size 2–5 μm; Whitacker, Clark & Daniels, Plainfield, NJ) for 30 min at 37°C in 5% CO_2. Cells were normally suspended in Hank's BSS containing 3% FCS. Thereafter, the cells and silica were centrifuged down at 1,500 rpm at 4°C, and the supernatant was used as a source of erythropoietic stimulating material.

Two mass cultures of silica-treated spleen supernatant were prepared. Batch 1 employed 20 C57Bl mouse spleens, which produced 450 ml, and batch 2 used 25 C57Bl mouse spleens and gave rise to 550 ml of supernatant. Control spleen suspensions were set up without silica addition. The supernatants from both batches were concentrated down to 50 ml using Sephadex G-25 medium grade (Pharmacia). The concentrate was dialysed against 4 changes of double-distilled water in 24 h and lyophilized.

Batch 1 was dissolved in 25 ml physiological saline and assayed in the polycythemic mouse bioassay. For this, virgin Swiss Webster mice (about 22 gm) with 5 mice/group were used after exposure to hypoxia at 0.4 atm for 19 h/day for 3 weeks. Animals having a hemocrit below 65% were not used. The mice were injected intraperitoneally with the samples on days 5 and 6 post-hypoxia, followed by intravenous injection of radioiron on day 7. The ^{59}Fe uptake into red blood cells was determined 72 h later. Batch 1 was also used to test if the activity was Ep by reaction with antierythropoietin (anti-Ep). Antiserum was prepared in rabbits by injection of crude human urinary Ep. One milliliter of the sample was assayed alone or after treatment with anti-Ep (capable of neutralizing 0.6 IU Ep) for 1 h. To prevent carry-over of anti-Ep into assay mice, the reaction was treated with goat anti-rabbit gamma globulin (GARGG) and the precipitate removed before injection. As control, GARGG was also added to the untreated sample, along with normal rabbit IgG.

Batch 2 of the lyophilized silica-treated spleen supernatant was dissolved in 2 ml of degassed phosphate buffer (see below). Of this, 1.5 ml were passed over a wheat germ lectin-Sepharose 6MB (Pharmacia) affinity chromatography column, which had first been equilibrated with 10 column volumes (1 column volume was equivalent to 2.6 ml) of PBS. All material that remained unbound to wheat germ lectin was eluted with 5 column volumes of PBS alone. Bound material was eluted with 8 column volumes of PBS containing 0.1 M N-acetyl glucosamine. All fractions of unbound or bound material were pooled, dialysed for 24 h against 4 changes of double-distilled water, and lyophilized. The lyophilized unbound and bound pooled fractions were then reconstituted with physiological saline in the same volume as was applied to the column; namely, 1.5 ml. The same fractionation procedure was used for step III Ep (Connaught Laboratories, Ontario, Canada), but in this case 2 ml containing 100 units was added to the column.

Protein determinations for all fractions were performed using a modification [18] of the Lowry procedure employing a standardized serum (Kontrollogen-L, Behringwerke, West Germany).

The phosphate-buffered saline (PBS) used for both chromatography and other experiments described in the Results was prepared according to Rabinowitz [8], but without Ca^{2+}, Mg^{2+}, EDTA, and glucose. The pH was 7.2.

Twelve-day fetal liver cell suspensions [10] were prepared by passing the dissected out organs through decreasing diameter syringe needles from 18 gauge to 22 gauge, and finally 25 gauge, in Hank's BSS containing 3% FCS. Six organs were suspended in 1 ml medium. Cells were counted as described above.

The erythroid colony-forming technique using 12-day fetal liver CFU-E as target cells for detecting erythropoietic stimulating activity in the supernatants was similar to that described previously [11]. Experiments were performed at least 2 times in duplicate wells containing 0.5 ml of the culture components. Colonies were enumerated using an inverted microscope. At optimal (75 mU Ep/ml) Ep concentrations, the number of fetal liver CFU-E colonies obtained is approximately 3,500 CFU-E/10^5 cells.

Regression analysis of the normalized in vitro dose-response curves has been described previously [11]. Probit analysis of the dose responses was performed using the BMD 03S program from the Health Sciences Computing Facility, UCLA. In addition, a chi-squared test was used to examine the degree of association of the points on the probit line.

RESULTS

Table I shows the effect of injecting 0.4 ml, 0.8 ml, and 1.6 ml of the C57Bl concentrated silica-treated spleen supernatant (batch 1) into polycythemic mice. Increasing volume of this supernatant results in a corresponding increase in the ^{59}Fe incorporation, thus providing evidence that the dose response so obtained is similar to that obtained when Ep is injected.

Controls – ie, suspensions incubated without silica (not shown in the diagram) – consistently gave ^{59}Fe incorporation readings below the 50 mU/ml level, the lowest concentration that can be detected using this in vivo polycythemic mouse bioassay.

Table II demonstrates that anti-Ep can almost completely neutralize the erythropoietic stimulating effect shown in Table I. GARGG is used to neutralize the effect of anti-Ep action after it has reacted with the sample (and Ep as control), and therefore to eliminate the effect of antiserum in the assay mouse.

TABLE I. Effect of the Concentrated Silica-Treated Spleen Supernatant (Batch 1) When Injected Into Polycythemic Mice*

Material assayed	% RBC ^{59}Fe incorporation (mean ± SEM)
Saline	0.3 ± 0.05
0.2 IU Ep	3.92 ± 0.42
0.4 IU Ep	8.14 ± 0.65
0.4 ml sample	2.06 ± 0.29
0.8 ml sample	4.14 ± 0.62
1.6 ml sample	10.43 ± 1.12

*Similar volumes of the supernatant obtained from cells incubated without silica produced values in ^{59}Fe incorporation not significantly different from the saline control.

An additional attempt to show that the ESF present in the silica-treated spleen supernatant was similar to Ep, was to purify the factor and compare the in vitro dose-response curve with similarly treated step III Ep. Figure 1 shows the dose-response curves for unfractionated step III Ep and the original, concentrated batch 2 supernatant. Also shown are the dose-response curves for the wheat germ lectin unbound and bound fractions for both step III and batch 2 supernatant. As shown in Figure 1, the majority of the protein material remains unbound to wheat germ lectin and is eluted using PBS alone. In the case of EP, some of the activity was eluted in this first peak and is probably due to overloading of the column. The second peak of eluted bound material contains most if

TABLE II. Effect of Reacting the Concentrated Silica-Treated Spleen Supernatant With Anti-Ep. Tested in the Polycythemic Mouse Bioassay

Material assayed	% RBC ^{59}Fe incorporation (mean ± SEM)
Saline	0.53 ± 0.08
0.4 IU Ep + GARGG + normal rabbit serum	9.32 ± 1.47
0.4 IU Ep + Anti-Ep IgG + GARGG	0.58 ± 0.10
1 ml sample + GARGG + normal rabbit serum	7.22 ± 1.30
1 ml sample + anti-Ep IgG + GARGG	0.23 ± 0.05

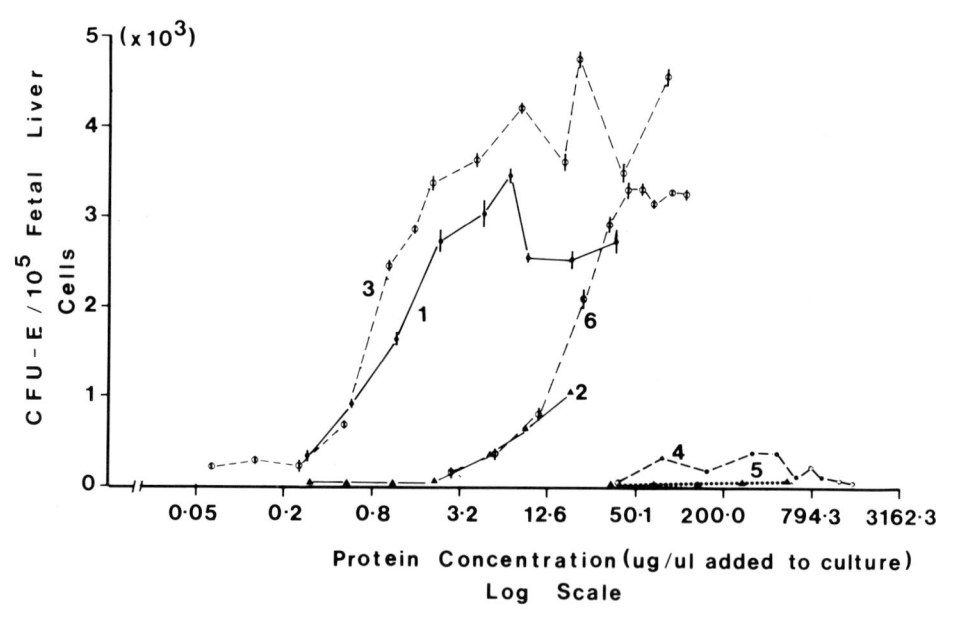

Fig. 1. Comparison of dose-response curves for step III Ep and the concentrated silica-treated spleen supernatant (batch 2) before and after fractionation on wheat germ lectin-Sepharose 6MB. Curve 1, original, unfractionated step III Ep (4.6 μg protein/μl); curve 2, fractionated step III Ep, unbound peak. (2.15 μg proteins/μl); curve 3, fractionated step III Ep, bound peak. (1.02 μg protein/μl); curve 4, original, concentrated supernatant (batch 2) (6.1 μg protein/μl); curve 5, fractionated, unbound peak (batch 2) (2.65 μg protein/μl); curve 6, fractionated, bound peak (batch 2) (0.44 μg protein/μl); two experiments performed.

not all the activity. With regard to the Ep fractionation, the dose response shows a modest additional stimulation of the fetal liver CFU-E, and a plateau at optimal concentrations is obtained. When batch 2 supernatant was passed through the column, almost all the activity was found in the wheat germ lectin bound material, and it also shows a plateau.

That the erythroid colony-forming technique can be used as a specific in vitro assay for Ep is shown in Figure 2. Here, the results for dose-response curves 1, 3, and 6 of Figure 1 have been transformed into percent CFU-E calculated from the maximum number of CFU-E obtained and plotted against the log protein concentration. In this form, a dose-response line for an unknown Ep preparation (in this case the silica-treated, concentrated, and fractionated spleen supernatant) must produce a dose response parallel to that obtained for a known Ep preparation — ie, step III Ep. Examination of the regression coefficients (Fig. 2) indicates similarity between those for both Ep preparations and the fractionated supernatant. Regression analysis substantiates this within 95% confidence limits. The horizontal displacement to the right for the silica-treated spleen supernatant dose response indicates an Ep preparation of lower potency than that for step III Ep. A concentration 12.8 times higher than that for step III Ep is required in order to stimulate 50% of the potential CFU-E present — ie, the essential dose 50 (ED_{50}) obtained by probit analysis.

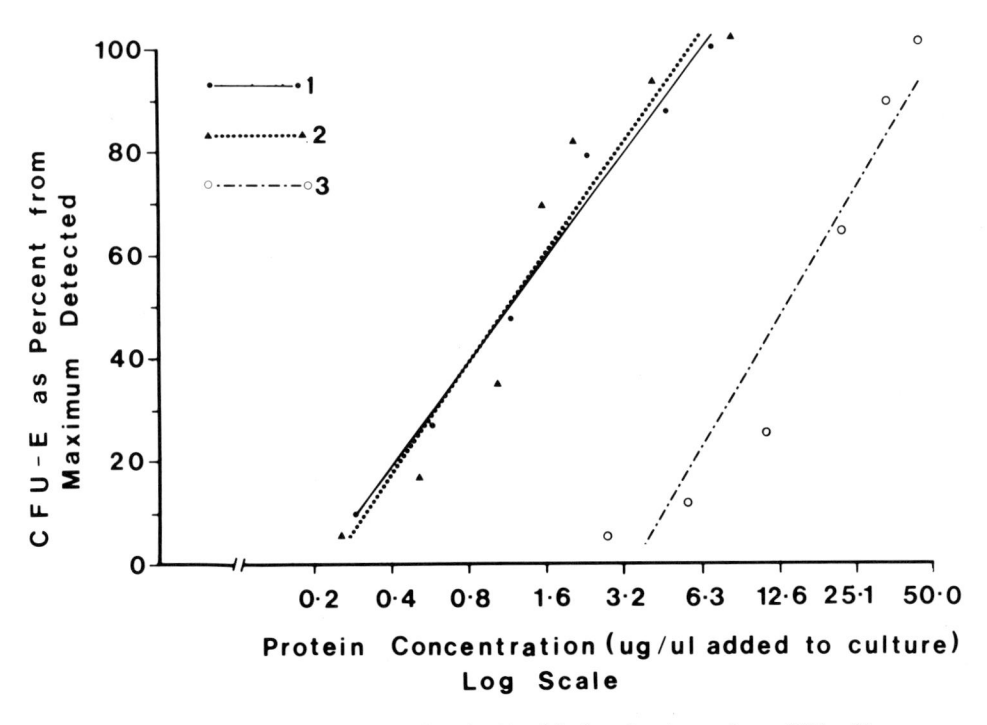

Fig. 2. Comparison of linear dose-response lines for 1) original, unfractionated step III Ep. Linear regression: y = 45.6 + 67.4(x); r = 0.99; P < 0.001. Probit line: 4.9 + 2.3 (x); P < 0.001; 2) Fractionated step III Ep, bound peak. Linear regression: y = 45.9 + 72.1 (x); r = 0.97; P < 0.05. Probit line: 4.9 + 2.3 (x); P < 0.05; 3) fractionated, bound peak (batch 2). Linear regression: y = −45.2 + 74.5 (x); r = 0.96, P < 0.05. Probit line: 1.5 + 3.1 (x); P < 0.05; two experiments performed.

Preincubation of macrophage-containing cell suspensions with silica is routinely performed using 3% fetal calf serum added to the suspension culture. Since this is a very low concentration of fetal calf serum, it was of interest to know whether the reaction would occur in serum-free conditions or whether certain cations were necessary. To investigate this, CBA/Ca spleen cell suspensions were incubated either 1) without fetal calf serum or with different concentrations of fetal calf serum or 2) with 2 mM Ca^{2+} ($CaCl_2$) and with 2 mM Ca^{2+} plus different concentrations of fetal calf serum added to PBS.

Table III shows the effect of incubating spleen cell suspensions in PBS alone or with different concentrations of fetal calf serum and/or 2 mM Ca^{2+}. When 1% fetal calf serum is added to PBS, an enhancement in CFU-E growth is observed compared to that incubated in the absence of fetal calf serum (330 ± 20 compared to 220 ± 20). Increasing the concentration of fetal calf serum from 1% to 20% does not affect the inital enhancement. When 2 mM Ca^{2+} is added to PBS in the absence of fetal calf serum, a similar increase in the number of CFU-E is obtained over that found in the presence of silica in PBS alone (350 ± 20 compared to 220 ± 20). The increase in CFU-E observed with 2 mM Ca^{2+} alone (350 ± 20) is the same as that obtained when fetal calf serum concentrations from 1% to 20% are added to the PBS. The addition of fetal calf serum to PBS containing 2 mM Ca^{2+} slightly increases the number of CFU-E over the number obtained for fetal calf serum on 2 mM Ca^{2+} alone. Increasing the serum concentration from 1% to 20% does not have an effect on the activity released into the supernatant after the spleen cells have been treated with silica.

DISCUSSION

A previous report demonstrated that an erythropoietic stimulating factor (ESF) could be shown to be released into the extracellular fluid when fetal liver and adult bone marrow and spleen cell suspensions were preincubated with silica [13]. In this communication we have cited evidence that the ESF is almost definitely erythropoietin (Ep), with the in vivo polycythemic mouse bioassay, its neutralization by anti-Ep and its similar effect to Ep in vitro as substantiation for our hypothesis. In addition, the ESF activity released by silica from macrophages is dependent on the erythropoietic status of the animal, indicating that the ESF activity can be manipulated in a similar manner to that of plasma Ep [13]. Moreover, release of ESF is actinomycin D sensitive both in normal and anemic mice [14], a result similar to that obtained by Schooley and Mahlman [15] in their investigations on extrarenal Ep production in lead-poisoned rats. Thus all data point to the macrophage-released ESF being Ep.

The in vitro dose-response effects of macrophage-released ESF deserve an additional comment. Recent work by Spivak [16] showed that Ep could be separated from other protein components using affinity chromatography on wheat germ lectin covalently bound to Sepharose 6MB. Both wheat germ lectin and Ep have N-acetyl glucosamine residues, so that the lectin binds Ep and can be eluted with this sugar. The results presented in Figure 1 agree very well with those of Spivak et al [16] and therefore provide an easy method of removing many of the toxic substances from commercially obtained Ep and human urinary Ep. The fact that the ESF also binds to wheat germ lectin indicates that the molecule also contains the required sugar moiety. It does not, however, indicate that the substance is Ep. We [11] and others [4] have shown that when raw data from CFU-E dose-response curves are normalized and transformed as indicated in Figure 2, different Ep

TABLE III. The Effect of Fetal Calf Serum or 2 mM Ca^{2+} Alone or in Combination on the Release of Ep From Silica-Treated Spleen Cells

	12-Day fetal liver CFU-E/10^5 cells[a]			
	Fetal calf serum alone		Fetal calf serum + 2 mM Ca^{2+}	
Serum concentration (%)	Control ± SEM	After silica ± SEM	Control ± SEM	After silica ± SEM
0	110 ± 10	220 ± 20	155 ± 15	350 ± 20
1	140 ± 10	330 ± 20	160 ± 20	500 ± 10
2.5	155 ± 5	365 ± 15	145 ± 15	455 ± 27
5	160 ± 10	375 ± 5	160 ± 20	455 ± 5
10	185 ± 15	355 ± 15	150 ± 20	425 ± 25
20	145 ± 15	370 ± 10	165 ± 15	445 ± 25

[a]Results are expressed as the mean ± SEM of 4 replicates.

preparations acting on the same target cell must result in parallel linear dose-response lines. These dose-response lines can be horizontally displaced, showing that the Ep preparations tested have different relative potencies or specific activities. This is the case when normal or wheat germ lectin-purified step III Ep is compared to the concentrated, wheat germ lectin-purified silica-treated spleen supernatant. This again indicates that the silica-treated spleen supernatant contains Ep.

Release of Ep from macrophages by treatment with silica takes place in the presence of serum. That fetal calf serum is necessary in the preincubation procedure is shown in Table III. Also shown is that a fetal calf serum concentration as low as 1% is required, increasing concentrations having no additional effect. In addition, fetal calf serum can be replaced by 2 mM Ca^{2+}. Addition of 1% fetal calf serum caused a slight increase in released activity. However, increasing this concentration to 20% again did not result in any significant increase (or decrease) in released activity. The possibility therefore arises of obtaining a native Ep preparation that does not contain the protein contamination usually present in both commercial and human urinary Ep.

The fact that macrophages can release Ep indicates an important role for this cell in erythropoietic differentiation. This comes from the presence in both the fetal liver and adult bone marrow of the so-called blood islands. The blood islands consist of a central macrophage-like cell with differentiating erythroblasts surrounding it. We [10] have postulated that the release of only minute amounts of Ep from macrophages could be high amounts for the differentiating cells in the immediate vicinity or even in association with these centrally lying cells. It is known that macrophages can release a wide range of different factors into the medium. Thus, when macrophages (and perhaps other cell types) are present in hemopoietic cultures either with or without the addition of known stimulating factors (eg, CSF, Ep), it is necessary to know whether these cells are releasing similar factors or other factors, which could result in the production of cells from one or several different pathways of differentiation.

An important implication raised by macrophage-released Ep is that it would provide an explanation for the spleen being a favorable environment for differentiating erythroblasts when the animal is subjected to erythropoietic stimulus, particularly hypoxia. Earlier work by Rambach et al [9] and Zangheri et al [19] showed that spleen extracts could cause an increase in erythropoiesis. This work led to the postulation that the spleen could be an important extrarenal site for Ep production. Gruber et al [3] showed that fetal liver macrophages are responsible for Ep production during hepatic erythropoiesis since the normal Ep production site, the kidney, has not developed this function at this stage of development. Almost all investigations concerned with the apparent production of Ep from extrarenal sites have laid emphasis on the adult liver. Considering the fact that early hepatic erythropoiesis plays such an important role in the development of the erythropoietic system, it is not surprising that a remnant of this role be carried over into adult life. The same argument can be applied to the spleen, since erythropoiesis is observed in the spleen before being fully transferred to the bone marrow [2, 5, 11].

Besides the study of Gruber et al [3], Peschle et al [7] showed that the extrarenal Ep source in the adult liver was a function of the reticuloendothelial system. Since crystalline silica is specifically cytotoxic for macrophages [1, 6], release of Ep can be shown to be a function of this cell not only in the fetal liver, bone marrow, and spleen [13] but also in the lung, peritoneum, and adult liver [14]. The finding that all macrophage-containing cell suspensions tested can release Ep activity would tend to argue for a ubiquitous production and/or storage phenomenon. Recent experiments in our laboratory utilizing gas-permeable Teflon foils with hydrophobic surfaces on which spleen cells are grown, indicate that Ep can in fact be produced by these cells for at least 4 weeks in culture [Rich, Heit, and Kubanek, in preparation].

ACKNOWLEDGMENTS

This work was supported by the Deutsche Forschungsgemeinschaft (Sonderforschungsbereich 112/project A2) and the Volkswagen Foundation, as well as by VA Research funds and grant AM 24027 from NIAMDD, NIH.

REFERENCES

1. Allison AC, Harrington SS, Birbeck M: J Exp Med 124:141, 1966.
2. Cole RJ, Regen T, White SL, Cheek EM: J Embryol Exp Morphol 34:575, 1975.
3. Gruber DF, Zucali JR, Mirand EA: Exp Hematol 5:392, 1977.
4. Gregory CJ, Tepperman AD, McCulloch EA, Till JE: J Cell Physiol 84:1, 1974.
5. Metcalf D, Moore MAS: "Hemopoietic Cells." Amsterdam: North Holland Publishing Company, 1970.
6. Pearsall NN, Weiser R: J Reticuloendothel Soc 5:107, 1968.
7. Peschle C, Marone G, Genovese A, Rappaport IA, Condorelli M: Blood 47:325, 1976.
8. Rabinowitz Y: Blood 23:811, 1964.
9. Rambach WA, Alt HL, Cooper JAD: Proc Soc Exp Biol Med 108:793, 1961.
10. Rich IN, Kubanek B: J Embryol Exp Morphol 50:57, 1979.
11. Rich IN, Kubanek B: J Embryol Exp Morphol 58:143, 1980.
12. Rich IN, Anselstetter V, Heit W, Kubanek B: J Supramol Struct Suppl 4, 1980.
13. Rich IN, Heit W, Kubanek B: Blut 41:29, 1980.
14. Rich IN, Anselstetter V, Heit W, Kubanek B: In Lucarelli B, Gale RP, Fliedner TM (eds): "Proceeding of the First International Symposium on Fetal Liver Transplantation." Amsterdam: North Holland Publishing Company, 1979, p 95.
15. Schooley JC, Mahlman LJ: Blood 43:425, 1974.
16. Spivak JL, Small D, Shaper JH, Hollenberg MD: Blood 52:1178, 1978.
17. Stephenson JR, Axelrad AA, McLeod DL, Shreeve MM: Proc Natl Acad Sci USA 68:1542, 1971.
18. Wang CS, Smith RL: Anal Biochem 63:414, 1975.
19. Zangheri EO, Fora-De-Moraes F, Lopez OI, Marios I: Experientia 29:706, 1973.

Purification and Characterization of Multiplication-Stimulating Activity (MSA) Carrier Protein

Daniel J. Knauer, Fred W. Wagner, and Gary L. Smith

School of Life Sciences, and Laboratory of Agricultural Biochemistry, University of Nebraska, Lincoln, Nebraska 68583

> The rat liver cell line, BRL-3A, is known to produce a family of polypeptides referred to as multiplication-stimulating-activity (MSA). Serum-free conditioned medium from this cell line is a rich source for the purification of these somatomedin-like molecules. Somatomedins in serum, as well as MSA produced by BRL-3A cells in culture, exist primarily as a high molecular weight complex bound to specific carrier proteins. This study describes the purification of the MSA carrier protein (MCP) from conditioned medium using affinity chromatographic procedures. The purified carrier protein is shown to specifically bind labeled MSA and generates a complex with an apparent molecular weight of 60,000–70,000 daltons. Characterization of the carrier protein indicates that it consists of two different noncovalently linked protein chains with apparent molecular weights of 30,000 and 31,500 daltons. The availability of a pure carrier protein should provide a unique opportunity to investigate the functional significance of the carrier protein in the biological activity of the somatomedins.

Key words: MSA, somatomedin, carrier protein

The somatomedins are a family of low molecular weight polypeptide hormones (mol wt $\simeq$ 7,500) that are the proposed mediators of the peripheral action of growth hormone on skeletal tissue [1, 2]. As a group, the somatomedins also have weak insulin-like activity and promote the growth of a variety of cell types in culture [1–5]. Several small polypeptides that fit the current definition of somatomedins have been purified from human serum. These include somatomedins A and C and the insulin-like growth factors IGF I and IGF II [3, 6–8]. A unique property of the somatomedins is that they circulate in plasma bound to high molecular weight protein(s) referred to as somatomedin carrier proteins (SM carrier proteins) [9–11]. While the role of the large carrier protein(s) in the biological activity of the somatomedins is still speculative, it has been established that the presence of at least one form of SM carrier protein(s) is also growth hormone dependent,

Received April 4, 1980; accepted October 9, 1980.

and association of somatomedin with this carrier greatly prolongs the circulating half-life of somatomedin activity in vivo [12, 13].

The somatomedin activity in rat serum is also complexed to a somatomedin carrier protein(s), analogous to the situation that exists for human somatomedins [3]. Although it is at this time difficult to predict structural similarities between human and rat SM carrier protein(s), highly purified Somatomedin A, IGF I, and IGF II all bind to rat SM carrier protein in a highly specific manner [14]. The precise chemical nature of the interactions of these molecules with rat SM carrier protein(s) is not known, but it is probably a weak noncovalent interaction since treatment at low pH is routinely used to dissociate the molecules [9–14].

In addition to the human somatomedins described above, purified multiplication-stimulating-activity (MSA) has also been shown to bind to SM carrier proteins in rat serum [14]. MSA refers to a family of low molecular weight somatomedin-like polypeptides purified from serum-free medium conditioned by the growth of a line of buffalo rat liver cells (BRL-3A) [15–18]. MSA appears to be structurally and biochemically similar to the human somatomedins. This similarity is strongly supported by the demonstration that somatomedin A, IGF I, and IGF II compete effectively with MSA for cell surface mitogenic receptors on a variety of cell types and tissues [4, 18], as well as for the same SM carrier proteins in rat serum [14].

At present, the molecular events involved and the physiological significance of the interactions between somatomedins and their serum carriers is a complex and largely unresolved issue. When highly purified ^{125}I-MSA is added to whole, untreated rat serum at neutral pH, two molecular weight forms of carrier protein can be demonstrated by Sephadex G-200 gel filtration [13, 14]. Only the low molecular weight form can be demonstrated in serum exposed to acid conditions (1 M acetic acid) and subsequently returned to neutral pH prior to the addition of ^{125}I-MSA. All attempts to generate the high molecular weight form from the low molecular weight form have been unsuccessful. That the relationship between the high and low molecular forms may be physiologically significant is indicated by experiments showing that only the low molecular weight form is demonstrable in serum from rats that have been hypophysectomized [13, 14].

In an attempt to better understand the molecular basis of the interactions between somatomedins and their carrier protein(s), we have taken advantage of the fact that BRL-3A cells produce SM carrier protein as well as MSA under serum-free conditions in culture and have utilized this as a source from which to purify a rat SM carrier protein. In this report, we demonstrate the purification of MSA carrier protein (MCP) from serum-free BRL-3A-conditioned medium and show that purified MCP specifically binds ^{125}I-MSA and generates a ^{125}I-MSA—carrier protein complex with a molecular weight similar in size to the low molecular weight form seen after the addition of ^{125}I-MSA to rat serum [13, 14]. We feel that this represents an important step in understanding the biochemistry and biological significance of SM carrier protein(s).

MATERIALS AND METHODS

MSA Purification and Separation From Carrier Protein

Serum-free conditioned medium was prepared from roller bottle cultures of BRL-3A rat liver cells as described by Dulak and Shing [19] MSA was purified from the conditioned medium using established procedures [16–18]. Briefly, 4-liter batches of conditioned medium were subjected to ion exchange chromatography using Dowex 50W-X8 resin in

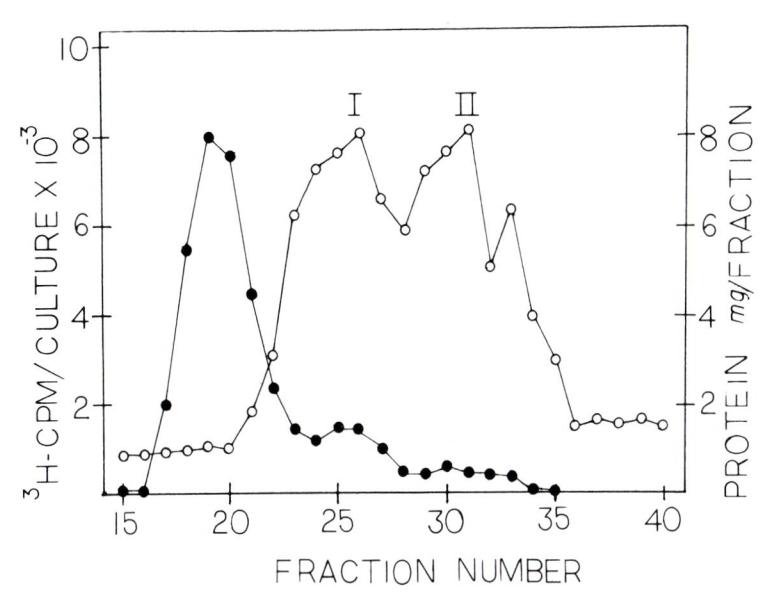

Fig. 1. Sephadex G-50 chromatography of Dowex-50 MSA in 1 N acetic acid. Fifty milligrams were chromatographed over a 2.5 × 100 cm column at 4°C. Ten-milliliter fractions were collected and assayed for protein content (●) and for multiplication-stimulation activity (○).

the Na$^+$ form. MSA absorbs to the resin at neutral pH (apparently indirectly via its carrier protein [19]) and is eluted at pH 11. This Dowex-50 MSA was then chromatographed on Sephadex G-50 in 1 N acetic acid. Most of the protein elutes from the G-50 column in the void-volume region, including the MSA carrier protein(s) that have been dissociated from the MSA by the acid treatment (Fig. 1). The pooled void-volume fractions from this G-50 step served as the starting material for the subsequent isolation of the carrier protein. MSA elutes from the G-50 column in two peaks of activity as measured by the stimulation of labeled-thymidine incorporation into quiescent cultures of chicken embryo fibroblasts. Both peaks have comparable activity in this assay and together consist of a family of seven or eight biologically active polypeptides. The MSA affinity column (to be described later) was prepared using a mixture of these two peaks of MSA since it was not feasible to purify enough of a single molecular weight species of MSA for this purpose.

For studies characterizing the binding properties of the carrier protein using radioactively labeled MSA, a single species of MSA was purified. The first peak of MSA activity to elute from the G-50 column was pooled and subjected to preparative scale disc polyacrylamide gel electrophoresis in acetic acid:urea at pH 2.7 as previously described [18]. This MSA preparation consisted of a single protein-staining band in acid and alkaline acrylamide systems (mol wt 9,000) and was active in stimulating DNA synthesis in chicken embryo fibroblasts at concentrations of 10–20 ng/ml and was maximally active at concentrations of 150–200 ng/ml. This preparation of purified MSA probably corresponds to MSA polypeptide II-1 of the designation system of Moses et al [20].

MSA Radioiodination

Six μg of highly purified MSA (mol wt 9,000) was radioiodinated to a high specific activity (~80 Ci/g) by a modification of the chloramine-T procedure [21]. The reaction mixture was chromatographed over a 1 × 50 cm column of Sephadex G-50 resin (fine) equilibrated at 4°C in 1 M acetic acid containing 1 mg/ml of bovine serum albumin (BSA). One-milliliter fractions were collected and assayed for specific binding to MSA receptors on chick embryo fibroblasts. Fractions containing ^{125}I-MSA that showed the highest degree of specific binding were pooled, dispensed in 0.2 ml aliquots and stored at −20°C. ^{125}I-MSA was stable for approximately 4 weeks.

Affinity Chromatography

The activated N-hydroxysuccinimide ester of Sepharose 4B was prepared as described by Parikh et al [22]. To 8 ml of packed, activated Sepharose 4B equilibrated in 0.1 M phosphate buffer, pH 7.2 at 4°C, was added an equal volume of the same buffer containing 10 mg of a mixture of MSA polypeptides from the G-50 column (see above). The reaction mixture was gently stirred overnight at 4°C. The resin was then extensively washed with distilled deionized water and then incubated for 2 h at room temperature with 0.1 M glycine, pH 9.0, in 0.1 M bicarbonate buffer. This was done to mask residual reactive ester groups. The resin was then washed with several volumes of 0.1 M acetate buffer (pH 3.5), followed by several volumes of distilled deionized water, and was finally equilibrated in Mg^{2+}- and Ca^{2+}-free Dulbecco's phosphate buffered saline (PBS, pH 7.4). This yielded a product consisting of Sepharose 4B containing a succinylaminoethyl arm to which MSA was covalently linked presumably via the α amino group. The slurry was transferred to a 10 ml plastic syringe yielding a column with a packed bed volume of approximately 6 ml.

Purification of MSA Carrier Protein

The beginning material used as the source for the purification of MSA carrier protein (MCP) from rat liver cell conditioned medium was a pool of the fractions comprising the void volume eluted from the Sephadex G-50 gel filtration step in the MSA purification scheme described above. It is during this step that MSA is separated from the high molecular carriers by gel filtration under acidic conditions. Fractions comprising the void volume were pooled, concentrated by lyophilization and rechromatographed over an identical G-50 column in 1 N acetic acid to insure complete separation of the MSA. Again the fractions comprising the void volume were pooled, dried by lyophilization, and then resuspended in PBS, pH 7.4, to a final concentration of 500 μg/ml. Ten milliliters of this solution was applied to the affinity column at 22°C at a flow rate of 5 ml/h. The protein content of the column effluent was continuously monitored by measuring the absorbance at 280 nm. Two-milliliter fractions were collected. After sample loading, the column was washed with PBS, pH 7.4, until the effluent was free of protein. The column was then eluted with 0.1 M acetic acid, pH 3.5, at a flow rate of 15 ml/h. Fractions comprising the eluted protein peak were pooled, dialyzed extensively against 1 M acetic acid, and concentrated by lyophilization. The protein residue was resuspended in 1 M acetic acid and chromatographed over a 1 × 50 cm Sephadex G-50 column as previously described. One-milliliter fractions were collected and analyzed by acetic acid-urea (4M:8M) polyacrylamide gel electrophoresis. Fractions free of MSA peptides, (V_0 fractions) were pooled as purified MCP.

Electrophoresis

Samples subjected to electrophoresis in acetic acid-urea (4M:8M), were first dried by lyophilization and resuspended in 0.05 ml of acetic acid-urea (4M:8M) containing 0.01% methylene blue as tracking dye. The samples were applied to 0.5 cm × 8.0 cm tube gels of the following composition: 4M acetic acid:8M urea, 7.5% acrylamide (w/v), 0.25% methylene bisacrylamide (w/v), 0.5% N,N,N',N'-tetramethylene diamine (TEMED) (v/v) and 0.21% ammonium persulfate (w/v). Electrophoresis was carried out toward the cathode for 4–6 h at 22°C. Gels were stained in a solution of 0.05% coomassie blue in methanol-acetic acid-H$_2$O (5:1:5). The gels were destained in 7.5% acetic acid (v/v) containing 20% methanol (v/v).

When protein bands were to be eluted from gels and assayed for biological activity, a pH 4.4 electrophoretic system was used in 8 M urea. Samples were lyophilized and resuspended in 0.05 ml of acetic acid-urea (0.5 M:8 M) with 0.01% methylene blue as tracking dye. Samples were applied to 0.5 × 8.0 cm tube gels of the following composition: 7.5% acrylamide (w/v), 8 M urea, 0.75 M acetic acid (pH 4.4), 0.2% methylene bisacrylamide (w/v), 0.5% TEMED (v/v), and 1.25 µg riboflavin. An unstained gel was sliced into 2 mm slices and the slices subjected to electrophoretic elution. Samples derived from each slice were dialyzed, lyophilized, and resuspended in 400 µl of PBS, pH 7.4, containing 5 mg/ml fatty acid-free BSA. The ability of the eluted protein to specifically bind ^{125}I-MSA was determined using the fatty acid-free BSA activated charcoal assay described by Moses et al [14].

Electrophoresis in 10% SDS-polyacrylamide tube gels (0.5 × 8.0 cm) was carried out as described by Weber et al [23]. Samples were denatured and reduced in buffer containing 2-mercaptoethanol by heating for 10 min at 65°C prior to electrophoresis. Where appropriate, molecular weight standards were included. The gels were stained and destained as described for acetic acid-urea gels.

Binding of ^{125}I-MSA to Purified MCP and Rat Serum

^{125}I-MSA (7 × 10^5 CPM) was incubated with 0.5 ml of rat serum, acid dialyzed rat serum, Dowex-50-treated rat serum or 5.0 µg of purified MCP (from BRL-3A conditioned medium) in 0.5 ml of 0.1 M Hepes-buffered DMEM (pH 7.4) containing 10 mg/ml of BSA. The reaction mixtures were incubated for 5 h at 22°C. To determine binding specificity, 10 µg of unlabeled MSA was added to a parallel reaction mixture. At the completion of the binding reaction, the mixtures were immediately diluted 1:1 with cold PBS, pH 7.4, and fractionated by chromatography over Sephadex G-200. The BSA used in the reaction mixtures had no detectable ^{125}I-MSA binding activity (data not shown).

Sephadex G-200 chromatography was performed using a 2.5 × 40 cm column of Sephadex G-200 (fine) equilibrated in PBS, pH 7.4. The column was poured and eluted under 12 cm constant head pressure in the descending mode. Fractions of 2.25 ml were collected at a flow rate of 5 ml/h. To calibrate the column with protein markers, 0.5 ml of calf serum diluted 1:1 with PBS, pH 7.4, was filtered over the column. Aliquots of the fractions were assayed for protein content by the method of Lowry et al [24]. When ^{125}I-MSA binding reaction mixtures were passed over the column, aliquots (0.1 ml–0.25 ml) of the fractions were assayed for ^{125}I-CPM by liquid scintillation spectrometry. Approximately 70% of the applied radioactivity was recovered.

Competitive Binding Assay for MSA

Specific binding of ^{125}I-MSA to purified MCP for Scatchard analysis [25] and for determination of binding activity in protein samples eluted from gel slices was performed using fatty acid-free BSA activated charcoal to separate bound from free MSA as described by Moses et al [14]. For the Scatchard analysis, 40 ng of purified MCP was incubated with 0.5 ng of ^{125}I-MSA and various concentrations of unlabeled MSA for 3 h in 0.4 ml of PBS

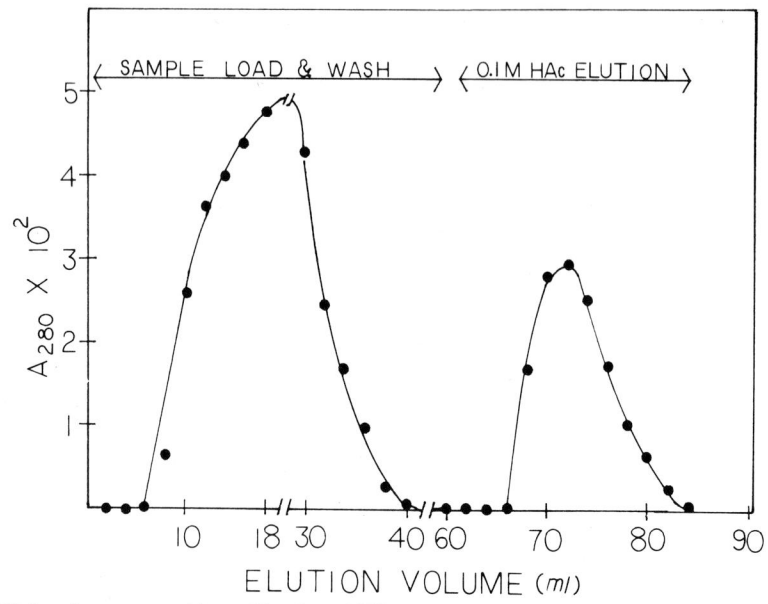

Fig. 2. Affinity chromatographic purification of MSA carrier protein. Five milligrams of a mixture of high molecular weight proteins, obtained from the void-volume region following chromatography of Dowex-50 MSA on Sephadex G-50 in 1 N acetic acid, were applied to a column of Sepharose 4B resin containing covalently immobilized MSA. Following extensive washing with PBS to remove unbound proteins, the column was eluted with 0.1 M acetic acid. Protein content in the eluant was continuously monitored by absorbance at 280 nm.

containing 5 mg/ml BSA. At the end of the incubation period, bound MSA was separated from free MSA and the supernatants assayed for radioactivity. Binding of ^{125}I-MSA to BRL-3A2 conditioned medium (a generous gift from M. Rechler) was also performed in a similar manner substituting 100 μl of medium in place of the purified MCP.

Binding of ^{125}I-MSA to samples eluted from gel slices was performed by resuspending protein samples in 0.4 ml of PBS containing 5 mg/ml BSA. Four assay tubes were used for each gel slice sample and contained 0.1 ml of the protein sample, 0.3 ml assay buffer, and 1×10^5 CPM of ^{125}I-MSA. Two of the tubes also contained 1 μg of unlabeled MSA for determination of specific binding. Separation of bound from free MSA was done as described above.

RESULTS

MSA polypeptides in BRL-3A conditioned medium exist in a high molecular weight form bound to a specific carrier protein [14]. During the purification of MSA, conditioned medium is first subjected to ion exchange chromatography using Dowex 50W-X8 resin in the Na$^+$ form. MSA is known to absorb to the resin at neutral pH and is then eluted at pH 11. Dulak and Shing [19] first suggested that MSA does not bind to the resin directly but does so via more basic proteins in the medium, presumably the carrier. After the Dowex-50 step, the MSA preparation is chromatographed on Sephadex G-50 in 1 N acetic acid. The detectable MSA activity elutes from the column as two peaks in the low molecular weight region clearly separated from the bulk of the protein, which appears near the void volume of the column (Fig. 1). This procedure dissociates the MSA-carrier protein complex and the MSA carrier protein (MCP) now elutes along with the majority of the protein in the void-volume region. BRL-3A conditioned medium thus provides a rich source of MCP as well as MSA polypeptides.

Because of the simple pH dependence of the binding of MSA peptides to their carrier protein(s), it was reasoned that MSA peptides immobilized on a solid support should provide a highly specific and simple method with which to purify BRL-3A-produced MSA carrier protein (MCP). The N-hydroxysuccinimide active ester of Sepharose 4B was prepared as described by Parikh et al [22]. A mixture of MSA polypeptides (peaks I and II, Fig. 1) were reacted with the activated Sepharose 4B to prepare the affinity column. For details of this procedure refer to Materials and Methods. The void-volume proteins obtained from Sephadex G-50 chromatography of Dowex-50 MSA (Fig. 1) were pooled and concentrated by lyophilization for use as the source for the purification of carrier protein. The Sephadex G-50 step was repeated to insure removal of the MSA polypeptides and the protein preparation was resuspended in PBS at a concentration of 500 µg/ml. Ten milliliters of this solution were loaded onto the affinity column at a flow rate of 5 ml/h. Protein content in the eluant was monitored by absorbance at 280 nm. After sample loading, the column was washed extensively with PBS, pH 7.4, until the protein content in the eluant was negligible (Fig. 2). The column was then eluted with 0.1 M acetic acid, pH 3.5, at a flow rate of 15 ml/h. Two-milliliter fractions were collected. One major peak of protein was released from the column during the acidic elution (Fig. 2). The fractions comprising this peak (68–86 ml total elution volume) were pooled, dialyzed against 1,000 volumes of 2% acetic acid (v/v) to remove salt, and concentrated by lyophilization. The residue was resuspended in 1 ml of 1 M acetic acid and chromatographed over a 1 × 50 cm column of Sephadex G-50 in 1 N acetic acid at 4°C (data not shown). This step was necessary to remove MSA peptides that were released during the acidic elution of the affinity column. The void-volume fractions were analyzed by acetic acid-urea (4M:8M) polyacrylamide gel electrophoresis. Fractions free of MSA polypeptide contamination were pooled as purified carrier protein (data not shown). To test the purity of this preparation and to examine its presence throughout the purification procedure, 5 µg were electrophoresed in the same acetic acid-urea system (Fig. 3). Only one protein band was observed for the purified carrier protein preparation. There were no detectable MSA peptides present. Shown for reference in Figure 3 are protein samples of BRL-3A conditioned medium (40 µg), and Dowex-50 MSA (40 µg). No additional protein bands were evident when as much as 30 µg of the purified MCP preparation was applied to similar gels.

To show that the purified protein band appearing in the gels corresponded to binding activity, a similar gel was sliced prior to staining, and protein was extracted from the slices by electrophoretic elution. Samples were then assayed for binding capacity using the fatty acid-free BSA activated charcoal procedure for the separation of bound and free MSA described by Moses et al [14]. As can be seen in Figure 4, binding activity in the gel coincided with the observable stained protein band in an identical gel. This proves that the protein purified by the affinity chromatographic procedure is indeed the MSA carrier protein.

To determine the relationship of purified MCP to the MSA or somatomedin binding proteins in rat serum, the experiments detailed in Figures 5 and 6 were performed. MSA labeled with iodine 125 was incubated with normal rat serum, acid dialyzed rat serum, Dowex-50-treated rat serum, or purified MCP. Identical reaction mixtures contained excess unlabeled MSA for determination of specific binding. After 5 h of incubation, the mixtures

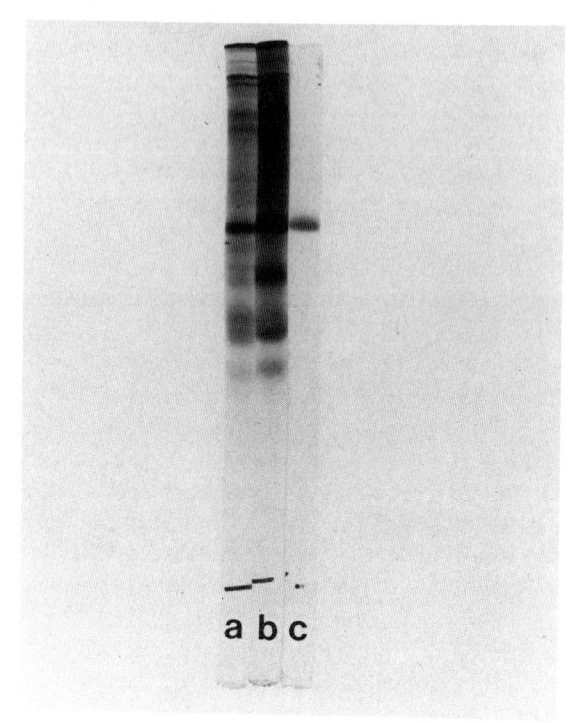

Fig. 3. Electrophoresis of purified carrier protein in acetic acid-urea gels. Carrier protein, 5 μg, purified by affinity chromatography (c) was subjected to polyacrylamide gel electrophoresis as described in Materials and Methods. For comparison, 40 μg samples of whole conditioned medium proteins (a) and Dowex-50 MSA (b) are shown.

were individually applied to the same 2.5 × 40 cm column of Sephadex G-200 equilibrated in PBS at 4°C. Aliquots of the 2.25 ml fractions were assayed for radioactivity. The results shown in Figure 5 confirm those previously reported by others [13, 14]. MSA binds to normal rat serum and after chromatography on Sephadex G-200 at neutral pH, radioactivity appears in five peaks. Binding to peaks 2 and 3 is specific and can be eliminated in the presence of an excess of unlabeled MSA. Binding to peak 1 is nonspecific, peak 4 is free ^{125}I-MSA and peak 5 is free ^{125}I. Acid dialysis of normal rat serum eliminates binding activity in the region of peak 2 (Fig. 5B).

When MCP purified by affinity chromatography was incubated with ^{125}I-MSA and the mixture chromatographed on Sephadex G-200, specific binding was detected in a region corresponding in size to the peak 3 binding activity in normal and acid-dialyzed rat serum (Fig. 6A). This complex elutes in a position similar in size to that of calf serum albumen and has an approximate molecular weight of 66,000–68,000 daltons.

Dowex-50 chromatographic treatment of normal rat serum followed by acid dialysis also destroys binding in the region of peak 2 (Fig. 6B). These results show that the complex generated when purified MCP is incubated with ^{125}I-MSA appears to be equivalent to the lower molecular weight (peak 3) complex found in normal or acid-dialyzed rat serum [13, 14].

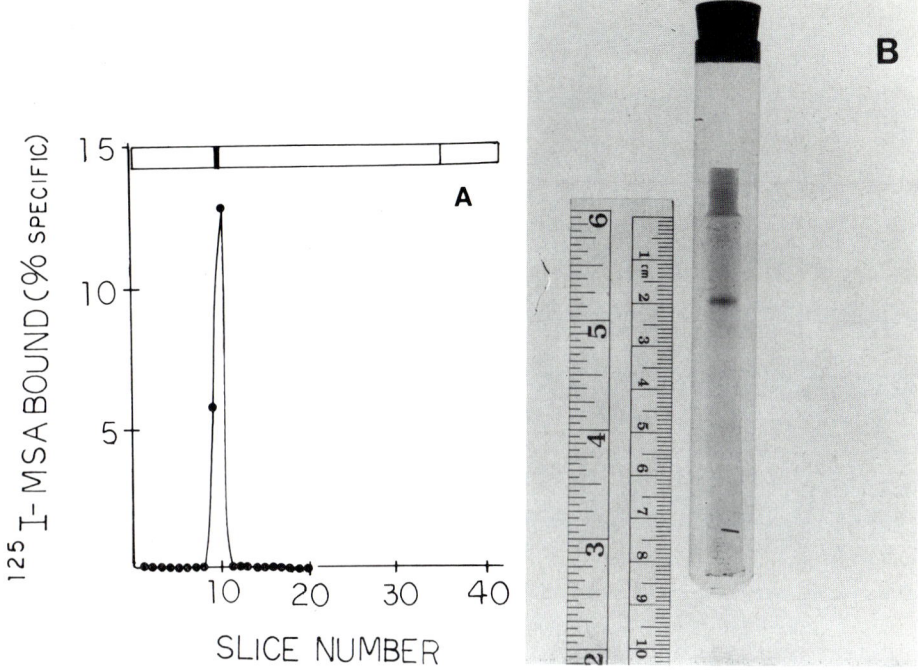

Fig. 4. Localization of MCP activity in acetic acid-urea gels. Samples of purified MCP (5 μg) were subjected to electrophoresis in acetic acid-urea gels at pH 4.4. The lower gel consisted of 7.5% acrylamide and a 1 cm upper gel was applied with an acrylamide concentration of 2.5% acrylamide. Following electrophoresis, one gel was sliced into 2 mm slices, and the slices were individually subjected to electrophoretic elution. Eluted samples were then assayed for carrier protein activity in the fatty acid-free BSA activated charcoal binding assay (A). An identical gel was stained for visualization of protein bands (B). A schematic diagram of the stained gel appears over the profile of carrier protein activity in panel A.

The approximate molecular weight of purified MCP obtained by affinity chromatography was determined using SDS polyacrylamide gel electrophoresis according to the method of Weber et al [23]. Samples were prepared for electrophoresis in the presence and absence of 2-mercaptoethanol. Under reducing conditions, the purified MSA carrier protein ran as two closely migrating bands with apparent molecular weights of 30,000 and 31,500 daltons (Fig. 7B). The faster-migrating band stained much lighter than the slower band. In the absence of 2-mercaptoethanol, a single sharp band with an apparent molecular weight of 31,500 was observed (Fig. 7A). It should be noted that these molecular weight determinations may be subject to error due to the possible glycoprotein nature of these proteins. This may also explain the difference in staining intensity of the two bands appearing under reducing conditions.

To determine the binding affinity of purified MCP to MSA, a competitive binding assay was performed using the fatty acid-free BSA activated charcoal procedure described by Moses et al [14] for the separation of bound and free MSA (Fig. 8). Scatchard analysis of the results yielded an affinity constant of $0.23 \times 10^9 M^{-1}$. In addition, the affinity constant was determined for the binding of ^{125}I MSA to BRL 3A2 conditioned medium.

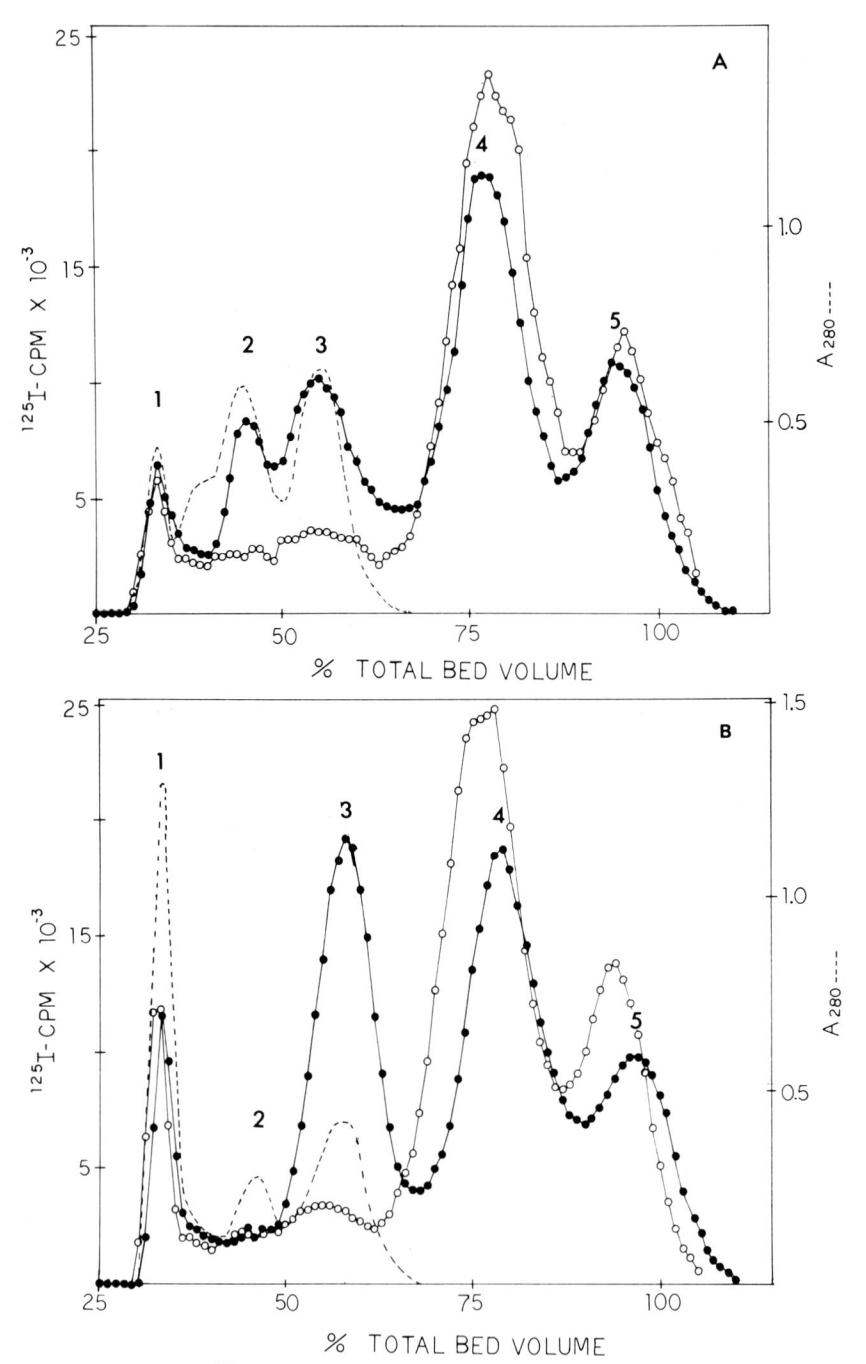

Fig. 5. Elution profile of ^{125}I-MSA and normal rat serum or acid-dialyzed rat serum on Sephadex G-200. ^{125}I-MSA was incubated with normal rat serum (A) or acid-dialyzed rat serum (B) for 5 h at 22°C. The reaction mixtures were applied to a 2.5 × 40 cm column of Sephadex G-200 equilibrated in PBS at 4°C. Fractions of 2.25 ml were collected and assayed for radioactivity (●). Identical mixtures contained 10 μg of unlabeled MSA for determination of specific binding (○). The dotted line represented the elution profile of 0.5 ml of calf serum monitored by absorbance at 280 nm, which was applied prior to each run to calibrate the column.

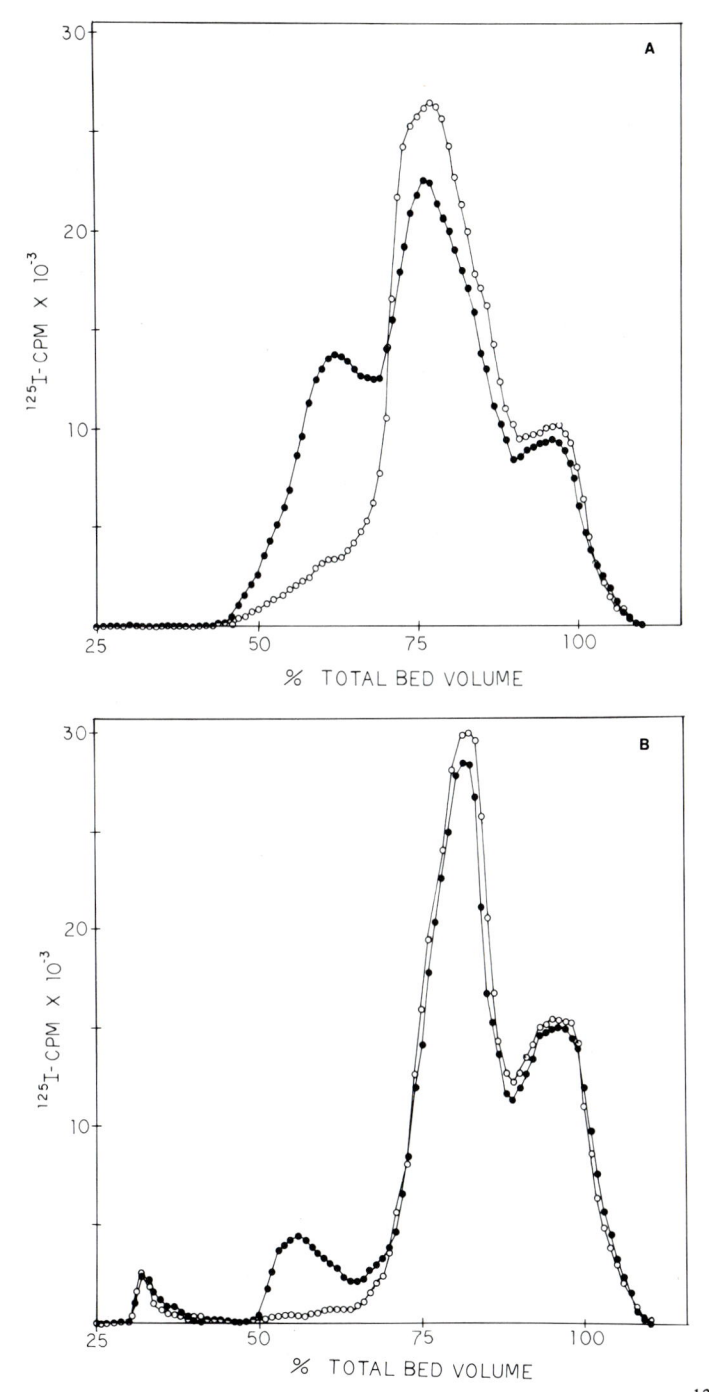

Fig. 6. Elution profile of ^{125}I-MSA and purified MCP or Dowex-50-treated rat serum. ^{125}I-MSA was incubated with purified MCP (A) or a Dowex-50 preparation of normal rat serum (B) for 5 h at 22°C. The reaction mixtures were then applied to the same Sephadex G-200 column described in Figure 5. Fractions were assayed for radioactivity (●). Identical reaction mixtures contained 10 µg of unlabeled MSA for determination of specific binding (○).

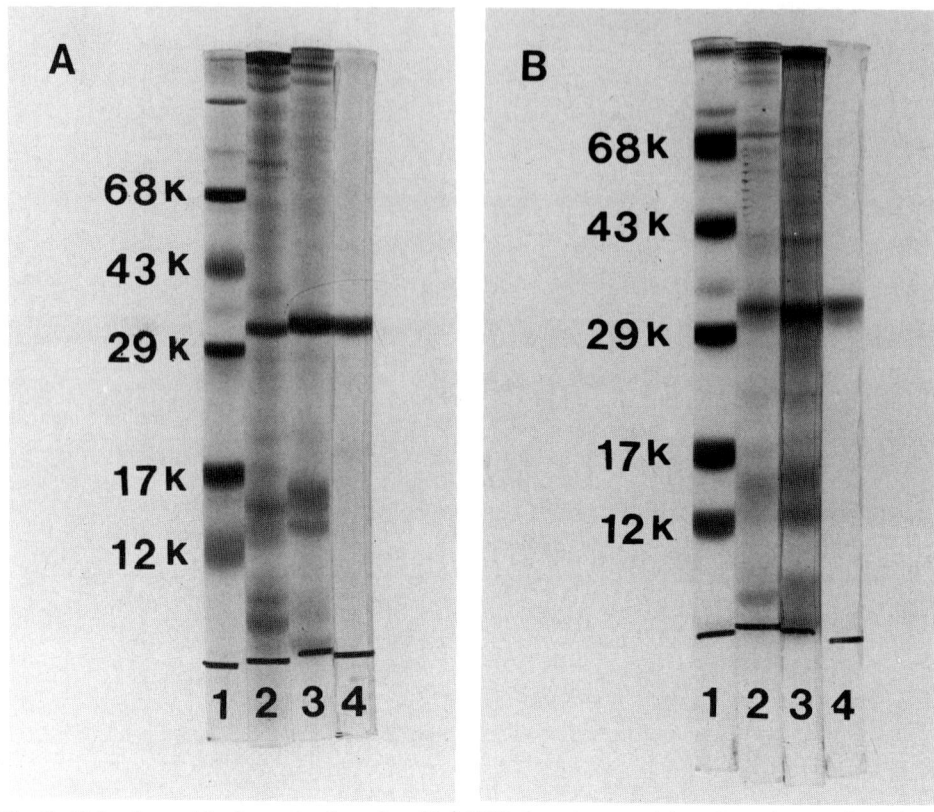

Fig. 7. Molecular weight determination of purified MCP. Samples containing (1) a mixture of standard molecular weight markers, (2) 30 μg of whole BRL-3A conditioned medium proteins, (3) 30 μg of Dowex-50 MSA, and (4) 5 μg of MCP purified by affinity chromatography were electrophoresed on 10% SDS polyacrylamide gels in the absence (A) or presence (B) of 2-mercaptoethanol.

BRL-3A2 cells are a subclone of BRL-3A cells that produce MSA carrier protein but do not produce MSA [14]. This allows a direct comparison of purified MCP to carrier protein in its native state in conditioned medium to determine if the MCP has been altered in any way by the purification procedure. The affinity constant for BRL-3A2 medium was determined to be $0.3 \times 10^9 M^{-1}$. These values are not significantly different. It was not possible to use BRL-3A medium for this purpose since MSA is produced in apparent excess, and all of the MCP is complexed with MSA. When ^{125}I-MSA is incubated with BRL-3A conditioned medium followed by fractionation on neutral Sephadex G-200, all of the radioactivity appears as free MSA (data not shown and personal communication with M. Rechler).

DISCUSSION

In this report, we have described the purification of the MSA carrier protein from BRL-3A rat liver cell condition medium. The purification scheme involved ion exchange chromatography using Dowex 50W-X8 resin in the Na+ form. MSA is known to absorb to the resin at neutral pH and is then eluted at pH 11. Based on the amino acid composition of

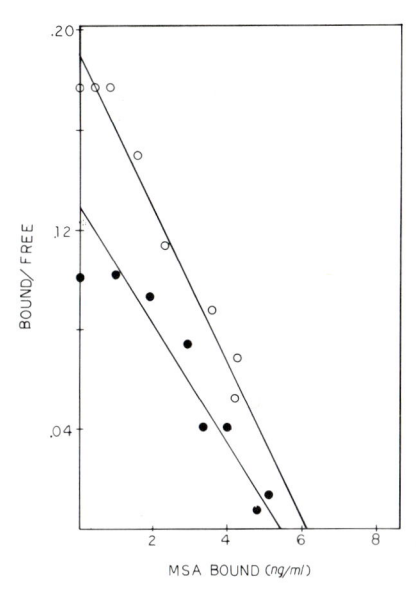

Fig. 8. Scatchard analysis of MSA binding to purified MCP or to BRL-3A2 conditioned medium
Equal amounts of MCP (40 ng, ●) or conditioned medium (100 μl, ○) were incubated with 500 pg of ^{125}I-MSA and increasing concentrations of unlabeled MSA (5-800 ng/ml) for 3 h in 0.4 ml of PBS containing 5 mg/ml fatty acid-free BSA. Bound MSA was then separated from free MSA by the fatty acid-free BSA activated charcoal procedure. The slopes of the lines were determined by linear regression analysis.

purified MSA, Dulak and Shing [19] previously suggested that MSA probably does not bind to the Dowex column directly but does so via more basic proteins in the conditioned medium. We now know that the carrier protein accounts for this behavior of MSA. Next the carrier protein-MSA complex is dissociated under acidic conditions and the components separated by chromatography on Sephadex G-50 in 1 N acetic acid. The carrier proteins, appearing in the void volume of the column can then be purified by specific binding to a column containing covalently bound MSA. It should be noted that the procedures used for the preparation of the affinity column involved the synthesis of the N-hydroxysuccinimide ester derivative of Sepharose 4B as described by Parikh et al [22]. This is a simple and mild technique to immobilize proteins essential via the α amino group to yield a ligand that is separated from the support by lengthy hydrocarbon extensions. This facilitates binding that might be hindered sterically in other methods of immobilization. In addition, since coupling is via primary amino groups, the possibility of inducing protein configuration changes is minimal.

    The carrier protein purified by these procedures migrates as a single band in acetic acid-urea and in SDS polyacrylamide gel electrophoresis systems with an apparent molecular weight of approximately 31,500 daltons. In the SDS system under reducing conditions (2-mercaptoethanol), a second minor band appears with an apparent molecular weight of 30,000 daltons. Thus, the carrier protein appears to be composed of two different protein chains. The possibility that the components we have purified consist of a single chain with heterogeneity in the degree of glycosylation cannot be excluded at this time. This would require the separation and analysis of both species for amino acid and carbohydrate content. The molecular weight determinations may be subject to error since the carrier protein may be a glycoprotein. This data is consistent with a report by Fryklund et al [26] that a somatomedin binding protein purified from human serum is a glycoprotein and is composed of two dissimilar protein chains with molecular weights between 35,000 and 45,000 daltons. Unfortunately, these investigators reported rather low total binding of somatomedin A by their purified preparation and further details were not given.

The MSA carrier protein purified by affinity chromatography bound labeled MSA and formed a complex that eluted from Sephadex G-200 with a molecular weight in excess of 60,000 daltons. This binding was specific and could be inhibited in the presence of unlabeled MSA. It is tempting to speculate that the complex is composed of one each of the carrier protein subunits and one MSA molecule; however, this must await more careful study. The size of the MSA-carrier protein complex seen in this report corresponds to the low molecular weight complex observed upon the addition of labeled somatomedin, NSILA or MSA to rat serum [14]. Moses et al [13, 14] showed that when labeled MSA is incubated with rat serum and then fractionated on Sephadex G-200 at neutral pH, radioactivity could also be detected in five peaks. Only binding to peaks II (mol wt $>$ 150,000) and III (mol wt $>$ 50,000) as also shown in this report were specific and could be inhibited by the presence of an excess of unlabeled MSA. The high molecular weight complex (peak 2) shows a striking sensitivity to acid treatment. When serum is treated with 1 N acetic acid and returned to neutral pH prior to mixing with the labeled MSA, binding in the region of peak 2 disappears [13, 14]. Attempts to regenerate the high molecular weight form following dissociation of the complex with acid have been unsuccessful [14]. While the relationship between these two forms of carrier is not yet clear, it has been suggested that the high molecular weight form is the functionally significant one and the lower molecular weight form may be a precursor [12–14]. First, Kaufmann et al [11] have shown that when labeled NSILA is injected into normal rats, radioactivity first appears in the low molecular weight complex followed by conversion into the high molecular weight form. In addition, Moses et al [13] demonstrated that the appearance of radioactivity in the high molecular weight complex is growth hormone dependent and does not occur in hypophysectomized rats. Chronic treatment with growth hormone restored the normal binding pattern.

The protein band observed here in acetic acid-urea gels was shown to be the carrier protein by detection of binding activity in extracts from gel slices that corresponded to the stained band in identical gels. Using these procedures approximately 300–500 μg of purified MCP were obtained from one liter of BRL-3A conditioned medium. Purified MCP was shown to be similar to native carrier protein in BRL-3A2 conditioned medium in that they had similar affinity constants. These values are similar to those previously reported by others [14] and encourage that assertion that purified MCP has not been altered significantly by the isolation techniques used.

In a separate report [27], we have shown that purified MCP will inhibit the biological activity of MSA on chicken embryo fibroblasts. The stimulation of glucose transport and thymidine incorporation by MSA are inhibited 70% and 96%, respectively, when equimolar quantities of MSA and MCP are present in the medium. MCP had no effect on insulin stimulation of these events in the same cells. These results suggest that the MSA-carrier protein complex described in this report is inactive on cells and that the free hormone must be released to exert its activity on responsive cells. The availability of purified components will facilitate studies to determine the nature of the high molecular weight complex (peak 3) in serum and the biological significance of these carrier protein complexes in the mechanism of somatomedin action. The isolation in pure form of the MSA carrier protein from BRL-3A cell conditioned medium is an important step in characterizing these carrier protein complexes and will provide the opportunity to understand their functional significance. The fact that the purified carrier protein described in this report, as well as the carrier protein from the BRL-3A2 clone that does not produce MSA [14], does not spontaneously form the high molecular weight complex when mixed with MSA, should not diminish the importance of this endeavor. The opportunity to utilize these purified components to search for the conditions or factors necessary for the conversion to the high molecular weight form of the complex now exists. The acid-sensitive, growth-hormone-dependent component of the larger complex does not appear to be the carrier protein itself since Moses et al [14] have shown that impure carrier protein preparations from serum and BRL-3A CM exposed to 1 N acetic acid have the same affinity for labeled MSA as the native carrier protein in BRL-3A2 CM.

ACKNOWLEDGMENTS

These studies were supported by USPHS grant #CA 17620 from the National Cancer Institute. We thank Ms. Russette Lyons for the excellent technical assistance and Dr. M. Rechler for his generous gift of BRL-3A2 conditioned medium.

REFERENCES

1. Daughaday WH, Hall K, Raben MS, Salmon WD, Van den Brande JL, Van Wyk JJ: Nature 235:107, 1972.
2. Brinsmead MW, Liggins GC: In Scarpelli EM, Cosmi EV (eds): "Reviews in Perinatal Medicine." New York: Raven Press, 1979, pp 207–242.
3. Van Wyk JJ, Furlanetto RW, Plet AS, D'Ercole AJ, Underwood LE: Natl Cancer Inst Monogr 48:141, 1978.
4. Rechler MM, Fryklund L, Nissley SP, Hall K, Podskalny JM, Skottner A, Moses AC: Eur J Biochem 82:25, 1978.
5. Rinderknecht E, Humbel RE: Proc Natl Acad Sci USA 75:2365, 1976.
6. Fryklund L, Skottner A, Sievertsson H, Hall K: In Percile E, Müller EE (eds): "Growth Hormones and Related Peptides." Amsterdam: Excerpts Medica, 1976, pp 156–179.
7. Rinderknecht E, Humbel RE: J Biol Chem 253:2769, 1978.
8. Westermark B, Wasteson A: Adv Metab Disord 8:85, 1975.
9. Zapf J, Waldvogel M, Froesch ER: Biochem and Biophys 168:638, 1975.
10. Hintz RL, Liu F: J Clin Endocrinol Metab 45:988, 1977.
11. Kaufmann U, Zapf J, Torretti B, Froesch ER: J Clin Endocrinol Metab 44:160, 1977.
12. Cohen KL, Nissley SP: Acta Endocrinol (Copenh) 83:243, 1976.
13. Moses AC, Nissley SP, Cohen KL, Rechler MM: Nature 263:137, 1976.
14. Moses AC, Nissley SP, Passamani J, White RM, Rechler MM: Endocrinology 104:536, 1979.
15. Dulak NC, Temin HM: J Cell Physiol 81:153, 1973.
16. Dulak NC, Temin HM: J Cell Physiol 81:161, 1973.
17. Smith GL Temin HM: J Cell Physiol 84:181, 1974
18. Nissley SP, Rechler MM: Natl Cancer Inst Monogr 48:167, 1978.
19. Dulak NC, Shing YW: J Cell Physiol 90:127, 1977.
20. Moses AC, Nissley SP, Short PA, Rechler MM, Podskalny JM: Eur J Biochem 103:387 1980.
21. Rechler MM, Podskalny JM, Nissley SP: J Biol Chem 252:3898, 1977.
22. Parikh I, March S, Cautrecasas P: In Jakoby WB, Wilchek M (eds): "Methods in Enzymology." New York: Academic Press, 1974, vol 34, pp 77–102.
23. Weber K, Pringle JR, Osborn M: In Hirs CHW, Timasheff SN (eds): "Methods in Enzymology." New York: Academic Press, 1972, vol 26, pp 3–27.
24. Lowry OH, Roseborough NH, Fan AL, Randall RJ: J Biol Chem 195:265, 1951.
25. Scatchard G: Ann NY Acad Sci 51:660, 1949.
26. Fryklund L, Scottner A, Forsman A, Castenson S: In Sato GH, Ross R (eds): "Hormones and Cell Culture." Cold Spring Harbor Laboratory 1979, pp 49–59.
27. Knauer DJ, Smith GL: Proc Natl Acad Sci USA (in press).

Multiplication Stimulating Activity (MSA) From the BRL 3A Rat Liver Cell Line: Relation to Human Somatomedins and Insulin

Matthew M. Rechler, S. Peter Nissley, George L. King, Alan C. Moses, Ellen E. Van Obberghen-Schilling, Joyce A. Romanus, Alfred B. Knight, Patricia A. Short, and Robert M. White

Section on Biochemistry of Cell Regulation, Laboratory of Biochemical Pharmacology (M.M.R., E.E.V.O.-S, J.A.R., A.B.K.), Section on Cellular and Molecular Physiology, Diabetes Branch (G.L.K.), National Institute of Arthritis, Metabolism and Digestive Diseases; Endocrine Section, Metabolism Branch, National Cancer Institute (S.P.N., A.C.M., P.A.S., R.M.W.), National Institutes of Health, Bethesda, Maryland 20205

> The properties of multiplication stimulating activity (MSA), an insulin-like growth factor (somatomedin) purified from culture medium conditioned by the BRL 3A rat liver cell line are summarized. The relationship of MSA to somatomedins purified from human and rat plasma are considered. MSA appears to be the predominant somatomedin in fetal rat serum, but a minor component of adult rat somatomedin. In vitro biological effects of MSA and insulin in adipocytes, fibroblasts and chondrocytes are examined to determine whether they are mediated by insulin receptors or insulin-like growth factor receptors. The possible relationship of a primary defect of insulin receptors observed in fibroblasts from a patient with the rare genetic disorder, leprechaunism, to intrauterine growth retardation is discussed.

Key words: insulin-like growth factor, somatomedin, multiplication stimulating activity, insulin, receptors: insulin; insulin-like growth factor, fibroblasts, DNA synthesis, amino acid transport, glucose, leprechaunism, liver cells

Alan C. Moses is now at 2000 Washington Street, Newton LF, MA 02162.

Received June 6, 1980; accepted September 15, 1980.

Abbreviations: MSA, multiplication stimulating activity; IGF, insulin-like growth factor; NSILA-s, acid ethanol soluble nonsuppressible insulin-like activity; BRL, Buffalo rat liver; NGF, nerve growth factor; EGF, epidermal growth factor; FGF, fibroblast growth factor; PDGF, platelet-derived growth factor; SDS, sodium dodecyl sulfate; hypox, hypophysectomized; GH, growth hormone; LDL, low-density lipoprotein; AIB, α-aminoisobutyric acid; TAT, tyrosine aminotransferase; LH, luteinizing hormone; FSH, follicle stimulating hormone; PGE_1, prostaglandin E_1.

MULTIPLICATION STIMULATING ACTIVITY (MSA): AN INSULIN-LIKE GROWTH FACTOR FROM A RAT LIVER CELL LINE

Definitions

Multiplication-stimulating activity (MSA) is the name given by Dulak and Temin [1] to a family of closely related polypeptides synthesized by a particular line of cells (BRL 3A) derived from normal rat liver. Our laboratories have purified two forms of MSA to homogeneity and have characterized their chemical, biological, and immunological properties and their interactions with specific cell surface receptors and serum carrier proteins [2–5]. It became apparent that, as postulated by Dulak and Temin, MSA from BRL 3A cells (BRL-MSA) closely resembled peptides present in human plasma known as somatomedins [6, 7] and nonsuppressible insulin-like activity [8, 9]. Direct comparative studies of MSA and homogeneous preparations of the human peptides (insulin-like growth factor (IGF)I, IGF-II, somatomedin A, and somatomedin C) have established that these five peptides are a family of closely related molecules [10–12].
We shall refer to this group of peptides as insulin-like growth factors because the two members of the group whose amino acid sequences have been defined, IGF-I and IGF-II, show remarkable sequence identity with insulin [13, 14]. The human peptides appear to differ in the extent to which their plasma levels are regulated by growth hormone, suggesting that only some members of the group may be mediators of growth hormone action or somatomedins [6, 7, 9, 15].

The properties common to the 5 insulin-like growth factors are summarized in Table I. Two features deserve special comment: 1) The extraordinary structural similarity — IGF-I and IGF-II are identical at 62% of their amino acid positions — and 2) preservation of the molecular features most related to biological function — recognition by cell surface receptors and by serum carrier proteins.

The designation insulin-related growth factor also has been used recently by Bradshaw and Niall [16, 17] to refer to a partially overlapping set of polypeptides, IGF-I, IGF-II, NGF, and relaxin [18, 19], to call attention to structural similarities and a presumed evolutionary relationship. It should be noted that IGF-I is much more closely related to IGF-II ($\cong$75% identical residues in the regions corresponding to insulin A and B chains) than to either NGF (14% identity) or relaxin (26% identity). Moreover, as summarized in Table II, IGF-I, NGF, and relaxin are synthesized by different tissues and have different biological activities in different target organs. Although relaxin stimulated thymidine incorporation in human fibroblasts [23] like insulin and IGFs [2, 10], relaxin from Dr. C. Schwabe [21] did not stimulate thymidine incorporation in chick embryo fibroblasts [S.P. Nissley, unpublished results]. Whereas IGF-I and insulin reacted with IGF receptors and insulin receptors, NGF and relaxin did not cross-react with these receptors, nor did insulin and IGF-I cross-react with relaxin and NGF receptors (Table III). In addition, NGF did not react with antibodies to MSA [32]. Thus, our classification of insulin-like growth factors emphasizes a functional relationship (probably based on close sequence homology), whereas the alternate classification of insulin-related growth factors that includes NGF and relaxin [16, 17], emphasizes a more distant chemical and possible evolutionary relationship.

Properties of MSA Purified From BRL 3A Conditioned Medium at NIH

The BRL 3A cell line was established in culture by Hayden Coon in 1968 by primary cloning from the liver of a 5-week-old normal rat [33]. It had the interesting prop-

TABLE I. Common Properties of the Insulin-Like Growth Factors (IGF-I, IGF-II, Somatomedin A, Somatomedin C, MSA)*

A. Chemical
   1. Single-chain polypeptides
   2. Molecular weight 7,000–9,000
   3. Acid soluble
   4. Acid and heat stable
   5. Inactivated by thiols[a]
   6. Amino acid sequence homology:
      a. IGF-I and IGF-II identical at 45 of 73 positions (62%)
      b. Somatomedin C identical to IGF-I at 22 of 25 positions tested; however, 2 fragments (8 and 9 residues) in somatomedin C are not present in IGF-I
B. Biological
   1. Adipose tissue: insulin-like activity (eg, glucose oxidation, lipogenesis); 0.2–2.0% as potent as insulin; not inhibited by antibodies to insulin
   2. Cartilage: sulfate incorporation into glycosaminoglycans; DNA, RNA, protein synthesis; hypox rat costal cartilage, chick pelvic leaflet
   3. Chick embryo fibroblasts in culture: DNA synthesis, cell multiplication
C. Occurrence
   Present in human plasma (IGF-I, IGF-II, somatomedin A, somatomedin C), rat plasma (rat somatomedin, MSA-like peptides), and BRL 3A conditioned medium (MSA) in association with higher molecular weight carrier proteins. Binding protein-IGF complexes dissociate at acid pH.
D. Recognition by specific receptors, binding proteins, and antibodies
   1. Receptors
      a. Insulin receptors: weak cross-reaction
      b. Insulin-like growth factor receptors
   2. Carrier proteins from rat and human serum
   3. Antibodies

*Based on references 2, 6, 7, and 9.
[a]Not reported for somatomedin A.

TABLE II. Properties of Peptides Structurally Related to Insulin

	Insulin [20]	IGF [9]	Relaxin [16, 21]	NGF [22]
1. Tissue source	Pancreas (β-cell)	Plasma Liver Liver-derived cells	Ovary (pregnant corpus luteum)	Submaxillary gland (male mouse)
2. Target tissues	Fat Muscle Liver	Cartilage Muscle	Pubic symphysis Uterus	Sensory and sympathetic neurons
3. Biological activities	Glucose transport and metabolism Amino acid transport Lipogenesis Gluconeogenesis Cell growth	Pleiotypic anabolic effects in cartilage, fibroblasts, muscle	Dilate cervix Inhibit uterine contractions Relax pubic ligaments	Development and survival of neurons

TABLE III. Cross-Reactivity of Peptides Structurally Related to Insulin in Receptor Assays

	Insulin	IGF	Relaxin	NGF
[125]I-insulin	+	+ (weak)	− [23]	− [28]
^{125}I-MSA (IGF)	+ (some)	+	− a,b	− c
[125]I-relaxin	− [23]d	+ (weak) [23a]	+ [23]	NTe
^{125}I-NGF	+ [29, 30]f, − [28, 31]	−	−	+

[a]Relaxin [18]: rat liver plasma membranes [M.M. Rechler, unpublished results].
[b]Relaxin [19]: chick embryo fibroblasts [M. M. Rechler, unpublished results].
[c]Chick embryo fibroblasts [24]; human skin fibroblasts [25]; rat liver plasma membranes [M. M. Rechler, unpublished results]; human placental membranes [26]; NRK fibroblasts [27].
[d]Weak inhibition by proinsulin in human skin fibroblasts [23] and rat uterus [23a].
[e]NT = not tested.
[f]<30% inhibition: chick embryo: dorsal root and sympathetic ganglia [29]; dissociated heart and brain [30].

erty that, following plating at low density in medium containing serum, BRL 3A cells would replicate at a nearly normal rate for several generations in serum-free medium without protein or hormonal supplements. This cell line was used by Dulak and Temin [1] for the initial isolation of MSA. We obtained BRL 3A cells from Dr. Temin in 1973. The BRL 3A cell line was submitted to the American Type Culture Collection in 1978 for distribution. Its ATCC designation is CRL 1442.

The starting material for MSA isolation is serum-free medium conditioned by confluent cultures of BRL 3A cells for 3–4 days. The purification scheme in effect in our laboratories from 1976 to 1980 has recently been published [34]. In brief, conditioned medium is chromatographed on Dowex 50 X-8. The pH 11 eluate containing MSA is chromatographed on Sephadex G-75. Fractions containing MSA activity (ie, capable of stimulating the incorporation of [^{3}H]thymidine into the DNA of serum-starved tertiary passage chick embryo fibroblasts) were further analyzed by disc gel electrophoresis at pH 2.7 in 9 M urea. Based on the electrophoretic pattern, fractions from the G-75 column have been grouped into 3 broad regions: peak I, peak II, and peak III. Peak I contains a single MSA species, mol wt 16,300, R_f 0.36; peak II contains 4 closely migrating species, mol wt 8,700, R_f 0.41–0.49; peak III contains 2 closely migrating species, mol wt 7,100, R_f 0.59–0.62.

Two species of MSA have been isolated in apparently homogeneous form from pools of the appropriate Sephadex G-75 fractions by preparative acrylamide gel electrophoresis under conditions similar to those of the analytical gel electrophoresis. These peptides have been designated MSA II-1 (mol wt 8,700) and MSA III-2 (mol wt 7,100).* MSA II-1 gives a single protein band following electrophoresis at two pHs and has a single COOH-terminal amino acid (glycine) [34]. Although MSA II-1 has a lower specific insulin-like activity than IGF-I in stimulating glucose oxidation and lipogenesis in adipocytes (50 mU/mg versus 144–170 mU/mg), this may represent the intrinsic potency of MSA II-1 rather than heterogeneity of the preparation [24, 35, 36]. MSA II-1 has been radioiodinated [24] and used as ligand in receptor [24] assays, binding protein [38, 39] assays, and some immunoassays [32]. MSA III-2 has been radioiodinated and used as ligand in some immunoassays [32], since it appears to bind preferentially to a different antibody population in one immune rabbit serum.

*These molecular weight assignments are based on Sepharose 6B-guanidine chromatography. They are considered provisional pending determination of the amino acid sequence of MSA II-1 and MSA III-2. For this reason, MSA potencies are expressed in ng/ml rather than in molar concentrations.

Sephadex G-75 peak II pool has been substituted for MSA II-1 in many biological studies and as standard in competitive binding assays. The rationale for this substitution is as follows. The G-75 peak II pool contains only 4 protein species on analytical gel electrophoresis [34]. The mixture of peak II peptides and MSA II-1 possess identical activity in the chick fibroblast bioassay (Fig. 1), radioimmunoassay (Fig. 1), and rat liver membrane receptor assay (unpublished results). The amino acid compositions of Sephadex peak II MSA, homogeneous MSA II-1, and a mixture of MSA II-2, 3, 4 obtained following preparative electrophoresis are indistinguishable [34]. The molecular weights of MSA II-1 and MSA II-2, 3, 4 determined by gel filtration on Sepharose 6B in 6 M guanidine HCl (in the absence of reducing agents) were identical, 8,700 [34]. In fact, the only discernible difference (aside from minor differences in electrophoretic mobility) between MSA II-1 and MSA II-2, 3, 4 occurred following incubation with 6 M guanidine HCl in the presence of reducing agents: The molecular weight of MSA II-1 was unaffected, whereas a mixture of MSA II-2, 3, 4 peptides was partially converted to lower molecular weight forms [34]. We interpret these results to indicate that some forms of MSA II-2, 3, 4 contain interruptions in the polypeptide chain but that 1) this break does not occur at a site critical for biological activity or recognition by receptors, antibody, or binding protein, and 2) in the absence of reducing agents, these cleaved molecules are held together in the native conformation by disulfide bonds.

Considerable evidence suggests that the sequence of MSA III-2 is contained within MSA II-1 [32, 34]. No amino acids are significantly more abundant in MSA III-2 than in

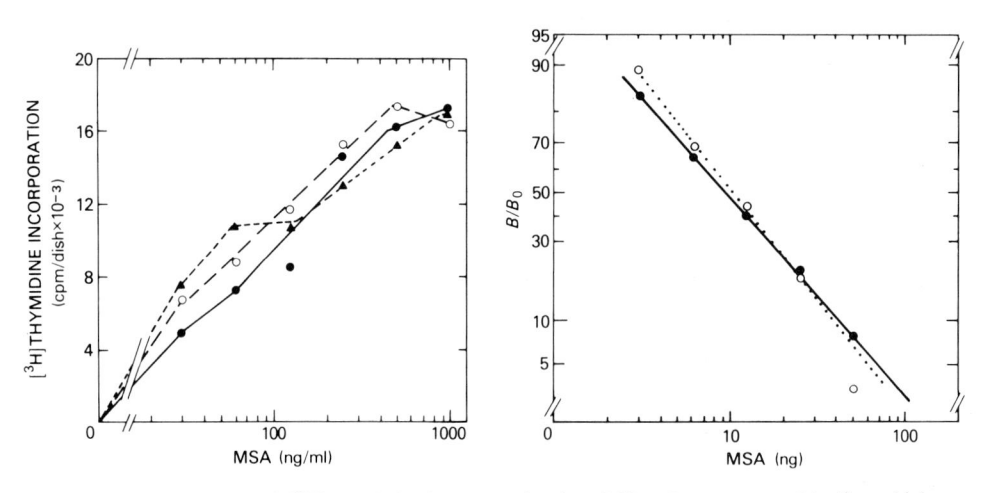

Fig. 1. Left: stimulation of [^{3}H]thymidine incorporation into DNA of serum-starved tertiary chick embryo fibroblasts by different MSA preparations: Sephadex G-75 peak II (●), MSA II-1 (○), and a mixture of MSA II-2, -3, and -4 (▲) obtained by preparative electrophoresis. MSA preparations were purified as previously described [32]. Confluent fibroblast cultures in 60 mm dishes were plated in Temin's modified Eagle's medium containing 0.4% calf serum. After 3–5 days, peptides were added in medium without serum for 12 h. The cultures were then pulsed for 1 h with [^{3}H]thymidine, and the incorporation of radioactivity into acid-precipitable material was quantitated as previously described [40]. Redrawn from [34] with permission.

Right: reactivity of MSA peptides in MSA radioimmunoassay. The immunoassay was performed as previously described [32], with MSA antiserum at 1:1,000 dilution, ^{125}I-labeled MSA II-1, and precipitation of antibody-bound radioactivity with polyethylene glycol. Dilutions of MSA II-1 (●) and a mixture of MSA II-3 and -4 (○) were tested. B_0 is the radioactivity bound to antibody in the absence of unlabeled MSA; B is the radioactivity bound at each concentration of unlabeled MSA. Results are plotted on a logit-log scale. Redrawn from [32] with permission.

CCDD:A:415

MSA II-1. Antibodies raised in rabbits to Sephadex G-75 MSA peak II (and devoid of MSA III-2) recognize MSA III-2. In fact, in a radioimmunoassay using ^{125}I-labeled MSA III-2, parallel dose-response curves are obtained with MSA III-2, MSA II-1, and peak I MSA, suggesting the presence of a common antigenic determinant. However, using ^{125}I-labeled MSA II-1 and the same antiserum, MSA III-2 competes poorly, with a dose-response curve markedly nonparallel to the MSA II-1 standard. MSA III-2 does not react at all with a different antiserum to peak II MSA [J. A. Romanus and M. M. Rechler, unpublished results]. This suggests that these populations of antibodies to peak II MSA recognize determinants that have been lost or altered in MSA III-2.

MSA III-2 reacts with MSA receptors and with the MSA binding protein of rat serum. It is approximately 3-fold more potent than MSA II-1 against the rat liver membrane receptor [4], chick embryo fibroblast receptor [unpublished results, M.M. Rechler and J.A. Romanus], somatomedin binding protein [4], and the insulin receptor of placental membranes [41]. Peak III MSA is more potent than peak II MSA in stimulating DNA synthesis (Fig. 2). When ^{125}I-MSA II-1 was incubated with chick embryo fibroblasts for 18 h under conditions of the thymidine incorporation bioassay, no conversion to MSA III-2 could be detected by gel filtration on Sephadex G-75 in 1 M acetic acid [Y. Yang and M. M. Rechler, unpublished results]. Reduction of MSA II-3 generated a species electrophoretically indistinguishable from MSA III-2 [34].

MSA peak I has not been purified to homogeneity and has been less extensively studied. Peak I possesses MSA biological activity (Fig. 2), and following electrophoresis, this activity corresponds to the major protein band in the preparation [34]. Peak I MSA reacts with the MSA receptor of liver membranes [34], weakly reacts with rat serum

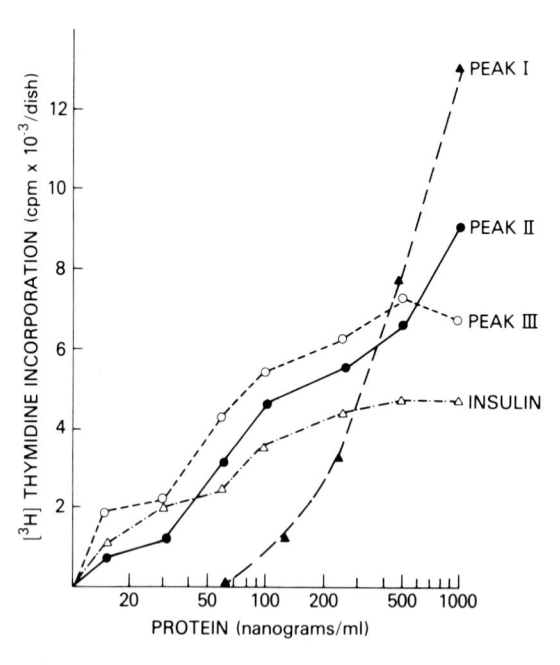

Fig. 2. Activity of different Sephadex G-75 purified MSA fractions and insulin in the chick embryo fibroblast bioassay. MSA fractions were purified as described elsewhere [34]. Porcine insulin was purchased from Eli Lilly Co. Conditions for [^{3}H]thymidine incorporation were as described in Figure 1 (left).

binding protein [4], and gives a displacement curve parallel to MSA III-2 and MSA II-1 in the radioimmunoassay using ^{125}I-MSA III-2 [32]. Peak I MSA may represent a precursor form.

Assays for MSA

Several assays for MSA are commonly in use in our laboratory. Their sensitivity and specificity for MSA species, human somatomedins, rat serum somatomedins, and insulin are summarized in Table IV. No one assay measures a single MSA-related polypeptide without cross-reaction of the other species. Interpretation of results with complex mixtures should be made with caution until the relative contributions of different reacting species can be resolved.

Relation of NIH-MSA to Other Known MSA Preparations From the BRL 3A Cell Line

Dulak MSA. MSA purified by Dowex chromatography and successive preparative electrophoreses in SDS and acetic acid-urea as described by Dulak and Shing [47] was kindly provided by Dr. Dulak. Dulak-MSA was equipotent with NIH-MSA peak II in the thymidine incorporation bioassay, rat liver membrane receptor assay, and radioimmunoassay [S. P. Nissley, M. M. Rechler, and A. C. Moses, unpublished results]. Gel electrophoresis at both acid and alkaline pHs [34] identified the major component in Dulak MSA as II-1, with other peak II peptides (but no non-MSA peptides) represented.

Collaborative Research. Partially purified MSA (after Sephadex G-50 gel filtration in 1 M acetic acid) was provided by R. Forand and tested in April 1978. Two components were found to have 50–100% the potency of NIH-MSA peak II in the chick embryo fibroblast bioassay [S.P. Nissley, unpublished results]. We have not examined the electrophoretic purity of this preparation.

New Zealand MSA. Brinsmead and Liggins [26] have purified MSA by Dowex 50 and Sephadex G-50 chromatography. In comparative studies, NIH-MSA and New Zealand-MSA were within a factor of 2 in potency in 1) the placental membrane receptor assay using ^{125}I-MSA [26] and 2) the chick embryo fibroblast thymidine incorporation bioassay [S.P. Nissley, unpublished results]. We have not performed electrophoretic studies with the New Zealand MSA preparation.

Relationship of MSA to Homogeneous Human Insulin-Like Growth Factors

MSA has been extensively compared with somatomedin A [10] and with IGF-I and IGF-II [11, 12, 24]. In competitive binding assays using multiple receptor systems or rat serum binding proteins, results obtained with a matrix of ^{125}I-labeled peptides and unlabeled peptides uniformly indicated substantial cross-reactivity [10, 11], although minor differences were appreciated in the relative potencies of different peptide preparations. For example, the results in Figure 3 suggest that ^{125}I-MSA and ^{125}I-IGF-I may, under appropriate conditions, bind to the same receptor in human fibroblasts, identified by the relative potencies of the competing unlabeled peptides. The 4 peptides also manifested similar biological activities; for example, they stimulated lipogenesis in rat fat pad [35] and DNA synthesis in chick embryo fibroblasts [10, 35].

Immunological differences between MSA and the human somatomedins are much greater than the differences in biological activity and in reactivity with receptors and binding proteins. MSA II-1 did not react in immunoassays for IGF-I [9], IGF-II [9], and somatomedin A [48] at concentrations indicating potencies <2%. Somewhat greater

TABLE IV. Assays for MSA in Use in Our Laboratories

Assay	Bioassay: [^{3}H]thymidine incorporation, chick embryo fibroblasts	Receptor assay: rat liver plasma membranes	Radioimmunoassay [32]: ^{125}I-MSA III-2	Radioimmunoassay [32]: ^{125}I-MSA II-1	Competitive binding assay: binding proteins[a]
1) Sensitivity (ED$_{50}$)	≈ 30 ng/ml	≈ 8 ng/ml	≈ 4 ng/ml	≈ 13 ng/ml	≈ 10 ng/ml[b]
2) Specificity MSA species	III > II > I (Fig. 2)	III > II > I [4]	III > II > I	II > I; III not parallel	III > II > I [4]
Insulin	+ (Fig. 2) [42]	- [42]	-	-	-
Proinsulin	+ [43][c]	-[d]	NT[e]	-	NT
IGF-I, IGF-II	+ [45]	+ [12][f]	+g	Not parallel, weak	+ [11, 39]
Somatomedin A	+ [10]	+ [11][f]	+	NT	+ [39]
Somatomedin C	+ [44]	+ [10]	NT	NT	±d
Rat somatomedin	+ (Table V)	+ (Fig. 4)	- (Table V)	NT	+ (Table V)
3) Rat serum	Hypox[h] < normal [4]	- (Table V)	Hypox < normal [4]	NT	Fetal high [46]
	Hypox < GH-treated hypox [4]	Fetal high [46]	Fetal >> maternal [46]		
	Maternal ≅ fetal [4]				

[a]Normal rat serum, GIBCO.
[b]Sensitivity when bound and free ^{125}I-MSA were separated using charcoal activated with fatty acid-free albumin. Approximately 5–10-fold lower sensitivity was observed with other albumin preparations [39].
[c]Proinsulin 20% as potent as insulin.
[d]Unpublished results, M. M. Rechler.
[e]NT = Not tested.
[f]IGF-II most potent.
[g]MSA II-1:IGF-II:IGF-I = 100:10:3 (parallel).
[h]hypox = hypophysectomized; GH = growth hormone.

HUMAN FIBROBLASTS

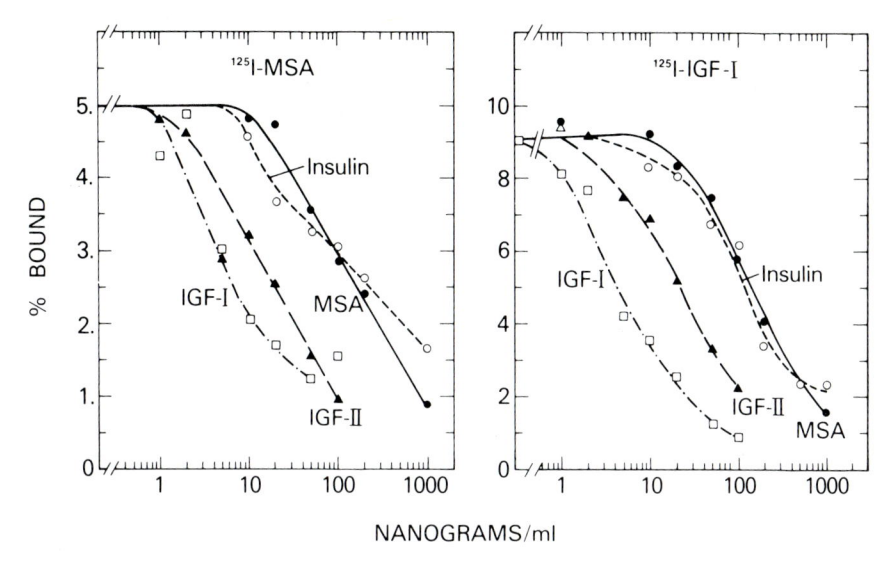

Fig. 3. Competition for binding of ^{125}I-labeled MSA (left) and ^{125}I-labeled IGF-I (right) to human skin fibroblasts in culture by unlabeled IGF-I and IGF-II (a kind gift of Dr. R. E. Humbel), Sephadex G-75 peak II MSA, and porcine insulin. Conditions of the binding assay have been previously described [11, 25]. Percent of input radioactivity bound to $1.3-1.5 \times 10^6$ suspended fibroblasts following 2 h incubation at 15°C is shown.

The results presented for ^{125}I-IGF-I are observed consistently, with the exception that the MSA dose-response curve is typically to the left of the insulin curve. The curve presented for unlabeled IGF-I inhibition of ^{125}I-MSA binding is atypical in one respect. In other experiments, unlabeled IGF-I only inhibited specific binding by $\cong 50\%$, reaching a plateau [11]. The latter result has been interpreted as indicating that ^{125}I-MSA binds to two sites in human fibroblasts, only one of which interacts with IGF-I. Apparently cell culture conditions affect the relative proportions of the two sites; in the present experiment, ^{125}I-MSA binds predominantly to the IGF-I inhibitable site, which presumably is identical to the ^{125}I-IGF-I binding site.

cross-reactivity was observed in the MSA radioimmunoassay using ^{125}I-MSA III-2; IGF-II and IGF-I were 10% and 3% as potent as MSA II-1 [32].

Recently, homogeneous preparations of somatomedin C [49] have been kindly provided by Dr. Van Wyk and compared with MSA. ^{125}I-somatomedin C appeared to bind to the same receptor as ^{125}I-MSA in chick embryo and human fibroblast cultures, but not in liver membranes or cultured liver cells. ^{125}I-somatomedin C (and ^{125}I-MSA) binding to chick embryo fibroblasts and human fibroblasts was inhibited by unlabeled MSA, somatomedin A, and insulin [M. M. Rechler and J. M. Podskalny, unpublished results]. Binding of ^{125}I-somatomedin C to BRL 3A2 cells (a subclone of BRL 3A) also was inhibited by insulin (nonparallel), as well as by MSA and somatomedin A [unpublished results]. By contrast, binding of ^{125}I-MSA and ^{125}I-IGF-II to BRL 3A2 cells was not inhibited by insulin [2, 11], whereas ^{125}I-IGF-I and ^{125}I-somatomedin A were similar to ^{125}I-somatomedin C in this respect [10, 11]. Like ^{125}I-somatomedin A [10] and ^{125}I-IGF-I [11], ^{125}I-somatomedin C bound poorly to purified rat liver plasma membranes [unpublished results]. However, unlabeled somatomedin C, MSA, and somatomedin A potently competed for ^{125}I-MSA binding to liver membranes (Fig. 4). This suggests differences between the iodo- and non-iodopeptides. In the placental membrane assay using ^{125}I-somatomedin C, MSA III-2 and MSA II-1

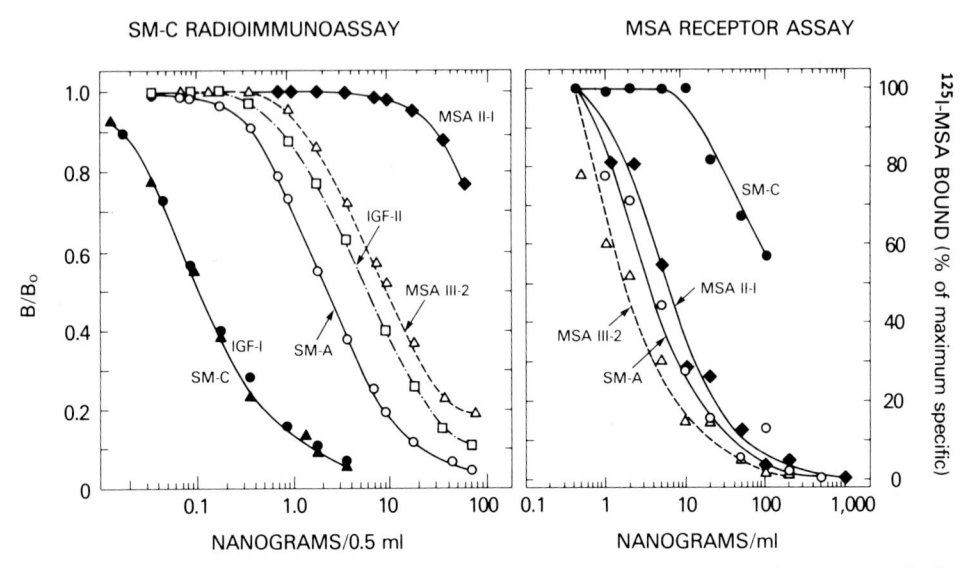

Fig. 4. Left: cross-reactivity of MSA III-2 and MSA II-1 in the radioimmunoassay for somatomedin C, using rabbit anti-human somatomedin C [50]. The ordinate represents binding (B) of ^{125}I-somatomedin C, expressed relative to maximum tracer binding in the absence of competing unlabeled peptide (B_0). Redrawn from [41] with permission.

Right: cross-reactivity of somatomedin C (●) and somatomedin A (○) in the radioreceptor assay using rat liver plasma membranes and ^{125}I-labeled MSA II-1. Assay conditions were as previously described [10, 24]. The same preparations of somatomedin C, MSA III-2, and MSA II-1 were used as in the left-hand panel. Homogeneous somatomedin A (SPE 2451) was a kind gift of Dr. Linda Fryklund (AB KABI, Stockholm). The percent of ^{125}I-MSA bound (equal to 10.7% of input radioactivity) is plotted against ng/ml of unlabeled peptide in the incubation.

were 21% and 5% as potent as unlabeled somatomedin C, respectively [41]. Finally, MSA III-2 was 1% as potent as somatomedin C in the somatomedin C radioimmunoassay (Fig. 4); IGF-II and somatomedin A were 2- and 4-fold more potent than MSA III-2, whereas MSA II-1 was 10-fold less potent [41].

This completes the demonstration that BRL-MSA is closely related to each of the defined members of the family of human somatomedins or insulin-like growth factors. Undoubtedly, this relatedness is based on major structural identities. IGF-I and IGF-II are identical at >60% of their residues [13, 14]. Partial sequence data on somatomedin C indicate that it, too, is quite similar to IGF-I [49], and IGF-I and somatomedin C are indistinguishable in radioreceptor and radioimmunoassays [41]. Residues at 22 of 25 sites identified in somatomedin C (9 NH_2-terminal residues and 3 internal tryptic peptides) correspond to residues in IGF-I [49]. However, two additional tryptic peptides containing 8 and 9 residues, respectively, do not match known IGF-I sequences, suggesting that the two molecules are similar but not identical [49].

Although it is not yet possible to decide which human somatomedin/IGF MSA resembles most closely, MSA and iodinated MSA serve as useful surrogates for the group of human somatomedins. The IGF/somatomedins and the cell surface receptors with which these peptides interact, however, are heterogeneous. To define completely how somatomedins interact with a given target tissue will require use of a panel of purified IGFs, including MSA. Hopefully, the subtle differences in reactivity will soon become understood in terms of the chemical and structural differences among this family of similar peptides.

TABLE V. Reactivity of Partially Purified Rat Somatomedin in Bioassays and Competitive Binding Assays for MSA

Assay	Potency of rat somatomedins relative to MSA peak II (%)[a]
1) Bioassay: [^{3}H]thymidine incorporation, chick embryo fibroblasts	≈60
2) Receptor assay: rat liver membranes	<0.1[b]
3) Receptor assay: chick embryo fibroblasts	≈10[c]
4) Competitive binding assay using rat serum binding proteins	≈3
5) Radioimmunoassay[d]	<0.25[d]

[a]Rat somatomedin was purified 40,000-fold from the serum of rats bearing the growth hormone-secreting tumor MStT/W15 by hollow fiber ultrafiltration, gel filtration, isoelectric focusing, and cellulose thin-layer chromatography (2 times) [53]. Rat somatomedin protein content was estimated from amino acid analysis. Its activity was 86 somatomedin units/mg protein in the hypox rat costal cartilage bioassay, and ≈46 milliunits insulin-like activity/mg protein in the placental membrane receptor assay using IGF-I tracer and partially purified NSILA standard [I. Mariz and W. H. Daughaday, personal communication]. In contrast to less purified preparations, this preparation was not contaminated by peptides resembling the C_{3A} component of complement, a serum protein that copurified with rat somatomedin in early stages of purification [52]. The purity of this preparation is not known. Experiments shown were performed with the same rat somatomedin preparation (Amicon 10–16).
[b]Inhibition of ≈25% only at highest concentration tested.
[c]Estimated. Rat somatomedin was 3.2% as potent as MSA III-2 in this experiment. In other experiments, MSA III-2 is approximately 3-fold more potent than MSA II-1.
[d]Radioimmunoassay used MSA III-2 tracer. No inhibition by rat somatomedin at highest concentration tested (400 ng).

Relation of MSA to Somatomedin in Rat Serum

Whether MSA synthesized by the BRL 3A rat liver cell line corresponds to a somatomedin that circulates in rat plasma cannot be definitively answered until the amino acid sequences of MSA and the rat somatomedins are available. Present information suggests that MSA represents a minor component of the somatomedins in adult rat plasma, but it may be the predominant somatomedin in fetal rats.

1) Radioimmunoassay highly specific for MSA indicated that 50–100 ng/ml of MSA equivalent are present in normal adult rat serum. By contrast, bioassay (thymidine incorporation in chick embryo fibroblasts) yields values of approximately 5 µg/ml MSA in rat serum [4]. Our provisional conclusion is that rat serum contains other low molecular weight, acid-soluble, biologically active, growth hormone-dependent peptides — presumably somatomedins — that do not react with antibodies to MSA. That is, MSA (by immunologic criteria) would represent a minor fraction of the somatomedin activity in normal adult rat serum. Two qualifications of this interpretation should be made: First, all of the biologic activity in rat serum may not result from somatomedins, and second, non-somatomedin components in rat serum may act synergistically with MSA to increase its growth-stimulating effect on chick embryo fibroblasts. For example, Cohen and Nissley have described enhancement of the mitogenic effect of MSA by Cohn Fraction VI of human serum [51].

2) A partially purified preparation of rat somatomedin has been obtained from the serum of rats bearing a growth hormone-producing tumor [52, 53]. Preparations have been made available for comparison with MSA by W. H. Daughaday and I. Mariz. As summarized in Table V, rat somatomedin had considerable activity in the MSA bioassay,

weak but distinct reactivity with chick fibroblast MSA receptors and rat serum somatomedin binding protein, and negligible reactivity with rat liver membrane MSA receptors and MSA antibodies.

Thus, the rat somatomedin preparation tested shows significant reactivity in three MSA assays, indicating that it contains one or more polypeptides related to MSA. The lack of reactivity in the MSA immunoassay and liver membrane receptor assay suggests that these related peptides, however, are not identical with BRL-MSA. The relationship of this rat somatomedin preparation obtained from a tumor-bearing animal to the somatomedins present in normal rat serum remains to be established.

MSA RECEPTORS AND INSULIN RECEPTORS: MEDIATION OF BIOLOGICAL RESPONSES

Introduction

The underlying assumption and working hypothesis during the last decade of research on polypeptide hormone receptors is that the hormones bind to specific receptors on the cell surface, and that a series of biochemical events are triggered by the binding reaction that result in a biological response [54]. This simple statement has been complicated by the realization that some polypeptides are taken into the cell (eg, EGF), often with their receptors [55], and that receptors for peptide hormones are present on membranes of internal organelles [56]. The functional significance of these latter observations is not yet fully understood: whether they are part of the transmission of the signal, or, alternatively, a means to terminate the signal, to regulate receptor number, or a stage in the synthesis of new membrane receptors. As a first approximation, it seems reasonable that binding to cell surface receptors is the first step in peptide hormone action, and that the specificity of the binding reaction should mirror that of biological potency at least semi-quantitatively. We shall briefly review the evidence that separate specific receptors occur on most cells for the insulin-like growth factors and for insulin and that these receptors (like the hormones themselves) have certain similarities; namely, they cross-react with the same spectrum of peptides. From the specificity of binding and biological responses, we shall attempt to deduce which receptor mediates particular biological effects.

An operational definition of an insulin receptor is that it binds ^{125}I-labeled insulin and that the relative abilities of other peptides to bind to this receptor (measured as competition for binding of radioligand) reflect the insulin-like biological potency of these molecules [54]. A panel of insulins of different biological potencies might include insulins of different species (pig, fish, chicken, guinea pig), chemically or enzymatically modified insulins (desalanine-desasparagine insulin, desoctapeptide insulin), and insulin precursors (proinsulin). If ^{125}I-insulin binds to the biologically relevant site, and if all that is required of the hormone to initiate a biological response is to bind to the receptor, then one would expect that the order and approximate potencies of these peptides should be about the same for biological effect and inhibition of binding. Given the complexities of the assay systems and the many unknown components, it would seem fortuitous if the potencies for biological effect and binding correspond quantitatively. Strict correspondence might not occur if 1) the receptor population binding a radioligand were heterogeneous; receptors may have different affinities, the same affinity but different specificity, or the same affinity and specificity but be connected to different biological functions (functional compartmentalization); 2) spare receptors were present; 3) hormones were degraded at different rates; 4) steps subsequent to binding and involved in signal transmission (eg, internalization, binding to intracellular sites) required active participation of the hormone.

422:CCDD:A

The insulin-like growth factors have weak insulin-like activity in adipocytes that is not inhibited by antibodies to insulin [8]. Hintz et al [57] initially demonstrated that somatomedins weakly inhibited binding of ^{125}I-insulin to the insulin receptors of placental membranes. This observation has been extended to the other human somatomedins and MSA [2, 9, 58, 59]. In general, the insulin-like growth factors are 100–1,000-fold less potent than insulin in biological and binding assays for insulin.

Specific receptors for insulin-like growth factors soon were demonstrated in placenta [58, 59], chick embryo fibroblasts [24, 60], rat liver plasma membranes [42, 61, 62], and thereafter in a broad range of tissues and cultured cells [reviewed in 2, 12]. These receptors interact with all other insulin-like growth factors (Figs. 3, 4). They do not react with chemically unrelated peptides (eg, EGF, hGH, glucagon) or functionally distinct peptides (eg, relaxin and NGF; see Table III). IGF receptors differ in their reactivity with insulin (and proinsulin). The MSA receptor in rat liver plasma membranes is <0.1% as reactive with insulin and proinsulin as with MSA [24, 42] (Table IV).* By contrast, insulin reacts with the MSA receptor of chick embryo fibroblasts quite effectively, being $\cong 50\%$ as potent as MSA [24]. Insulin reacts with the IGF-I/MSA receptor of cultured human fibroblasts 2–10% as potently as unlabeled IGF-I and is comparable in potency to unlabeled MSA (Fig. 3). Intermediate reactivity was observed in placenta; insulin is 0.1–1% as potent as the insulin-like growth factors in inhibiting radiolabeled IGF binding to placental membranes [26, 58, 59].

The MSA receptors of human fibroblasts [25] appear to share immunologic determinants with insulin receptors, indicating a possible structural relationship between the two receptors [12, 64]. This was studied using IgG purified from the serum of a patient with the type B syndrome of extreme insulin-resistant diabetes mellitus and acanthosis nigricans [65, 66], in which circulating antibodies to insulin receptors have been demonstrated. Antireceptor IgG inhibits binding of ^{125}I-insulin to insulin receptors of rat liver membranes and human fibroblasts (Fig. 5). The inhibition of binding is quite specific: IgG from controls did not inhibit insulin binding, and the patient's IgG did not inhibit the binding of the chemically unrelated polypeptides EGF and glucagon. Concentrations of antireceptor IgG that inhibited binding of ^{125}I-insulin by 75–95% in liver membranes inhibited ^{125}I-MSA binding by only 5% and 18%, respectively (Fig. 5). By contrast, high concentrations of antireceptor IgG inhibited both ^{125}I-insulin and ^{125}I-MSA binding to human fibroblasts by >95%. Approximately 5–20-fold higher concentrations of IgG are required to produce inhibition of ^{125}I-MSA binding equivalent to that of ^{125}I-insulin [12, 64]. It is not known whether the same molecules in a heterogeneous population of antibodies inhibit binding to both receptors. If this were the case, these results would suggest that the insulin-sensitive MSA receptor of human fibroblasts shares antigenic determinants with insulin receptors in a species-specific and possibly tissue-specific fashion.

There is little evidence to suggest that ^{125}I-MSA is internalized by chick embryo fibroblasts. By contrast, after EGF [55] and LDL [69] bind to surface receptors in human fibroblasts at higher temperatures (37°C, but not 4°C), they are taken into the cell in endocytotic vesicles and degraded in lysosomes. When ^{125}I-MSA binds to chick embryo fibro-

---

*Megyesi et al initially reported inhibition of binding of ^{125}I-labeled NSILA-s ($\cong 15\%$ pure, containing IGF-I and IGF-II) to rat liver plasma membranes by insulin and proinsulin [63]. Subsequently, they presented more complete dose-response curves showing no inhibition of ^{125}I-NSILA-s (or ^{125}I-MSA) binding to the same liver membrane preparations by insulin at high concentrations [62]. We have not observed inhibition of binding of ^{125}I-MSA, ^{125}I-IGF-I, or ^{125}I-IGF-II binding to the same Neville liver membrane preparations by insulin or proinsulin [2, 11, 24, 42].

CCDD:A:423

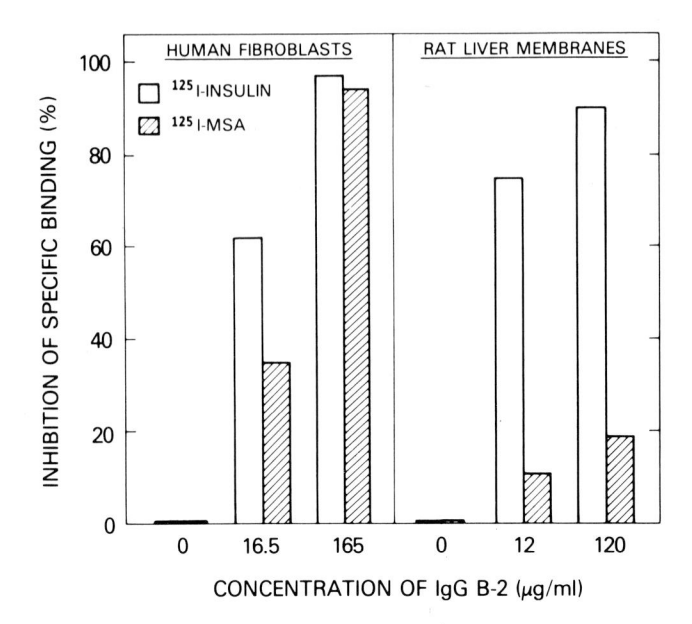

Fig. 5. Inhibition of binding to insulin receptors and MSA receptors by anti-receptor IgG B-2. IgG from patient B-2 [65, 66] with circulating antibodies to insulin receptor was purified by precipitation with 33% ammonium sulfate and DEAE-cellulose chromatography as previously described [67]. Percent inhibition of specific binding of ^{125}I-insulin (open bar) or ^{125}I-MSA (hatched bar) to human fibroblast cultures (left panel), or rat liver plasma membranes (right panel) by the indicated concentration of IgG B-2 is shown.

Binding assays were performed as previously described. Incubation conditions were as follows: a) human fibroblasts, ^{125}I-insulin, incubated 2 h at 15°C [68]; b) human fibroblasts, ^{125}I-MSA, incubated 3 h at 15°C [25]; c) ^{125}I-insulin and ^{125}I-MSA, incubated with rat liver membranes for 17 h at 2°C in 0.05 M Tris-HCl, pH 7.5, containing 10 mg/ml bovine serum albumin [42, 61].

blasts under standard binding conditions (3 h, 22°C) or at 37°C for 1 h, however, there is little evidence for degradation: ^{125}I-MSA remaining in the incubation medium and cell-associated ^{125}I-MSA extracted with Triton-urea-acetic acid remain physically intact [24]. No significant differences were observed in the rate or extent of dissociation of ^{125}I-MSA bound to chick embryo fibroblasts under diverse conditions [22°C, 3 h (steady state): 22°C, 1 h; 4°C, 19 h; 37°C, 1 h] and dissociated at 37°C (faster) or 22°C (slower). These results suggest that the major fraction of ^{125}I-MSA bound to chick embryo fibroblasts is not internalized and degraded or rendered inaccessible.

The properties of MSA and insulin most highly conserved are those determining biological activity and receptor binding. Somatomedin binding proteins in rat serum do not interact with insulin [39]. Rat insulin does not react with rabbit antibodies to MSA [32]. MSA does not react in radioimmunoassay for guinea pig and porcine insulin [J. Rosenzweig, unpublished results], nor does it react with monoclonal antibodies to rat insulin [70; J. Schroer, A. B. Knight, M. M. Rechler, unpublished observations].

In summary, insulin and the IGFs have separate receptors, but they frequently interact with the receptor for the other peptide. The receptors in some cases may share structural features recognized by anti-receptor antibodies. Insulin and the IGFs also have a similar spectrum of biological activities. We shall next turn to identifying which biological function is mediated by which receptor.

Mitogenic Effects

Growth effects in fibroblasts. MSA stimulates the incorporation of [^{3}H]thymidine into DNA in chick embryo fibroblasts [42, 71], human skin fibroblasts [10, 25] and Balb c/3T3 mouse fibroblasts [71]. Chick embryo fibroblasts are most sensitive: half-maximal stimulation occurred at $\cong 30$ ng/ml. Human fibroblasts and 3T3 cells require 10-fold higher concentrations. High concentrations of MSA ($\geqslant 1$ μg/ml) increase DNA content and cell number in chick fibroblasts [10, 43] and 3T3 cells [71]. Cell multiplication has not been demonstrated in human fibroblasts. Even in chick embryo fibroblasts, the doubling time in medium containing MSA is only about one-third that in serum-supplemented medium [10].

These results, obtained in serum-free, hormone-free medium (Eagle's Minimum Essential Medium supplemented with bovine serum albumin), tend to indicate that MSA is a less potent mitogen in some cells than platelet-derived growth factor (PDGF) [72, 73], fibroblast growth factor (FGF) [74], and epidermal growth factor [55]. However, recent studies have indicated that plasma factors (such as somatomedins) are not themselves sufficient to induce DNA synthesis in Balb c/3T3 cells [75]. Rather, a two-stage sequence is required in which cells are first made "competent" to respond to plasma factors (by addition of PDGF or FGF), and then are induced to "progress" to S phase by plasma factors that include somatomedin C and MSA [76]. In addition, Gospodarowicz [77] has observed that a variety of cell types that grow in serum but not in plasma when plated on plastic culture dishes, grow equally well in medium containing plasma or serum when maintained on an extracellular matrix. In the light of these results, the mitogenic role of MSA and other somatomedins should be reevaluated in cells made competent by addition of PDGF or by plating on extracellular matrix. Plasma factors may indeed be important mitogens in vivo.

Hypothesis: MSA receptors mediate growth effects of insulin and MSA in fibroblasts. Radiolabeled MSA binds to a receptor in chick embryo fibroblasts with properties that led us to consider it the probable receptor mediating the growth response [2, 10, 24, 42, 43]. The relative potencies of MSA:insulin:proinsulin for inhibiting binding to this receptor and for stimulating thymidine incorporation were 100:50:20. The absolute concentrations of MSA inhibiting binding and stimulating thymidine incorporation were similar [25]. Maximally effective concentrations of MSA, insulin, and proinsulin did not stimulate thymidine incorporation additively [43]. The human insulin-like growth factors also stimulate DNA synthesis in chick embryo fibroblasts: IGF-I and IGF-II were active at 10-fold lower concentrations than MSA [35, 45, 78]; somatomedin A [10] was equipotent with MSA. The human IGFs appear to interact with the receptor identified by ^{125}I-MSA binding, with relative potencies appropriate to their biological potency [10, 11]. Together, these results suggested that the MSA receptor in chick embryo fibroblasts might mediate the mitogenic effectors of MSA, the human IGFs, insulin, and proinsulin.

Human fibroblasts possess an MSA receptor with the same relative affinities for MSA, insulin, and proinsulin as the chick fibroblast MSA receptor [25] (Fig. 3). These peptides stimulate thymidine incorporation into DNA [25, 79, 80] (Fig. 6). The shape of the dose-response curve with insulin was less steep than the curve obtained with MSA for unexplained reasons, making it difficult to determine relative potencies. Proinsulin was $\cong 50\%$ as potent as insulin and gave a parallel dose-response curve to insulin [81]. Although the process is probably less straightforward than in chick fibroblasts, we have pro-

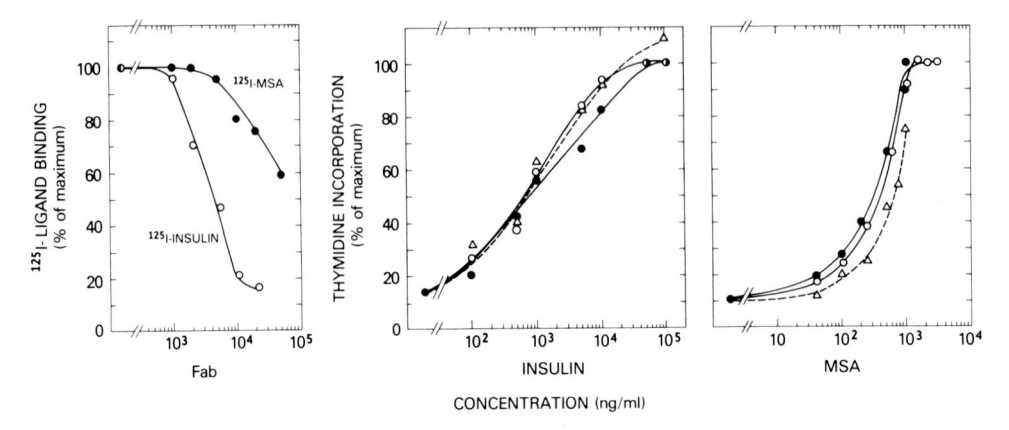

Fig. 6. Left: inhibition of ^{125}I-labeled porcine insulin and ^{125}I-labeled MSA II-1 binding to human fibroblast cultures by Fab fragments of antireceptor IgG. The Fab fragments were prepared from the serum of patient B-2 with the syndrome of extreme insulin resistance and acanthosis nigricans [65, 66]. The IgG fraction, containing antibodies to the insulin receptor, was purified by Protein A–Sepharose chromatography [80]. The IgG was digested with papain and chromatographed on Protein A–Sepharose. Undigested IgG and Fc fragments bind to Protein A; Fab fragments do not [80].

^{125}I-insulin (○) or ^{125}I-MSA (●) and the indicated concentrations of Fab fragments were incubated for 2 h at 15°C with 5 × 10^6 fibroblasts in 0.5 ml of Hepes binding buffer, pH 8.0 [25, 80]. Cell-associated radioactivity was determined following microfuge centrifugation as previously described [25].

Inhibition of ^{125}I-MSA binding to human fibroblasts by DEAE-purified IgG from patient B-2 has previously been observed [64]. Interestingly, other MSA receptors (eg, chick embryo fibroblasts, rat liver membranes) were not inhibited by the same IgG preparations [M. M. Rechler, J. M. Podskalny, C. R. Kahn, S. P. Nissley, manuscript in preparation]. Redrawn from [80] with permission.

Right and center: effect of antireceptor Fab fragments on insulin-stimulated (center) and MSA-stimulated (right) [^{3}H]thymidine incorporation in human fibroblasts. Confluent fibroblast cultures in serum-free medium were incubated with the indicated concentrations of insulin (center) or Sephadex G-75 MSA II (right) in the presence of 0 (●), 4 μg/ml (○), or 10 μg/ml (△) of Fab. After 18 h incubation, the cultures were pulsed with [^{3}H]thymidine for 30 min, and incorporation of radioactivity into acid-precipitable DNA was determined as previously described [79]. Neither antireceptor Fab fragment nor antireceptor IgG stimulated thymidine incorporation when added separately [80]. Redrawn from [80] with permission.

posed that the mitogenic effects of insulin, MSA, and human insulin-like growth factors in human fibroblasts are mediated by MSA/IGF receptors rather than by insulin receptors [2, 25, 79].*

Demonstration that the growth effects of insulin and MSA in fibroblasts are not mediated by the insulin receptor. Fab fragments prepared from IgG from the serum of a patient with insulin resistance resulting from circulating antibodies to insulin receptors are potential antagonists of insulin receptors and insulin receptor-mediated biological effects [80]. Dose-dependent inhibition of ^{125}I-insulin binding to insulin receptors on human fibroblasts by increasing concentrations of antireceptor Fab fragments is observed (Fig. 6). Blockade of MSA receptors required higher concentrations of Fab (Fig. 6), making it possible to select Fab concentrations (4 and 10 μg/ml) at which insulin binding to insulin receptors

---

*Additional support for this hypothesis has come from the observation that insulin isolated from casiragua, a hystricomorph [82], is 20-fold more potent in stimulating thymidine incorporation in human fibroblasts than glucose oxidation in adipocytes [81].

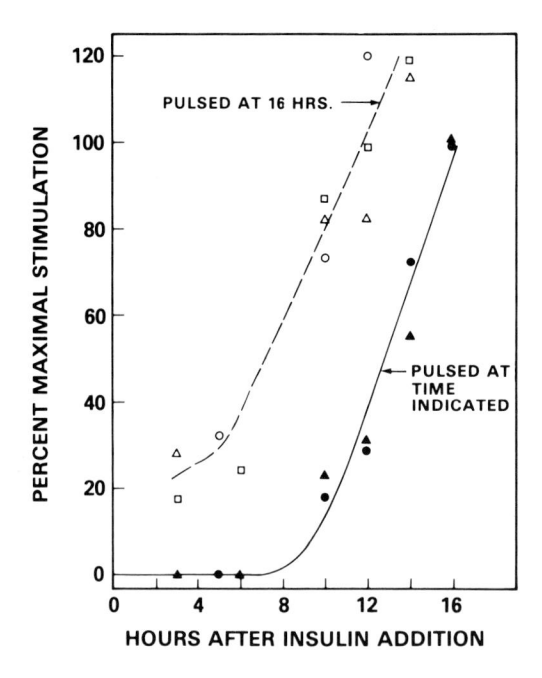

Fig. 7. Incorporation of [^{3}H]thymidine into the DNA of human fibroblasts following incubation with insulin for different durations. Confluent, serum-starved, quiescent fibroblast cultures received 20 µg/ml of porcine insulin at time zero. Some cultures were pulsed with [^{3}H]thymidine for 30 min beginning at the times (3–16 h) after insulin addition indicated (●, ▲). Parallel cultures (○, △, □) received insulin (20 µg/ml) for the indicated times (3–13 h), following which fresh serum-free medium (without insulin) was added. The incubations were continued until 16 h after insulin addition, at which time these cultures were pulsed for 30 min with [^{3}H]thymidine as previously described [79].

was effectively blocked, but MSA receptors were negligibly affected. The ability of insulin and MSA to stimulate thymidine incorporation in human fibroblasts was examined in the presence and absence of these concentrations of Fab (Fig. 6). No significant inhibition (reflected in a rightward shift of the dose-response curve or decreased amplitude of the response) was observed.

Provided that insulin receptors were blocked by anti-receptor Fab fragments under conditions of the thymidine incorporation experiment and for the time required for insulin to exert its mitogenic effects, these results would suggest that blockade of insulin receptors has no effect on insulin- and MSA-stimulated thymidine incorporation; that is, these effects are not mediated by insulin receptors. As seen in Figure 7, insulin must be continuously present for 12–14 h for maximal stimulation of thymidine incorporation to occur. That is, human fibroblasts are not committed to DNA synthesis until approximately 4 h before the start of DNA synthesis. Similar results have been reported in chick embryo fibroblasts [83]. Since low concentrations of ^{125}I-insulin tracer (<1 ng/ml) are degraded by fibroblasts during incubation at 37°C in culture media, it is difficult to verify directly that antireceptor Fab fragments continually block insulin binding under the prolonged incubation conditions of the thymidine incorporation assay required for commitment to DNA synthesis. Independent evidence suggests that this may be the case. Antireceptor IgG inhibits insulin binding but retains some insulin-like biological activity. Under thymidine incorporation bioassay conditions, antireceptor IgG produced a small but significant increase in the protein content of the fibroblast cultures, similar in magnitude to that produced by incubation with insulin or MSA [80]. Thus, it seems likely that insulin receptors were blocked under the experimental conditions and that the growth effects of insulin and MSA in human fibroblasts are mediated by receptors other than insulin receptors, possibly MSA/IGF receptors.

It has not been possible to perform the analogous experiments in chick embryo fibroblasts. Antireceptor Fab fragments did not inhibit MSA binding to the MSA receptor of chick fibroblasts [G. L. King, unpublished results]. However, insulin binds poorly to chick embryo fibroblasts [24], and these insulin receptors are incompletely inhibited by antireceptor Fab fragments (50–60% inhibition by 25 μg/ml Fab [G. L. King, unpublished results]) and antireceptor IgG (50% inhibition by 400 μg/ml IgG [M. M. Rechler, unpublished results]).

Mitogenic effects in other cell types. Two recent studies suggest that our hypothesis that insulin acts via IGF receptors to stimulate growth, originally formulated for cultured fibroblasts, may not hold for all cultured cells. Koontz [84] described the stimulation of thymidine incorporation in H4 cells derived from a minimal deviation Reuber hepatoma by low concentrations of insulin; half-maximal stimulation occurred at 50–100 pM. The sensitive dose-response curve for insulin suggested an insulin receptor-mediated response. Proinsulin was $\cong$1% as effective as insulin, consistent with this interpretation. Anti-insulin receptor IgG prepared from the serum of a patient described by Baldwin et al [85] inhibited binding of radiolabeled insulin to the insulin receptors of H4 cells. Although anti-receptor IgG B-2 did not stimulate thymidine incorporation in fibroblasts [80], the antireceptor IgG used by Koontz stimulated thymidine incorporation in H4 cells. Together, these results suggest that insulin receptors may mediate growth effects in hepatoma cells.

The different effects of anti-insulin receptor IgG on thymidine incorporation – stimulation in H4 cells, no stimulation in fibroblasts – may result from the fact that the IgG preparations were from different patients or that different target cells were used. For example, Kahn et al [86] reported that serum from 3 patients with antibodies to insulin receptors either inhibited (1 patient) or stimulated (2 patients, different dose-response curves) glucose oxidation in rat adipocytes. Of several liver-derived membranes or cells that we have examined for MSA binding, only one (BRL 61t) has an MSA receptor that is inhibited by insulin [33; M.M. Rechler, unpublished results]. Insulin does not cross-react (at 10^3 molar excess) with the MSA receptor of the Neville preparation of liver plasma membranes [10, 24] from Sprague-Dawley rats [2, 10, 24, 42],* HTC hepatoma cells [87], BRL 3A2 or BRL 3A cells [2, 10, 24, 33; M. M. Rechler and J. M. Podskalny, unpublished observations]. This suggests that if insulin acts as a mitogen in liver, it must do so via its own receptor.

A second system in which insulin appears to act as a mitogen via the insulin receptor occurs in a series of variant cell lines of Cloudman S91 melanoma cells [88]. Insulin at low concentrations inhibits the growth of S91 cells. Variant cell lines resistant to this inhibitory effect were selected. One of these lines was, in fact, dependent upon insulin for growth. Insulin stimulated cell multiplication in the dependent line at low concentrations (ED_{50} 0.1 nM); MSA was approximately 0.2% as potent. These results were similar to the relative potencies of insulin and MSA as inhibitors of ^{125}I-insulin binding to the insulin receptors of the insulin-dependent melanoma line: K_D 1–2 nM, MSA $\cong$1% as potent as insulin. The simplest interpretation of these results is that insulin utilizes insulin receptors to stimulate the growth of the insulin-dependent melanoma line.

*See footnote on page 13.

By contrast, Florini et al [89] have presented evidence that MSA is a mitogen for rat myoblasts in culture and presumably utilizes MSA/IGF receptors. Multiplication was stimulated significantly by MSA at 10^{-7} M, but not by insulin at 10^{-6} M.

Sato and his colleagues have established hormonal supplements of defined culture medium that allow a variety of cell types to replicate in the absence of serum [90]. One component required for growth of all cells tested is insulin, typically added at high concentrations (2–5 μg/ml). At these concentrations, insulin might be acting via an MSA/IGF receptor rather than via an insulin receptor. It would be of interest to evaluate whether MSA can substitute for insulin in some of these cell types and to determine the reactivity of their MSA receptors with insulin.

Glucose Oxidation: Adipocytes

The original name for the insulin-like growth factors, nonsuppressible insulin-like activity, arose from the observation that human plasma contained insulin-like biological activity in an in vitro bioassay in fat that could not be neutralized (suppressed) by antibodies to insulin [8]. Partially purified NSILA also was active in vivo, inducing hypoglycemia in adrenalectomized rats [91]. Purified IGF-I, IGF-II, and MSA have insulin-like activity in adipose tissue in vitro; for example, stimulation of glucose oxidation and lipogenesis [35, 36]. The purified insulin-like growth factors compete for ^{125}I-insulin binding to insulin receptors of adipocytes and other tissues (Fig. 8) [2, 24]. In addition, radioiodine-labeled IGF-I, IGF-II and MSA also bind to IGF receptors in adipocytes [92]. This is illustrated using ^{125}I-MSA (Fig. 8). Insulin did not inhibit ^{125}I-MSA binding to adiocytes but instead increased binding approximately 50% in a dose-dependent fashion (Fig. 8). Similar enhancement of ^{125}I-labeled NSILA [92] and IGF-II [9, 78] binding to adipocytes and ^{125}I-MSA binding to rat myoblasts [J. Florini, personal communication] by insulin have been observed. King et al [93] have presented evidence that the enhancement of ^{125}I-MSA binding to the MSA receptor by insulin requires participation of the insulin receptor and has a concentration dependence similar to insulin stimulation of glucose oxidation.

As in human fibroblasts, Fab fragments of antireceptor IgG offer a tool with which to inhibit adipocyte insulin receptors selectively, without significantly affecting MSA receptors. Antireceptor Fab fragments at 1, 3, and 5 μg/ml inhibited ^{125}I-insulin binding to adipocytes by 10%, 25%, and 80%, respectively; ^{125}I-MSA binding was not significantly decreased (Fig. 9).

Although antireceptor IgG from NIH patient B-2 stimulates glucose oxidation, the Fab fragments prepared from this IgG have no stimulatory activity and are pure antagonists of insulin action [80, 94] (Fig. 9). Increasing concentrations of Fab progressively shifted the dose-response curve for insulin stimulation of glucose oxidation to the right. This is the result expected if the Fab fragments were competitive inhibitors of insulin action. (The alternative possibility that the Fab fragments were noncompetitive inhibitors cannot rigorously be excluded at the concentrations of Fab that it was practicable to test.) Assuming that only 10–20% of insulin receptors need be occupied to produce a maximal response in adipocytes ("spare receptors"), either a competitive inhibitor or a non-competitive inhibitor at low concentration would shift the dose response curve to the right. At higher concentrations, however, a noncompetitive inhibitor but not a competitive inhibitor also would decrease the magnitude of the response.

The effect of antireceptor Fab fragment on MSA-stimulated glucose oxidation was next examined. If MSA acted through the IGF/MSA receptor, which was not blocked by antireceptor Fab, the dose-response curve would be unchanged by the presence of anti-

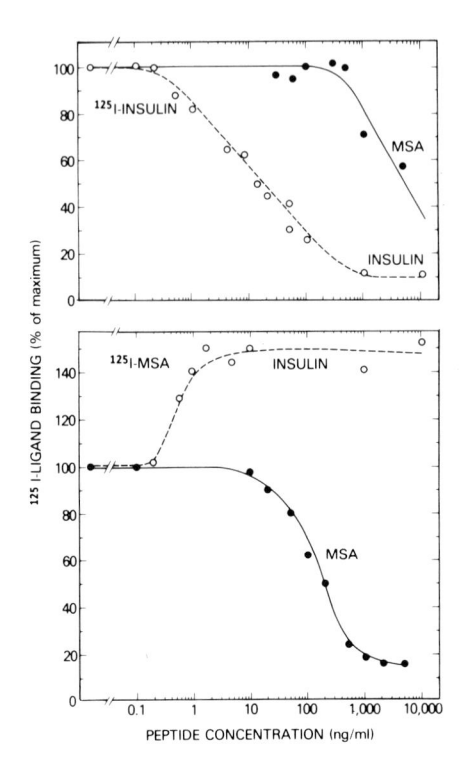

Fig. 8. Binding of ^{125}I-insulin (top) and ^{125}I-MSA (bottom) to isolated rat adipocytes. Adipocytes were prepared from epididymal fat pads of 100–160 gm Sprague-Dawley rats by collagenase treatment [80]. Adipocytes (1.2×10^5 cells), ^{125}I-ligand, and unlabeled insulin (○) or MSA (Sephadex G-75 peak II) (●) at the indicated concentrations were incubated in 0.5 ml pH 7.4 Krebs-Ringer buffer containing 20 mg/ml bovine albumin. Steady-state conditions were used: for ^{125}I-insulin, 20 min at 37°C; for ^{125}I-MSA, 40 min at 24°C. Adipocyte-associated radioactivity was recovered in the layer above dinonylphthalate following microfuge centrifugation. Percent of maximum binding is plotted; maximum binding represents 4% of input tracer radioactivity for both experiments. Redrawn from [80] with permission.

receptor Fab. If, however, MSA acted through the insulin receptor to stimulate glucose oxidation, blockade of the insulin receptor by Fab fragments should shift the dose-response curve to the right, as observed with insulin. As seen in Figure 9 (right panel), the latter result was obtained, indicating that some if not all of MSA's activity in adipocytes occurs via the insulin receptor. It should be noted that the effects of antireceptor Fab are specific for agents that stimulate glucose oxidation via the insulin receptor; stimulation by vitamin K_5 and spermine, which act by different mechanisms, is not inhibited [80]. The function of the MSA/IGF receptor in adipocytes, as well as the significance of the interaction of this receptor with insulin and/or the insulin receptor, remains to be elucidated.

Glucose Incorporation and Amino Acid Transport in Human Fibroblasts

The preceding sections have focused on two extremes of insulin action: an acute metabolic effect in a traditional target tissue (glucose oxidation in adipocytes) and a chronic growth effect (DNA synthesis in cultured fibroblasts). We now examine two

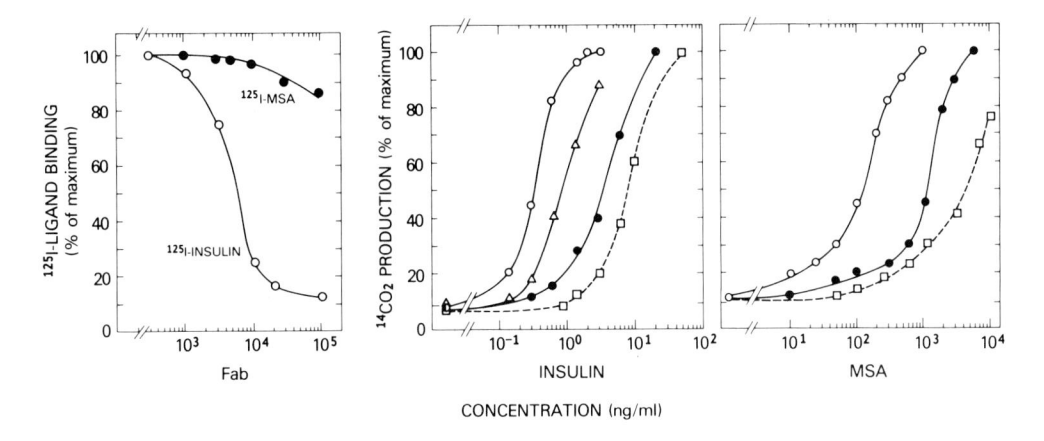

Fig. 9. Left: inhibition of ^{125}I-insulin and ^{125}I-MSA binding to isolated rat adipocytes by Fab fragments of antireceptor IgG. Fab fragments were prepared as described in Figure 6. Preparation of adipocytes and binding conditions were as described in Figure 8. Percent of maximum binding is plotted for different Fab concentrations.

Right and center: effect of antireceptor Fab fragments on insulin-stimulated (center) and MSA-stimulated (right) glucose oxidation in rat adipocytes. Glucose oxidation was quantitated by the conversion of (U)^{14}C-glucose to ^{14}CO$_2$ during 1 h of incubation as previously described [92]. Insulin (center) and MSA (right) were added at the indicated concentrations in the presence of 0 (○), 3 µg/ml (△), 5 µg/ml (●), or 10 µg/ml (□) of antireceptor Fab fragment. Redrawn from [80] with permission.

acute metabolic effects in human fibroblasts: transport of the non-metabolizable amino acid, α-aminoisobutyric acid (AIB) and glucose uptake. Our interest in cultured fibroblasts derives from the fact that they represent one of a limited number of cell types that can be isolated from patients with disorders of target-organ resistance to insulin or IGFs and propagated for 50–60 generations in culture, thereby allowing expression of their genotype free from influences of the physiological environment at the time the initial cells were removed by biopsy from the patient. Human fibroblasts possess both classic insulin receptors [68] and an MSA/IGF receptor that interacts with insulin [25] (Fig. 3). We recently have reported a profound reduction of insulin binding to insulin receptors on fibroblasts cultured from an infant with leprechaunism, severe insulin resistance, and intrauterine growth retardation [95]. Prior to evaluating the functional consequences of this defective insulin binding, we have attempted first to define the biological functions mediated by insulin in fibroblasts and to determine whether insulin produces these responses via an insulin receptor or an MSA/IGF receptor.

Glucose incorporation. Howard et al [96] have demonstrated that insulin at low concentrations stimulated total glucose incorporation into human fibroblast cultures. Stimulation is seen after 30 min of exposure to insulin; maximal stimulation requires 2 h of incubation. Fibroblasts incubated with [^{14}C]glucose for 20 min show increased incorporation in all compartments in insulin-treated cells: glycogen, acid-soluble nucleotides, lactic acid, etc. Using the procedures described by Howard et al [96] with some modifications, we have confirmed that insulin induces a 60–100% increase in [^{14}C]glucose incorporation. Half-maximal stimulation was observed with ≈ 5 ng/ml of insulin [97]. Proinsulin stimulated incorporation to the same extent, but required 50-fold higher concentrations (Table VI) [97]. These results suggested that insulin and proinsulin probably were acting via the insulin receptor.

MSA also was a potent stimulator of glucose incorporation in human fibroblasts (Table VI). Surprisingly, and in contrast to its relative impotence in adipocytes, MSA was 12% as potent as insulin; that is, it was 6 times more potent than proinsulin. Since MSA interacts more weakly with the human fibroblast insulin receptor than proinsulin (<1% as potent as insulin compared to 5%) [24, 68], this result provisionally suggests that the stimulation of glucose incorporation by MSA cannot be explained completely by its interaction with insulin receptors, and it probably represents an activity mediated by the MSA/IGF receptor.

Similar reasoning led Meuli and his colleagues to conclude that insulin and NSILA-s elicited the same biological effects (3-O-methyl-glucose transport, glucose uptake, lactate production) in the perfused rat heart by interacting with different receptors [98, 99]. The preparation of NSILA-s used was 1/60th as potent as insulin in adipocytes, but 1/2–1/5th as potent as insulin in rat heart muscle.

Similar results have been reported recently by Poggi et al in mouse soleus muscle [100]. IGF-I stimulated glucose metabolism (2-deoxyglucose uptake, glycolysis, glycogen synthesis) 4–9% as potently as insulin. Since insulin did not bind to the IGF-I receptor in this tissue, these results were interpreted as indicating that insulin acts via the insulin receptor and that IGF-I acts at least in part via the IGF receptor [100].

Amino acid transport. Neutral α-amino acids may be transported by several specific transport systems in animal tissues [101]. Although the non-metabolizable amino acid, AIB, may be transported by 3 different systems [101], stimulation of AIB transport in hepatocytes by insulin appears to involve selective stimulation of the A-transport system [102]. Of the 3 systems that potentially transport AIB, only the Na^+-dependent A-system is inhibited by N-methyl-AIB.

The ability of insulin, proinsulin, and MSA to stimulate uptake of [^{14}C]methyl-AIB via the A-transport system has been examined in human fibroblasts [97] (Table VI). Experimental procedures are similar to those described by Hollenberg [103, 104]. Insulin stimulated methyl-AIB uptake approximately 2-fold [97]. Half-maximal stimulation occurred at approximately 5 ng/ml [97], similar to values previously reported by others [104, 105]. Proinsulin was approximately 5% as potent as insulin [97] (Table VI), consistent with published results [104] and with the potency of proinsulin as an inhibitor of ^{125}I-insulin binding to insulin receptors [68].

As seen for glucose incorporation in human fibroblasts, MSA was considerably more potent in stimulating methyl-AIB transport (~50% as potent as insulin) than would have been anticipated from its interaction with insulin receptors [97] (Table VI). Presumably it acts through the MSA/IGF receptor. In preliminary experiments, IGF-I also stimulated both methyl-AIB uptake and glucose incorporation with a potency greater than would be expected if it acted predominantly via the insulin receptor [M. M. Rechler, A. B. Knight, unpublished results]. Hollenberg and Fryklund [106] previously reported that somatomedin A stimulated AIB transport in human fibroblasts with a high potency, suggesting that it acted via a somatomedin receptor.

As summarized in Table VII, insulin and MSA are capable of stimulating AIB transport in different experimental systems by different mechanisms. In human fibroblasts and HTC hepatoma cells, insulin and proinsulin appear to utilize insulin receptors, whereas MSA and other IGFs appear to utilize the MSA/IGF receptor (Table VII). By contrast, in

TABLE VI. Relative Potencies of Insulin, MSA, and Proinsulin for Different Biological Responses and Receptors

	Insulin[a]	Proinsulin[a]	MSA[a]
1) Insulin receptors: adipocytes [80], human fibroblasts [68], IM-9 lymphoblasts [24], rat liver membranes [54] [b]	100	5	0.1–1.0
2) MSA receptors: chick fibroblasts, human fibroblasts [24, 25, 43]	100	40	200
3) Adipocytes: glucose oxidation [80]	100	5	0.2
4) Chick fibroblasts: thymidine incorporation[c]	100	40	200
5) Human fibroblasts			
Glucose incorporation [97][d]	100	2	12
Methyl AIB transport [97][e]	100	5	50
6) Hepatoma (HTC) cells: Tyrosine aminotransferase induction [87][f]	100	13	28
7) Perfused rat heart: glucose transport and metabolism [98, 99]	100	NT[g]	20–40[h]
8) Muscle, mouse soleus: glucose uptake and metabolism [100]	100	NT	4–9[i]
9) Chondrocytes (Swarm rat chondrosarcoma): proteoglycan synthesis [100a]	100	3	1

[a] Porcine insulin, porcine proinsulin, Sephadex G-75 peak II MSA.
[b] M. M. Rechler, unpublished results.
[c] Proinsulin also is 40% as potent as insulin in stimulating thymidine incorporation in human fibroblasts [81]. As discussed in the text, the dose-response curve for MSA is nonparallel (steeper), so that relative potency is difficult to assess.
[d] A. B. Knight, unpublished results.
[e] M. M. Rechler, unpublished results.
[f] Relative potencies based on 4 paired experiments each for insulin-MSA and insulin-proinsulin.
[g] NT = not tested.
[h] Performed with partially purified NSILA-s, 1/60th as potent as insulin in adipocytes.
[i] Performed with IGF-I.

chick embryo heart, MSA, insulin, and proinsulin all appear to stimulate AIB transport via the MSA receptor and not the insulin receptor [107]. (It should be emphasized that this mechanism requires that the MSA/IGF receptor shows substantial cross-reactivity with insulin and proinsulin.) This summary clearly illustrates the necessity of determining the relative biological potencies of both classes of peptides, as well as defining the properties of the receptors with which these peptides interact, before reaching conclusions about the receptor that mediates a given biological response.

HTC Rat Hepatoma Cells: Induction of Tyrosine Aminotransferase (TAT)

Insulin causes a 2-fold increase in the amount of TAT in HTC cells in which TAT has been induced by dexamethasone, by decreasing the rate of TAT degradation [112, 113]. Half-maximal stimulation is achieved with $\cong 30$ ng/ml of insulin [87]. Proinsulin and MSA are 9% and 28% as potent as insulin, respectively [87]. HTC cells possess both an insulin receptor and an MSA receptor [87]. ^{125}I-insulin binding is inhibited 50% by $\cong 33$ ng/ml of insulin [87]. Proinsulin inhibits insulin binding with a potency slightly lower than its biological potency in HTC cells [87]. The MSA receptor of HTC cells is relatively insensitive (50% inhibition of ^{125}I-MSA binding by 200 ng/ml unlabeled MSA) and does not interact with insulin (no inhibition of ^{125}I-MSA binding by 10 μg/ml insulin [87]. These results

TABLE VII. Predominant Receptors Mediating the Stimulation of AIB Transport by Insulin and Insulin-Like Growth Factors in Different Experimental Systems

Experimental system	Peptide	Receptor used	Reference
1) Human fibroblasts	Insulin	Insulin	[97, 104, 105][a]
	Proinsulin	Insulin	[97, 104][a]
	MSA	MSA/IGF	[97][a]
	IGF-I	MSA/IGF	[97][a]
	Somatomedin A	MSA/IGF	[106]
2) Chick embryo fibroblasts	MSA	MSA/IGF (?)[b]	[83]
3) Chick embryonic heart	MSA	MSA/IGF[c]	[107]
	Insulin	MSA/IGF[c]	[107]
	Proinsulin	MSA/IGF[c]	[107]
4) Rat thymocytes	Insulin	Insulin[d]	[108]
5) HTC cells	Insulin	Insulin	[87]
	Proinsulin	Insulin	[87]
	MSA	MSA[e]	[87]
6) Rat myoblasts	MSA	MSA/IGF (?)[f]	[109]
	Insulin	(?)[f]	[109]
	Somatomedin A	(?)[f]	[110]
7) Rat hepatocytes	Insulin	Insulin (?)[g]	[102, 111]

[a]M.M. Rechler, unpublished results [107a].
[b]Insulin not tested.
[c]The relative potencies for stimulation of [^{14}C]methyl-AIB transport and inhibition of ^{125}I-MSA binding were identical: MSA:insulin:proinsulin = 2–3:1:0.2–0.3.
[d]The dose-response curve for insulin stimulation of AIB transport was complex. Assignment was based on the fact that one K_D for binding and biological response were similar. Studies with insulin analogues or insulin-like growth factors were not performed.
[e]The MSA receptor of HTC cells does not interact with insulin [87].
[f]MSA and insulin stimulation of AIB transport in L6 rat myoblasts required high concentrations ($ED_{50} \cong 1 \mu g/ml$; maximal response at $\cong 10 \mu g/ml$) [109]. Concentration dependence was similar for the two peptides. Stimulation by somatomedin A [110] was observed at high concentration (1 $\mu g/ml$); a dose-response curve was not presented.
[g]ED_{50} for insulin 12–18 ng/ml [102], 40 ng/ml [111]. Insulin-like growth factors not tested.

suggest that insulin and proinsulin act predominantly by an insulin receptor-mediated mechanism. The greater relative potency of MSA suggests that it exerts at least part of its biological effect via the MSA receptor.

Proteoglycan Synthesis: Chondrocytes

The existence of growth hormone-dependent substances in rat serum that presumably mediate the action of growth hormone on cartilage was first appreciated using as in vitro bioassay the stimulation of [^{35}S]sulfate incorporation into glycosaminoglycans in costal cartilage from hypophysectomized (hypox) rats [114]. Sulfate incorporation assays in hypox rat costal cartilage or chick pelvic leaflets guided the purification of somatomedin C [7] and somatomedin A [6] from human serum. Moreover, cartilage assays on serum samples have been the classic biological assay for somatomedin activity in different physiological and pathological states [115]. Over the years, it has been appreciated that supraphysiological concentrations of insulin are capable of mimicking the actions of somatomedins in cartilage [116, 117], presumably by weak cross-reaction with somatomedin receptors [57]. The studies to be described present evidence for an alternative mechanism; namely, that insulin and MSA stimulate proteoglycan synthesis in chondrocytes prepared from the Swarm rat chondrosarcoma by an insulin receptor-mediated mechanism [100a].

Proteoglycan synthesis in rat chondrosarcoma chondrocytes has been extensively studied by Hascall and co-workers [118]. Insulin and MSA stimulated proteoglycan synthesis 2–3-fold; the proteoglycan was normal in size, chondroitin sulfate content, and ability to aggregate [119]. Half-maximal response was observed with 1 ng/ml of insulin, 30 ng/ml of proinsulin, and 100 ng/ml of MSA [100a, 119]. Chondrocytes were shown to possess a typical insulin receptor [100a] and an MSA receptor that did not react with insulin at high concentrations [100a]. These results strongly suggest that insulin stimulates proteoglycan synthesis through an insulin receptor; MSA could act through the insulin receptor and/or the MSA receptor.

Miscellaneous Biological Actions of MSA

Casein synthesis: mouse mammary epithelium in culture. Insulin at high concentrations acts synergistically with glucocorticoids and prolactin to stimulate synthesis of milk proteins by mouse mammary explants [120]. Casein synthesis by explants from midpregnant mice was evaluated by immunoprecipitation after 48 h of incubation with [^{3}H]amino acids, as previously described [121]. In the presence of hydrocortisone (1 μg/ml) and prolactin (5 μg/ml), insulin (5 μg/ml) and MSA (5 μg/ml) stimulated casein synthesis, whereas no stimulation was observed with MSA at 1 μg/ml [T. Oka, C. Hori, S. P. Nissley, M. M. Rechler, unpublished results]. Synthesis with 5 μg/ml MSA was $\cong 40\%$ of that obtained with insulin at the same concentration. In other experiments, significant stimulation has been observed with 50 ng/ml of insulin [T. Oka, unpublished results]. These results suggest that MSA is 1–5% as potent as insulin in stimulating casein synthesis, a finding consistent with both peptides acting via the insulin receptor.

Intracellular cyclic AMP: chick embryo fibroblasts. MSA induced a rapid, concentration-dependent inhibition of PGE$_1$-stimulated cAMP accumulation in chick embryo fibroblasts [122]. Maximal inhibition was achieved with 100 ng/ml. Inhibition of PGE$_1$-stimulated and fluoride-stimulated adenylate cyclase was observed in membranes prepared from these cells [122].

Promote meiosis: Xenopus oocytes. Insulin can induce meiotic division in Xenopus laevis oocytes at high concentrations (ED$_{50}$ 2 μM) [123]. In prelimary experiments, MSA also is a meiosis-promoting agent of stage 5 Xenopus oocytes in vitro [M. El-Etr and E. E. Baulieu, personal communication]. MSA is effective at lower concentrations than insulin ($\cong$10-fold). Evaluation of insulin and MSA receptors is in progress.

Ornithine decarboxylase: porcine granulosa cells. MSA, like a variety of other hormones, including LH, FSH, EGF, cAMP analogs, and prostaglandins of the E series, stimulates ornithine decarboxylase activity in isolated porcine granulosa cells maintained under defined conditions in vitro [124]. The dose-response curve is quite sensitive, with ED$_{50}$ $\approx$ 10 ng/ml MSA [124]. MSA has previously been shown to stimulate ornithine decarboxylase in 3T3 mouse fibroblasts [71].

ROLE OF INSULIN AND INSULIN-LIKE GROWTH FACTORS IN FETAL GROWTH

The hormonal determinants of intrauterine growth are poorly understood. There is little to indicate a significant role for pituitary growth hormone, since fetuses lacking fetal growth hormone or maternal growth hormone are normal in size [125]. Insulin has been implicated by association in the pathogenesis of the large babies born to diabetic mothers [126]. The converse, insulin deficiency, has been reported in rare cases of pancreatic agenesis, and is associated with cessation of fetal growth in the third trimester, exceedingly small birthweight, and early neonatal death [127–129]. The insulin-like growth factors have received relatively little attention as fetal growth factors. In this section, we shall consider evidence implicating MSA (specifically among insulin-like growth factors) as a fetal growth factor in the rat, evidence implicating insulin receptors (and perhaps IGF receptors) in human intrauterine growth retardation in a patient with leprechaunism, and possible mechanisms for the large size of infants of diabetic mothers.

MSA as Fetal Growth Factor in the Rat

In 1979 we reported that explants of near-term fetal rat liver, placed in organ culture for up to 4 days in serum-free completely defined medium, synthesized significant quantities of peptides indistinguishable from BRL-MSA in biological, chemical, and immunological properties, and in interactions with receptors and binding proteins [36]. This observation was particularly intriguing because the liver explant system is known to faithfully recreate the developmental state, pattern of enzyme regulation, and pattern of fetal protein synthesis appropriate to the living fetus at the same gestational age [36]. This strongly suggested that fetal rat liver at near term synthesizes significant quantities of MSA-like peptides.

Moses et al [46] next examined serum from rats at different gestational ages, employing the MSA assays described in Table IV. Fetal rat serum of 16–20 days gestational age was found to contain 1–4 μg/ml of MSA by radioimmunoassay, liver membrane receptor assay, and competitive binding protein assay. Maternal serum contained 20–100 times lower amounts of MSA by immunoassay. After birth, radioimmunoassayable MSA levels began to fall to the low adult levels after 5–10 days [46].

Total somatomedin activity, measured in the chick embryo fibroblast thymidine incorporation bioassay, was similar to the level of immunoreactive MSA in fetal rat serum [46]. However, somatomedin levels determined by this bioassay were higher in maternal serum than in fetal serum, and they were substantially greater than the levels of immunoreactive MSA in maternal rat serum. These results suggest that MSA may be the predominant fetal somatomedin in the rat, but that other somatomedins, which appear after birth, represent the major postnatal species. Studies are in progress to evaluate the association of fetal somatomedin with serum binding proteins, and to determine MSA and binding protein levels in states of pathological intrauterine growth.

Leprechaunism: Genetic Disorder of Decreased Insulin Binding and Intrauterine Growth Retardation

Leprechaunism is a rare (<30 cases), heterogeneous, presumably genetic syndrome characterized by low birthweight, absent subcutaneous fat, poorly developed muscles, unusual facies, hirsutism, and frequently by hyperinsulinemia and insulin resistance [reviewed in 130]. A small subgroup of leprechaun patients with insulin resistance has been studied more intensively. Available results suggest a variety of molecular defects, including primary (cellular) decrease in insulin receptors, secondary (humoral) decrease in insulin receptors, or post-receptor defects. In patients with a post-receptor defect, insulin binding to insulin receptors would be normal, but the biological response to insulin would be decreased. Two patients have been described with a possible defect at this level: one patient from Arkansas studied by Kobayashi et al [131] had normal insulin binding to monocytes and fibroblasts, but abnormal glucose transport*; one patient from North Carolina had normal insulin binding, but insulin function in the patient's cells was not examined [132]. Two siblings studied in Arkansas [128, 133] exhibited abnormal insulin binding to their circulating cells (erythrocytes). Binding studies to fibroblasts cultured from these patients have not yet been reported, so it is not possible to determine whether the receptor defect is secondary to humoral factors (eg, insulin, antireceptor antibodies) or to a primary cellular defect. Finally, we have had the opportunity to study fibroblasts from a leprechaun patient seen in Winnipeg, Canada [130], which exhibit a profound selective defect of insulin binding to insulin receptors [95, 134]. Since the abnormality in this patient was expressed in fibroblasts propagated for multiple generations in culture, we consider this the first demonstration of a primary, cellular, genetic defect of insulin receptors in man. Studies defining the defect in this patient will be described below.†

Binding of ^{125}I-insulin to fibroblasts from our leprechaun patient was consistently ≤20% of binding to age- and sex-matched control fibroblasts (determined per cell or per microgram of DNA, after correction for nonspecific or nonsaturable binding) [95]. Although the experimental observation is clearcut, the validity of the conclusion critically depends on eliminating possible extraneous influences that might spuriously generate these results. These include the following:

1) Cell handling. Patient and control cultures were established in Winnipeg by identical procedures and shipped to NIH at similar low passage number at the same time. Subsequent handling of stock cultures and cultures prior to experiments were performed in parallel (ie, same media, split ratio, time). Experiments were performed with patient and control cells at the same (±2) passage number. To eliminate possible variations in binding with cell growth rate, only confluent, quiescent cultures were used. Patients and control fibroblasts grew at a similar rate [J.A. Romanus, E.E. Schilling, M.M. Rechler, unpublished observations]. Cultures were examined for Mycoplasma contamination by culture techniques [95] and found negative.

2) Binding to suspended cells or fibroblast monolayers. We have routinely suspended fibroblasts by gentle and controlled trypsinization, for greater flexibility in optimizing binding assay conditions (temperature, cell concentration) [24, 25, 68, 136]. The defect in insulin binding to patient's fibroblasts was observed in suspension or in situ.

3) Control fibroblasts were representative of fibroblasts from other normal individuals. The extent and dose response of insulin binding to the control fibroblasts obtained from Winnipeg and paired with the leprechaun patient's fibroblasts did not differ significantly from a broader experience with fibroblasts cultured from forearm skin of adult donors [136, 137] and studied by the same procedures in our laboratory. The control fibroblasts from Winnipeg were established from foreskin; the patient's fibroblasts were derived from skin of the upper arm. The site from which the skin was derived does not appear to be a major variable, since similar control results were obtained with foreskin, skin from

---

*Insulin binding to erythrocytes from this patient appears to be decreased [131a; M. J. Elders, personal communication], suggesting that the defect may be at the level of the insulin receptor.

†Recently, a patient cared for in Minnesota was found to have decreased insulin binding to permanent cultures of B lymphocytes [135], suggesting a cellular defect possibly similar to that of the Winnipeg patient [95].

the arm of an adult, or skin from a 3-month-old female [95].

4) Degradation of ^{125}I-insulin. Following incubation with patient's fibroblasts, ^{125}I-insulin recovered from the supernate is physically intact [95] and fully capable of binding to IM-9 lymphoblasts [J. A. Romanus, unpublished results].

5) Incubation conditions. Binding of ^{125}I-insulin to patient's fibroblasts was reduced to the same extent when binding studies were performed at different times of incubation, at different temperatures, and at different pHs [95; E. E. Schilling, unpublished results].

6) Cell size. Patient's fibroblasts appear to have slightly less surface area than control fibroblasts (Coulter channelyzer, assuming suspended cells are spheres), but the magnitude of this decrease is too small to account for the reduction of insulin binding [95]. The appearance of patient's and control fibroblasts on scanning electron microscopy is similar [B. Rentier and E. E. Schilling, unpublished observations]. A more sensitive control for cell surface area is the binding of EGF, a chemically unrelated peptide, to patient and control fibroblasts. EGF binding is complicated by the fact that EGF and its receptors are internalized by fibroblasts, taken up by lysosomes where EGF (and possibly the receptor as well) is degraded [55]. Studies were performed in the presence of 10 mM NH_4Cl to inhibit lysosomal uptake and degradation (suggested by Dr. Stanley Cohen) in order to achieve a steady state of binding, and in the absence of NH_4Cl. Under both conditions, EGF binding to fibroblasts from the leprechaun patient and control was indistinguishable [95; E.E. Schilling, unpublished results]. These results indicate that the abnormality in the leprechaun patient's fibroblasts is selective, and not a general cell-surface abnormality or simple reduction in cell surface area.

Recent studies provide further fascinating insights into the nature of the genetic abnormality in the leprechaun patient's cells, and into the coupling of insulin receptors to insulin action [97, 138].

First, binding of ^{125}I-IGF-I to the IGF-I receptor (Fig. 3) of patient and control fibroblasts was examined [138]. Binding to the leprechaun patient's fibroblasts (per cell) was decreased by 78 ± 9% compared to control fibroblasts in 5 paired experiments, a reduction comparable in magnitude to that observed for insulin receptors [97, 138]. Since insulin receptors and IGF receptors react with the same peptides, and since antibodies to insulin receptors recognize determinants on the MSA/IGF receptor of human fibroblasts, it is possible that insulin receptors and IGF receptors share a common structural domain, and that this component may be altered by mutation in our leprechaun patient's cells. Alternatively, insulin receptors and IGF receptors may be present in the plasma membrane adjacent to common membrane components (? transducers), that may be altered by the leprechaun mutation and decrease binding to both receptors. Needless to say, this dual receptor defect compels us to withhold final judgment on whether the intrauterine growth retardation in leprechaunism results from a failure of insulin action via insulin receptors or IGF receptors, or a lack of IGF effect via the IGF receptor.

Second, two acute metabolic functions of insulin and MSA have been examined in the leprechaun patient's and control fibroblasts: glucose incorporation and AIB transport [97]. As discussed above, insulin (and proinsulin) appear to stimulate both responses with similar sensitivity ($\cong$5 ng/ml) and specificity, compatible with these effects being mediated by insulin receptors. MSA, however, appears to stimulate both effects via IGF/MSA receptors (Table VI). In the leprechaun patient's fibroblasts, stimulation of glucose incorporation by insulin and MSA is greatly impaired [97]. By contrast, stimulation of AIB trans-

HORMONAL DETERMINANTS OF IN UTERO GROWTH

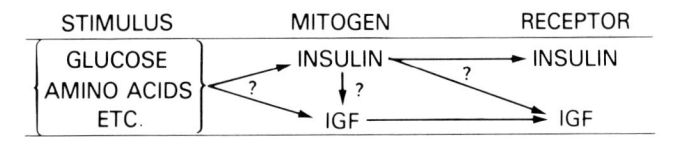

STIMULUS	MITOGEN	RECEPTOR
GLUCOSE AMINO ACIDS ETC.	INSULIN ↓ ? IGF	INSULIN IGF

Fig. 10. Schematic representation of possible mechanisms of insulin and IGF regulation of intrauterine growth.

port by insulin and MSA is normal in magnitude in the patient's fibroblasts and slightly shifted in sensitivity [97, 107a]. One possible interpretation of these results would be the existence of functional subclasses of insulin receptors and IGF receptors or multicomponent receptors [107a]: Certain receptors (components) are coupled to effectors mediating glucose incorporation (and are abnormal in the patient's cells); other receptors (components) are coupled to effectors mediating amino acid-transport (normal in the leprechaun patient's cells). Moreover, receptors for insulin and IGF that mediate the same function (glucose incorporation or amino acid transport) appear to be coupled to the same effectors. If the coupling molecules are in proximity to the insulin and IGF binding sites, mutational alteration of those molecules coupling both binding sites to glucose uptake might be affected in our leprechaun patient's fibroblasts. These putative coupling molecules would differ from those linking insulin receptors and IGF receptors to AIB transport.

Infant of the Diabetic Mother: Pathogenesis of Large Babies

Infants of diabetic mothers have increased weight and length at birth compared to infants of normal mothers [126]. This fetal macrosomia is thought to arise by the following pathogenetic sequence: maternal hyperglycemia results in fetal hyperglycemia, which induces fetal hyperinsulinism [126]. Supporting evidence links elevated fetal insulin with large infant size [126]. Questions deserving further exploration include whether insulin acts via insulin receptors (which has been tacitly assumed), or whether the insulin-like growth factors and IGF receptors may be involved [134]. Some of the mechanistic alternatives are diagrammed in Figure 10. 1) If insulin were the responsible mitogen, it might act via insulin or IGF receptors (provided that the IGF receptors in the relevant tissues interact with insulin). 2) Alternatively, insulin might act indirectly by stimulating IGF synthesis and release. 3) IGF synthesis might increase in response to the same stimuli as those that trigger increased insulin synthesis (eg, amino acids and/or glucose). 4) IGF would act as an in utero mitogen via IGF receptors. Precedent exists for each of these mechanistic alternatives. Given the complex interrelationship of insulin and IGFs, and insulin and IGF receptors, in their biological actions and involvement in the intrauterine growth retardation of leprechaunism, we submit that these mechanisms should be considered as possible factors in the pathogenesis of the large infants of diabetic mothers.

SUMMARY AND CONCLUSIONS

1) The insulin-like growth factors (IGF-I, IGF-II, somatomedin A, somatomedin C, MSA), constitute a closely related group of polypeptides with similar chemical structure, biological activities, target organs, and recognition by antibodies, carrier proteins, and re-

ceptors. By contrast, the insulin-related growth factors (NGF, relaxin) have a more distant structural resemblance and differ in site of synthesis, biological responses, target tissues, and recognition by receptors.

2) Properties of one of the insulin-like growth factors, MSA, are described in detail. Two MSA species have been obtained in homogeneous form. The relationship among MSA species is inferred from chemical, physical, and immunological studies.

3) Comparative studies have previously established the close resemblance of MSA, synthesized by a cultured rat liver cell line to human somatomedin A, IGF-I, and IGF-II. Studies are presented that indicate the similarity between MSA and human somatomedin C in receptor and immunoassays.

A preparation of rat somatomedin partially purified from rat serum cross-reacts in some, but not all, assays for MSA, indicating that these polypeptides are not identical.

4) Most tissues possess separate receptors for insulin and for the insulin-like growth factors. Cross-reaction of insulin with IGF/MSA receptors, and of IGFs with insulin receptors is frequently observed. Because of the overlapping specificity, insulin and the IGFs frequently exhibit the same biological effects. Comparison of the relative potencies of IGF and insulin in a given bioassay with the properties of IGF and insulin receptors in that tissue provides preliminary clues as to which receptor mediates the activity of a given peptide. More direct analysis has been possible in some instances by using selective blockade of insulin receptors (and insulin receptor-mediated function) with Fab fragments of human antibodies to insulin receptors. Results of these studies include the following a) The metabolic effects of IGFs and insulin in adipocytes are mediated by insulin receptors; b) stimulation of amino acid transport may occur via both receptors (eg, HTC cells, human fibroblasts), or insulin and IGF (MSA) both may utilize the MSA/IGF receptor (embryonic chick heart); c) insulin stimulates proteoglycan synthesis in rat chondrosarcoma chondrocytes via the insulin receptor; d) mitogenic effects of insulin and MSA in human fibroblasts may be mediated by the MSA receptor, but are not mediated by the insulin receptor. However, insulin may exert mitogenic effects in other cell types via insulin receptors.

5) MSA-like polypeptides probably constitute a small portion of the total somatomedin activity of adult rat plasma, but they may represent the predominant somatomedin in fetal rat plasma. Fetal rat serum contains 20-100 times higher levels of immunoreactive MSA than does maternal or adult rat serum, suggesting that MSA may have a special role in fetal growth in the rat.

6) Insulin acting through insulin receptors also may be a critical determinant of in utero growth. Support for this view derives from demonstration of a profound reduction of insulin binding to fibroblasts from an infant with leprechaunism, intrauterine growth retardation, and marked insulin resistance (without consistent glucose intolerance). Poor third trimester growth, absent fat stores, and reduced muscle mass at birth may reflect the lack of insulin effect in fetal development that results from this heritable deficiency of insulin binding.

ACKNOWLEDGMENTS

These studies reflect the contributions and cooperation of many colleagues and collaborators, including L. Fryklund, J. Zapf, E. R. Froesch, R. E. Humbel, K. Hall, C. R. Kahn, H. Eisen, J. M. Podskalny, O. Z. Higa, C. B. Bruni, Y. Yang, V. Hascall, R. Stevens, T. Foley, J. Heaton, T. D. Gelehrter, A. M. Rosenberg, C. Grunfeld, J. J. Van Wyk, L. E.

Underwood, W. H. Daughaday, and T.-Y. Liu. We thank J. Koontz and R. Kahn for permission to discuss their results prior to publication. We thank J. Miller, J. Smith, and J. Todd for assistance in preparation of the manuscript.

REFERENCES

1. Dulak NC, Temin HM: J Cell Physiol 81:153, 1973.
2. Nissley SP, Rechler MM: Natl Cancer Inst Monogr 48:167, 1978.
3. Moses AC, Nissley SP, Rechler MM, Short PA, Podskalny JM: In Giordano G, Van Wyk JJ, Minuto F (eds): "Somatomedins and Growth." London: Academic Press, 1979, vol 23, pp 45–59.
4. Nissley SP, Rechler MM, Moses AC, Eisen HJ, Higa OZ, Short PA, Fennoy I, Bruni CB, White RM: In "Hormones and Cell Culture." Cold Spring Harbor, New York: Cold Spring Harbor Laboratory, 1979, vol 6, pp 79–94.
5. Nissley SP, Rechler MM: Proceedings, 6th International Congress of Endocrinology, Melbourne, Australia, February, 1980 (in press).
6. Hall K, Fryklund L: In Gray CH, James VHT (eds): "Hormones and Blood." London: Academic Press, 1979, vol 1, pp 255–278.
7. Van Wyk JJ, Underwood LE: In Litwack G (ed): "Biochemical Actions of Hormones." New York: Academic Press, 1978, vol V, pp 101–148.
8. Oelz O, Froesch ER, Bünzli HF, Humbel RE, Ritschard WJ: In Steiner DF, Freinkel N (eds): "Handbook of Physiology." Baltimore: Williams and Wilkins, 1972, vol 1, sec 7, pp 685–702.
9. Zapf J, Rinderknecht E, Humbel RE, Froesch ER: Metabolism 27:1803, 1978.
10. Rechler MM, Fryklund L, Nissley SP, Hall K, Podskalny JM, Skottner A, Moses AC: Eur J Biochem 82:5, 1978.
11. Rechler MM, Zapf J, Nissley SP, Froesch ER, Moses AC, Podskalny JM, Schilling EE, Humbel RE: Endocrinology 107:1451, 1980.
12. Rechler MM: In Waldhäusl WK (ed): "Diabetes 1979: Proceedings of the 10th Congress of the International Diabetes Federation, Vienna, Austria, September 9–14, 1979." Amsterdam: Excerpta Medica, 1980, pp 266–271.
13. Rinderknecht E, Humbel RE: J Biol Chem 253:2769, 1978.
14. Rinderknecht E, Humbel RE: FEBS Lett 89:283, 1978.
15. Zapf J, Walter H, Morell B: In "International Congress Series No. 481." Amsterdam: Excerpta Medica, 1979, abstr 684.
16. Bradshaw RA, Niall HD: TIBS 3:274, 1978.
17. Bradshaw RA: J Supramol Struct Suppl 4 (abstr 310) 124, 1980.
18. Schwabe C, McDonald JK, Steinetz BG: Biochem Biophys Res Commun 75:503, 1977.
19. James R, Niall H, Kwok S, Bryant-Greenwood G: Nature 267:544, 1977.
20. Kahn CR: TIBS 4:263, 1979.
21. Schwabe C, Steinetz B, Weiss G, Segaloff A, McDonald JK, O'Bryne E, Hochman J, Carriere B, Goldsmith L: Recent Prog Horm Res 34:123, 1978.
22. Mobley WC, Server AC, Ishii DN, Riopelle RJ, Shooter EM: N Engl J Med 297:1096, 1977.
23. McMurtry JP, Floersheim GL, Bryant-Greenwood GD: J Reprod Fertil 58:43, 1980.
23a. Mercado-Simmen RC, Bryant-Greenwood GD, Greenwood FC: J Biol Chem 255:3617, 1980.
24. Rechler MM, Podskalny JM, Nissley SP: J Biol Chem 252:3989, 1977.
25. Rechler MM, Nissley SP, Podskalny JM, Moses AC, Fryklund L: J Clin Endocrinol Metab 44:820, 1977.
26. Brinsmead MW, Liggins GC: Aust J Exp Biol Med Sci 56:513, 1978.
27. DeLarco JE, Todaro GJ: Nature 272:356, 1978.
28. Banerjee SP, Snyder SH, Cuatrecasas P, Greene LA: Proc Natl Acad Sci USA 70:2519, 1973.
29. Frazier WA, Boyd LF, Bradshaw RA: J Biol Chem 249:5513, 1974.
30. Frazier WA, Boyd LF, Pulliam MW, Szutowicz A, Bradshaw RA: J Biol Chem 249:5918, 1974.
31. Herrup K, Shooter EM: Proc Natl Acad Sci USA 70:3884, 1973.
32. Moses AC, Nissley SP, Short PA, Rechler MM: Eur J Biochem 103:401, 1980.
33. Nissley SP, Short PA, Rechler MM, Podskalny JM, Coon HG: Cell 11:441, 1977.
34. Moses AC, Nissley SP, Short PA, Rechler MM, Podskalny JM: Eur J Biochem 103:387, 1980.
35. Zapf J, Schoenle E, Froesch ER: In Kastrup KW, Nielsen JH (eds): "Proceedings 11th FEBS Meeting, Copenhagen." New York: Pergamon Press, vol 48, 1977, pp 59–64.

CCDD:A:441

36. Rechler MM, Eisen HJ, Higa OZ, Nissley SP, Moses AC, Schilling EE, Fennoy I, Bruni CB, Phillips LS, Baird KL: J Biol Chem 254:7942, 1979.
38. Moses AC, Nissley SP, Cohen KL, Rechler MM: Nature 263:137, 1976.
39. Moses AC, Nissley SP, Passamani J, White RM, Rechler MM: Endocrinology 104:536, 1979.
40. Cohen KL, Short PA, Nissley SP: Endocrinology 96:193, 1975.
41. Van Wyk JJ, Svoboda ME, Underwood LE: J Clin Endocrinol Metab 50:206, 1980.
42. Rechler MM, Podskalny JM, Nissley SP: Nature 259:134, 1976.
43. Nissley SP, Rechler MM, Moses AC, Short PA, Podskalny JM: Endocrinology 101:708, 1977.
44. Van Wyk JJ, Underwood LE, Baseman JB, Hintz RL, Clemmons DR, Marshall RN: Adv Metabolic Disorders 8:128, 1975.
45. Rinderknecht E, Humbel RE: Proc Natl Acad Sci USA 73:2365, 1976.
46. Moses AC, Nissley SP, Short PA, Rechler MM, White RM, Knight AB, Higa OZ: Proc Natl Acad Sci USA 77:3649, 1980.
47. Dulak NC, Shing YW: J Cell Physiol 90:127, 1977.
48. Hall K, Brandt J, Enberg G, Fryklund L: J Clin Endocrinol Metab 48:271, 1979.
49. Svoboda ME, Van Wyk JJ, Klapper DG, Fellows RE, Grissom FE, Schlueter RJ: Biochemistry 19: 790, 1980.
50. Furlanetto RW, Underwood LE, Van Wyk JJ, D'Ercole AJ: J Clin Invest 60:648, 1977.
51. Cohen KL, Nissley SP: Horm Metab Res 12:164, 1980.
52. Daughaday WH, Mariz IK, Daniels JS, Jacobs JW, Rubin JS, Bradshaw RA: In Giordano G, Van Wyk JJ, Minuto F (eds): "Somatomedins and Growth." London: Academic Press, 1979, pp 25–29.
53. Daughaday WH, Jacobs JW, Mariz IK, Bradshaw RA: In Japan Medical Res Foundation (eds): "Growth and Growth Factors," Proc Int Symposium, June 4–5, 1979. Tokyo: University of Tokyo Press, 1980, pp 85–102.
54. Kahn CR: Methods Membrane Res 3:81, 1975.
55. Carpenter G, Cohen S: Annu Rev Biochem 48:193, 1979.
56. Kahn CR: J Cell Biol 70:261, 1976.
57. Hintz RL, Clemmons DR, Underwood LE, Van Wyk JJ: Proc Natl Acad Sci USA 69:2351, 1972.
58. Marshall RN, Underwood LE, Voina SJ, Foushee DB, Van Wyk JJ: J Clin Endocrinol Metab 39: 283, 1974.
59. Takano K, Hall K, Fryklund L, Holmgren A, Sievertsson H, Uthne K: Acta Endocrinol 80:14, 1975.
60. Zapf J, Mäder M, Waldvogel M, Schalch DS, Froesch ER: Arch Biochem Biophys 168:630, 1975.
61. Megyesi K, Kahn CR, Roth J, Froesch ER, Humbel RE, Zapf J, Neville DM Jr.: Biochem Biophys Res Commun 52:307, 1974.
62. Megyesi K, Kahn CR, Roth J, Neville DM, Nissley SP, Humbel RE, Froesch ER: J Biol Chem 250: 8990, 1975.
63. Megyesi K, Kahn CR, Roth J, Gorden P: J Clin Endocrinol Metab 38:931, 1974.
64. Rechler MM, Podskalny JM, Jarrett DB, Nissley SP, Moses AC, Flier JS, Kahn CR: 59th Annual Meeting of The Endocrine Society, abstr 159, 1977.
65. Flier JS, Kahn CR, Roth J, Bar RS: Science 190:63, 1975.
66. Kahn CR, Flier JS, Bar RS, Archer JA, Gorden P, Martin MM, Roth J: N Engl J Med 294:739, 1976.
67. Flier JS, Kahn CR, Jarret DB, Roth J: J Clin Invest 58:1442, 1976.
68. Rechler MM, Podskalny JM: Diabetes 25:250, 1976.
69. Goldstein JL, Brown MS: Annu Rev Biochem 46:897, 1977.
70. Schroer JA: In Fellows, RE (ed): "Monoclonal Antibodies in Endocrine Research." New York: Raven Press (in press).
71. Nissley SP, Passamani J, Short P: J Cell Physiol 89:393, 1976.
72. Antoniades HN, Scher CD, Stiles CD: Proc Natl Acad Sci USA 76:1809, 1979.
73. Heldin C-H, Westermark B, Wasteson A: Proc Natl Acad Sci USA 76:3722, 1979.
74. Gospodarowicz D: J Biol Chem 250:2515, 1975.
75. Pledger WJ, Stiles CD, Antoniades HN, Scher CD: Proc Natl Acad Sci USA 75:2839, 1978.
76. Stiles CD, Capone GT, Scher CD, Antoniades HN, Van Wyk JJ, Pledger WJ: Proc Natl Acad Sci USA 76:1279, 1979.
77. Gospodarowicz D: J Supramol Struct Supplement 4:125, 1980 (abstr 312).
78. Zapf J, Schoenle E, Froesch ER: Eur J Biochem 87:285, 1978.

79. Rechler MM, Podskalny JM, Goldfine ID, Wells CA: J Clin Endocrinol Metab 39:512, 1974.
80. King GL, Kahn CR, Rechler MM, Nissley SP: J Clin Invest 66:130, 1980.
81. King GL, Kahn CR: Diabetes 29(Suppl 2):14A, 1980 (abstr 53).
82. Horuk R, Goodwin P, O'Connor K, Neville RWJ, Lazarus NR, Stone D: Nature 279:439, 1979.
83. Smith GL, Temin HM: J Cell Physiol 84:181, 1974.
84. Koontz J: J Supramol Struct Suppl 4:171, 1980 (abstr 448).
85. Baldwin D Jr, Terris S, Steiner DF: Diabetes 28:392, 1979 (abstr 191).
86. Kahn CR, Baird K, Flier JS, Jarrett DB: J Clin Invest 60:1094, 1977.
87. Heaton JH, Schilling EE, Gelehrter TD, Rechler MM, Spencer CJ, Nissley SP: Biochim Biophys Acta 632:192, 1980.
88. Kahn R, Murray M, Pawelek J: J Cell Physiol 103:109, 1980.
89. Florini JR, Nicholson ML, Dulak NC: Endocrinology 101:32, 1977.
90. Bottenstein J, Hayashi I, Hutchings S, Masui H, Mather J, McClure DB, Ohasa S, Rizzino A, Sato G, Serrero G, Wolfe R, Wu R: In Jakoby WB, Paston IH (eds): "Methods in Enzymology." New York: Academic Press, 1979, pp 94–109.
91. Oelz O, Jakob A, Froesch ER: Eur J Clin Invest 1:48, 1970.
92. Schoenle E, Zapf J, Froesch ER: FEBS Lett 67:175, 1976.
93. King GL, Kahn CR, Rechler MM: In Program, 62nd Annual Meeting, The Endocrine Society, Washington, D.C., June, 1980.
94. Kahn CR, Baird KL, Jarrett DB, Flier JS: Proc Natl Acad Sci USA 75:4209, 1978.
95. Schilling EE, Rechler MM, Grunfeld C, Rosenberg AM: Proc Natl Acad Sci USA 76:5877, 1979.
96. Howard BV, Mott DM, Fields RM, Bennett PH: J Cell Physiol 101:129, 1979.
97. Knight AB, Schilling EE, Romanus JA, Rechler MM: Diabetes 29 (Suppl 2): 5A, 1980 (abstr 20)
98. Meuli C, Froesch ER: Arch Biochem Biophys 177:31, 1976.
99. Meuli C, Froesch ER: Biochem Biophys Res Commun 75:689, 1977.
100. Poggi C, Le Marchand-Brustel Y, Zapf J, Froesch ER, Freychet P: Endocrinology 105:723, 1979.
100a. Foley TP, Stevens RL, Nissley SP, Hascall VC, King GL: 62nd Annual Meeting of the Endocrine Society, abstract No. 784, June, 1980.
101. Christensen HN: Fed Proc 32:19, 1973.
102. Le Cam A, Freychet P: Diabetologia 15:117, 1978.
103. Hollenberg MD and Cuatrecasas P: J Biol Chem 250:3845, 1975.
104. Hollenberg MD: Life Sci 18:521, 1976.
105. Martin MS and Pohl SL: J Biol Chem 254:9976, 1979.
106. Hollenberg MD and Fryklund L: Life Sci 21:943, 1977.
107. Wheeler FB, Santora II AC, Elsas II LJ: Endocrinology 107:195, 1980.
107a. Knight AB, Rechler MM, Romanus JA, Van Obberghen-Shilling EE, Nissley SP; Proc Natl Acad Sci USA (in press).
108. Goldfine ID, Gardner JD, Neville DM Jr: J Biol Chem 247:6919, 1972.
109. Merrill GF, Florini JR, Dulak NC: J Cell Physiol 93:173, 1977.
110. Ewton DZ, Florini JR: Endocrinology 106:577, 1980.
111. Kletzien RF, Pariza MW, Becker JE, Potter VR, Butcher FR: J Biol Chem 251:3014, 1976.
112. Gelehrter TD, Tomkins GM: Proc Natl Acad Sci USA 66:391, 1970.
113. Spencer CJ, Heaton JH, Gelehrter TD, Richardson KI, Garwin JL: J Biol Chem 253:7677, 1978.
114. Salmon WD Jr, Daughaday WH: J Lab Clin Med 49:825, 1957.
115. Phillips LS, Daughaday WH: In James VHT (ed): "Endocrinology: Proceedings of the V International Congress of Endocrinology." Excerpta Medica International Congress Series No. 402, 1977, p 150.
116. Salmon WD, Duvall MR, Thompson EY: Endocrinology 82:493, 1968.
117. McCumbee WD, Lebovitz HE: Endocrinology 106:905, 1980.
118. Kimura JH, Hardingham TE, Hascall VC, Solursh M: J Biol Chem 254:2600, 1979.
119. Stevens RL, Hascall VC, Nissley SP, Rechler MM: Fed Proc 39:2120, 1980.
120. Topper YJ: Recent Prog Horm Res 26:287, 1970.
121. Hori C, Oka T: Proc Natl Acad Sci USA 76:2823, 1979.
122. Anderson WB, Wilson J, Rechler MM, Nissley SP: Exp Cell Res 120:47, 1979.
123. El-Etr M, Schorderet-Slatkine S, Baulieu EE: Science 205:1397, 1979.
124. Veldhuis JD, Hammond JM: Endocrinol Res Commun 6:299, 1979.
125. Turner RC, Cohen NM: Dev Med Child Neurol 16:371, 1974.
126. Pederson J: "The Pregnant Diabetic and Her Newborn," Ed 2. Baltimore: Williams and Wilkins

CCDD:A:443

Co, 1977.
127. Dodge JA, Laurence KM: Arch Dis Child 52:411, 1977.
128. Hill DE: Semin Perinatol 2:319, 1978.
129. Lemons JA, Ridenour R, Orsini EN: Pediatrics 64:255, 1979.
130. Rosenberg AM, Haworth JC, DeGroot GW, Trevenen CL, Rechler MM: Am J Dis Child 134:170, 1980.
131. Kobayashi M, Olefsky JM, Elders J, Mako ME, Given BD, Schedwie HK, Fiser RH, Hintz RL, Horner JA, Rubenstein AH: Proc Natl Acad Sci USA 75:3469, 1978.
131a. Herzberg VL, Boughter JM, Carlisle SK, Elders MJ, Schedewie HK, Hill DE: Pediatr Res 14:573, 1980 (abstr 886).
132. D'Ercole AJ, Underwood LE, Groelke J, Plet A: J Clin Endocrinol Metab 48:495, 1979.
133. Jacobs RF, Nestrud RM, Beard AG, Fiser RH, Fawcett DD, Hill DE, Morris MD, Elders MJ: Clin Res 26:71A, 1978 (abstr).
134. Rechler MM, Schilling EE, King GS, Fraioli F, Rosenberg AM, Higa OZ, Podskalny JM, Grunfeld C, Nissley SP, Kahn CR: Pepeu G, Kuhar MJ, Enna SJ (eds): "Receptors for Insulin and Insulin-like Growth Factors in Disease." New York: Raven Press, 1980, pp 489–497.
135. Taylor SI, Podskalny JM, Samuels B, Roth J, Brasel DE, Pokora T, Engel RR: Clin Res 28:408A, 1980.
136. Bar RS, Levis WR, Rechler MM, Harrison LC, Siebert C, Podskalny J, Roth J, Muggeo M: N Engl J Med 298:1164, 1978.
137. Muggeo M, Kahn CR, Bar RS, Rechler MM, Flier JS, Roth J: J Clin Endocrinol Metab 49:110, 1979.
138. Van Obberghen-Schilling EE, Romanus JA, Rechler MM: 16th Annual Meeting of the European Association for the Study of Diabetes, September, 1980.

Transforming Growth Factors (TGFs): Properties and Possible Mechanisms of Action

George J. Todaro, Joseph E. De Larco, Charlotte Fryling, Patricia A. Johnson, and Michael B. Sporn

Laboratory of Viral Carcinogenesis (G.J.T., J.E.D.L., C.F., P.A.J.) and Laboratory of Chemoprevention (M.B.S.), National Cancer Institute, National Institutes of Health, Bethesda, Maryland 20205

Transforming growth factors (TGFs) are growth-promoting polypeptides that cause phenotypic transformation and anchorage-independent growth of normal cells. They have been isolated from several human and animal carcinoma and sarcoma cells. One TGF is sarcoma growth factor (SGF) which is released by murine sarcoma virus-transformed cells. The TGFs interact with epidermal growth factor (EGF) cell membrane receptors. TGFs are not detectable in culture fluids from cells which contain high numbers of free EGF cell membrane receptors. SGF acts as a tumor promoter in cell culture systems and its effect on the transformed phenotype is blocked by retinoids (vitamin A and synthetic analogs). The production of TGFs by transformed cells and the responses of normal cells to the addition of TGFs to the culture medium raise the possibility that cells "autostimulate" their own growth by releasing factors that rebind at the cell surface. The term "autocrine secretion" has been proposed for this type of situation where a cell secretes a hormone-like substance for which it has external cell membrane receptors. The autocrine concept may provide a partial explanation for some aspects of tumor cell progression.

Key words: transforming growth factor, sarcoma growth factor, epidermal growth factor, membrane receptor, tumor promoter, retinoid, growth factors, transformation

A growth-promoting transforming polypeptide is characterized by the following properties: It is a strong mitogen which causes loss of density-dependent inhibition of cell growth in monolayer culture; it causes morphologic transformation of normal cells and anchorage-independent growth (a property in cell culture that correlates best with tumorigenicity in vivo) [1, 2]. Polypeptides that cause phenotypic transformation of indicator cells and meet the above criteria for a transforming protein have been isolated from a number of human and animal carcinoma and sarcoma cells. These polypeptides have been termed transforming growth factors (TGFs) [3]. The first TGF to be recognized as such was sarcoma growth factor (SGF) [4].

Received and accepted January 7, 1981.

0275-3723/81/1503-0287$04.50 © 1981 Alan R. Liss, Inc.

It was observed that murine sarcoma virus (MSV)-transformed cells are characterized by a loss of measurable cell surface receptors for the growth-stimulating polypeptide epidermal growth factor (EGF) [5, 6]. The apparent loss of cell surface receptors occurs in both fibroblastic and epithelioid cells transformed by MSV and can be demonstrated with cells derived from various species [6, 7]. The effect is seen with transforming RNA viruses but not with DNA virus transformation nor with most chemical carcinogen-induced transformation. Over the years we have accumulated cells transformed by a variety of agents, including DNA viruses such as simian virus 40 (SV40) and polyoma, RNA viruses such as murine and avian sarcoma viruses, chemical carcinogens, and radiation, as well as cells which have become transformed spontaneously during passage in cell culture. These have been obtained from Swiss/3T3, Balb/3T3, and other mouse and rat cell systems. In collaboration with Stanley Cohen, these transformed cells were tested for their ability to bind ^{125}I-labeled EGF [5]. Of 47 independently isolated, chemically transformed cells, five show a pattern like the MSV-transformed cells, ie, almost complete loss of EGF receptors with normal levels of other receptors maintained. The chemically transformed cells without detectable EGF receptors have not yet been further characterized for the growth factors they may be producing. They represent a minority of chemically transformed cells that, with respect to this phenotype, behave like the MSV-transformed cells. The basis for this finding appears to be the production by the sarcoma virus-transformed cells of a family of growth factors called "sarcoma growth factors" (SGFs) [4]. Sufficient quantities are released into serum-free medium of Moloney MSV-infected mouse 3T3 cells to allow for their partial purification and characterization [4].

The growth factors that are produced by the sarcoma virus-transformed cells are a family of heat- and acid-stable transforming polypeptides. Addition of these SGFs to the culture medium of normal cells results in rapid and reversible changes. They cause normal rat fibroblasts to grow and form large colonies in soft agar (induction of anchorage-independent cell growth). They also have a pronounced morphologic effect on normal fibroblasts, converting them to transformed cells that pile up and are virtually indistinguishable from those genetically transformed by sarcoma viruses (Fig. 1). Thus, these polypeptides have the property of reversibly conferring the transformed phenotype on normal cells in vitro, and, in this sense, can tentatively be considered proximate effectors of the malignant phenotype [4]. The SGFs are specific for murine or feline sarcoma virus-transformed cells in that supernatants from untransformed cells or DNA tumor virus-transformed cells do not contain detectable quantities of these factors [4].

One of these SGFs has been further purified and shown to specifically bind to EGF membrane receptors [8]. The ability to bind to and be eluted from EGF receptors provides an important purification step in the isolation and characterization of EGF-like growth factors. SGF binding to EGF receptors can be completely blocked by mouse salivary gland EGF. The chemical properties of radiolabeled SGF that has been purified using this method give further support to the idea that SGF and EGF are distinctly different molecules. The SGFs have been shown to compete with EGF for available membrane receptors, yet they do not crossreact with antibodies to EGF, and their biological activity is distinct from that of EGF. Cells lacking EGF receptors are unable to respond to the growth-stimulating effects of this partially purified SGF. We concluded, therefore, that SGF released by MSV-transformed cells elicits its biologic effects via specific interaction with EGF membrane receptors.

Polypeptides which are characterized by their ability to confer a transformed phenotype on an untransformed indicator cell have also been isolated directly from tumor cells

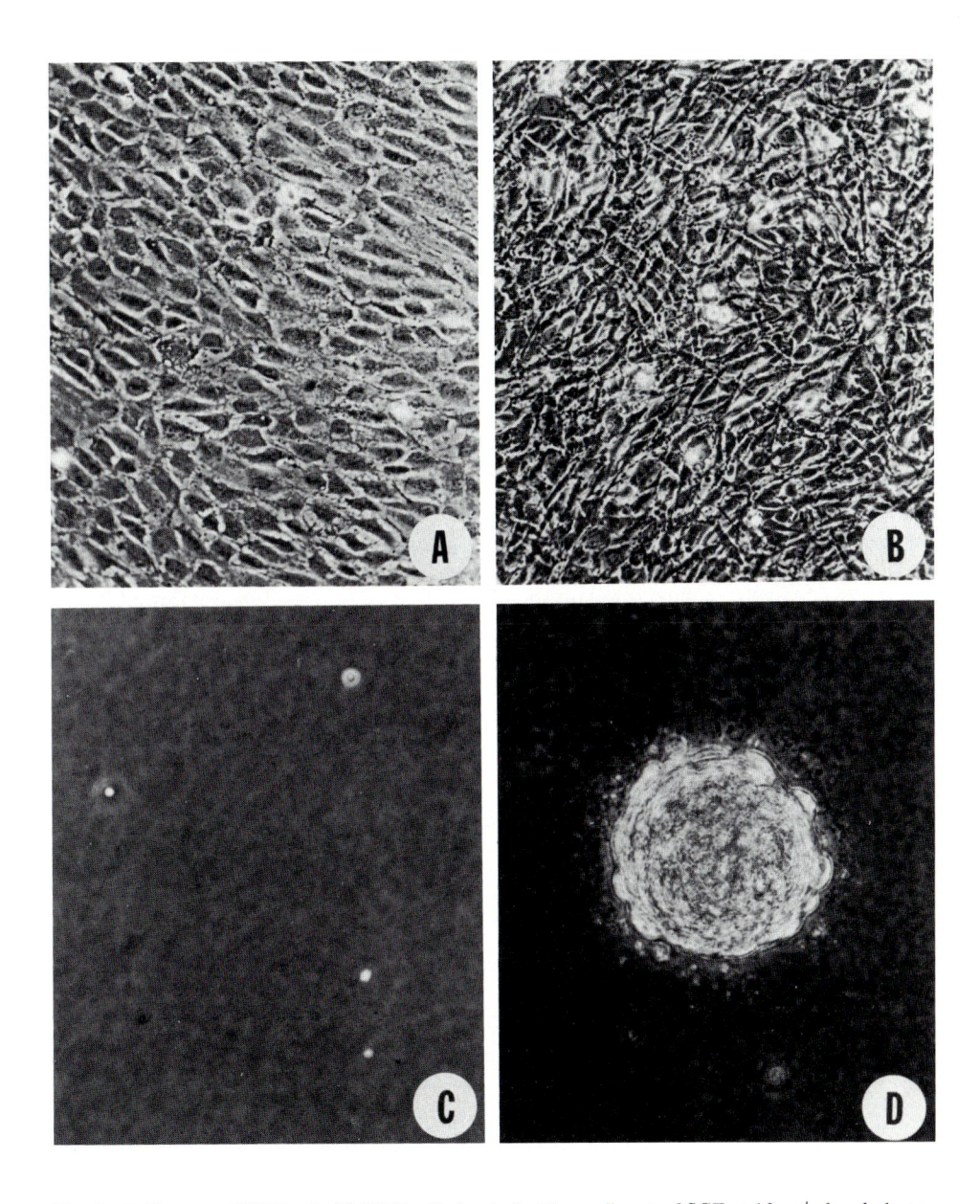

Fig. 1. A) Untreated NRK cells. B) NRK cells treated with an aliquot of SGF at 10 μg/ml and photographed six days later. The cells have grown to considerably higher cell density and display a morphology similar to that of virus-transformed cells. Magnification: A and B, 125×. C) Untreated NRK cells plated in 0.3% soft agar. D) NRK cells plated in 0.3% soft agar, treated with an aliquot of SGF at 10 μg/ml and photographed two weeks after treatment. The untreated cultures show primarily single cells with two or three cell colonies, but none of larger size. In the treated cultures, many colonies contained well over 500 cells. Magnification: C and D, 250×.

growing both in culture and in the animal using an acid-ethanol extraction procedure [9]. The properties of these intracellular polypeptides from both virally and chemically transformed cells are similar to those described for the SGFs isolated from the conditioned medium of sarcoma virus-transformed mouse 3T3 cells, suggesting the definition of a new class of transforming growth factors common to tumor cells of different origin. Thus, the TGFs represent a new class of polypeptides common to cells transformed either by chemicals or by sarcoma viruses and possess biological activity distinct from that of EGF.

Murine sarcoma virus-transformed cells lack available receptors for EGF. We have shown that this altered phenotype is the result of the endogenous production of growth factors by the MSV-transformed cells themselves. There is no evidence that SGF acts as a complete carcinogen itself, producing permanent cell transformation; its properties resemble classical chemical promoters of carcinogenesis, like 12-O-tetradecanoylphorbol-13-acetate (TPA) [10–12], the highly active component of croton oil. While TPA is an exogenous plant derivative acting on an animal or a cell, SGF is an endogenous, virally induced growth promoter.

Retinoids (vitamin A and synthetic analogs) [13] block the action in vivo of exogenous and endogenous promoters, preventing carcinogens from producing new tumors, but do not reverse the growth of many established tumors [13–16]. Retinoids prevent cancer of the lung [13, 17], skin [15], bladder [18], and mammary gland [19] in experimental animals, block cell transformation induced by chemicals [20] and radiation [20, 21] in culture, and reverse the anchorage-independent growth of transformed mouse fibroblasts [22]. If SGF is part of the natural tumor-promoting system and retinoids are part of the natural defense against that system, then one should be able to demonstrate a direct antagonism in cell culture.

We have used a subclone (536-7) of a rat fibroblast cell clone (NRK 49F) [6] that showed pronounced morphologic transformation and anchorage-independent growth when treated with SGF, forming multiple cell layers and crisscrossing each other in an apparently random fashion. Although the effects are all reversible, these treated cells resemble MSV-transformed cells in their phenotype [4]. The cells that were treated with both SGF and retinoids did not have a disordered growth pattern. Retinyl acetate, at 6 ng/ml, almost abolished the growth-stimulatory effect of SGF, as determined by the final cell density reached by the monolayer cultures ten days after the experiment began. The effect of retinoids on SGF-induced morphologic alterations was evident within a few days after treatment.

The retinoid concentrations ($1-2 \times 10^{-8}$ M) neither reversed the phenotype of virally transformed cells, nor blocked cell transformation produced by transforming viruses, such as the Moloney strain of MSV or SV40. Mouse 3T3 cells and rat fibroblasts were tested for susceptibility to transformation by MSV and by SV40. Neither retinyl acetate nor retinoic acid, up to 2×10^{-6} M, could be demonstrated to block either the initiation or the maintenance of virally induced transformation when efficient transforming viruses, like MSV or SV40, were used. In the same experiment, however, the SGF-induced morphologic transformation was inhibited. Retinyl acetate did not inhibit normal cell growth or the cloning efficiency of the rat fibroblast cell clones in petri dishes, but did have a pronounced effect on the final cell ("saturation") density of cells treated with SGF. SGF-induced cell growth was blocked and normal growth properties were essentially retained.

Table I shows that at concentrations well below those that show any evidence of toxicity, retinoids prevent SGF-induced colony formation in soft agar. Colonies that did form were smaller and contained fewer cells than those treated with SGF alone. Retinoic

TABLE I. Effect of SGF and Various Retinoids on the Colony-Forming Ability of Rat Fibroblasts Plated in Soft Agar

Treatment	Colonies/plate		
	Expt. 1	Expt. 2	Expt. 3
Untreated controls	0	0	0
+ Retinyl acetate (1.9×10^{-8} M)	0	0	0
+ Retinoic acid (2.0×10^{-8} M)	NT[a]	0	NT
+ Retinylidene dimedone (1.5×10^{-8} M)	NT	NT	0
+ Retinyl methyl ether (2.0×10^{-8} M)	NT	NT	0
SGF-treated (10 µg/ml)	44.5	39.0	49.5
+ Retinyl acetate (1.9×10^{-8} M)	2.5	1.5	8.0
+ Retinoic acid (2.0×10^{-8} M)	NT	3.2	NT
+ Retinylidene dimedone (1.5×10^{-8} M)	NT	NT	0.5
+ Retinyl methyl ether (2.0×10^{-8} M)	NT	NT	14.5

On day 0, 1×10^5 rat fibroblast cells, clone 536-7, were treated in monolayer cultures using DMEM with 1% fetal calf serum. On day 2, they were seeded at 1×10^4 cells per plate in 0.3% soft agar containing the additions shown as previously described [8]. All cells not treated with SGF (whether treated with retinoid or not) remained as single cells with occasional (<10%) small colonies of 2–4 cells. Colonies with greater than 20 cells after two weeks in agar were scored as positive.
[a]NT = Not tested.

acid, retinyl acetate, retinyl methyl ether, and retinylidene dimedone were all effective. Cells not treated with SGF and plated in agar remained as single cells with occasional 2–4 cell colonies. The clone used (536-7) has a spontaneous transformation rate, as determined by agar colony growth, of less than 1 in 10^6 cells plated. The preparation of SGF used, at 10 µg/ml, produced 40–50 large colonies per 10^4 treated cells and many smaller colonies with between four and 20 cells as well. The inhibiting effect of the retinoids was less evident or absent when more active SGF preparations, or higher concentrations of SGF were used. Retinylidene dimedone, of the compounds tested, was the most efficient inhibitor of SGF-induced phenotypic transformation (Table I, experiment 3). As a control against selective toxicity of retinoids to transformed cells, as compared to normal cells, MSV-transformed mouse and rat cells that grow well in agar without adding SGF were plated in soft agar in the presence of retinyl acetate at 2×10^{-6} M. No reduction in colony-forming ability was seen.

These experiments establish that, in the system used here, retinoids block the transforming effect of the polypeptide hormone, SGF. Only one concentration of each retinoid, well below the level that shows any cell toxicity, was used, and both the growth promoter and the antagonists were added to the cells at the same time. This system is now available for further studies where the concentrations, duration of treatment, and the nature of the interaction between each of the three components (promoter, antagonist, and responding cell clone) can be varied in a systematic manner.

CCDD:A:449

The general transformation model we are proposing has these features: Viruses and chemical carcinogens act by inducing cells to produce normally repressed or inactive growth-promoting factors. These factors, which may be endogenous or exogenous to given cells, could be important in embryonic development, but if inappropriately expressed later in life could lead to transformation. Tumor viruses either provide transforming genes directly or activate cellular genes; chemical carcinogens do only the latter. These growth-promoting and transforming factors may be produced during early embryogenesis and then "switched off." The endogenous viruses, with their capacity to recombine with cellular genes, have the ability to transfer information between cells and presumably within a cell, like bacterial insertion sequences. They may well be vehicles that allow expression of the endogenous growth-promoter structural genes. In this model, the promoters, be they endogenous (SGF) or exogenous (TPA), act as proximal effectors of transformation.

The virogene-oncogene hypothesis [23] points out the possibly erroneous assumption that virally induced tumors would have to arise through external infection by emphasizing that virus-coded or virus-associated genes are already present in several animal species. These genes, rather than the environmentally transmissible agents, are more likely to be involved in the origin of natural cancers. The tumor viruses, although unnatural in that they had often been selected for producing rapid disease, have provided extremely powerful tools to dissect out and understand the molecular mechanisms involved. Genetically transmitted viral genes and transforming genes are now accepted as being part of the normal genetic makeup of many organisms and of being activated by agents such as chemical carcinogens, hormones, and radiation [24, 25]. In parallel with this is the frequently made assertion that chemical carcinogenesis and environmental carcinogenesis, or even industrial carcinogenesis, are almost interchangeable with one another. The finding that SGF, produced by animal cells themselves, is an extremely potent promoter in cell culture systems, suggests that endogenous growth promoters may be significant factors in naturally occurring cancers.

Since it was established that mouse sarcoma virus-transformed cells produced TGFs, we decided to screen human tumor cells for similar endogenous factors related to EGF and SGF. The human tumor cells tested for production of factors analogous to SGF were chosen for study because they had no apparent EGF receptors and readily form colonies in soft agar. Normal embryonal lung fibroblasts, unable to grow in soft agar, and A431 epidermoid carcinoma cells, which have a very high number of EGF receptors and grow poorly in soft agar, were used as controls.

Figure 2 shows the results of experiments comparing the five cultures for their ability to form colonies in soft agar. The cells were grown in monolayer cultures, harvested and seeded at varying densities into medium with 0.3% agar. Colonies were scored at five and 10 days. Colonies with more than 10 cells were counted as positive. The results shown in Figure 2 were obtained at five days; the later reading showed no additional positive cells. The cell line 9812 (a bronchogenic carcinoma) formed progressively growing colonies even when relatively low numbers of cells were seeded. A431 cells only showed colony growth when high cell inocula were used. This suggests that a critical concentration of diffusible factors from these cells is required for anchorage-independent growth.

Cells which are potential producers of factors that stimulate growth in soft agar (eg, human tumor cells) were seeded in one layer of agar at 1×10^6 cells per plate and overlaid with indicator cells (eg, rat fibroblasts) at 1×10^4 cells per plate. The indicator cells formed colonies when certain human tumor cells were seeded in the other layer. A673 (human rhabdomyosarcoma), 9812 and A2058 (human metastatic melanoma) cells elicited the

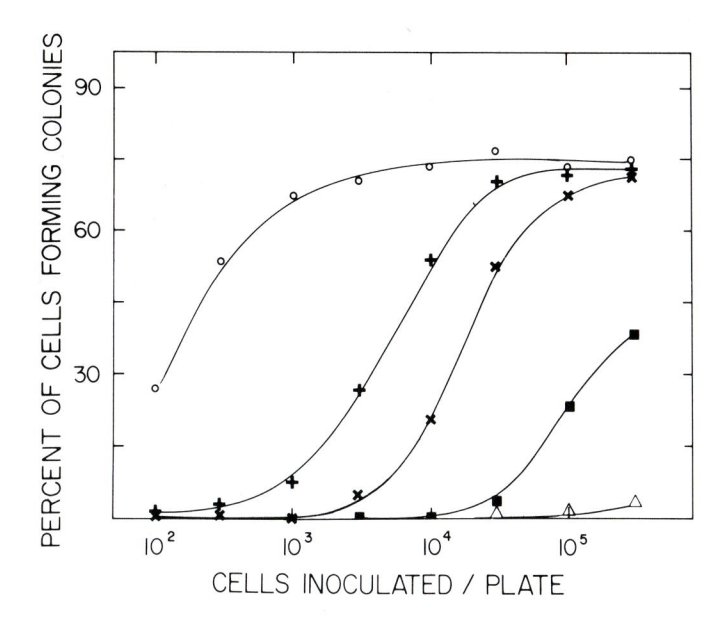

Fig. 2. Soft agar colony formation as a function of cell density. Soft agar assays were set up in 60-mm tissue culture dishes (Falcon #3002) by applying a base layer of 0.5% soft agar (Difco, Noble) and a 2-ml layer of 0.3% agar containing the appropriate cell number. HEL 299 (△); A431 (■); 9812 (○); A673 (+); A2058 (X).

greatest response and released as much agar growth-stimulating activity as did a comparable number of MSV-transformed mouse 3T3 cells.

Figure 3 shows the results of experiments in which serum-free supernates from A673 cells were collected, concentrated, and run over a Bio-Gel P-100 column in 1 M acetic acid. Individual fractions were tested for protein concentration, ability to stimulate cells to form colonies in soft agar, and ability to compete with ^{125}I-labeled EGF [26]. The majority of the protein is in the void volume of the column. A major peak of soft agar growth-stimulating activity was found in the included volume with maximal activity in fraction 54. When the same fractions were tested for competition with ^{125}I-EGF binding, one major peak was again found, with maximal activity also in fraction 54. Aliquots were tested for stimulation of cell division in serum-depleted cultures of mouse 3T3 cells, rat NRK cells, and human skin fibroblasts; in all cases, the major growth-stimulating activity was found in fraction 54. Fractions 51 to 57 were pooled, concentrated by lyophilization, and used for further studies.

The identical procedure was used to test for growth-stimulating factors and EGF-competing peptides from the supernates of four other human cell cultures. Figure 4 shows that the two highly transformed tumor cell lines, 9812 and A2058, release a growth-stimulating and EGF-competing activity with an apparent molecular weight of 20,000–23,000 daltons (Fig. 4B,C). A2058 cells release a second factor with an apparent molecular weight of 6,000–7,000 daltons. Figure 4A shows that the supernate from normal human fibroblast cells did not release a detectable growth stimulating activity and had no significant

CCDD:A:451

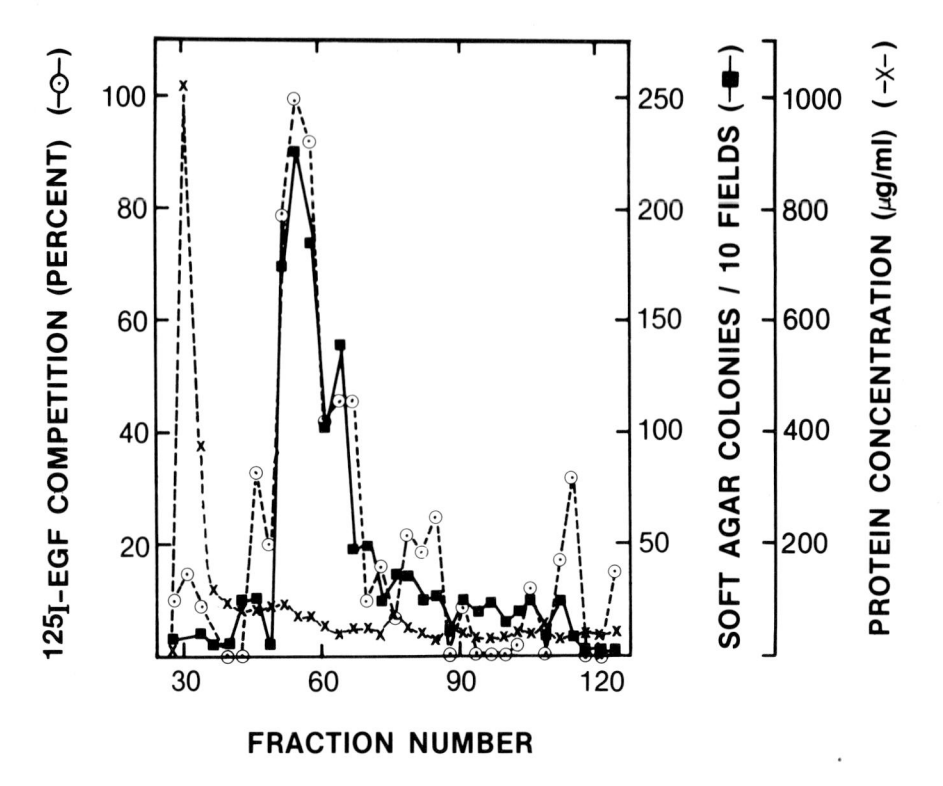

Fig. 3. Biological activity and protein determination of P-100 column fractions of concentrated conditioned media from A673 cells. EGF competition was performed as described. Nonspecific binding, determined by an addition of a 500-fold excess of unlabeled EGF, was approximately 200 counts per minute (cpm). Specific binding was approximately 1,200 cpm. Percent competition was determined after correcting for nonspecific binding. Soft agar assays were performed as described. Protein concentration was determined by the method of Lowry et al [48].

EGF-competing activity. A431 cells showed a smaller peak of growth-stimulating activity with an apparent molecular weight of 21,000 daltons; no EGF-competing activity was found.

Figure 5A shows a dose-response curve measuring soft agar growth as a function of protein concentration. The pooled, peak fractions from A673 cells are compared with those from normal human fibroblasts. There was a 50- to 100-fold difference in soft agar growth-stimulating activity.

The relative sensitivities of three different assays for growth-stimulating activity are compared in Figure 5B. The data are presented as the percentage of the maximal response. Induction of DNA synthesis as tested with serum-depleted rat fibroblast monolayer cultures was slightly more sensitive than the soft agar growth assay; EGF-competing ability was the least sensitive. The latter two assays were used in further studies, since they have greater specificity. Each of the TGF activities was destroyed by trypsin or dithiothreitol but was stable at 100°C for two minutes and to repeated lyophilization from 1 M acetic acid.

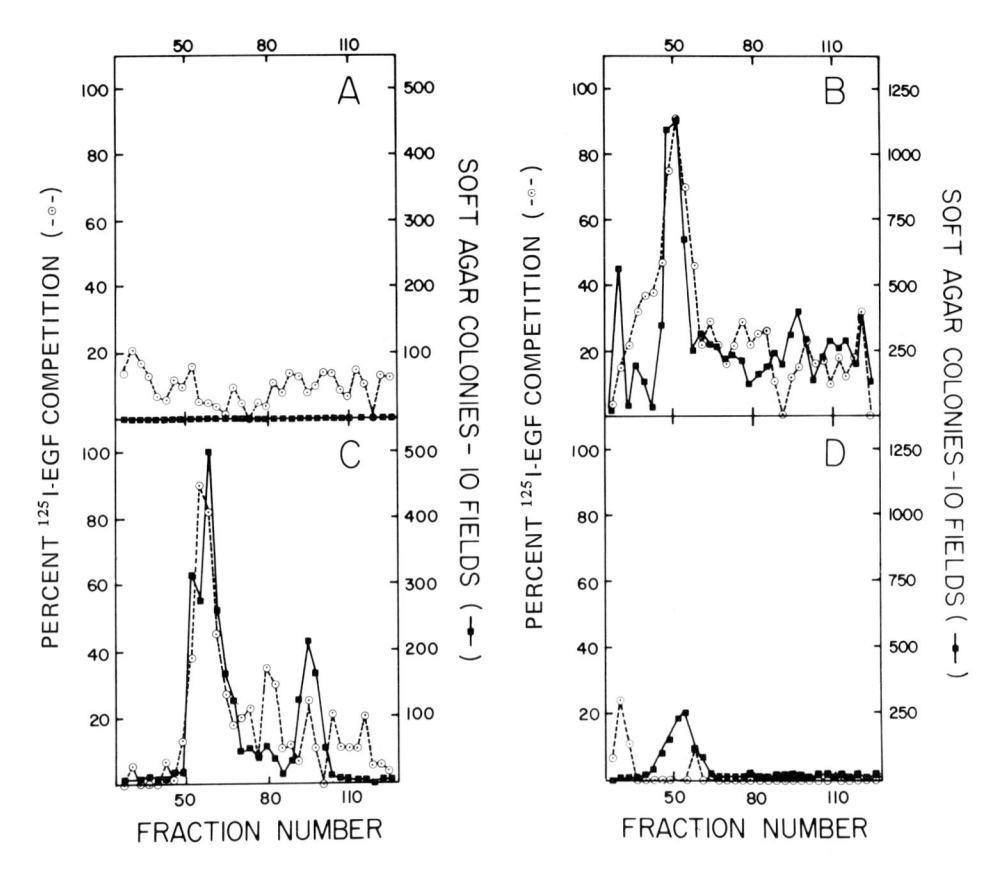

Fig. 4. Biological activity in P-100 column fractions of serum-free conditioned media from four human cell lines in culture. EGF competitions and soft agar assays were performed as described. A) HEL 299; B) 9812; C) A2058; D) A431.

Table II shows that the growth-stimulatory factor(s) released by the human tumor cells induce anchorage-independent growth of normal human fibroblasts. Two cell strains were tested; passage eight of HEL 299 (a human embryonic lung cell line) and the fourteenth passage of HsF (a skin strain from a normal human adult). A673 cells were tested at 10 μg/ml and 1 μg/ml. 1×10^4 cells were seeded per plate and 1,000 single cells were followed for two weeks. Those that grew to colonies containing 10 cells were scored as positive. The percentages of HEL 299 and HsF single cells that gave rise to colonies were 4.2% and 3.1%, respectively, using 10 μg/ml of P-100 purified TGF. In contrast, 23.6% of the rat fibroblast cells showed a pronounced response even at 1 μg/ml. TGF also induced soft agar growth of a mouse epithelial cell line MMC-1 [27] (data not shown).

In order to test whether human tumor cell lines could also respond to TGFs, cells such as A431, that untreated could not form colonies in agar unless inoculated at high density, were used. Carcinoma cell growth in agar also depends on "conditioning" factors, such as TGFs, which partially replace the requirement for high cell density. The results

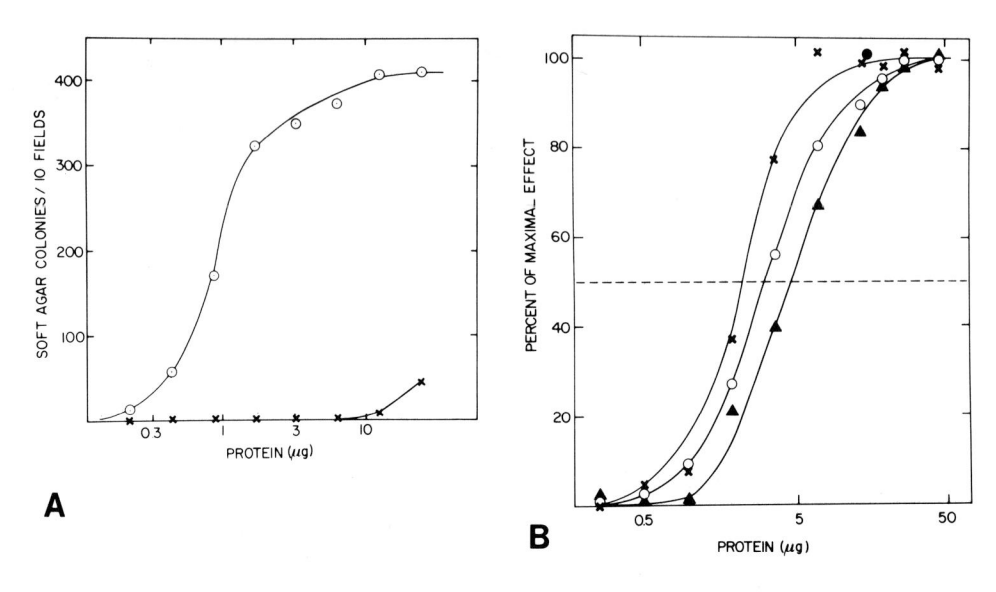

Fig. 5. A) Soft agar colony formation as a function of protein concentration. Bio-Gel P-100 column fractions from the 20,000–23,000 dalton region were pooled and lyophilized. Aliquots in 0.1 M acetic acid were added with the cells in the soft agar overlay. A673 (○); HEL 299 (×). B) Plot of the percent of the maximal effect as a function of protein. ^{3}H-Thymidine incorporation, EGF competition, and soft agar growth assays were performed as described. Maximal response was seen at 25–50 μg of protein; 73,500 cpm and 1,800 cpm, respectively, were incorporated for the ^{3}H-thymidine assays and control plates; 430 colonies per 10 fields where the control plates had none for the soft agar growth assay; 96% inhibition of ^{125}I-EGF binding.

TABLE II. Stimulation of Growth in Agar of Human Diploid Fibroblasts and Human Tumor Cells by TGF

Cell	Type	Control	+ TGF (10 μg/ml)	+ TGF (1 μg/ml)
HEL299	Embryonic lung fibroblast	1	42	3
HsF	Adult skin fibroblast	2	31	2
A431	Epidermoid carcinoma	3	31	8
TE85	Osteosarcoma	1	75	14
NRK (clone 49F)	Rat kidney fibroblasts	0	236	37

Colonies >10 cells/1,000 cells

were more striking when the human osteosarcoma line TE85, which can be further transformed by MSV and certain chemical carcinogens [28], was used as an indicator cell. These results demonstrate that normal and tumor cells respond to TGFs in the same manner as rat fibroblasts. The results, then, are not dependent on an unusual property of a particular indicator cell.

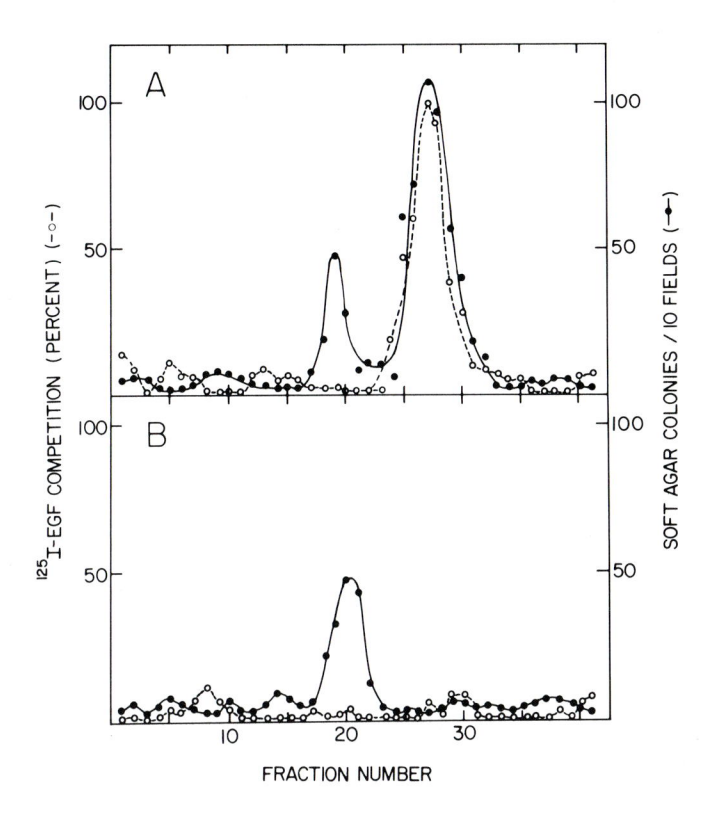

Fig. 6. Chromatography of biological activities in the peak region of Bio-Gel P-100 columns rechromatographed on a carboxymethyl cellulose column. A) A673; B) A431.

The active fractions from P-100 columns of A673 and A431 cells were pooled, concentrated, and applied to carboxymethyl cellulose columns. Two peaks of agar growth-stimulating activity were obtained from A673 cells; only the major activity was associated with the peak of EGF-competing activity. Dose-response curves from each peak show an activity detectable when concentrations of 10 to 20 ng/ml are added to soft agar. The comparable fraction from supernates of cultures of the normal human fibroblast showed no activity. Fractions derived from A431 cells showed (Fig. 6) only the less active, earlier eluting peak which is not associated with EGF-competing activity. We conclude that A431 cells which grow poorly in agar and have a high level of EGF receptors produce a factor capable of stimulating anchorage-independent growth of cells through a mechanism independent of the EGF receptor system. The highly transformed A673 cells, however, make at least two different factors. One interacts with the EGF receptor system and accounts for over 90% of the total activity in the fraction. The other is independent of the EGF receptor system and may be analogous to the factor produced by the A431 cells.

These results demonstrate that human tumor cells produce a growth factor(s) capable of inducing transformation in normal indicator cells. It has many properties in common with the factor from mouse and rat sarcoma virus-transformed cells. The major activity, although considerably larger than SGF, is closely associated with EGF-competing activity. We have found that a chemically transformed mouse 3T3 cell line produces growth-stimulating factor(s) active in the soft agar growth assay (unpublished experiments). Production of these factors then, is not restricted to RNA tumor virus-transformed cells, sarcoma cells or rodent cells but, rather, may be a more general expression of the transformed phenotype. In assays comparing growth stimulation of mouse, rat, and human fibroblasts in monolayer cultures there is no evidence for species specificity of the factors produced by human cells. Conclusions as to whether the carcinoma, sarcoma, and melanoma cells are producing an identical factor(s) await further chemical purification. The present experiments show that anchorage-independent growth of tumor and normal cells is stimulated by these growth factors. Their production by transformed cells and the responses of their normal counterparts raise the possibility that cells "auto-stimulate" their growth by releasing factors that rebind at the cell surface [29]. Experiments demonstrating that growth in soft agar of tumor cells depends on the number of cells seeded per unit area argue that diffusible substances released by cells stimulate neighboring cells. Those cells that grow best in soft agar are the most efficient producers of transforming peptides. Additional cell lines will have to be tested under different conditions before conclusions can be drawn as to the significance of this association.

Roberts et al [9] described a procedure for purifying TGFs. The peptides are stable in acidic 70% alcohol. Intracellular growth factors have been extracted from cultured MSV-transformed mouse cells and from tumor cells in athymic mice. The major peptide with soft agar growth-stimulating activity has an apparent molecular weight of 6,700 daltons. The peak of EGF-competing activity is in the same fraction. A transplantable, transitional cell, mouse bladder carcinoma had agar growth-stimulating activity for rat fibroblasts. Ozanne et al [30] described a transforming factor from Kirsten sarcoma virus-transformed rat fibroblasts with properties like SGF and TGFs and report a similar activity in a spontaneously transformed rat cell line. The effect of the transforming factor on morphologic transformation can be blocked by actinomycin D early after treatment, suggesting that new RNA is produced prior to the change in phenotype of the indicator cells. Inhibitors of protein synthesis also produce a rapid reversion in the phenotype of the treated cells [30].

If release of the factor and rebinding to EGF receptors is essential for growth stimulation, tumor cell growth could be interrupted by exogenous agents, perhaps analogues that interact with the receptors but do not confer the ability to proliferate under anchorage-independent conditions [3]. Anchorage-independent growth is a cell culture property closely associated with the transformed state in vivo [31, 32]. These peptides, then, are potent proximal effectors of cell transformation. Their continued production appears to play a role in maintaining the transformed phenotype. This can be directly demonstrated in temperature-sensitive mutant transformants of rodent cells [30, 33], but has not yet been shown for factors produced by human tumor cells. The approach described here offers a sensitive assay for growth-stimulatory factors associated with maintaining the transformed state. Purification of such factors may lead to the development of specific immunologic assays for their production by tumor cells and their presence in body fluids. The factors may be analogous to peptide growth factors expressed early in normal embryonic development [33]. This is supported by experiments by Nexo et al [34]. In the mouse embryo (days 11 to 18) there is 5–10 times more EGF-like material than mouse EGF. Why the factors produced by transformed cells are so potent in stimulating anchorage-independent growth while EGF is not effective is unclear but suggests the possibility that there may be more "transforming" variants of the normally expressed growth factors produced in adult life.

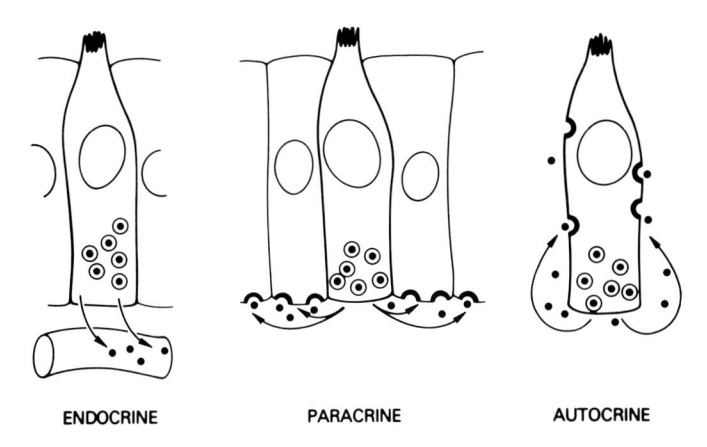

Fig. 7. Diagrammatic representation of autocrine, paracrine, and endocrine secretion (adapted from Dockray [21]). Regulatory chemical messengers are shown in latent form within the cell. The thickened, semicircular regions of the cell membrane represent receptor sites.

We suggest that these factors, like SGFs, are EGF-related peptides as insulin and somatomedins are related [35] and appear to have evolved from common ancestral proteins [36]. Further purification of these and other growth factors from human tumor cells is needed to define their relationship to other biologically active peptides that cells produce. We are also testing the possibility that certain tumor cells may also produce factors related to the phorbol ester family of growth promoters.

The growth of normal cells is largely controlled by the interplay between several polypeptide hormones and hormone-like growth factors that are present in tissue fluids [37]. Many new polypeptide growth factors have recently been identified in blood, serum, tissue fluids, and cellular extracts [38, 39]. Malignant cells, however, are not subject to all the same growth controls as are normal cells. In general, malignant cells require less of these exogenous growth factors than do their normal counterparts for optimal growth and multiplication, and it has been suggested that "transformed or malignant cells escape from normal growth controls by requiring less of [such] hormones or growth factors" [37]. For example, chick fibroblasts transformed by Rous sarcoma virus require less of an insulin-like growth factor for cell multiplication than do normal chick fibroblasts [40], and murine 3T3 cells transformed by SV40 virus require much less serum for multiplication and growth than do their nontransformed, contact-inhibited counterparts [41].

Furthermore, to account for the previously mentioned "lesser requirements of transformed cells for exogenous growth factors" [40, 41], one might suggest two additional properties: The transforming polypeptide should be produced by the putative transformed cell itself [29], and the putative transformed cell should have its own functional cellular receptors for this polypeptide, allowing phenotypic expression of the peptide by the same cell that produced it. The term "autocrine secretion" has been proposed [42] for this type of self-stimulation, whereby a cell secretes a hormone-like substance for which the cell itself has functional external receptors (Fig. 7). With this model of autocrine secretion, the classic "lesser requirement of malignant cells for exogenous growth factors" can be simply

CCDD:A:457

explained: The endogenous production of growth-promoting polypeptides by the transformed cells lessens its own requirement for an exogenous supply of similar growth factors [29, 43].

The autocrine concept that we have outlined provides a simple conceptual model for certain aspects of malignant transformation, in suggesting that one of the ways in which cells become transformed is by endogenous production of growth factors for which they have their own receptors and to which they are capable of responding [29]. This internal production of growth-promoting polypeptides would serve as a constant stimulus for continued cell division, thereby releasing the peptide-producing cells from some of their normal exogenous physiologic controls. This molecular and cellular concept of malignant transformation may be placed in the broader context of developments in both neuroendocrinology and gastrointestinal endocrinology, in which the concept of "peptide humoral regulation" [44] is assuming increasing importance. As Grossman has noted, "We are coming to recognize that the substances that had been called gastrointestinal hormones are members of a broader group of regulatory chemical messengers that are produced by neural, endocrine, and paracrine cells in many parts of the body" [45]. Increasing attention is now being given to the study of paracrine control mechanisms, which involve the local diffusion of a peptide or other regulatory molecule to its target through the extracellular space but not via the bloodstream. In primitive organisms, such as coelenterates, that have no circulatory or glandular endocrine system, paracrine secretion is a principal form of humoral regulation of cells [46]. This type of paracrine humoral regulation of growth and differentition must also be important in the very early premammalian vertebrate embryo, which also has no circulatory or glandular endocrine system to sustain it.

The idea that malignant cells have some relation to early embryonic cells is an old one. Study of primitive mechanisms of "peptide humoral regulation" has already been shown to be of major importance for understanding comparative aspects of neuroendocrinology and gastrointestinal endocrinology [44]. On a purely deductive basis, autocrine secretion should be viewed as an even more primitive use of "regulatory chemical messengers" than either endocrine or paracrine secretion. Autocrine mechanisms for self-stimulation would confer obvious selective growth advantages on very early embryonic cells and could help to account for the explosive growth and multiplication of cells that occur during the earliest stages of embryogenesis, when a critical mass of cells that will survive as an organism must be established very rapidly. For example, a functioning circulatory system is established in the developing chick embryo within 48 hours of the start of incubation, long before the development of many aspects of endocrine function. It is also obvious that autocrine mechanisms are potentially very dangerous to the survival of the organism if they are not closely regulated as soon as they are no longer needed. We are suggesting that malignant transformation of cells may result from inappropriate later expression of autocrine growth factors that were required by cells during normal early embryogenesis [3, 29, 47]. The recent isolation and characterization of defined polypeptide transforming growth factors, which appear to function by such autocrine mechanisms, suggests that malignant transformation may be controlled some time in the future by means of specific inhibitors of the action of these peptides.

REFERENCES

1. Kahn P, Shin S-I: J Cell Biol 82:1, 1979.
2. Cifone MA, Fidler IJ: Proc Natl Acad Sci USA 77:1039, 1980.

3. Sporn MB, Newton DL, Roberts AB, De Larco JE, Todaro GJ: In Sartorelli AC, Bertino JR, Lazo JS (eds): "Molecular Actions and Targets for Cancer Chemotherapeutic Agents." New York: Academic (in press).
4. De Larco JE, Todaro GJ: Proc Natl Acad Sci USA 75:4001, 1978.
5. Todaro GJ, De Larco JE, Cohen S: Nature 264:26, 1976.
6. De Larco JE, Todaro GJ: J Cell Physiol 94:335, 1978.
7. De Larco JE, Todaro GJ: Cell 8:365, 1976.
8. De Larco JE, Todaro GJ: Symp Quant Biol 44:643, 1980.
9. Roberts AB, Lamb LC, Newton DL, Sporn MB, De Larco JE, Todaro GJ: Proc Natl Acad Sci USA 77:3494, 1980.
10. Boutwell RK: Crit Revs Toxicol 2:419, 1974.
11. Slaga TJ, Sivak A, Boutwell RK (eds): "Mechanisms of Tumor Promotion and Carcinogenesis." New York: Raven, 1978.
12. Weinstein IB, Wigler M: Nature 270:659, 1977.
13. Sporn MB, Dunlop NM, Newton DL, Smith JM: Fed Proc 35:1332, 1976.
14. Bollag W: Cancer Chemother Rep 55:53, 1971.
15. Bollag W: Eur J Cancer 8:689, 1972.
16. Verma AK, Boutwell RK: Cancer Res 37:2196, 1977.
17. Saffiotti U, Montesano R, Sellakumar AR, Borg SA: Cancer 20:857, 1967.
18. Sporn MB, Squire RA, Brown CC, Smith JM, Wenk ML, Springer S: Science 195:487, 1977.
19. Moon RC, Grubbs CJ, Sporn MB, Goodman DG: Nature 267:620, 1977.
20. Merriman RL, Bertram JS: Cancer Res 39:1661, 1979.
21. Harisiadis L, Miller RC, Hall EJ, Borek C: Nature 274:486, 1978.
22. Dion LD, Blalock JE, Gifford GE: Exp Cell Res 117:15, 1978.
23. Huebner RJ, Todaro GJ: Proc Natl Acad Sci USA 64:1087, 1969.
24. Aaronson SA, Stephenson JR: Biochim Biophys Acta 458:323, 1976.
25. Todaro GJ, Callahan R, Sherr CJ, Benveniste RE, De Larco JE: In Stevens JG, Todaro GJ, Fox CF (eds): "Persistent Viruses." ICN–UCLA Symposia on Molecular Biology. New York: Academic, vol 11, 1978, pp 133–145.
26. De Larco JE, Reynolds R, Carlberg K, Engle C, Todaro GJ: J Biol Chem 255:3685, 1980.
27. Keski-Oja J, De Larco JE, Rapp UR, Todaro GJ: J Cell Physiol 104:41, 1980.
28. Cho HY, Rhim JS: Science 205:691, 1979.
29. Todaro GJ, De Larco JE: Cancer Res 38:4147, 1978.
30. Ozanne B, Fulton J, Kaplan PL: J Cell Physiol 105:163, 1980.
31. Shin S, Freedman VH, Risser R, Pollack R: Proc Natl Acad Sci USA 72:4435, 1975.
32. Montesano R, Drevon C, Kuroki T, Saint Vincent L, Handelman S, Sanford KK, DeFeo D, Weinsten IB: J Natl Cancer Inst 59:1651, 1977.
33. Todaro GJ, De Larco JE: In Jimenez de Asua L, Levi-Montalcini R, Shields R, Iacobelli S (eds): "Control Mechanisms in Animal Cells: Specific Growth Factors." New York: Raven, vol 1, 1980, pp 223–243.
34. Nexo E, Hollenberg MD, Figueroa A, Pratt RM: Proc Natl Acad Sci USA 77:2782, 1980.
35. Rinderknecht E, Humbel RE: J Biol Chem 253:2769, 1978.
36. Niall HD: In Goodman G, Meienhofer J (ed): "Peptides: Proceedings of the Fifth American Peptide Symposium." New York: Halsted, 1977, pp 127–135.
37. Holley RW: Nature 258:487, 1975.
38. Sato G, Ross R (eds): "Cold Spring Harbor Conference on Cell Proliferation." New York: Cold Spring Harbor Press, vol 6, 1979.
39. Gospodarowicz D, Moran JS: Ann Rev Biochem 45:531, 1976.
40. Temin HM: J Cell Physiol 69:377, 1967.
41. Holley RW, Kiernan JA: Proc Natl Acad Sci USA 60:300, 1968.
42. Sporn MB, Todaro GJ: N Engl J Med 303:878, 1980.
43. Pastan I: Adv Metab Dis 8:7, 1975.
44. Zimmerman EG: Fed Proc 38:2286, 1979.
45. Grossman MI: Fed Proc 38:2341, 1979.
46. Dockray GJ: Fed Proc 38:2295, 1979.
47. Todaro GJ, Heubner RJ: Proc Natl Acad Sci USA 69:1009, 1972.
48. Lowry OH, Rosebrough NJ, Farr AL, Randall RJ: J Biol Chem 193:265, 1951.

CCDD:A:459

Normal and Neoplastic Lymphocyte Maturation

I.L. Weissman, M.S. McGrath, E. Pillemer, N. Hollander, R.V. Rouse, L. Jerabek, S.K. Stevens, R.G. Scollay, and E.C. Butcher

Laboratory of Experimental Oncology, Stanford University School of Medicine, Stanford, California 94305

Lymphocytes are cells that are responsible for processes of specific antigen recognition and for those aspects of the immune response that characterize adaptive immunity. In this respect adaptive immunity can be characterized as antigen-induced immune *memory* and *effector* functions as compared to native immunity — the nonspecific phagocytic and humoral protective elements in lower vertebrates. In vertebrates both B and T lymphocytes apparently express self-synthesized receptors that 1) are involved in the recognition of antigens, and 2) mediate the interactions between various important cells in the hematolymphoid system. There are three major subclasses of T lymphocytes — those involved with helper/inducer functions, those involved with suppressor functions, and those involved in direct cytotoxicity of antigenic target cells [1, 2].

IMMUNOGENETICS OF T CELL RECOGNITION AND FUNCTION

A major advance in the understanding of T lymphocyte classes, T lymphocyte function, and T lymphocyte recognition developed through the combination of immunogenetics and cellular immunology by Cantor and Boyse and by Zinkernagel and Doherty [3, 4]. The general case could be stated as follows: Both killer and suppressor T lymphocytes are cells that express high concentrations of the T cell markers Lyt-2,3, whereas inducer lymphocytes express high concentrations of the marker Lyt-1, with no apparent expression of Lyt-2,3. Inducer lymphocytes recognize allogeneic cells expressing MHC I region markers, or recognize foreign antigens in the context of self I region markers, whereas cytotoxic T cells recognize allogeneic cells via MHC K/D markers, or foreign antigens in the context of self K/D surface glycoproteins [3, 4]. Thus a triad of markers and properties seem to go together — cell function, cell surface Lyt expression, and the portion of the MHC region recognized by antigen-reactive T cells. Exceptions to this triad have been sought extensively, and current evidence indicates an absolute correlation between Lyt type and the type of MHC gene product the T cell recognizes (I or K/D), rather than a correlation between function and either of the two markers described above.* This implies that the Lyt molecules

*Swain S. Federation Proceedings 39:13, 3110, 1980.

Received January 16, 1981; accepted January 21, 1981.

0275-3723/81/1503-0303$03.50 © 1981 Alan R. Liss, Inc.

are involved somehow in the recognition process rather than the functional property of the cell that bears them, and that the generality of the triad reveals the most probable associations, presumably via selective processes of antigen-independent and/or antigen-dependent T cell proliferation and survival.

There is a second important principle involved in T lymphocyte recognition of antigen: Most T lymphocytes are G_0 small cells which are only brought into cycle upon appropriate recognition of antigen. The following is a model of antigen-induced T cell blastogenesis and T cell cooperation: Small G_0 Lyt 1 cells, which recognize I region antigens on the surface of the antigen-presenting or stimulating cell, enlarge to become cycling large lymphoblasts, and release lymphokines such as T cell growth factor (TCGF, also called Interleukin 2 [IL2]), a substance which maintains proliferation in several other subclasses of T lymphocytes, including some (if not all) Lyt-2,3 cells [5–7]. Small G_0 Lyt-2 cells recognize H-2-K/D antigens, enlarge, and begin to express surface receptors for TCGF-IL2. Their continued proliferation may be growth-factor-dependent and antigen-independent [5–7]. Thus antigen recognition is necessary for the *initiation* of Lyt-2 cell division, but is not required for *maintenance* of proliferation by these cells. Lyt-2 effector lymphoblasts of the killer cell series next use their *antigen recognition apparatus* to bind to antigenic target cells, which they subsequently lyse [1]. We have demonstrated that anti-Lyt-2 monoclonal antibodies block the two antigen recognition phases of Lyt-2 cell function – the initial blastogenesis (but not TCGF-dependent proliferation) and the recognition phase necessary for cytolysis of antigenic target cells [8]. Similarly, anti-Lyt-1 may affect the recognition of antigen by Lyt-1 cells, although in this case the monoclonal anti-Lyt-1 antibodies augment rather than inhibit antigen-dependent blastogenesis by these cells [9].

DEVELOPMENT OF IMMUNOCOMPETENT SUBCLASSES OF T LYMPHOCYTES OCCURS IN THE THYMUS

Several experiments indicate not only that the thymus is the site of maturation of immunocompetent T cells from immunologically incompetent hematopoietic precursors [2], but also that specific recognition of self-MHC markers utilized by T lymphocytes for corecognition of antigen and antigen-bearing stimulator/target cells is also a property that is developed, or selected for, within the thymus [10, 11]. The precursors of thymocytes reside originally in the yolk sac [12], but are produced throughout life within the bone marrow from a relatively infrequent subset of precursors [13]. Within the thymus, the primitive lymphoblasts reside mainly under the capsule in the thymic outer cortex, and these provide precursors of all Lyt-defined subpopulations of thymocytes and peripheral T cells [14, 15, 16]. The most frequent progeny of these outer cortical lymphoblasts are cortical small T lymphocytes, which are mainly immunologically incompetent, most of which appear to die in situ within 3–5 days after their appearance [17]. A small proportion of immunocompetent virgin T lymphocytes either enter the medulla, or emigrate to peripheral lymphoid organs [15, 17]. About 70% of emigrating cells in the mouse throughout life are Lyt-1$^+$, Lyt-2$^-$,3$^-$, the other 30% or so expressing high levels of Lyt-2,3 and low levels of Lyt-1 [16, 18]. Study of the outer cortical thymocytes suggests that these surface marker-defined lymphocyte subclasses may be separate lineages throughout their thymic existence, from the primitive lymphoblasts in the outer cortex through the small lymphocytes in the deep cortex and the virgin T cells in the medulla and in the periphery [16].

While in the thymus (for their 3–5-day maturation/selection sojourn) thymic lymphocytes come into contact with at least three distinctive types of relatively sessile cells. In the outer cortex a rare subpopulation of thymic epithelial cells may be marked by transcapsular staining with fluorescein isothiocyanate [16, and a personal communication from B. Kyewski]. The cells so marked have been described elsewhere as "nurse" cells because they are in tight association with thymic lymphoblasts when prepared in suspension [19]. These nurse cells express the MHC markers of both K/D and I regions. The relationship of these nurse cells to the dendritic epithelial cells scattered throughout the cortex needs further study. As developing thymic lymphocytes filter down through the thymic cortex (as G_0 small thymocytes), they pass through a dendritic array of membranous processes from *cortical thymic epithelial cells;* these cortical dendritic epithelial cells express high levels of I-region products of the MHC [20]. Figure 1 shows such a field of cells stained with the monoclonal anti I-A^k antibody. Because these cells express, for the most part, only low-to-undetectable levels of allotypic determinants of the H-2K and H-2D regions, they may be unique among all other cells in the body in terms of this peculiar expression of I-region MHC gene products. It has been reported [21] that this class of cells contains populations of cells that produce and secrete thymopoietin, a polypeptide hormone believed to be involved in thymic lymphocyte maturation and/or proliferation. Thymic lymphocytes passing through the interstices of these dendritic processes may collect in large pools traversing the corticomedullary junction, and come into contact in that region with macrophages expressing both I-region and K/D-region MHC antigens [22]. In the medulla are more *spatulate epithelial cells,* also expressing both I-region and K/D-region MHC markers. It should be noted that infrequent dendritic epithelial cells in the thymic cortex also express allotypic determinants detected by anti-J-region antibodies [20]. The emigration site of lymphocytes from the thymus is believed to be in the region of the corticomedullary junction; and a possible graveyard for the thymocytes that die appears to be in the medulla in structures called Hassall's corpuscles. Those thymic lymphocytes destined to die appear to have an aberrant expression of H-2K region markers [24]. We have proposed elsewhere that subclasses of thymic epithelial cells are involved in the maturation and proliferation of, and selection for lymphocytes expressing receptors directed to their expressed MHC markers, and that the frequency of emigrating Lyt-1 vs Lyt 1,2,3 lymphocytes mirrors somewhat the abundance of MHC I vs K/D markers on thymic epithelial cells [20].

NEWLY DEVELOPING T AND B LYMPHOCYTES EXPRESS SURFACE RECEPTORS THAT GOVERN THEIR MIGRATION AND LOCALIZATION IN THE PERIPHERY

Emergent virgin B and T lymphocytes enter the blood stream and quickly migrate to peripheral lymphoid organs in a process involving 1) recognition of and binding to the surface of a specialized subclass of postcapillary endothelial cells, called high endothelial venule cells (HEV); and 2) subsequent migration through the vessel wall into surrounding lymphoid tissue [25–27]. The expression of receptors for these highly specialized endothelial cells is a property of mature lymphocytes that is not shared by their immature bone marrow and thymocyte precursors, or by nonlymphoid blood cells [28–30]. The in vivo process of lymphocyte binding to HEV, and subsequent transvascular migration, is outlined in Figure 2A. Figure 2B demonstrates that this cell-cell interaction can be studied in vitro in a model system, described originally by Stamper and Woodruff [28], in which lymphocytes bind specifically to HEV during incubation on frozen sections of murine lymph nodes or Peyer's patches. By quantitating the binding ability of various lymphocyte

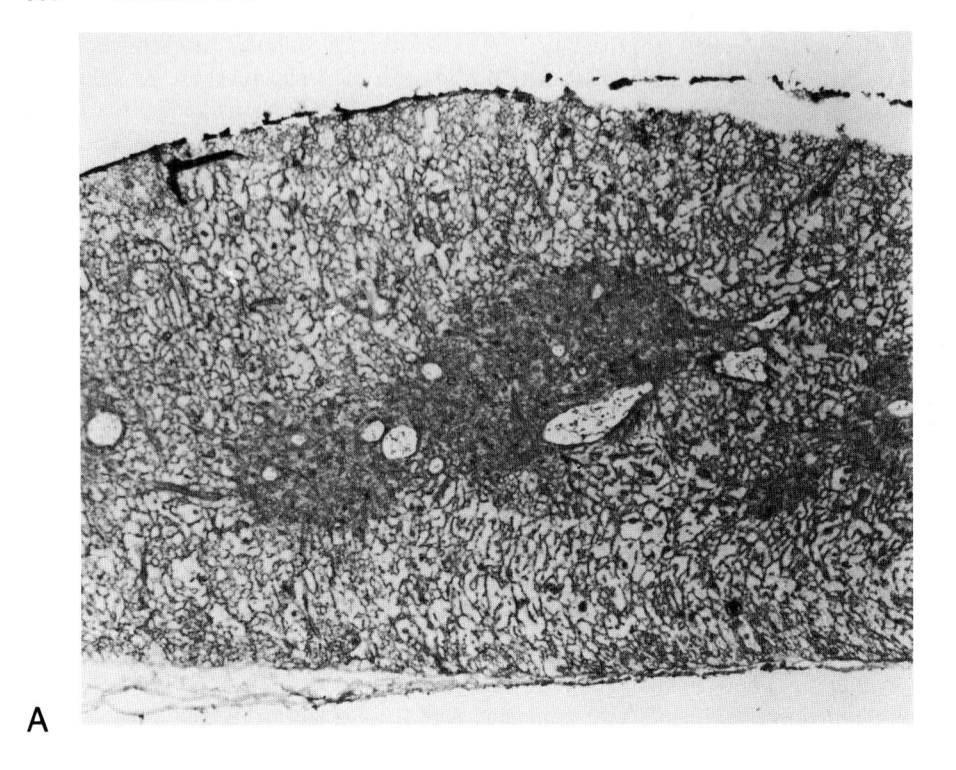

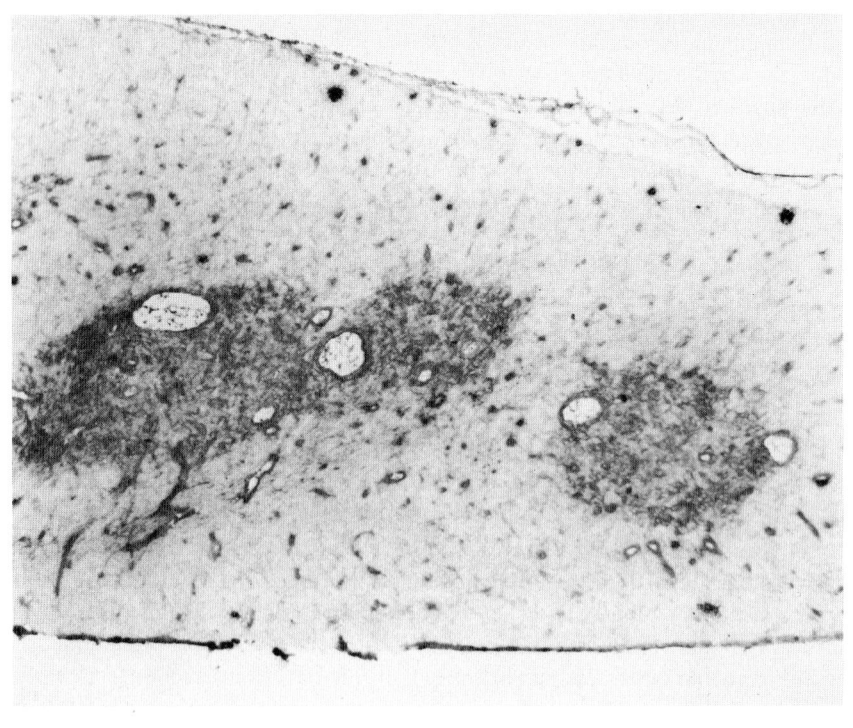

Fig. 1. C57B1/6-H-2^k mouse thymus stained with A) monoclonal anti-I-A^k and B) monoclonal anti-H-2-K^k. Both antibodies produce confluent medullary staining, but only I-A is expressed on dendritic cortical cells.

464:CCDD:A

postcapillary venule

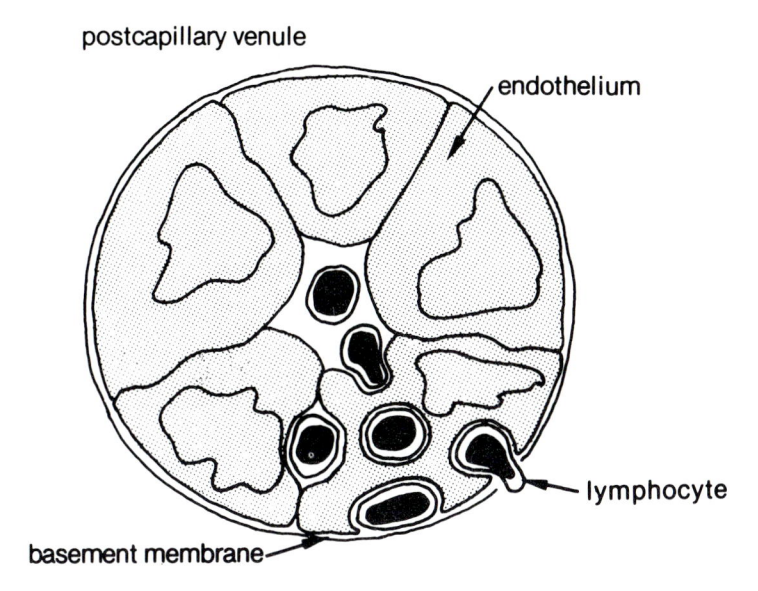

Fig. 2A. Diagrammatic cross section of a lymph node postcapillary venule. Small dark cells are lymphocytes adhering to and migrating across the specialized endothelial walls of these vessels. (Redrawn from J.L. Gowans, Hosp Pract 3(3):34, 1968.

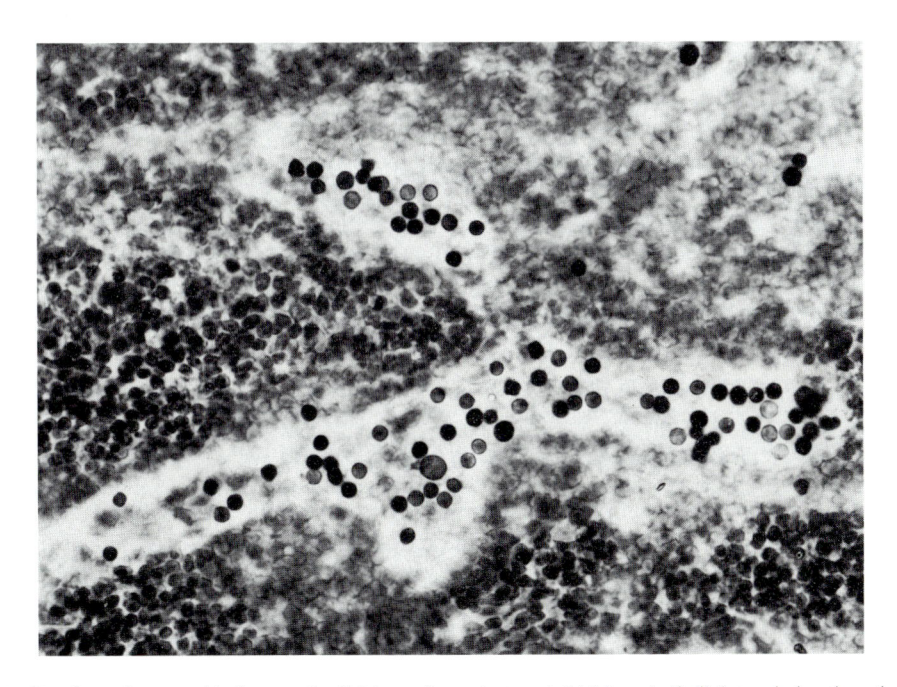

Fig. 2B. Lymphocytes (dark, round cells) bound to a transected high-endothelial venule in a lymph node frozen section after in vitro incubation.

populations in this in vitro system [29, 31], we have gained some insight into the evolution of this cell-cell interaction, and some understanding of its importance in directing lymphocyte traffic. Figure 3 demonstrates that the ability of lymphocytes from various

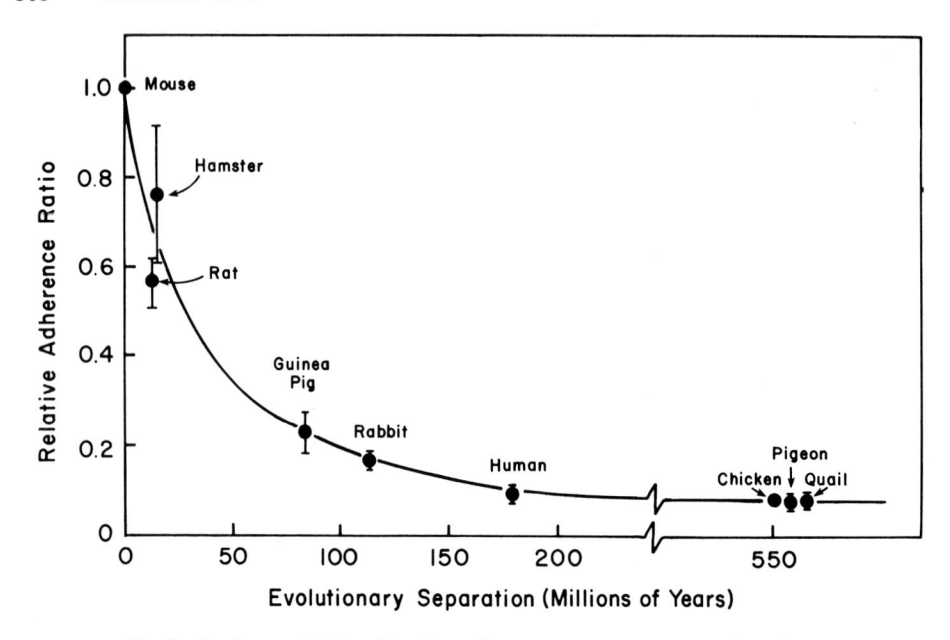

Fig. 3. Species specificity of in vitro adherence to mouse mesenteric node HEV.

vertebrate species to bind to mouse HEV declines exponentially with the evolutionary distance separating the lymphocyte donor from the mouse host. We have interpreted this as evidence for continuous coevolution of the lymphocyte and endothelial cell recognition elements [31, 32]. Not only is there a progressive evolutionary divergence between species, but there are also detectable variations in HEV-binding properties of lymphocytes within each species: We have defined one major difference in the HEV-binding characteristics of BALB/c and C57BL lymphocytes, and have mapped this trait to a single region on chromosome 7 [31]. Whether this genetically defined difference is due to an allotypic variant of the lymphocyte recognition molecule(s) for HEV is currently unknown. In addition to these interspecies and interstrain differences, evolutionary pressures have resulted in the existence of more than one set of complementary lymphocyte and HEV recognition receptors or elements *within* each lineage, apparently to direct lymphocyte migration through particular regions of the body. For instance, selective lymphocyte migration through the gut-associated Peyer's patches or through the gut-independent peripheral lymph nodes is well documented [33–40]. We have demonstrated the existence of distinct recognition determinants on HEV in Peyer's patches and peripheral lymph nodes, and have argued that the selective expression of receptors for these endothelial determinants by various lymphocyte populations is a major factor controlling their circulation through these sites [30–39]. Our experiments have defined two distinct kinds of organ-specific homing by lymphocytes: 1) Normal small lymphocyte populations demonstrate limited organ specificity of migration based on lymphocyte class. For instance, peripheral node high endothelial cells bind T cells in preference to B cells, whereas Peyer's patch HEV cells select about twice as many B cells as T cells. These relative homing preferences of B and T cell populations are observed regardless of their organ of origin. It is interesting that the class-specific binding preferences of peripheral node and Peyer's patch HEV are closely correlated with (and thus may well determine) the fractional representation of B and T

RELATIVE ADHERENCE RATIOS vs MNL

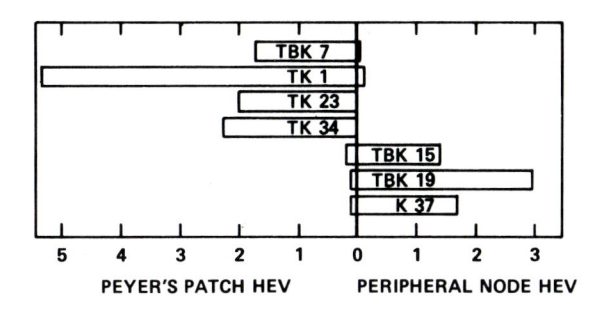

[Bar chart showing relative adherence ratios. Left axis: Peyer's Patch HEV (5, 4, 3, 2, 1, 0). Right axis: Peripheral Node HEV (0, 1, 2, 3). Bars from top to bottom: TBK 7, TK 1, TK 23, TK 34 (extending left toward Peyer's Patch HEV); TBK 15, TBK 19, K 37 (extending right toward Peripheral Node HEV).]

Fig. 4. Organ specificity of in vitro adherence of lymphomas to HEV in peripheral (axillary and brachial) nodes and Peyer's patches. The adherence of normal syngeneic mesenteric node lymphocytes to HEV in either organ is defined as unity.

lymphocyte populations in the two organs (S.K. Stevens, I.L. Weissman, and E.C. Butcher, in preparation). 2) Some lymphocyte subpopulations may demonstrate nearly absolute organ specificity of homing, or of HEV recognition, as exemplified by certain neoplastic T cells and their (presumably clonal) progeny, which bind exclusively or nearly exclusively to either Peyer's patch HEV, or peripheral node HEV (Fig. 4) [31, 41]. These T cell lymphomas are of thymic origin, and apparently are committed to organ-specific homing prior to their exit from the thymus [30]. This observation suggests either a) that normal thymocytes, from which these lymphomas derive, may themselves be precommitted to future expression of a particular HEV receptor, or b) that the process of neoplastic transformation and blastogenesis may nonspecifically induce organ-specific homing receptors, perhaps in imitation of a normal differentiative process associated with antigen-induced blastogenesis in the periphery. Finally, the exciting possibility exists that the metastatic potential of these tumors may be determined in part by their particular lymphoid organ HEV receptors.

RECEPTOR-MEDIATED PROLIFERATION OF NORMAL AND NEOPLASTIC T LYMPHOCYTES

In the first section, we outlined some of our evidence that proliferation of T lymphocytes (and perhaps B lymphocytes) appears to be governed largely by recognition of antigen by antigen-specific cell surface receptors linked to cell surface receptor-mitogen complexes. Several years ago we proposed the hypothesis that lymphomagenesis by oncogenic retroviruses was controlled by misregulation of or uncontrolled stimulation of these cell surface receptor-mitogen complexes [42–44].

Briefly, we proposed that oncogenic retroviruses bind to that subset of normal lymphocytes bearing antigen-specific receptors directed against retroviral envelope moieties. The retroviruses thus bound might infect the lymphocyte leading to cellular production of more retrovirus envelope structures. In turn, this could lead to receptor-mitogen complex stimulation of progeny cells into several (and perhaps endless) rounds of division. Using a sensitive virus-binding assay we showed that, indeed, each T lymphoma induced by a particular retrovirus bears surface receptors specific for the virus that induces it (Fig. 5). At least 20 T and 1 B lymphoma bear receptors specific for the viruses they produce and

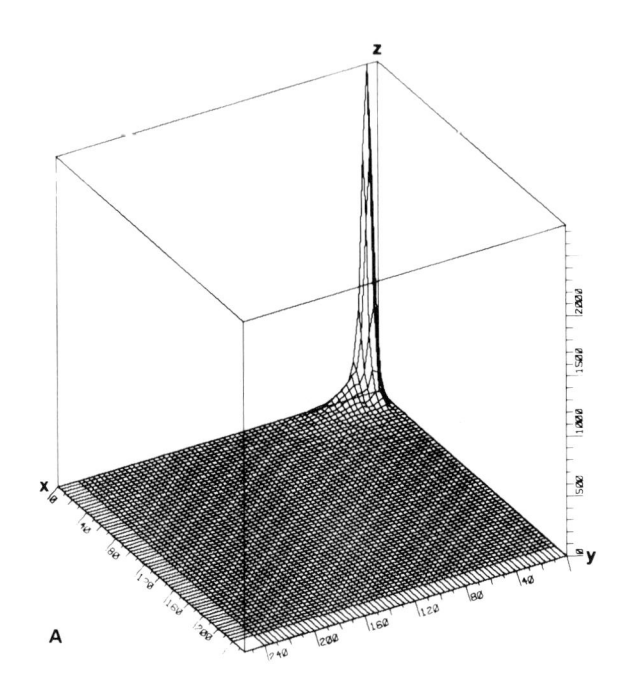

310 Weissman et al

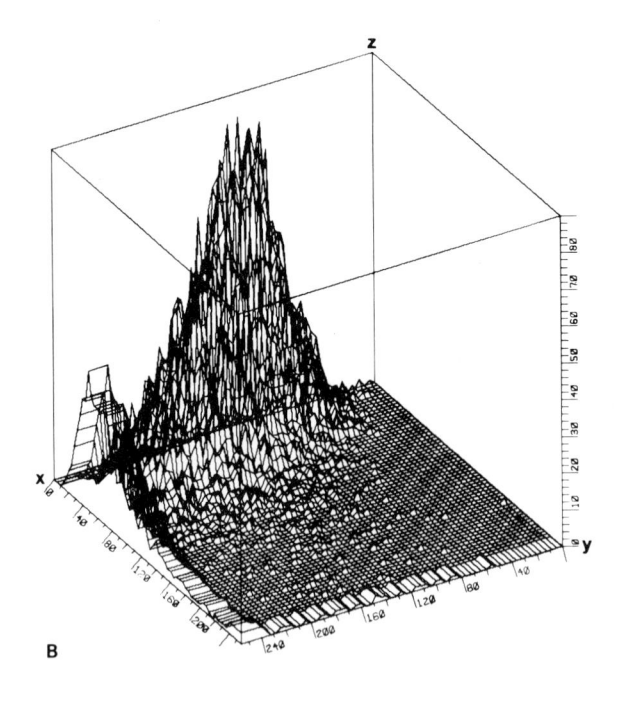

A

B

468:CCDD:A

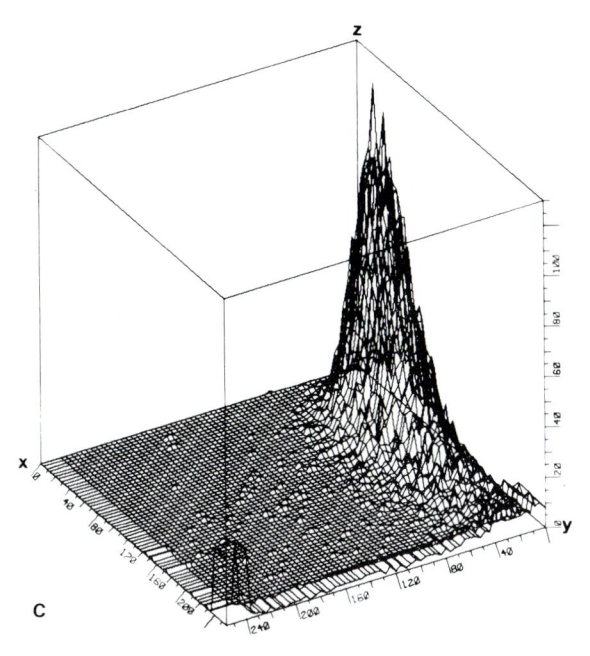

Fig. 5. Two-color FACS analysis: MCF-247 and RadLV/VL$_3$ binding specificites. These three-dimensional perspective plots show rhodaminated MCF-247 and fluoresceinated RadLV/VL$_3$ binding to three different target cell populations. MCF-247-binding increases along the X axis, fluoresceinated RadLV/VL$_3$-binding increases along the Y axis, and the frequency of cells binding a particular level of fluorescence increases along the Z axis. In each case, 10^5 cells were analyzed, and the fluorescein and rhodamine backgrounds were equal to 30 units. A) 10^6 One-week-old AKR/J thymocytes were incubated simultaneously with 0.2 A$_{260}$ unit of fluoresceinated RadLV/VL$_3$ and 0.2 A$_{260}$ unit of rhodaminated MCF-247. Virus-binding assay was performed as previously described [44], and two-color binding was assessed by the FACS. B) 10^6 KKT-1 cells were assayed for simultaneous binding of fluoresceinated RadLV/VL$_3$ and rhodaminated MCF-247 as in (A). C) 10^6 BL/VL$_3$ cells were assayed for simultaneous binding of fluoresceinated RadLV/VL$_3$ and rhodaminated MCF-247 as in (A). Cells with fluorescence greater than 250 units, representing < 5% of all cells analyzed, were collected and combined in the 250-unit computer channel, as represented by the ridges along the distal edges of the three-dimensional plots. All data were plotted by a Calcomp plotter interfaced with a PDP-11 computer using the MAINSAIL computer program designed by W. Moore, Stanford University.

which presumably induced them [45–47]. This receptor-binding assay is extremely sensitive, and with the use of the fluorescence-activated cell sorter (FACS) we have been able to detect and transfer incipient lymphoma cells within the thymus of otherwise apparently healthy hosts; morphologically identical non-virus-binding thymic lymphocytes in the same thymus are not capable of transferring autonomously replicating lymphoma cells [44].

An important postulate of this hypothesis is that interference with receptor-virus interaction should lead to the inhibition of proliferation by these T lymphoma cells. Accordingly, we raised several types of monoclonal antibodies directed against lymphoma cell surface markers, and found four that blocked the binding of virus to lymphoma cell surfaces, whereas several others bound to the lymphoma cell surface but did not block virus binding. Those monoclonal antibodies that block virus binding also block proliferation of T lymphoma cells, either in the first G$_1$ cell cycle phase after addition of antibodies, or in the second round of cell division [48] (Fig. 6). Inhibition of lymphoma cell prolifera-

CCDD:A:469

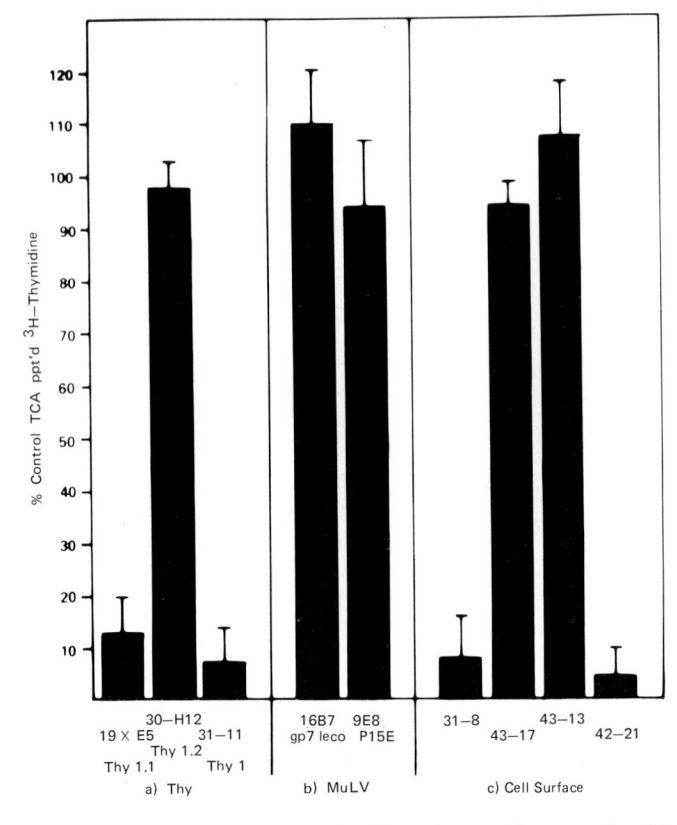

Fig. 6. Lymphoma cell growth inhibition assay. KKT-2 lymphoma cells were pelleted from subconfluent cultures and were resuspended in serum-free growth medium (MEM, GIBCO) for 2 h at 37°C at a density of 10^5 cells/ml. After repelleting, the cells were resuspended in cold tissue culture medium (MEM, 5% FCS) at a density of 2×10^4 cells/ml. Two and one-half milliliters (5×10^4 cells) were placed in the bottom of 25 cm^2 Corning tissue culture flasks and monoclonal antibody was added for a 1:12 final dilution. After growing for 14 h at 37°C, 0.1 ml of medium containing 10 μCi [^{3}H]-thymidine (NEN) was added to each culture for 2 h. Labeled cells were washed, 5% TCA was precipitated, and percent growth inhibition was calculated using cells without antibody as equal to 100% growth, KKT-2 cells were tested for growth inhibition with a) anti-Thy 1 (19 X E5, 30–H12, 31–11), b) anti-MuLV (16B7, 9E8), and c) anti-cell-surface antibodies (31–8, 43–17, 43–13, 42–21). The above data represent five experiments ± standard deviations. Monoclonal antibody 42–21 has subsequently been shown to react with Thy-1 determinants.

tion by these monoclonal antibodies could be prevented by prebinding to the lymphoma cell the oncogenic retrovirus that induced the tumor (Fig. 7). Analysis of the molecules detected by these monoclonal antibodies reveals that they are not the actual virus receptors, but are molecules (such as Thy-1) that are in great abundance on the T lymphoma cell surface. Inhibition of virus binding in these instances is probably due to steric, rather than competitive, inhibition. We proposed that the monoclonal antibodies act via inhibition of a receptor-mitogen complex that is normally activated and reactivated by binding of intact (but not necessarily infectious) retrovirus. If this is a parallel to the antigen-induced activation of T cell proliferation noted above, then it is possible that these retroviruses bind to and activate via either the antigen-specific T cell receptors or via some cell-surface entity involved in the receptor-mitogen complex.

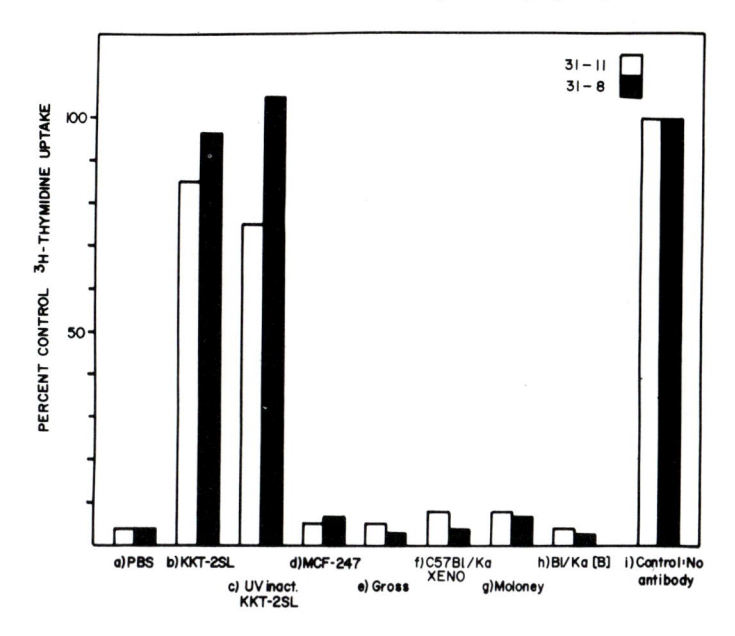

Fig. 7. The standard KKT-2 cell growth inhibition assay as outlined in Figure 6 was carried out with 31–11 and 31–8 antibodies at a 1:250 dilution on 5×10^4 KKT-2 cells after preincubation with purified retroviruses. One one-hundredth A_{260} unit of Sepharose-4B-purified virus [50] in 0.3 ml of PBS was incubated with KKT-2 cells for 60 min at room temperature prior to addition of inhibitory antibodies. This amount of virus represents a receptor saturation level as previously determined [45]. The origin of each retrovirus population has also been previously described [43–45]. Sepharose-4B-purified KKT-2-SL virus was also UV inactivated and used to inhibit antibody-induced KKT-2 cell growth inhibition. Five milliliters of virus in PBS (1 A_{260} unit/ml) was irradiated for 145 sec at 4,000 erg/mm^2 prior to use.

ACKNOWLEDGMENTS

Dr. I.L. Weissman received research support from the National Cancer Institute, contract CP 91011; National Institutes of Health, grant AI 09072; and American Cancer Society, grant IM56. Michael S. McGrath received support from a Cancer Biology Training grant, CA 09302, and he and Eric Pillmer were supported through the Medical Scientist Training Program by National Institutes of Health Training grant GM-07365. Nurit Hollander is the recipient of an American Cancer Society Senior Research Fellowship. Robert V. Rouse was supported by National Institutes of Health grant CA 05838. Susan K. Stevens was supported by the Stanford Alumni Medical Scholars Program. Eugene Butcher received support through an American Cancer Society Senior Fellowship.

REFERENCES

1. Hood L, Weissman IL, Wood W: In Benjamin, Cummings (ed): "Immunology," Chap 1, 1978.
2. Cantor H, Weissman IL: Prog Allergy 20:1, 1976.
3. Cantor H, Boyse EA: Immunol Rev 33:105, 1977.
4. Doherty PC, Blanden RV, Zinkernagel RM: Transplant Rev 29:89, 1976.
5. Morgan DA, Ruscetti␣␣␣W, Gallo␣␣C: Science 193:1077, 1976.

6. Gillis S, Ferm MM, Ou W, Smith KA: J Immunol 120:2027, 1978.
7. Watson J, Gillis S, Marbrook J, Mochizuki D, Smith KA: J Exp Med 150:849, 1979.
8. Hollander N, Pillemer E, Weissman IL: J Exp Med 152:674, 1980.
9. Hollander N, Pillemer E, Weissman IL: Proc Natl Acad Sci USA (in press).
10. Zinkernagel RM, Callahan GN, Althage A, Cooper S, Klein PA, Klein J: J Exp Med 147:882, 1978.
11. Fink PJ, Bevan MJ: J Exp Med 148:766, 1978.
12. Weissman I, Papaioannou V, Gardner R: In "Differentiation of Normal and Neoplastic Hematopoietic Cells." Cold Spring Harbor Laboratory, 1978, pp 33–47.
13. Le Douarin NM, Jotereau FV, Houssaint E, Belo M: Ann Immunol 127:849, 1976.
14. Weissman IL, Small M, Fathman CG, Herzenberg LA: Fed Proc 34:141, 1975.
15. Weissman I: J Exp Med 137:504, 1973.
16. Scollay R, Weissman I: J Immunol 124:2841, 1980.
17. Scollay R, Butcher E, Weissman IL: Eur J Immunol 10:210, 1980.
18. Scollay R, Kochen M, Butcher E, Weissman I: Nature 276:79, 1978.
19. Wekerle H, Ketelsen U-P, Ernst M: J Exp Med 151:925, 1980.
20. Rouse RV, van Ewijk W, Jones PP, Weissman IL: J Immunol 122:2508, 1979.
21. Goldstein G: Ann NY Acad Sci 249:177, 1975.
22. van Ewijk W, Rouse RV, Weissman IL: J Histochem Cytochem 28:1089, 1980.
23. Rouse RV, Weissman IL: Ciba Foundation Symposium, 84, Excerpta Medica (in press).
24. Scollay R, Jacobs S, Jerabek L, Butcher E, Weissman I: J Immunol 124:2845, 1980.
25. Gowans JL, Knight EJ: Proc R Soc Ser B 159:257, 1964.
26. Marchesi VT, Gowans JL: Proc R Soc Ser B 159:283, 1964.
27. Scollay R, Butcher E, Weissman IL: Eur J Immunol 10:210, 1980.
28. Stamper HB, Woodruff JJ: J Immunol 119:772, 1977.
29. Butcher E, Scollay R, Weissman I: Adv Exp Med Biol 114:64, 1979.
30. Butcher E, Scollay R, Weissman I: J Immunol 123:1996, 1979.
31. Butcher EC, Weissman IL: Ciba Foundation Symposium 71:265, Excerpta Medica, 1980.
32. Butcher E, Scollay R, Weissman I: Nature 280:496, 1979.
33. Griscelli C, Vassalli P, McCluskey RT: J Exp Med 130:1427, 1979.
34. Scollay R, Hopkins J, Hall J: Nature 260:528, 1976.
35. Guy-Grand D, Griscelli C, Vassalli P: J Exp Med 148:1661, 1978.
36. Smith ME, Martin AR, Ford WL: Monogr Allergy 16:203, 1980.
37. McWilliams M, Phillips-Quagliata JM, Lamm ME: J Immunol 115:54, 1975.
38. Hall JG, Hopkins J, Orlans E: Eur J Immunol 7:30, 1977.
39. Scollay R, Hopkins J, Hall J: Nature 260:528, 1976.
40. Cahill RNP, Poskitt DC, Frost H, Trnka A: J Exp Med 145:420, 1977.
41. Butcher EC, Scollay RG, Weissman IL: Eur J Immunol 10:556, 1980.
42. Weissman IL, Baird S: In Koprowski H (ed): "Life Sciences Research Report 7, Neoplastic Transformation: Mechanism and Consequences." Dahlem Konferenzen, 1977, p 135.
43. McGrath MS, Weissman IL: In "Hematopoietic Mechanisms." Cold Spring Harbor Laboratory, 1978, pp 577–589.
44. McGrath MS, Weissman IL: Cell 17:65, 1979.
45. McGrath MS, Lieberman M, Decleve A, Kaplan HS, Weissman IL: J Virology 28:819, 1978.
46. McGrath MS, Pillemer E, Kooistra D, Weissman IL: Contemp Top Immunobiol 11:157, 1980.
47. McGrath MS, Weissman IL: In preparation, 1981.
48. McGrath MS, Pillemer E, Weissman IL: Nature 285:259, 1980.
49. McGrath MS, Jerabek L, Pillemer E, Steinberg RA, Weissman IL: In Neth R (ed): "Modern Trends in Human Leukemia IV." In press.
50. McGrath MS, Witte O, Pincus T, Weissman IL: J Virol 25:923, 1978.

Control of Hemopoietic Cell Proliferation and Differentiation

Donald Metcalf

Cancer Research Unit, Walter and Eliza Hall Institute of Medical Research, Royal Melbourne Hospital, Victoria 3050, Australia

Hemopoietic populations offer by far the best available model systems for analyzing many aspects of the fundamental mechanisms controlling cell proliferation and differentiation. Semisolid cloning systems of high plating efficiency exist for normal and leukemic hemopoietic populations, and in these culture systems both proliferation and differentiation occur under defined culture conditions. Furthermore, the various differentiated end cells differ so extremely in morphology, membrane markers, and functional activity that a wide variety of parameters is available for monitoring differentiation. For the erythroid and granulocyte-macrophage systems, specific progenitor cells can be stimulated to generate large clones of differentiating progeny using the purified glycoprotein regulators erythropoietin, GM-CSF, and M-CSF. Exploitation of these systems will provide information on the events leading to proliferation and differentiation that should have general relevance for many other cell systems.

Key words: granulocyte-macrophage, colony-stimulating factor, myeloid leukemia

The formation of mature, usually short-lived, blood cells (hemopoiesis) occurs continuously during fetal and adult life. This massive cellular proliferation results in the production of 7 major types of blood cell, each with distinctive morphology and highly specialised functional activity. Progress in analyzing the precise mechanisms controlling these complex processes has been catalysed by the introduction of a series of semisolid culture systems supporting the clonal proliferation and differentiation of the various specialised hemopoietic precursors.

This subject has recently been reviewed extensively [1, 2]. The present discussion will be restricted, therefore, to some of the principles emerging from an analysis of hemopoiesis, since preliminary information from other cell systems suggests the likelihood that these principles may be generally applicable to other mammalian tissues.

CLONAL ORIGIN AND DIVERSITY OF HEMOPOIETIC PRECURSORS

The clonal generation of blood cells appears in principle to parallel the clonal generation of mature end cells in many solid tissues — eg, skin, gut, epithelium [3]. Hemopoiesis is sustained throughout adult life by the proliferative activity of the limited number of

Received and accepted March 4, 1980.

© 1981 Alan R. Liss, Inc., 150 Fifth Avenue, New York, NY 10011

multipotential hemopoietic stem cells that originate in early embryonic life in the yolk sac and migrate into the developing embryo (the multipotential primordial germ cells and melanoblast precursor cells have a similar developmental and migratory history [4]). Hemopoietic populations are expanded in fetal and adult life by the dual capacity of the hemopoietic stem cells both for self-replicative division (in the mouse for well in excess of 100 divisions) and for generating progenitor cells that each can produce up to 5,000–10,000 progeny cells. Progenitor cells differ from multipotential stem cells in having little or no capacity for self-replication and in having entered a restricted (usually single) pathway of hemopoietic differentiation.

At any one time only a small fraction of stem cells is active in either self-generation or generating progenitor cells, and there is some evidence to suggest that steady-state hemopoiesis in adult life may be sustained at any given time by the proliferative activity of a single original stem cell and its progeny [4]. This may be the basis for the age-related stratification demonstrable within cells of the stem cell compartment, with subsets of stem cells exhibiting a progressively reduced capacity for self-generation [5]. Although clonal dominance by one hemopoietic stem cell and its progeny has not been formally proved in normal adults, this sequential, "cascade-type" of cell production seems possible from available data. In patients with acute or chronic myeloid leukemia prior to treatment, a common clonal origin from the same cell of both the leukemic and co-existing phenotypically normal erythroid and megakaryocytic populations has been well documented, and this therefore may represent only a minimal deviation from the situation existing in normal adult life.

Until recently, the only method available for monitoring multipotential hemopoietic stem cells was based on the capacity of these cells to produce discrete hemopoietic colonies in the spleen of irradiated syngeneic recipients [6], and thus the alternative name used for such cells is, colony-forming units-spleen (CFU-S). Methods are now available for the growth of mixed hemopoietic colonies in semisolid culture [7–12], and analysis has indicated that the cells forming such colonies are able to generate both CFU-S (self-generation) [9, 10] and extensive populations of progenitor and differentiating hemopoietic cells of the erythroid, neutrophil, macrophage, eosinophil, megakaryocyte, and B-lymphoid series [9, 11, 12; Johnson, G.R., personal communication]. Further studies on these in vitro colonies may permit an analysis of two of the most crucial steps in hemopoiesis: the control of stem cell self-generation and the generation by stem cells of committed progenitor cells of various types.

Progenitor cells are more numerous than the stem cells from which they are generated; eg, in the mouse marrow CFU-S levels are $30/10^5$ cells, granulocyte-macrophage progenitor cells $300/10^5$ cells, and most progenitor cells are in active cell cycle at any one time. Progenitor cells are detectable by their capacity to form colonies of differentiating cells in semisolid culture [1], hence the term colony-forming cells (eg, granulocyte-macrophage colony-forming cells, GM-CFC).

In the well-analyzed granulocyte-macrophage (GM) system it is clear that extensive heterogeneity exists within the various progenitor cell populations [1], and preliminary evidence exists for a similar heterogeneity in erythroid, eosinophil, and megakaryocyte progenitor populations. GM progenitors in the mouse have been shown to vary widely in the following attributes:

1) Proliferative capacity – colony size at 7 days can vary from 50 to 5,000 cells
2) Ability to generate progeny of various types – some form only granulocytic

progeny (G-CFC), some only macrophage progeny (M-CFC), while most in the mouse are bipotential, forming both granulocytes and macrophages (GM-CFC).

3) Responsiveness to regulatory factors — quantitative differences exist in responsiveness to a single regulatory factor, and qualitative differences exist in responsiveness or unresponsiveness to related GM regulators

4) Size, adherence, density, and expression of surface antigens

To date there is no evidence to suggest either that progenitor cells can "dedifferentiate" and again become multipotential cells or that one progenitor cell can form a progenitor cell of another class — eg, a GM progenitor become an erythroid or eosinophil progenitor. Within the GM progenitor population, it is possible that some of the heterogeneity may be based on parent–progeny relationships; eg, many M-CFC with a restricted capacity for proliferation could be the immediate progeny of GM-CFC. However, the proposed sequence G-CFC → GM-CFC → M-CFC [13] seems unlikely to apply to all GM progenitors, since certain G-CFC never generate colonies containing macrophages [14]. A striking feature of the heterogeneity observed between GM progenitor cells is that the progeny of such cells continue to exhibit the same characteristics as the originating progenitor cell — eg, responsiveness or unresponsiveness to various regulatory molecules [1]. This leads to the conclusion that end cell heterogeneity — eg, differences between subsets of apparently similar polymorphs or macrophages — must exist and be ultimately demonstrable in terms of heterogeneity of end cell function. This concept is well established for T- and B-lymphoid cells but to date has not been extensively investigated for other blood cells.

While the heterogeneity of progenitor cells and their generation of clonally distinctive progeny seems to introduce unnecessary complexity to hemopoiesis, a system of this design is highly flexible and lends itself readily to "fine tuning" in response to fluctuating demands either for differing subsets of end cells or for varying numbers of end cells.

REGULATORY CONTROL OF HEMOPOIESIS

Because colonies of differentiating progeny can be generated in vitro by progenitor cells when stimulated by defined specific regulatory factors, it has been concluded that blood cell production and differentiation from the progenitor cell stage onwards are controlled exclusively by the interaction of various humoral regulatory factors. Conversely, data from analysis of spleen colony formation in vivo have led to the conclusion that self-replication of stem cells and the formation of progenitor cells are processes regulated by cell–cell interactions between stem cells and specialised microenvironmental cells in the hemopoietic tissues (Fig. 1).

Particular importance has been placed on the development of sharply demarcated second hemopoietic populations in expanding spleen colonies as indicating spread into an adjacent microenvironmental niche that is programmed to induce a second, different, type of hemopoiesis [15] — eg, the development of a GM area in an expanding colony of erythropoietic cells (Fig. 2, 1). This has been interpreted as indicating that one or more of the CFU-S in the expanding erythropoietic colony must be committed by the adjacent niche to granulopoiesis.

There has been some dispute in the literature as to whether the microenvironmental control of a developing hemopoietic colony in the spleen necessarily requires intimate cell contact between the stem cell and microenvironmental cells (Fig. 3, 1) or whether micro-

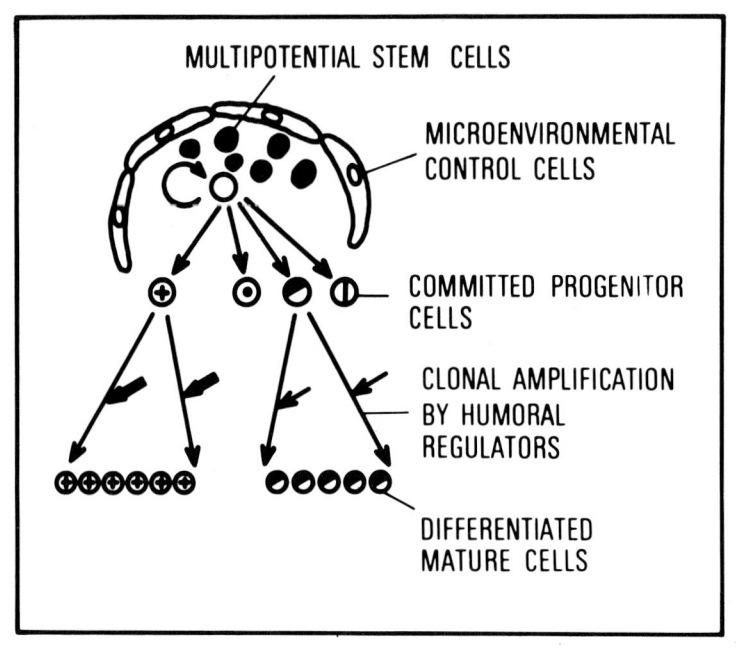

Fig. 1. The conventional view of control of hemopoiesis. Self-replication of multipotential stem cells and their commitment to specific progenitor cells are regarded as being under local microenvironmental control. The subsequent clonal proliferation of differentiating cells from progenitor cells is regarded as being regulated by specific proliferative humoral factors modulated by a variety of inhibitory factors.

environmental cells scattered throughout the tissues could direct differentiation by generating gradients of short-range regulatory factors (Fig. 3, 2). Extensive observations on colony formation in vitro have shown that different, clonally pure populations of hemopoietic colonies can develop side by side in cultures containing a mixture of regulatory factors (Fig. 3, 3). This indicates clearly that once a cell has been committed to a particular pathway of differentiation, the subsequent generation of progeny of the same class proceeds irrespective of what irrelevant regulatory molecules may also impinge on the dividing colony cells. For this reason, much of the growth of hemopoietic colonies in the spleen certainly does not require either continuing contact with microenvironmental cells or special focussing of differentiation-inducing gradients.

More recent information suggests that the sequential two-tiered system postulated as controlling hemopoiesis (Fig. 1) is likely to be a misleading oversimplification. Progenitor cells are present in the peripheral blood, as are regulatory factors, yet the production of differentiating hemopoietic cells occurs only in certain tissues — eg, the bone marrow and spleen. At the very least, this implies that microenvironmental processes must play a permissive role and may well be involved actively in regulating terminal proliferative events in various hemopoietic families. While production of regulatory factors by microenvironmental cells has been documented, some of these factors (eg, GM-CSF controlling GM proliferation) are also produced by many non-hemopoietic organs, so the microenvironmental cells in hemopoietic organs must be presumed to produce additional regulatory factors vital for localising hemopoiesis to these sites.

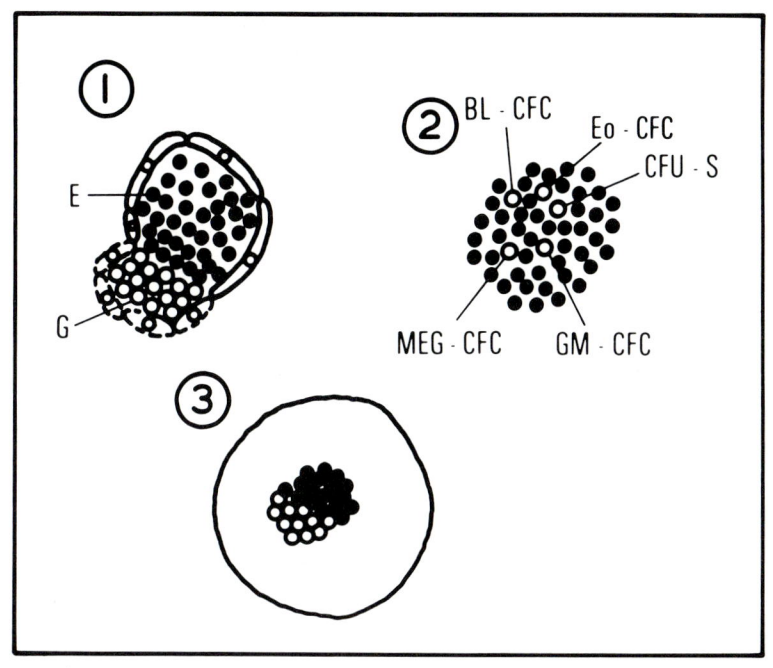

Fig. 2. Sharp demarcation lines between erythroid and granulocytic elements in spleen colonies (1) have been used to support the concept of specific microenvironmental niches, but similar sharp demarcations are common in mixed colonies grown in vitro (3) where no microenvironmental cells are present. The homogeneity of spleen colonies is also more apparent than real, since morphologically "pure" erythroid colonies (2) commonly contain progenitor cells (CFC) of other lineages – eg, granulocyte-macrophage (GM), eosinophil (EO), megakaryocyte (MEG), and B-lymphocyte (BL) – in addition to multipotential stem cells (CFU-S).

Similarly, the ability to grow a mixed colony in a semisolid culture containing only a single cell [7, 16] implies that hemopoietic stem cell proliferative activity does not necessarily require cell—cell interaction with microenvironmental cells. Cell contact control may possibly be a more efficient process, but it clearly is not mandatory.

It is not uncommon in mixed colonies grown in vitro to see distinct zones of erythroid and nonerythroid cells that are as clearly demarcated as in expanding spleen colonies in vivo [9] (Fig. 2, 3). This indicates that demarcation does not require any special cell—cell influence by microenvironmental cells arrayed in niches. Furthermore, clonal analysis of spleen colonies that, on cytological analysis, appear to be composed wholly of hemoglobin-synthesizing erythroid cells has revealed the frequent presence in such colonies not only of CFU-S but also of committed progenitors for GM, eosinophil, megakaryocyte, and B-lymphoid cells [14, 17, 18]. This casts considerable doubt on the concept of line-specific niches, since clearly in such colonies a variety of different commitment steps must have occurred.

While the simple two-stage control concept – microenvironmental (to generate progenitor cells) and humoral (to generate differentiated end cells) – is clearly inadequate in view of recent data, it is useful to keep both types of control system in mind, even though both possibly operate in concert and on all stages of hemopoiesis.

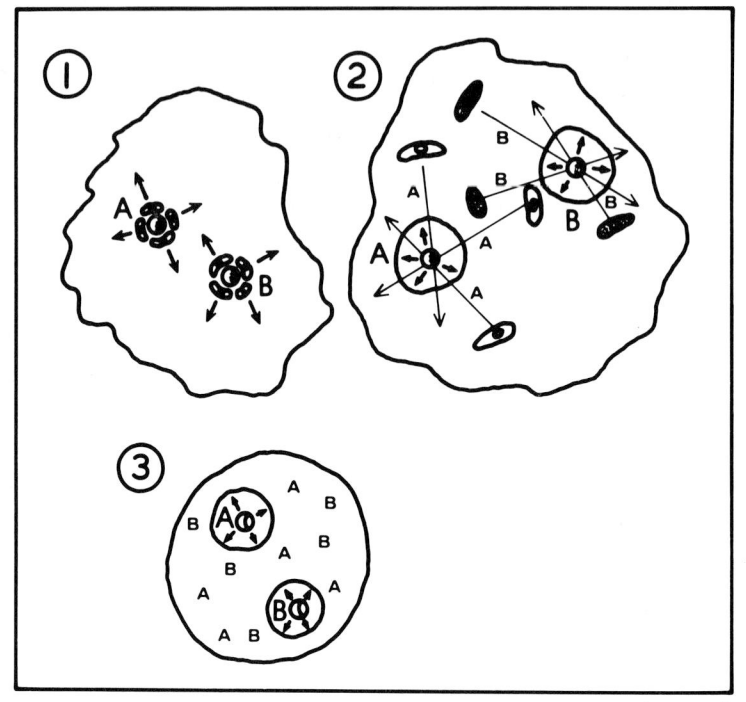

Fig. 3. Microenvironmental control of developing spleen colonies may require intimate cell contact with specific microenvironmental cells (1). However, microenvironmental cells could regulate colony development at a distance by focussed short-range gradients of regulatory molecules (2). Following commitment to a defined hemopoietic pathway, clonally pure populations of progeny cells can develop, as seen in vitro (3), regardless of the co-existence of other specific regulatory factors.

HUMORAL REGULATORY FACTORS

Because the humoral regulatory factors have been technically easier to detect and analyze in vitro, the remainder of the discussion will deal with aspects of their mode of action.

Specific regulatory factors have now been described that are able to stimulate the formation of colonies of differentiating erythroid, granulocyte-macrophage, eosinophil, and megakaryocyte cells in vitro. In the mouse all are neuraminic acid-containing glycoproteins, and three have been obtained in chemically pure form — erythropoietin [19], GM-CSF [20], and M-CSF [21].

These regulators exhibit certain common features in their mode of action which can best be discussed by a consideration in detail of the most extensively analyzed — the control of granulocyte and macrophage formation by the granulocyte-macrophage colony stimulating factor (GM-CSF) (syn. colony-stimulating activity, CSA; or macrophage-granulocyte-inducer, MGI).

Two major forms of GM-CSF are produced by mouse cells — a 23,000 mol wt form (GM-CSF) stimulating both granulocyte and macrophage formation and a 70,000 mol wt form (M-CSF) preferentially, but not exclusively, stimulating macrophage formation. A less well characterized G-CSF may exist and exclusively stimulate granulocyte formation.

GM-CSF has been purified from mouse lung conditioned medium [20] and M-CSF from L-cell conditioned medium [21]. Both are neuraminic acid-containing glycoproteins and both are biologically active in vitro at 10^{-11} M concentrations. GM-CSF and M-CSF are normally bioassayed in vitro by their capacity to stimulate GM progenitor cells to proliferate and form colonies of granulocytes and/or macrophages. Differentiation of colony granulocytes and macrophages occurs progressively during colony growth, leading to the formation of polymorphs and macrophages with essentially all of the morphological features, membrane markers, and functional activities exhibited by mature cells of these types in vivo.

GM colonies are clones and can be grown using purified GM-CSF in cultures containing only a single micromanipulated progenitor cell [7, 16, 22, 23] or in cultures with fully defined, serum-free medium [24]. In the absence of added GM-CSF, no proliferation of GM cells is possible in vitro, and in fact GM-CFC rapidly die or lose their capacity for proliferation. Actively growing GM colonies cease proliferation and die if transferred to cultures lacking GM-CSF [25, 26]. From this it can be deduced that GM-CSF has a direct action on GM target cells, is required continuously, and that proliferation and differentiation of GM cells can be achieved by the action of a single molecular regulator. This action of GM-CSF is highly specific, as the molecule is incapable of stimulating the formation in vitro of any other type of hemopoietic colony or the proliferation of other cell types.

The biological activity of the 23,000 molecular weight GM-CSF is of special interest, since at high concentrations granulocyte colony formation occurs preferentially, and at low concentrations preferential macrophage colony formation occurs [27]. Since most of the target GM progenitor cells in the mouse are bipotential, this indicates that control of the formation of two distinct end cells — polymorphs and macrophages — from a common progenitor cell can be achieved simply by altering the concentration of a single regulatory molecule. A naturally occurring example of this relationship is seen in the disease cyclic neutropenia, in which matching inverse cycles of polymorph and monocyte production occur, paralleled by cyclic fluctuations in serum and urine GM-CSF levels [28, 29]. The situation is complicated, however, by the heterogeneity of progenitor cells, since subsets of GM progenitors exist that are capable only of granulocyte formation, and these require high concentrations of GM-CSF before being stimulated to proliferate. Conversely, subsets of GM progenitor cells exist that are apparently preprogrammed only to generate macrophage progeny, and these are highly responsive to stimulation by low GM-CSF concentrations [30]. Thus the observed relationship between GM-CSF concentration and granulocyte or macrophage production appears to be based on the operation of two distinct processes — selective stimulation of preprogrammed subsets of precursor cells and a directed differentiation of biopotential precursors [31] (Fig. 4).

Dose-response relationships exist between GM-CSF concentration and both the number and growth rate of GM colonies developing. The characteristic sigmoid dose–response curve for colony numbers versus GM-CSF concentration reflects the heterogeneity of target colony-forming cells; the most unresponsive colony-forming cells require GM-CSF concentrations 10–20-fold higher than those stimulating the most responsive cells. While the formation of up to 5,000–10,000 cells from a single colony-forming cell following stimulation by GM-CSF is dramatic evidence of the proliferative effects of the regulator, analysis of the actual mechanisms involved is incomplete. Specific binding of ^{125}I-labeled M-CSF has been reported to target populations [32], but subsequent intracellular events are not known. Although cyclic nucleotides have been reported both to increase and inhibit GM-CSF-stimulated colony formation [33], it is not clear from these experiments

TWO MECHANISMS CONTROLLING THE ALTERNATE FORMATION OF G OR M CELLS

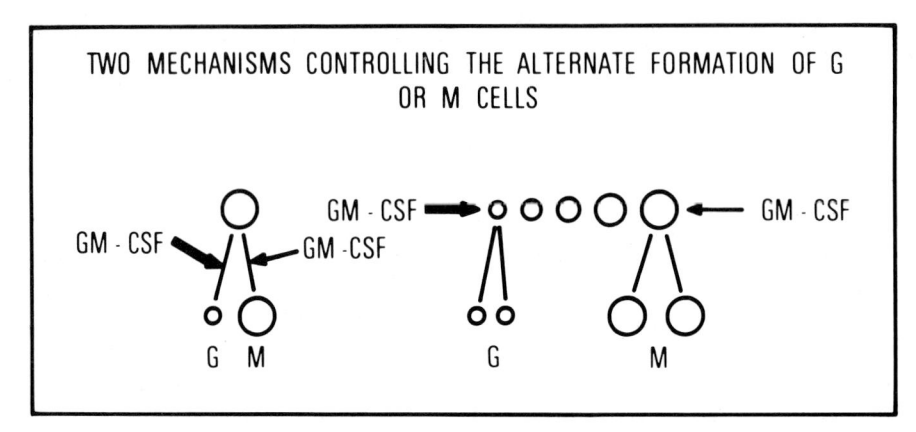

Fig. 4. High concentrations of GM-CSF preferentially stimulate granulocyte colony formation; low concentrations, macrophage colony formation. Two mechanisms are responsible: a concentration-dependent influence of GM-CSF on bipotential GM progenitors, and selective activation of subsets of GM progenitors preprogrammed to form either G or M progeny. The small G-forming progenitors require higher concentrations of GM-CSF to be stimulated to proliferate than do M-forming progenitors.

whether binding of GM-CSF to target cells necessarily is followed by activation of the cyclic AMP system as the mediator of GM-CSF action.

If colony-forming cells are not cycling, they can be induced to enter S-phase within 3 hours by exposure to GM-CSF [34]. Similarly, mean cell cycle times of dividing colony cells are shortened progressively by increasing concentrations of GM-CSF [35; Metcalf, D., unpublished data]. However, the greatly increased size achieved by colonies growing in the presence of high concentrations of GM-CSF may not depend entirely on shortening of cell cycle times. Incomplete evidence indicates that the proportion of cells capable of division (ie, myeloblasts, myelocytes) in a developing G colony may remain high in the presence of high concentrations of GM-CSF, thus delaying cessation of colony growth due to terminal differentiation to non-dividing end cells. This needs further examination, since this postulated effect of GM-CSF could constitute a temporary arrest of differentiation, whereas other evidence indicates that GM-CSF enforces or accentuates the expression of GM differentiation. Thus, incubation of polymorphs with GM-CSF leads to increase RNA and protein synthesis [27, 36], and incubation of macrophages with GM-CSF or M-CSF leads to increased phagocytic activity [37], increased prostaglandin E synthesis [38], and plasminogen activator synthesis [39].

The dual action of GM-CSF on cell proliferation and end cell functional activity (Fig. 5) has obvious parallels with the action of erythropoietin, which stimulates both the proliferation of erythropoietic cells and hemoglobin synthesis in such cells. It is possible that similar dual activity may be exhibited by the glycoprotein regulators controlling eosinophil and megakaryocyte production, and this is worthy of investigation.

Recent experiments have indicated that GM-CSF acting in concert with erythropoietin can increase the number of erythroid colonies developing in cultures of marrow or fetal liver cells [40, 41]. Analysis of this phenomenon indicated that GM-CSF, but not M-CSF, has the capacity to stimulate directly the initial proliferation of at least some multipotential, erythroid, eosinophil, and megakaryocyte precursor cells for up to five cell divisions without altering the capacity of these cells to form mixed, erythroid, eosinophil,

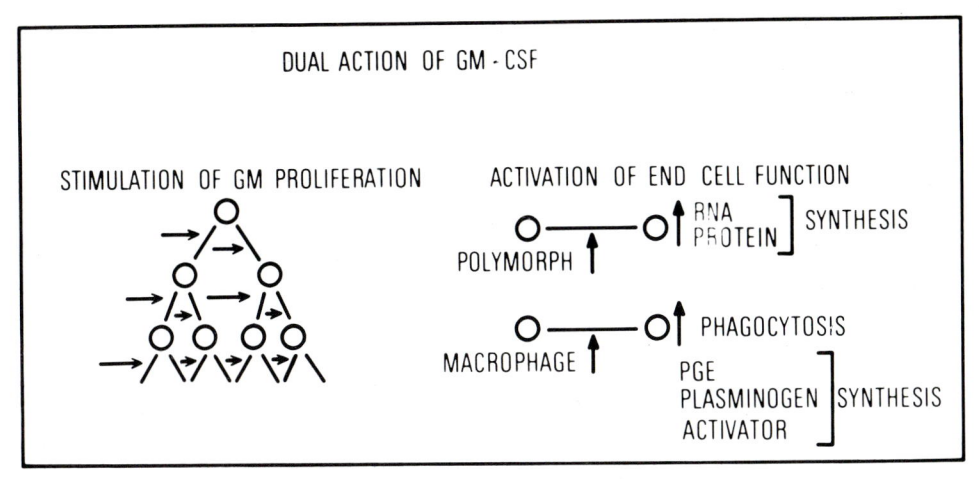

Fig. 5. Dual action of GM-CSF on granulocyte-macrophage populations: (a) stimulation of clonal proliferation, and (b) stimulation of functional activity of mature end cells.

or megakaryocyte colonies when the appropriate CSF for these latter cell types is subsequently added to the cultures [16, 41, 42]. The failure of GM-CSF alone to stimulate the formation of these latter types of colonies stems from the fact that GM-CSF is unable to stimulate the terminal cellular proliferation and specific differentiation necessary in such colonies to allow the formation of identifiable colonies.

The capacity of GM-CSF to stimulate the proliferation for five successive divisions of a cell subsequently forming a large colony composed only of hemoglobin-synthesizing erythroid cells is an intriguing example of the capacity of an inappropriate regulator to stimulate cell division without modifying the genetic programming of the cell to proceed along a particular pathway of differentiation — eg, erythropoiesis. The various hemopoietic populations are, of course, closely related ancestrally, and the initial responsiveness of a multipotential cell to GM-CSF may simply be lost when progeny cells differentiate to pathways other than GM formation — eg, by the loss of the appropriate membrane receptor for GM-CSF by maturing precursors of erythroid, eosinophil, and megakaryocytic cells.

One characteristic of GM-CSF that is shared by comparable regulators such as nerve growth factor and epidermal growth factor is that GM-CSF production occurs in many (perhaps all) tissues and is not restricted, as in the case of classical hormones, to a single organ. Thus GM-CSF is extractable from all organs in concentrations higher than are present in the serum [43, 44] and is actively synthesized in vitro by a wide variety of mouse organs [45, 46]. It is difficult to assess the synthetic capacity of individual normal cell types free of other contaminating cells, but at least four different cell types — macrophages, lymphocytes, fibroblasts, and endothelial cells — have been shown to synthesize GM-CSF [see review, 31]. Since these cells are present in all tissues, it so far has been difficult to determine whether other cell types also can synthesize GM-CSF, although this seems likely. It is uncertain whether the production of other hemopoietic regulators is also multifocal. Because of the apparent conviction of early workers on erythropoietin that this regulator was exclusively of renal origin, few detailed studies were undertaken to determine whether other tissues might also produce erythropoietin. However, recent studies have at least identified Kupffer cells in the liver as an additional site of erythropoietin production

[47], and it seems likely that other production sites will be documented. No studies have been reported on the in vivo production sites of MEG-CSF and EO-CSF, but lymphoid cells are capable of synthesizing both, and in humans, EO-CSF is also produced by the placenta, marrow cells, and peripheral blood monocytes [31].

The multifocal origin of GM-CSF presents complexities in determining the relative importance of various tissues in GM-CSF production, and thus in the control of GM formation in vivo. Some evidence has been produced to indicate the importance of local production of GM-CSF within the bone marrow, particularly during GM population regeneration following irradiation damage [48, 49], but it is conceivable that, in other situations, other tissues may become more important. For example, in infections with endotoxemia, the high circulating GM-CSF levels possibly originate mainly from endothelial and macrophage populations widely distributed throughout the body. Similarly, in patients with lung tumors who exhibit a granulocytosis, the tumor cells themselves have been shown to produce GM-CSF [50]. Since varying granulocyte levels in these patients correlate well with the varying capacity of individual tumors to secrete GM-CSF [51], the tumors in these patients may constitute a major source of GM-CSF.

Initial biochemical studies suggested that GM-CSF produced by different tissues might vary widely in biochemical nature although sharing considerable antigenic cross-reactivity [44, 52]. However, more extensive analysis of GM-CSF produced by 17 different adult mouse tissues has indicated that, following neuraminidase treatment and examination under dissociating conditions, all possess similar biological activity, charge, and molecular weight (23,000) [46]. The 70,000 mol wt. M-CSF is produced by continuous lines of L-cells, and apparently similar molecules are produced by the yolk sac [53] and pregnant uterus [54; Burgess, A.W., Nicola, N.A., Metcalf, D., unpublished data]. It is of interest that most GM progenitor cells in the fetus are M-progenitors, and the local production of M-CSF in the fetus and adjacent tissues may represent a matching regulatory system for these hemopoietic precursors.

CONTROL OF LEUKEMIC POPULATIONS

Leukemic cells from patients with myeloid leukemia are able to proliferate in semisolid cultures and form colonies or clusters [see reviews 1, 2]. Karyotypic analysis of colony and cluster cells grown from suitable patients has confirmed that the proliferating cells are members of the leukemic clone. In the case of colonies grown from patients with chronic myeloid leukemia, the colony cells exhibit relatively normal differentiation to granulocytes and macrophages. However, maturation defects are common in cluster cells grown from patients with AML, suggesting more extreme internal derangement of the AML leukemic cells.

The most striking finding of the in vitro clonal analysis of leukemic populations grown from patients with acute and chronic myeloid leukemia has been the demonstration that all such populations remain absolutely dependent on GM-CSF for proliferative stimulation in vitro. When marrow or blood cells from such patients are prefractionated to remove endogenous GM-CSF-producing cells, leukemic cells are absolutely unable to proliferate in vitro in the absence of added GM-CSF [1, 2]. While some differences exist between the dose-response curves to GM-CSF for normal and leukemic cells, these are minor in degree [55, 56].

In two mouse myeloid leukemias extensively analysed in vitro (the M1 and WEHI-3 leukemias) a similar situation has been observed. Although with continuous culture in vitro the leukemic populations have become progressively less dependent for proliferation on exogenous GM-CSF, the leukemic populations exhibit extensive maturation to relatively normal polymorphs and macrophages when cultured in the presence of CSF (MGI)-containing materials or in the presence of purified GM-CSF [57, 58].

Of more importance, clonal analysis of mouse leukemic colonies grown in the presence of GM-CSF has indicated that when GM-CSF forces cellular differentiation in leukemic colonies, stem cell (colony-forming cell) self-replication is markedly suppressed [58, 59]. This is reflected in vivo where injection of leukemic populations, previously forced to differentiate in vitro, is followed by a prolongation of survival time or in some cases failure to develop transplanted leukemia [60–62].

Differentiation-resistant subclones of M1 leukemia have been documented, but it is of interest that combined treatment of such cells with chemotherapeutic agents plus differentiation-inducing preparations leads to the development of differentiation [63].

Certain CSF-containing preparations are outstandingly effective in inducing differentiation in mouse myeloid leukemic cells (eg, ascitic fluid from tumor-bearing mice or serum from endotoxin-injected mice). Since the activity of these materials is disproportionate to their content of GM-CSF, the possibility exists that a special subset of GM-CSF molecules is involved or that a second differentiation-inducing factor is present that acts either independently or in collaboration with GM-CSF [59, 64].

The capacity of leukemic cells to respond to control systems enforcing differentiation in normal cells raises the obvious possibility of controlling the proliferation of a leukemic population by enforced differentiation. This possibility is the more intriguing in view of the ability of various chemotherapeutic agents to induce responsiveness in other unresponsive leukemic subclones. It might be wondered, therefore, whether the apparently normal GM populations reappearing in patients with AML following induction of a remission represent leukemic populations now exhibiting normal differentiation. Karyotypic and in vitro clonal analyses have clearly shown, however, that in most patients in complete remission, the normal GM populations are indeed of normal origin and have been regenerated from normal stem and progenitor populations previously suppressed by the leukemic populations.

If the intervention of regulator-enforced differentiation occurs during remissions, as now seems likely, it would appear that differentiation induction contributes to extinction of the previously dominant leukemic clone by suppressing leukemic stem cell self-replication and forcing differentiation of surviving leukemic cells to post-mitotic cells (Fig. 6).

UNRESOLVED PROBLEMS

Three hemopoietic regulators (erythropoietin, GM-CSF, and M-CSF) have been purified and are available for a more detailed analysis of cellular events during the expression of differentiation in proliferating hemopoietic populations. The amounts available from existing source materials are relatively small (microgram amounts), which prevents sequencing or similar studies. There would be value in attempting molecular cloning of the messenger RNA for these regulators in an effort to produce more reasonable quantities of these regulators following insertion of complementary DNA into bacterial plasmids.

Methods are available for producing the corresponding regulators for eosinophil, megakaryocytic, and erythroid cells by the use either of mitogen-stimulated lymphoid cells or WEHI-3 leukemic cells, but available protein separative procedures have so far failed to obtain the three regulators in pure form due to the close similarity of the molecules

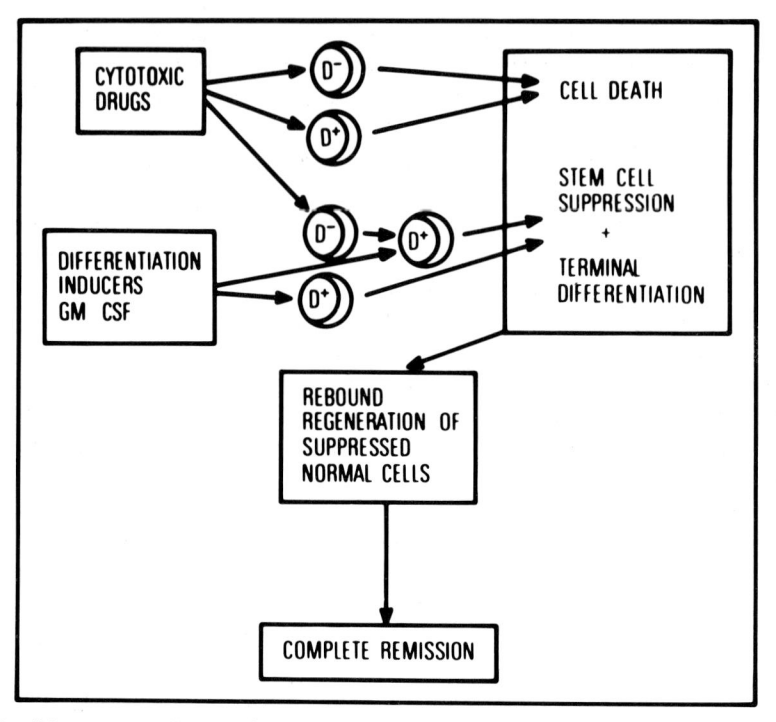

Fig. 6. Possible sequence of events during the induction of a complete remission in a patient with acute myeloid leukemia. Differentiation-enforcing factors, including GM-CSF, restrict leukemic stem cell self-replication and suppress the leukemic clone by forcing responsive (D^+) cells to differentiate to post-mitotic cells. Cytotoxic drugs not only cause death of leukemic cells but also induce unresponsive (D^-) cells to transform to D^+ cells. Suppression of the leukemic clone allows rebound regeneration of previously suppressed normal hemopoietic cells, possibly by eliminating production by leukemic cells of factors inhibitory for normal cells [65].

(23,000 mol wt glycoproteins) [66]. Neither production source is outstanding, and it may be worthwhile investigating alternative cellular sources — eg, hybridomas produced by fusion of activated T-lymphocytes [67] as a richer source of starting material, possibly containing only one regulator.

No studies to date have thrown any light on the mechanisms controlling the self-generation of stem cells. Although sustained production of stem cells can be achieved using underlayers of marrow cells (the Dexter culture system [68]), this system is too complex to analyze at present. Possibly, cloning of the various stromal populations present in such underlayers may provide a first approach to simplifying the system. Equally unknown are the mechanisms influencing the alternative behavior of such stem cells — self-generation or the generation of committed stem cells. Despite evidence that GM-CSF can influence proliferation of cells in this compartment, GM-CSF does not appear to be able to influence commitment, and no other agent so far tested appears to influence this step.

Commitment of multipotential stem cells to a variety of precursor cells does occur during mixed colony formation in vitro [9] when the cultures are stimulated by spleen-conditioned medium, and further fractionation of this material may isolate regulator molecules capable of influencing both multipotential cell self-replication and commitment.

The nature of the control systems determining self-generation or commitment of multipotential cells is the most fundamental unanswered question in hemopoiesis. Since leukemia is essentially a derangement of this process, eventual understanding of the nature of leukemic cells also depends on understanding this deceptively simple biological event. A puzzling aspect of leukemia is that an analysis of differentiation patterns in leukemic populations, either by morphology or assessing progenitor cell frequency in culture, suggests that in no two leukemias is the deranged pattern of differentiation precisely identical [69, 70]. Furthermore, if insertion of viral-derived DNA into the host cell genome is instrumental in initiating the leukemic derangement, it is equally disturbing that the location of viral DNA insertion appears not to be constant [70], which implies that leukemia can result from a wide variety of insertions. Both observations are difficult to reconcile with the natural assumptions that control of differentiation in a hemopoietic cell should reside in a precise portion of the genome and that the initiation of differentiation is likely to be an all-or-nothing event.

ACKNOWLEDGMENTS

This work was supported by the Carden Fellowship Fund of the Anti-Cancer Council of Victoria and the National Cancer Institute, grants ROI-CA-25972 and ROI-CA-22556.

REFERENCES

1. Metcalf D: "Hemopoietic Colonies." New York: Springer, 1977.
2. Metcalf D: Clin Haematol 8:263, 1979.
3. Potten CS, Schofield R, Lajtha LG: Biochim Biophys Acta 560:281, 1979.
4. Metcalf D, Moore MAS: "Haemopoietic Cells." Amsterdam: North-Holland, 1971.
5. Rosendaal M, Hodgson GS, Bradley TR: Cell Tissue Kinet 12:17, 1979.
6. Till JE, McCulloch EA: Radiat Res 14:213, 1961.
7. Johnson GR, Metcalf D: Proc Natl Acad Sci USA 74:3879, 1977.
8. Johnson GR, Metcalf D: J Cell Physiol 94:243, 1978.
9. Metcalf D, Johnson GR, Mandel TE: J Cell Physiol 98:401, 1979.
10. Humphries RK, Jacky PB, Dill FJ, Eaves AC, Eaves CJ: Nature 279:718, 1979.
11. Hara H, Ogawa M: Am J Hematol 4:23, 1978.
12. Fauser AA, Messner HA: Blood 53:1023, 1979.
13. Bol S, Williams N: J Cell Physiol 102:233, 1980.
14. Metcalf D: In Baum SJ, Ledney GD (eds): "Experimental Hematology 1978" New York: Springer, 1978, pp 35–46.
15. Wolf NS, Trentin JJ: J Exp Med 127:205, 1968.
16. Metcalf D, Johnson GR, Burgess AW: Blood 55:138, 1980.
17. Metcalf D, Johnson GR: In Battisto JR, Streilein JW (eds): "Immuno-Aspects of the Spleen." Amsterdam: Elsevier/North-Holland Biomedical Press, 1976, pp 27–34.
18. Lala PK, Johnson GR: J Exp Med 148:1408, 1978.
19. Miyake T, Kung CK-H, Goldwasser E: J Biol Chem 252:5558, 1977.
20. Burgess AW, Camakaris J, Metcalf D: J Biol Chem 252:1998, 1977.
21. Stanley ER, Heard PM: J Biol Chem 252:4305, 1977.
22. Moore MAS, Williams N, Metcalf D: J Cell Physiol 79:283, 1972.
23. Johnson GR, Metcalf D: In Le Douarin (ed): "Cell Lineage, Stem Cells and Cell Differentiation." INSERM Symposium No 10. Amsterdam: Elsevier/North-Holland Biomedical Press, 1979, pp 199–213.
24. Guilbert LJ, Iscove NN: Nature 263:594, 1976.
25. Metcalf D, Foster R: Proc Soc Exp Biol Med 126:758, 1967.
26. Paran M, Sachs L: J Cell Physiol 72:247, 1968.

27. Burgess AW, Metcalf D: In Baum SJ, Ledney GR (eds): "Experimental Hematology Today." New York: Springer, 1977, pp 135–146.
28. Dale DC, Alling DW, Wolf SM: J Clin Invest 51:2197, 1972.
29. Moore MAS, Spitzer G, Metcalf D, Penington DG: Br J Haematol 27:47, 1974.
30. Metcalf D, MacDonald HR: J Cell Physiol 85:643, 1975.
31. Metcalf D: In Baserga R (ed): "Tissue Growth Factors." New York: Springer (in press).
32. Stanley ER: Proc Natl Acad Sci USA 76:2969, 1979.
33. Kurland JI, Hadden JW, Moore MAS: Cancer Res 37:4534, 1977.
34. Moore MAS, Williams N: In Robinson WA (ed): "Hemopoiesis in Culture." Washington: DHEW Publication No (NIH) 74-205, 1973, pp 16–26.
35. Metcalf D, Moore MAS: In Wolstenholme GEW, O'Connor M (eds): "Haemopoietic Stem Cells." Amsterdam: Elsevier-Excerpta Medica North-Holland, 1973, pp 157–175.
36. Burgess AW, Metcalf D: J Cell Physiol 90:471, 1977.
37. Handman E, Burgess AW: J Immunol 122:1134, 1979.
38. Kurland JI, Broxmeyer HE, Pelus LM, Bockman RS, Moore MAS: Blood 52:388, 1978.
39. Lin H-S, Gordon S: J Exp Med 150:231, 1979.
40. Van Zant G, Goldwasser E: In Clarkson B, Marks PA, Till JE (eds): "Differentiation of Normal and Neoplastic Hematopoietic Cells." Cold Spring Harbor, New York: Cold Spring Harbor Laboratory, 1978, pp 165–177.
41. Metcalf D, Johnson GR: J Cell Physiol 99:159, 1979.
42. Metcalf D, Burgess AW, Johnson GR: In Baum SJ (ed): "Experimental Hematology 1979." New York: Springer (in press).
43. Sheridan JW, Stanley ER: J Cell Physiol 78:451, 1971.
44. Sheridan JW, Metcalf D: J Cell Physiol 80:129, 1972.
45. Bradley TR. Sumner MA: Aust J Exp Biol Med Sci 46:607, 1968.
46. Nicola NA, Burgess AW, Metcalf D: J Biol Chem 254:5290, 1979.
47. Zucali JR, Mirand EA: In Murphy MH (ed): "In Vitro Aspects of Erythropoiesis." New York: Springer, 1978, pp 218–224.
48. Chan SH, Metcalf D: Blood 40:646, 1972.
49. Chan SH, Metcalf D: Cell Tiss Kinet 6:185, 1973.
50. Asano S, Urabe A, Okabe T, Sato N, Kondo Y, Ueyama Y, Chiba S, Ohsawa N, Kosaka K: Blood 49:845, 1977.
51. Asano S, Sato N, Mori M, Ohsawa N, Kosaka K, Ueyama Y: Br J Cancer (in press).
52. Shadduck RK, Metcalf D: J Cell Physiol 86:247, 1975.
53. Johnson GR, Burgess AW: J Cell Biol 77:35, 1978.
54. Bradley TR, Stanley ER, Sumner MA: Aust J Exp Biol Med Sci 49:595, 1971.
55. Metcalf D, Moore MAS, Sheridan JW, Spitzer G: Blood 43:847, 1974.
56. Francis GE, Berney JJ, Chipping PM, Hoffbrand AV: Br J Haematol 41:545, 1979.
57. Sachs L: Nature 274:535, 1978.
58. Metcalf D: Int J Cancer 24:616, 1979.
59. Metcalf D: Int J Cancer 25:225, 1980.
60. Ichikawa Y: J Cell Physiol 76:175, 1970.
61. Fibach E, Sachs L: J Cell Physiol 86:221, 1975.
62. Honma Y, Kasukabe T, Hozumi M: J Natl Cancer Inst 61:837, 1978.
63. Hayashi M, Okabe J, Hozumi M: Gann 70:235, 1979.
64. Hozumi M, Umezawa T, Takenaga K, Ohno T, Shikita M, Yamana I: Cancer Res 39:5127, 1979.
65. Broxmeyer HE, Jacobsen N, Kurland J, Mendelsohn N, Moore MAS: J Natl Cancer Inst 60:497, 1978.
66. Burgess AW, Metcalf D, Russell SHM, Nicola NA: Biochem J 185:301, 1980.
67. Howard M, Burgess A, McPhee D, Metcalf D: Cell 18:993, 1979.
68. Dexter TM, Allen TD, Lajtha LG: J Cell Physiol 91:335, 1977.
69. Metcalf D, Nakamura K, Wiadrowski M: Aust J Exp Biol Med Sci 43:413, 1965.
70. Moore MAS: In Cleton FJ, Crowther D, Malpas JB (eds): "Advances in Acute Leukemia." Amsterdam: North-Holland, 1975, pp 161–227.
71. Rowe WP: In Zabriskie JB (ed): Harvey Lectures Series 71. New York: Academic Press, 1978, pp 173–192.

Erythropoietin and Red Cell Differentiation

Eugene Goldwasser

Department of Biochemistry, The University of Chicago, Chicago, Illinois 60637

Detailed study of the molecular and cellular mechanisms of red cell differentiation is facilitated by the fact that the glycoprotein, erythropoietin (epo), has an unambiguous primary effect as the inducer of erythropoiesis. In adult animals, under normal conditions, it has been well established that production of mature erythrocytes does not occur in the absence of epo. In cultures of adult bone marrow cells, the synthesis of hemoglobin and the appearance of erythroid colonies and of bursts are dependent on addition of epo to the medium. This system may, therefore, be viewed as a prototype for the study of the induction of cell differentiation in general and, more specifically, of blood cell differentiation. This paper will deal with the properties of epo, its mode of action, and the nature of its target cells.

PROPERTIES OF EPO

Epo has been purified from two sources: plasma from anemic sheep [1] and urine from patients with hypoplastic anemia [2]. The amount of sheep plasma epo obtained was far too small for extensive study of its chemical properties. What little is known is summarized in Table I, in comparison with the still incomplete information on human urinary epo. One surprising difference between the two is seen in their potency (units per mg of protein*). Human epo has about 7- to 8-fold greater potency than sheep plasma epo when assayed in mice or rats. Since it has already been shown that human and sheep epo differ immunologically [3], the difference in activity could represent an intrinsic difference in the properties of the two substances. Alternatively, since both were assayed in heterologous animals, the difference in potencies may be due to the assay method; it is possible that sheep epo assayed in sheep and human epo assayed in man would have the same potency. Since potency must be measured with reference to a standard, and since the present International Reference Preparation is a crude fraction of human urinary epo, this problem may not be solved for a very long time.

The two epos also differ somewhat in molecular size and in carbohydrate content. The carbohydrate composition of sheep epo was determined analytically; for human epo, we are still in the process of determining the composition, and the total carbohydrate calculated as the residual material not accounted for by amino acids is very tentative.

---

*A unit of epo is arbitrarily defined as one-tenth the activity contained in an amupule of the International Reference Preparation.

Received and accepted April 7, 1980.

© 1981 Alan R. Liss, Inc., 150 Fifth Avenue, New York, NY 10011

TABLE I. Properties of Purified Erythropoietin

	Sheep	Human
Potency (units/mg protein)	9,200	70,400
Sedimentation coefficient	4.1S	3.6S
Molecular weight	46,000	39,000
$A^{1\%}$ (λ max)	7.4	8.5
Percent carbohydrate	24	$\cong 59$
Sialic acid	16–18 residues/mole	–
Mannose	11–12 residues/mole	–
Galactose	18 residues/mole	–
N-acetyl glucosamine	11–12 residues/mole	–
Glucose	6 residues/mole	–

TABLE II. Amino Acid Compositions of α and β Erythropoietin

	Residues per mole	
	α	β
asp	11	11
thr	9	9
ser	9	9
glu	17	17
pro	8	8
gly	9	9
ala	16	15
val	9	8
met	2	2
ile	7	6
leu	21	19
tyr	4	4
phe	4	3
his	2	3
lys	6	6
arg	10	9
cySH	3	3
trp	2	2

In the last step of the purification of human epo, the hydroxylapatite column yielded two fractions (α and β) of identical potency and molecular size, but with slightly different electrophoretic mobilities [2]. The amino acid compositions of the α and β forms are essentially the same (Table II). We do not yet know the molecular basis for the difference between α and β epo; it could be either in the carbohydrate portion or in amidation of the glutamates and aspartates. The former possibility is now being studied. The amino acid compositions show no peculiarities; there are the expected high proportions of hydrophobic residues anticipated from our observation that epo tends to bind to glass surfaces, so that, at very low concentrations, it is lost by adsorption.

Human epo loses biological activity when it is iodinated (Table III). The amount of activity remaining after iodination with either chloramine-T [4] or "Iodogen" [5] is less than, or approximately equal to, the amount of unlabeled epo calculated from the degree of labeling and the number of tyrosine residues. These data suggest that tyrosine is in, or closely associated with, the active site of epo, and that substitution of one or more iodines

TABLE III. Iodination of Erythropoietin

	Specific activity of erythropoietin (μCi/μg)	Atoms of I/molecule	% of original biological activity found	% of unlabeled erythropoietin
Chloramine-T, DMSO	2.19	0.049	68	95
Iodogen	12.6	0.276	37	76
Bolton-Hunter	71	0.71	52	48

TABLE IV. Effects of Oxidation and Iodine Substitution on Erythropoietin Activity

Time of exposure to Iodogen (sec)	% of original activity	
	Without iodide	With iodide
20	86	75
40	85	48
55	94	36

in the ring results in loss of biological activity. The conditions of iodination (eg, exposure to a strong oxidant) are not responsible for inactivation; this is indicated in Table IV, which shows that oxidation in the absence of iodine does not result in loss of activity. There is a report in the literature that iodination of epo with lactoperoxidase does not cause inactivation [6]. Since the epo preparations used in these earlier experiments were 0.5% and 0.1% pure, and since, in our hands, the lactoperoxidase method results in labeling equivalent to 0.06 atoms of iodine per molecule of epo (this is equivalent to 99.4% of the epo being unlabeled), it is probable that the finding of active epo after iodination was due to the fact that a negligible fraction of the epo had any iodine substitution. The data in Table III show that acylation of free amino groups by use of the Bolton-Hunter [7] reagent also causes loss of biological activity, implicating the α amino and/or lysyl residues in the active site. Although these findings are disappointing because they make it impossible to do cell-binding experiments with epo iodinated by either of these methods, iodination does not interfere with the reaction between epo and its antibody. We have, therefore, been able to develop a radioimmunoassay for epo by using the biologically inactive iodinated material labeled by the "Iodogen" or Bolton-Hunter method [8].

An alternative technique of labeling epo is based on the method of Van Lenten and Ashwell [9], in which the terminal two carbon atoms of the sialic acid residues are removed by periodate oxidation and the resulting adehydic carbon termini are reduced with NaB^3H_4. This method yields an epo derivative with 12 residues tritiated and with no loss of biological activity. Unfortunately, even the highest-specific-activity borohydride does not yield epo with enough radioactivity for binding studies. We calculate that cellular binding studies will involve measuring of femtomolar amounts of labeled epo.

At least 1.5 of the 3 sulfhydryl groups of epo can be acylated with ^{14}C-N-ethyl maleimide (NEM) with full retention of activity, indicating that all three cysteines are reduced and that two of them are not part of the active site. Neither ^{14}C- nor ^3H-NEM is sufficiently radioactive for cell-binding experiments. With N-(7-dimethylamino-4-methyl coumarinyl) maleimide (DACM) [10], it is possible to derivatize the sulfhydryls, yielding fluorescent epo with complete activity. DACM-epo has been used very recently for the study of cell binding by fluorescent microscopy. This method, with the equipment available to us, is not suitable for quantitative measurements, but it can yield useful qualitative information.

MODE OF ACTION OF EPO

In view of recent findings that some of the effects ascribed to epo were really due to the ubiquitous contaminant, bacterial endotoxin, we prepared epo free of endotoxin by a modification of a method suggested by Dr. R. Shadduck. Pure epo was centrifuged over a cushion of endotoxin-free human albumin at 35,000 rpm for 17 hours in a Beckman SW50L rotor; the solution above the albumin cushion contained 70–100% of the epo activity and was free of endotoxin, according to the limulus lysate test. This endotoxin-free epo still caused increased hemoglobin synthesis in suspension cultures, erythroid burst formation in semi-solid medium, increased RNA synthesis by marrow cells, and the formation of the marrow cytoplasmic protein, which we consider to be the mediator between the surface and the nucleus of epo-responsive cells.

The problem of the molecular mode of action of epo is made difficult by the fact that several distinctly different types of cells are affected by it. Much of the biochemical study of epo action in this and other laboratories has been done on late forms, and we do not yet know whether similar effects will be seen in more primitive cells when it becomes possible to study them.

In very recent studies of binding of DACM-epo to cells, we found that a small fraction (about 2%) of normal mouse marrow cells bind epo (become fluorescent) within a few minutes of exposure. In contrast, mouse thymus cells show no binding of DACM-epo at all. With marrow, we have seen binding to some small cells, not yet identified, and to large, early megakaryocyte-like cells. The bound DACM-epo becomes trypsin-resistant after a relatively short time, suggesting that it may be internalized.

In the late-differentiated cells which predominate in normal marrow, the earliest action of epo appears to be on the appearance of a cytoplasmic protein that can cause an increase in nuclear transcription when assayed in isolated nuclei [11]. This can occur in the absence of measurable protein synthesis, as can the early effect of epo on transcription [12], indicating that the cytoplasmic protein may be modified in some manner from a preexisting form, or that it may be released from a complex. The active cytoplasmic protein does not appear to have any epo activity when assayed using fresh marrow cells, but it must be emphasized that the assay is rather insensitive; it would not detect fewer than 4×10^{12} molecules. It is possible that epo reacts with cellular receptors and becomes internalized and that the complex of epo, receptor, and possibly other cell surface constituents is the active cytoplasmic factor that causes nuclear transcription. It is also possible that the interaction of epo at the outer cell surface initiates transmembrane signaling, which results in the release or alteration of a protein on the inner surface of the membrane, and that this protein is the mediator of epo action on the nucleus. This mechanism would not require internalization of epo for its action. When sufficiently active radiolabeled epo is available, it will be possible to test these ideas rigorously. In the meantime, we are continuing the arduous, frustrating, and at present unrewarding task of trying to characterize the marrow cytoplasmic factor and to learn more about its mechanism of action. Unfortunately, very little can be added now to the statements made here two years ago on this subject [13] – ie, the cytoplasmic protein appears to act on RNA polymerase I + III, on polymerase II, and on the availability of template DNA in whole nuclei, without having an effect on broken nuclei.

TABLE V. Effect of Erythropoietin on Synthesis of Globin mRNA

Time (hours)		Globin mRNA[a] synthesized	+epo/−epo
2.5	−epo	0.52	0.9
	+epo	0.49	
10	−epo	0.23	1.8
	+epo	0.41	
20	−epo	0.08	4.1
	+epo	0.33	

[a]In arbitrary units; assayed with the Ehlich's ascites cell lysate method.

In whole marrow cells (late forms) exposed to epo, stimulated synthesis of at least two classes of heterogeneous nuclear RNA occurs within 15–30 minutes [13]. The function of these transcripts is not yet known. The question of precisely when globin message is synthesized is another unanswered one. In fetal liver cell cultures, it has been reported that an increase in translatable globin mRNA appears about 6 hours after addition of epo and that RNA hybridizable to globin cDNA also appears at this time [15]. Unfortunately, no data between zero and 6 hours were provided. In marrow cells, a 9S–10S RNA is synthesized as early as one hour after epo [16], but by 2.5 hours there is still no epo-stimulated increase in translatable globin message, suggesting that the early 9S–10S RNA is not globin message (Table V). After 10 hours of exposure to epo, there is an 80% increase in translatable globin RNA [17]; this interval is similar to the time at which globin synthesis starts, and may suggest that regulation is primarily at the transcriptive step.

We believe that, in the late erythroid cells at least, the events resulting in massive hemoglobin synthesis are all causally related to the initial intracellular events. We have studied the temporal relationship between hemoglobin synthesis and the synthesis of erythrocyte-characteristic membrane proteins [18]. In these experiments we used marrow from rats made polycythemic by exposure to 0.5 atmosphere for 14 days, followed by 5 days at normal pressure. These animals form very few red cells, probably because the surplus of red cells in the circulation causes a lowering of the epo titer and consequent suppression of erythropoiesis.

Marrow cells from such rats, when cultured in the presence of epo, take about 90 hours for maximal (about 20 times control) hemoglobin synthesis. This observation is in marked contrast to our finding with marrow from mice, in which maximal hemoglobin synthesis in cultures of "primitive" cells — ie, hemoglobin synthesis by cells derived from BFU-E — occurs at 8 days. The difference may be related to the curious fact that erythroid bursts cannot be grown in cultures of rat marrow cells.

We examined the effect of epo on the synthesis of red cell membrane proteins by marrow cells from suppressed rats and found that stimulated synthesis of the major integral membrane protein, band 3, occurred at about the same time as stimulated hemoglobin formation (90–114 hours after addition of epo). In contrast, stimulated synthesis of one of the major glycoproteins of the red cell membrane, glycophorin, is maximal between 30 and 66 hours after epo. We found, in addition, that epo stimulated the synthesis of some membrane proteins not found in mature red cells and that it had no detectable effect on the synthesis of spectrin, the other major glycoprotein of the red cell membrane. These observations suggest that epo causes the "switch-on" of the synthesis of some "temporary" membrane proteins that may characterize intermediate erythroid cells. These are either replaced later or modified and do not appear, as such, in the final erythrocyte. Another con stituent, glycophorin, is synthesized long before hemoglobin and band 3, and it persists through the final stages of differentiation. It could, therefore, be considered an early marker for erythroid cells. Spectrin, however, appears in this system to be a non-changing membrane constituent that probably is present throughout the erythroid pathway from the most primitive epo-responsive cells to the end-product red cell.

THE NATURE OF THE EPO-RESPONSE CELL

As mentioned earlier, work in this field is made difficult by the lack of a single, well-defined, epo-responsive cell type. The epo-responsive cells probably represent a population containing a continuum of cells that range from the most primitive of the blood cell precursors (in our view, as summarized below, this includes the hemopoietic pluripotent stem cell) through the earliest burst-forming cell, intermediate forms, the colony-forming cell, and possibly some defined erythroid cells. The only cells in the erythroid series for which there is evidence of no epo effect are normoblasts, reticulocytes, and the end-product erythrocyte. It is important in this respect to emphasize the difference between those cells that can be affected by epo, but are already on the erythroid pathway, and those that are epo-responsive, but may not be exclusively on the erythroid pathway. It is quite possible that similar considerations hold for the granulocyte/macrophage pathway — ie, that CSF (if it is a granulopoietin) can act on the whole continuum of cells from the pluripotent to almost the end stage. Action on the latter type of cells, as in the erythroid pathway, would be on a "committed" cell, whereas action on the earliest forms may involve pluripotent cells.

We have published evidence consistent with the view that epo can act on pluripotent stem cells [19] and that epo and CSF can act on the same cell [20, 21]. In summary form, these data show 1) that epo added to mouse marrow cell cultures causes an increase in spleen colony-forming cells; 2) that epo causes a decrease in the number of cells responsible for the eventual formation of non-erythroid spleen colonies; 3) that epo can cause a decrease in the number of CSF-induced granulocyte/macrophage colonies in vitro; 4) that CSF can cause a decrease in the number of erythroid bursts, but does not affect colonies composed of CFU-E progeny; 5) that CSF can cause a decrease in epo-induced hemoglobin synthesis; 6) that the "competitive" effects of CSF and epo are dose-dependent (eg, the suppressive effect of CSF can be decreased by increasing the concentration of epo and increased by increasing the concentration of CSF); 7) that pure lung-cell CSF at a very low concentration (10–25 units/ml) in the presence of epo causes a substantial potentiation of the epo effect; and 8) that a low concentration of epo in the presence of CSF causes a moderate increase in the number of granulocyte/macrophage colonies. The mechanism of these positive effects of one inducer, at low concentration, on the action of the other is not understood, but the data indicate clearly that the inducers can act on the same cell, which is in agreement with evidence cited earlier. A similar sort of effect has recently been described by Metcalf et al [21].

The most widely accepted working model of hemopoiesis postulates the locus of action of inducers such as epo to be committed, unipotent stem cells. These have the option of either not differentiating further or of having their progeny become only erythroid cells (or, with different unipotent cells and either granulopoietin or thrombopoietin, become neutrophils or platelets). This view is not consistent with the evidence presented above, which suggests that epo and CSF have a common target cell. We have proposed an alternative model [23], the features of which are 1) pluripotent stem cells are predominantly in G_0, but are capable of cycling (this is well documented); 2) when in cycle, these cells de-

TABLE VI. Features of Alternative Models of Blood Cell Differentiation

Noninstructional	Instructional
Pathway of differentiation determined by unknown process that converts pluripotent stem cells to "irreversibly committed unipotent stem cells" (ICUSC)	Pathway of differentiation determined by combination of time pluripotent stem cell has proper receptor and concentration of inducer[a]
Existence of ICUSC early	No ICUSC early; cells become irreversibly committed as differentiation proceeds
Inducers act on ICUSC to cause cell division; mitogenic action primary	Inducers act on pluripotent (3 or 4) stem cells to instruct which pathway to take; mitogenic action secondary
Majority view	Minority view
Not proven	Not proven

[a]"Inducer" means erythropoietin (epo), granulopoietin (gpo), or thrombopoietin (tpo).

velop, transiently, receptors for epo, granulopoietin (gpo), and thrombopoietin (tpo); 3) when the correct concentration of inducer coincides with the appearance of receptors, the cell is induced to go down the pathway defined by receptor and inducer; 4) if there is no inducer, or if it is present at too low a concentration, the cell will, in the process of cycling, lose the receptor and the capacity for induction; if it does not encounter a sufficient amount of any inducer, it will revert to the G_0 state. In this model, cells are inducible only during distinct periods of the cell cycle — ie, the state of commitment is reversible and is defined by presence of receptors. In Table VI the features of this model are contrasted with those of the "three-tier" model.

Almost certainly this model will prove to be wrong in many, if not all, of its details. At the moment, it can account for our data and for data obtained in vivo that show competitive effects operating within the three main pathways of blood cell differentiation.

There is one profoundly disturbing observation of an epo effect that is not consistent with either model. It has been shown that, in vitro, addition of pure epo to the medium results in the appearance of megakaryocyte colonies [24, 25]. At present, these findings can only prove a challenge to us to elaborate a plausible mechanism for this effect. All of us, of course, will ponder this challenge and try to devise hypotheses and experimental strategies to test them, in the hope that this unexpected development will open up new approaches to the understanding of blood cell differentiation.

ACKNOWLEDGMENTS

The work reported here was done jointly with Drs. G. Van Zant, J. B. Sherwood, B. D. Tong, R. Pine, T. Weiss, G. Inana, and C. K.-H. Kung under the following grants from The National Institutes of Health: CA 18375, HL 21676, HL 16005, CA 19265; and from DOE contract EY-76-C-02-0069.

REFERENCES

1. Goldwasser E, Kung CK-H: Proc Natl Acad Sci USA 68:697, 1971.
2. Miyake T, Kung CK-H, Goldwasser E: J Biol Chem 252:5558, 1977.
3. Schooley JC, Mahlman LJ: In Gordon AS, Condorelli M, Peschle C (eds): "Regulation of Erythropoiesis." Milan: I. L. Ponte, 1972, p 167.

4. Fraker PJ, Speck JC Jr: Biochem Biophys Res Commun 80:849, 1978.
5. Hunter WM, Greewood FC: Nature 194:495, 1962.
6. Murphy MJ Jr: Biochem J 159:287, 1976.
7. Bolton AE, Hunter WM: Biochem J 133:529, 1973.
8. Sherwood JB, Goldwasser E: Blood 54:885, 1979.
9. Van Lenten L, Ashwell G: J Biol Chem 246:1889, 1970.
10. Machida M, Ushijima N, Machida MI, Kanoka Y: Chem Pharm Bull 23:1385, 1975.
11. Chang SC-S, Goldwasser E: Dev Biol 34:246, 1973.
12. Gross M, Goldwasser E: Biochim Biophys Acta 287:514, 1972.
13. Goldwasser E, Inana G: In Golde D, Cline M, Metcalf D, Fox CF (eds): "Hematopoietic Cell Differentiation." New York: Academic Press, 1978, p 15.
14. Gross M, Goldwasser E: Biochemistry 8:1795, 1969.
15. Terada M, Ramirez F, Cantor L, Maniatis GM, Bank A, Rifkind RA, Marks PA: In Nakao K, Fisher JW, Takaku F (eds): "Erythropoiesis." University of Tokyo Press, 1975, p 23.
16. Gross M, Goldwasser E: J Biol Chem 246:2480, 1971.
17. Pine R, Goldwasser E: Unpublished data.
18. Tong BD, Goldwasser E: Unpublished data.
19. Van Zant G, Goldwasser E: J Cell Physiol 90:241, 1977.
20. Van Zant G, Goldwasser E: Blood 53:946, 1979.
21. Van Zant G, Goldwasser E: In Baum SJ, Ledney GD (eds): "Experimental Hematology Today." Berlin: Springer-Verlag, 1979, p 63.
22. Metcalf D, Johnson GR, Burgess AW: Blood 55:138, 1980.
23. Goldwasser E: Fed Proc 34:2285, 1976.
24. Vainchenker W, Bouguet J, Guichard J, Breton-Gorius J: Blood 54:940, 1979.
25. Axelrad A, McCleod D: Personal communication.

On Deciding Which Factors Regulate Cell Growth

Arthur B. Pardee, Paul V. Cherington, and Estela E. Medrano

Laboratory of Tumor Biology, Sidney Farber Cancer Institute, and Department of Pharmacology, Harvard Medical School, Boston, Massachusetts 02115

Many conditions affect the growth of animal cells in culture. Some, such as essential nutrients, are necessary for growth as precursors of macromolecules. Other substances appear to have regulatory functions. Their presence or absence determines whether the cell will continue to grow or move into a resting, quiescent state. In particular certain factors (usually provided by serum) are thought to have regulatory roles. To decide which factors are regulatory is a difficult problem. We propose here that no single criterion is sufficient to permit a decision and that the best course at present is to apply several criteria. Three such criteria are proposed. To the degree that the tests are satisfied we can tentatively decide whether or not a given factor is regulatory.

Key words: cell culture, growth control, serum factors, protein synthesis, cell cycle

Cells in culture require many soluble factors to permit their rapid growth. These factors include ions, sugars, amino acids, vitamins, and other low molecular weight compounds generally provided in tissue culture media. Also needed are hormones, peptides and proteins that are usually provided by serum. Absence of any of these factors prevents rapid growth. Cells deprived of certain factors go into a quiescent state from which they may emerge if they are provided with complete media [1]. The factors necessary for growth may be divided between those that are required as nutrients for example, and those that somehow regulate growth, eg, that under physiological conditions determine continued growth vs quiescence. Deciding which factors are nutrients and which are regulatory is not a simple problem. Thus, essential amino acids such as isoleucine are required for growth but they also have been suggested to be growth regulatory [2].

The problem addressed in this article is "What criteria can we at present use to decide whether a given substance is or is not a regulatory factor?" Our approach to this question is similar to one suggested by E.A. Adelberg about a quarter century ago. He asked how one decides which enzymes are involved in a given metabolic pathway as contrasted to those enzymes which only seem to be involved because they act on potential intermediates [3].

Received and accepted May 5, 1980.

© 1981 Alan R. Liss, Inc., 150 Fifth Avenue, New York, NY 10011

Adelberg proposed that no one criterion was adequate at the present state of knowledge, and that one had to apply several criteria. If all were satisfied, one might conclude that the enzyme in question was likely to be involved. Similarly, for growth control we suggest that three criteria can be used. We illustrate this proposal with results from our recent studies.

FIRST CRITERION: THE "ARREST TEST"

Regulatory factors are mainly functional in and required for traverse of the G_1 phase of the cell cycle (Fig. 1). In general, cells that are at rest in vivo have a 2N DNA content and are described as being in G_1 or G_0 [1]. Similarly, in culture mammalian cells come to rest with a 2N DNA content under most physiological conditions that are suboptimal for growth, eg, at high density, in the absence of a substratum, or with insufficient serum factors (Fig. 2). We have proposed earlier that cells stop in G_1 because they are unable to complete some crucial biochemical events under adverse conditions. These events are required for cells to initiate DNA synthesis [4]. Thus, in the absence of a truly regulatory factor cells should arrest in G_0/G_1, but removal of a nutrient could cause arrest in other parts of the cycle.

As an illustration of this test we refer to cytofluorimetric patterns of CHEF/18 [5] cells (Fig. 3). These cells grow in a defined medium lacking serum and containing epidermal growth factor (EGF), fibroblast growth factor (FGF), insulin, and transferrin [6]. In this medium the population has a typical distribution of cells in the G_1, S, and G_2 + M phases of the cycle. If these cells are placed for 12 hours in medium lacking EGF and insulin their DNA histograms reveal arrest in the G_1 part of the cycle. This specific arrest suggests that these two factors are involved in growth regulation. By contrast, after transferrin is removed the cell number no longer increases but the population remains distributed around the cycle. Thus, according to this test, transferrin is not growth regulatory although it is essential for growth.

That deletion of an individual factor preferentially arrests cells in G_0/G_1 does not prove or disprove that this factor also limits growth under other conditions, for example at confluence or after serum starvation. Addition of EGF and insulin, but not nutrients, stimulates the growth of both serum starved or confluent 3T3 cells [7, 8] indicating that the peptide growth factors, but not nutrients, are regulatory for both conditions.

THE REGULATED CELL CYCLE

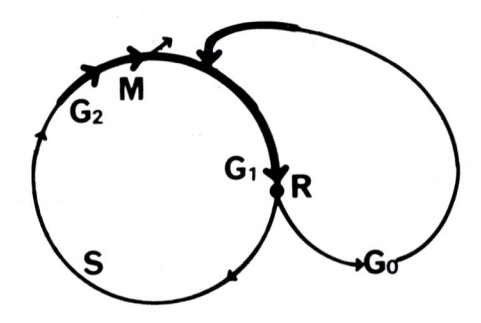

Fig. 1. The cell cycle. Heavy line indicates regions of the cell cycle where regulatory events may occur.

Nutrients have been proposed to be growth regulatory [2] because their rates of transport are reduced when cells become quiescent [9]. For instance, lower transport of phosphate or uridine by untransformed cells was correlated with G_0/G_1 arrest [10]. However, these observations do not rule out an alternative explanation: that cells shut down their underutilized active transport processes as a consequence of growth arrest. Indeed, a reduced rate of uridine transport, which is not essential for growth, is observed when cells come to confluence. Transport of glucose and of phosphate are also not regulatory [11, 12].

SECOND CRITERION: THE TRANSFORMATION TEST

This test is essentially based on mutation. We propose that when a mutation causes growth control to become relaxed or lost so that cells do not arrest as they did previously, a true regulatory factor should no longer have the same role. If removal of the factor does have the same effect, still stopping the mutated cells in G_0, this factor is either not regulatory or its function is not modified by this mutation (though it might be modified by another mutation).

We already have a plentiful supply of cells whose growth control is relaxed or eliminated; they are transformed cells [1]. These cells generally have a diminished requirement for serum: they continue to grow at much lower serum concentrations than normal cells require. Some, but not others, depending on the specific transformation, can be arrested with a 2N DNA content under conditions more stringent than needed to arrest untransformed cells [6, 13]. These cells generally become arrested at high density owing to their using up nutrients rather than from lack of serum factors [14].

As an example of selective mutational loss of a growth factor requirement, we can consider effects of transformation on the CHEF/18 line [6, 26]. In totally defined medium two transformed derivatives of CHEF/18, CHEF/16, and T30-4, still required insulin and transferrin, but both lost their EGF requirements, as did some transformed BHK lines [6].

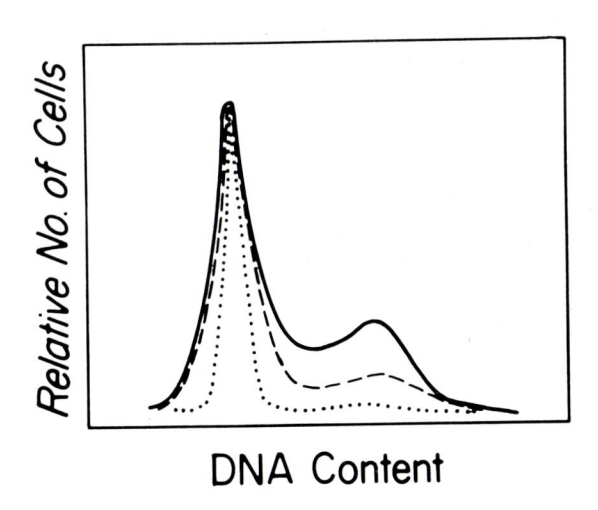

Fig. 2. Cell cycle distributions of Balb/c 3T3 cells. Balb/c 3T3 fibroblasts were plated into 60-mm dishes in medium containing 10% fetal calf serum at an initial density of 3×10^3 cells/cm^2. (———) cells processed for cytofluorometric analysis 24 hr after plating (exponentially growing). (- - - -) cells processed after 72 hr (at high density). (. . . .) cells shifted to 0.5% calf serum and processed for cytofluorometric analysis after 48 hr (serum-starved cells).

By this test, EGF is a regulatory factor. Insulin is not eliminated as a regulatory factor; the test shows only that these particular transformations did not diminish its role.

Transferrin is an example of a non-regulatory factor. This was indicated by examining the growth requirements of the polyoma virus transformed BHK-21 cells, which by all criteria seem to be totally transformed. These cells continued to grow slowly in the absence of serum, but cells did not accumulate at G_1 (Table I). They still required transferrin but not EGF, FGF, or insulin for rapid growth. Similarly, we have observed that SV40-transformed Swiss 3T3 cell (SV3T3) growth in the presence of 0.5% serum is greatly enhanced by transferrin (results not shown), confirming a previously reported result [15] that SV3T3 cells require transferrin, as well as biotin, for growth in serum-free medium. Thus transferrin is needed, but according to these tests it is not regulatory.

THIRD CRITERION: AVAILABILITY TEST

A true regulatory factor should be present under physiological conditions at a reasonable concentration. This criterion rules out some substances that affect growth and fulfill the first two criteria, but which are not found in vivo in the animal in question. Examples that come to mind are caffeine [16], cycloheximide (CHM) [17], lectins [18], or phorbol myristate acetate [19].

Such compounds may not be natural regulators, but they can provide clues as to mechanisms of regulation. For example, we consider the effects of CHM [17] which satisfies the first two criteria. At low CHM concentrations normal cells are arrested with a G_0/G_1 DNA content. Transformed cells are not arrested in G_1 by CHM as readily as are normal cells [20]. The small effects of CHM on four differently transformed 3T3 lines are shown in Figure 4. CHM diminishes the growth rates of all these lines similarly, but cells remain distributed through the cell cycle [20]. These results with transformed cells strongly support our original conclusion that CHM specifically affects the major growth regulatory event that is important in transformation [17]. We have proposed that CHM's specific effect is to limit the accumulation of a labile protein that has a half life of a few hours. This protein is proposed to be required for transit past the restriction point in G_1 in preparation for DNA synthesis.

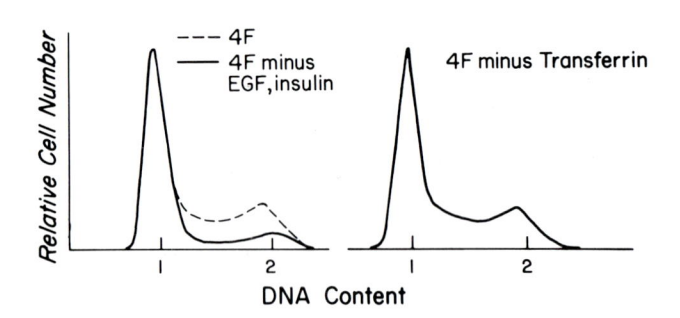

Fig. 3. Cell cycle distributions of CHEF/18 deprived of various growth factors. CHEF/18 cells were plated into 60-mm dishes in medium containing 10% fetal calf serum at an initial density of 3×10^3 cells/cm^2. After one day the cells were extensively washed and fed with media containing EGF, insulin, FGF, and transferrin (4F) or with media lacking one or more of these factors. After 12 hr, plates were sampled for cytofluorometric analysis.

Limiting the concentrations of amino acids [21] might cause cells to be arrested in G_1 by the same mechanism as CHM, ie, by limiting the rate of protein synthesis. Although growth of yeast is not dependent on intercellular communication and hormone-like factors (except for mating hormone) [22], limiting an amino acid essential for an auxotrophic yeast does restrict its growth [23], as CHM also does [24]. This basic mechanism might persist in animal cells and have superimposed on it the effects of growth factors such as EGF and insulin.

TABLE I. Transferrin is Non-Regulatory, but Essential for Rapid Growth of Polyoma-Transformed BHK Cells

Condition	Cells/cm^2 ($\times 10^{-3} \pm$ SEM)	%G$_1$
4F	99 ± 1	45
4F minus transferrin	58 ± 7	47
4F minus EGF, FGF, insulin	116 ± 7	47
no factors or serum added	38 ± 6	50

Polyoma-virus-transformed BHK cells were plated at 3×10^3 cells/cm^2 into 60-mm dishes in medium containing 10% serum. After one day the medium in each plate was replaced with serum-free media containing EGF, insulin, FGF, and transferrin (4F) or 4F with the indicated factors omitted. After one day, samples were prepared for cytofluorometric analysis. Analysis of pyBHK cells after two days under similar conditions gives the same result. Duplicate cell counts were made at the time of the shift and two days later. Cell density at the shift to experimental conditions was $12 \pm 0.5 \times 10^3$ cells/cm^2.

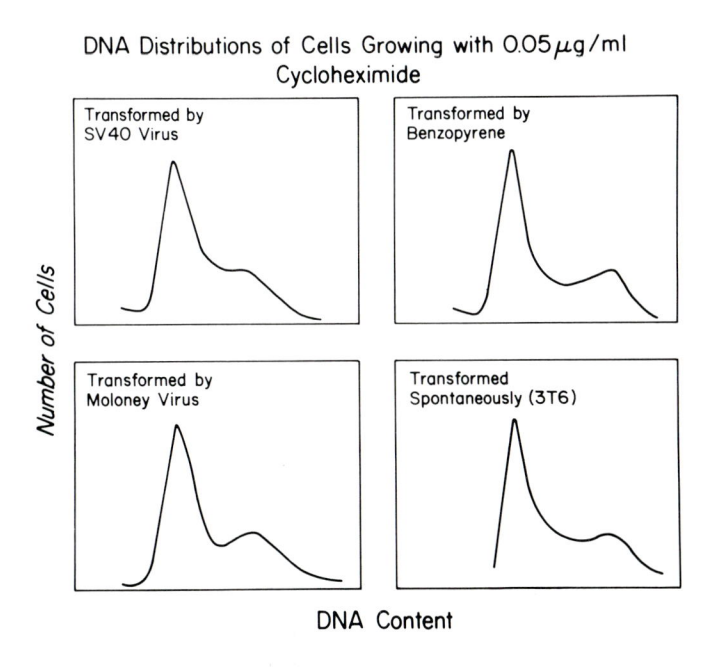

Fig. 4. Effects of CHM on various transformed 3T3 cells. Cells were plated into 60-mm dishes in medium containing 10% calf serum at an initial density of 3×10^3 cells/cm^2. After 24 hr, 0.05 μg/ml CHM was added and cells processed for cytofluorometric analysis 48 or 72 hr later.

THE CHOICE BETWEEN POSSIBLE REGULATORY FACTORS

More than one factor sometimes fulfills the three criteria just proposed. For example, somatomedins appear to replace insulin for permitting growth of Balb/c 3T3 cells [25]. Inability of serum from hypophysectomized animals to support cell growth unless somatomedin C is added suggests that somatomedin C may be a physiologically important factor. The low concentration at which it functions in culture also indicates that somatomedin C is physiologically important. Indeed, the fact that insulin is needed in culture at concentrations of micrograms/ml as compared to its much lower (nanogram/ml) concentrations in vivo, and the nanogram/ml concentrations required for other factors, has suggested that insulin might substitute for the actual growth factor for these cells.

We further illustrate the dilemma of choice between factors with some of our recent observations. Thrombin supports growth in defined medium of CHEF/18 cells even better than does EGF. Other proteases, including trypsin and urokinase, did not stimulate growth of these cells. Thrombin almost completely eliminated the cells' need for EGF, and EGF could not support as rapid growth as did thrombin [26]. Both proteins functioned optimally at similar concentrations (Fig. 5). Transformation eliminated the requirements for both factors (Table II). Although both thrombin and EGF are present in vivo, the role of thrombin in wound repair (blood clotting) makes it a reasonable candidate for regulating fibroblast growth, because fibroblasts fill a wound following clotting. Which is the true regulatory factor?

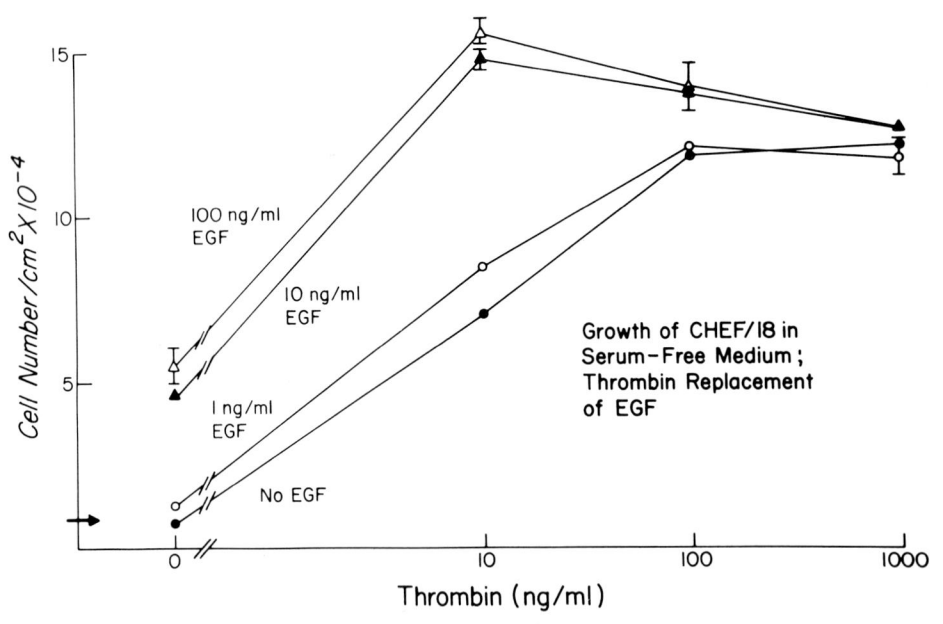

Fig. 5. Growth of CHEF/18 in serum-free medium: thrombin replacement of EGF. CHEF/18 cells were plated at 3×10^3 cells/cm^2 into 35-mm dishes in medium containing 10% serum and one day later shifted after extensive washing into media containing FGF, insulin, and transferrin with no EGF (●), 1 ng/ml EGF (○), 10 ng/ml EGF (▲), or 100 ng/ml EGF (△). Thrombin concentrations were varied as indicated. Duplicate cell counts were made at the time of the shift (arrow) and again three days later. SEM (bar) is noted only where the SEM is larger than the symbol.

TABLE II. Transformed CHEF Cells Have Lost Requirements for EGF and Thrombin, but Not for Insulin or Transferrin

Condition	Population doublings in three days ($\pm$ SEM)		
	CHEF/18	CHEF/16	T30-4
4F + thrombin (10 ng/ml)	3.3 ± 0.1	4.0 ± 0.1	3.9 ± 0.1
minus EGF	1.8 ± < 0.1	3.9 ± 0.1	4.0 ± 0.1
minus thrombin	2.4 ± < 0.1	3.7 ± 0.2	3.9 ± 0.1
minus insulin	1.3 ± < 0.1	1.9 ± < 0.1	2.9 ± 0.1
minus transferrin	1.1 ± 0.1	2.1 ± 0.3	(decreased cell number)

Each cell line was plated at 3×10^3 cells/cm^2 into 35-mm dishes in medium containing 10% serum. After one day, the medium in each plate was replaced with serum-free medium containing 4F (10 ng/ml EGF, 10 µg/ml insulin, 10 ng/ml FGF, 5 µg/ml transferrin), and thrombin (10 ng/ml), or with media lacking one of the above factors. Duplicate cell counts were made at the time of the shift into experimental conditions and three days later. The number of population doublings under each condition was calculated as follows:

$$\text{Number of doublings} = \frac{\log \frac{\text{cell number 3 days after shift}}{\text{cell number at shift}}}{\log 2}.$$

Among all the compounds of nature some could accidently interact with a cellular process unrelated to its native action. For example, factors involved with other aspects of growth or development, such as EGF or insulin, can stimulate the growth of cells in culture that are otherwise stimulated by thrombin or somatomedin C. Compounds derived from unrelated organisms such as plants, eg, lectins or phorbol esters, can also interact with mammalian cells and perhaps mimic endogenous factors (like opiates with enkephalins). These compounds that mimic regulatory factors could replace the endogenous factor by acting at the cell surface. Thus, the presence of receptors is not proof that the compound has a physiological role. Accidentally a compound of similar structure may be involved. These factors could act intracellularly on events normally stimulated as a secondary response to the endogenous factor.

Clearly, there are still some dilemmas regarding the nature of the true growth factors. We need other criteria. As possible tests genetic defects can be indicative: defective platelet formation has consequences that can be attributed to deficiencies in the platelet-derived growth factor [27]. The in vivo function of all hormones was pinpointed by organ ablation surgically or loss owing to genetic diseases [28]. Surgical removal of the source of a growth factor, such as hypophysectomy to eliminate somatomedin C production, can be used together with supplementation of a test system with the factor in question [25]. The possibility also exists of using specific antibodies against a growth factor for in vivo testing of a factor's role, as reported some time ago for nerve growth factor [29].

ACKNOWLEDGMENTS

We are grateful to Gail Morreo for technical assistance and to Eileen Fingerman for typing. This work was supported by U.S. Public Health Service grant GM 24571 (to ABP), a U.S. Public Health Service International Research Fellowship to EEM, and a National Research Service Award Predoctoral Training grant 5 T32 GM 07306 to PVC.

REFERENCES

1. Baserga R: "Multiplication and Division in Mammalian Cells." New York: Marcel Dekker, 1976.
2. Holley R: Proc Natl Acad Sci USA 69:2840, 1972.
3. Adelberg EA: Bacteriol Rev 17:253, 1953.
4. Pardee AB: Proc Natl Acad Sci USA 71:1286, 1974.
5. Sager R, Kovac PE: Somatic Cell Genet 4:375, 1978.
6. Cherington PV, Smith BL, Pardee AB: Proc Natl Acad Sci USA 76:3937, 1979.
7. Rossow PW, Pardee AB: Unpublished data.
8. Mierzejewski K, Rozengurt E: Biochem Biophys Res Commun 83:874, 1978.
9. Cunningham DD, Pardee AB: Proc Natl Acad Sci USA 64:1049, 1970.
10. Dubrow R, Pardee AB, Pollack R: J Cell Physiol 95:203, 1978.
11. Barsh GS, Greenberg DB, Cunningham DD: J Cell Physiol 92:115, 1977.
12. Naiditch WP, Cunningham DD: J Cell Physiol 92:319, 1977.
13. Dubrow R, Riddle VGH, Pardee AB: Cancer Res 39:2718, 1979.
14. Moses HL, Proper JA, Volkenant ME, Wells DJ, Getz MJ: Cancer Res 38:2807, 1978.
15. Young DV, Cox FW III, Chipman S, Hartman SC: Exp Cell Res 118:410, 1979.
16. Pardee AB, James LJ: Proc Natl Acad Sci USA 72:4994, 1975.
17. Rossow PW, Riddle VGH, Pardee AB: Proc Natl Acad Sci USA 76:4446, 1979.
18. Gunther GR, Wang JL, Edelman GM: Exp Cell Res 98:15, 1976.
19. Slaga TJ, Sivak A, Boutwell RK (eds): "Mechanism of Tumor Promotion and Cocarcinogensis." New York: Raven Press, 1978.
20. Medrano EE, Pardee AB: Proc Natl Acad Sci USA 77:4123, 1980.
21. Ley KD, Tobey RA: J Cell Biol 47:453, 1970.
22. Hartwell LH: Bacteriol Rev 38:164, 1974.
23. Shilo B, Simchen G, Pardee AB: J Cell Physiol 97:177, 1978.
24. Shilo B, Riddle VGH, Pardee AB: Exp Cell Res 123:221, 1979.
25. Stiles CD, Capone GT, Scher CD, Antoniades HN, VanWyk JJ, Pledger WJ: Proc Natl Acad Sci USA 76:1279, 1979.
26. Cherington PV, Pardee AB: J Cell Physiol (in press).
27. Gerrard JM, Phillips DR, Rao GHR, Plow EF, Walz DW, Ross R, Harker LA, White JG: Blood 54:241A, 1979.
28. Tepperman J: "Metabolic and Endocrine Physiology." Chicago: Year Book Publ, 1973.
29. Levi-Montalcini R, Booker B: Proc Natl Acad Sci USA 46:384, 1960.

The Platelet-Derived Growth Factor: A Perspective

Russell Ross

University of Washington School of Medicine, Seattle, Washington 98195

The observation that most diploid cells require serum to multiply in culture is almost as old as the culture technique itself. Plasma clots were used to grow cells during the early period of development of the methodology of cell culture; however, it was not long before serum was added to the culture medium and a requirement for serum could be demonstrated.

A number of approaches have been used to identify the factors present in serum responsible for its mitogenic effect. Several years ago it became clear that the principal mitogenic component in serum is a relatively low molecular weight protein present in platelets termed the "platelet-derived growth factor" (PDGF). When serum is formed, platelets are exposed to thrombin, aggregate, and undergo what is commonly termed the "platelet release reaction." During this release response, many constituents are secreted from the platelet, including PDGF [1,2].

The corollary to these findings was the observation that it is possible to maintain cells in a quiescent and reasonably healthy state in culture for relatively long periods (6–8 weeks or longer) in 5% serum, if the serum is derived from cell-free plasma. From a teleological point of view, cell-free plasma may provide an environment that is somewhat more analogous to the environment to be found among cells and tissues of most adult organisms. Serum, however, represents a fluid environment that would be found in vivo only at sites where injury and blood coagulation have occurred. Most cells are constantly exposed to a filtrate of plasma, not too dissimilar from the plasma-derived serum that is now being used to maintain cells in a quiescent state in culture. Such quiescent cells can then be studied under conditions that permit an analysis of their response to hormones and many other stimuli [3].

Several laboratories have isolated PDGF to varying stages of purity. The group of Heldin, Wasteson and Westermark have purified the protein to homogeneity, and have provided an amino acid composition that substantiates the cationic character of the protein [4]. The PDGF has a number of distinctive characteristics. Its pH is 9.8, it is quite hydrophobic, and it is extraordinarily stable not only to heat but to a number of denaturing agents, including urea and guanidine. The protein is susceptible to disulfide bond breakage, since its activity is destroyed upon exposure to agents such as dithiothreitol. It is present in the alpha granules of the platelet; in pure form, approximately 4 ng/ml has the equivalent growth properties of 5% whole blood serum [5].

Received and accepted September 11, 1980.

© 1981 Alan R. Liss, Inc., 150 Fifth Avenue, New York, NY 10011

THE EFFECTS OF PDGF UPON CELLS

One of the principal effects of PDGF upon cells is initiation of DNA synthesis; in the appropriate environment this leads to cell division and proliferation. Pledger et al [6] have suggested that PDGF and plasma each play separate roles. They interpret their data to suggest that the mitogen commits the cells to DNA synthesis, and that plasma components play a role in the progression of cells through the cell cycle. The situation is probably more complex, since Vogel et al [7] and Heldin et al [personal communication] found that cells will go through one or more doublings after exposure to PDGF in a serum-free culture environment. Furthermore, Vogel et al [7] have shown that the amount of PDGF per cell is an important determinant of how many cells respond to the mitogenic activity of this protein. Although it has been shown by numerous studies that PDGF somehow acts in a coordinate way with components in plasma to induce not only DNA synthesis but also cell doubling, the mode of action of the components in plasma and the identity of all of those components that are important in these events have not been completely ascertained.

Experiments are under way to determine whether PDGF acts in a fashion similar to other mitogens, such as epidermal growth factor. Preliminary studies in our laboratory suggest that PDGF binds to a high-affinity receptor presumably located at the cell surface. The mitogen appears to bind to its receptor in a saturable fashion and can be competed for by unlabeled PDGF. A number of intracellular mechanisms that radically alter the metabolism of the cell are probably set into motion within a short time after the factor binds to its receptor.

Within 3 h after exposure to PDGF, cells such as 3T3 cells or smooth muscle cells undergo a marked increase in the rate of endocytosis of molecules such as sucrose or horseradish peroxidase [8]. This increased rate of endocytosis appears to be associated with the stimulus for DNA synthesis, since cells that are not susceptible to PDGF, such as arterial endothelial cells, also undergo an increase in the rate of endocytosis if they are stimulated by other means to make DNA and divide. Numerous other events occur within 6 h or less after exposure of susceptible cells to PDGF. These include an increase in cholesterol synthesis [9], increase of binding of LDL to high-affinity receptors located at the surface of the cell [10], and increased RNA and protein synthesis. The first 2 activities presumably relate to the utilization of increased cholesterol for synthesis of new membranes. In the case of cholesterol synthesis, or the breakdown of LDL into cholesterol and cholesteryl ester, the cells appear to utilize LDL as an exogenous source of cholesterol, presumably for new membrane formation.

In a recent series of unpublished studies, Habenicht et al [11] have demonstrated that a number of changes occur quite early after exposing 3T3 cells to PDGF. These changes relate largely to alterations in lipid metabolism. They include a loss of radioactive arachidonic acid, glycerol, phosphorus, and myoinositol from prelabeled phosphatidylinositol; a loss of radioactive arachidonic acid from prelabeled phosphatidalcholine; incorporation of radioactive arachidonic acid into phosphotidylinositol, phosphatidylethanolamine, and diglyceride; as well as incorporation of precursors into other phospholipids. One of the earliest changes observed was that diglyceride was formed from glycerol or arachidonic acid within 60 sec following exposure of the cells to the mitogen. These studies suggest that one of the early effects of the mitogen is the activation of a phospholipase C to catalyze the hydrolysis of phosphotidylinositol. The results also suggest that the diglyceride that was released is also hydrolized by a diglyceride lipase to form monoglyceride and free arachidonic acid. The phospholipase C may also be active toward a number of phospholipids in cells after activation by PDGF.

504:CCDD:A

RELATION OF PDGF TO OTHER GROWTH FACTORS

A large number of mitogens have been described during the last 5 years, including FGF, EGF, factors formed by transformed cells, and recently, factors formed by endothelial cells and by macrophages. Many of these appear to have a similar spectrum of responsive cells; however, the relationships among them are unclear. PDGF appears to be different from FGF and EGF, and possibly from the factors formed by endothelium and macrophages as well.

THE ROLE OF THE PLATELET-DERIVED GROWTH FACTOR IN VIVO

When it was first discovered, it was surprising that platelets contained within their granules a mitogen so potent that a few nanograms will stimulate 10,000 cells to synthesize DNA. A number of possibilities are being explored concerning the biological role of this factor in vivo.

Several pathologic processes involve platelet activation associated with tissue injury and blood coagulation. These include the processes of wound repair and atherogenesis, as well as chronic inflammatory disease. Cells such as endothelium and macrophages could provide growth factors in disease processes such as arthritis and chronic granulomatous disease.

A sequence of cellular events has been clearly delineated during the process of wound repair. The first cells to enter the wound are those intimately associated with blood coagulation, the platelets. These are followed in sequence by polymorphonuclear neutrophylic leukocytes, followed by monocytes that become macrophages, and subsequently by fibroblasts and capillaries. It is not clear whether mitogens such as PDGF provide the initial stimulus for the succeeding events. Experiments to answer this question are under way.

There is considerable evidence to suggest that the lesions of atherosclerosis arise by some form of injury to the endothelial cells that line the artery wall. Several different forms of experimentally induced endothelial injury have been shown to be followed by adherence of platelets at sites of injury, together with degranulation of platelets and release of their constituents. Together with plasma constituents, the platelet factors that gain entry at sites of injury could provide the stimulus that leads to the intimal smooth muscle proliferative response that is the hallmark of the lesions of atherosclerosis [12].

Functional platelets are necessary to reduce the lesions of experimental atherosclerosis. Data have come from laboratories in which lesions have been induced by mechanically injuring the endothelium with an intra-arterial catheter, or chemically by homocystinemia or chronic hypercholesterolemia. In each of these instances, if platelet function is inhibited by pharmacological agents [13], or if the animals are made thrombocytopenic with an antiplatelet serum [14, 15], then the proliferative response does not occur. Hypercholesterolemia will induce extensive smooth muscle proliferative lesions of atherosclerosis in normal swine. However, if animals that genetically lack the capacity for platelets to interact at sites of injury and release their contents are made hypercholesterolemic, then the proliferative lesions of atherosclerosis cannot be induced [16].

Each of these approaches has demonstrated that platelet function is required to promote the cellular proliferative response in vivo. No data are available to determine whether PDGF is the agent responsible for these events. Once an antibody to the factor becomes available and sufficient amounts of pure labeled factor are at hand, it should be possible to provide answers to this question.

Several recent papers contain data that suggest that the differences observed between plasma and serum are not due to the absence of PDGF in plasma. Rather, they suggest that the lack of response of cells, such as smooth muscle, to plasma, is due to the plastic substrate upon which the cells have been plated [17, 18]. These workers go on to suggest that substratum-induced elaboration of an extracellular matrix is necessary for cells to be capable of responding to plasma factors that are otherwise sufficient for growth stimulation. Unfortunately, the data presented in these investigations raise more questions than they answer. The investigators use Triton X-100 treatment to remove the cells from the matrix, and Triton X-100 treatment, as used in these investigations, disrupts the cell membrane, leaving the bulk of the cells in place [19–21]. Preliminary observations in our laboratory have also shown that the cells are not removed with the treatment, even with much more extensive washing; hence, numerous factors may be available to interact with cells under such conditions, including PDGF, making the data most difficult to interpret.

REFERENCES

1. Ross R, Glomset J, Kariya B, Harker L: Proc Natl Acad Sci 71:1207–1210, 1974.
2. Kohler N, Lipton A: Exp Cell Res 87: 297–301, 1974.
3. Ross R, Nist C, Kariya B, Rivest M, Raines E, Callis J: J Cell Physiol 97: 497–508, 1978.
4. Heldin C-H, Westermark B, Wasteson A: Proc Natl Acad Sci USA (in press).
5. Ross R, Vogel A: Cell 14: 203–210, 1978.
6. Pledger WJ, Stiles CD, Antoniades HN, Scher CD: Proc Natl Acad Sci USA 74:4481–4485, 1977.
7. Vogel A, Ross R, Raines E: J Cell Biol 85: 377–385, 1980.
8. Davies PF, Ross R: J Cell Biol 79: 663–671, 1978.
9. Habenicht A, Glomset J, Ross R: J Biol Chem 255:5134–5140, 1980.
10. Chait A, Ross R, Albers J, Bierman E: Proc Natl Acad Sci USA 77:4084–4088, 1980.
11. Habenicht AJR, Glomset JA, King W, Nist C, Ross R: J Biol Chem (in press).
12. Ross R, Glomset J: N Engl J Med 295:369–377, 420–425, 1976.
13. Harker LA, Ross R, Slichter SJ, Scott CR: J Clin Invest 58: 731–741, 1976.
14. Moore S, Friedman RJ, Singal DP, Gauldie J, Blajchman MS, Roberts RS: J Intl Soc Thromb Hemostasis 35: 70–81, 1976.
15. Friedman RJ, Stemerman MB, Wenz B, Moore S, Gauldie J, Gent M, Tiell MI, Spaet TH: J Clin Invest 60: 1191–1201, 1977.
16. Fuster V, Bowie EJW, Lewis JC, Fass DN, Owen CA Jr, Brown AL: J Clin Invest 61: 722–730, 1978.
17. Gospodarowicz D, Ill CR: Proc Natl Acad Sci USA 77: 83: 2726–2730, 1980.
18. Gospodarowicz D, Delgado D, Vlodavsky I: Proc Natl Acad Sci USA 77: 83: 4094–4098.
19. Yu J, Fischman DA, Steck TL: J Supramol Struct 1: 233–248, 1973.
20. Brown S, Levinson W, Spudich JA: J Supramol Struct 5: 119–130, 1976.
21. Ben-Ze'ev A, Duerr A, Solomon F, Penman S: Cell 17:859–865, 1979.

Plasma Components in Growth and Development: Workshop Report

S. P. Nissley and W. J. Pledger

National Institutes of Health, Bethesda, Maryland 20205 (S.P.N.) and University of North Carolina, Chapel Hill, North Carolina 27514 (W.J.P.)

The workshop on serum growth factors focused on two areas: the insulin-like growth factors (IGF) or somatomedins (SM) and their relationship to insulin, and the competence-progression model for control of the cell cycle.

Judson Van Wyk briefly discussed the current status of the somatomedins (IGF-I, IGF-II, SM-C, SM-A, and MSA) and their relationship to one another. Rinderknecht's and Humbel's amino acid sequence data for IGF-I and IGF-II demonstrated a 38–48% sequence homology with insulin in the A and B domains of the IGF molecules. Recently, Van Wyk's laboratory has shown that IGF-I and SM-C behave identically in three different assay systems: a radioreceptor assay using human placental membranes and ^{125}I-insulin, a radioreceptor assay using placental membranes and radiolabeled SM-C, and a radioimmunoassay for SM-C. Thus, IGF-I and SM-C appear to be functionally identical. The identity of IGF-I and SM-C is supported by recently published amino acid sequence data for SM-C, although there are some sequence differences that remain to be resolved before chemical identity has been assured. Although there are no amino acid sequence data for SM-A, the published amino acid composition points to SM-A being distinct from IGF-II and IGF-I (SM-C).

Multiplication stimulating activity (MSA) is of rat origin, in contrast to the other insulin-like growth factors, which have been purified from human plasma. Moses and his colleagues have recently shown that MSA is elevated in fetal rat serum and then drops to low levels by day 25 of extrauterine life, suggesting that MSA may be a fetal growth factor in the rat.

It is well known that insulin at high concentrations will partially replace the serum requirement for the growth of some cells in tissue culture. Indeed, insulin is a constituent of many of the defined media developed by Sato's laboratory for the growth of diverse cells in tissue cultures. In these instances, is insulin simply functioning as a somatomedin analog by interacting with the somatomedin receptor, or is insulin exerting its mitogenic effect by interacting with a classic insulin receptor? Matthew Rechler summarized data supporting the view that, in chick embryo fibroblasts, insulin is acting as a somatomedin analog. Thus, the relative potency of MSA and insulin in competing for binding of ^{125}I-MSA to chick embryo fibroblasts is the same as the relative potency of MSA and insulin in stimulating [^{3}H]thymidine incorporation into DNA. James Florini has also presented evidence [abstract 319] that insulin is acting like a somatomedin analog in supporting the

Received and accepted April 29, 1980.

© 1981 Alan R. Liss, Inc., 150 Fifth Avenue, New York, NY 10011

growth of the L_6 myoblast. However, insulin does not bind to all somatomedin receptors. For example, in purified rat liver membranes, insulin does not displace ^{125}I-MSA from its receptor. John Koontz discussed data [abstract 448] showing that, in a rat hepatoma cell line, insulin stimulates DNA synthesis and cell multiplication at physiologic concentrations. In addition, an antibody directed against the insulin receptor duplicates insulin's mitogenic effect on these cells. These important experiments suggest that, in this rat hepatoma cell line, insulin is acting as a mitogen by interacting with the insulin receptor.

In serum or plasma, the somatomedins are complexed to larger carrier proteins. Knauer and Smith [abstract 546] have purified a binding protein from the BRL-3A rat liver cell line, the cell line that produces MSA. Knauer discussed data showing that the binding protein inhibits the MSA-stimulated increase in [^{3}H] thymidine incorporation into DNA in chick embryo fibroblasts, suggesting that the somatomedin-binding protein complex is biologically inactive.

A recent model [Proc Natl Acad Sci USA 74:4481; 75:2839; 76:1279), relating the role of plasma-derived growth factors in the control of cellular proliferation of Balb/c-3T3, was presented. Density-dependent, growth-inhibited fibroblasts become competent when exposed to platelet-derived growth factor or fibroblast growth factor. Competent cells undergo progression through the cell cycle in the presence of plasma-supplemented medium. Incompetent cells do not undergo progression when exposed to plasma. Platelet-derived growth factor and plasma-derived growth factors function synergistically to stimulate cell proliferation. This means that, in combination, PDGF and plasma stimulate cell proliferation at concentrations that used singly do not stimulate DNA synthesis in quiescent fibroblastic cells.

The factors in plasma control an ordered sequence of events that leads to the commitment of competent cells to DNA synthesis. The commitment of a cell to DNA synthesis immediately precedes S phase. Factors present in somatomedin C-deficient rat plasma from hypophysectomized rats allow competent cells to progress to the "V" point. This V point growth arrest leaves arrested cells temporally 6 h before the S phase. Addition of somatomedin C, MSA, or high concentrations of insulin to cells arrested at the V point in plasma from hypophysectomized rats allows the arrested cells to begin DNA synthesis after a lag of 6 h. Amino acid deficiency also arrested progressing cells at the V point.

Another plasma-dependent arrest point, "W," immediately precedes S phase and is believed to be the commitment point. Both Drs. C. D. Stiles and W. J. Pledger have found that the addition of EGF, transferin, and somatomedin C to competent Balb/c-3T3 cells resulted in DNA synthesis. The addition of soybean trypsin inhibitor apparently maximized the effect of this hormonal supplemented serum-free medium for maximal progression from competence to the entry into S phase.

During the discussion of this model, Dr. Ake Wasteson pointed out that at present no information is available to indicate how long PDGF resides on or in the stimulated cell. However, since the purification and ^{125}I-iodination of PDGF has been reported, such information concerning the binding of PDGF to the cell and its subsequent processing should be forthcoming. The discussion also pointed out that the requirements for serum factors when cells are grown on a matrix as compared to growth on plastic will be interesting. It is of interest to note that transformation of fibroblasts abrogates the PDGF requirement. In addition the abortive transformation of density-dependent growth-inhibited Balb/-c-3T3 cells with SV40 has been shown to alter both PDGF and plasma-derived growth factor requirements necessary to stimulate DNA synthesis.

Dr. John Groelke discussed his data concerning the stimulation of quiescent chick fibroblasts [abstract 532]. Serum-deprived chick embryo fibroblasts initiate DNA synthesis after a 6 h lag when placed in media containing serum or insulin. However, a 6–8 h exposure to cycloheximide renders the quiescent cells nonresponsive to insulin and results in a 12 h lag before initiation of DNA synthesis after exposure to serum. The addition of one of several α-keto-acids restores sensitivity to insulin stimulation in the cycloheximide treated cells.

Dr. Harold Moses presented data from collaborative work with Dr. Bob Robinson. They found that AKR mouse embryo cells arrest at G_0 (12 h G_0/G_1 transit time) when density-dependent, growth-inhibited, or serum deprived. However, when these cells are grown in medium deficient for cystine or tryptophan, the cells become growth arrested at a point between G_0 and S phase. Cells at this arrest point have an 8 h lag before DNA synthesis, when the deficient amino acid is replaced. Unlike the normal AKR cells, 3-methylcholanthrene-transformed cells, grown in standard medium, growth arrest at the arrest point that precedes DNA synthesis by an 8 h lag and not G_0 (12 h lag). These transformed cells apparently deplete the growth medium of cystine or tryptophan.

Lymphocyte Triggering: Workshop Report

C.W. Parker and J.D. Watson

Department of Microbiology, University of California, Irvine, California 92717

Although lymphocytes normally exist in a growth-arrested state in peripheral lymphoid organs, their interactions with antigen leads to DNA synthesis, cell division, and maturation to either antibody-forming (B) cells of effector T cells. Lymphocyte triggering appears to be regulated by two membrane-mediated events: the first is the binding of antigen to surface immunoglobulin receptors, and the second by a helper T cell signal. It is generally recognized the helper signal may be mediated by a soluble factor released by helper T cells. The small number of antigen-specific lymphocytes in any given cell population and the difficulty associated with selection and enrichment for these cells have long been problems in the biochemical analysis of lymphocyte triggering. A large number of agents have been found to exert mitogenic activity on lymphocyte populations, and as many of these result in polyclonal activation of cells, these agents have been used to probe the triggering process. For those agents that interact at the cell surface to activate lymphocytes, there must exist a mechanism for translating the binding process into a biochemical signal. The discussion in this workshop was generally confined to the following issues: 1) What effector molecules activate lymphocyte? 2) What intracellular biochemical changes take place following interaction of these effectors with cells?

Lymphocyte activating agents appear to fall into several groups. Agents such as anti-immunoglobulin reagents and lipopolysaccharides (LPS) activate B cells alone, whereas the T cell mitogens Con A and PHA appear to require the presence of a growth factor now termed Interleukin 2 (IL-2).

Several lymphocyte-activating agents were discussed in detail. Evidence for a regulatory role of surface immunoglobulin on B lymphocytes in the activation process comes directly form the observation that a number of anti-Ig antibodies will induce some of these cells to proliferate. Mouse splenic B lymphocytes are known to proliferate in the presence of a number of anti-immunoglobulin reagents. Dr. David Parker has shown that the divalent $F(ab')_2$ fragment of purified rabbit anti-mouse IgM stimulates B lymphocytes to divide in the absence of immunoglobulin secretion. Removal of the Fc portion of such reagents appears important, as many anti-mouse Ig antibodies do not appear to be mitogenic, possibly because of inhibitory effects arising from the interaction of the Fc region with non-B cell types present. Mouse B cell activated by $F(ab')_2$ fragments of rabbit anti-mouse IgM can be driven to immunoglobulin secretion by the addition of supernatant factors derived from con A-activated spleen cells. The majority of murine splenic B cells ex-

Received and accepted May 12, 1980.

© 1981 Alan R. Liss, Inc., 150 Fifth Avenue, New York, NY 10011

press surface IgM and IgD molecules. Dr. Ian Zitron has developed a monoclonal cell hybrid (68A2.C6) that produces alloanti-delta antibody. B lymphocytes from mice of the Ig^a and Ig^c allotype groups are stimulated to proliferate by this monoclonal anti-delta antibody. The demonstration of B cell triggering with a monoclonal antibody directed against a surface IgD determinant is the first demonstration of a triggering function for IgD.

In contrast to the mitogenic properties of the F(ab')$_2$ fragments of anti-Ig reagents, Dr. E. Morgan described the generation of an Fc fragment derived from human IgG1, which induces murine B lymphocytes to divide and differentiate to antibody-forming cells. Macrophages appear to cleave the Fc fragment into a 14,000 molecular weight subfragment, which in turn functions as the B cell mitogen. For the stimulation of polyclonal antibody production by Fc, T cells are required. The requirement for T cells can be substituted for by the addition of supernatant factors release by con A-activated spleen cells.

These data suggest that B lymphocyte proliferation and antibody synthesis occurs through two signals. One signal is associated with proliferation and is generated by the interaction of anti-IG or Fc mitogenic fragments with B lymphocytes; a second signal is derived from a supernatant factor.

Dr. Diane Hacobs discussed the activation of mouse B cells by LPS. The C3H/HeJ mouse strains has a defect at the *Lps* locus that renders B cells unresponsive to LPS. A number of studies indicate that the defect is associated with the triggering mechanism. A radiolabeled LPS of very high specific activity has been used to analyze LPS binding to spleen cells, and little difference is seen in the binding of this probe to cells from high and low mouse strains.

The mitogenic effects of con A and PHA on T lymphocytes also appear to be due to the delivery of two signals. The interaction of these plant lectins with T lymphocytes does not lead directly to cell proliferation. These mitogens stimulate the production of a proliferative factor termed Interleukin 2 (IL-2). Resting T lymphocytes do not appear to express cellular receptors for IL-2. Following activation with Con A or PHA, T cells rapidly express a receptor for IL-2, and thereafter, the presence of sufficient concentrations of IL-2 is sufficient to cause proliferation of the activated T cells.

Lymphocytes can also be activated by treatment with neuraminidase-galactose oxidase (NAGO). Dr. A. Novogrodsky has shown that such oxidizing mitogens function by generating aldehyde moieties on cell surface glycoproteins. NAGO-treated cells also produce a growth factor, and it remains to be seen whether the factor is IL-2.

Many changes in cellular metabolism have been detected following the addition of mitogenic agents to lymphocytes. These vary from membrane changes in the turnover of phosphatidylinositol, changes that affect ion flux, amino acid and sugar transport, and cyclic nucleotide metabolism leading to protein phosphorylation, increases in cytoplasmic RNA, protein and fatty acid synthesis, and finally increases in nuclear synthetic activity. Any agent that interacts at the surface membrane of a growth-arrested cell and results in the initiation of DNA synthesis and cell division must activate a complex set of metabolic reactions. The problem faced in all eukaryotic systems is to determine how mitogens or other activating agents initiate such changes. The involvement of calcium and changes in the intracellular levels of cyclic AMP and cyclic GMP as pleiotypic regulators in the triggering process was discussed. Whether a single intracellular regulator can control all of the diverse alterations in cellular metabolism associated with lymphocyte activation remains to be established. Moreover, it seems possible that some of the metabolic alterations represent epiphenomena not truly essential to the activation process. This is particularly likely when the mitogen is a lectin. The workshop ended with a general recognition of the problems faced in analyzing the biochemical complexity of the triggering process and the need to find intracellular markers of the activation sequence.

Purification of Murine Helper T cell-Replacing Factors

James Watson, Diane Mochizuki, and Steven Gillis

Department of Microbiology, University of California, Irvine, California 92717 (J.W., D.M.) and Fred Hutchinson Cancer Research Center, Seattle, Washington 98104 (S.G.)

Humoral factors secreted by T lymphocytes and macrophages appear to play a role in cell communication leading to the triggering of immune responses. A factor has been purified from the culture supernatants of concanavalin A(Con A)-activated murine spleen cells with lymphokine activity in four assay systems. 1) stimulation of antibody responses to erythrocyte antigens in BALB/c.nu spleen cultures, 2) amplification of production of cytotoxic T cells in thymocyte cultures, 3) stimulation of mitogenic responses to Con A in thymocyte cultures where the cell density is too low to support responses to Con A alone, and 4) maintenance of continuous T cell growth. The biologic activity has been sequentially purified by salt precipitation, gel filtration, chromatography, ion-exchange chromatography, and isoelectric focusing (IEF). These four lymphokine activities appear to be due to one class of molecules, termed Interleukin 2 (IL-2), with an apparent molecular weight of 30,000 daltons. A T cell lymphoma, LBRM-33 has been found to produce high titers of IL-2 activity upon mitogen stimulation. LBRM-derived IL-2 possesses similar chemical and biological properties to splenic-derived IL-2. The mode of action of IL-2 appears to be the clonal expansion of antigen- or mitogen-activated T cells. While mitogen-activated LBRM cells secrete IL-2, which appears to exert helper T cell-replacing activity via the expansion of pre-T cells in nude spleens, mitogen-activated spleen cells appear to secrete both IL-2 and another class of T cell-replacing factor (TRF). These findings raise the question of whether there exist multiple classes of factors with TRF activity. The identification of these molecules may lead to an understanding of the mechanism that regulates the induction of antigen-sensitive cells.

Key words: lymphokines, interleukin 2, T cell-replacing factors

Antigen-sensitive lymphocytes are composed of two distinct cell populations that differ in both phenotype and function. The cells, found as resting cells in the peripheral lymphoid organs, have differentiated to a stage where they require a specific stimulus to drive them to their effector-cell states, which represent the terminal stages in an extensive developmental sequence. Antigen-sensitive cells are triggered by the recognition of antigen in association with a cooperating cell system. The cellular components of the cooperating

Received and accepted June 9, 1980.

© 1981 Alan R. Liss, Inc., 150 Fifth Avenue, New York, NY 10011

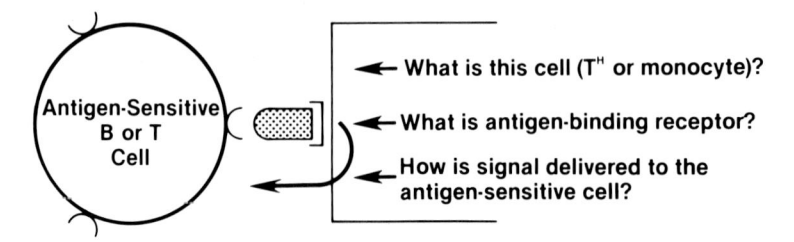

Fig. 1. The role of the cooperating cell system in the triggering of antigen-sensitive cells.

system and the mechanism by which signals are delivered to antigen-sensitive cells are the focus of a variety of immunological studies.

Three major questions examine the involvement of the cooperating cell system in the activation of B and T cells (Figure 1). First, what is the cell type that delivers the inducing signal to antigen-sensitive cells? Second, what is the chemical nature of the antigen-binding receptor of T cells? Third, what is the signal that results following the binding of antigen to helper cells?

A common approach to analyzing these problems is to delete one cellular component of the cooperating system and attempt to determine how the function can be replaced. This approach has resulted in the finding that lymphocyte culture supernatants contain a variety of factors that stimulate lymphocyte maturation in T cell or macrophage-depleted culture assay systems.

Factors have been found in murine, human, and rat lymphocyte culture supernatants that can replace helper T cell function in the induction of in vitro antibody production or cytotoxic T cell responses [1–4]. Some of the factors appear to be antigen-specific [5], while other factors lack antigen-specificity in their mode of action [6]. The relationship between specific and nonspecific factors is unknown, but it is possible that the specific factors are involved in antigen-binding processes, while nonspecific factors are involved in intercellular communication processes (Fig. 1).

Two classes of nonspecific factors that exhibit helper T cell-replacing properties have been distinguished on the basis of biochemical and biological criteria [7]. One of these factors can be obtained from the culture supernatants of human and murine macrophages and possess a molecular weight in the range of 15,000 [8]. This macrophage-derived factor, originally termed lymphocyte-activating factor, is now termed Interleukin I or IL-1 [7]. A second factor obtained from murine lymphocyte culture supernatants possesses a molecular weight in the range of 30,000 and has been termed Interleukin 2 or IL-2 [7]. IL-2 is defined as the class of molecules that stimulates the continuous proliferation of T cells in culture [7]. We focus here on the biochemical and biological properties of IL-2 and, in particular, consider its helper T cell-replacing activity in the induction of in vitro antibody responses to erythrocyte antigens.

There are a number of difficulties in purifying IL-2 from spleen culture supernatants. First, the small amount of IL-2 obtained from activated murine, rat, or human sources makes it difficult to purify the quantities necessary for further molecular characterization. Second, mitogen-activated lymphocyte culture supernatants contain multiple lymphokine activities. Therefore, until a factor has been purified to homogeneity, it remains difficult to definitively associate biological activities with distinct molecular species. We have recently described a cultured murine lymphoma cell line, LBRM-33, which can be activated by phytohemagglutinin (PHA) or concanavalin A (Con A) to secrete IL-2 activity [3, 4]. In this paper, we describe the purification of IL-2 from LBRM cells. In comparing the TRF activity of LBRM-derived and splenic-derived IL-2, we observe that splenic IL-2 preparations may contain a class of TRF not found in LBRM supernatants.

MATERIALS AND METHODS

LBRM-33 cells, originally derived from a B10.BR splenic lymphoma, were grown in culture as described elsewhere [3, 4], utilizing LBRM-33 clone 5A4. To produce IL-2 for purification, LBRM cells were seeded in Falcon T75 tissue culture flasks in medium supplemented with 10% fetal calf serum and were grown to confluency. Culture medium was removed and replaced with serum-free medium containing 1% PHA (Grand Island Biological Company, New York) for 20 h. Culture supernatants were then removed, and the cells removed by centrifugation at 10,000g for 15 min. The resulting supernatant was then concentrated by ammonium sulfate precipitation [1, 2].

Biochemical Purification of IL-2

All chromatography was performed at 4°C using sterile buffers. Concentrated supernatants containing IL-2 activity were dialyzed against 0.05 M NaCl-Hepes buffer (pH 7.2) and then fractionated using DEAE-Sephacel as detailed elsewhere [1, 2]. The ionic strength of the fractions collected was determined using a conductivity meter and then assayed for activity. Column fractions containing biological activity were pooled and fractionated by gel-filtration chromatography using a 2 X 90 cm column of AcA54 (LKB Produkter, Sweden) equilibrated in 0.9% saline as detailed elsewhere [1, 2].

Flat-bed IEF was performed in a horizontal layer of Sephadex [2]. Samples were prepared in 100 ml solution containing 1% glycine, 0.1% aspartic acid, and 2% ampholines (pH 3–10, LKB Produkter, Sweden). The sample was mixed with 4 g Ultradex (LKB Produkter) and the gel suspension spread in a gel tray. The tray was subjected to electrophoresis on a cooling tray for 20–24 h under a constant current of 7mA, during which time the voltage increased from 100–1,000 V. After electrophoresis, the gel was sectioned into 30 portions, and the pH and IL-2 activity in each portion was determined as detailed elsewhere [1, 2, 4].

Biological Assays for IL-2 Activities

T cell growth assay. CTLL-2 cells, a continuous IL-2-dependent line of cytotoxic T cells, and HT-1, a continuous IL-2-dependent line of helper T cells were used to perform T cell growth assays as detailed elsewhere [1, 2].

TRF activity and antibody synthesis. Antibody synthesis to heterologous erythrocyte antigens was measured in a microculture system using BALB/c.nu spleen cells as detailed elsewhere [2]. In all assays, 120 microcultures were used to measure IL-2 activity at each concentration [1, 2].

RESULTS

Purification of IL-2 From LBRM Cells

Culture supernatants from PHA-activated LBRM cells were sequentially subjected to a variety of separation techniques. After each purification step the fractions were tested

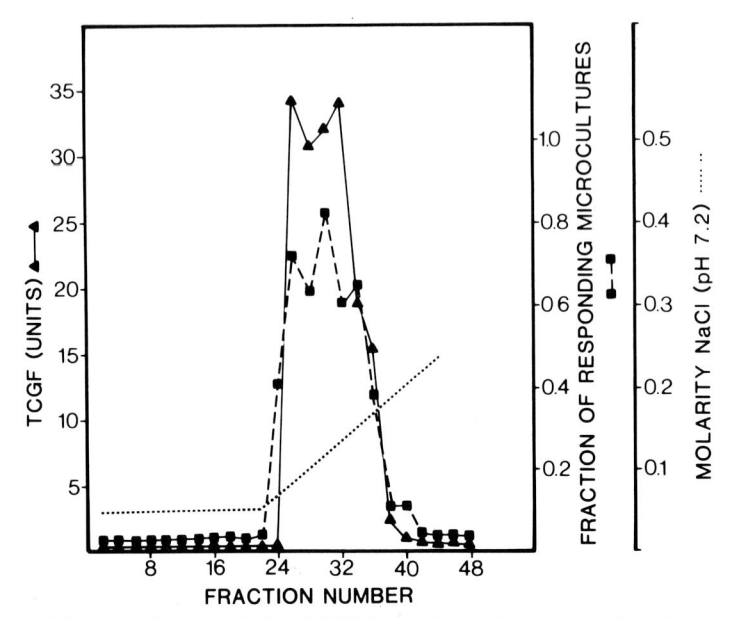

Fig. 2. Lymphokine activities assayed after DEAE ion-exchange chromatography of concentrated medium conditioned by PHA-stimulated LBRM cells. TCGF activity: ▲—▲. GRBC-specific antibody responses in BALB/c.nu spleen microcultures: ■—■.

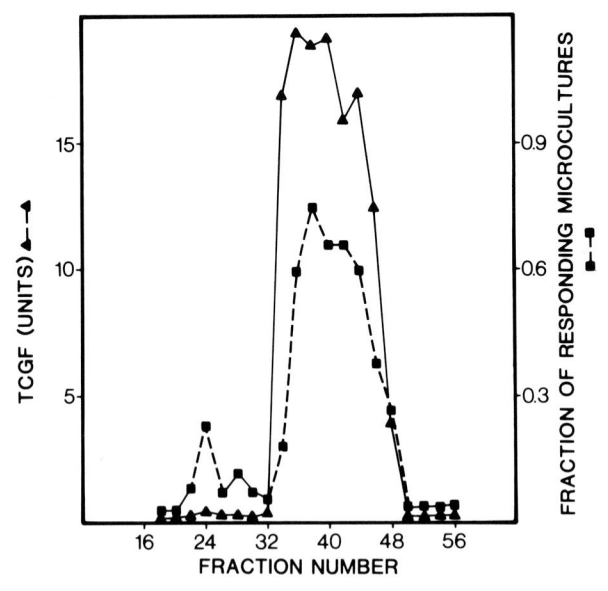

Fig. 3. Lymphokine activities obtained after AcA54 gel-exclusion chromatography of pooled IL-2-containing fractions separation by AcA54 gel-exclusion chromatography. TCGF activity: ▲—▲. GRBC-specific antibody responses in BALB/c.nu spleen microcultures: ■—■. The markers used to calibrate the AcA54 columm were bovine serum albumin, ovalbumin, and cytochrome c.

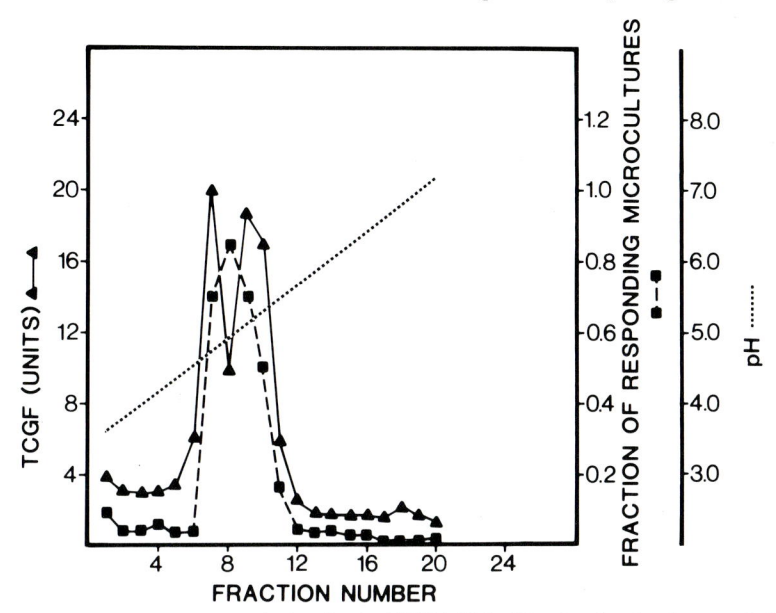

Fig. 4. Lymphokine activities assayed after flat-bed IEF (pH 3–10) of active fractions pooled after DEAE ion-exchange chromatography. TCGF activity: ▲—▲. GRBC-specific antibody responses in BALB/c.nu spleen microcultures: ■—■.

in two assay systems as detailed in Materials and Methods: 1) T cell growth assay and 2) the in vitro stimulation of antibody responses to GRBC in BALB/c.nu spleen microcultures.

Supernatants prepared from PHA-activated LBRM cells were concentrated and then fractionated by ion-exchange chromatography. The data presented in Figure 2 summarize the assays performed on each column fraction in each of the lymphocyte response systems. The IL-2 activity was tested for T cell growth activity on CTLL-2 cells eluted from DEAE-Sephacel between 0.06 and 0.2 M NaCL (pH 7.2) (Fig. 2). The activity that stimulated antibody responses in nude-mouse spleen cultures (Fig. 2) co-eluted with the IL-2 activity. The elution of these activities from a DEAE ion-exchange resin is identical to that obtained with murine spleen-derived IL-2 [1].

When the LBRM IL-2 activity in the column fractions obtained by ion-exchange chromatography (Fig. 2) were pooled and fractionated by gel filtration using AcA54, the IL-2 activity was observed to elute in the size range of 30,000 daltons (Fig. 3). Furthermore, the factor activities responsible for the stimulation of GRBC-specific antibody responses in nude-mouse spleen cultures (Fig. 3) appeared to be inseparable from the T cell growth-promoting activity of IL-2. Again, the size and properties of this LBRM-derived IL-2 are identical to that previously reported for splenic IL-2 [1].

The IL-2 activity in column fractions obtained by AcA54 gel-filtration chromatography (Fig. 3) was then pooled and examined by IEF. IL-2 activity was observed to separate into two major peaks of activity with isoelectric values of approximately 4.5 and 5.0, respectively (Fig. 4). The factor activity responsible for the stimulation of antibody responses in nude spleen cultures was again inseparable from the IL-2 activity (Fig. 4). We have observed some variability in different experiments in the isoelectric values of these two components. As observed with splenic IL-2, the pI values range from 4.3–4.7 and 4.9–5.2, respectively, for each component [1].

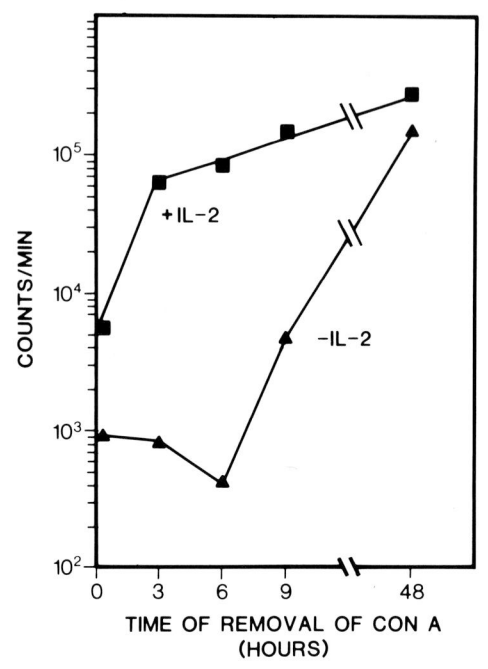

Fig. 5. Effect of removal of Con A from thymocyte cultures on the mitogenic response to Con A. C57BL/6J thymocytes were cultured at a density of 2×10^6 cells/ml with 2 μg/ml of Con A, and incubated at 37°C. At 0, 3, 6, 9, and 48 h, 0.1 M α-methyl mannoside was added to the cultures indicated; the cells were then harvested by centrifugation and recultured in fresh medium lacking Con A. Thymocytes recultured with 10 units/ml LBRM-derived IL-2: ■ — ■. Thymocytes recultured in the absence of added IL-2: ▲ — ▲. All cultures were radioactively labeled with 0.25 μCi/ml of ^{3}H-thymidine [1] between 48 and 56 h after the initial culturing, and the radioactive incorporation determined.

We stress that following the IEF purification step we detected one major difference between preparations of spleen- and LBRM-derived IL-2. While splenic and LBRM IL-2 stimulate antibody responses in nude spleen cultures, we have previously observed that splenic IL-2 possesses a second class of molecules with antibody-inducing activity. This second class of molecules is absent in the LBRM IL-2 preparations and is distinguished by the ability to induce antibody synthesis to heterologous erythrocyte antigens in nude-mouse spleen cultures, while lacking any T cell growth-factor activity [1]. This material, detected in spleen supernatants, showed considerable charge heterogeneity, with pI values ranging from 3.0 to 4.2 [1], and again we emphasize LBRM cells do not secrete this class of lymphokine (Fig. 4).

Mode of Action

The results presented here and elsewhere [1–4] suggest that IL-2 possesses a common mode of action in each of the lymphocyte assay systems. Activated T cells, but not resting T cells, readily absorb IL-2 [10]. The IL-2-induced proliferative effects on cloned cytotoxic and helper T cell lines also imply that IL-2 interacts directly with T cells [1–4, 9–11]. However, the striking feature of IL-2 activity in the induction of antibody synthesis or CTL responses is that there is a strict requirement for antigen to observe cellu-

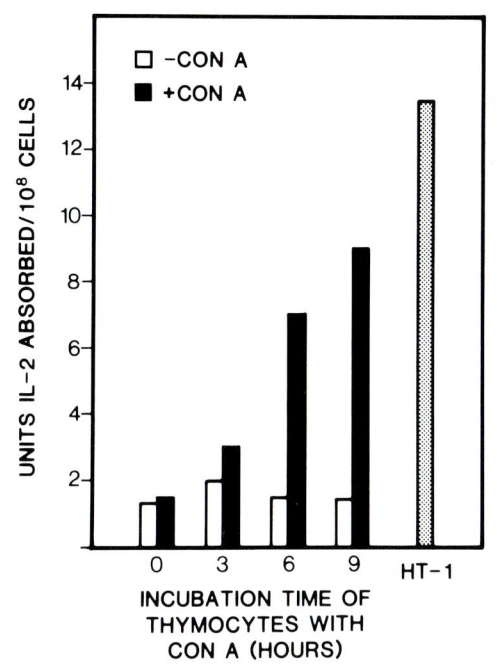

Fig. 6. The absorption of IL-2 activity by thymocytes treated with Con A. C57BL/6J thymocytes were cultured at a density of 2×10^6 cells/ml in the presence of 2 μg/ml of Con A (solid bars), or in the absence of Con A (open bars). After 0, 3, 6, or 9h, 0.1 M α-methyl mannoside was added to the cultures indicated, and the cells harvested by centrifugation. For each absorption performed, 10^8 untreated or Con A-treated thymocytes were incubated with 20 units of LBRM-derived IL-2 in 0.5 ml fresh medium at 37°C for 2 h. As a control, 10^8 HT-1 cells [1] were incubated in the same conditions with 20 units of IL-2. After the incubation period, cells were removed by centrifugation and the IL-2 activity remaining in the supernatant was determined by a titration in the T-cell growth assay [1].

lar responses [1, 2]. We have suggested that following an interaction with antigen or mitogen, T cells or their precursors in thymocyte or nude-mouse spleen cultures respond by expressing receptors for IL-2 [1–4, 9, 10]. The subsequent clonal expansion of antigen- or mitogen-activated T cells from each of the effector classes (helpers, killer, or suppressors), requires only the presence of IL-2 [9–11].

The experiments presented in Figures 5 and 6 support the above suggestions. Thymocytes were incubated for varying periods of time with mitogenic concentrations of Concanavalin A (2 μg/ml). At various times, cultures were treated with 0.1 M α-methyl mannoside to remove Con A, collected by centrifugation, and then recultured in the presence or absence of 10 units of IEF-purified LBRM-derived IL-2. This IL-2 preparation contained no PHA. After 48 h in culture, cells were labeled with radioactive thymidine. The results show that when thymocytes are cultured with 2 μg/ml of Con A for 3 h, they are rendered responsive to the proliferative activity of IL-2 (Fig. 5). In contrast, when Con A is removed from thymocyte cultures in the initial 9-h period, in the absence of added IL-2, virtually no mitogenic response results (Fig. 5). While these results support the contention that mitogens render T cells responsive to the proliferative effects of IL-2 by stimulating the expression of receptors, it should be noted that in the activation of thymocytes Con A has two roles: to stimulate changes in responsive cells that result in receptor expression and to stimulate the production of IL-2 from another cellular source.

To determine whether the acquisition of responsiveness to IL-2 is due to the expression of receptors on thymocytes (Fig. 5), a parallel experiment was performed. Thymocytes were incubated in the presence of 2 µg/ml of Con A for 3, 6, or 9 h, washed with 0.1 M α-methyl mannoside, and then tested for the ability to absorb IL-2 (Fig. 6). Thymocytes incubated in the absence of Con A failed to absorb significant amounts of IL-2 activity. However, thymocytes incubated for 3, 6, or 9 h with Con A showed an increasing ability to absorb IL-2 activity. After 9 h of incubation with Con A, thymocytes absorbed amounts of IL-2 that approached those absorbed by a cloned T cell line HT-1 (Fig. 6).

DISCUSSION

A class of nonspecific factor, IL-2 has been shown to be present in the culture supernatants of activated murine splenocytes. IL-2 is active in vitro in four assay systems: 1) stimulation of antibody responses to erythrocyte antigens in BALB/c.nu spleen cultures, 2) amplification of production of cytotoxic T cells, 3) stimulation of Con A-induced mitogenic responses in thymocyte cultures at limiting cell densities, and 4) maintenance of proliferation of antigen-activated T cells. Although it is not possible to approach the problem of a physiological role for IL-2, the specificity of its T cell growth-promoting activities has become a valuable tool for the selection and generation of cloned lines of antigen-specific T cells [9–13].

Mitogen-activated lymphocyte populations are the source of many diverse lymphokine activities including IL-2, and the small quantities produced of each of these lymphokines has caused a major problem in the purification and assignment of biologic activities to distinct classes of lymphocyte regulatory molecules. In the production of the lymphokine IL-2, we have circumvented the problems of small amounts and multiple-factor classes by utilizing LBRM-5A4 cloned lymphoma cells as a source of IL-2 activity. Our results indicate that LBRM IL-2 and splenic IL-2 are similar in chemical and biological activities following purification through ion-exchange chromatography, gel-filtration chromatography, and isoelectric focusing. The IL-2 activity (T cell growth stimulation) from LBRM cells was inseparable from activity responsible for the generation of antibody-forming cells in nude-mouse spleen cultures (Figs. 2–4). Since the source of LBRM IL-2 is a cloned T-cell line, it is likely that we are detecting a single class of molecules with stimulatory activity in the two assay systems. We have previously shown that murine, rat, and human lymphocytes can all be induced to secrete IL-2 activity, defined by T cell growth-promoting activity [1, 2]. Since IL-2 from these sources exhibit identical biologic activities in murine lymphocyte response assays, we have suggested that murine, rat, and human IL-2 exert stimulatory effects utilizing a common mechanism via the proliferation of activated T cells [1–4]. The results presented here with LBRM IL-2 support this concept.

The induction of IL-2 activity in murine spleen cell supernatants has been shown to require the presence of two cell types, a T cell and an Ia-positive, non-T cell [14, 15]. Formally, it has been difficult to assign one cell type or the other as the cellular origin of IL-2. The finding that a cloned T cell line LBRM is capable of producing IL-2 indistinguishable from splenic IL-2 indicates that the T cell is a cellular origin of IL-2 activity. The LBRM T-cell line has apparently lost the requirement for interaction with the non-T cell population for induced IL-2 secretion.

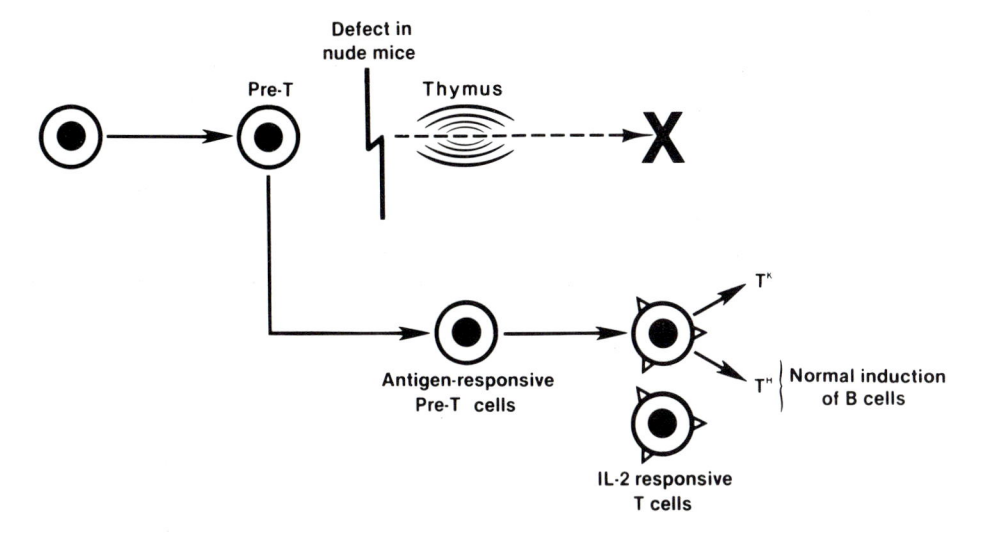

Fig. 7. A possible mode of action for IL-2 in spleen cultures derived from athymic (nude) mice. It is suggested that antigen-responsive pre-T cells accumulate in the spleens of nude mice and, in the presence of antigen and IL-2, can be clonally expanded to form functional effector cells. For the induction of antibody synthesis to red blood cell antigens in nude-mouse spleen cultures, IL-2 may stimulate the proliferation of antigen-specific helper T cells. Activation of the B cells would then involve the normal induction mechanism.

We would focus here on the T cell-replacing properties of IL-2 in the induction of antibody responses to erythrocyte antigens. The data presented in Figures 5 and 6 support our previous hypothesis that IL-2 exerts its effects directly on activated T cells. Since both antigen and IL-2 are required to stimulate antibody responses (Figs. 2–4), it appears that in nude spleen cultures antigen and IL-2 are stimulating the clonal expansion of helper T cells. The implication is that the nude-mouse spleen cell populations contain antigen-responsive helper T cells (Fig. 7). If antigen and IL-2 are acting as a stimulus to drive these cells to maturity, the induction of antibody responses in these cultures may result from the normal cooperative interactions of helper T cells and antigen-sensitive B cells (Fig. 7).

There remain, however, a number of unexplained observations. In splenic-derived IL-2 preparations, we have reported a class of molecules similar in size and charge to IL-2, lacking T cell growth-promoting activity, but possessing T cell-replacing activity in the induction of antibody synthesis [1]. This second class of molecules is not observed in LBRM-derived IL-2 preparations (Figs. 2–4). The other lymphokine associated with murine IL-2 preparations [1] may therefore have three, rather than four, T cell-replacing activities and be termed TRF [7].

In a previous study, an attempt was made to analyze the spleen cells producing TRF activity in a limiting dilution analysis [6, 16]. The surprising observation was that under conditions of limiting dilution where TRF appeared to be derived from single cells, the TRF showed a segregation of biologic activities as revealed by an ability to stimulate antibody responses to one erythrocyte antigen, but not another [6, 16]. These results indicated that TRF produced in Con A-treated spleen cultures may have antigenic specificity, an unexpected result if the activity detected was solely due to IL-2. If supernatants from Con A-treated spleen cells contain two factors, a T cell growth factor, or IL-2, and a T cell-replacing factor (TRF), the biological activities we have discussed may result from different modes of action. IL-2 may act directly on T cells in the induction of antibody responses to stimulate the clonal proliferation of any T cell that has been activated by antigen. IL-2 would thus stimulate growth of T cells in culture by exerting an antigen-nonspecific or hormonal effect on T cells. In contrast, TRF may replace T cells in the induction of antigen-sensitive cells. TRF may exhibit an antigen-specific activity, as revealed by the limiting dilution analysis [6, 16], and may be secreted by T cells but bind cytophilically to another cell type in the cooperating system to provide helper activity [6]. It is clear that only the molecular characterization of these various lymphokine activities will allow the delineation of their biologic activities.

ACKNOWLEDGMENTS

This work was supported by Research Career Development award AI-00182 and grant AI-13383 from the National Institute of Allergy and Infectious Diseases, and grant 1-469 from the National Foundation to James Watson. Steven Gillis is a fellow of the Leukemia Society of America and is supported by a grant from the National Foundation (1-724) and NCI grant CA 28419.

REFERENCES

1. Watson J, Gillis S, Marbrook J, Mochizuki D, Smith KA: J Expt Med 150:849, 1979.
2. Gillis S, Smith KA, Watson J: J Immunol 124:1875, 1980.
3. Gillis S, Scheid M, Watson J: J Immunol 125:2570, 1980.
4. Mochizuki D, Watson J Gillis S: J Immunol 125:2579, 1980.
5. Howie S, Feldmann M: Eur J Immunol 7:417, 1977.
6. Watson J, Mochizuki D, Thoman M: (1979) In Bach F, Vitetta E, Bonavita B (eds): "Biochemistry of Lymphocyte Communication in T and B Lymphocytes: Recognition and Function," Fifth ICN-UCLA Symposium on Molecular Biology. New York: Academic pp 123–142.
7. Aarden LA et al: J Immunol 123:2928, 1979.
8. Mizel SB: J Immunol 122:2167, 1979.
9. Gillis S, Ferm MM, Ou W, Smith KA, J Immunol 120:2027, 1978.
10. Smith KA, Gillis S, Ruscetti FW, Baker PE, McKenzie D: Ann NY Acad Sci 332:423, 1979.
11. Watson J: J Expt Med 150:1510, 1979.
12. Gillis S, Smith KA: Nature (Lond) 268:154, 1977.
13. Gillis S, Union NA, Baker PE, Smith KA: J Expt Med 149:460, 1978.
14. Swain S, Dutton RW: J Immunol 124:437, 1980.
15. Wagner HH, Rollinghoff M: J Expt Med 148:1523, 1977.
16. Lefkovits I, Waldmann H: Immunol 32:915, 1977.

Insulin Receptor Synthesis and Turnover in Differentiating 3T3-L1 Preadipocytes

M. Daniel Lane, Brent C. Reed, and Peter R. Clements

Department of Physiological Chemistry, The Johns Hopkins University School of Medicine, Baltimore, Maryland 21205

3T3-L1 "preadipocytes" can be induced to differentiate in culture into cells having the morphological and biochemical characteristics of adipocytes. The binding of ^{125}I-insulin to the cell-surface of differentiated and undifferentiated 3T3-L1 cells and nondifferentiating 3T3-C2 cells was compared. In the absence of agents which induce adipocyte conversion, ie, insulin or insulin plus methylisobutylxanthine (MIX) and dexamethasone (DEX), 3T3-L1 cells fail to express the adipocyte phenotype and maintain a constant number of insulin binding sites. Induction of adipocyte conversion with 3T3-L1 cells in the presence of insulin causes apparent down-regulation of insulin receptors followed by a 12–15-fold increase in receptor number which parallels differentiation. Approximately 170,000 insulin binding sites per cell are expressed when > 75% of the cells have differentiated. The rise of insulin receptor level is differentiation-dependent. 3T3-C2 cells, which do not differentiate in the presence of insulin or insulin plus MIX and DEX, exhibit only insulin-induced down-regulation of insulin receptors. The increase of insulin receptor level in 3T3-L1 cells is receptor-specific since the levels of epidermal growth factor receptor or choleragen receptor, respectively, remain constant or decrease substantially.

A heavy isotope, density-shift technique was used to analyze insulin receptor synthesis and turnover in cells labeled with "heavy" (^{2}H, ^{13}C, and ^{15}N) amino acids. Solubilized newly-synthesized "heavy" and old "light" receptors were separated by isopycnic banding on CsCl gradients and quantitated. The size of the soluble receptor isolated after isopycnic banding in CsCl gradients is approximately 400,000 daltons. Mixing of "light" and "heavy" membranes prior to extraction of receptor revealed no change in "light" or "heavy" receptor isopycnic banding densities. Thus, no detectable interchange of subunits occurs between receptor molecules during extraction or equilibrium centrifugation.

Insulin receptor synthesis and turnover, studied by the density-shift technique showed that the rise of receptor level during differentiation results primarily from an increased rate of receptor synthesis. The rate of insulin receptor degradation was not significantly altered. The t½ for degradation of the insulin receptor in differentiated 3T3-L1 cells in culture was 6–7 hours in

Abbreviations: EGF, epidermal growth factor; MIX, methylisobutylxanthine; DEX, dexamethasone.

Received and accepted May 6, 1980.

© 1981 Alan R. Liss, Inc., 150 Fifth Avenue, New York, NY 10011

the presence of insulin. Removal of insulin from the medium did not materially affect the rate of receptor degradation. Inhibition of protein synthesis with cycloheximide causes a lengthening of the $t_{1/2}$ for insulin receptor degradation to 26 hours. Thus, the synthesis of a short-lived protein appears to be required for a critical step in the pathway of insulin receptor degradation.

Key words: insulin receptor, I. receptor synthesis, I. receptor degradation, differentiation, adipocytes, preadipocytes, tunicamycin, clycloheximide

Insulin serves the vital role in animal cells of regulating the synthetic phase of energy metabolism, in particular, glucose and amino acid uptake, glycogen synthesis, lipogenesis, and protein synthesis [1, 2]. Circulating insulin functions as an indicator of the global energy status of the organism and, together with glucagon, regulates nutrient flux into energy storage versus catabolic pathways. The first point at which a cell can respond to a change in insulin level is through the binding interaction of the hormone with its specific receptor in the plasma membrane. The cell itself can alter its response to insulin. This can be accomplished either by adjusting the number of receptors in the plasma membrane, as in insulin-induced receptor "down-regulation" [3, 4], or by affecting the coupling of the receptor to its biological response system [5]. The mechanisms by which insulin receptor level and coupling are regulated have not been elucidated.

We have undertaken to investigate the regulation of insulin receptor metabolism with the 3T3-L1 preadipocyte cell line [6, 7] which expresses a high level of the receptor upon differentiation [8–11]. 3T3-L1 cells were cloned from 3T3 mouse embryo fibroblasts and undergo differentiation in culture to acquire the morphological and biochemical characteristics of adipocytes [6–9, 11–16]. Several days after reaching confluence, 3T3-L1 cells lose their fibroblastic morphology, round up, and deposit large amounts of cytoplasmic triglyceride [6, 7]. When transplanted subcutaneously into Balb C (athymic) nude mice, 3T3 preadipocytes differentiate and develop into tissue indistinguishable from normal adipose [14].

During differentiation of 3T3-L1 cells in culture the expression of adipocyte characteristics is associated with alterations in the binding of and response to hormones which control lipogenesis and lipolysis in mature adipocytes [5, 8–11]. Simultaneously, 3T3-L1 preadipocytes acquire enhanced lipogenic capacity which is correlated with a coordinate rise in the activities of the enzymes of de novo fatty acid synthesis [12, 15, 17] and triacylglycerol synthesis [13]. The coordinate nature of these enzyme responses suggests that the differentiation process causes concurrent induction of the entire group of lipogenic enzymes.

In this paper it will be shown that a marked and specific rise in number of cell surface insulin receptors accompanies the differentiation of 3T3-L1 preadipocytes into "adipocytes". The mechanism by which the level of insulin receptor is increased was studied using a recently-developed isotope density-shift technique [9, 18, 19] to quantitate receptor synthesis and turnover. It will be concluded that the differentiation-induced rise in receptor level results from an increased rate of receptor synthesis with little or no changes in the rate of receptor activation or degradation. Part of this work has recently been published [8–10].

RESULTS AND DISCUSSION

Induction of Differentiation of 3T3-L1 Cells Into Adipocytes

In the studies to be described, expression of the differentiated adipocyte phenotype was induced by either of two methods:

Method I: 3T3-L1 preadipocytes and 3T3-C2 control cells are plated at a density of 1300 cells per cm^2 and cultured in Dulbecco's Modified Eagles Medium supplemented with 10% calf serum, insulin, d-biotin, and pantothenate [12]. Deposition of cytoplasmic triglyceride in 3T3-L1 cells often begins about 4–5 days after confluence is achieved in which case by the 8th and 15th days post-confluence, 20%–25% and > 80%, respectively, of cells in the monolayer have accumulated triglyceride. Frequently, however, differentiation induced by this method is delayed. Under identical conditions, 3T3-C2 cells do not express the adipocyte phenotype and, thus, serve as a nondifferentiating control.

Method II: This method utilized the fact that differentiation is markedly and consistently accelerated by treating 3T3-L1 cells with methylisobutylxanthine [20] and dexamethasone [11]. Cells are brought to confluence as described in Method I except that fetal calf serum replaces calf serum. Two days after reaching confluence, cell monolayers are treated for three days with a combination of methylisobutylxanthine and dexamethasone in insulin-containing medium [9, 17, 21]. Under these conditions the deposition of cytoplasmic triglyceride begins within a day after removing methylisobutylxanthine and dexamethasone. 3T3-L1 control cells, which do not receive insulin, methylisobutylxanthine and dexamethasone do not express the adipocyte phenotype for up to 10 days post-confluence.

Coordinate Expression of the Enzymes of the Lipogenic Pathway

The most striking phenyotypic change during differentiation of 3T3-L1 preadipocytes is the massive deposition of cytoplasmic triglyceride. The rapid rate of formation and storage of triglyceride [6, 7, 12, 13] indicates a major reprogramming of cellular lipid metabolism. Triglyceride accumulation is closely associated with an increased rate of de novo lipogenesis as measured by [^{14}C]acetate incorporation into cellular triglyceride [7, 12]. These results are consistent with the coordinate rise observed in the activities of four key enzymes of fatty acid synthesis, ie, ATP-citrate lyase [12], acetyl-CoA carboxylase [12], fatty acid synthetase [12], and pyruvate carboxylase [15], which are, in part, responsible for the observed elevation of de novo triglyceride synthesis. Differentiation of 3T3-L1 cells also leads to an enhanced lipoprotein lipase activity [16] providing these cells with the capacity (like normal adipocytes) to utilize fatty acids derived from exogenous lipoprotein triglyceride. Fatty acids, derived both from de novo synthesis and from exogenous lipoprotein, are converted to triglyceride by the microsomal enzymes of triacylglycerol biosynthesis. During preadipocyte differentiation there is also a coordinate increase in the activities of the enzymes of this pathway, ie, fatty acid:CoA ligase, glycerol-3-phosphate acyltransferase, lysophosphatidic acid acyltransferase, and diacylglycerol acyltransferase [13].

A comparison of the extents of amplification (differentiated/undifferentiated) of the individual enzymes in the lipogenic pathway from glucose during adipocyte conversion reveals an interesting pattern (Table I). Those enzymes (Group III) that catalyze reactions through which all lipogenic precursors must pass, ie, fatty acids derived both from the de novo synthetic pathway and from the action of lipoprotein lipase on extracellular lipo-

TABLE I. Differential Amplification of Sequential Enzyme Activities During Differentiation of 3T3-L1 Preadipocytes

Enzyme group		Pathway	Metabolic role(s)	Amplification of enzyme activity Differentiated/undifferentiated
I	PFK Ald CS	glucose ↓ citrate	– energy metabolism – provide lipogenic precursors	3–5-fold
II	CBX_{pyr} CBX_{ac} CL FAS	↓ fatty acid	fatty acid synthesis	15–20-fold
III	FAL GAT LPA-AT DGAT	↓ triglyceride	esterification of fatty acids from: – de novo synthesis – exogenous origin	40–100-fold

Abbreviations: PFK, phosphofructokinase; Ald, aldolase; CS, citrate synthase; CBX_{pyr}, pyruvate carboxylase; CBX_{ac}, acetyl-CoA carboxylase; CL, ATP-citrate lyase; FAS, fatty acid synthetase; FAL, fatty acid:CoA ligase; GAT, glycerophosphate acyl transferase; LPA-AT, lysophosphatidic acid acyl transferase; and DGAT, α,β-diglyceride acyl transferase.
Differentiation was induced by Method I. Results are from [12, 13, 15].

proteins, are amplified to the greatest extent, ie, 40–100-fold (Table I). On the other hand, enzymes of de novo fatty acid synthesis (Group II, Table I), which provide only part of the fatty acid precursors for triglyceride synthesis, are amplified only 15–20-fold. Finally, enzymes that have major metabolic functions in addition to lipogenesis, eg, the glycolytic enzymes and citrate synthase and amplified to even a lower extent, 3–5-fold (Group I, Table I). This differential amplification implies that the lipogenic program is fine-tuned such that the coordinate changes of the three groups of enzymes in this complex pathway are differentially modulated.

It has now been established that the increased activities of acetyl-CoA carboxylase [12], fatty acid synthetase [17, 21], and pyruvate carboxylase [22] in differentiated 3T3-L1 adipocytes arise from increased levels of enzyme protein. Recently, we determined [17, 21] that the differentiation-induced increase in the level of one of these enzymes, ie, fatty acid synthetase, results entirely from a dramatic rise in the rate of synthesis of this enzyme with no alteration in its rate of degradation. The fact that the activities of acetyl-CoA carboxylase, pyruvate carboxylase, and all of the other lipogenic enzymes rise coordinately with fatty acid synthetase [13] supports the hypothesis that there is coordinate induction of the synthesis of all of the enzymes of the lipogenic pathway.

Differentiation-Induced Rise of Cell-Surface Insulin Receptor Number

During differentiation induced by insulin (Method I) 3T3-L1 preadipocytes exhibit a dramatic increase in cell-surface insulin binding capacity (Fig. 1). Maximal binding capaci-

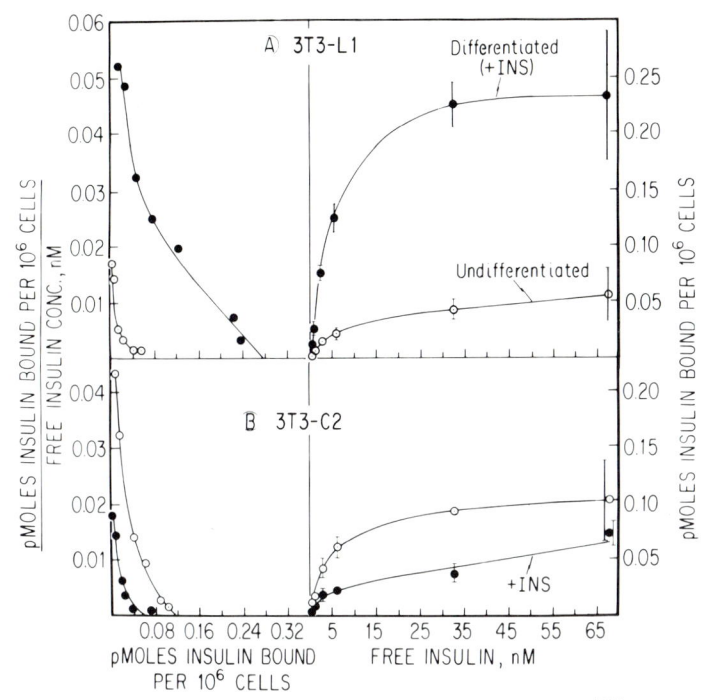

Fig. 1. Effect of differentiation induced by insulin (Method I) on cell-surface ^{125}I-insulin binding by 3T3-L1 preadipocytes and non-differentiating 3T3-C2 cells. Confluent monolayers of 3T3-L1 (A) and 3T3-C2 (B) cells were maintained in medium with (-●-) or without (-○-) insulin. By 29 days post-confluence > 75% of the 3T3-L1 cells and < 1% of the 3T3-C2 cells had expressed the adipocyte phenotype. Cell-surface ^{125}I-insulin binding capacity was determined on cell monolayers as previously described [8]. Results are from [8].

ties, estimated from Scatchard plots of ^{125}I-insulin binding data (Fig. 1A), indicate that 3T3-L1 adipocytes express 170,000 receptor sites per cell while their undifferentiated counterparts, not exposed to insulin, express only 25,000 receptor sites per cell. Since treatment of 3T3-L1 cells with insulin to induced differentiation causes an apparent initial receptor "down-regulation" from 25,000 to an estimated 13,000 sites per cell, the magnitude of the increase of cell-surface insulin receptor number during adipocyte conversion, ie, 13,000 to 170,000 sites per cell, is 13-fold.

This marked rise of insulin receptor number appears to be differentiation-specific and not due to a nonspecific effect of the chronic insulin treatment used to induce differentiation. This is indicated by the fact that the nondifferentiating 3T3-C2 cell line exhibits an inverse response to chronic insulin treatment. These cells undergo insulin-induced insulin receptor "down-regulation" (Fig. 1B), and do not show the subsequent rise in receptor number observed with 3T3-L1 cells.

To follow the kinetics of insulin receptor changes during adipocyte conversion, ^{125}I-insulin binding at an insulin concentration of 2 nM was measured at intervals following the induction of differentiation with insulin. It is assumed that changes in insulin binding reflect alterations primarily in receptor number and not affinity of the receptor for

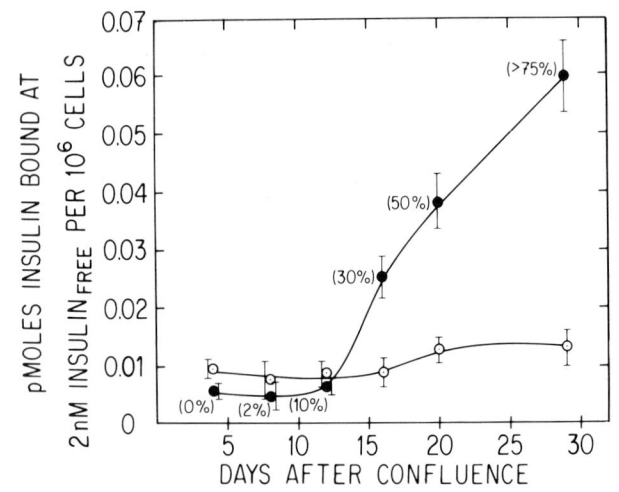

Fig. 2. Kinetics of expression of cell-surface ^{125}I-insulin binding capacity during differentiation of 3T3-L1 preadipocytes. Cell monolayers were carried from confluence with (-●-) or without (-○-) insulin (Method I). Cell-surface binding of ^{125}I-insulin was determined as in Figure 1. Values in parenthesis are the percentage of cells exhibiting triglyceride accumulation; in all other cases, < 1% of the cells accumulated triglyceride. Results are from [8].

TABLE II. Cell-Surface Receptor Changes During Differentiation of 3T3-L1 Preadipocytes

	Sites per cell		
Receptor	Undifferentiated	Differentiated	Change due to differentiation
Insulin	25,000	170,000	6.8-fold increase
	13,000[a]	170,000	13-fold increase
Epidermal growth factor	40,000	35,000	no change
Choleragen	5.0×10^6	3.3×10^5	15-fold decrease
Ganglioside G_{M1}	45×10^6	6.1×10^6	7.4-fold decrease

[a]Estimated number of sites per cell following 24-hr exposure of cell monolayers to insulin-containing medium.
The binding of ^{125}I-insulin and ^{125}I-epidermal growth factor to cell-monolayers was determined as described in [8]. The binding of ^{125}I-choleragen to cell monolayers and ganglioside G_{M1} content of total cell membranes was determined as described in [23].

insulin. This assumption appears to be justified by the similarity of the shapes of the binding isotherms on Scatchard plots of data from differentiated and undifferentiated cells (Fig. 1). Furthermore, the relative amounts of insulin bound at 2 nM insulin by differentiated and undifferentiated cells closely parallels their relative maximal binding capacities estimated from Scatchard plots. As illustrated in Figure 2, 3T3-L1 cells maintained in culture in the absence of insulin did not express the adipocyte phenotype and maintained a relatively constant insulin receptor level. Cells maintained in the presence of insulin initially exhibited 50% down-regulation of receptor level followed by a dramatic increase in receptor number beginning at about Day 12 (Fig. 2). This rise of receptor level paralleled the change in the percentage of cells expressing the adipocyte phenotype (values in parenthesis in Fig. 2) and finally achieved a receptor level 12-fold higher than the initial down-regulated level. Hence, insulin receptor number, like the activities of the key lipogenic enzymes [12, 13, 15], increases coordinately with expression of the adipocyte phenotype.

Specificity of Changes in Insulin Receptor Level: Effect of Differentiation on Epidermal Growth Factor and Choleragen Receptor Levels

Since a substantial increase in cell surface area (~ 2.5-fold) accompanies expression of the adipocyte phenotype [8], the possibility was considered that a nonspecific rise of the cell's membrane content might account for the observed increase in insulin receptor level. This possibility was ruled out by the patterns of expression exhibited by two other receptors, ie, the epidermal growth factor (EGF) and choleragen receptors. As illustrated in Table II, EGF receptor number per cell remained constant during differentiation of 3T3-L1 cells. In contrast, during the course of differentiation, 3T3-L1 cells undergo a loss of choleragen receptors as evidenced by a decreased capacity to bind ^{125}I-choleragen, a reduced fluorescent staining of bound choleragen and a decrease in membrane ganglioside G_{M1} content, the putative choleragen receptor (Table II and [23]). The decrease in choleragen receptor during the 3T3-L1 fibroblast to adipocyte differentiation is consistent with the finding that adipocytes possess fewer choleragen receptors than fibroblasts [24–28]. Thus, the rise of receptor level resulting from differentiation is not simply related to an increase in the surface area of exposed plasma membrane. It is evident that induction of the differentiation program of 3T3-L1 cells results in receptor-specific changes in which the number of insulin receptors per cell increases; the number of EGF receptors remains constant and the number of choleragen receptors decreases.

Possible Mechanisms for the Differentiation-Induced Rise in Cell-Surface Insulin Receptor Level

Several mechanisms that could account for the rise in the level of cell-surface insulin receptors are illustrated in Figure 3. Based on current knowledge of the pathway of plasma membrane receptor synthesis, it is assumed that the receptor polypeptides are translated in the rough endoplasmic reticulum, translocated to the Golgi, and finally, delivered via a Golgi-derived vesicle for insertion into the plasma membrane (Fig. 3A). Thus, one mechanism by which the steady-state level of insulin receptor may be elevated is by an increased rate of synthesis of the receptor. Alternatively, a reduced rate of receptor degradation, with or without a change in synthetic rate, could also produce an elevated insulin receptor level in the steady-state. A reduced rate of any step in the pathway of insulin receptor turnover (Fig. 3A) could elicit this change provided the step becomes rate-limiting. Insulin receptor turnover in some cell types is thought to involve migration of the receptor to the coated pit regions of the cell surface, internalization by endocytosis, fusion of the endocytotic vesicle with a lysosome and then hydrolytic degradation within the secondary lysosome [29, 30]. Finally, the differentiation-linked rise in the active cell-surface receptor level could be caused either by recruitment to the cell surface of spare active receptors from an intracellular pool (Fig. 3B) or by the activation of inactive receptors in a precursor pool (Fig. 3C). Since the insulin receptor is a glycoprotein [31, 32], activation of the receptor polypeptide may involve glycosylation or perhaps yet another processing event. Preliminary evidence suggesting that glycosylation is required for the formation of active insulin receptor is discussed below.

Fig. 3. Possible mechanisms for the differentiation-induced rise in cell-surface insulin receptor level. I refers to insulin, R to inactive insulin receptor, and R̲ to active insulin receptor which is capable of binding insulin. In A, dark and light receptor symbols refer to newly-synthesized "heavy" and previously-synthesized "light" receptors, respectively.

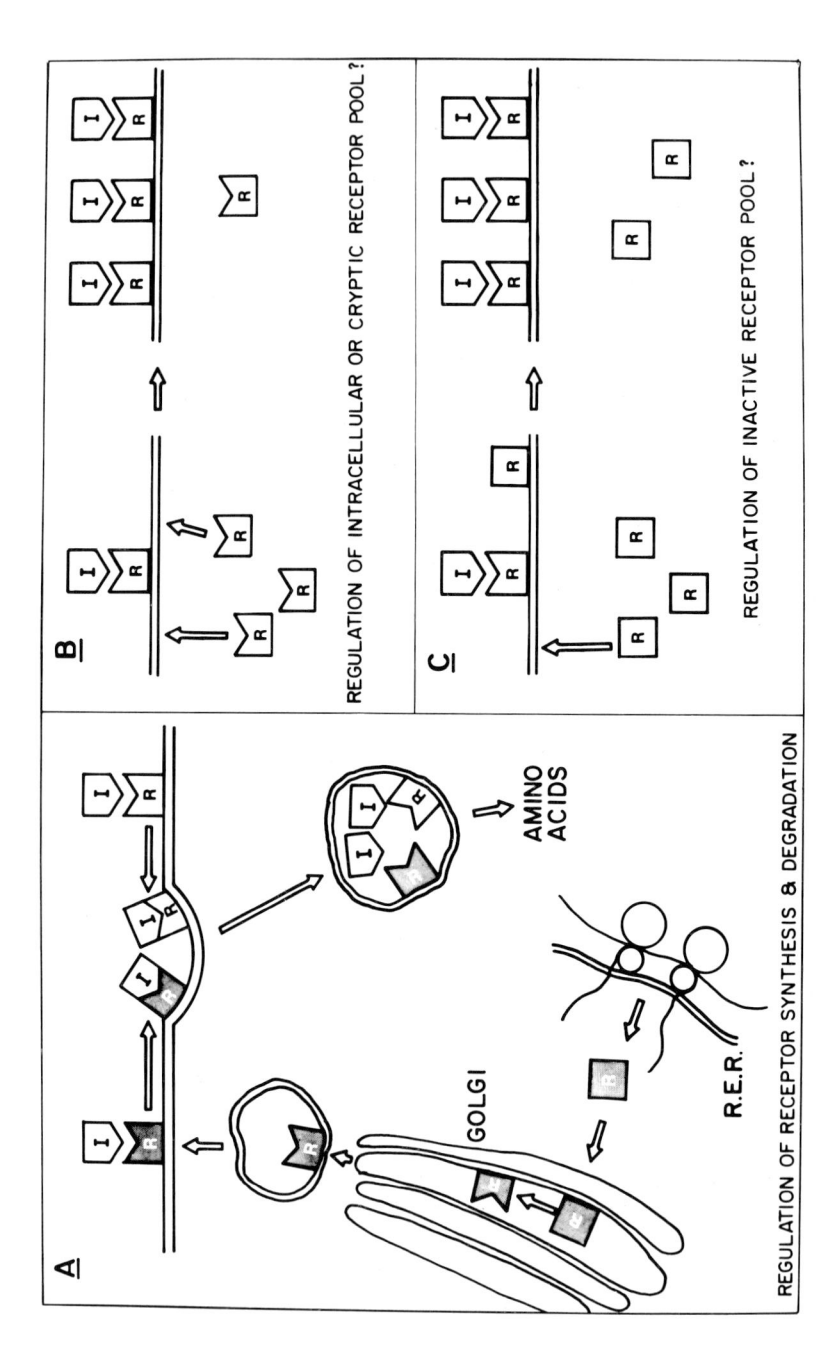

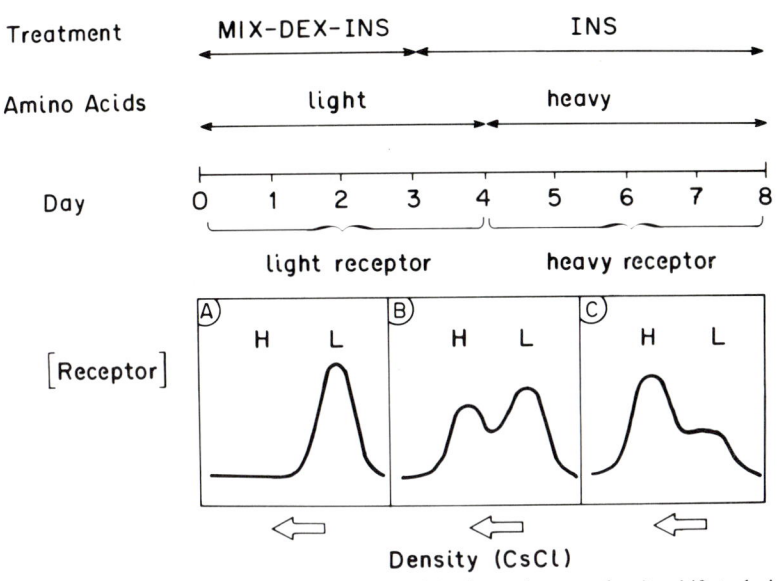

Fig. 4. Diagrammatic illustration of the application of the heavy isotope, density-shift technique to follow the synthesis and degradation of cellular insulin receptors. Light and heavy amino acids refer to [^{1}H, ^{12}C, ^{14}N] and [> 95% ^{2}H, ^{13}C, ^{15}N] amino acids, respectively, present in culture medium. A, B, and C represent hypothetical isopycnic CsCl density gradients of detergent-solubilized cellular insulin receptors on Days 4, 5, and 8, respectively. See text for explanation.

Density-Shift Method for the Analysis of Receptor Synthesis and Turnover

To determine which mechanism(s) (Fig. 3) is responsible for the differentiation-induced rise in level of cell-surface insulin receptor, it was necessary to develop a method with which rates of synthesis and degradation of insulin receptor could be measured unequivocally. This was accomplished by modifying the heavy isotope "density-shift" method of Fambrough and Devreotes [18, 19]. In principle, cells are shifted from a medium containing "light" (^{1}H, ^{12}C, and ^{14}N) amino acids to a medium containing "heavy" (> 95% ^{2}H, ^{13}C, and ^{15}N) amino acids [9, 10]. Following the shift the rate of formation of newly-synthesized "heavy" insulin receptor and the rate of degradation of old "light" insulin receptor are followed concomitantly [9, 10]. The application of this approach to the analysis of insulin receptor metabolism in 3T3-L1 cells is illustrated diagramatically in Figure 4. Differentiation is induced by treating confluent cell monolayers with methylisobutylxanthine (MIX), dexamethasone (DEX), and insulin for three days (Method II). On Day 3, MIX and DEX are removed with a medium change, but treatment with insulin is continued. During the first 2–3 days, the cell population undergoes several rounds of division at which point expression of the adipocyte phenotype begins. By Day 6, > 90% of the cells in the monolayer exhibit cytoplasmic triglyceride accumulation. To measure receptor synthesis and turnover the cells can be shifted to the heavy amino acid-containing medium at any point during differentiation, for example at Day 4 in Figure 4. Thus, soluble receptor extracted prior to Day 4, would be composed entirely of light amino acids and, when banded isopycnically in a CsCl density gradient, would generate a single receptor peak as shown in Figure 4A. Gradients with receptor extracted from cells on Day 5 should contain both a heavy peak of newly-synthesized receptor and a reduced light peak of receptor

Cell monolayers after incubation with "light" [^{12}C, ^{14}N, ^{1}H] or "heavy" [> 95% ^{13}C, ^{15}N, ^{2}H] amino acids

- homogenize
- centrifuge, > 120,000g, 1 hr

membrane pellet   supernate

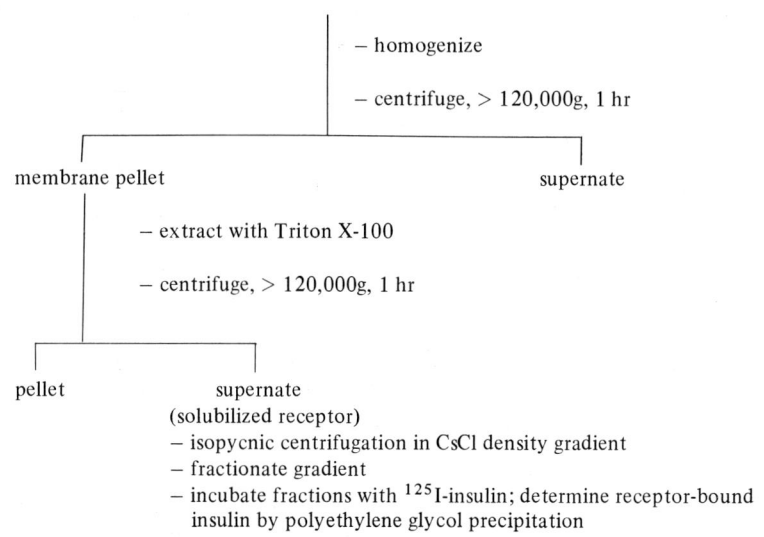

- extract with Triton X-100
- centrifuge, > 120,000g, 1 hr

pellet   supernate
(solubilized receptor)
- isopycnic centrifugation in CsCl density gradient
- fractionate gradient
- incubate fractions with ^{125}I-insulin; determine receptor-bound insulin by polyethylene glycol precipitation

Fig. 5. Protocol for extraction, isopycnic banding, and quantitation of insulin receptors from cells maintained in light or heavy medium. See text and [9] for procedural details.

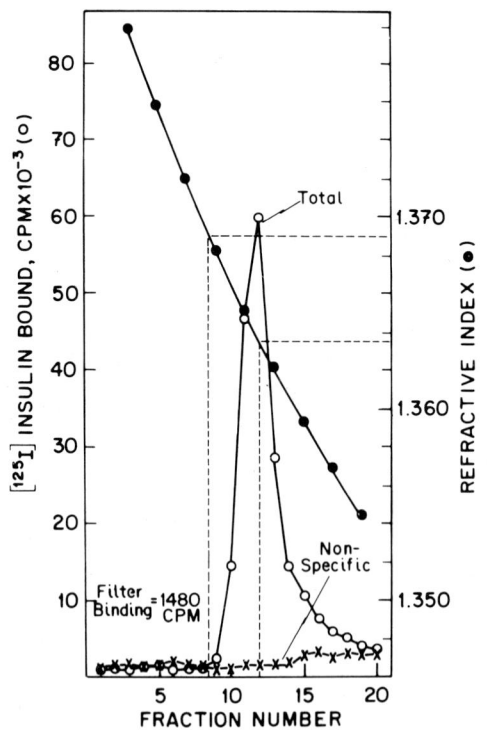

Fig. 6. Isopycnic banding of Triton X-100-solubilized insulin receptors in a CsCl density gradient. Receptors extracted from 2.2×10^6 3T3-L1 cells differentiated in light medium (Method II, Fig. 4) were subjected to equilibrium centrifugation and analysis as outlined in Fig. 5 and [9]. Specific activity of ^{125}I-insulin, 9.8×10^5 cpm per pmol. From Reed and Lane [9].

synthesized prior to the shift to heavy medium (Fig. 4B). Gradient profiles of soluble receptor obtained after longer time periods, eg, Figure 4C, should show increasing amounts of heavy receptor and decreasing amounts of light receptor. By quantitating the change in each class of receptor from differentiated and undifferentiated cells with time after the switch to "heavy" medium, the effect of differentiation on rate of receptor synthesis and degradation can be assessed.

The protocol for extraction and isopycnic banding of heavy and light receptor is presented in Figure 5. After incubation with medium containing light or heavy amino acids, the cell monolayer is washed successively to remove unlabeled insulin [9]. The cells are scraped from the dish, homogenized in hypotonic buffer and centrifuged for one hr at $> 100,000g$. Insulin receptor in the membrane pellet is solubilized by extraction with 4% Triton X-100. Following centrifugation for one hr at $> 100,000g$, the supernatant fraction containing both solubilized cell-surface and intracellular membrane receptor is mixed with CsCl and banded isopycnically during an 18-hour centrifugation [9]. To locate and quantitate light and heavy receptor, the gradient is fractionated and the ^{125}I-insulin binding activity in each fraction is measured by the polyethylene glycol precipitation method [33]. Nonspecific ^{125}I-insulin binding is also measured in duplicate fractions containing excess (4.6 μM) unlabeled insulin. A typical CsCl gradient profile of insulin receptor from differentiated 3T3-L1 cells cultured in light medium is shown in Figure 6. A single sharp peak of specific insulin-binding activity is observed at a refractive index (η_D^{25}) of 1.3635. It will be shown subsequently that heavy receptor from cells cultured in heavy medium bands at a refractive index of 1.3690.

To ascertain the state of aggregation of the isopycnically-banded receptor, the peak fractions of insulin-binding activity from a similar CsCl gradient were pooled and subjected to gel filtration on a calibrated column of cross-linked Sepharose 4B-CL. The column was equilibrated and eluted with 2.59M CsCl (in 50 mM Tris·Cl buffer, pH 7.4, 4°C) which is the concentration of CsCl at the position in the gradient where the receptor bands (Fig. 6). Under these conditions the receptor exhibits an elution volume which corresponds to a molecular weight of a globular protein of approximately 700,000. This value approximates those reported by others [34–36] for insulin receptor solubilized with nonionic detergents. While the insulin receptor is believed to be an oligomeric structure, its subunit composition has not yet been firmly established. Evidence presented below indicates that no detectable subunit interchange between light and heavy oligomeric receptor units occurs during the solubilization and isopycnic banding procedures.

Before the density-shift approach could be applied to the measurement of insulin receptor synthesis and turnover during differentiation, it was necessary to verify that the rise of extractable receptor, determined by the isopycnic banding method, parallels the rise of cell-surface receptors measured with intact cell monolayers (Figs. 1, 2). It is evident from the CsCl gradient profiles shown in Figure 7 that receptor peak area, which estimates insulin receptor level, increases markedly during adipocyte conversion. Also of importance is the fact that the position of the receptor peak in the gradient, hence the density of the receptor, does not change during differentiation of 3T3-L1 cells in "light" medium (Fig. 7).

Kinetic plots of the integrated peak areas of total extractable (soluble light) insulin receptor (Fig. 7) and cell-surface insulin receptor are compared in Figure 8. The 10–12-fold rise in insulin binding to soluble light receptor is closely correlated with the increase of cell-surface receptor. These changes parallel the expression of the adipocyte phenotype as measured by the percentage of adipocytes in the cell population (values in parenthesis, Fig. 8). In other experiments (not shown) non-differentiating 3T3-C2 cells and untreated

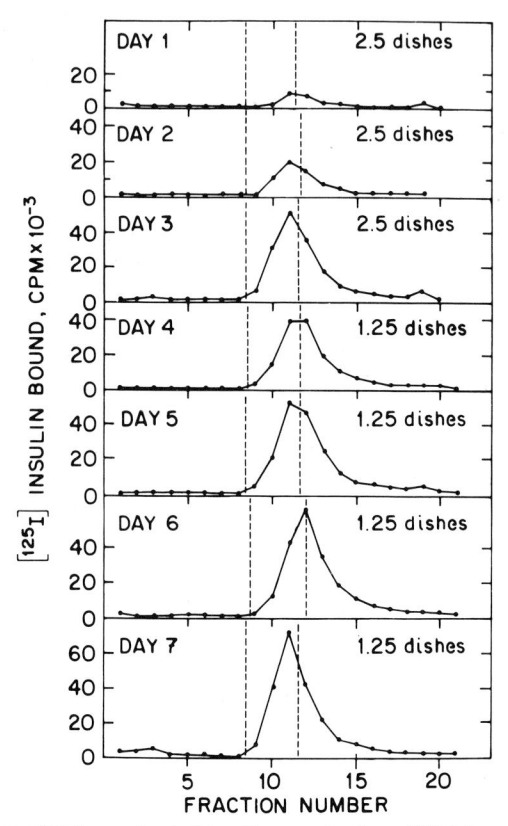

Fig. 7. Isopycnic banding of Triton-extracted insulin receptor from 3T3-L1 preadipocytes during the course of differentiation (Method II, Fig. 4). Receptors from 1.25 or 2.5 dishes of cells were applied to a CsCl gradient and analyzed as in Figure 5. From Reed and Lane [9].

3T3-L1 cells neither exhibit a rise in extractable insulin receptor nor express the adipocyte phenotype. The fact that the kinetics of appearance of cell-surface and total extractable receptors in 3T3-L1 cells are similar during differentiation is inconsistent with the mobilization to the cell surface of an active pool of intracellular receptors (Fig. 3B) as the mechanism by which surface insulin receptors are increased during differentiation. Such a large intracellular pool of active receptor which could account for the observed 10–12-fold rise in surface receptor should have been readily detected as extractable receptor prior to differentiation. This was not observed.

Insulin Receptor Synthesis and Degradation in Differentiated 3T3-L1 Adipocytes

To assess the effect of labeling with heavy amino acids on the expression of insulin receptors and to determine whether the subsequent increase of receptor level results from an increased rate of receptor protein synthesis, 3T3-L1 cells were shifted from light to heavy medium on Day 3 following the induction of differentiation with MIX, DEX and insulin. The CsCl density-gradient profiles of receptors extracted from these cells or from cells maintained in light medium were compared for the next several days. Within 16 hours

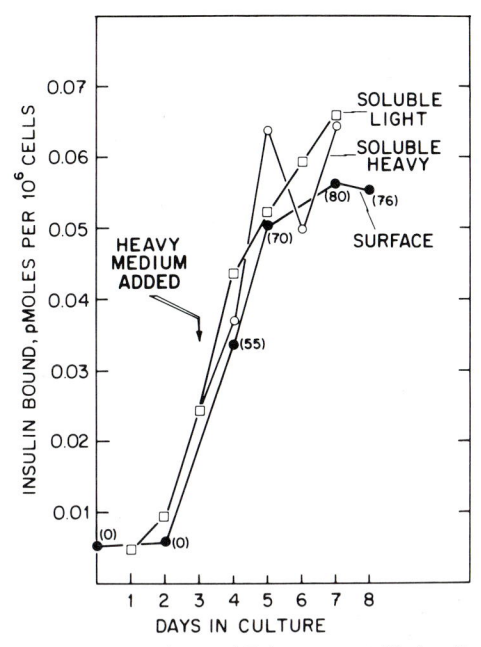

Fig. 8. Kinetics of the rise in level of surface and Triton-extractable insulin receptor during differentiation of 3T3-L1 preadipocytes in light or heavy medium. Cell surface insulin binding determined as in Figure 1. Heavy medium was added to some cultures on Day 3. Soluble light and heavy insulin receptor levels were determined by the CsCl density-gradient isopycnic banding method. Values for soluble receptor were calculated from the results in Figures 7 and 9. The values in parentheses are percentages of cells exhibiting triglyceride accumulation. Results are from Reed and Lane [9].

after the switch to heavy medium only one major peak of insulin binding activity is observed and this peak is centered in a denser region of the gradient than that from cells maintained in light medium (Fig. 9). Furthermore, the position of the newly-synthesized heavy receptor remains constant as the cells continue to differentiate in heavy medium. Importantly, the rate of expression of the insulin receptor, as measured by the isopycnic banding method, does not appear to be materially affected when the cells are exposed to the heavy medium during differentiation (Fig. 8). These results argue against the activation of an inactive receptor pool (Fig. 3C) as a mechanism for the differentiation-induced elevation of insulin receptor level unless such a pool of receptors is turning over rapidly, ie, with a $t_{1/2}$ of four hr or less. Activation of a relatively stable pool of light receptors, present prior to differentiation, should have led to an increase of receptor localized in the light region of the gradient. Instead, the newly-expressed receptor appeared exclusively in the denser region of the gradient (Fig. 9).

The rapid shift, ie, within 16 hours (Fig. 9), from light to heavy receptor suggested that either a rapid turnover of the insulin receptor had occurred or that the old light and newly-synthesized heavy subunits exchange rapidly between oligomeric receptor molecules. In either case, a single denser receptor species would result. However, when cells cultured continuously from Day 3–7 in light or in heavy medium were mixed prior to extraction of

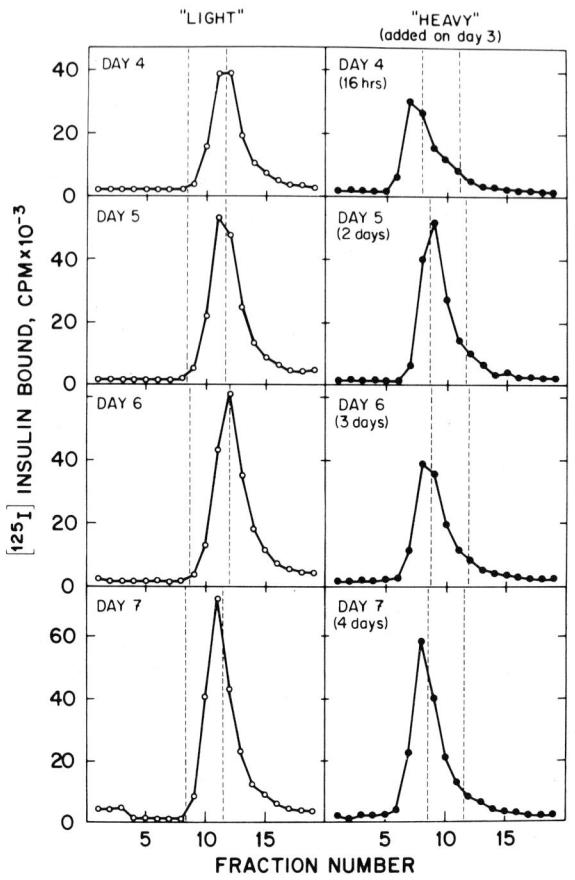

Fig. 9. Isopycnic banding of Triton-extracted insulin receptor from 3T3-L1 preadipocytes during differentiation (Method II) in light or heavy medium. On Day 3 (Fig. 4) some cells were shifted from light to heavy medium. Cells from 1.25 culture dishes were subjected to density gradient centrifugation and gradient fractions subjected to ^{125}I-insulin (9.9×10^5 cpm per pmol) binding as described in Figure 5 and [9]. From Reed and Lane [9].

receptor (Fig. 10) or their respective extracts were mixed (results not shown), the resulting isopycnic gradients resolved the heavy and light receptors with no change in their respective density positions or peak areas as compared with the same receptors run separately. The gradient pattern of insulin-binding activity of the mixed receptors (Fig. 10C) closely approximates the composite pattern (not shown) obtained when the light (Fig. 10A) and heavy (Fig. 10B) peaks are summed. These results indicate that newly-synthesized heavy and old light receptor are resolved by CsCl gradient centrifugation and that little or no subunit interchange between receptors occurs.

To determine the exact half-life of the insulin receptor in differentiated 3T3-L1 adipocytes, an experiment was conducted with shorter sampling intervals following the shift to heavy medium (Fig. 11). Accordingly, 3T3-L1 cells were shifted to heavy medium

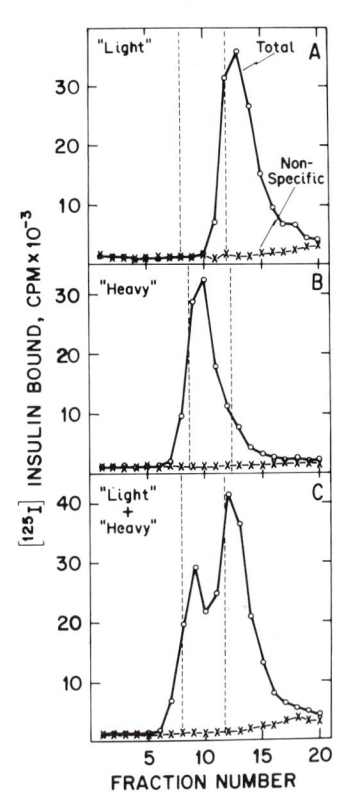

Fig. 10. Isopycnic banding of insulin receptors extracted after mixing membranes from differentiated 3T3-L1 cells (Method II) maintained in light or heavy medium. Receptors, extracted from light membranes(A), heavy membranes (B), and a 1:1 mixture of light and heavy membranes (C), were subjected to equilibrium centrifugation in CsCl gradients and analysis as described in Figure 5. From Reed and Lane (submitted).

on Day 7 after initiating MIX, DEX, and insulin treatment, a point at which the cells have expressed nearly maximal insulin receptor levels. Within three hr after the shift to heavy medium, active heavy receptor was detected (Fig. 11). As time progressed, the heavy receptor peak, which represents newly-synthesized receptor, increased while the light receptor peak, which represents receptor synthesized prior to the shift, decreased concomittantly. By 27.5 hr following the shift, the light peak had virtually disappeared indicating nearly complete turnover of old receptor.

Since the leading edge of the light receptor peak does not contribute to ^{125}I-insulin binding at the peak of heavy receptor (eg, fractions 15–16 at 0 time, Fig. 11), the heavy receptor peak height may be used to calculate the total binding contribution of incompletely resolved heavy receptor. A linear correlation ($r^2 = 0.98$) was observed between peak area and peak height from an analysis of 28 CsCl density gradients containing only single peaks, but variable amounts, of light or heavy receptor [9]. Knowing the total ^{125}I-insulin binding capacity on a gradient and the contribution of heavy receptor, the corresponding value

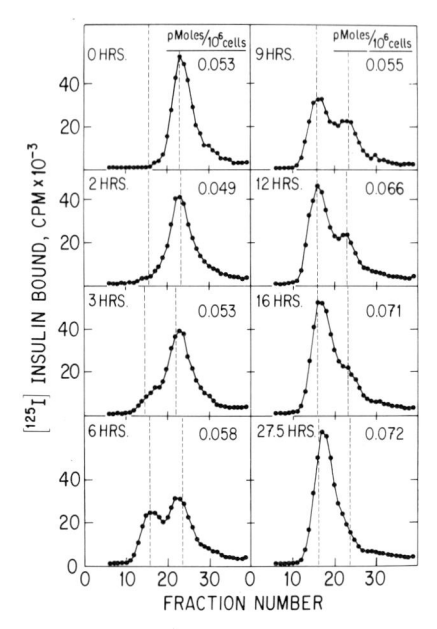

Fig. 11. Isopycnic banding of Triton-extracted insulin receptor from differentiated (Method II) 3T3-L1 adipocytes (2.5 culture dishes per gradient) shifted from light to heavy medium on Day 6. Density-gradient centrifugation and assay for [^{125}I] insulin (8.1 × 10^5 cpm per pmole) binding activity were described in Figure 5. Total insulin binding capacity in each gradient is expressed as pmoles of insulin bound per 10^6 cells. From Reed and Lane [9].

for light receptor was calculated by difference for each gradient in Fig. 11. With this information progress curves were constructed (Fig. 12) for the rate of accumulation of new heavy receptor and the rate of degradation of old light receptor. Following a lag period, active heavy receptor is rapidly formed and by 27.5 hr has replaced 95% of the old light receptor. The 1.5–2.0-hr lag in the appearance of active heavy receptor, after the shift to heavy medium, appears to represent the time required for processing or assembly of receptor polypeptide into a functional form capable of binding insulin.

The degradation of old light receptor followed first-order kinetics (inset, Fig. 12) and was unexpectedly rapid. The $t_{1/2}$ for receptor degradation in this experiment was 6.7 hr. Other experiments using the heavy isotope density-shift method (Krupp, M. and Lane, M.D., unpublished results) reveal that the $t_{1/2}$ for insulin receptor degradation by chick liver cells in primary monolayer culture without insulin is also 6–7 hr. Previous investigations by others on insulin receptor decay in 3T3-L1 cells and other cell lines [5, 37–39], measured in the presence of cycloheximide, indicated a far longer $t_{1/2}$ of 20–40 hr. This discrepancy suggested that inhibition of protein synthesis extends the life-time of the receptor.

To determine whether cycloheximide affects the turnover of the insulin receptor, differentiated 3T3-L1 adipocytes were exposed to the inhibitor for one hr prior to and after the shift to heavy medium. Under these conditions cycloheximide inhibited [^{3}H]-leucine incorporation into protein by > 90% (results not shown). Gradient profiles of receptor extracted at intervals after the shift show that cyloheximide totally inhibits the

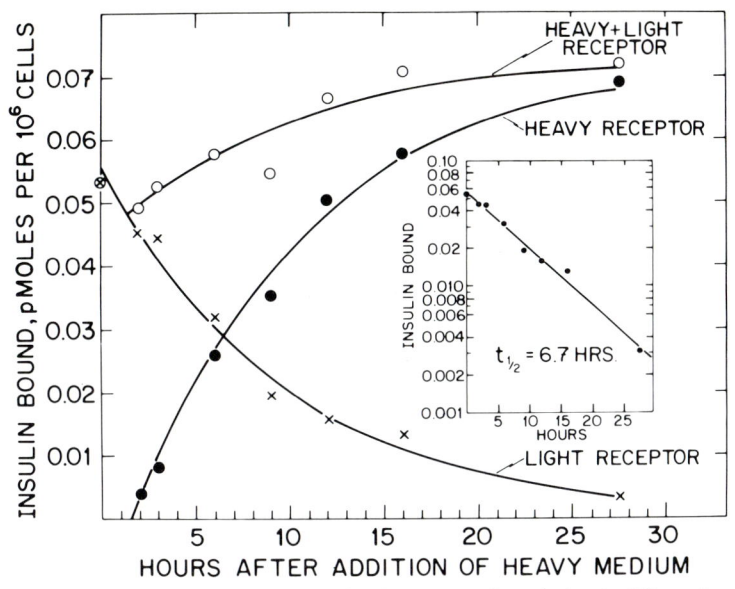

Fig. 12. Kinetics of heavy receptor synthesis and light receptor degradation in differentiated 3T3-L1 adipocytes. Inset: semilog plot of light receptor degradation. The results were calculated from the data in Figure 11. The line through the data points for light receptor represents a least-squares fit ($r^2 = 0.987$) to the equation,

$$R_t = \frac{k_s}{k_d}(1 - e^{-k_d t}) + R_0 e^{-k_d t},$$

when k_s for synthesis of light receptor in heavy medium equals 0. Values of 0.103 hr^{-1} for k_d, and 6.7 hr for receptor half-life were obtained. The line through the data points for heavy receptor represents the values of R_t generated by the Equation assuming t_0 equals 1.5 hr, $k_s = 0.0075$ pmoles of insulin binding activity (at 2 nM) per hr per 10^6 cells (approximated by the limiting slope of heavy receptor synthesis at t_0), $k_d = 0.103$ hr^{-1}, and that heavy receptor $R_0 = 0$ at t_0. The line through the data points for total receptor was generated using the above values for k_s and k_d, and $R_0 = 0.048$ pmoles insulin bound per 10^6 cells, calculated at $t_0 = 1.5$ hours, from the least-squares fit of the data for light receptor decay. From Reed and Lane [9].

formation of heavy receptor (Fig. 13) and lengthens the $t_{1/2}$ for receptor degradation from 7.5–26 hr (Fig. 14). These results suggest that cycloheximide blocks the synthesis of a short-lived proteins(s) required for the normal degradation of the insulin receptor. Inhibition of any step in the degradation process, eg, receptor clustering, endocytosis, or lysosomal hydrolysis, would increase receptor half-life.

Tunicamycin inhibits the synthesis of dolichol-linked N-acetylglucosamine, thereby blocking glycosylation of asparaginyl residues of glycoproteins [40, 41]. Treatment of 3T3-L1 cells with Tunicamycin completely blocks the appearance of active heavy insulin receptor on CsCl density gradients when differentiated cells are maintained in heavy medium (results not shown). General protein synthesis, on the other hand, remains substantially intact, ie, 60–70% of untreated controls. These findings were not entirely unexpected since the insulin receptor is thought to be a glycoprotein and it was shown

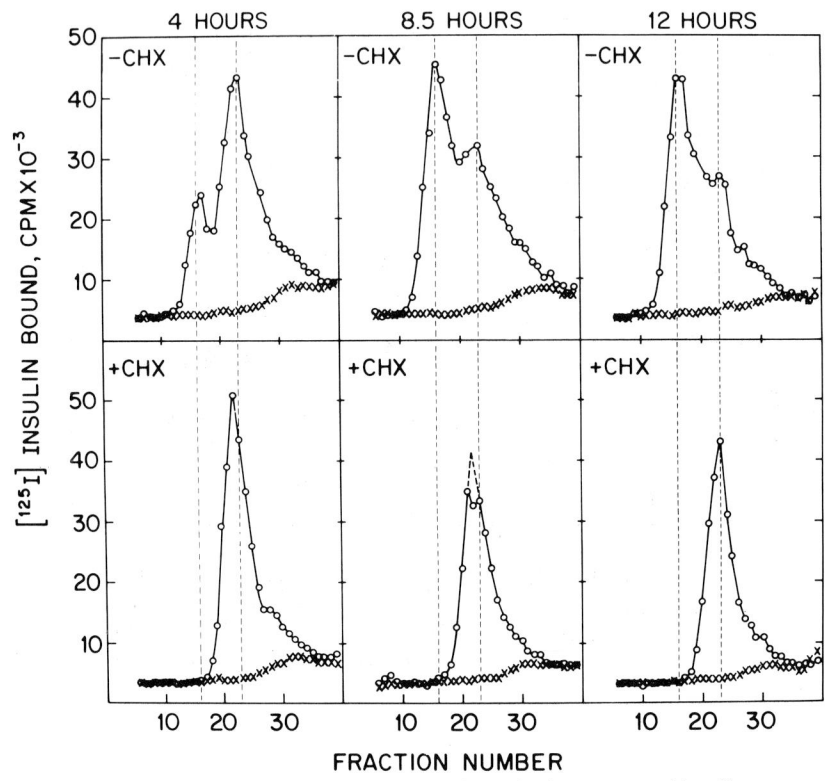

Fig. 13. Effect of cycloheximide on the isopycnic banding of Triton-extracted insulin receptors from differentiated 3T3-L1 cells at 4, 8.5, and 12 hr after shifting from light to heavy medium. Cells to be treated with cyloheximide (5μg/ml) were exposed to the inhibitor 1 hr prior to the shift to heavy medium. The procedures were otherwise identical to those described in Figure 11. From Reed and Lane (submitted).

earlier [42] that treatment with β-galactosidase and neuraminidase caused the loss of insulin binding activity. Furthermore, it has been recently shown [43] that Tunicamycin increases the rate of loss ($t_{1/2}$ = 9 hr) of surface insulin binding capacity by 3T3-L1 cells. Taken together these results suggest that glycosylation of receptor polypeptide(s) is a requisite processing step in the formation of active insulin receptor.

Comparison of Apparent Rates of Insulin Receptor Synthesis and Degradation by Differentiated 3T3-L1 and Undifferentiated 3T3-C2 Cells

To determine whether an altered rate of insulin receptor synthesis or degradation is responsible for the differentiation-linked rise in receptor level (Figs. 1, 2), these rates were compared in differentiated 3T3-L1 and undifferentiated 3T3-C2 cells using the heavy isotope density-shift method. As shown in Table III, receptors extracted from differentiated 3T3-L1 cells exhibited 16-fold greater insulin-binding capacity at 2 nM insulin than did receptors extracted from undifferentiated 3T3-C2 cells. Following differentiation (Day 7 after initiating induction with MIX, DEX, and insulin), the apparent rate of receptor synthesis by 3T3-L1 adipocytes, estimated from the initial rate of formation of heavy receptor

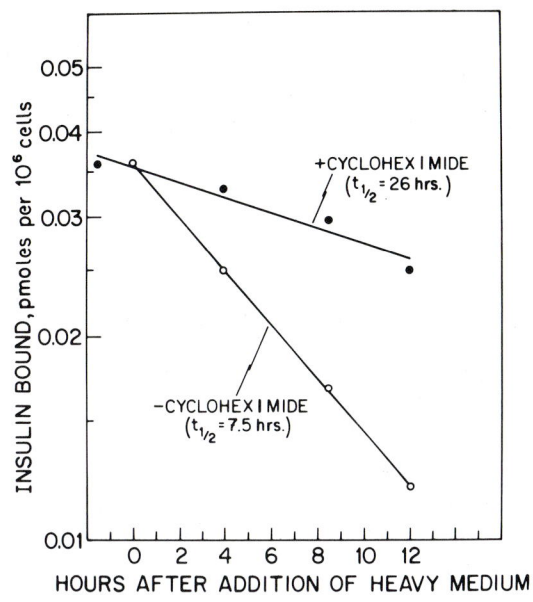

Fig. 14. Effect of cycloheximide on the rate of degradation of Triton-extractable insulin receptor from differentiated 3T3-L1 cells. The data from Figure 13 were analyzed and plotted as described in Figure 12 (inset). From Reed and Lane (submitted).

TABLE III. Effect of Differentiation on the Synthesis and Turnover of Insulin Receptors

Cell line	State	Sites per cell[a]	Rate of synthesis (sites/cell/hr)	$t_{1/2}$ degradation (hr)
3T3-L1	differentiated	42,000	4,200	6.7
3T3-C2	undifferentiated	2,600	300	7.5–10

[a] estimated at 2 nM [insulin].
Differentiation of 3T3-L1 cells was induced by Method II. Total ^{125}I-insulin binding capacity at 2nM insulin was determined on isopycnically-banded receptor from 3T3-L1 adipocytes on Day 7 following induction and from untreated non-differentiating 3T3-C2 cells after a comparable period in culture. Rates of receptor synthesis and degradation in 3T3-L1 cells were calculated from the results in Figure 12 and from comparable results with 3T3-C2 cells.

(Fig. 12), was 14-fold greater than that by 3T3-C2 cells at a comparable time post-confluence (results not shown). Since the half-lives of the insulin receptor are 6.7 hr for 3T3-L1 adipocytes and 7.5–10 hr for 3T3-C2 cells (Table III), it is evident that the differential in receptor level is due to the markedly greater rate of synthesis by 3T3-L1 adipocytes. The somewhat longer $t_{1/2}$ of receptor decay for 3T3-C2 cells would, if not compensated by a lower synthetic rate, lead to a higher receptor level relative to 3T3-L1 cells. Thus, it appears that the basis for the far higher level of insulin receptors in differentiated 3T3-L1 cells compared to their undifferentiated counterparts is a higher rate of receptor synthesis (Table III).

ACKNOWLEDGMENTS

This work was supported by research grants from the National Institutes of Health (AM-14574) and the American Heart Association (#78822).

B.C.R. is the recipient of postdoctoral Fellowship (AM-05506) from the National Institutes of Health.

REFERENCES

1. Fritz IB. In Litwack G (ed): "Biochemical Actions of Hormones." Vol 2:165. New York: Academic Press, 1972.
2. Manchester KL. In Litwack G (ed): "Biochemical Actions of Hormones." Vol 1:267. New York: Academic Press, 1970.
3. Catt KJ, Harwood JP, Aguilera G, Dufau ML: Nature (London) 280:109, 1979.
4. Tell G, Haour F, Saez J: Metabolism 27:1566, 1978.
5. Karlsson FA, Grunfeld C, Kahn CR, Roth J: Endocrinology 104:1383, 1979.
6. Green H, Kehinde O: Cell 1:113, 1974.
7. Green H, Kehinde O: Cell 5:19, 1975.
8. Reed BC, Kaufman SH, Mackall JC, Student AK, Lane MD: Proc Natl Acad Sci USA 74:4876, 1977.
9. Reed BC, Lane MD: Proc Natl Acad Sci USA 77:285, 1980.
10. Reed BC, Lane MD: Adv Enzyme Regul 18:97, 1980.
11. Rubin CS, Hirsch A, Fung C, Rosen OM: J Biol Chem 253:7570, 1978.
12. Mackall JC, Student AK, Polakis SE, Lane MD: J Biol Chem 251:6462, 1976.
13. Coleman RA, Reed BC, Mackall JC, Student AK, Lane MD, Bell RM: J Biol Chem 253:7256, 1978.
14. Green H, Kehinde O: J Cell Biol 101:169, 1979.
15. Mackall JC, Lane MD: Biochem Biophys Res Commun 79:720, 1977.
16. Wise LS, Green H: Cell 13:233, 1978.
17. Student AK, Hsu RY, Lane MD: J Biol Chem 255:4745, 1980.
18. Fambrough DM, Devreotes PN: In Cook JS (ed): "Biogenesis and Turnover of Membrane Macromolecules." New York: Raven, 1976, p 124.
19. Devreotes PN, Gardner JM, Fambrough DM: Cell 10:365, 1977.
20. Russell TR, Ho RJ: Proc Natl Acad Sci USA 73:4516, 1976.
21. Student AK, Lane MD: In Ailhaud G (ed): "Symposium on Cellular and Molecular Aspects of Obesity." Vol 87:35, Paris: INSERM, 1979.
22. Freytag SO, Utter MF: Proc Natl Acad Sci USA 77:1321, 1980.
23. Reed BC, Moss J, Fishman PH, Lane MD: J Biol Chem 255:1711, 1980.
24. Mora PT, Brady RO, Bradley RM, McFarlan VW: Proc Natl Acad Sci USA 63:1290, 1969.
25. Yogeeswaran G, Sheinin R, Wherret JR, Murray RK: J Biol Chem 247:5146, 1972.
26. Fishman PH, Brady RO: Science 194:906, 1976.
27. Pacuszka T, Moss J, Fishman PH: J Biol Chem 253:5103, 1978.
28. Ohashi M: Lipids 14:52, 1979.
29. Goldstein J, Anderson RGW, Brown MS: Nature 279:679, 1979.
30. Davies PJA, Davies DR, Levitjki A, Maxfield FR, Milhand P, Willingham MC, Pastan IH: Nature 283:162, 1980.
31. Jacobs S, Schechter Y, Bissel K, Cuatrecasas P: Biochem Biophys Res Commun 77:981, 1977.
32. Cautrecasas P, Tell GPE: Proc Natl Acad Sci USA 70:485, 1973.
33. Cautrecasas P: Proc Natl Acad Sci USA 69:318, 1972.
34. Gavin JR III, Mann DL, Buell DN, Roth J: Biochem Biophys Res Commun 49:870, 1971.
35. Cautrecasas P: J Biol Chem 247:1980, 1972.
36. Lang V, Kahn CR, Chrambach A: Endocrinology 106:507, 1980.
37. Rubin CS, Rosen OM: Fed Proc 38:306, 1979.
38. Ginsburg BH, Kahn CR, Roth J, DeMeyts P: Biochem Biophys Res Commun 73:1068, 1976.
39. Kosamakos FC, Roth J: Endocrine Society, 58th Annual Meeting, Abstr 26, 1976.
40. Takatsuki A, Kohno K, Tamura G: Agric Biol Chem 41:2089, 1975.
41. Tkacz JS, Lampen JO: Biochem Biophys Res Commun 65:248, 1975.
42. Cuatrecasas P: Ann Rev Biochem 43:169, 1974.
43. Rosen OM, Chia GH, Fung C, Rubin CS: J Cell Physiol 99:37, 1979.

Cell Surface Receptors: Workshop Report

H. R. Herschman and J. F. Perdue

University of California, Los Angeles, California 90024 (H.R.H.) and Lady Davis Institute for Medical Research, Montreal, Quebec, Canada H3T 1E2 (J.F.P.)

The workshop on cell surface receptors was limited to the discussion of two areas: 1) evidence for the presence of multiple cell surface receptors that bind specific ligands with different affinities and 2) the requirement for the continued presence of plasma-membrane-bound growth factors to elicit the mitogenic response. A second workshop, "Coated Pits, Vesicles, and Intracellular Protein Transport" (R. Bradshaw, S. Cohen, and T. Roth), continued with discussions of the mechanism and role of receptor-ligand internalization.

EVIDENCE FOR MULTIPLE RECEPTORS

Several reports in the literature concluded from measurements of epidermal growth factor (EGF) binding as a function of concentration that receptors for this growth factor exist as a homogeneous class of high-affinity sites on mouse and human cells, ie, a linear slope was obtained when bound/free vs bound/iodo-EGF was plotted by the method of Scatchard. Using a similar method of analysis, other investigators reported curvilinear plots of the binding data. Curvilinearity has been ascribed to 1) the presence of 2 or more populations of specific receptors each having a distinctive affinity for the ligand, 2) evidence for negative cooperativity, ie, the occupation of free receptors accelerating the rate of dissociation of previously bound ligand by an alteration in affinity, or 3) evidence for selective down-regulation of a receptor at a high ligand concentration. A stimulating exchange of views arguing for the presence of single vs multiple classes of receptor sites for EGF and the other growth factors was provoked in large measure by the observations reported at this meeting by Iwata, Williams, and Fox (J Supramol Struct, Suppl 9, 162a, 1980) that Scatchard plots of EGF binding to numerous human and mouse clonal cell lines as well as cell strains were curvilinear. These investigators also reported that they obtained curvilinear plots for EGF binding to isolated plasma membrane preparations and to detergent-extracted soluble membrane receptors. In addition, measurements of the rate constants for EGF dissociation from membranes and intact cells gave values that were consistent with the presence of high- and low-affinity EGF binding components. From the

Received and accepted May 5, 1980.

© 1981 Alan R. Liss, Inc., 150 Fifth Avenue, New York, NY 10011

latter results, it was argued that the curvilinearity of the plotted data for the binding of EGF to the intact cells cannot be explained by negative cooperative, ie, site-site interactions. Alternative explanations for these observations ranging from different gene products for high- and low-affinity receptors to differences in physical state of a single gene product were discussed. The observations of Iwata et al were considered in the light of the studies of Aharonov et al (J Biol Chem 253:3970, 1978), who observed linear Scatchard plots of EGF binding to 3T3 cells when the studies were carried out at 4°C. In studies of variants of 3T3 cells, Pruss and Herschman reported the loss of all EGF binding, suggesting a single gene product for the EGF receptor. Cohen (Vanderbilt University, Nashville, TN) reported observing linear Scatchard plots as did C. King, from the laboratory of Pedro Cuatrecasas (Wellcome Research Laboratories, Research Triangle Park, NC), when the studies were carried out at 37°C. Cohen qualified these past observations, however. He cautioned that, because of the possibility of site-site interactions and the influence of other proteins and membrane perturbants on the affinity of EGF binding to its receptors, he would only have confidence in the results of studies carried out with soluble receptors. He cited, as an example, that he now observes curvilinear Scatchard plots of EGF binding to its receptors using a Triton extract of A431 cells. It was also pointed out by Kermit Carroway (University of Oklahoma) that, in studies of the enzyme 5'-nucleotidase, 2 distinctive K_m's were observed for intact cells. However, when a solubilized membrane preparation was examined, a single K_m was obtained. In further support of a single-receptor mode, N. Shimizu (Arizona University, Tucson, AZ) described studies of chromosome and isozyme marker analysis of clones of mouse L cells fused with the human diploid fibroblast, WI38. The mouse cells were deficient both in hypoxanthenephosphoribosyl transferase activity and the ability to bind ^{125}I-labeled EGF. Using chromosome and isozyme marker analysis, Shimizu concluded that EGF binding was associated with human chromosome number 7. In response to this presentation, however, M. Rechler (National Institute of Arthritis and Metabolic Diseases, Bethesda, MD) pointed out that the results of this analysis were inconclusive since the size of the genetic material that was being lost, ie, whole chromosomes, was so great as to preclude any conclusive statement concerning single vs multiple EGF receptor sites. Several investigators argued for the presence of at least 2 distinctive binding components of cells. J. Baker (University of California, Irvine, CA) described studies with human fibroblasts in which cells grown in the presence of glucocorticoids bound additional EGF. Analysis of binding data by the method of Scatchard indicated that a homogeneous class of low-affinity sites were complemented by the appearance of a second class of high-affinity binding sites. M. Rechler summarized studies with various insulin-like growth factors, ie, IGF I and II, somatomedins A and C, etc, on placental, adipocyte, and liver plasma membranes, as well as on cultured cells. In many of these studies, iodinated ligands including insulin bound to distinctive receptors and, in some cases, to common receptors but with altered affinities. It has yet to be established, however, if there are unique receptors that bind the same ligand but with different affinities. B. Yankner from Eric Shooter's laboratory (Stanford University, Stanford, CA) described the characteristics of binding of nerve growth factor (NGF) to the clonal pheochromocytoma cell line, PC-12. During the first 4–6 h of incubation of the cells with NGF, a rapidly dissociable binding site, ie, low affinity, was selectively down-regulated; a membrane component that bound the ligand tightly, however, was maintained at a steady state. Under conditions in which NGF induced neurite outgrowth, the membrane component that bound the ligand tightly increased in concentration; these events were temporally correlated. Yankner argued that NGF was recognized by 2 classes of binding sites. R. Bradshaw (Washington University, St. Louis,

MO) observed nonsaturable binding of NGF to cultured dorsal root ganglion and isolated membrane from these cells. His studies indicated, however, that the curvilinear binding plots were explained by negative cooperative interactions of a homogeneous class of receptors, rather than by the existence of 2 specific receptors for NGF. During the various exchanges, expressions of caution were voiced on the failure, in certain of these studies, to take into consideration the effect of growth-factor concentration, environmental factors, etc, on specific ligand binding. J. Gordon (University of Colorado, Denver, CO) described studies on concanavalin-A (Con-A) binding to fixed and unfixed erythrocytes in which both positive and negative cooperativity could be demonstrated. On further analysis, the alteration in cooperativity could be accounted for by the state of ligand interaction, ie, from tetramers to monomers, as well as by the numbers of cells interacting with the plant lectin. For example, by changing cell number and holding the con-A concentration constant, he demonstrated changes in the K_d, total binding capacity (R_o), and the degree of negative and positive cooperativity.

Two abstracts presented at this meeting by authors who did not participate in this workshop contained data which may be pertinent to this discussion. Magun and his colleagues (U. Arizona, Tucson, AZ), in studies of the effects of the tumor promoter, tetradecanoyl phorbol acetate (TPA), on iodo-EGF binding to rat cells, observed a dramatic decrease in the binding of ligand to the high-affinity receptor class. In control cells, a curvilinear Scatchard plot was obtained. This decrease in binding was attributed to a TPA-induced change in affinity of the receptor for EGF, rather than a specific loss of a population of receptors. Similar observations on alteration of EGF-receptor affinity by TPA have been made by Shoyab et al (Nature 279:387, 1979) and Brown et al (Biochem Biophys Res Comm 86:1037, 1979). These studies contrast with the earlier one by Lee and Weinstein (Science 202:315, 1978) in which TPA treatment of HeLa cells was ascribed to a specific loss of receptor number with no change in the receptor affinity. In studies carried out with succinylated con-A, Ballmer and Burger (University of Basel, Switzerland) correlated an inhibition of EGF-induced mitogenesis in 3T6 cells by succinylated con-A with a decreased binding of iodo-EGF. Analysis of the binding of iodo-EGF indicated lectin treatment caused a 10-fold reduction in the affinity of the ligand with its receptor; the R_o remained unchanged.

Some recently published studies (not discussed in this workshop) on the effects of anti-EGF and anti-insulin IgG on the binding and biological activity of EGF and insulin in responsive cells, have been made that may have a strong bearing on the interpretation of 2 classes of affinity sites. Schechter et al, Nature 278:835, 1979) demonstrated that the addition of anti-EGF IgG to mitogenically inactive, cyanogen bromide-treated EGF or to suboptimally mitogenic concentrations of EGF enhanced this growth factor's biological activity. For example, full restoration of mitogenic activity could be demonstrated when a 1:8,000–1:10,000 dilution of anti-EGF IgG was incubated with 3T3 monolayers in the presence of 5 ng/ml of treated EGF. Monovalent Fab' fragments of anti-EGF were ineffective. This antibody was also effective in stimulating 3T3 cells to replicate with submitogenic concentrations of EGF. Accompanying its ability to reactivate and to enhance the biological activity of modified and unmodified EGF, respectively, the anti-EGF induced macro-clustering of fluorescent-labeled EGF on the surface of the receptive cells, but at significantly higher antibody concentration than that which induced mitogenesis. In a study of insulin binding to liver plasma membrane and to 3T3 cells, Schechter et al (PNAS 76:2720, 1979) observed curvilinear Scatchard plots. The addition of anti-insulin IgG but not Fab' fragments stimulated specific insulin binding to the liver membranes.

This was accompanied by the conversion of the low-affinity receptor sites to high-affinity sites with no alteration in the total binding capacity. Kinetic binding analysis indicated that the rate constant for association was increased approximately 15 times. In the presence of antibody, there was also a decrease in the rate of dissociation of membrane-bound insulin. Using similar methodology, suboptimal concentrations of iodinated insulin bound to a greater extent to 3T3-C2 fibroblasts treated with the bivalent anti-insulin antibody but not the Fab' fragments. Accompanying this greater binding, there was also a significant stimulation of thymidine incorporation into DNA. These observations were unique to liver plasma membranes and the cultured cells, since iodo-insulin binding was actually inhibited by anti-insulin IgG when tested on plasma membranes derived from fat cells. The latter cells already have a large number of very high-affinity insulin binding sites that exist in macroclusters. Anti-EGF IgG did not affect binding of iodo-EGF to the 3T3-C2 fibroblasts. However, similar to insulin's binding to fat cell membranes EGF already demonstrates a 20-fold greater affinity than insulin for binding to specific receptors on these cells. These observations, when viewed in the context of evidence for 2 classes of binding sites on cultured cells, raise the possibility that alteration in the degree of molecular interaction of plasma membrane receptors influence both the affinity with which they bind their ligand and the induction of the biological response.

In conclusion, although no definitive statements could be made with respect to the existence of multiple vs single receptors for the various growth factors, the participants and audience (including those actively engaged in this aspect of research) gained a new appreciation of the complexity of the systems involved and the need for caution in interpreting the results of such studies. Moreover, since there is still so little known about the mechanism whereby receptor occupancy leads to a subsequent mitogenic response (ie, the roles of persistence of the ligand at the plasma membrane, aggregation of receptors, processing of ligand and/or the receptor, the final cytoplasmic target for processed or unprocessed ligand and/or receptor, etc). The nature of initial specific binding events at the cell membrane needs continued re-evaluation in the light of future observations on distal processes in the mitogenic response.

THE ROLE OF PERSISTENT ASSOCIATION OF GROWTH FACTORS WITH PLASMA MEMBRANE RECEPTORS

The most extensive study of the requirement for persistence of EGF in 3T3 mitogenesis is by Aharonov et al (J Biol Chem 253:3970, 1978). These investigators observed a lack of correlation between the kinetics for EGF-induced receptor loss, ie, down-regulation, and the stimulation of TdR incorporation into 3T3 cells. The latter response required the continuous presence of EGF over a 3- to 4-day period, whereas down-regulation was completed within 4 h. Studies by Schechter et al (PNAS 75:5788, 1978) also indicated that the stimulation of DNA synthesis by EGF required the persistent occupation of a relatively small fraction of the total measurable externally located receptors. The enhancement of DNA synthesis observed 24 h after the addition of EGF to serum-deprived fibroblasts was completely blocked following the addition of anti-EGF IgG at any time during the first 8–10 h of incubation.

The question of persistence as it applied to the mitogenic events associated with platelet-derived and fibroblast growth factors (PDGF and FGF) was also discussed. C. Scher (Sidney Farber, Boston, MA) commented on the observations made by Pledger et al (Proc Natl Acad Sci USA 74:4481, 1977) that the addition of PDGF for 2 h followed by its removal and the addition of platelet-poor plasma allowed quiescent cells to replicate. The concept of the commitment of cells to replicate vs their progression through G_1 to S emanated from these studies. However, as pointed out by Scher, there presently is no evidence that the absorbed growth factors were actually removed from the cells following the rinsing of the plates with buffer and fresh culture medium. Finally, the possible existence of receptors that have even greater affinities for growth factors than are detectable using standard binding assays and Scatchard plot analysis must be kept in mind during our consideration of mechanisms regulating growth factor-induced mitogenesis. With respect to the concept of persistence, there presently is too little data to determine the necessity of maintaining small levels of plasma membrane-occupied receptor for either PDGF or FGF for cell replication. In contrast, data from Cuatrecasas', Cohen's, and Herschman's laboratories suggest that EGF must be present continuously for maximal mitogenic response.

The concept of ligand-induced receptor alterations as a mechanism for inducing a biological response was discussed in much more detail at the workshop and, in some respects, offers a more plausible and exciting mechanism for understanding growth factor-stimulated replication of cells. Rechler (NIH) described studies previously carried out by the laboratories of Kahn and Cuatrecasas in which anti-insulin-receptor antibody (ARA) obtained from insulin-refractory diabetic patients or made against isolated rat-liver receptors, respectively, displayed many of the insulin-like activities ascribed to this hormone, including the stimulation of hexose transport and its metabolism. Both groups established the necessity for a bivalent antibody for the biological response since Fab' fragments of ARA inhibited insulin's interaction with specific adipocyte receptors, but did not stimulate either glucose transport or its metabolism. Although insulin binds to specific receptors on mitogenically responsive cells with high affinity, it also binds to receptors that bind insulin-like growth factors but with a much lower affinity. The occupancy of what has been termed by Rechler growth factor sites resulted in the stimulation of TdR incorporation into DNA and cell replication. The addition of ARA to quiescent avian and human fibroblasts failed to stimulate replication, although it did stimulate glucose transport. This observation is consistent with the hypothesis that the growth factor receptor is different from the insulin receptor. In a recent development reported at this meeting, J. Koontz (University of Colorado, Denver, CO) has been able to initiate a growth response in the H35 rat hepatoma cell line with ARA. This cell line was unique in that insulin at nM levels stimulated hexose transport and cell replication. Purified IgG from ARA, but not IgG from normal patients, displaced iodinated insulin and, in the same concentration ranges, stimulated the incorporation of TdR into DNA. These observations are unique and complement to a large extent those of Kahn and Cuatrecasas, since they can be interpreted to indicate that mitogenic response (like what has been observed for insulin's stimulation of hexose transport) does not require internalization and processing of the hormone but is achieved by perturbation and/or modulation of specific cell-surface receptors. If this interpretation has merit, it suggests that the internalization of growth factors is not causal for replication. Further studies with antibody against other growth-factor receptors are necessary to truly evaluate this model. For a continued discussion of this issue, readers should see the workshop on "Coated Pits, Vesicles, and Intracellular Protein Transport."

Coated Pits and Vesicles in Intracellular Protein Transport: Workshop Report

Ralph A. Bradshaw, Stanley N. Cohen, and Thomas F. Roth

Department of Biological Chemistry, Washington University School of Medicine, St. Louis, Missouri 63110 (R.A.B.), Departments of Genetics and Medicine, Stanford University School of Medicine, Stanford, California 94305 (S.N.C.), and Department of Biological Sciences, University of Maryland, Baltimore County, Catonsville, Maryland 21228 (T.F.R.)

It has become evident that the specific endocytosis of a variety of polypeptides is mediated by specialized regions of the plasma membrane called coated pits. Since their first description by Roth and Porter in 1962, their postulated ubiquitous role in specific protein transport has been demonstrated in a wide variety of systems in which a wide variety of polypeptides have been demonstrated to be bound to and moved by this organelle.

As recently reviewed by Goldstein, Brown, and Anderson [Nature], the list of ligands known to enter cells by this mechanism is already large and is becoming larger each year. It is equally clear, however, that not all ligands that are transported by this mechanism share the same fate. In fact, there are clear differences in the eventual intracellular sites to which these ligands go.

Viewed simplistically, the fate of the coated pit is not complex. After forming, it pinches off as a coated vesicle which very rapidly loses its clathrin-containing coat. Now naked, the vesicle has several choices as to its fate, choices apparently dictated by the cell type and by the particular ligand bound to its receptor.

Although the number of routes are thus far not many or, seemingly, complicated, the means by which the choice is made is unknown. The routes can be simplistically divided based upon their initial intracellular targets — eg, storage vesicles (yolk proteins), basolateral plasmamembranes (IgG and IgA), multivesicular bodies (MVB) or lysosomes (LDL, EGF, etc), nuclear membrane (NGF), and Golgi (ER proteins for export and recycling of PM). Although this selection is not large, it is a choice that results in profoundly different effects on the cell.

For example, in either the transcellular routes of IgG, as in transplacental transport or IgA transport across the liver parenchymal cell, the molecule Ig is moved specifically and in an intact form. This is also the case for yolk proteins where the various proteins go to storage vacuoles for later use during early embryogenesis. These two routes, in which no apparent degradation occurs, stand in clear contradistinction to the vesicles that are targeted for the lysosomes. In such instances, exemplified by the ligand LDL, the fate seems assured and degradation occurs. It is not clear in most examples to what degree and for which components of the vesicles degradation takes place. It may well be that some ligands are delivered to the lysosomes for degradation with concomitant release of a biologically active portion, as exemplified by cholesterol release in the case of LDL. In other instances the lysosome could be a way station for specific processing of the ligand or receptor. The eventual fate of the undegraded material would then be determined by its further movement to a secondary site in the cells. Such a mechanism could be responsible for the process-

Received and accepted June 20, 1980.

© 1981 Alan R. Liss, Inc., 150 Fifth Avenue, New York, NY 10011

ing and removal of nonspecifically internalized materials. Perhaps a mechanism like this delivers molecules such as NGF to the nuclear region to initiate the events in stimulation.

Membrane retrieval via coated pits after synaptic vesicle fusion at the nerve terminal or after pancreatic secretion may be a process than can either involve or bypass the lysosomal shunt as the vesicles proceed to the Golgi. In such instances, the vesicles would be delivering membrane to the Golgi for reutilization in the secretory process. This process would provide a necessary conservation of an energy-rich structure in which there is no obvious need for specialized removal of specific membrane molecules before reuse.

The routes described above involve coated pit-derived, smooth-surfaced vesicles that arise at the cell surface. Equally important are the coated pits that appear to arise from the ER. These were postulated many years ago by Dr. G. Palade to shuttle products of the ER to the Golgi. Recently the work of Dr. J. Rothman has lent credence to this role for the coated vesicles. That such a subset of morphologically smaller coated vesicles, and their derived smooth vesicles, exists in this shuttle is entirely consistent with the general role for coated vesicles postulated by Roth and Porter. If such structures are shown to contain specific receptors for the ER synthesized proteins, it would go far toward explaining how different proteins are moved from the ER at different rates after synthesis.

It is unknown whether a coated pit mechanism may also operate from the surface of the Golgi to deliver specific proteins directly to other organelles such as the lysosome or to the plasma membrane.

ITEMS DISCUSSED

After the brief preceding summary of the roles that coated vesicles have in cellular processes, several topics were discussed that are related to the initial events in receptor ligand interactions and internalization. In particular, questions were raised as to whether the receptor ligand complex must cluster to initiate internalization and what is known about the mechanism of such clustering. A related discussion also ensued in an attempt to determine whether the biological effect is initiated by the receptor ligand interaction at the surface or if it occurs after internalization; or if both may be needed for the complete biological response and receptor regulation. Because of the limited time available, the only other topic addressed was internalization and intracellular routes. Because the large group of people attending the workshop was specialized, most of the discussion centered on the role these events may have in EGF, NGF, and insulin action.

CLUSTERING

Clustering appears to be a requisite event leading to the internalization of the receptor-ligand complex, and in some instances the biological response. Dr. S. Cohen (Vanderbilt) indicated that EGF clustering in A431 tissue culture cells occurs within the first half minute after binding, followed by internalization in coated pits by 2.5 min, and its subsequent appearance in multivesicular bodies in 30 mins. These events, illustrated by electron micrographs taken by Drs. Haigler, McKanna, and Cohen, elegantly demonstrated the endocytosis route using an EGF-ferritin derivative. After the initial complexation events, in which all of the sites appear to have the ferritin associated with the various membranes, the 1 h time point showed the ferritin free from the membranes of the MVB-lysosomes and in clumps in the MVB. This was interpreted as an initial event in the degradation of the EGF-ferritin-receptor complex. The morphology clearly suggested that the ligand triggered the formation of coated pits and the subsequent internalization. When the experiment was done in the presence of primary amines, the EGF-ferritin remained bound to the MVB membranes and was not degraded; internalization was not prevented.

Of note was Dr. Haigler's observation that some of the ferritin found in the MVB at 30 min was on small vesicles but was reoriented to face outwards, suggesting a vesicle-fusion process that inverts some vesicles as they enter the MVB.

Dr. Cohen noted that the ^{125}I on the EGF has a half-life in the cell of approximately 20 min but that the fate of the ferritin was not known, probably because of the short incubation time (1 h) prior to fixation.

Dr. V. Marchesi (Yale) raised the issue of whether the clustering might be due to a small percentage of aggregates in the preparation. Dr. Haigler suggested that this was not the case because there was conjugation of EGF solely through the alpha amino group and subsequent purification of the conjugates on sizing columns. Also he noted that ferritin- and fluorescent-^{125}I-EGF conjugates all gave identical localization and kinetics, although at different levels of resolution. He pointed out, however, that he was not able to detect clusters of fluorescent-EGF on the cell surface due to the limits of resolution of the method. He could detect larger clusters when they accumulated in the larger vesicles, the MVBs and the lysosomes.

In a related series of experiments, Dr. H. Herschman (UCLA) and Dr. R. Fine (Boston University) explored the internalization routes by using sucrose gradient fractionation of the cellular organelles. Using anti-clathrin to precipitate fractions containing externally oriented clathrin, they demonstrated a flow of ^{125}I-EGF from the cell surface that peaks in an anti-clathrin precipitable peak in 3–4 min, emptying in 7 mins. Subsequently, the lysosomes fill maximally at 15 min. This suggests a kinetic flow through the coated vesicles and into a non-CV compartment prior to lysosomal fusion. Interestingly, when ammonium chloride is added, it permits these various compartments to fill with the same kinetics, but each compartment stays full. After 1 h 20% of the EGF remains at the surface, and the clathrin-precipitable fractions are still loaded, as are the lysosomes. No degradation of the ^{125}I-EGF was noted. As Dr. Herschman said, "It is as though you turned off the conveyor belt and trapped everything along the way."

Dr. Cohen noted that the rapid uptake of the EGF receptor occurs only if bound to ligand; otherwise, the half clearance of receptors from the surface takes 6–7 h.

Dr. P. Cuatrecasas (Burroughs-Wellcome) spoke more directly to the question of whether clustering is essential for the biological rather than the morphological response of internalization. He noted that if EGF is applied to 3T3 cells at concentrations 100- to 1,000-fold lower than that normally needed for the biological response, no effect was noted. If, however, these treated cells were then exposed to anti-EGF, they elicited a "substantial enhancement of DNA synthesis." Monovalent Fab fragments did not elicit the response. Since Dr. Cohen had previously shown that anti-EGF can prevent the biological response, Dr. Cuatrecasas was asked why his anti-EGF elicited a response. He believed that the difference was due to the amount of antibody added. If too much is used it is effectively monovalent and thus does not elicit clustering. Of note was his observation that anti-EGF applied to the EGF receptor complex reduces the affinity of the EGF for the receptor and promotes EGF release from the membrane. The same phenomenon has also been observed when insulin is bound to cells at substimulating concentrations and anti-insulin is then applied.

Dr. Cuatrecasas also noted that when they cleaved EGF at its single methionine with CNBr, they obtained an EGF derivative of two chains that was held together by a disulfide bridge. This EGF binds to cells to some extent, but there is no biological activity. In the fluorescent microscope it appears evenly distributed on the cell surface. When these cells are then treated with anti-EGF, there is clustering and the cells are biologically activated. The most obvious interpretation for both of these experiments is that some minimal clustering is needed for a biological response. However, it does not address the question of whether the subsequent internalization is also a necessary condition for initiating the biological response. No data presented during the workshop related directly to this latter question.

Dr. R. Bradshaw (St. Louis) raised the question of whether there is any case in which clustering is not involved in internalization. The people present could think of no well-studied example. However, Dr. Haigler noted that you can get bulk pinocytosis of non-clustered EGF from A431 cells, where perhaps as much as 20% of the surface EGF is internalized in the first 30 sec after warming to $37°C$. This early stimulation of bulk pinocytosis was also observed when phorbol esters were applied to the cells.

As a corollary to the above question, Dr. Cuatrecasas noted that clustering can occur without internalization. In his work on encephalins, beta-endorphin localized on neuroblastoma hybrids showed good clustering that was permanent and independent of the biological activity. Interestingly, if you reduce and alkylate, there is no clustering, but you do not affect the ability of the encephalins to affect adenylate cyclase. But of course that is not a long-term biological effect.

Dr. Herschman raised a related question about whether there was any evidence at the EM level for what happens when there is coordinate up or down regulation of one receptor by a ligand of a different specificity. No example was known to the group, although the cointernalization of multiple fluorescently labeled ligands as reported by Pastan's laboratory was cited as a possible case. However, Dr. F. Fox (UCLA) noted that the kinetics of down-regulation of insulin and EGF were quite disparate, suggesting that the process was not normally coordinate. However, the presence on the cell of three ligands might possibly stimulate an early bulk phase pinocytosis similar to that observed by Dr. Haigler. Clearly, the experiment needs to be redone at the EM level to determine the precise mechanism by which the multiple ligands are internalized.

Dr. Cohen raised the issue of triggered vs preformed pits as sites at which receptor ligand complexes would be found. "Does clustering occur first and is clathrin then 'attracted' to the cluster, or are there preformed pits to which the ligand-receptor complexes go?" He stated that he "previously believed the fluorescent experiments in Pastan's laboratory showing the simultaneous labeling of alpha-2-macroglobulin, insulin, and EGF clusters demonstrated that they went to the same pit." However, he was "not sure any longer whether pits or large vesicles were seen under the fluorescent microscope."

Dr. H. Green (MIT) wondered whether clustering was requisite to initiate the transepithelial route for IgG transport from the colostrum in the rat gut to the circulation of the newborn rat. In answer, Dr. T. Roth (University of Maryland, Baltimore County) cited the work of Dr. Rodewald (Virginia) as being the best study of this problem. However, since the cell surface between the microvillae is fully covered with the IgG that will be endocytosed, clusters in the usual sense of the word are not observed.

Dr. Green and Dr. Marchesi asked what brings clathrin and its related proteins to the site of formation of the pits and whether the ligand stimulates pits. Dr. Roth noted that at present little is known as to what triggers coat formation at a particular site, although this question is being investigated in several laboratories. He then asked Dr. J. Salisbury (Einstein) to discuss his group's work on IgM capping in a human lymphoblastoma cell line.

Salisbury noted that low levels of anti-IgM stimulate capping and subsequent endocytosis by coated pits. In his system there is a threefold stimulation of coated pits in the area where endocytosis occurs. When stelazine, a somewhat nonspecific inhibitor of calmodulin, is added to these cells, uptake is inhibited about 50%, and the ferritin-labeled sites do not appear to be associated with coated material. According to Dr. Salisbury, this "suggests a possible role for calmodulin in either the conversion of a submembranous clathrin to form the coat or in recruitment of cytoplasmic pools of clathrin to the membrane." Dr. Roth noted that this thought was consistent with his laboratories report that calmodulin was one of the proteins that can be found specifically associated with coated vesicles when they are isolated in the presence of calcium.

Dr. Marchesi raised the issue of the physiological relevance of clustering. He asked whether the number of coated pits was the same in cells exposed to multivalent ligands as in unexposed cells. Since they were not the same in IgM or EGF treated cells, he asked, "how can you escape from the conclusion that you are bombarding the cell membrane with these potentially multivalent ligands?" The assembled group thought that most ligands were stimulating the cell in this manner and that this, in fact, was triggering a response in a manner similar to that by which the anti-EGF triggered the biological response in the presence of subtriggering levels of bound EGF.

INTERNALIZATION

Dr. Roth then raised the issue of coated pit internalization as an effector route versus the concept of the involvement of coated pits as "elegant garbage dumps" to clear receptor-ligand complexes from the surface. "Wouldn't classical down-regulation of receptors involve such a mechanism of coated pit internalization?" Dr. Cohen believed this was certainly true for EGF. Dr. Roth thought the superb studies of Goldstein, Brown, and Anderson clearly showed the coated pit mechanism to be involved in both down-regulation of the LDL receptor complex and, once the LDL was in the lysosome, in the biological response as well.

The question was raised as to the presence of other examples of biological effects after internalization. In answer, the example of the NGF response in pheochromocytoma cells was offered. In this instance, there is a rapid down-regulation and degradation of the receptor-ligand complex that is slower than, but similar to, EGF. Thereafter, there is an accumulation of NGF on receptors at the nuclear membrane at the time that lysosomal degradation is stopping. Interestingly, if the lysosomal inhibitor chloroquine is added to the cells, NGF arrives in the vicinity of the nucleus within a few minutes. If the surface NGF is removed by trypsin treatment, the decay of the biological response parallels the decay of NGF from the nucleus with a 33 h half-life.

A similar effect was observed using EGF on bovine corneal cells. If chloroquine was added at time zero, within 4–6 h some 20–30% of the accumulated EGF reached the nucleus. This nuclear EGF is antibody precipitable and comigrates with native EGF. It was not bound to DNA or histone. However, the investigators did not know whether it was nonspecifically associated, only that it was found there at the 6 h time point and thus correlated roughly with the 10 h lag in the EGF effect. They hypothesize that it does not go to the lysosomal system but goes via vesicles directly to the nuclear membrane. Native EGF will not bind to the nuclear membrane.

In contrast, Dr. Cohen cited the results of Drs. P. Gordon and L. Orci, who did not observe ^{125}I-EGF in the nuclear region by autoradiography in either the presence or absence of ammonia. However, they only monitored the experiment for 1 h, unlike the 4–6 h needed in the previously cited experiment. He also noted that in his laboratory, no ferritin-EGF was found in the nuclear region in the presence of chloroquine, but here too the experiment was only carried out for 1 h. It was volunteered that chloroquine increases the total cell-associated EGF about 12-fold.

Dr. Bradshaw mentioned that Dr. R. Savage (Temple University), at the International Endocrine Meeting in Melbourne in February, reported EGF binding sites in the nuclei of hepatocytes. Dr. Bradshaw was then asked by Dr. Cuatrecasas if any NGF that was transported retrogradely in the axon was then exocytosed. Dr. Bradshaw stated that Dr. H. Thoenen (Max-Planck, Munich) found none and that the uptake of the NGF is several orders of magnitude greater than anything else (although anything can be taken up nonspecifically). With fluorescent and ferritin labels, they find that most of the NGF becomes associated with the lysosomes, some with the ER, and no significant amount with the nucleus, Golgi, or other organelles. In contrast, Dr. Bradshaw found in his experiments, in which subcellular fractionation was used after retrograde transport, that a small but significant fraction of the NGF is found in the nucleus.

PC-12 neuroblastoma cells appear to move very large amounts of NGF into the nucleus. Dr. Bruce Yankner (Stanford) commented that the reason most people do not see NGF in the nucleus may be because NGF on the nuclear receptor has a half-life of about 33 hours. If this is the case, then neurons cultured in NGF would have their nuclear receptors filled and thus would be able to bind ^{125}I-NGF only as the nuclear receptors emptied. When either he or Dr. P. Calissano (LBC, Rome) did 48 h uptake studies with labeled NGF on sympathetic neurons or PC-12 cells, they found NGF around the perimeter of the nucleus. He went on to mention that there are striking similarities between the intracellular insulin receptor reported by Dr. Goldfine (San Francisco) and that reported by Dr. Bradshaw's and Dr. Shooter's laboratories. Both receptors are on the nuclear membrane, both are detergent resistant, and both have biphasic Scatchard plots. Here again he felt that the divergence of results with regard to those working on the insulin receptor was due to the use of too short a labeling period (0.5 h). Dr. Goldfine used 3 h and was able to demonstrate grains over the nuclear membrane and the RER. He concluded that "shorter periods do not saturate the lysosomal system, and thus you do not get the material into a biologically significant pathway. Perhaps this is why it takes so long for peptide hormones to act." He and his associates propose that vesicles fuse with the nuclear membrane, thus freeing their content into the inner space of the nuclear membrane contiguous to the chromatin. Triton would most likely leave the receptor–inner membrane–chromatin complex intact.

Dr. Roth took exception to the idea of having to saturate the lysosomal system prior to having a biological effect, primarily because of the large amount of ligand that probably would be needed to saturate the lysosomal system. The biological effect in those systems is triggered by small amounts of ligand. Dr. B Yankner responded that perhaps it was not saturation per se that was needed but rather some other event. If you add chloroquine, you get NGF into the nucleus earlier. However, the chloroquine effect seems to argue against a role for the lysosome in the processing of the NGF.

The question was then raised as to whether it was important to move the NGF or the receptor to the nucleus. For if you consider the chloroquine effect as a blockade of lysosomal degradation, this would inhibit the removal of labeled ligand from the receptor-ligand complex. Thus, the labeled ligand would not normally accompany the receptor to the vicinity of the nucleus. However, in the chloroquine case, the label would be with the receptor as it moves to the nucleus and thus would permit the receptor to be visualized.

The NGF people agreed that such a hypothesis was consistent with the data. In fact, if you incubate cells with NGF for long periods there is a large increase in the number of detectable nuclear receptors with the high-affinity site in PC 12 cells.

RECYCLING

Recycling of receptors is a phenomenon that may be a specialized case included in the general problem of membrane recycling or retreival. At present there are varying amounts of evidence to support the idea that LDL, the asialoglycoprotein, and the vitellogenin receptors are recycled. In contrast, Dr. Bradshaw pointed out, there is a larger list of receptors for which recycling does not occur: insulin, EGF, NGF, prolactin, and growth hormone, among others. He suggested that the reason these are not recycled may be because the receptors are used in the biological response, whereas the recycled receptors are "merely" carriers to bring material into the cytoplasm and thus are fit for reutilization.

Epidermal Growth Factor-Receptor-Protein Kinase Interactions

Stanley Cohen, Graham Carpenter, and Lloyd King, Jr.

Departments of Biochemistry (S.C., G.C.) and Medicine (Dermatology) (G.C., L.K.), Vanderbilt University School of Medicine, and Veterans Administration Hospital (L.K.), Nashville, Tennessee 37232

Membranes prepared from A-431 human epidermoid carcinoma cells retained the ability to bind ^{125}I-labeled epidermal growth factor (EGF) in a specific manner. In the presence of $[\gamma\text{-}^{32}P]$ATP and Mn^{2+} or Mg^{2+}, this membrane preparation was capable of phosphorylating specific endogenous membrane proteins, as well as exogenously added histone. The binding of EGF to membranes in vitro resulted in a several-fold stimulation of the phosphorylation reaction. The phosphorylation reaction was not dependent on cyclic AMP or cyclic GMP.

These findings suggest that one of the biochemical consequences of the binding of EGF to membranes is a rapid activation of a cyclic AMP-independent phosphorylating system. The activation of the membrane-associated protein kinase by EGF appears to be a reversible pheonomenon.

The membrane preparation could be solubilized by a number of nonionic detergents with the retention of both ^{125}I-labeled EGF binding activity and EGF-enhanced phosphorylation of specific membrane proteins. The solubilized membrane preparation was purified by affinity chromatography. The purified preparation retained both EGF-binding activity and EGF-enhanced phosphorylation activity. Analysis of the affinity-purified preparation by SDS gel electrophoresis indicated the presence of one major protein band of molecular weight 150,000 and several trace bands.

The evidence suggests that the major 150,000 protein band is the receptor for EGF and is a substrate of the phosphorylation reaction. The copurification of EGF-binding activity and EGF-stimulated phosphorylation activity suggests an inherent close relationship.

The EGF-stimulated phosphorylation reaction appears to be specific for tyrosine residues and in this regard resembles the *src* protein kinase.

Key words: phosphotyrosine, membrane phosphorylation, EGF-receptor purification

Epidermal growth factor (EGF) binds to specific plasma membrane receptors in target cells and initiates and maintains a complex program of biochemical and morphological events leading to cell growth and multiplication [see review, 1]. Subsequent to the initial binding of EGF to the plasma membrane, the hormone, probably together with its receptor, rapidly is internalized in endocytotic vesicles and eventually degraded in lysosomes [1–10]. It is reasonable to assume that the observed biochemical and morpho-

Received and accepted May 20, 1980.

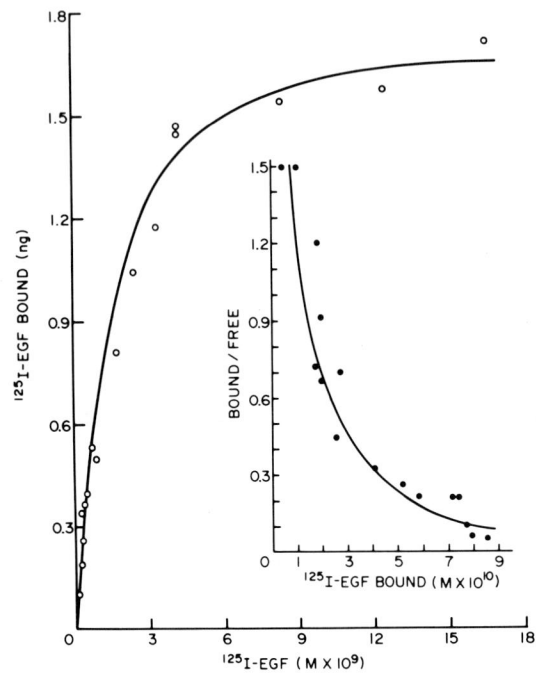

Fig. 1. Effect of increasing ^{125}I-EGF concentrations on binding to A-431 membranes. Indicated concentrations of ^{125}I-EGF (specific activity 32, 950 cpm/ng) were added to standard reaction mixtures that contained the A-431 membrane preparation (12 μg of protein). The reaction mixtures were incubated at 25°C for 30 min, filtered, washed, and counted for radioactivity. [Data from 13].

logical alterations induced by EGF result from the generation, amplification, and propagation of a series of "signals." To understand the nature of these signals and the mechanisms by which they are amplified and propagated, it would appear necessary to obtain cell-free systems that are responsive in vitro to the addition of EGF. For this purpose, we have chosen to study membrane preparations from the human epidermoid carcinoma cell line A-431, which has been shown to contain an extraordinarily high concentration of membrane receptors for EGF, approximately 2 to 3 × 10^6 receptors per cell [6, 11]. Since membranes contain endogenous protein kinases, and since phosphorylation and dephosphorylation reactions participate in the regulation of many metabolic pathways, we initiated this study on the possible role of EGF as a control element in these processes.

In this brief report we review the work from our laboratory, which indicates that one of the biochemical consequences of the binding of EGF to membranes is the rapid activation of a cyclic AMP-independent phosphorylating system that is capable of phosphorylating not only the membrane receptor for EGF but also certain exogenously added substrates.

BINDING OF ^{125}I-LABELED EGF TO A-431 MEMBRANES

Membranes may be prepared from A-431 cells that retain the ability to bind ^{125}I-labeled EGF in a specific manner [12, 13]. The effect of increasing concentrations of ^{125}I-EGF on the extent of the binding reaction at 25°C is shown in Figure 1. The binding

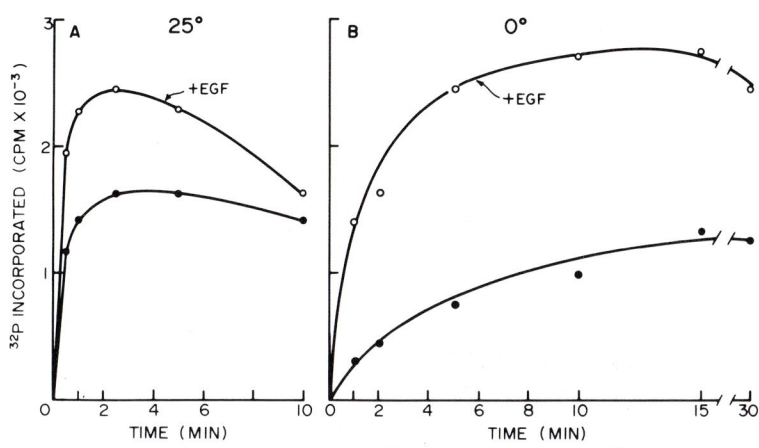

Fig. 2. Stimulation by EGF of the incorporation of [^{32}P] phosphate from [γ-^{32}P] ATP into cell membranes. The standard phosphorylation assay was performed as described elsewhere [13], but with temperature varied. A. all incubations carried out at 25°C. B. all incubations carried out at 0°C. [Data from 13].

is maximal at approximately 1.6×10^{-8} M total EGF and is half-maximal in the presence of 1.2×10^{-9} M total EGF. At these concentrations, 5% and 25%, respectively, of the total added ^{125}I-EGF were bound. At the lowest concentration of ^{125}I-EGF (8×10^{-11} M) tested, 60% of the total hormone was bound. A plot of these binding data by the Scatchard procedure is curvilinear (inset, Fig. 1), perhaps reflecting the presence of heterogeneous classes of binding sites or negative cooperativity in the binding reaction.

MEMBRANE PHOSPHORYLATION IN THE PRESENCE OR ABSENCE OF EGF

The incubation of A-431 membranes with [γ-^{32}P] ATP resulted in the incorporation of radioactivity into trichloroacetic acid-insoluble material at 25°C and 0°C (Fig. 2A and B). The addition of EGF to the reaction mixture at 0°C stimulated 3- to 4-fold the phosphorylation of endogenous membrane-associated proteins. The data in Figure 2B show that at 0°C in the presence of EGF the initial rate of phosphorylation was increased from 16.7 to 50 pmoles of [^{32}P]-phosphate incorporated per min per mg of protein. Membrane phosphorylation reached a maximal level by 10 min at 0°C. The addition of EGF to the reaction mixture at 25°C also resulted in a stimulation of membrane phosphorylation (Fig. 2A), with a maximal level by 2.5 min. The decline in the level of phosphorylation after 2.5 min at 25°C might be the result of increased activity of phosphatases in the membrane preparation; therefore, all subsequent experiments were performed at 0°C. Depletion of ATP during the incubation by reactions such as hydrolysis by ATPase was not a major factor influencing the extent of phosphorylation.

PROPERTIES OF THE EGF-STIMULATED MEMBRANE PHOSPHORYLATION SYSTEM

The stimulation of membrane phosphorylation by EGF was specific for EGF and its biologically active derivatives [13], such as fluorescein-EGF, EGF$_{1\ 51}$, EGF$_{1\ 48}$, and hu-

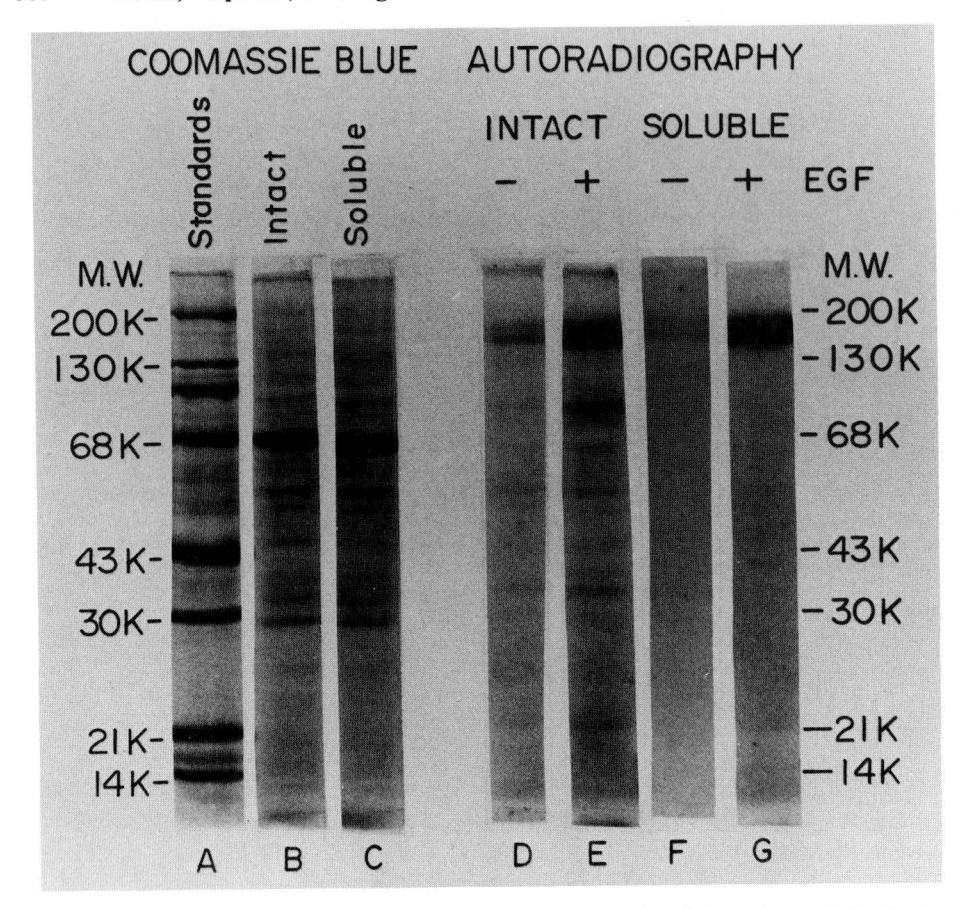

Fig. 3. Electrophoresis and autoradiography of intact and solubilized A-431 membranes following incubation with [γ-^{32}P]ATP in the presence and absence of EGF. A-431 membranes (668 μg protein) were solubilized in 200 ml of 20 mM Hepes buffer, pH 7.4, containing 1% Triton X-100 and 10% glycerol as described elsewhere [15]. After centrifugation aliquots (15 μl) of the solubilized membrane and comparable amounts of intact membrane were subjected to the standard phosphorylation procedures in the presence and absence of EGF for 5 min at 0° as described elsewhere [15]. The reaction was stopped by the addition of 50 μl of Laemmli SDS sample buffer. The mixtures were boiled for 5 min and subjected to SDS gel electrophoresis, Coomassie blue staining, and autoradiography. A. standards. B and C. Coomassie blue stain of membrane components from intact (B) and solubilized (C) preparations. D–G. autoradiography of membrane components phosphorylated in absence (D and F) and presence (E and G) of EGF. D and E. intact membranes. F and G. solubilized membranes [Data from 15].

man EGF. The basal level of A-431 membrane phosphorylation was not affected by any of the following hormones: insulin, glucagon, thyrotropin, follicle stimulating hormone, luteinizing hormone, prolactin, or growth hormone.

Both the basal and EGF-stimulated phosphorylation reactions required the presence of Mg^{2+}, Mn^{2+}, or Co^{2+}. The optimum Mg^{2+} concentration for the reaction was approximately 50 mM, while that for Mn^{2+} was considerably lower. In the presence of 1 mM Mn^{2+}, the addition of 2 mM EDTA completely inhibited both the basal and EGF-stimulated phosphorylation reactions. In the absence of added metal ions, only traces of membrane phosphorylation were detectable. No effects were noted by the addition of Ca^{2+} (1 to 5 mM), Cd^{2+} (5 mM), or putrescine (10mM).

Neither cyclic AMP nor cyclic GMP appears to be involved in the phosphorylation of A-431 membranes under the conditions employed in these studies. Addition of these cyclic nucleotides (10^{-4} to 10^{-6} M) to the membrane phosphorylation system did not stimulate the incorporation of ^{32}P into the total membrane preparation or individual membrane proteins [14] either in the presence or absence of EGF. The phosphorylation system of the membrane can ultilize [γ-^{32}P]GTP in both the basal and EGF-stimulated reactions; the same membrane proteins appear to be phosphorylated in the presence of labeled ATP or GTP [14].

The net increase in the phosphorylation of membrane-associated proteins by EGF was ascribed to an enhancement of phorphorylation and not to an inhibition of dephosphorylation. This conclusion was drawn from [γ-^{32}P]ATP pulse-chase experiments [13, 15], in which the presence of EGF had no effect on the rate of membrane dephosphorylation. The A-431 membrane preparation was capable of EGF-enhanced phosphorylation of not only endogenous membrane proteins (see later section) [14] but also certain exogenously added substrates such as histones.

The activation of the membrane protein kinase system by EGF appears to be a reversible process [15]; exposure of EGF-activated membranes to anti-EGF serum resulted in a return of the phosphorylating capacity of the preparation to basal levels. Addition of EGF to the washed antiserum-inactivated membrane preparation resulted in a re-activation of the phosphorylating system.

The nature of the A-431 membrane components phosphorylated by [γ-^{32}P]ATP in the presence and absence of EGF was examined by SDS gel electrophoresis [14, 15], and the results are depicted in Figure 3 (lanes A, B, D, and E). Coomassie blue staining (lane B) indicates the presence of many components. (The heavily stained band, which migrates at 68,000, is BSA used in the reaction mixtures.) Also presented in Figure 3 are autoradiographs (lanes D and E) of intact A-431 membranes phosphorylated in the presence or absence of EGF and separated by SDS electrophoresis. The data show 1) EGF increases the phosphorylation of many membrane components, particularly those having molecular weights of approximately 170,000, 150,000, and 80,000; 2) proteins phosphorylated in the presence of EGF are also phosphorylated, but to a lesser extent, in the absence of the growth factor; and 3) there is no apparent correlation between the intensity of Coomassie blue staining and the amount of phosphorylation.

EGF-responsive phosphorylation was also detected in membranes from normal human fibroblasts [14] and from term human placenta [14, 16], although the enhancement by EGF was lower than that observed with A-431 membranes.

SOLUBILIZATION OF RECEPTOR AND EGF-ENHANCED PHOSPHORYLATION ACTIVITY

The solubilization of the ^{125}I-EGF binding activity of A-431 membranes by the use of Triton X-100 has been reported [17]. We examined detergent-solubilized membrane preparations for protein kinase activity and responsiveness to EGF. In preliminary experiments it was found that over 50% of the endogenous basal kinase activity could be extracted with 1% Trition X-100; the inclusion of 10% glycerol in the detergent solution increased the yield to 70–90%. Attempts to stimulate the basal phosphorylation by the addition of EGF to solubilized preparations under the usual assay conditions (all incubations carried out at 0°) were not successful. However, if EGF was allowed

TABLE I. Comparison of Effects of EGF on Phosphorylation in Intact and Solubilized Membranes*

Preparation	^{32}P incorporated	
	−EGF	+EGF
	(cpm)	
Intact membranes	1050	2200
Solubilized (1% Trition)	510	1310
Solubilized (1% Trition, 10% glycerol)	670	1790

*Aliquots of A-431 membranes (400 μg protein) were suspended in 200 μl of either a) 20 mM Hepes, pH 7.4, b) 20 mM Hepes containing 1% Trition X-100, or c) 20 mM Hepes containing 1% Trition X-100 and 10% glycerol. The suspensions were allowed to stand at room temperature for 20 min, and the mixtures containing detergent were centrifuged at 100,000g for 60 min. Aliquots (15 μl) of the intact membranes or detergent-solubilized supernatant were preincubated with EGF for 10 min at room temperature prior to the assay for phosphorylation. (Data from [15]).

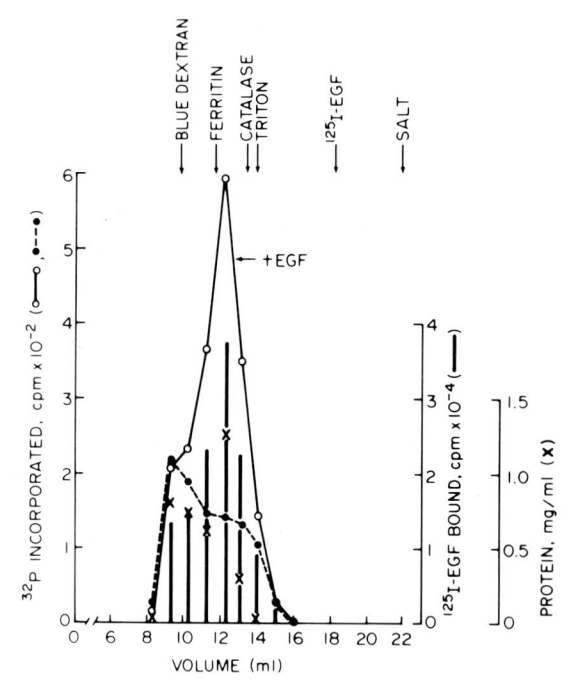

Fig. 4. EGF stimulated phosphorylation and ^{125}I-labeled EGF binding in Triton solubilized membranes fractionated by gel filtration. Membranes (6 mg protein) were solubilized in 450 μl of 20 mM Hepes buffer, pH 7.4, containing 1% Triton X-100 and 10% glycerol as described elsewhere [15]. After centrifugation, 350 μl of the solubilized preparation were applied to a Sephacryl S-300 column (0.9 × 31.5 cm) equilibrated with the membrane solubilizing buffer. One ml fractions were collected at a flow rate of 0.15 ml/min. Aliquots (35 μl) of the fractions were assayed for their capacity to bind ^{125}I-EGF as described elsewhere [15]. Aliquots (15 μl) of the same fractions were assayed for their endogenous phosphorylation capacity in the presence and absence of EGF by the standard method described elsewhere [15]. The vertical bars indicate the ^{125}I-EGF specifically bound in the aliqouts taken from each fraction; X, protein content; ○──○, ^{32}P incorporated in the presence of EGF; ●──●, ^{32}P incorporated in the absence of EGF. [Data from 15].

to react with the solubilized membrane preparation at room temperature prior to chilling the reaction tubes to 0°C for the phosphorylation assay, considerable activation of endogenous phosphorylation by EGF was detected [15].

A comparison of the effects of EGF on phosphorylation using intact and solubilized membranes is shown in Table I. Much (64%) of the basal activity and almost all (over 90%) of the EGF responsive activity were recoverable after solubilization. Figure 3 compares membrane proteins phosphorylated in the presence and absence of EGF, using both intact and solubilized membrane preparations, by SDS acrylamide electrophoresis and autoradiography. The results show that the major phosphorylated component of both preparations is a doublet whose components have molecular weights of approximately 170 and 150 K. The phosphorylation of this doublet was clearly stimulated in both intact and solubilized membranes by the addition of EGF. The phosphorylation of other membrane components, however, was less apparent in the solubilized preparations. While added histones were phosphorylated in both membrane preparations, the solubilized preparation was much less efficient in this regard as compared to intact membranes (data not shown). As expected, the Coomassie blue patterns of the intact membranes and solubilized preparation were similar (Fig. 3).

Attempts were made to fractionate the phosphorylating and EGF-binding activity of the solubilized membranes by gel filtration using Sephacryl S-300. The results of these efforts are shown in Figure 4. From the data we conclude that 1) most of the ^{125}I-EGF-binding activity and EGF-stimulated phosphorylation activity elute together in a volume that suggests a molecular size larger than a Triton X-100 micelle (mol wt 90,000) or catalase (mol wt 232,000) but smaller than ferritin (mol wt 440,000); 2) most of the EGF-independent phosphorylating activity elutes with the excluded peak; and 3) most of the solubilized membrane protein distributes from the excluded peak to the position of Triton micelles. Since both the EGF-binding and EGF-stimulated phosphorylation activities in the Triton extracts penetrated the Sephacryl matrix, these experiments provided additional evidence that "solubilization" had indeed occurred.

AFFINITY PURIFICATION OF EGF RECEPTOR-KINASE COMPLEX

In view of the complex composition of the Triton extracts and the limited quantites of membrane available, we attempted to purify and resolve the EGF receptor and EGF-sensitive phosphorylation activities by affinity chromatography with EGF covalently linked to Affi-Gel. Since high-affinity ligand binding and elution were involved, we chose to effect the separation by a "batch" procedure where temperature and time could be closely controlled. In brief, Triton extracts of A-431 membranes were stirred with EGF-Affi-Gel, and the gel was washed exhaustively and then incubated with a high concentration of EGF to effect elution. Since the EGF present in the eluate precluded ^{125}I-EGF binding studies, only phosphorylation activity was examined. Elution of the EGF-Affi-Gel with EGF resulted in the recovery of about 37% of the original activity.

The elution of material from the EGF-Affi-Gel by EGF presumably was dependent upon competition for the solubilized adsorbed receptor. Since assaying for receptor activity in the presence of such large quantities of EGF was not possible, we examined alternate procedures to elute the receptor and/or kinase from the affinity gel. It was found that dilute solutions of ammonia or ethanolamine were capable of eluting both the EGF receptor and EGF-dependent phosphorylating activity from the EGF-Affi-Gel beads. In brief, Trition extracts of A-431 membranes were stirred with EGF-Affi-Gel, and the gel was washed exhaustively and incubated with 5 mM ethanolamine pH 9.7 to effect elution. The resulting eluate was neutralized and assayed for ^{125}I-labeled EGF binding and for endogenous phosphorylation capacity in the presence and absence of EGF. The details of the procedure and a typical result are shown in Table II.

Approximately 39% of the original ^{125}I-EGF binding activity was recovered in the ethanolamine eluate. With regard to the endogenous phosphorylation activity eluted from the gel, it should be noted that a stimulatory effect of EGF was detectable. The recovery of phosphorylating activity in the ethanolamine eluate (Table II, third column) appears low (14%) relative to the recovery of binding activity (39%). This may reflect the fact that not all of the kinase activity is associated with the receptor or that some of the substrates for phosphorylation are also removed by the purification procedure.

In the following experiments we compared 1) the composition of the original solubilized membrane preparation with the material eluted from the EGF-Affi-Gel (by SDS gel electrophoresis and Coomassie blue staining) and 2) the nature of the components phosphorylated in the original extract with those phosphorylated in the material eluted from EGF-Affi-Gel (by SDS gel electrophoresis and autoradiography). The elutions were carried out both by the EGF procedure and by the ethanolamine procedure.

We conclude from the Coomassie blue data shown in Figure 5 that 1) the composition of the non-adsorbed material in the Triton extracts (lane B) is almost identical to that of

TABLE II. EGF-Affi-Gel Purification of Receptor and Protein Kinase Activity by Elution With Ethanolamine*

Sample	^{32}P incorporated −EGF (cpm)	^{32}P incorporated +EGF (cpm)	^{125}I-EGF binding (cpm)
Original extract	710	2540	370,000
Non-adsorbed supernate	730	1030	80,000
Ethanolamine eluate	150	360	144,000

*Membranes were solubilized and treated with EGF-Affi-Gel beads as described elsewhere [15]. The beads were centrifuged for 1 min in an Eppendorf centrifuge, the non-adsorbed supernatant fluid was collected, and the gel was washed 3 times with 1 ml of cold buffer (10% glycerol, 1% Triton X-100, adjusted to pH 7.2 with NaOH). To the final gel pellet (approximately 150 μl) were added 150 μl of elution buffer (10% glycerol, 1% Triton X-100, 5 mM ethanolamine, final pH 9.7). The mixture was stirred for 30 min at 0°C. After centrifugation, 15 μl aliquots of the ethanolamine eluate were assayed for endogenous phosphorylation in the presence and absence of EGF, and 30 μl aliquots of the eluate were assayed for specific ^{125}I-EGF binding. Simultaneously, comparable aliquots of the original Triton extract and non-adsorbed first supernatant fluid also were assayed. The standard phosphorylation and binding assays were performed as described elsewhere in [15].

Fig. 5. Electrophoresis and autoradiography of solubilized A-431 membranes purified by affinity chromatography. Membranes (10 mg protein) were solubilized in 800 μl of 20 mM Hepes buffer, pH 7.4, containing 1% Triton X-100 and 10% glycerol. After centrifugation, aliquots (200–400 μl) were adsorbed to EGF-Affi-Gel beads, washed, and the adsorbed material was eluted by either EGF or ethanolamine. Aliquots of the original extract, the non-adsorbed material, and the EGF or ethanolamine eluates were subjected to the standard phosphorylation procedures in the presence and absence of EGF for 10 min at 0°C. The reactions were stopped and analyzed by SDS gel electrophoresis, Coomassie blue staining, and autoradiography. All procedures are described elsewhere [15]. A–D. Coomassie blue strain of original extract (A), non-adsorbed material (B), EGF eluate (C), and ethanolamine eluate (D). E–K. autoradiography of membrane components phosphorylated in absence (E, G, J) and presence (F, H, I, K) of EGF. E and F. original extract. G and H. non-adsorbed material. I. EGF eluate. J and K. ethanolamine eluate. [Data from 15].

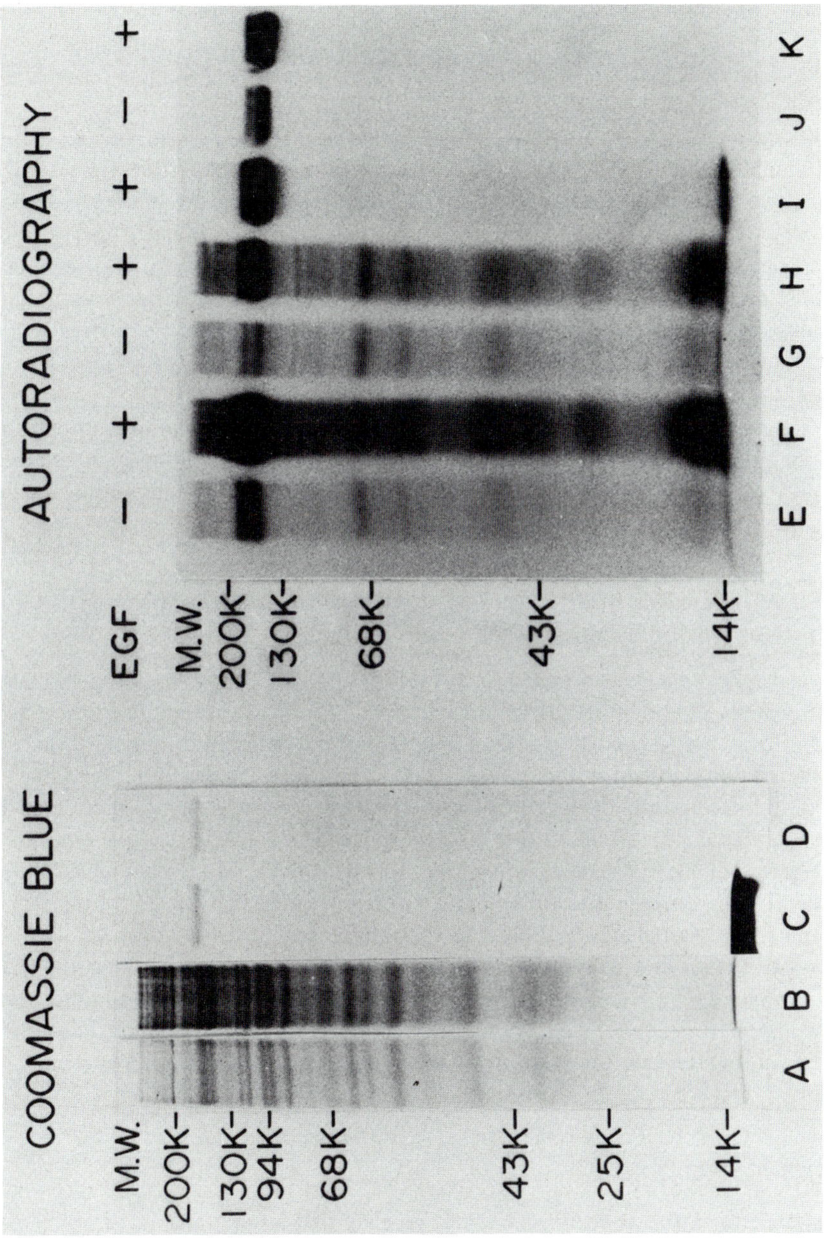
COOMASSIE BLUE
M.W.
200K–
130K–
94K–
68K–
43K–
25K–
14K–
A B C D
AUTORADIOGRAPHY
EGF – + – + + – +
M.W.
200K–
130K–
68K–
43K–
14K–
E F G H I J K

the original extract (lane A); 2) elution of the gel with either EGF or ethanolamine results in identical patterns (lanes C and D) consisting of one major 150 K band. Trace quantities of other bands (not apparent on the photograph) were detectable on the original gels at the 170 K, 130 K, and 50–60 K regions. Thus, we achieved by this method a very considerable purification. Due to the difficulties in measuring the trace quantities of protein available, especially in the presence of Triton, precise purification ratios were not obtained. However, amino acid analysis of the original extract and the ethanolamine eluate (after acid hydrolysis) indicated that approximately 1% of the original protein was recovered.

We conclude from the autoradiography data shown in Figure 5 (lanes E–K) that under the conditions employed, the major phosphorylated component in the original Triton extract was a doublet in the 150 K–170 K region (lanes E and F). In the unadsorbed Triton supernate, the patterns were similar, except that the EGF-stimulated 150–170 K components had decreased (lanes G and H). In both the EGF and ethanolamine eluates, only this doublet was detected, with the major radioactive band (150 K) corresponding with the major Coomassie blue staining material. The stimulatory effect of adding EGF to the ethanolamine eluate during the phosphorylation was clearly seen (compare lanes J and K).

To summarize, the ethanolamine eluate, which by SDS electrophoresis contains one major protein band and a few trace bands, possesses both ^{125}I-EGF binding capacity and EGF-stimulated phosphorylation activity. The major 150 K protein band is a substrate of the phosphorylation reaction.

IDENTIFICATION OF TYROSINE AS THE PHOSPHORYLATED AMINO ACID IN THE EGF-ENHANCED PHOSPHORYLATION REACTION

We have previously preported [13] that the A-431 membrane kinase system phosphorylated mainly threonine residues and in this, and other regards, resembled the *src*-gene kinase. When we became aware [Dr. David Baltimore, personal communication; 18, 19] that certain transforming virus-associated protein gene kinases phosphorylated tyrosine residues (originally mistakenly identified as threonine due to co-migration of the two phosphorylated amino acids in the electrophoretic system employed), we reinvestigated the nature of the EGF-protein kinase reaction [Ushiro and Cohen, unpublished experiments]. The original identification of threonine phosphate as the major phosphorylated amino acid in partial acid hydrolysates of A-431 membranes is in error; we have now identified the product as tyrosine phosphate based on co-migration with authentic tyrosine phosphate after electrophoresis at pH 1.7 and 3.5, chromatography in three solvents, and chromatography of a dansylated derivative. Pronase digestation of the membrane preparation after phosphorylation in the presence of EGF also yields tyrosine phosphate.

CONCLUDING REMARKS

The experiments reviewed in this report indicate that a functional relationship exists between the EGF receptor and membrane kinase activity.

The co-purification of EGF-binding activity and the EGF-enhanced phosphorylation activity by EGF-affinity chromatography suggests that the two activities are inherently associated in the membrane. The results also suggest that the 150 K Coomassie blue band detected in the purified preparation is the receptor, which is phosphorylated in the presence of EGF. Whether one of the minor bands detectable on the SDS gel is actually the kinase or whether the receptor itself has kinase activity is not yet resolved.

Several groups have attempted to identify the EGF receptor by covalent crosslinking to ^{125}I-labeled EGF. Hock et al [20] report receptor molecular weights of 180,000 and 160,000 in placenta. Wrann and Fox [21] and Das et al [8, 9] report molecular weights for the receptor of 175,000 in A-431 cells and 190,000 in 3T3 cells, respectively.

Many alternative hypotheses may be considered in attempting to understand the biochemical mechanisms, and the biological consequences, of the EGF-enhanced phosphorylation reaction [see review, 1], but definitive experiments are not yet available.

A number of similarities between the EGF-enhanced protein kinase activity and the *src* protein kinase have been previously noted [13, 22]. The finding that, in addition, both kinase systems are capable of phosphorylating tyrosine residues raises the interesting question of whether transforming viruses utilize and amplify a normal (hormone-receptor) cellular regulatory process.

ACKNOWLEDGMENTS

This work was supported by United States Public Health Service grants HD-00700 (S.C.) and AM 26518 (L.K.); research funds from the Veterans Administration (L.K.); a Basil O'Connor grant from the March of Dimes – Birth Defects Foundation (G.C.); American Cancer Society grant BC-294 (G.C.); and a National Cancer Institute grant, CA 24071 (G.C.). S.C. is an American Cancer Society Research Professor.

REFERENCES

1. Carpenter G, Cohen S: Annu Rev Biochem 48:193, 1979.
2. Carpenter G, Cohen S: J Cell Biol 71:159, 1976.
3. Gorden P, Carpentier J, Cohen S, Orci L: Proc Natl Acad Sci USA 75:5025, 1978.
4. Shechter Y, Schlessinger J, Jacobs S, Chang KJ, Cuatrecasas P: Proc Natl Acad Sci USA 75:2135, 1978.
5. Schlessinger J, Shecter Y, Willingham MC, Pastan I: Proc Natl Acad Sci USA 75:2659, 1978.
6. Haigler H, Ash JF, Singer SJ, Cohen S: Proc Natl Acad Sci USA 75:3317, 1978.
7. Aharonov A, Pruss RM, Herschman HR: J Biol Chem 253:3970, 1978.
8. Das M, Miyakawa T, Fox CF, Pruss RM, Aharonov A, Herschman HR: Proc Natl Acad Sci USA 74:2790, 1977.
9. Das M, Fox FC: Proc Natl Acad Sci USA 75:2644, 1978.
10. McKanna JA, Haigler HT, Cohen S: Proc Natl Acad Sci USA 76:5689, 1979.
11. Fabricant RN, DeLarco JE, Todaro GJ: Proc Natl Acad Sci USA 74:565, 1977.
12. Carpenter G, King L Jr, Cohen S: Nature 276:409, 1978.
13. Carpenter G, King L Jr, Cohen S: J Biol Chem 254:4884, 1979.
14. King LE Jr, Carpenter G, Coehn S: Biochemistry 19:1524, 1980.
15. Cohen S, Carpenter G, King L Jr: J Biol Chem 255:4834, 1980.
16. Carpenter G, Poliner L, King L Jr.: Mol Cell Endocrino 18:189, 1980.
17. Carpenter G: Life Sci 24:1691, 1979.
18. Eckhart W, Hutchinson MA, Hunter T: Cell 18:925, 1979.
19. Hunter T, Sefton BM: Proc Natl Acad Sci USA 77:1311, 1980.
20. Hock RA, Nexo E, Hollenberg MD: Nature 277:403, 1979.
21. Wrann MN, Fox CF: J Biol Chem 254:8083, 1979.
22. Chinkers M, McKanna JA, Cohen S: J Cell Biol 83:260, 1979.

Role of Proteases in Growth and Development: Workshop Report

J.F. Perdue and D. Rifkin

Lady Davis Institute for Medical Research, Montreal, Quebec, Canada H3T 1E2 (J.F.P.) and New York University Medical School, New York, New York 10016 (D.R.)

It has been known for a number of years that the growth of confluent cells can be stimulated by the addition of serum and factors present in and isolated from serum — ie, insulin, insulin-like growth factors, platelet derived growth factor, etc. Proteases also have mitogenic activity on cultured cells, although the history of these investigations is of much more recent origin. Sefton and Rubin [Nature 227:843, 1970] demonstrated that quiescent cultures of chick embryo fibroblasts could be stimulated to replicate with microgram quantities of trypsin. Pronase and ficin also had some activity. Studies from Buchanan's laboratory revealed that thrombin was a potent mitogen for avian fibroblasts [Proc Natl Acad Sci USA 72:131, 1975]. These observations have been confirmed and extended to human and mouse embryo fibroblasts by Pohjanpelto [J Cell Physiol 91:387, 1977] and Cunningham and his colleagues [J Cell Physiol 95:13, 1978]. The latter group of investigators characterized the binding of iodothrombin to mouse cells and established that the ligand need not be internalized to achieve a mitogenic event. There are about 200,000 sites per cell that bind the ligand with a K_d of < 1 nM. Studies by Perdue [J Cell Biol 83:55, 1979] indicate that avian and rat embryo fibroblasts and human neonatal foreskin fibroblasts have about 20,000–50,000 sites per cell that bind thrombin with a similar K_d.

Using photoactivatable thrombin, Glen and Cunningham [Nature 278:711, 1979] and Carney et al [J Biol Chem 254:6244, 1979] identified the thrombin-binding site(s) on chick and mouse embryo fibroblasts as having molecular weights of about 43 and 50K daltons, respectively. In a recent publication Baker et al [Nature 278:743, 1979] reported that iodothrombin will form SDS and heat-stable complexes with membrane components of human neonatal foreskin fibroblasts in the absence of photoactivation. The cellular component in the complex was characterized as having a molecular weight of 38 Kdaltons.

During his plenary session presentation, Cunningham indicated that Joff Baker and David Low from his laboratory have identified a cellular component designated protease-nexin (PN) that is released over a two- to three-week period into the medium of cultured human foreskin fibroblasts; PN can combine with exogenously added ^{125}I-urokinase and ^{125}I-α-thrombin to form SDS and heat-stable complexes. Moreover, the protease–PN complex binds to and is internalized by the human cells with rates indistinguishable from those observed for thrombin binding to chicken and mouse cells. Since it was suggested that the majority of the previously described ^{125}I-thrombin binding to human cells could be accounted for by the binding of ^{125}I-thrombin–PN complexes, and that a similar mechanism was operating, although at a much reduced level, on avian and other mammalian cells, it

Received and accepted May 14, 1980.

© 1981 Alan R. Liss, Inc., 150 Fifth Avenue, New York, NY 10011

was deemed important by the workshop organizers that the background and conditions employed by Dr. Baker in the PN experiments be discussed and the contribution of ^{125}I-thrombin–PN to the calculation of R_o (total binding capacity) and molecular weight of the thrombin binding site on the various cells be assessed.

Dr. Baker described the experiments and results that led to the conclusion that PN is secreted by human and possibly other cells, and that when it is complexed with serine proteases, rebinds to a specific site on the cells of origin. When nanogram quantities of ^{125}I-labeled thrombin were incubated with human fibroblasts at 37°C in serum-free conditioned media, a steady-state level of specific binding to cells was achieved by 90–120 min. This binding required esterase active thrombin; ie, diisopropylfluorophosphate-treated protease or hirudin-treated protease did not bind. As evidenced by SDS-PAGE of detergent-extracted cells, the specifically bound thrombin existed as a stable complex with an estimated molecular weight of 68 Kdaltons. Incubating the extract either with 1 M hydroxylamine or at a pH of 12 dissociated the complex into its respective components — ie, the 30 Kdalton B chain of thrombin and the 38 Kdalton PN. The stability to SDS and heat and sensitivity to hydroxylamine and alkaline pH sensitivity of the extracted thrombin–PN complex are properties similar to those displayed by serine protease inhibitors, including anti-thrombin III (AT3). AT3 is a 65K plasma glycoprotein that binds trypsin, thrombin, etc, and inactivates the protease by forming a stable serine-ester linkage at the catalytic site. AT3 also has a heparin-binding site, which when occupied increases the rate of thrombin inactivation by several hundredfold. It is likely that the thrombin in thrombin–PN complexes is inactivated, although proof of this will await purification of PN. Dr. Baker described experiments that showed that PN is different from AT3.

As much as 4 ng/ml/day of PN was released continuously over a period of 10 days into the medium of the cultured fibroblasts. In addition, incubating this conditioned medium with ^{125}I-labeled thrombin or urokinase resulted in the formation of an SDS and heat-stable complex of 68 or 72 Kdaltons, respectively. Similar to AT3, heparin binds to a specific site on PN and accelerates the rate of complex formation between PN and the added protease. This complex then re-bound to the cultured cells at a site distinct from that which bound iodothrombin but with similar kinetics. The PN binding site on cells may be identical with or overlap the region on PN which binds heparin, since this mucopolysaccharide (but not chondroitin sulfate) inhibited thrombin– or urokinase–PN complex binding. Heparin had little effect on the binding of free thrombin to the cultured cells, nor did it prevent protease-induced mitogenesis. The latter result was interpreted to indicate that the formation of a protease–PN complex and its binding and internalization had little to do with protease-induced mitogenesis. In his discussion of studies with other cells, Dr. Baker indicated that PN-like components are secreted by human fibrosarcomas and chick embryo, mouse embryo, and Chinese hamster lung cells but at markedly reduced levels — ie, < 5% of the level observed for human foreskin cells. However, if these cells were conditioned by depriving them of serum for extended periods, Dr. Baker believed that the greatest quantity of iodoprotein binding would then be via the PN complex formation and rebinding mechanisms. Dr. Baker concluded by stating that their laboratory does not view these new results as invalidating the earlier estimation of receptor molecular weights for chick and mouse embryo fibroblasts or on the estimation of 40,000 and 200,000 sites, respectively, for these cells. During the general discussion of this work, Dr. Baker indicated he had no explanation for J.F. Perdue's observation that trypsin did not compete with ^{125}I-thrombin for binding to substratum-attached human foreskin fibroblasts during incubation at 24°C. He has observed that trypsin does compete with thrombin for complex formation when the incubations are carried out in conditioned media at 37°C.

Dr. V. Hatcher mentioned his observation that serum-deprived cultures of human fibroblasts secrete into the medium uncomplexed plasminogen activator (PA) and other neutral proteases, as well as inhibitors that may have an affinity for trypsin. There was puzzlement expressed as to why free-PA would be detected in the presence of large quantities of PN. It is conceivable, however, that differences in specificity could account for these discrepancies since Baker's studies were carried out with human ^{125}I-labelled urokinase of 35 Kdaltons, whereas secreted PA may be the 55 Kdalton type.

One question that was not dealt with but should be considered in future discussions was: Is the secretion of PN under conditions of serum deprivation a natural response? Serum contains many factors, one of which has a survival value. Culturing cells in the complete absence of this factor(s) for extended periods in most cases results in their death. In this respect, the human foreskin fibroblast appears heartier than most other cell types. One must then wonder whether such cells in situ would secrete PN molecules, since they may be exposed continuously to AT3, α-trypsin inhibitor, α_2-macroglobulin, etc. Accepting these qualifications, a consensus was reached that the identification of cell-secreted PN was an important observation that may provide insights into the mechanisms whereby cells regulate their responsiveness and sensitivity to the environmental intrusion of serine proteases like plasmin, trypsin, thrombin, PA, and elastase, among others. At present, the contribution of PN to protease-induced mitogenesis is unknown. The role of additional protease-mediated mechanisms of mitogenesis, possibly by the modification of the binding site on cells through limited proteolysis, should be considered and evaluated in future investigations to explain this phenomenon.

Cell-Mediated Lympholysis in *H-2K/D* Identical Congenic Strain Combinations

Dorothee Wernet and Jan Klein

Abteilung Immungenetik, Max-Planck-Institut für Biologie, 7400 Tübingen, Federal Republic of Germany

Eight *H-2K/D* identical congenic strain combinations were tested in cell-mediated lympholysis. The reaction pattern of selected mouse strains revealed unidirectional reactivity to *Tla*-linked antigens in six strain combinations, whereas in two strain combinations the loci coding for the target antigen could only tentatively be linked to *H-2*. The *Tla*-linked loci defined by the different strain combinations appear to be identical with the *H-2T* (*Qa-1*) locus. Typing of B10.W lines for the *H-2T* (*Qa-1*) locus revealed the presence of alleles that appear to be the same as those present in inbred strains, as well as alleles that are different from the inbred-defined alleles but that share some determinants with the latter. The *H-2T* (*Qa-1*) system thus appears to consist of a limited number of alleles, some of which are closely related and presumably similar in their genetic structure.

Key words: CML, *K/D* identical strains, *H-2T*, B10.W lines

Cell-mediated lympholysis (CML) is an in vitro response of cytolytic lymphocytes to molecules bound to target-cell membranes [for a review, see reference 1]. The stimulating molecules can be of two kinds: molecules that are recognized alone, without the need for simultaneous recognition of other membrane elements (unrestricted CML), and molecules recognized in the context of other molecules (restricted CML). The targets of the unrestricted CML are class I and class II molecules encoded by the major histocompatibility complex (MHC) of the species. The target of the restricted CML can be a variety of substances, ranging from simple haptens to virus-specific and minor histocompatibility antigens [2-4]; the restricting molecules are encoded by class I MHC loci.

Recently, in this laboratory a locus (*H-2T*) was discovered that resides outside the conventionally defined murine MHC (*H-2*) but codes for molecules capable of stimulating unrestricted CML [5]. Simultaneously, but independently, Forman and Flaherty described another locus (*Qa-2*) [6], which mapped into the same region and displayed much the same properties as *H-2T* but was clearly distinct from *H-2T*. Subsequently, other investigators, working with different combinations than those used for the definition of *H-2T*, have described loci that apparently are identical with *H-2T* [7, 8].

Received and accepted April 2, 1980.

© 1981 Alan R. Liss, Inc., 150 Fifth Avenue, New York, NY 10011

TABLE I. Strain Combinations Used in the Cell-Mediated Lympholysis Experiments (A) and the *H-2* Haplotypes* of the Strains Used (B)

A
1) B10.HTT anti-B10.S(7R)
2) B10.S(7R) anti-B10.HTT
3) B10.HTT anti-B10.S(9R)
4) B10.S(9R) anti-B10.HTT
5) B10.A(3R) anti-B10.D2(R107)
6) B10.D2(R107) anti-B10.A(3R)
7) B10.A(1R) anti-B10.AM
8) B10.AM anti-B10.A(1R)

B

| | *H-2* Regions |||||||| |
	K	A	B	J	E	C	S	D	Tla
B10.HTT	s	s	s	s	k	k	k	d	c
B10.S(7R)	s	s	s	s	s	s	s	d	a
B10.S(9R)	s	s	°	k	k	d	d	d	a
B10.A(3R)	b	b	b	b	k	d	d	d	a
B10.D2(R107)	b	b	b	b	b	b	b	d	c
B10.A(1R)	k	k	k	k	k	d	d	b	b
B10.AM	k	k	k	k	k	k	k	b	b

*H-2 haplotypes according to Klein et al [11].

To determine whether CML-stimulating loci other than *H-2T* and *Qa-2* exist in the region (on the right-hand side of *H-2D*), we initiated a systematic screening of congenic mouse strain combinations identical at class I loci but possibly differing in regions of chromosome 17 immediately adjacent to the outer limits of the *H-2* complex (ie, regions at the *K* end and the *D* end of *H-2*). In this communication, we report some preliminary results of this screening. In addition, we also present data pertaining to the polymorphism of the *H-2T* locus.

MATERIALS AND METHODS

The congenic strain combinations tested are listed in Table IA; their *H-2* haplotype composition is given in part B of Table I. In each combination, the responding strain was immunized in vivo by lymphoid tissue inoculation, the mice were killed at various times after immunization, and their lymphocytes were restimulated in vitro in the CML assay described elsewhere [9]. The effector cells were tested against ConA and LPS target cells of the immunizing strain, and when the reaction was positive, they were also tested against a panel of 10 or more strains selected to provide information about the *H-2* linkage of the putative loci coding for the CML-target antigens. For each testing, three effector-to-target cell ratios were used (80:1, 40:1, and 20:1).

One combination, B10.HTT anti-B10.S(7R), was selected to test a panel of 30 B10.W congenic lines carrying *H-2* haplotypes and adjacent segments of undetermined length derived from wild mice. All of these lines have the rest of the genome derived from strain C57BL/10Sn (= B10). These lines and their mode of derivation have been described elsewhere [10].

TABLE II. Reactivity Pattern in the Cell-Mediated Lympholysis of Six Strain Combinations

	Reactivity of effector cells					
Target cells	B10.HTT anti-B10.S(7R)[a]	B10.HTT anti-B10.S(9R)	B10.S(9R) anti-B10.HTT	B10.A(3R) anti-B10.D2(R107)	B10.A(1R) anti-B10.AM	B10.AM anti-B10.A(1R)
A	+	+	–	–	–	+
A-Tla^b	–	–	+	+	–	+
B10.A	+	+	–	–	NT	NT
B10	–	–	(–)[b]	(–)[b]	–	+
B6	–	–	+	NT	NT	NT
B6-Tla^a	+	+	–	NT	NT	NT
B6.K1	–	–	+	NT	NT	NT
B6.K2	–	–	+	NT	NT	NT
C3H.SW	–	NT	NT	NT	–	NT
BALB/c	–	+	+	+	–	+
DBA/2	–	–	+	+	–	+
A.CA	–	–	–	NT	–	–
BALB.HTG	NT	NT	NT	NT	–	NT
C3H	–	–	–	NT	+	–
CBA	NT	–	+	NT	+	–
B10.BR	+	NT	–	NT	–	NT
AKR	–	NT	NT	NT	–	NT
BALB.K	NT	–	+	NT	+	NT
DBA/1	–	–	+	NT	–	–
A.SW	–	–	+	NT	–	–
129	NT	–	+	NT	–	–

[a]These data have been published previously [9].
[b]Sometimes a low rather than completely negative reaction with B10 was seen.
NT = not tested.

RESULTS

Altogether 16 strain combinations were tested, of which eight were examined in more detail. The eight combinations can be divided into three groups: combinations 1 to 6, in which positive reactions were obtained against antigens encoded by loci mapping close to the Tla locus; combination 7, which shows K^k-restricted reactivity; and combination 8, in which the locus coding for the target cell antigen could not be mapped. In none of these combinations was the reactivity directed against antigens encoded by loci within the H-2 complex.

Of the first six strain combinations, two reacted, under the conditions used, only unidirectionally: B10.HTT reacted against B10.S(7R) (this reaction has been described in detail elsewhere [9]), but B10.S(7R) did not react against B10.HTT; similarly B10.A(3R) reacted against B10.D2(R107), but B10.D2(R107) did not react against B10.A(3R). B10.HTT anti-B10.S(7R) reacted positively with Tla^a strains and negatively with strains carrying other Tla alleles [9], whereas B10.A(3R) anti-B10.D2(R107) reacted positively with Tla^b-carrying strains (eg, A-Tla^b) and negatively with Tla^a strains (eg, A, see Table II). The B10.HTT anti-B10.S(9R) effector cells reacted similarly to the B10.HTT anti-B10.S(7R) cells, with the exception of BALB/c, which was positive with the former and negative with the latter cells (Table II). B10.S(9R) anti-B10.HTT effector cells showed, with a few exceptions, an opposite reactivity pattern to that seen with B10.HTT anti-B10.S(9R) cells (Table II).

TABLE III. Reactivity Pattern in CML Typing of B10.HTT Anti-B10.S(7R) on 30 B10.W Strains

BUA 1	−	KPA 128	−
BUA 16	−	KPA 132	−
BUA 19	+	LIB 18	−
CAA 2	+	LIB 55	+
CAS 1	−	MOL 1	+
CAS 2	+	SAA 48	−
CHA 2	−	SNA 57	−
CHR 51	−	SNA 70	−
DRB 62	+/−	STA 10	+
GAA 20	−	STA 12	+/−
GAA 37	−	STA 39	+/−
KEA 2	−	STA 62	−
KEA 5	−	STC 77	−
KPA 42	−	STC 90	−
KPA 44	−	WOA 105	−

+ = highly positive; +/− = intermediate reaction; − = negative.

One combination, B10.HTT anti-B10.S(7R), was used to type B10.W mouse strains [10]. Of the 30 B10.W lines tested, six gave a positive reaction, three showed only an intermediate reaction, and all the other strains were negative (Table III). In two of the six positive reactions, cold target inhibition studies showed discrepancies with the results obtained by direct typing (data not shown).

The B10.A(1R) anti-B10.AM effector cells reacted not only with the B10.AM cells but also with some, though not all, H-2^k-bearing strains. All the positive strains, including our B10.AM line, carry the agouti gene, whereas the negative H-2^k strains are agouti negative. Therefore, it is likely that the cells react with a minor histocompatibility antigen controlled by a locus closely linked to the agouti gene; the reactivity is probably restricted by an H-$2K^k$ allele.

The reactivity pattern of the B10.AM anti-B10.A(1R) cells remains to be explained.

DISCUSSION

The data described in this communication, in combination with the data of Kastner and Rich [8] and Fischer-Lindahl [7], indicate that H-2T or H-2T-like CML can be induced in several different strain combinations. The minor discrepancies observed between the strain distribution pattern of positive reactions as defined by some of these new combinations and the original combination can be explained by the phenotypic complexity of the detected target molecules rather than by the genetic complexity of the controlling regions. Although the existence of a tight cluster of CML loci in the *H-2T* region cannot be excluded, there is no need to resort to the cluster hypothesis unless direct evidence for genetic complexity is obtained.

The fact that strain combination after strain combination defines what appears to be the same locus makes it unlikely that the *T* region contains a large number of easily separable CML loci (unless some of the loci are monomorphic in the strains tested). The *H-2T* locus itself is probably identical with the serologically detectable *Qa-1* locus [12]. The reason it was given a new designation was that the *Qa-1* locus was originally mapped between

H-2D and *Qa-2*, whereas the *H-2T* clearly mapped between *Qa-2* and *Tla*. Later it was learned, however, that the original mapping of *Qa-1* was in error, and the *Qa-1* locus had to be relocated to the same region as the *H-2T*. Taking into account some of the present uncertainties about allele assignment by both the serological and the CML assays, one finds a good correlation between the strain distribution patterns of the *Qa-1* and *H-2T* loci. Furthermore, recent evidence presented by Kastner et al [8] indicates that the H-2T-like CML can be inhibited by anti-Qa-1 sera. It would therefore be most desirable that the different symbols now in existence for the locus controlling this *T*-region CML be dropped and the *Qa-1* symbol adapted on the basis of priority.

The *Qa-1*-related CML is not a simple biallelic system. The first indication of the existence of a third allele was provided by our typing of the *H-2^f* strains, A.CA and B10.M [5], which showed intermediate reactivity in comparison with the defining alleles. In this communication, we provide additional evidence for the existence of intermediate alleles in several of the B10.W lines. The simplest interpretation of these intermediate alleles is that they code for molecules sharing some determinants with products of other alleles. This interpretation is supported by preliminary cold-target inhibition studies. There is also evidence that some of the intermediate alleles may not be the same. However, the frequent occurrence among the B10.W lines of alleles that type identical to alleles found in the inbred strains suggests that the polymorphism of the *Qa-1* locus is considerably lower than that of the class I loci (and also of some of the class II loci). The apparent sharing of determinants even between what appear to be different *Qa-1* alleles is in agreement with this conclusion. (In the CML directed against the polymorphic class I loci, sharing of determinants could be demonstrated only between closely related alleles derived by recent mutations.) Thus, the picture of the *Qa-1* system emerging from the limited data available is one of a not too large number of alleles, some of which are apparently closely related as reflected by sharing of antigenic determinants between the molecules controlled by these alleles.

ACKNOWLEDGMENTS

We gratefully acknowledge the expert technical assistance of Mrs. Gudrun Labib. This work was supported by the Forschergruppe "Leukämieforschung" of the Deutsche Forschungsgemeinschaft.

REFERENCES

1. Klein J: Springer Seminars in Immunopathology 1:31, 1978.
2. Shearer GM, Rehn R, Garbarino CA: J Exp Med 141:1348, 1975.
3. Zinkernagel RM, Doherty PC: J Exp Med 141:1427, 1975.
4. Bevan MJ: J Exp Med 142:1349, 1975.
5. Klein J, Chiang CL: Immunogenetics 6:235, 1978.
6. Forman J, Flaherty L: Immunogenetics 6:227, 1978.
7. Fischer-Lindahl K: Immunogenetics 8:71, 1979.
8. Kastner DL, Rich RR, Shen F-W: J Immunol 123:1232, 1979.
9. Wernet D, Klein J: Immunogenetics 8:361, 1979.
10. Klein J: Science 168:1362, 1970.
11. Klein J, Flaherty L, Van de Berg JL, Shreffler DC: Immunogenetics 6:489, 1978.
12. Stanton TH, Boyse EA: Immunogenetics 3:525, 1976.

Cellular Interaction and the Environment in Lymphocyte Development: The Roles of Antigen, Histocompatibility, and Growth Factors in T cell-Dependent B cell Stimulation

Fritz Melchers, Jan Andersson, Waldemar Lernhardt, and Max H. Schreier

Basel Institute for Immunology, CH-4058 Basel, Switzerland (F.M., W.L., M.H.S.) and Biomedicum, University of Uppsala, Uppsala, Sweden (J.A.)

T cell-dependent B cell stimulation depends on cellular interactions between macrophages, T cells, and B cells. T cells recognize antigen in the context of Ia determinants on macrophages. This leads to the activation of helper T cells and to the production of helper factor(s) either by the activated helper T cells or by the macrophages. Cloned lines of helper T cells produce factors ("help") for B cell replication and Ig secretion in the presence of histocompatible macrophages and of specific antigen. These factors stimulate histocompatible, as well as histoincompatible, mitogen-activated B cell blasts polyclonally. Thus, neither antigen nor histocompatibility, but antigen-unspecific, soluble factors are required to stimulate an activated B cell blast through the next round of division.

Small resting B cells, however, are not stimulated to replication by these factors, but only to polyclonal, H-2-unrestricted maturation to Ig-secreting cells. Replication (and Ig secretion) of small resting B cells, however, is only induced when antigen-specific small B cells bind their specific antigen via surface Ig molecules and interact with histocompatible helper T cells. The preference of the resting B cells for such collaboration with T cell help is mapped to the K-end of the H-2 locus and probably constitutes the antigen expressed on B cells by the I region. It appears that a resting B cell is excited by the binding of specific antigen to surface Ig and by the interaction of its surface Ia with helper T cells. After this dual recognition the excited B cell can be stimulated by the antigen-unspecific factor(s) generated by the interaction of helper T cells, adherent cells and antigen to initiate growth. Immune induction of a B cell thus involves three controlling elements: Ig, Ia, and B cell growth factor receptors.

Key words: helper T cells, B cell growth, Ia-restriction, helper factors, resting B cells

The majority of B lymphocytes in the immune system are resting cells. Activation of these B cells from their resting, G_0-state of the cell cycle can generally be achieved in two ways. One way involves helper T cells specific for an antigen, adherent cells (also called accessory cells, macrophages, dendritic cells), and specific antigen and results in replication and maturation to immunoglobulin (Ig) secretion of those few B cell clones producing antigen-specific Ig (Fig. 1, mode A or B) [1–6]. In this antigen-specific, T cell-dependent

Received and accepted May 12, 1980.

© 1981 Alan R. Liss, Inc., 150 Fifth Avenue, New York, NY 10011

A SRC-specific stimulation

C Polyclonal stimulation

B TNP-specific stimulation

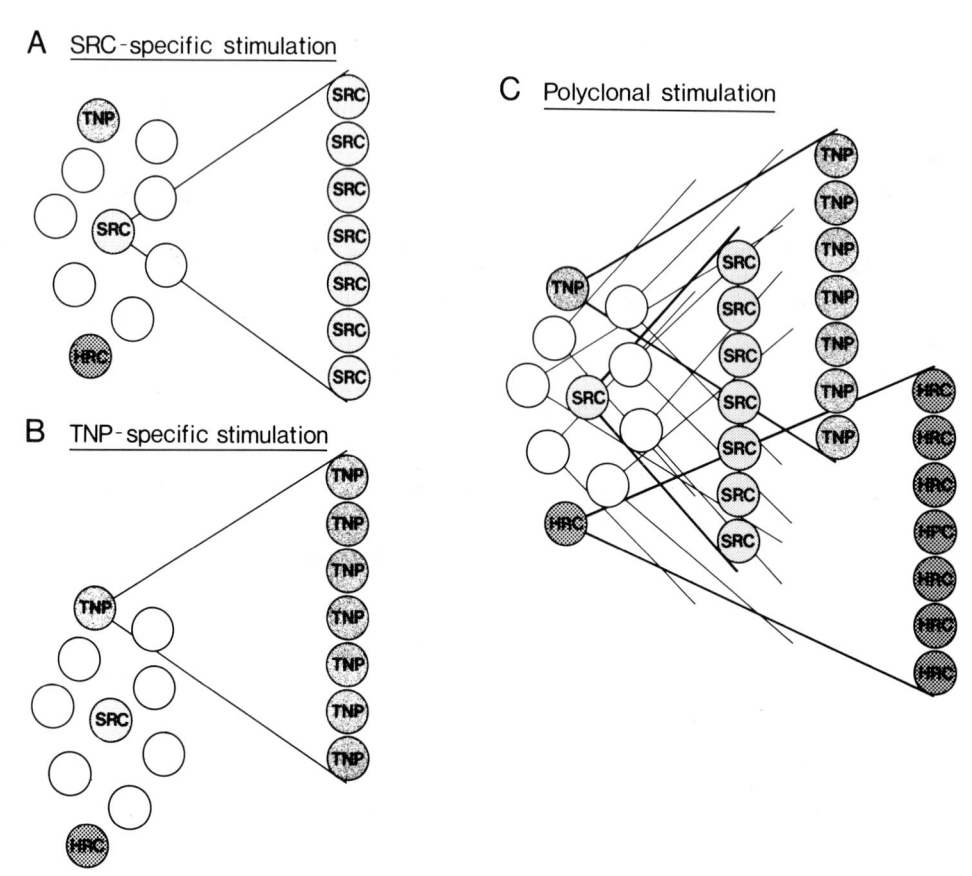

Fig. 1. Modes of B cell stimulation. The repertoire of B lymphocytes includes those specific for SRC, HRC, and TNP. Specific stimulation, either dependent or independent of helper T lymphocytes – eg, by SRC-specific helper T cells and SRC as an antigen [16–19] (see below), or by TNP-LPS at low concentrations [8] – results in replication and maturation of the antigen-specific clones only (modes A and B). Polyclonal stimulation of the same B cell population by mitogens, such as LPS at high concentrations, results in replication and maturation of all (or a large part of) B cells (mode C).

B cell stimulation all signals except antigen are endogenously supplied by the host. The other way of B cell stimulation is T cell-independent – ie, will also occur in T cell-depleted lymphocyte populations. One very effective way of antigen-specific, T cell-independent B cell stimulation is done by antigens that are covalently bound to mitogens (Fig. 1, mode A or B) [7, 8]. Mitogens, such as lipopolysaccharide (LPS) [9] or lipoprotein (LPP), [10] are substances produced by bacteria – ie, are normally exogenous to the organism. At high concentrations these substances will stimulate B cells polyclonally and irrespective of the histocompatibility (H-2) haplotype of the lymphocytes (Fig. 1, mode C).

At least three structures on B cells appear to play a role in antigen-dependent B cell stimulation: 1) Ig, which binds antigen. One B cell expresses only one antigen-binding capacity by a given set of variable (v) region of heavy (H) and light (L) chains. 2) Ia antigens, encoded by I- A and I-E/C region genes of the H-2 complex, which are involved in histocompatible interactions with (either) macrophages and/or helper T cells. 3) Receptors for factors (mitogens) involved in the regulation of replication and maturation to Ig secretion.

While T cell-dependent, antigen-specific B cell stimulation appears to use these three structures in order to activate resting B cells and stimulate them through successive rounds of divisions and to Ig secretion, mitogenic B cell activation, at high concentrations of mitogen, circumvents the binding step to Ig and any restrictions imposed by the H-2 antigens expressed on B cells. Mitogenic stimulation of B cells is a convenient way to transform B cells into an activated state. It can also be used to monitor the extent of reactiveness of the B cell compartment and to assess its repertoire of B cells specific for a given antigen. Thus, in most strains of mice, at two or three months of age, one-third of all B cells are reactive to LPS and LPP, and most of them are reactive to both [11, 12]. Since genetic analyses [13, 14] of LPS and LPP responsiveness argue for two separate reactivities and, therefore, receptors, this indicates that one B cell can be regulated by more than one growth-regulating substance/receptor interaction. Within the mitogen-activated B cell population 10^{-3} cells are specific for SRC or HRC, and 10^{-2} are specific for TNP-SRC [15]. Frequency analysis of specific B cell clones activated under a given protocol among the total activated B cell population, therefore, is a quantitative measure to assess the extent of the specificity of activation.

This report summarizes our current knowledge of the requirements of the activation of resting B cells from the resting state, and of the stimulation of activated B cells through successive cell cycles and to maturation to Ig secretion. The details of our experiments relevant to T cell-dependent and mitogen-dependent B cell stimulation have been published elsewhere [16–19].

QUANTITATION OF IN VITRO B-LYMPHOCYTE RESPONSES

The response of a given lymphocyte population to stimulation can best be quantitated by determining the frequency of cells that can be activated to clonal replication and maturation to Ig secretion by antigen or mitogen [11, 15, 20]. Under ideal culture conditions all inducible cells should be stimulated, and each such cell should do so as a single cell in a culture. Such conditions [21] then allow limiting dilution analyses in which the frequencies of reactive cells can be calculated by Poisson's distribution. These analyses also require that single clones of cells of varying sizes can be detected. This is possible since stimulated B cells secrete Ig. The Ig secreted into the culture supernatants can be quantitated by radioimmune assays that are specific for a given antigen, or class, or type of Ig. Single Ig-secreting cells in a clone can be detected by the hemolytic plaque assay, either specific for an antigen on red cells or modified to detect all cells secreting Ig of a given class or type [22].

Conditions of in vitro cultures of normal murine B cells have been improved over recent years. Changes in the composition of Dulbecco's medium and the addition [23] of 2-mercaptoethanol [24] allow normal murine lymphocytes to survive for many days in high numbers and, upon stimulation, to replicate for many divisions and to mature to Ig secretion. Replacement of serum by albumin, transferrin, and soybean lipid [23, 25] excludes possible stimulatory or suppressive effects. Finally, addition of thymus cells as "filler" or "feeder" cells allows the dilution of reactive B cells to one cell per culture, inducible by stimulation to clonal replication and Ig secretion.

Mitogen-Dependent B cell Stimulation

In such tissue culture conditions one of three splenic B cells of a variety of mouse strains, all at 6 weeks to 3 months of age, will respond to LPS or to LPP by replication and maturation to Ig secretion. Activation from the resting state is asynchronous and is complete within the first 18 h [W. Lernhardt and F. Melchers, in preparation]. Very little, if any, recruitment into the cell cycle occurs after that time. The activated B cells divide every 18 h [21]. Remarkable synchrony of a given clone is observed for at least 5 divisions. Every cell in the clone divides, and every dividing cell secretes Ig [21]. In fact, Ig secretion is measurable after 2 to 4 h of activation, in time well before the onset of DNA synthesis and the first mitosis [26]. LPS and LPP stimulate largely overlapping B cell populations [12]. Both mitogens effect mainly maturation to IgM secretion. Approximately 10% of all stimulated B cell clones will switch to IgG_{1+2} secretion, and that may occur at any cell cycle [27]. Mitogen-stimulated B cells do not replicate forever. After 10–15 divisions clonal expansion comes to a halt and cannot be restarted by a renewed supply of fresh medium and/or mitogen. Activated B cells require the mitogen early in the G_1 phase of every cell cycle to continue the cell cycle and to complete mitosis [W. Lernhardt and F. Melchers, in preparation]. Although no further divisions take place in the absence of the stimulating mitogen, Ig secretion continues in the activated cells for days [12]. It is, in fact, so far not possible to revert in vitro mitogen-activated B cell blasts back to the state of a resting small B cell. While under normal in vitro conditions of mitogenic activation of B cells, replication and Ig secretion always occur simultaneously, the two processes can be uncoupled under certain conditions. First, in a replicating B cell clone the rate of Ig synthesis and secretion increases with every division over the rate of all cellular protein synthesis. This apparently disfavors further replication so that the clone stops dividing. Second, when DNA synthesis is blocked by hydroxyurea or cytosine arabinoside, induction of small B cells to Ig secretion still occurs [28]. Finally, when small B cells are exposed to anti-Ig antibodies prior to mitogenic stimulation the mitogen-stimulated clones will divide but will not secrete Ig or show any measurable increase of Ig synthesis [29].

HELPER T-LYMPHOCYTE LINES AND CLONES IN CULTURE

In vivo-activated helper T cells can be propagated in vitro in the presence of specific antigen and irradiated adherent cells, usually in the form of spleen cells. Upon prolonged restimulation in vitro they can be cloned by limiting dilution in the presence of antigen, irradiated adherent cells, and T cell growth factors (TCGF). Such clones of specific helper T cells have been obtained with sheep or horse erythrocytes (SRC or HRC) as antigens [30]. They continue to replicate in culture for months if they are provided either with TCGF or with antigen plus adherent cells as probable sources of TCGF production. The lines and clones of erythrocyte-specific helper T cells that have been employed in the work summarized below help B cells to replicate and to secrete Ig [30]. The production of this help, as measured by the capacity of these helper T cells to induce B cells to antigen-specific clonal replication and Ig secretion, is antigen-specific and H-2 restricted [30] (Table I). The restriction of this production of help maps to the I-region of the H-2 complex.

TABLE I. Antigen-Specific, H-2-Restricted Production of T cell Help

Helper T cell line or clone[a] (C57BL/6J, b b b[b]) specificity	PFC response[c]	Response of C57BL/6J (b b b)[b] B cells[c] in the presence of helper T cells[c] and irradiated adherent cells[c] of:											
		b b b haplotype[c]		b d d haplotypes[b]			b b d b haplotypes						
		No antigen	HRC	SRC	No antigen	HRC	SRC	No antigen	HRC	SRC
Anti-SRC	Anti-SRC	–	–	+[d]	–	–	–	–	–	+[d]
	Anti-HRC	–	–	–	–	–	–	–	–	–
Anti-HRC	Anti-SRC	–	–	–	–	–	–	–	–	–
	Anti-HRC	–	+[d]	–	–	–	–	–	+[d]	–

[a]Long-term cultured T cells grown in the presence of irradiated, histocompatible adherent cells and antigen, or TCGF [30].

[b]b b b denotes the b-haplotype of the K-, I-A, and D-region of the H-2 complex of the mouse characteristic of C57BL/6J mice. b b b are the corresponding haplotypes of B10.A (5R) mice; b d d, those of B10.A (4R) mice.

[c]Measured in cultures containing 2.5×10^4 helper T cells/ml, 2.5×10^6 erythrocytes/ml and 1×10^6/ml 3,300 rad-irradiated spleen cells as source of adherent cells with 5×10^5 nu/nu spleen cells or anti thy 1-treated spleen cells/ml as source of B cells at day 6 or 7 of culture, as IgM-secreting, direct PFC.

[d]At day 7 of culture 5×10^3 – 5×10^4 IgM-secreting, antigen-specific PFC were found in cultures of 5×10^5 initially cultured spleen cells (–) denotes no response – ie, less than 50 antigen-specific PFC in cultures of 5×10^5 initially cultured spleen cells.

EFFECT OF T CELL HELP ON ACTIVATED B CELL BLASTS

Antigen-activated T cell help was generated by co-cultivation of antigen, specific helper T cells, and irradiated histocompatible adherent cells (see Table I). Activated B cell blasts were produced from spleen cells of a variety of mouse strains differing in their H-2 haplotype by a 48 h incubation period with the B cell mitogen LPS, followed by enrichment of the B-blasts through velocity sedimentation. Such activated B-blasts, as well as "background"-activated B-blasts of normal spleens were all restimulated by antigen-activated T cell help to successive rounds of division and to increased maturation to IgM secretion. The restimulation was polyclonal and occurred irrespective of the H-2 haplotype of the B-blasts, indicating that neither antigen nor H-2 compatibility – ie, the occupance of neither Ig nor H-2 antigens such as K,D, or Ia – was required to stimulate the blasts. It was concluded that, most likely, B cell replication and maturation factors (BCMF) were stimulating the B-blasts. Therefore, antigen-specific and H-2-restricted induction of T cell help resulted in the production of antigen-unspecific B cell help [16, 30].

This antigen-unspecific "help" for activated B-blasts could be found in the supernatant media of 2-day incubations of helper T cells, antigen, and irradiated adherent cells. It stimulated activated B-blasts to polyclonal replication and maturation in an H-2-unrestricted fashion [18] (summarized in Table II).

EFFECTS OF T CELL HELP ON SMALL RESTING B CELLS

The effect of the same T cell help-conditioned media on small resting B cells was markedly different [18] (Table II). The factors, in the absence of any added antigen, did not induce replication as judged by the uptake of radioactive thymidine or by cell counting. Small cells, however, changed their morphological appearance. Within 24 h of incubation with the T cell help-conditioned media, large, blast-like cells appeared in large numbers. Within 4–5 days approximately one-third of the splenic B cells of a variety of mouse strains developed in this way. The large, non-replicating blasts purified by velocity sedimentation were shown to secrete IgM and IgG. Their repertoire was polyclonal. Therefore, soluble factors of T cell help (ie, BRMF) induce maturation without replication in small resting B cells in a polyclonal and H-2-unrestricted fashion. The capacity of the T cell help-factors (BRMF)-induced non-replicating B blasts to replicate is lost as they mature to Ig secretion. Within 1 to 2 days the frequency of cells initiating clonal replication after stimulation by LPS or by T cell help is reduced by 90–99%. BRMF therefore appears to uncouple maturation to Ig secretion from replication in small B cells. Maturation of small B cells to Ig secretion, in fact, appears to antagonize their capacity to replicate.

What, then, constitutes an antigen-specific, H-2-restricted [5, 6] response of B cells to T cell-dependent antigens such as erythrocytes? This is revealed when small resting B cells are exposed to antigen-activated T cell help – ie, to antigen, specific helper T cells, and irradiated adherent cells [17] (Table II). Clonal replication of only the antigen-specific B cell clones is observed in frequencies expected from the repertoire determinations of the B cell compartment, with LPS as a polyclonal stimulator. Many other B cells with specificities for other than the erythrocyte antigens still only mature in a polyclonal fashion to Ig secretion without replicating. When a second, non-cross-reacting erythrocyte antigen that is not stimulating the helper T cells is added (ie, SRC to an HRC-specific helper T cell/HRC/adherent cell mixture, and vice versa), B cells with specificities for that second ("bystander") antigen will now replicate and mature to Ig secretion, again in frequencies expected from previous repertoire determinations.

TABLE II. Summary of the Action of Antigen-Activated T cell Help and of Soluble Factors Released by Such Help on Activated B cell Blasts and on Small Resting B cells

B cell source (mouse strain)	H-2 haplotype K, I-A, D	Activated B cell-blasts — Antigen-activated T cell help haplotype b b b — Replication	Maturation	Soluble factors of antigen-activated T cell help — Replication	Maturation	Small resting B cells — Antigen-activated T cell help haplotype b b b — Replication	Maturation	Soluble factor of antigen-activated T cell help — Replication	Maturation
C57BL/6J	b b b	Polyclonal	Polyclonal	Polyclonal	Polyclonal	Antigen-specific only	Antigen-specific plus polyclonal	No response	Polyclonal
B10.A (4R)	b d d	Polyclonal	Polyclonal	Polyclonal	Polyclonal	No response	Polyclonal	No response	Polyclonal
B10.A (5R)	b b d	Polyclonal	Polyclonal	Polyclonal	Polyclonal	Antigen-specific only	Antigen-specific plus polyclonal	No response	Polyclonal
C3H	k k k	Polyclonal	Polyclonal	Polyclonal	Polyclonal	No response	Polyclonal	No response	Polyclonal

Antigen-specific replication is, however, only initiated with small resting B cells if these small cells are histocompatible, probably in the I-A region of the H-2 complex, with the H-2 (I-A) environment to which the helper T cells have been adapted during their differentiation from precursor cells. Histo*in*compatible small B cells are not stimulated to antigen-specific replication [17] (Table II).

CONCLUSIONS

From these studies the following picture of T-B-lymphocyte collaboration in antigen-specific immune responses emerges (Fig. 2). Antigen-specific helper T cells recognize antigen on the presenting macrophage. Activation of the helper T cells, and consequently

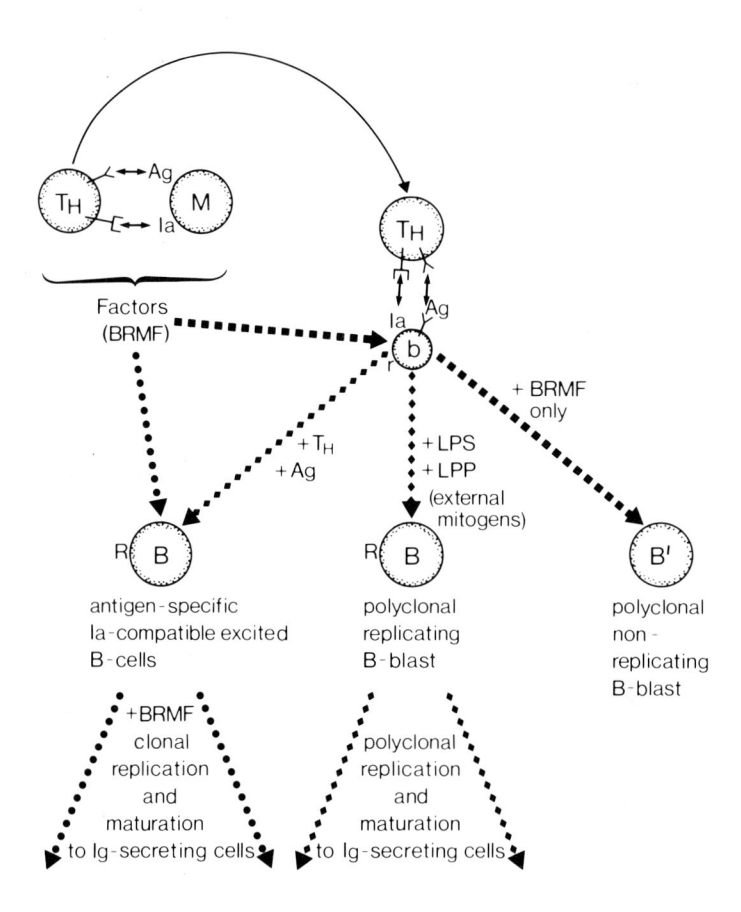

Fig. 2. Scheme of helper T cell–macrophage–B cell cooperation. Helper T cells (T$_H$) recognize antigen (Ag) in the context of Ia antigens on macrophages (M). This interaction produces B cell replication and maturation factors (BRMF). In a second interaction the same antigen-specific, H-2 (Ia)-restricted helper T cell interacts with antigen (Ag) bound by Ig molecules on B cells, and with Ia antigens on B cells. This excites the small, resting B cell (b) so that receptors for BRMF (r) change their conformation to R. Action on b by BRMF via r induces polyclonal maturation without replication. Action on the excited B cell via R induces maturation and replication. Small, resting B cells (b) can also be excited to an activated state (B) by mitogens such as LPS or LPP. LPS, LPP and BRMF will, then, continue the polyclonal replicating and maturation of such blasts.

"help" for B cells will occur, however, only if the helper T cell is also adapted to recognize the particular haplotype of Ia-antigen expressed on the macrophage. The dual recognition of antigen and Ia-determinants, then, leads to the production of soluble B cell replication and maturation-factors (BRMF) by either the helper T cells or the macrophages.

These factors act differently on activated B-blasts and on resting small B cells. Maturation without replication — in fact, exhaustive terminal differentiation of small B cells — is effected by the action of BRMF if neither Ig nor Ia molecules on the B cells are occupied. The occupance of Ig by antigen alone is insufficient to stimulate the antigen-specific B cells to replication. Such small B cells have to be doubly occupied by antigen and by histocompatible interaction structures of the T cells, fitting the Ia-antigen on the B cells, before they become excited. Excitation leads to a state similar to that of a mitogen-activated B-blasts, in which BRMF can now initiate the cell cycle. BRMF is then needed at the beginning of each successive cell cycle to expand the B cell clone. Once a resting B cell has become excited, neither antigen nor histocompatible interactions — ie, neither occupance of Ig nor of Ia — is needed to stimulate the activated B cell through the next rounds of division. Antigen-unspecific BRMF, acting like growth hormones or mitogen, are only then needed.

Three elements, therefore, control B lymphocyte replication and maturation: Ig, Ia, and receptors for BRMF. Ig and Ia control the resting state of the cell and thereby safeguard B cells of the immune system against antigen-unspecific, histoincompatible stimulation. Double occupancy of Ig and Ia renders the cell susceptible to BRMF. If BRMF has previously been produced by the interaction of antigen, helper T cells, and macrophages, antigen-specific, H-2-restricted B cell clones will then replicate.

REFERENCES

1. Claman HN, Chaperon EA, Triplett RF: Proc Soc Exp Biol Med 122:1167, 1966.
2. Miller JFAP, Mitchell GF: Nature 216:659, 1967.
3. Mosier DE: Science 158:1573, 1967.
4. Rosenthal AS, Shevach EM: J Exp Med 138:1194, 1973.
5. Katz DH, Hamaoka T, Dorf ME, Benecerraf B: Proc Natl Acad Sci USA 70:2624, 1973.
6. Sprent J: Immunol Rev 42:108, 1978.
7. Andersson J, Sjöberg O, Möller G: Transplant Rev 11:131, 1972.
8. Coutinho A, Möller G: Adv Immunol 21:114, 1975.
9. Andersson J, Sjöberg O, Möller G: Eur J Immunol 2:349, 1972.
10. Melchers F, Braun V, Galanos C: J Exp Med 142:473, 1975.
11. Andersson J, Coutinho A, Melchers F: J Exp Med 145:1511, 1977.
12. Andersson J, Coutinho A, Melchers F: J Exp Med 149:553, 1979.
13. Watson J, Riblet R: J Exp Med 140:1147, 1974.
14. Watson J, Riblet R: J Immunol 114:1462, 1975.
15. Andersson J, Coutinho A, Melchers F: J Exp Med 145:1520, 1977.
16. Schreier MH, Andersson J, Lernhardt W, Melchers F: J Exp Med 151:194, 1980.
17. Andersson J, Schreier MH, Melchers F: Proc Natl Acad Sci USA 77:1612, 1980.
18. Melchers F, Andersson J, Lernhardt W, Schreier MH: Eur J Immunol 10:679, 1980.
19. Melchers F, Andersson J, Lernhardt W, Schreier MH: Immunol Rev 52:89, 1980.
20. Lefkovits I, Waldmann H: "Limiting Dilution analysis of Cells in the Immune System." Cambridge, London, New York, Melbourne: Cambridge University Press, 1979.
21. Andersson J, Coutinho A, Lernhardt W, Melchers F: Cell 10:27, 1977.
22. Gronowicz E, Coutinho A, Melchers F: Eur J Immunol 6:588, 1976.
23. Iscove NN, Melchers F: J Exp Med 147:923, 1978.
24. Click RE, Benck L, Alter BJ: Cell Immunol 3:156, 1972.
25. Guilbert LJ, Iscove NN: Nature 263:594, 1976.
26. Melchers F, Andersson J: Biochemistry 13:4645, 1974.

27. Andersson J, Coutinho A, Melchers F: J Exp Med 147:1744, 1978.
28. Melchers F, Andersson J: Eur J Immunol 4:687, 1974.
29. Andersson J, Bullock WW, Melchers F: Eur J Immunol 4:715, 1974.
30. Schreier MH, Tees R: Allergy Appl Immunol 61:227, 1979.

Density Inhibition of Growth: Workshop Report

Robert W. Tucker and Charles W. Boone

The Johns Hopkins Oncology Center, Baltimore, Maryland 21205 (R.W.T.) and National Cancer Institute, Bethesda, Maryland 20205 (C.W.B.)

It is a commonly observed phenomenon in tissue culture that non-neoplastic fibroblasts become growth-arrested when crowded. Explanations for this phenomenon, termed "density inhibition of growth" (or proliferation), have included growth factor depletion [1], accumulation of toxic factors or chalones [2], and inaccessibility of growth factors or nutrients [3]. In analogy with contact inhibition of movement [4], many workers have assumed that density inhibition of growth was due to cell–cell contact. However, recent evidence has shown that changes in cell spreading or cell attachment can mimic density inhibition of growth in the absence of cell–cell contact [5]. These same studies have also linked the mechanism of density inhibition of growth to anchorage and serum dependence of growth. Additionally, there are in vivo correlates of these phenomena that should be considered.

IN VITRO DENSITY INHIBITION, ANCHORAGE DEPENDENCE, AND SERUM GROWTH REQUIREMENT

Recently, a new approach in studying density inhibition of growth has allowed the separation of cell–cell contact from other factors [5]. Cells placed on a polymer [poly(HEMA)] spread out to an extent dependent on the thickness of the polymer. The amount of cell spreading can be easily quantitated by either the height of the cell [5] or the area of the substratum covered by the cell [6]. When non-neoplastic cells are held at different degrees of cell spreading, those that are 70–100% maximally spread are capable of synthesizing DNA. Round, poorly attached cells do not synthesize DNA and do not grow. In contrast, neoplastic fibroblasts grow even when rounded up on thick poly(HEMA) films [6, 7]. The use of these polymer films allows one to culture cells with different amounts of rounding, but without the cell–cell contact characteristic of a crowded confluent monolayer. When sparse cells on poly(HEMA) and confluent cells on plastic cover the same area of substratum, both show the same amount of DNA synthesis [5]. Thus, the rounded-up configuration of the cells in a confluent monolayer is sufficient to explain the cessation of DNA synthesis. The forces that inhibit cell spreading are generated by the surrounding cells and have recently been shown in this conference to contribute to the movement of

Received and accepted May 9, 1980.

cells into a wound of a confluent monolayer [8]. The movement itself has not been shown to be necessary for subsequent DNA synthesis [9], but the increased amount of cell spreading at the edges of the wound is sufficient to explain DNA synthesis and growth [5]. From this point of view, density inhibition of growth is the direct result of deprivation of anchorage for cells in the crowded monolayer. Similarly, in the assays for anchorage-independent growth, both poly(HEMA) and agarose prevent cell attachment. The fundamental change in both density inhibition and anchorage dependence is in the degree of cell spreading.

The specific nature of the information signaled to the cell by changes in cell spreading has not yet been explored. Increased cell spreading might produce decreased sensitivity to local inhibitory factors or increased sensitivity to exogenous growth factors. Both of these general mechanisms would predict increased growth at constant serum concentrations as the cells spread more extensively. In fact, this is what happens; the sensitivity of the cell to serum stimulation of growth depends on the degree of cell spreading [6, 7]. That is, cell growth rate is proportional to the degree of cell spreading at constant serum concentrations; moreover, the higher the serum concentration, the greater is the change in growth rate for a given change in the area of spreading. In view of these results, one can now group density inhibition, anchorage dependence, and serum growth requirement as categories that all depend on the area of cell attachment or spreading.

There are many examples in this conference of the importance of cell spreading and cell attachment to an extracellular matrix in determining differentiation and growth. The considerations listed here suggest that one of the possible factors determining these matrix effects is the amount of cell spreading and/or the physiochemical nature of cellular attachments. Neoplastic or viral transformation changes the cell's dependence on cell spreading on poly(HEMA) [6] or other cellular matrixes, as examined in this conference for growth of SV40-transformed cells on monolayers of normal cells [10]. Future research will undoubtedly delineate the role of cell spreading and other interactions with the cellular matrices as possible controlling factors in cellular differentiation and growth in nonneoplastic cells. A corollary of such studies may be an increased understanding of the neoplastic process as an aberration of these normal controls.

IN VIVO COUNTERPARTS OF DENSITY INHIBITION, ANCHORAGE DEPENDENCE, AND SERUM GROWTH REQUIREMENT

Most of the characteristics shown by connective tissue cells (ie, fibroblasts, endothelial cells, pericytes, and histiocytes, among others) in tissue culture are an expression of their in vivo functions related to inflammation, wound repair, and foreign body rejection. This fact is not always taken into consideration by investigators who deal with the in vitro characteristics of cell behavior. Yet the interpretation of in vitro cell behavior within the framework of in vivo cell and tissue physiology undoubtedly would suggest significant new questions in many diverse areas of cell biology. Therefore, it seems desirable to examine thoroughly the in vivo counterparts of density inhibition of proliferation, anchorage dependence, and serum growth requirement.

Activation of the Normally Dormant Fibroblast

First, it should be emphasized that under normal circumstances the fibroblast in vivo exists as a relatively dormant cell, which ordinarily is cigar-shaped, nonproliferative, and nonmotile. During the process of wound repair, however, it is activated to become highly motile and rapidly proliferative. It is important to realize that the fibroblast in the tissue culture dish in vitro is such an activated cell, which is behaving as it would in vivo when reacting to a wound or the traumatic introduction into the tissues of a smooth-surfaced foreign body, for instance, a sheet of plastic or glass. Stated in a way that dramatizes the situation, "The fibroblast placed in vitro behaves as it would have in vivo if its host had been stabbed with a Petri dish which remained in the wound."

In Vivo Counterpart of Serum Growth Requirement

Serum, of course, is never present in the tissue fluids in vivo unless hemorrhage and clot formation have occurred. To continue the analogy introduced above, immediately after the placement of a smooth-surfaced foreign body into the tissues, hemorrhage and clot formation would lead to the production of serum-containing factors (eg, platelet-derived growth factor (PDGF) and thrombin) that stimulate fibroblasts to proliferate. Indeed, fibroblasts proliferate in vitro only very slowly in the presence of platelet-free plasma that is devoid of PDGF [11].

In Vivo Counterpart of Anchorage Dependence

In vitro, attachment to an adhesive surface is required to induce the fibroblast to occupy a larger area of attachment, become flat, motile, and rapidly proliferative. If adhesivity were also required for the activation of fibroblastic proliferation in vivo, one would expect that nonadhesive foreign bodies introduced in vivo would tend not to induce a fibroblastic reaction with capsule formation. This is indeed the case: non-adhesive prostheses made of silicone or chemically inert metals are well known to be less prone to induce a foreign body rejection reaction. Thus, the term "foreign surface activation" is an in vivo counterpart to the in vitro term "anchorage dependence."

In Vivo Counterpart of Density Inhibition of Proliferation

Generally speaking, when foreign material is traumatically embedded in the tissues in vivo, one or more of three types of response occurs in temporal sequence. First, polymorphonuclear leukocytes are attracted to the area by chemotactic factors associated with cell injury. The leukocytes liberate from their lysosomes a variety of hydrolases, which break down and solubilize most organic components of the foreign material. Second, macrophages also become activated to release hydrolases and phagocytose material that has resisted solubilization and happens to be in particulate form. Third, whatever is left of the foreign material that cannot be solubilized or phagocytosed is surrounded by proliferating fibroblasts, which form a multilayered capsule that "walls off" the foreign material. In the case of a smooth-surfaced foreign body made of plastic or glass, the formation of a capsule of growth-arrested fibroblasts is the most prominent reaction that occurs [12]. In close analogy, fibroblasts in vitro confronted with the smooth surface of the Petri dish, attempt to form a multilayer upon it as they would if it were embedded in tissue. Thus, density inhibition of proliferation occurs during the process of fibroblastic multilayering both in vitro and in vivo. There is a further analogy between the foreign body rejection reaction in vivo and fibroblastic behavior in vitro. In rodents, just as tumors readily appear in the capsule around a smooth-surfaced foreign body, so connective tissue cells in tissue culture spontaneously become tumorigenic for the host after a time. This analogy has recently been experimentally explored [13].

Table I summarizes the relationship between in vitro properties of serum growth requirement, anchorage dependence, and density inhibition of proliferation, and their

TABLE I.

In vitro (tissue culture)	In vivo (wound by smooth foreign body)
Density-dependent inhibition of proliferation. Growth slows as rapidly proliferating fibroblasts become crowded on plastic.	Intracapsular inhibition of proliferation. Growth slows as rapidly proliferating fibroblasts form a multilayered capsule around a smooth-surfaced foreign body.
Anchorage dependence. Cells "require" attachment to an adhesive solid surface to flatten, move, and proliferate.	Foreign surface activation. Cells are activated by contact with an adhesive smooth-surfaced foreign body to convert from a cigar-shaped, dormant cell to a flat, motile, proliferative cell.
Serum growth requirement. Cells require the presence of a serum growth factor in order to proliferate.	Serum stimulation. Serum produced by hemorrhage and clot formation contains factors that stimulate fibroblast growth during wound repair.

corresponding counterparts in vivo, which occur during and after the traumatic introduction of a smooth-surfaced foreign body into the tissues.

In summary, we have pointed out that of the many properties of cells in vitro now being revealed and characterized by the investigators at this conference, many significant insights could result if more attention were paid to relating in vitro cell behavior to the physiological roles and functions of cells and tissues in vivo.

REFERENCES

1. Holley RW: Nature 258:487, 1975.
2. Holley RW, Armour R, Baldwin JH: Proc Natl Acad Sci USA 75:1864, 1978.
3. O'Neill CH, Riddle PN, Jordan PW: Cell 16:909, 1979.
4. Abercrombie M, Heaysman JEM: Exp Cell Res 94:293, 1954.
5. Folkman J, Moscona A: Nature 273:345, 1978.
6. Tucker RW, Pardee AB, Folkman J: J Supramol Struct, Suppl 4:150, 1980.
7. Folkman J, Tucker RW: Dev Biol (in press).
8. Rosen P, Misfeldt DS: J Supramol Struct, Suppl 4:209, 1980.
9. Stoker M, Piggott D: Cell 3:207, 1974.
10. Robinson CC, Lehman JM: J Supramol Struct, Suppl 4:208, 1980.
11. Scher CD, Pledger WJ, Martin P, Antoniades HN, Stiles CD: J Cell Physiol 97:371, 1978.
12. Brand KG: In Becker FF (ed): "Cancer," Vol 1. New York: Plenum Press, 1975, pp 485–505.
13. Boone CW, Takeichi N, Eaton SA, Paranjpe M: Science 204:177, 1979.

Random Transitions and Cell Cycle Control

Robert F. Brooks

Cell Proliferation Laboratory, Imperial Cancer Research Fund Laboratories, London WC2A 3PX, England

Differences between the cycle times of sister cells are exponentially distributed, which means that these differences can be explained entirely by the existence of a single critical step in the cell cycle which occurs at random. Cycle times as a whole are not exponentially distributed, indicating an additional source of variation in the cell cycle. It follows that this additional variation must affect sister cells identically; ie, sister cell cycle times are correlated. This correlation and the overall distribution of cycle times can be predicted quantitatively by a model that was developed initially in order to explain certain problematic features of the response of quiescent cells to mitogenic stimulation — in particular, the significance of the lag that almost invariably occurs between stimulation and the onset of DNA synthesis. This model proposes that each cell cycle depends not on one but *two* random transitions, one of which (at reasonably high growth rates) occurs in the mother cell, its effects being inherited equally by the two daughter cells. The fundamental timing element in the cell cycle is proposed to be a lengthy process, called L, which accounts for most of the lag on mitogenic stimulation and also for the minimum cycle time in growing cultures. One of the random transitions is concerned with the initiation of L, whereas the other becomes possible on completion of L. The latter transition has two consequences: the first is the initiation of a sequence of events which includes S, G_2 and M; the second is the restoration of the state from which L may be initiated once more. As a result, L may begin (at random) at any stage of the conventional cycle, ie, S, G_2, M, or G_1.

There are marked similarities between the hypothetical process L and the biogenesis of mitotic centres — the structures responsible for organising the spindle poles.

Key words: transition probability, cell proliferation, mitogenesis, mitotic centres

The duration of G_1 varies not only between populations of like cells growing at different rates but also between individual cells within the same population growing in exponential steady state [1]. Indeed, essentially all of the variation in generation times can be attributed to the variation in G_1 times [2]. Furthermore, this variability — though widely considered to be no more than a nuisance — is in some way connected with the regulation of proliferation since alterations in proliferation rate are often brought about by changing the degree of variability, the minimum generation time changing very little [3].

In attempting to account for the variability of G_1 and at the same time explain how the cell cycle could be regulated, Smith and Martin [1, 4] were led to the idea that G_1 might contain a single, critical event (or transition), which took place at random. Regulation

Received and accepted May 20, 1980.

© 1981 Alan R. Liss, Inc., 150 Fifth Avenue, New York, NY 10011

of proliferation rate, they suggested, could be brought about by changing the probability of this transition in response to alteration of the environmental conditions.

Since the transition probability model was first formulated, a good deal of evidence has accumulated in its favour. The most rigorous is the distribution of differences between the cycle times of sister cells. This distribution (known as the β curve) is perfectly exponential, so far as can be judged, and the only simple way in which this could arise is if the different division times of sister cells were almost entirely the result of a single random transition [3, 5–7].

A Problem for the Transition Probability Theory

If the random transition were the sole cause of variation in cycle times, then the distribution of these (the α curve) should also be exponential. It is not, which implies an additional source of variation. However, as already mentioned, the differences between the cycle times of sister cells can be attributed almost entirely to one random transition. The additional variation must therefore affect sister cell pairs identically. This predicts that the generation times of sister cells should be correlated, which is indeed the case [5]. Nevertheless, the existence of another source of variation that affects pairs of sister cells identically is a puzzling conclusion which the transition probability theory does not address. An explanation of it has recently been proposed, though before proceeding it is necessary to discuss the response of quiescent cells to mitogens.

Mitogenic Stimulation

One of the most puzzling aspects of mitogenic stimulation is the significance of the relationship between mitogen dose and the number of cells caused to initiate DNA synthesis [eg, 8]. Why is it that only a fraction of a quiescent population appears to initiate DNA synthesis on stimulation with a particular concentration of serum, say, when all of the cells in a cloned population would be expected to bind growth factors to similar extent? Does this imply heterogeneity in the population despite cloning? This seems unlikely since the cells that respond to a low serum concentration do not give rise to progeny which continue to proliferate rapidly in low serum. Rather, their descendents have average generation times as long as the rest of the population (Brooks, unpublished observations). An alternative explanation was suggested on publication of the transition probability model, namely, that the concentration of serum determined the *probability* of initiating the cell cycle. If true, then the problem of commitment to initiate the cell cycle became a question of kinetics and, accordingly, a kinetic study of the response to serum was undertaken.

The results were rather surprising [9]. On adding back serum to serum-starved cultures, no cells entered S-phase until after a long lag. Thereafter, the cells left G_1 with approximately exponential (ie, first-order) kinetics. The apparent rate constant was a function of the serum concentration, though the duration of the initial lag was not.

Superficially, the approximately first-order kinetics and the serum dependence of the rate constant might seem to support the concept of a random transition in G_1. However, the very fact that first-order kinetics were obtained at all implied that the transition probability must have increased very abruptly. This could not have been anticipated. As a result, although the experiments provided qualitative support for the transition probability model, the experiments also raised other questions that could not be answered within its simple framework.

594:CCDD:A

The difficulty was to account for the abrupt increase in the rate of entry into S-phase at the end of a very long lag. The abruptness could be most readily explained if the random transition were located many hours before S-phase and if the addition of serum caused a more or less immediate increase in the transition probability. However, other observations, notably the shortness of G_1 in rapidly proliferating cells and the relatively rapid decline in the rate of initiating S-phase following a serum step-down, argued that the transition must be rather close to S-phase [10]. If so, and if there really were only one important random transition, then this raised the question of what sort of processes could be taking place following serum addition that would produce an abrupt increase in the transition probability at the *end* of a lag, the duration of which was independent of the intensity of stimulation (ie, serum concentration). Conventional explanations, such as the time required for some substance to reach a critical threshold, or a sequence of biochemical steps, could easily be ruled out [11]: the former does not account for the serum-independence of the lag, whereas the latter is incompatible with the apparently abrupt change in the transition probability. The second follows because the transition probability is not a discrete step but is a parameter that is continuous in time. As such it must reflect some steady-state activity of the cell. This in turn means that any intervening steps between serum factor binding and the mechanism controlling the transition probability must also be continuous, ongoing processes. However, the more steps there are (and many would be required to explain the length of the lag), the less abrupt would be the eventual change from one steady state to another.

A Two-Transition Model of the Cell Cycle

The key to the problem was the realization that the duration of the lag in a variety of cell types was rather similar to the minimum interdivision time found in rapidly growing cultures of the same type. It seemed worth considering the possibility that both intervals might be determined by the same underlying mechanism. Pursuing this, J. Smith, D. Bennett, and I arrived at a simple model for the lag which accounts qualitatively for the response of cells to manipulations of serum concentration, and which could be generalized to rapidly proliferating cells. Applied to steady-state cultures, the model predicts quantitatively the distribution of cycle times, and also the correlation between sibling generation times and the exponential distribution of differences between sibling generation times (the β curve). These ideas are developed in detail elsewhere [11, 12], so only a brief account is given here.

The model is outlined in Figure 1. The essential postulates are that the cell cycle contains not one but two random transitions which are linked by a lengthy process, called L, which determines most of the lag and is also equivalent to the minimum cycle time. Quiescent cells are considered to be located in a new indeterminate state, called Q, which they may leave at random by initiating process L. The probability of leaving Q increases quickly following mitogenic stimulation leading to the (asynchronous) initiation of L. This is the point at which Figure 1 begins. On completion of L, the cell enters another indeterminate state, called A (the same as in the Smith and Martin model [4]), which they also leave at random. The consequences of leaving A state are twofold: first, events leading to DNA synthesis and mitosis are initiated and second, the cell reenters Q. If the probability of leaving Q remains reasonably high (as in cells exposed continuously to mitogens), then in most cells L will be reinitiated before mitosis. Because the process L is considered to take longer than the duration of $S + G_2 + M$, it will nevertheless be completed after cell division in the G_1 phase of the subsequent generation. Accordingly, the two daughter cells

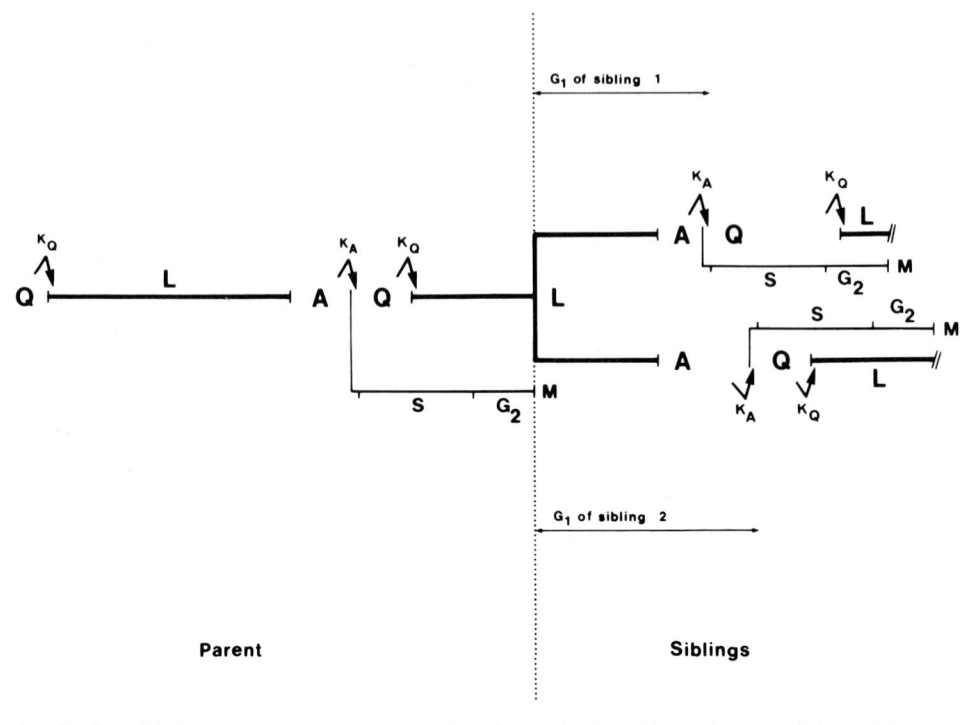

Fig. 1. A model for the mammalian cell cycle. The diagram begins with a quiescent cell located in an indeterminate state, called Q, just after the probability of leaving Q has been raised by mitogenic stimulation. On leaving Q, which occurs at random (rate constant k_Q), the cell embarks upon a lengthy process called L. This it must complete in order to reach another indeterminate state, called A. Exit from A state is also random (rate constant k_A) and leads both to reentry into Q and the initiation of a sequence of events that include the S, G_2, and M phases of the classical cell cycle. If k_Q is sufficiently high, then most cells will reinitiate L before mitosis. However, completion of L always takes place in the G_1 phase of the daughter cells. When L is initiated in the mother cell, the daughter cells reach the next A state at the same time, leading to a correlation between sibling generation times. (Random transitions are indicated by bent arrows.)

reach the next A state at the same time. Differences in the time of initiating S-phase (or mitosis, since the duration of S + G_2 is practically invariant [2]) will then depend only on the random exit from A state, which explains the exponential β curves. Different pairs of sister cells reach A state at different times, however, depending on when L was initiated in the mother cell. Hence, at reasonably high proliferation rates, the cycle times of sister cells are correlated. As the probability of leaving Q decreases, however, and proliferation rate falls, an increasing fraction of the population will fail to initiate L before division. The progeny of such cells (which make up the bulk of a so-called quiescent population) must then initiate L independently and the sibling correlation is lost. Preliminary experiments indicate that the sibling correlation does indeed disappear when cells become quiescent (Brooks, unpublished observations).

The minimum cycle time is given by a cell that spends zero time in A state and whose mother spent zero time in Q state. Thus the duration of L, which accounts for most of the lag on serum stimulation, is also equivalent to the minimum cycle time.

The response of cells to manipulations of serum concentration is explained if both transition probabilities are sensitive to serum factors. However, to explain why cells accumulate predominantly in Q state after serum deprivation, it is necessary to make the additional postulate that the probability of leaving Q falls much lower than that for leaving A state, or that the process L is to some extent "reversible" under adverse conditions, redepositing cells in Q.

Relation Between Q State and G_0

The two-transition model, then, postulates that quiescent cells exist in an indeterminate state called Q and the question arises as to why this term was preferred to the older one of G_0 [13, 14]. The latter is usually taken to signify a resting state outside of the normal cell cycle, in which cells are not engaged in activities relating to cell division. A quiescent cell in Q state would comply with this description, though the two-transition model proposes that Q is not *confined* to quiescent cells but is an integral part of all mammalian cell cycles. A quiescent cell in Q state differs from a "rapidly cycling" cell in Q state only in the probability of leaving Q, and manifest biochemical differences (eg, rates of protein synthesis) would reflect this rather than signal distinct cell cycle phases. In all instances, Q state begins at roughly the same time that DNA synthesis starts. With the onset of quiescence, the cell merely remains in Q for the rest of its cycle, its daughters being born in Q. In this way, Q may overlap with any stage of the classical cycle, ie, S, G_2, M, or G_1. Since it is not normally regarded as possible for a cell to be in G_0 at the same time as it is engaged in DNA replication or mitosis, it is evidently desirable that the two concepts be kept entirely separate.

Quantitative Predictions

Qualitatively, the two-transition model explains many seemingly unrelated features of cell proliferation. The model also gives good quantitative agreement with the cycle kinetics of cells growing in steady state.

We have shown elsewhere [12] that it is possible to calculate the values of the two-transition probabilities (or rate constants) in several independent ways. An equation for the distribution of cycle times has been derived and, using the calculated values of the rate constants, the expected distribution may be obtained. The agreement with the distribution actually obtained is extraordinarily good [12].

The most objective method of estimating the values of the rate constants is to make use of the relationship between the total variance in generation times and the observed correlation coefficient for sister cell cycle times [12]. However, correlation coefficients are particularly sensitive to sampling errors when the sample size is small. Also, both they and estimates of variance are likely to be meaningful only if both cells of *all* sister pairs are seen to divide — a situation only rarely encountered in time-lapse films because cells leave the field or fail to divide before the end of the film. The problem arises because incomplete observations tend to be biased toward cells with long cycle times. However, incomplete generation times can (and must) be included in the compilation of α and β curves up until the cell is lost from view. Procedures for doing this have been described previously [15]. Such curves may then be used to obtain graphical estimates of the two rate constants. This has been done for the experiment shown in Figure 2.

The data, which are for 3T3 cells, include three sister cell pairs for which one sister failed to divide during the period of observation. One of these pairs nevertheless gave the largest recorded difference in sibling generation times in the experiment. The cumulative frequency distributions of sister differences (β curve) and cycle times (α curve) are shown as step diagrams plotted on a linear ordinate (Fig. 2). The smooth line drawn through the β curve is a perfect exponential with the rate constant (k_A) determined by linear regression of log β_t against time (inset to Fig. 2). The line drawn through the first 85% of the α curve was chosen as a reasonable description of the data [see 12] and was used to calculate the frequency distribution of Smith and Martin's B phase (the time between A states) as described previously [12] from the relationship:

$$B_t = \alpha_t - (\dot{\alpha}_t / k_A)$$

where $\dot{\alpha}_t$ is the slope of the α curve at time t. According to the two-transition model, the distribution of B phase should be exponential (rate constant k_Q) plus a constant, T_L (the

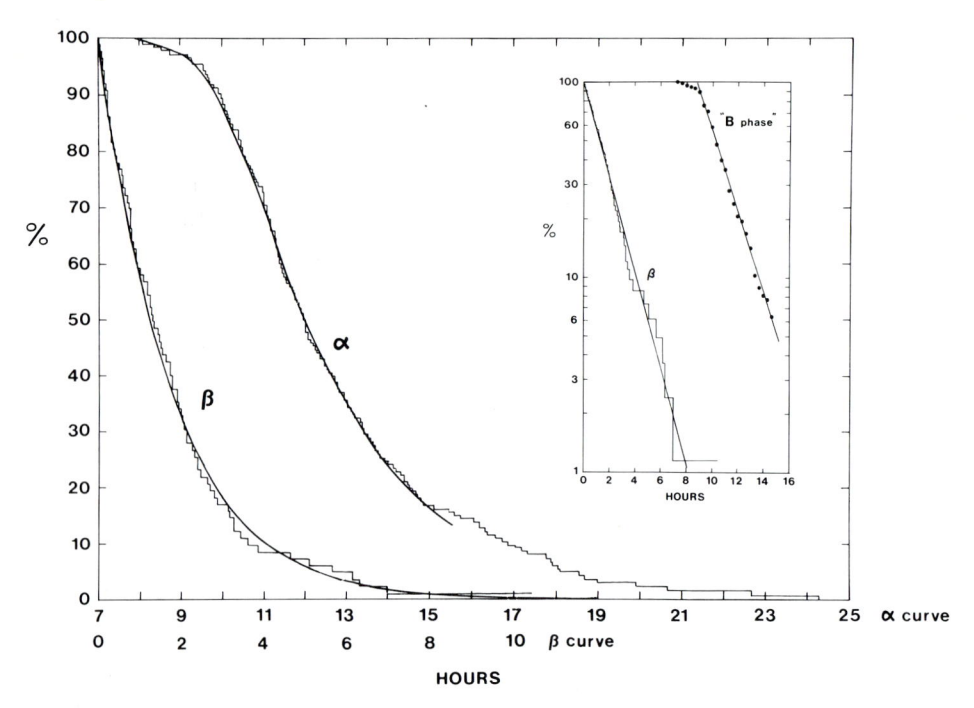

Fig. 2. The main figure shows α and β curves obtained from a time-lapse film of 3T3/4A/C5 cells [21] growing in Eagle's medium (Dulbecco's modification) supplemented with 15% calf serum, 25 μM inosine, and 3 μM thymidine. Proliferation rate was judged to be exponential from the lack of correlation between birth time and cycle time (r = −0.12, which was not significant at the 1% level). The α curve refers to the fraction of cells remaining undivided as a function of cell age. The sample size was 172. The β curve refers to the fraction of sister pairs whose generation times differ by an amount greater than or equal to the times given on the abscissa. The sample size was 86. Data points are represented by steps in the histogram.

The line drawn through the β curve is a perfect exponential distribution given by $\beta_t = e^{-k_A t}$ with the value of $k_A = 0.564$ hr^{-1} determined by linear regression of log β_t against t (inset). The corresponding correlation coefficient (r = −0.992) indicates that the data are well described by an exponential.

The line drawn through the α curve was chosen as a reasonable description of the data. This line was used to estimate the distribution of Smith and Martin's B phase [see 12], which is shown in the inset. Linear regression of log B_t against t for points > 9 hr yielded a line whose slope gave $k_Q = 0.480$ hr^{-1} (r = −0.998). The intercept gave $T_L = 8.8$ hr.

598:CCDD:A

duration of L). As shown in the inset (Fig. 2), this appears to be a good approximation, apart from a small amount of initial curvature. The value of k_Q was then estimated by linear regression of log B_t against t, ignoring the initial curvature (points earlier than 9 hours). The value of T_L was taken as the intercept of the regression line, ie, the value of t which made $B_t = 100\%$. The graphical estimates of k_A, k_Q, and T_L were then substituted into the equation for the α curve [12] which is:

$$\alpha_t = \frac{1}{k_A - k_Q} [k_A e^{-k_Q(t-T_L)} - k_Q e^{-k_A(t-T_L)}]$$

This generated the curve plotted in Figure 3 (linear ordinate) or the inset to Figure 3 (logarithmic ordinate). As can be seen, apart from small deviations at the top consistent with slight variability in T_L (evident only in the linear plot), the curve describes the experimental data rather well.

The correlation coefficient for sibling generation times is given by:

$$r_{ss} = \frac{1/k_Q^2}{1/k_Q^2 + 1/k_A^2} \qquad \text{(see [12])}.$$

With the graphical estimates of k_Q and k_A, this yields a predicted correlation of 0.58. The true correlation for the sample cannot be determined because the largest sister difference is given by a pair that contains an incomplete generation time. However, if it is assumed

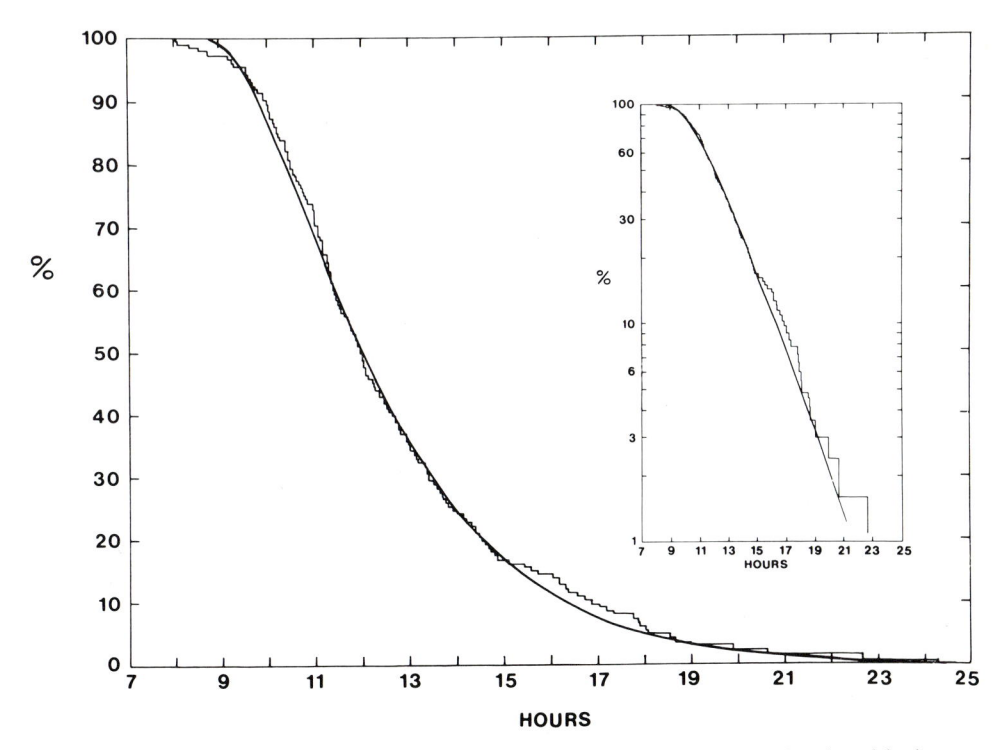

Fig. 3. The α curve from Figure 2 is shown replotted on either linear (main figure) or logarithmic ordinates (inset). The curve drawn through the data was calculated assuming two random transitions and invariant T_L as described in the text, using the values of k_A, k_Q, and T_L given in the legend to Figure 2.

that this cell divided in the very next frame after being lost, a maximum estimate of the correlation can be made. This was 0.61, which compares favourably with the predicted value of 0.58. Once again, we conclude that the two-transition model provides a good quantitative description of mammalian cell cycle kinetics.

Biological Counterpart of the Two-Transition Model

One of the most attractive aspects of the two-transition model is that a possible biological counterpart more or less suggests itself. For a variety of reasons, the properties of the process L are most readily understood in terms of the assembly of a complex structure [11, 12]. If so, then because L is frequently initiated in one cell cycle and completed in the next, it is evident that the hypothetical structure must duplicate during the cycle and identical copies must be segregated to the daughter cells. This, in turn, recalls the behaviour of the "mitotic centres" — the structures responsible for organising the spindle poles.

Working with sea urchin eggs, Mazia et al [16] concluded that at the time of mitosis, each spindle pole was a duplex structure consisting of a mature "centre" together with an immature "daughter" joined to it. Maturation in the sense of gaining functional independence was completed shortly after mitosis and involved separation of the mother and daughter centres. Shortly after this, the next duplication began, the new daughter centres reaching maturity in turn after the next mitosis. Thus, like our process L, the maturation of a mitotic centre is lengthy (equivalent to almost an entire cell cycle time) and spans cell division. Furthermore, the initiation of new centres and the separation of mother and daughter centres stand out as potential candidates for the two random transitions.

In metazoan cells, the mitotic centre represents the centriolar region of the cell, ie, the centrioles plus associated pericentriolar material. Centriole duplication, in mammalian cells, is known to occur close in time to the initiation of S-phase [17]. In budding yeast, the mitotic centre corresponds to a plaque of electron-dense material on the nuclear envelope. Interestingly, in this organism, genetic and ultrastructural studies have already implicated plaque duplication in the control of the cell cycle [18–20]. Whether a causal connection exists in metazoans between the biogenesis of mitotic centres and the control of the cell cycle remains to be established, as does the applicability of the two-transition model to yeasts. For the present, the remarkable parallels with a model proposed on purely kinetic grounds suggest that a closer inspection of mitotic centres is warranted.

ACKNOWLEDGMENTS

It is a pleasure to acknowledge that many of the ideas discussed here had origin in the innumerable conversations that took place with Jim Smith and Dot Bennett during the past few years. Without them, the analysis presented here would not have been possible. I also wish to express my gratitude to Peter Riddle and his assistants for making the time-lapse film and patiently extracting the data from it, and to Judy Marsden for help in analysing the data.

REFERENCES

1. Smith JA, Martin L: In Padilla GM, Cameron IL, Zimmerman A (eds): "Cell Cycle Controls." New York: Academic Press, 1974, pp 43–60.
2. Siskin JE, Morasca L: J Cell Biol 25(2):179, 1965.
3. Shields R, Smith JA: J Cell Physiol 91:345, 1977.

4. Smith JA, Martin L: Proc Natl Acad Sci USA 70:1263, 1973.
5. Minor PD, Smith JA: Nature 248:241, 1974.
6. Shields R: Nature 273:755, 1978.
7. Shields R: In Jimenez de Asua L, Levi-Montalcini R, Shields R, Iacobelli S (eds): "Control Mechanisms of Animal Cell Growth." New York: Raven Press, 1980, p 157.
8. Clarke GD, Stoker MGP, Ludlow A, Thornton M: Nature 227:798, 1970.
9. Brooks RF: J Cell Physiol 86:369, 1975.
10. Brooks RF: Nature 260:248, 1976.
11. Brooks RF: In John P (ed): "Society Experimental Biology Seminar Series." Cambridge, England: Cambridge University Press (in press).
12. Brooks RF, Bennett DC, Smith JA: Cell 19:493, 1980.
13. Quastler H: In Lamerton LF, Fry RJM (eds): "Cell Proliferation." Oxford: Blackwell Scientific Publications, 1963, p 18.
14. Lajtha LG: J Cell Comp Physiol 62, supplement 1:143, 1963.
15. Smith JA: Cell Biol Int Rep 1:283, 1977.
16. Mazia D, Harris PJ, Bibring T: J Biophys Biochem Cytol 7:1, 1960.
17. Robbins E, Jentzsch G, Micali A: J Cell Biol 36:329, 1968.
18. Byers B, Goetsch L: Cold Spring Harbor Symp Quant Biol 38:123, 1974.
19. Byers B, Goetsch L: J Bacteriol 124:511, 1975.
20. Hartwell LH: Bacteriol Rev 38:164, 1974.
21. Brooks RF: Cell 12:311, 1977.

Control of the Cell Cycle: Workshop Report

R. Baserga, R.F. Brooks, and A.B. Pardee

Temple University, Philadelphia, Pennsylvania 19140 (R.B.), Imperial Cancer Research Fund Laboratories, London WC2A 3PX, England (R.F.B.), and Sidney Farber Cancer Institute, Boston, Massachusetts 02115 (A.B.P.)

The first half of this workshop was chaired by R. Baserga. The Chairman made a brief introduction dealing with temperature-sensitive (ts) mutants of the cell cycle, defined as temperature-sensitive mutants of growth that arrest at the nonpermissive temperature in a specific phase of the cell cycle. These ts mutants of the cell cycle, although they have obtained a considerable amount of recognition in yeast, have for some reason, been somewhat neglected in mammalian cells. This is unfortunate since ts mutants clearly provide the evidence of the existence of specific genes whose expression is necessary for cell proliferation. On the other side, these mutants also opened the possibility of cloning these same genes. Cloning of genes responsible for certain precise steps in the cell cycle would result in the identification of such genes, as well as the possibility of studying, by molecular probes, the way in which they are expressed.

G. Jonak, Temple University, then showed evidence that the A gene of the small DNA oncogenic virus, SV40, when microinjected into G_1 ts mutants made quiescent by serum deprivation, caused these cells to enter S phase, even at the nonpermissive temperature. The studies were carried out by the technique of manual microinjection, in which the SV40 A gene, or fragments of it, were directly microinjected into the nuclei of quiescent cells. The cells were then incubated at the nonpermissive temperature, and although, under these conditions, they do not enter S phase when stimulated by serum, they did so when microinjected with SV40 A gene. These results are of considerable theoretical importance because they indicate that the large T-antigen of SV40 can induce cellular DNA replication in the absence of certain G_1 functions that are required by serum stimulated cells.

Gretchen Stein, University of Colorado, then showed that quiescent human diploid cells, when fused with other untransformed cells, inhibit the entry of the heterokaryons into S phase. On the contrary, SV40 transformed cells are dominant and induce DNA synthesis in quiscent human diploid cells. The conclusions of Gretchen Stein are that these experiments show that quiescent human diploid cells, especially senescent human diploid cells, contain an inhibitor that is dominant, except in the case of SV40 transformed cells, where the product of the SV40 genome becomes dominant over the cellular inhibitor.

R. Schlegel, Pennsylvania State University, showed evidence that nuclei of cells in S phase contain information for the reactivation of quiescent 3T3 nuclei in heterokaryon formation. The entry into S of the 3T3 nuclei is delayed, suggesting an effect on the trigger mechanism. The results open the question that a number of genes in the recipient cells have to be reactivated, even when the S phase cells contain all the necessary information for the transition from the resting stage to the growing stage.

Received and accepted April 14, 1980.

© 1981 Alan R. Liss, Inc., 150 Fifth Avenue, New York, NY 10011

R. Baserga then concluded by reminding the audience that although we are all fascinated by cell DNA replication, we should not forget that for a cell to divide, it also has to double its size — ie, double the amounts of RNA, proteins, and other substances. It is indeed possible that the two signals increasing cell size and cell DNA replication may be dissociated in mammalian cells.

The second half of the workshop dealt with the contentious issue of cell cycle models. The discussion was opened by R. Brooks (I.C.R.F., London) with an introduction to the revised version of the transition probability model (see Brooks, page 593). This claims to explain why there should be a lag, similar in duration to the minimum cycle time of proliferating cells, between the addition of a mitogenic stimulus to quiescent cultures and the subsequent start of DNA synthesis. The model also predicts quantitatively the distribution of cycle times in growing cultures and the correlation between the generation times of sibling cells. While Brooks did not wish to claim that this was the only model that could do all this, he did feel that it was the simplest and most comprehensive produced to date.

In the transition probability model, particular importance is attached to the observation that the *differences* between the cycle times of sister cells are exponentially distributed (the so-called β curve). L. Castor (Institute for Cancer Research, Fox Chase, Philadelphia) questioned the exponentiality of some of the published data on which this assertion is based, as well as the claim that the only possible explanation of such distributions is a single random transition. He showed that a model in which G_1 *rates* are normally distributed and coupled to a normal distribution of the duration of the rest of the cell cycle could also give β curves indistinguishable from those actually obtained, and could account for the overall distribution of cycle times and the correlation between sister cell cycle times. However, this required six parameters with values chosen by computer simulation in order to fit the data, compared to three parameters in the revised transition probability model, each of which is fixed by the measured values of the sister/sister correlation coefficient and the mean and variance of generation times.

A. Koch (University of Indiana) raised the question of how cells maintained their average size in the face of variable generation times. He argued that there must be a mechanism for preventing the variation in cell size from increasing indefinitely. Indeed, the coefficient of variation of size at division is roughly half that of age at division and this, he felt, was an indication that cell size was a more important determinant of cycle time than cell age. In the subsequent discussion, Brooks argued that this was also a property of cell cycle models which denied any relationship between the timing of cell division and cell size or age provided that cell growth during the cycle was not autocatalytic (ie, exponential). In such models, this follows because cell age at division is determined anew in each cell cycle, whereas cell size also depends on how big the cell was at the end of the preceding cycle. Koch conceded that this was the case for linear growth during the cycle, but doubted whether it was true for more complex growth patterns. He also felt, given available data, that exponential increase in mass was the most likely growth pattern — at least for bacteria.

Finally, S. Cooper (University of Michigan, Ann Arbor), in an impassioned and entertaining oration, argued that there was no such thing as a G_1-specific event. Rather, he believes that G_1 should be regarded as part of a larger period which began with the previous initiation of DNA replication and which is concerned with the time required to accumulate an "initiator" substance to a necessary threshold. In his view, whether one observes a G_1 period or not depends only on the rate of protein synthesis.

Subject Index

ACTH, 3, 4
adipocytes, 429-30, 524-26, 533-34, 536
aging, 54-61, 84
amino acid transport, 430-32
anchorage dependence, 589-92
animal cell proliferation, 1-6
antibody, polyclonal, 131, 133-39
anti-insulin-receptor antibody, 547
arachidonic acid, 89-96
arrest test, 496-97
atherosclerosis, 505
autocrine secretion, 457-458
autoradiography, 286, 327
availability test, 498-99

B cells
 developing, 463-67
 regulation of, 211-17
 stimulation of, 579-87
 see also lymphocyte(s)
biochemistry, 148-49
biotinyl epidermal growth factor, 287, 292, 294, 296
B-lymphoblastoid cell lines, 50
bone marrow cultures, 275-77
bovine vascular endothelial cells, 72, 74-75

cAMP, 435, 558
 and cell growth inhibition, 4
 and hormone-receptor complex, 195
 and lymphocyte triggering, 512
carcinogenesis, 450
carrier protein, 395-409
cartilage, 226
cartilage factors, 154-65
casein synthesis, 435
cell, effect of PDGF on, 504
cell-cell contact, 21-26

cell culture
 growth of, 495
 insulin-like, 103-4, 107
cell cycle
 and animal cell growth, 5
 clock, 246-48
 control, 496, 593-600, 603-4
 and PDGF, 141-50
cell density and
 bovine vascular endothelial cells, 72, 74-75
 extracellular matrix, 82
 growth factors, 23-24, 26
cell division, 121-22
cell growth
 factors controlling, 1-3, 495-501
 and serum spreading factor, 167-82
 see also growth control
cell inhibition and
 cartilage factors, 154
 heparin, 154
cell morphology, 167-82
cell shape and
 animal cell growth, 1, 3, 4
 growth factors, 77
cell spreading, 353-63
cell substratum, 225-39
cell surface
 and animal cell growth, 1, 3, 4
 and differentiation of vascular endothelial cell, 61, 76
 and lectins, 34, 36
 and polypeptide growth factor, 203
 proteins, 373-84
 receptors, 543-47
 receptors, and growth factors, 3-4
 receptors, for viral glycoproteins, 255-64
cellular clock, 241-54
cellular regulation, 50
chicken
 chondrocytes, 103-8

605

embryo skeletal muscle
  formation, 226-39
 oocytes, 315-22
 tissue, 29-36
cGMP, 512
choleragen receptors, 529
chondrocytes, 100-108, 434-35
chromatography, 111, 117,
  316-20
clustering, 550
coated pits, 549-50, 552, 553
coated vesicles, 315, 549-50, 553
collagen, 69, 74, 76, 84
colony-stimulating factor
 and erythropoietin, 10-18,
  492
 in granulopoiesis, 265-80
 and hemopoiesis, 474-84
competence, 142-43, 148-50, 194
covalent epidermal growth factor
  receptor complex, 292, 296-97
creatine kinase, 227-31, 238
C6 rat glioma cell, 172-74
cycloheximide, 498-99, 538-39

density inhibition, 589-92
density-shift method, 531, 533,
  534
dexamethasone, 531, 534, 537
  540
diabetic mother, 439
diffferentiation
 and extracellular matrix, 53
 of hemopoietic cell, 473-85
 of preadipocytes, 526-29
 red cell, 487-93
 skeletal muscle, 226-39
 of vascular endothelial cell,
  61-66
diffusion, 275
division delay, 248-50
DNA, content of, 496, 497
DNA, and erythropoietin, 490-
  91
DNA, and leukemia, 485
DNA, and program of senescence, 58
DNA replication, 603-4
DNA synthesis
 and animal cell growth, 4-6
 and cell spreading, 77

 and chondrocyte metabolism,
  103, 104
 and colony-stimulating factor,
  271
 and density inhibition, 589-
  90
 and epidermal growth factor,
  219-20, 222, 546
 and growth factors, 141-50,
  507-9
 and lymphocytes, 512
 and PDGF, 504, 505
 and plasma factors, 425
 rate of, 242, 252
 in 3T3 cells, 23-24
DNA virus, 446
down regulation, 204, 304, 524,
  553

electrophoresis and
 epidermal growth factor, 286
 insulin receptors, 327
 interleukin 1, 111, 117
 thrombin binding, 374, 376
endocrine system. See hormones
endocytosis, 204, 207, 549
enzymes, 525-26
epidermal growth factor
 and animal cell growth, 1,
  3-4
 and cell growth, 496, 499-
  501
 cell surface receptors, 255-64
 clustering, 550-53
 -competing activity, 455-56
 effect of differentiation on,
  529
 linkage of, to receptor, 283-
  301
 measurements, 543-47
 mechanism of action of, 206-
  7
 modulation of binding of,
  219-23
 persistence of, 546-47
 and protein kinase, 557-67
 relation of, to PDGF, 505
 and sarcomas, 446-48, 450
epidermal growth factor receptors, 455-56
 -complex binding, 165

proteolysis in, 303-12, 339-50
purification of, 563-66
erythropoiesis, 9
erythropoietic-enhancing activity,
    9-18
erythropoietic-stimulating factor,
    387-92
erythropoietin, 481-93
    mode of action, 490-92
    properties of, 487-90
    release of, from macrophages,
        387-94
    response of, 492-93
extracellular materials, 31
extracellular matrix and
    cell proliferation, 53-85
    skeletal muscle differen-
        tiation, 225-39

fatty acids, 525-26
Fc fragments, 131-39, 211-17
fetal growth, 435-39
fibroblast growth factor, 6
    and animal cell growth, 1
    in brain, 197
    and cell growth, 496
    and DNA synthesis, 425
    and extracellular matrix, 53,
        76, 83-84
    and growth, 194
    relation of, to PDGF, 505
    and vascular endothelial cell
        differentiation, 61-66
    and vascular endothelial cell
        proliferation, 54, 56-58
    and vascular smooth muscle
        growth, 81-82
fibroblasts
    amino acid transport in, 430-
        33
    cell cultures of, 219-23
    glucose incorporation in,
        430-32
    and multiplication-stimulating
        activity, 414, 423, 425-26
    neoplastic 353-63
    nonneoplastic, 353
    and thrombin binding, 374
    thrombin receptors, 119-29
fibronectin and
    cell proliferation, 53, 61-66,
        69, 74, 76

thrombin binding, 376-77,
    379-83
15-HPAA, 87, 89-96
fluorescent labeling, 121-29

glucose, 429-32
glycoproteins, 255-64, 382-83
glycosaminoglycans, 234, 238
Golgi cells, 549-50
granulocyte-macrophage system,
    474-84
granulopoiesis, 265-80
growth control
    models, 4-6
    and pinocytosis in 3T3 cells,
        21-26
    see also cell growth
growth factors
    additional polypeptide, 2, 3
    and animal cell growth, 1-6
    and cell cycle events, 141-50
    and fetal growth, 435-39
    human T-cell, 39-50
    interaction between, 2
    and pinocytosis in, 3T3 cells,
        21-26
    T cell, 39-50, 582
    see also specific growth
        factors
growth inhibitors
    and animal cell growth, 1, 4
    purification of, 3

hemopoiesis
    control of cells in, 473-85
    and hemopoietic progenitors,
        367-68
    and phytohemagglutinin,
        365-371
heparin, 154, 157-58, 164
H4 hepatoma cells, 325-36
hollow fiber diafiltration, 111-13,
    116
hormones
    and cell growth, 457
    and growth, 193
    and internalization, 204
    mechanism of action, 202
    and polypeptide growth
        factor, 195-97
    secondary, 193-207
    see also specific hormones

608 Subject Index

HPLC, 102, 105
HTC rat hepatoma cells, 433-34
H-2T locus, 573-74, 576-77
human cell line GTC, 9-18
humoral regulation factor, 478-82
hyaluronic acid, 226, 227-39
hydroperoxide, 87-96
hypercholesteremia, 505

immunogenetics, 461-62
immunoglobulin, 131-39, 211-17
indomethacin, 90
infants, large, 439
inhibition, lymphocyte activation, 91, 93-96
injury, 505
insulin, 199, 507
   activity, 439-40
   and cell growth, 496, 498
   and chondrocyte metabolism, 99, 107
   relation of multiplication-stimulating activity to, 417-22
   role of, in fetal growth, 435-39
   structure, 199, 201-2
insulin receptors
   and multiplication-stimulating activity receptor, 422-35
   photoaffinity labeling of, 325-36
   synthesis, 523-41
insulin-like growth factor, 197-99, 201-2, 507, 544
   activities, 439-40
   multiplication-stimulating activity in, 411, 422-23
   role of, in fetal growth, 435-39
interferon, 3
interleukin 1, 110
interleukin 2, 132, 138, 514-22
internalization, 553
isoelectric focusing and
   interleukin 1, 111, 113, 114, 116
   leukemia, 365-71
   T cell, 517, 520
intracellular protein transport, 549-55

lectin binding sites, 29-36
leprechaunism, 436-39
leukemia, 365-71, 482-85
lipogenesis, 524, 525-26
lipopolysaccharides, 580-82, 584
lipoprotein, 580-81, 582; *see also* low density lipoprotein
liver. *See* H4 hepatoma cells
low density lipoprotein, 26, 504, 549, 555
lymphocyte-activating factor. *See* interleukin 1
lymphocyte(s)
   activation, 87-96
   development, 579-87
   maturation, 461-70
   neoplasm, 576-70
   and proteases, 153
   triggering, 511-12
   see also B cells; T cells
lymphokines, 514, 518, 521-22
lympholysis, 573-77

macrophage, 216, 217
   -granulocyte system, 474-84
   and polyclonal antibody formation, 138
   release of erythropoietin from, 387-94
malignant cells, 456, 457-58
meiosis, 435
membrane
   phosphorylation, 559-63
   plasma, 335-36, 546-47
   proteases, 163-65
methylisobutylxanthine, 531, 534, 537, 540
mitogenesis, 594-95
mitosis, 600
models, 4-6
mother, diabetic, 439
mouse
   embryo cells, 124-29
   sarcoma virus, 446-49, 451, 454, 456-57
   type C virus, 255-64
multiple receptors, 543-46
multiplication-stimulating activity, 395-409, 411-40, 507
   biological actions of, 435
   as fetal growth factor, 436
   properties of, 440

relation of, to somatomedins, 419–22
relation of, to insulin, 411–35
multivesicular bodies, 549, 550
murine. *See* mouse
myelin basic protein, 197
myeloid leukemia, 482–85

neoplasms
    fibroblast, 353–63
    lymphocyte, 467–70
nerve growth factor, 544, 545
    activity, 199, 412, 440
    and animal cell growth, 2
    deprivation of, 194
    mechanism of action, 196, 205–6
    response, 553–55
    structure, 201–2
nutrients and
    cell growth, 1, 4, 497
    somatomedins, 148

oncogenic retroviruses, 467, 470
ornithine decarboxylase, 435
oscillation, limit cycle, 242–44

paracrine secretion, 458
pathology, 505
peptides, low-molecular-weight, 100–108
phase
    perturbation, 241–54
    response, 243
phosphotyrosine, 566–67
phosvitin, 315–22
photoaffinity labeling, 325–36
phytohemagglutinin, 365–71, 516, 517
pinocytosis, 21–26, 203
placenta, human, 339–50
plaque-forming cells, 95
plasma
    bioactivities in, 99
    and cell growth, 78–82, 85
    and chondrocyte metabolism, 100–108
    components, 507–9
    -derived serum, 23–25
plasma membrane
    liver, 335–36

receptors, and growth factor, 546–47
plasminogen activator, 163–64, 571
platelet-derived growth factor, 23, 26, 503–6, 508, 546
    cell cycle effects of, 141–50
    DNA synthesis, 425
    and extracellular matrix, 79, 81, 84
    and growth, 194
polyclonal antibody response, 213–15
polypeptide growth factors, 2, 3
    mechanism of action, 202–7
    properties of, 197–202
    relation of, to hormones, 195–97
    structure of, 193–207
preadipocytes, 124–41
progression, 143–50
proteases
    and epidermal growth factor, 346, 347, 350
    inhibition, 153–65
    membrane, 163–65
    -nexin, 569–71
    role of, in growth and development, 569–71
protein
    carrier, 395–409
    cell surface, 373–84
    kinase, 308–10, 557–67
    phosphorylation, 307, 310–12, 559
    synthesis, 498–99, 539
    synthesis, in smooth muscle cells, 157
    transport, 315–22, 549–55
proteoglycan synthesis, 434–35
proteolysis, 303–12, 339–50

rat liver cell line, 411–40
recycling, 555
red cell differentiation, 487–93
relaxin, 199, 201, 412, 440
reproductive tract, female, 199
retinoid, 448–49
RNA
    and erythropoietin, 490–91
    synthesis, 5, 91, 242, 252
    virus, 446

sarcoma growth factor, 446, 448–50, 456
serum, 355–63
serum-free medium, 167–82
serum growth factors
   and cell cycle events, 150–51
   and cell growth, 78–82, 85, 495, 496, 498
serum growth requirement, 589–92
serum spreading factor, 167–82
silica, 387–94
skeletal muscle differentiation, 226–39
smooth muscle cell
   and cartilage factor, 157–65
   and heparin, 154, 156, 157, 164
   and plasma, 506
somatomedins, 100, 107, 108, 194, 507
   activity, 395–96
   and cell growth, 500–501
   and fetal growth, 436
   and insulin, 500
   and multiplication-stimulating activity, 419–22, 424, 439–40
   in plasma, 151
   and progression, 145
   regulatory event, 148
SV40 virus, 176–77, 457
suppressor antibody, 188–89, 191

T cells
   activation of, 131–39
   and B cell stimulation, 579–87
   constant region, 185–86, 191
   developing, 463–67
   function, 461–62
   growth factor, 39–50
   helper, 582–87
   proliferation, 39–50
   receptor, 185–91
   recognition, 461–62
   -replacing factors, 514, 521–22
   see also lymphocyte(s)
3T3 cells, 176–77
   cycle regulation in, 141–50
   and epidermal growth factor, 545–46
   insulin receptors in, 524–41
   pinocytosis in, 21–26
thrombin, 569, 570
   and cell growth, 500
   cleavage of cell surface proteins by, 373–84
   receptors, 119–29
thymus, 462–63
transferrin, 496, 498
transformation test, 497–98
transforming growth factor, 445–58
transitions, 593–600
triglycerides, 525–26
trypsin, 307–8, 312
tumor promoter, 448, 450, 455
tunicamycin, 539–40
1242 mouse teratocarcinoma cells, 175–76
tyrosine aminotransferase, 433–34

ultrastructure, 148–49

vascular endothelial cells
   and aging, 54–61
   differentiation of, 61–66
   and extracellular matrix, 53–85
vascular smooth muscle cells, 78–82
virogene-oncogene hypothesis, 450
virus, mouse, 255–64

ERRATA

Mapping the Mitotic Clock by Phase Perturbation, R.R. Klevecz, G.A. King, and R.M. Shymko, pages 241–254.

On page 243, the reaction scheme representing the trimolecular model of Prigogine and Lefevre was incomplete. The complete reaction scheme is:

$$A \underset{k_{-1}}{\overset{k_1}{\rightleftharpoons}} X$$

$$B + X \underset{k_{-2}}{\overset{k_2}{\rightleftharpoons}} Y + D$$

$$2X + Y \underset{k_{-3}}{\overset{k_3}{\rightleftharpoons}} 3X$$

$$X \underset{k_{-4}}{\overset{k_4}{\rightleftharpoons}} E$$

On page 244, in the second sentence in the last paragraph, the word "suggests" should have no "s". The sentence should read: "However, our observations of subcyclic phenomena suggest that more than one loop is required per cycle."

PROGRESS IN CLINICAL AND BIOLOGICAL RESEARCH

Series Editors
Nathan Back
George J. Brewer

Vincent P. Eijsvoogel
Robert Grover
Kurt Hirschhorn

Seymour S. Kety
Sidney Udenfriend
Jonathan W. Uhr

Vol 1: **Erythrocyte Structure and Function,** George J. Brewer, *Editor*

Vol 2: **Preventability of Perinatal Injury,** Karlis Adamsons and Howard A. Fox, *Editors*

Vol 3: **Infections of the Fetus and the Newborn Infant,** Saul Krugman and Anne A. Gershon, *Editors*

Vol 4: **Conflicts in Childhood Cancer: An Evaluation of Current Management,** Lucius F. Sinks and John O. Godden, *Editors*

Vol 5: **Trace Components of Plasma: Isolation and Clinical Significance,** G.A. Jamieson and T.J. Greenwalt, *Editors*

Vol 6: **Prostatic Disease,** H. Marberger, H. Haschek, H.K.A. Schirmer, J.A.C. Colston, and E. Witkin, *Editors*

Vol 7: **Blood Pressure, Edema and Proteinuria in Pregnancy,** Emanuel A. Friedman, *Editor*

Vol 8: **Cell Surface Receptors,** Garth L. Nicolson, Michael A. Raftery, Martin Rodbell, and C. Fred Fox, *Editors*

Vol 9: **Membranes and Neoplasia: New Approaches and Strategies,** Vincent T. Marchesi, *Editor*

Vol 10: **Diabetes and Other Endocrine Disorders During Pregnancy and in the Newborn,** Maria I. New and Robert H. Fiser, *Editors*

Vol 11: **Clinical Uses of Frozen-Thawed Red Blood Cells,** John A. Griep, *Editor*

Vol 12: **Breast Cancer,** Albert C.W. Montague, Geary L. Stonesifer, Jr., and Edward F. Lewison, *Editors*

Vol 13: **The Granulocyte: Function and Clinical Utilization,** Tibor J. Greenwalt and G.A. Jamieson, *Editors*

Vol 14: **Zinc Metabolism: Current Aspects in Health and Disease,** George J. Brewer and Ananda S. Prasad, *Editors*

Vol 15: **Cellular Neurobiology,** Zach Hall, Regis Kelly, and C. Fred Fox, *Editors*

Vol 16: **HLA and Malignancy,** Gerald P. Murphy, *Editor*

Vol 17: **Cell Shape and Surface Architecture,** Jean Paul Revel, Ulf Henning, and C. Fred Fox, *Editors*

Vol 18: **Tay-Sachs Disease: Screening and Prevention,** Michael M. Kaback, *Editor*

Vol 19: **Blood Substitutes and Plasma Expanders,** G.A. Jamieson and T.J. Greenwalt, *Editors*

Vol 20: **Erythrocyte Membranes: Recent Clinical and Experimental Advances,** Walter C. Kruckeberg, John W. Eaton, and George J. Brewer, *Editors*

Vol 21: **The Red Cell,** George J. Brewer, *Editor*

Vol 22: **Molecular Aspects of Membrane Transport,** Dale Oxender and C. Fred Fox, *Editors*

Vol 23: **Cell Surface Carbohydrates and Biological Recognition,** Vincent T. Marchesi, Victor Ginsburg, Phillips W. Robbins, and C. Fred Fox, *Editors*

Vol 24: **Twin Research,** Proceedings of the 2nd International Congress on Twin Studies, Walter E. Nance, *Editor*
Published in 3 Volumes:
    Part A: Psychology and Methodology
    Part B: Biology and Epidemiology
    Part C: Clinical Studies

Vol 25: **Recent Advances in Clinical Oncology,** Tapan A. Hazra and Michael C. Beachley, *Editors*

Vol 26: **Origin and Natural History of Cell Lines,** Claudio Barigozzi, *Editor*

Vol 27: **Membrane Mechanisms of Drugs of Abuse,** Charles W. Sharp and Leo G. Abood, *Editors*

Vol 28: **The Blood Platelet in Transfusion Therapy,** G.A. Jamieson and Tibor J. Greenwalt, *Editors*

Vol 29: **Biomedical Applications of the Horseshoe Crab (Limulidae),** Elias Cohen, *Editor-in-Chief*

Vol 30: **Normal and Abnormal Red Cell Membranes,** Samuel E. Lux, Vincent T. Marchesi, and C. Fred Fox, *Editors*

Vol 31: **Transmembrane Signaling,** Mark Bitensky, R. John Collier, Donald F. Steiner, and C. Fred Fox, *Editors*

Vol 32: **Genetic Analysis of Common Diseases: Applications to Predictive Factors in Coronary Disease,** Charles F. Sing and Mark Skolnick, *Editors*

Vol 33: **Prostate Cancer and Hormone Receptors,** Gerald P. Murphy and Avery A. Sandberg, *Editors*

Vol 34: **The Management of Genetic Disorders,** Constantine J. Papadatos and Christos S. Bartsocas, *Editors*

Vol 35: **Antibiotics and Hospitals,** Carlo Grassi and Giuseppe Ostino, *Editors*

Vol 36: **Drug and Chemical Risks to the Fetus and Newborn,** Richard H. Schwarz and Sumner J. Yaffe, *Editors*

Vol 37: **Models for Prostate Cancer,** Gerald P. Murphy, *Editor*

Vol 38: **Ethics, Humanism, and Medicine,** Marc D. Basson, *Editor*

Vol 39: **Neurochemistry and Clinical Neurology,** Leontino Battistin, George Hashim, and Abel Lajtha, *Editors*

Vol 40: **Biological Recognition and Assembly,** David S. Eisenberg, James A. Lake, and C. Fred Fox, *Editors*

Vol 41: **Tumor Cell Surfaces and Malignancy,** Richard O. Hynes and C. Fred Fox, *Editors*

Vol 42: **Membranes, Receptors, and the Immune Response: 80 Years After Ehrlich's Side Chain Theory,** Edward P. Cohen and Heinz Köhler, *Editors*

Vol 43: **Immunobiology of the Erythrocyte,** S. Gerald Sandler, Jacob Nusbacher, and Moses S. Schanfield, *Editors*

Vol 44: **Perinatal Medicine Today,** Bruce K. Young, *Editor*

Vol 45: **Mammalian Genetics and Cancer: The Jackson Laboratory Fiftieth Anniversary Symposium,** Elizabeth S. Russell, *Editor*

Vol 46: **Etiology of Cleft Lip and Cleft Palate,** Michael Melnick, David Bixler, and Edward D. Shields, *Editors*

Vol 47: **New Developments With Human and Veterinary Vaccines,** A. Mizrahi, I. Hertman, M.A. Klingberg, and A. Kohn, *Editors*

Vol 48: **Cloning of Human Tumor Stem Cells,** Sydney E. Salmon, *Editor*

Vol 49: **Myelin: Chemistry and Biology,** George A. Hashim, *Editor*

Vol 50: **Rights and Responsibilities in Modern Medicine: The Second Volume in a Series on Ethics, Humanism, and Medicine,** Marc D. Basson, *Editor*

Vol 51: **The Function of Red Blood Cells: Erythrocyte Pathobiology,** Donald F. H. Wallach, *Editor*